MECHANICAL ASSEMBLIES

OXFORD SERIES ON ADVANCED MANUFACTURING

Series Editors

J. R. Crookall
Milton C. Shaw
Nam P. Suh

Series Titles

John Benbow and John Bridgewater, *Paste Flow and Extrusion*

John L. Burbidge, *Period Batch Control*

John L. Burbidge, *Production Flow Analysis for Planning Group Technology*

Shiro Kobayashi, S. Oh, and T. Altan, *Metal Forming and the Finite-Element Method*

Milton C. Shaw, *Metal Cutting Principles, 2nd ed.*

Milton C. Shaw, *Principles of Abrasive Processing*

Nam P. Suh, *The Principles of Design*

Daniel E. Whitney, *Mechanical Assemblies: Their Design, Manufacture, and Role in Product Development*

MECHANICAL ASSEMBLIES

Their Design, Manufacture, and Role in Product Development

Daniel E. Whitney
Massachusetts Institute of Technology

New York   Oxford
OXFORD UNIVERSITY PRESS
2004

Oxford University Press

Oxford   New York
Auckland   Bangkok   Buenos Aires   Cape Town   Chennai
Dar es Salaam   Delhi   Hong Kong   Istanbul   Karachi   Kolkata
Kuala Lumpur   Madrid   Melbourne   Mexico City   Mumbai   Nairobi
São Paulo   Shanghai   Taipei   Tokyo   Toronto

Published by Oxford University Press, Inc.
198 Madison Avenue, New York, New York 10016
www.oup.com

Library of Congress Cataloging-in-Publication Data

Whitney, Daniel E.
   Mechanical assemblies: their design, manufacture, and role in product development / by
Daniel E. Whitney.
     p. cm.
   Includes bibliographical references and index.
   ISBN 0-19-515782-6
    1. Production engineering.   2. Design, Industrial. I. Title.

TS171.4.W48 2004
658.5′752–dc22   2003066170

Printing number: 9 8 7 6 5 4 3 2 1

Printed in the United States of America
on acid-free paper

CONTENTS

PREFACE xix

1 WHAT IS ASSEMBLY AND WHY IS IT IMPORTANT?

1.A. Introduction 1

1.B. Some Examples 2
    1.B.1. Stapler Tutorial 2
    1.B.2. Assembly Implements a Business Strategy 6
    1.B.3. Many Parts from Many Suppliers Must Work Together 8
    1.B.4. Some Examples of Poor Assembly Design 9

1.C. Assembly in the Context of Product Development 9

1.D. Assembling a Product 11

1.E. History and Present Status of Assembly 12
    1.E.1. History 12
    1.E.2. Manual and Automatic Assembly 13
    1.E.3. Robotic Assembly 14
    1.E.4. Robotics as a Driver 15
    1.E.5. Current Status and Challenges in Assembly 16

1.F. Assemblies Are Systems 16

1.G. Chapter Summary 17

1.H. Problems and Thought Questions 17

1.I. Further Reading 18

2 ASSEMBLY REQUIREMENTS AND KEY CHARACTERISTICS

2.A. Prolog 19

2.B. Product Requirements and Top-Down Design 19

2.C. The Chain of Delivery of Quality 20

2.D. Key Characteristics 21

2.E. Variation Risk Management 22
    2.E.1. Key Characteristics Flowdown 23
    2.E.2. Ideal KC Process 25

2.F. Examples 26
    2.F.1. Optical Disk Drive 26
    2.F.2. Car Doors 27
2.G. Key Characteristics Conflict 29
2.H. Chapter Summary 31
2.I. Problems and Thought Questions 32
2.J. Further Reading 32

3 MATHEMATICAL AND FEATURE MODELS OF ASSEMBLIES

3.A. Introduction 34
3.B. Types of Assemblies 34
    3.B.1. Distributive Systems 34
    3.B.2. Mechanisms and Structures 35
    3.B.3. Types of Assembly Models 36
3.C. Matrix Transformations 36
    3.C.1. Motivation and Example 36
    3.C.2. Nominal Location Transforms 37
    3.C.3. Variation Transforms 42
3.D. Assembly Features and Feature-Based Design 42
    3.D.1. History 43
    3.D.2. Fabrication Features 43
    3.D.3. Assembly Features 44
    3.D.4. The Disappearing Fabrication Feature 44
3.E. Mathematical Models of Assemblies 45
    3.E.1. World Coordinate Models 45
    3.E.2. Surface-Constrained Models 46
    3.E.3. Connective Models 46
    3.E.4. Building a Connective Model of an Assembly by Placing Feature Frames
        on Parts and Joining Parts Using Features 47
    3.E.5. A Simple Data Model for Assemblies 51
3.F. Example Assembly Models 53
    3.F.1. Seeker Head 53
    3.F.2. Juicer 55
3.G. Chapter Summary 57
3.H. Problems and Thought Questions 57
3.I. Further Reading 60

4 CONSTRAINT IN ASSEMBLY

4.A. Introduction 62
4.B. The Stapler 63
4.C. Kinematic Design 63
    4.C.1. Principles of Statics 63
    4.C.2. Degrees of Freedom 65

4.C.3. How Kinematics Addresses Constraint 66
4.C.4. Kinematic Assemblies 68
4.C.5. Constraint Mistakes 68
4.C.6. "Good" Overconstrained Assemblies 71
4.C.7. Location, Constraint, and Stability 73
4.C.8. One-Sided and Two-Sided Constraints—Also Known as Force Closure and Form Closure 73
4.C.9. Force–Motion Ambiguity 75
4.C.10. Summary of Constraint Situations 75

4.D. Features as Carriers of Constraint 76

4.E. Use of Screw Theory to Represent and Analyze Constraint 77
4.E.1. History 77
4.E.2. Screw Theory Representations of Assembly Features 78

4.F. Design and Analysis of Assembly Features Using Screw Theory 86
4.F.1. Motion and Constraint Analysis 86
4.F.2. Basic Surface Contacts and Their Twist Matrices 87
4.F.3. Construction of Engineering Features and Their Twist Matrices 89
4.F.4. Use of Screw Theory to Describe Multiple Assembly Features That Join Two Parts 94
4.F.5. Graphical Technique for Conducting Twist Matrix Analyses 97
4.F.6. Graphical Technique for Conducting Constraint Analyses 98
4.F.7. Why Are the Motion and Constraint Analyses Different? 101

4.G. Advanced Constraint Analysis Technique 102

4.H. Comment 102

4.I. Chapter Summary 102

4.J. Problems and Thought Questions 103

4.K. Further Reading 106

4.L. Appendix: Feature Toolkit 107
4.L.1. Nomenclature for the Toolkit Features 107
4.L.2. Toolkit Features 107

5 DIMENSIONING AND TOLERANCING PARTS AND ASSEMBLIES

5.A. Introduction 112

5.B. History of Dimensional Accuracy in Manufacturing 113
5.B.1. The Rise of Accuracy and Interchangeability 113
5.B.2. Recent History of Parts Accuracy and Dimensioning and Tolerancing Practices 114
5.B.3. Remarks 116

5.C. KCs and Tolerance Flowdown from Assemblies to Parts: An Example 116

5.D. Geometric Dimensioning and Tolerancing 118
5.D.1. Dimensions on Drawings 118
5.D.2. Geometric Dimensioning and Tolerancing 118

5.E. Statistical and Worst-Case Tolerancing 123
5.E.1. Repeatable and Random Errors, Goalposting, and the Loss Function 124
5.E.2. Worst-Case Tolerancing 125
5.E.3. Statistical Process Control 126
5.E.4. Statistical Tolerancing 130

5.E.5. Summary of SPC and Statistical Tolerancing 133

5.E.6. Why Do Mean Shifts and Goalposting Occur? 133

5.E.7. Including Mean Shifts in Statistical Tolerancing 134

5.E.8. What If the Distribution Is Not Normal? 135

5.E.9. Remarks 136

5.F. Chapter Summary 136

5.G. Problems and Thought Questions 137

5.H. Further Reading 138

5.I. Appendix: Central Limit Theorem 138

5.J. Appendix: Basic Properties of Distributions of Random Variables 139

5.J.1. Mean of a Sum 139

5.J.2. Variance of a Sum 139

5.J.3. Average of a Sum 140

5.J.4. Variance of the Average of a Sum 140

6 MODELING AND MANAGING VARIATION BUILDUP IN ASSEMBLIES

6.A. Introduction 141

6.B. Nominal and Varied Models of Assemblies Represented by Chains of Frames 142

6.B.1. Calculation of Connective Assembly Model Variation Using Single Features 142

6.B.2. Calculation of Connective Assembly Model Variation Using Compound Features 143

6.C. Representation of GD&T Part Specifications as 4×4 Transforms 147

6.C.1. Representation of Individual Tolerance Zones as 4×4 Transforms 147

6.C.2. Worst-Case Representation of 4×4 Transform Errors 148

6.C.3. Statistical Representation of 4×4 Transform Errors 149

6.C.4. Remark: Constraint Inside a Part 152

6.D. Examples 152

6.D.1. Addition of Error Transforms to Nominal Transforms 152

6.D.2. Assembly Process Capability 152

6.D.3. Variation Buildup with Fixtures 155

6.D.4. Car Doors 157

6.E. Tolerance Allocation 162

6.E.1. Tolerance Allocation to Minimize Fabrication Costs 162

6.E.2. Tolerance Allocation to Achieve a Given C_{pk} at the Assembly Level and at the Fabrication Level 163

6.F. Variation Buildup in Sheet Metal Assemblies 165

6.F.1. Stress–Strain Considerations 165

6.F.2. Assembly Sequence Considerations 167

6.F.3. Adjustment Considerations 167

6.G. Variation Reduction Strategies 168

6.G.1. Selective Assembly 168

6.G.2. Functional Build and Build to Print 169

6.H. Chapter Summary 171

6.I. Problems and Thought Questions 173

6.J. Further Reading 176

6.K. Appendix: MATLAB Routines for Obeying and Approximating Rule #1 177

7 ASSEMBLY SEQUENCE ANALYSIS

7.A. Introduction 180

7.B. History of Assembly Sequence Analysis 181

7.C. The Assembly Sequence Design Process 183

      7.C.1. Summary of the Method 183

      7.C.2. Methods for Finding Feasible Sequences 184

      7.C.3. Methods of Finding Good Sequences from the
Feasible Sequences 186

      7.C.4. An Engineering-Based Process for Assembly Sequence Design 186

7.D. The Bourjault Method of Generating All Feasible Sequences 190

      7.D.1. First Question: $R(1;2,3,4)$ 191

      7.D.2. Second Question: $R(2;1,3,4)$ 191

      7.D.3. Third Question: $R(3;1,2,4)$ 191

      7.D.4. Fourth Question: $R(4;1,2,3)$ 191

      7.D.5. Reconciliation of the Answers 192

      7.D.6. Precedence Question Results 192

7.E. The Cutset Method 192

7.F. Checking the Stability of Subassemblies 193

7.G. Software for Deriving Assembly Sequences 194

      7.G.1. Draper Laboratory/MIT Liaison Sequence Method 194

      7.G.2. Sandia Laboratory Archimedes System 194

7.H. Examples 195

      7.H.1. Automobile Alternator 195

      7.H.2. Pump Impeller System 197

      7.H.3. Consumer Product Example 199

      7.H.4. Industrial Assembly Sequence Example 201

7.I. Chapter Summary 205

7.J. Problems and Thought Questions 205

7.K. Further Reading 206

7.L. Appendix: Statement of the Rules of the Bourjault Method 207

7.M. Appendix: Statistics on Number of Feasible Assembly Sequences a Product
Can Have and Its Relation to Liaisons Per Part for Several Products 208

8 THE DATUM FLOW CHAIN

8.A. Introduction 211

8.B. History and Related Work 213

8.C. Summary of the Method for Designing Assemblies 213

8.D. Definition of a DFC 215

      8.D.1. The DFC Is a Graph of Constraint Relationships 215

      8.D.2. Nominal Design and Variation Design 216

8.D.3. Assumptions for the DFC Method 216

8.D.4. The Role of Assembly Features in a DFC 216

8.E. Mates and Contacts 217

8.E.1. Examples of DFCs 217

8.E.2. Formal Definition of Mate and Contact 219

8.E.3. Discussion 219

8.F. Type 1 and Type 2 Assemblies Example 221

8.G. KC Conflict and Its Relation to Assembly Sequence and KC Priorities 224

8.H. Example Type 1 Assemblies 226

8.H.1. Fan Motor 226

8.H.2. Automobile Transmission 227

8.H.3. Cuisinart 231

8.H.4. Pump Impeller 232

8.H.5. Throttle Body 234

8.I. Example Type 2 Assemblies 235

8.I.1. Car Doors 235

8.I.2. Ford and GM Door Methods 236

8.I.3. Aircraft Final Body Join 240

8.J. Summary of Assembly Situations That Are Addressed by the DFC Method 243

8.J.1. Conventional Assembly Fitup Analysis 243

8.J.2. Assembly Capability Analysis 243

8.J.3. Assemblies Involving Fixtures or Adjustments 244

8.J.4. Selective Assembly 244

8.K. Assembly Precedence Constraints 244

8.L. DFCs, Tolerances, and Constraint 245

8.M. A Design Procedure for Assemblies 245

8.M.1. Nominal Design Phase 245

8.M.2. Variation Design Phase 247

8.N. Summary of Kinematic Assembly 247

8.O. Chapter Summary 248

8.P. Problems and Thought Questions 248

8.Q. Further Reading 250

8.R. Appendix: Generating Assembly Sequence Constraints That Obey the Contact Rule and the Constraint Rule 251

9 ASSEMBLY GROSS AND FINE MOTIONS

9.A. Prolog 253

9.B. Kinds of Assembly Motions 253

9.B.1. Gross Motions 253

9.B.2. Fine Motions 253

9.B.3. Gross and Fine Motions Compared 254

9.C. Force Feedback in Fine Motions 255

9.C.1. The Role of Force in Assembly Motions 255

9.C.2. Modeling Fine Motions, Applied Forces, and Moments 255

9.C.3. The Accommodation Force Feedback Algorithm 256

9.C.4. Mason's Compliant Motion Algorithm 258

9.C.5. Bandwidth of Fine Motions 259

9.C.6. The Remote Center Compliance 260

9.D. Chapter Summary 261

9.E. Problems and Thought Questions 261

9.F. Further Reading 261

9.G. Appendix 262

10 ASSEMBLY OF COMPLIANTLY SUPPORTED RIGID PARTS

10.A. Introduction 263

10.B. Types of Rigid Parts and Mating Conditions 263

10.C. Part Mating Theory for Round Parts with Clearance and Chamfers 264

10.C.1. Conditions for Successful Assembly 265

10.C.2. A Model for Compliant Support of Mating Parts 266

10.C.3. Kinematic Description of Part Motions During Assembly 269

10.C.4. Wedging and Jamming 270

10.C.5. Typical Insertion Force Histories 274

10.C.6. Comment on Chamfers 275

10.D. Chamferless Assembly 276

10.E. Screw Thread Mating 278

10.F. Gear Mating 280

10.G. Chapter Summary 282

10.H. Problems and Thought Questions 282

10.I. Further Reading 285

10.J. Appendix: Derivation of Part Mating Equations 285

10.J.1. Chamfer Crossing 285

10.J.2. One-Point Contact 286

10.J.3. Two-Point Contact 286

10.J.4. Insertion Forces 287

10.J.5. Computer Program 288

11 ASSEMBLY OF COMPLIANT PARTS

11.A. Introduction 293

11.A.1. Motivation 293

11.A.2. Example: Electrical Connectors 295

11.B. Design Criteria and Considerations 296

11.B.1. Design Considerations 296

11.B.2. Assumptions 297

11.B.3. General Force Considerations 297

11.C. Rigid Peg/Compliant Hole Case 299

11.C.1. General Force Analysis 299

11.D. Design of Chamfers 304

11.D.1. Introduction 304

11.D.2. Basic Model for Insertion Force 304

11.D.3. Solutions to Chamfer Design Problems 306

11.E. Correlation of Experimental and Theoretical Results 311

11.F. Chapter Summary 312

11.G. Problems and Thought Questions 313

11.H. Further Reading 314

11.I. Appendix: Derivation of Some Insertion Force Patterns 314

11.I.1. Radius Nose Rigid Peg, Radius Nose Compliant Wall 314

11.I.2. Straight Taper Rigid Peg, Cantilever Spring Hole 315

11.J. Appendix: Derivation of Minimum Insertion Work Chamfer Shape 316

12 ASSEMBLY IN THE LARGE: THE IMPACT OF ASSEMBLY ON PRODUCT DEVELOPMENT

12.A. Introduction 317

12.B. Concurrent Engineering 317

12.C. Product Design and Development Decisions Related to Assembly 319

12.C.1. Concept Generation 320

12.C.2. Architecture and KC Flowdown 320

12.C.3. Platform Strategy, Technology Plan, Supplier Strategy, and Reuse 321

12.D. Steps in Assembly in the Large 321

12.D.1. Business Context 321

12.D.2. Manufacturing Context 323

12.D.3. Assembly Process Requirements 323

12.D.4. Product Design Improvements 324

12.D.5. Summary 324

12.E. Chapter Summary 325

12.F. Problems and Thought Questions 325

12.G. Further Reading 326

13 HOW TO ANALYZE EXISTING PRODUCTS IN DETAIL

13.A. How to Take a Product Apart and Figure Out How It Works 327

13.B. How to Identify the Assembly Issues in a Product 328

13.B.1. Understand Each Part 329

13.B.2. Understand Each Assembly Step 329

13.B.3. Identify High-Risk Areas 330

13.B.4. Identify Necessary Experiments 330

13.B.5. Recommend Local Design Improvements 331

13.C. Examples 331

13.C.1. Electric Drill 331

13.C.2. Child's Toy 335

13.C.3. Statistics Gathered from a Canon Camera 338

13.C.4. Example Mystery Features 338

13.D. Chapter Summary 339

13.E. Problems and Thought Questions 339

13.F. Further Reading 340

14 PRODUCT ARCHITECTURE

14.A. Introduction 341

14.B. Definition and Role of Architecture in Product Development 341

    14.B.1. Definition of Product Architecture 341

    14.B.2. Where Do Architectures Come From? 342

    14.B.3. Architecture's Interaction with Development Processes and Organizational Structures 345

    14.B.4. Attributes of Architectures 345

14.C. Interaction of Architecture Decisions and Assembly in the Large 354

    14.C.1. Management of Variety and Change 354

    14.C.2. The DFC as an Architecture for Function Delivery in Assemblies 360

    14.C.3. Data Management 362

14.D. Examples 363

    14.D.1.   Sony Walkman 363

    14.D.2.   Fabrication- and Assembly-Driven Manufacturing at Denso—How Product and Assembly Process Design Influence How a Company Serves Its Customers 364

    14.D.3.   Airbus A380 and Boeing Sonic Cruiser 365

    14.D.4.   Airbus A380 Wing 366

    14.D.5.   Office Copiers 367

    14.D.6.   Unibody, Body-on-Frame, and Motor-on-Wheel Cars 368

    14.D.7.   Black and Decker Power Tools 369

    14.D.8.   Car Air–Fuel Intake Systems 370

    14.D.9.   Internal Combustion Engines 370

    14.D.10. Car Cockpit Module 371

    14.D.11. Power Line Splice 371

14.E. Chapter Summary 375

14.F. Problems and Thought Questions 375

14.G. Further Reading 376

15 DESIGN FOR ASSEMBLY AND OTHER "ILITIES"

15.A. Introduction 379

15.B. History 380

    15.B.1. DFM/DFA as Local Engineering Methods 380

    15.B.2. DFM/DFA as Product Development Integrators 381

    15.B.3. DFA as a Driver of Product Architecture 382

    15.B.4. The Effect on DFM/DFA Strategies of Time and Cost Distributions in Manufacturing 382

15.C. General Approach to DFM/DFA 383

15.D. Traditional DFM/DFA (DFx in the Small) 385

    15.D.1. The Boothroyd Method 385

    15.D.2. The Hitachi Assembleability Evaluation Method 388

    15.D.3. The Hitachi Assembly Reliability Method (AREM) 389

15.D.4. The Westinghouse DFA Calculator 391

15.D.5. The Toyota Ergonomic Evaluation Method 391

15.D.6. Sony DFA Methods 391

15.E. DFx in the Large 392

15.E.1. Product Structure 392

15.E.2. Use of Assembly Efficiency to Predict Assembly Reliability 401

15.E.3. Design for Disassembly Including Repair and Recycling (DfDRR) 403

15.E.4. Other Global Issues 406

15.F. Example DFA Analysis 407

15.F.1. Part Symmetry Classification 407

15.F.2. Gross Motions 408

15.F.3. Fine Motions 409

15.F.4. Gripping Features 409

15.F.5. Classification of Fasteners 409

15.F.6. Chamfers and Lead-ins 409

15.F.7. Fixture and Mating Features to Fixture 409

15.F.8. Assembly Aids in Fixture 411

15.F.9. Auxiliary Operations 411

15.F.10. Assembly Choreography 411

15.F.11. Assembly Time Estimation 413

15.F.12. Assembly Time Comparison 413

15.F.13. Assembly Efficiency Analysis 413

15.F.14. Design Improvements for the Staple Gun Design for Assembly 413

15.F.15. Lower-Cost Staple Gun 414

15.G. DFx's Place in Product Design 415

15.H. Chapter Summary 416

15.I. Problems and Thought Questions 417

15.J. Further Reading 417

16 ASSEMBLY SYSTEM DESIGN

16.A. Introduction 420

16.B. Basic Factors in System Design 420

16.B.1. Capacity Planning—Available Time and Required Number of Units/Year 421

16.B.2. Assembly Resource Choice 422

16.B.3. Assignment of Operations to Resources 423

16.B.4. Floor Layout 423

16.B.5. Workstation Design 424

16.B.6. Material Handling and Work Transport 424

16.B.7. Part Feeding and Presentation 424

16.B.8. Quality: Assurance, Mistake Prevention, and Detection 424

16.B.9. Economic Analysis 424

16.B.10. Documentation and Information Flow 425

16.B.11. Personnel Training and Participation 425

16.B.12. Intangibles 425

16.C. Available System Design Methods 425

16.D. Average Capacity Equations 426

16.E. Three Generic Resource Alternatives 428
    16.E.1. Characteristics of Manual Assembly 428
    16.E.2. Characteristics of Fixed Automation 429
    16.E.3. Characteristics of Flexible Automation 431

16.F. Assembly System Architectures 431
    16.F.1. Single Serial Line (Car or Airplane Final Assembly) 432
    16.F.2. Team Assembly 432
    16.F.3. Fishbone Serial Line with Subassembly Feeder Lines 432
    16.F.4. Loop Architecture 433
    16.F.5. U-Shaped Cell (Often Used with People) 434
    16.F.6. Cellular Assembly Line 434

16.G. Quality Assurance and Quality Control 435
    16.G.1. Approaches to Quality 435
    16.G.2. Elements of a Testing Strategy 436
    16.G.3. Effect of Assembly Faults on Assembly Cost and Assembly System Capacity 436

16.H. Buffers 440
    16.H.1. Motivation 440
    16.H.2. Theory 441
    16.H.3. Heuristic Buffer Design Technique 442
    16.H.4. Reality Check 442

16.I. The Toyota Production System 443
    16.I.1. From Taylor to Ford to Ohno 443
    16.I.2. Elements of the System 443
    16.I.3. Layout of Toyota Georgetown Plant 445
    16.I.4. Volvo's 21-Day Car 445

16.J. Discrete Event Simulation 447

16.K. Heuristic Manual Design Technique for Assembly Systems 449
    16.K.1. Choose Basic Assembly Technology 449
    16.K.2. Choose an Assembly Sequence 449
    16.K.3. Make a Process Flowchart 449
    16.K.4. Make a Process Gantt Chart 449
    16.K.5. Determine the Cycle Time 451
    16.K.6. Assign Chunks of Operations to Resources 451
    16.K.7. Arrange Workstations for Flow and Parts Replenishment 451
    16.K.8. Simulate System, Improve Design 452
    16.K.9. Perform Economic Analysis and Compare Alternatives 452

16.L. Analytical Design Technique 454
    16.L.1. Theory and Limitations 454
    16.L.2. Software 454
    16.L.3. Example 455
    16.L.4. Extensions 457

16.M. Example Lines from Industry: Sony 458

16.N. Example Lines from Industry: Denso 458
    16.N.1. Denso Panel Meter Machine (~1975) 458

16.N.2. Denso Alternator Line (~1986) 458

16.N.3. Denso Variable Capacity Line (~1996) 459

16.N.4. Denso Roving Robot Line for Starters (~1998) 460

16.N.5. Comment on Denso 460

16.O. Example Lines from Industry: Aircraft 461

16.P. Chapter Summary 463

16.Q. Problems and Thought Questions 463

16.R. Further Reading 464

17 ASSEMBLY WORKSTATION DESIGN ISSUES

17.A. Introduction 465

17.A.1. Assembly Equals Reduction in Location Uncertainty 465

17.B. What Happens in an Assembly Workstation 466

17.C. Major Issues in Assembly Workstation Design 467

17.C.1. Get Done Within the Allowed Cycle, Which Is Usually Short 467

17.C.2. Meet All the Assembly Requirements 468

17.C.3. Avoid the Six Common Mistakes 468

17.D. Workstation Layout 469

17.E. Some Important Decisions 470

17.E.1. Choice of Assembly "Resource" 470

17.E.2. Part Presentation 470

17.F. Other Important Decisions 476

17.F.1. Allocation of Degrees of Freedom 476

17.F.2. Combinations of Fabrication and Part Arrangement with Assembly 476

17.G. Assembly Station Error Analysis 476

17.H. Design Methods 477

17.H.1. Simulation Software and Other Computer Aids 477

17.H.2. Algorithmic Approach 478

17.I. Examples 481

17.I.1. Sony Phenix 10 Assembly Station 481

17.I.2. Window Fan 483

17.I.3. Staple Gun 483

17.I.4. Making Stacks 484

17.I.5. Igniter 484

17.J. Chapter Summary 488

17.K. Problems and Thought Questions 488

17.L. Further Reading 488

18 ECONOMIC ANALYSIS OF ASSEMBLY SYSTEMS

18.A. Introduction 489

18.B. Kinds of Cost 489

18.B.1. Fixed Cost 489

18.B.2. Variable Cost 490

18.B.3. Materials Cost 490

18.B.4. Administrative Cost 490

18.B.5. Direct Cost 490

18.B.6. Indirect Cost 490

18.B.7. Distribution of Costs in the Supply Chain 490

18.B.8. Cash Flows 492

18.B.9. Summary 493

18.C. The Time Value of Money 493

18.D. Interest Rate, Risk, and Cost of Capital 493

18.E. Combining Fixed and Variable Costs 494

18.F. Cost Models of Different Assembly Resources 495

18.F.1. Unit Cost Model for Manual Assembly 495

18.F.2. Unit Cost Model for Fixed Automation 495

18.F.3. Unit Cost Model for Flexible Automation 496

18.F.4. Remarks 497

18.F.5. How SelectEquip Calculates Assembly Cost 499

18.F.6. Is Labor Really a Variable Cost? 499

18.G. Comparing Different Investment Alternatives 499

18.G.1. Discounting to Present Value 500

18.G.2. Payback Period Method 501

18.G.3. Internal Rate of Return Method 501

18.G.4. Net Present Value Method 501

18.G.5. Example IRoR Calculation 501

18.G.6. Example Net Present Value Calculation 501

18.G.7. Remarks 504

18.H. Chapter Summary 504

18.I. Problems and Thought Questions 505

18.J. Further Reading 505

INDEX 507

PREFACE

AIMS OF THIS BOOK

The overt aim of this book is to present a systematic approach to the design and production of mechanical assemblies. It should be of interest to engineering professionals in the manufacturing industries as well as to post-baccalaureate students of mechanical, manufacturing, and industrial engineering. Readers who are interested in logistical issues, supply chain management, product architecture, mass customization, management of variety, and product family strategies should find value here because these strategies are enabled during assembly design and are implemented on the assembly floor.

The approach is grounded in the fundamental engineering sciences, including statics, kinematics, geometry, and statistics. These principles are applied to realistic examples from industrial practice and my professional experience as well as examples drawn from student projects.[1]

It treats assembly on two levels. *Assembly in the small* deals with putting two parts together. These are the basic processes of assembly, much as raising a chip is a fundamental process of machining. *Assembly in the large* deals with design of assemblies so that they deliver their required performance, as well as design and evaluation of assembly processes, workstations, and systems.

The sequence of chapters follows the three themes in the book's title: design of assemblies, manufacture of assemblies, and the larger role of assemblies in product development.

---

[1]Many of my curious experiences in professional practice are included in footnotes or used as quotes at the beginning of many chapters.

Assembly is the capstone process in discrete parts product manufacturing. Yet there is no book that covers these themes. This is very surprising because there are many books about the design and manufacture of machine elements like shafts and gears. But these items do not do anything by themselves. *Only assemblies of parts actually do anything,* except for a few one-part products like baseball bats and beer can openers. Assemblies are really the things that are manufactured, not parts. Customers appreciate the things products do, not the parts they are made of.

The lack of books on assemblies is reflected in many companies where it is easy to find job descriptions corresponding to the design of individual parts but hard to find job descriptions corresponding to design of assemblies. As one engineer told me, "The customer looks at the gap between the door and the fender. But it's an empty space and we don't assign anyone to manage empty spaces."

There are also many books about tolerances and statistical process control for the manufacture of individual parts, but little or nothing about assembly process capability or the design of assembly equipment to meet a particular level of capability, however it is defined. There are, in addition, many fine books about balancing assembly lines and predicting their throughput, given that there is a competently designed assembly ready to be assembled.

But what is a competently designed assembly and how would we know one if we saw one? This book is directed at that question.

A deeper aim of the book is to show how to apply principles from system engineering to design of assemblies. This is done by exploiting the many similarities between systems in general and assemblies in particular. Students who learn about parts but not about assemblies never get a high-level view of how parts work together to create function, and thus they do not know how to design parts that are intended to contribute to a function in conjunction with other parts. For this reason, they design parts as individual items and are satisfied when they think they have done their individual job well. They are as disconnected from the product they are designing as is the assembly line worker who installs the same part for thirty years without knowing what product is being produced. Products and companies can fail for lack of anyone who understands how everything is supposed to work together.

The systems focus of the book is part of a trend at MIT to complement traditional engineering science with integrative themes that unite engineering with economic, managerial, and social topics.

OUTLINE OF THIS BOOK

Chapter 1 provides a discussion about what an assembly is and why it is important. Chapters 2 through 8 deal with the design of assemblies, including

- a requirements-driven approach to designing assemblies that is based on mathematical and engineering principles,
- a theory of kinematic assemblies[2] that shows how to specify and tolerance assemblies so that they deliver geometrically defined customer requirements,
- the method of key characteristics for defining the important dimensions of an assembly, and
- the datum flow chain technique for designing assemblies to achieve their key characteristics.

Chapters 9 through 11 deal with the basic processes of assembly, including

- how to describe the motions that parts undergo during assembly operations and
- what the conditions are under which a part mating attempt will or will not be successful.

Chapters 12 through 18 extend the scope of inquiry to include manufacturing methods and systems and the role of assembly in product development. Important topics in these chapters include

- assembly in the large, a view of how product function and business issues each can be viewed through the prism of assembly,
- how to analyze an existing assembly and perform a design for assembly (DFA) analysis,
- an exploration of product architecture, including its relationships to business strategy and design for assembly,
- design of assembly systems and workstations, and
- economic analysis of assembly systems.

A compact disc accompanies this book. The CD-ROM contains an additional chapter, Chapter 19, which is a complete case study that applies the book's methods to an aircraft structural subassembly. In addition, the CD-ROM contains supporting material such as chapter appendixes, student class project reports, a professional consulting report, software, and MATLAB routines that duplicate examples and methods in Chapters 3, 4, 5, 6, 16, 18, and 19.

HOW THE COURSE HAS BEEN TAUGHT

The material in the book has been presented to MIT graduate students for several years. The explicit prerequisites include linear algebra (to help the students with the matrix math) and applied mechanics (to provide a background in statics and statically determinate structures). There is no prerequisite for a knowledge of probability and statistics, even though the treatment of tolerancing makes use of those ideas and presents the basics in passing. Nevertheless, one student emphasized to me the huge paradigmatic difference between the usual way of teaching design (there is one answer) and the fact that we live in a stochastic world where designs and objects are really members of histograms. Until he took this course, he had seen only the former, never the latter.

Implicit prerequisites that make it easier for students to grasp the concepts include some experience in mechanical design, some work in industry, and an ability to make reasonably realistic perspective or isometric sketches of mechanical parts and simple assemblies.

Raw ability to manipulate equations or computer simulations will not be enough to either teach or learn this material.

The class taught by me meets twice a week for 1.5 hours, for a total of 25 class sessions. Each session focuses on

---

[2] As explained more completely in Chapter 4, a kinematic assembly is one that can be assembled without applying force or storing energy in the parts.

one chapter, although several chapters, such as those covering constraint, variation, datum flow chain, and product architecture are conceptually challenging and require two or three class sessions each. Homework assignments provide practice with the concepts. In some cases, considerable class time is devoted to discussing the homework. In addition to class sessions and traditional homework, students form groups with four to six members and do a semester-long project.

Students with work experience enjoy telling the class how course material compares with corresponding methods at their current or previous employers. I and my students value contributions from the class, which are encouraged throughout the semester. Some of these contributions have enriched my knowledge and have made their way into the book.

Throughout the book, portions of student project work are used as examples to illustrate the concepts as well as to showcase the accomplishments of the students and encourage others to emulate them.

POSSIBLE TEACHING APPROACHES

My MIT classes consist of both traditional mechanical engineering students and students pursuing MBAs with an engineering emphasis. Since the engineering content, such as part mating physics and tolerance chains, appeals to the engineering students while the business content, such as product architecture and supply chains, appeals to the MBAs, each group grumbles a bit about being taught the other group's favored material. I strive to convince each group that the other's favorite material is important for them to understand, because that provides the integrated system-level view.

Nevertheless, teachers using this book may wish to partition the material cleanly into engineering focus and management focus semesters or quarters. To aid this, here are a few paths through the chapters for various emphases (all paths start with the Preface and Chapter 1, which are therefore not listed):

- Engineering design focus: Chapters 2–8, 10, 11, 13, 15
- Industrial/manufacturing engineering focus: Chapters 5–7, 9, 15–18
- Engineering management focus: Chapters 12, 14, 18, 19

- Bottom-up sequence from parts to systems: Chapters 9, 13, 10, 11, 2–8, 12, 14–19.

In the bottom-up sequence, which I use, not all chapters are taught each semester and not all get equal time or emphasis.

ACKNOWLEDGMENTS

I have benefited during preparation of this book, and throughout my career, from many people, to whom I am deeply grateful. If there are any errors in this book, they are mine and not those of any person who contributed material or ideas. I also apologize if anyone has been omitted from the following list.

Charles Stark Draper Laboratory colleagues: Mr. James L. Nevins, Dr. Thomas L. De Fazio, Mr. Alexander C. Edsall, Mr. Richard E. Gustavson, Mr. Richard W. Metzinger, and Mr. Donald S. Seltzer. Our work together over more than twenty years formed my understanding and appreciation of assembly as an intellectual focus and provided the backbone of many of the book's chapters. Some of these chapters are updates of chapters in our earlier book *Concurrent Design of Products and Processes,* New York, McGraw-Hill, 1989. I also wish to thank current and former Draper colleagues Dr. J. Edward Barton, Prof. Samuel H. Drake, Mr. Richard R. Hildebrant, Mr. Michael P. Hutchins, Dr. Daniel Killoran, Mr. Anthony S. Kondoleon, Mr. Steven C. Luby, Prof. Thomas J. Peters, Mr. Raymond Roderick, Mr. Jonathan M. Rourke, Dr. Sergio N. Simunovic, Mr. Thomas M. Stepien, the late Mr. Paul C. Watson, and Mr. E. Albert Woodin for their contributions to our collective work.

MIT colleagues and programs: Mr. Martin Anderson, Dr. Don P. Clausing, Dr. George L. Roth, Professors Edward F. Crawley, Steven D. Eppinger, Charles H. Fine, Daniel Frey, David C. Gossard, Stephen C. Graves, Christopher L. Magee, Joel Moses, Daniel Roos, Warren P. Seering, Alex H. Slocum, Nam P. Suh, James M. Utterback, and David Wallace; the Center for Innovation in Product Development, the International Motor Vehicle Program, the Leaders for Manufacturing Program, the System Design and Management Program, and the Ford–MIT Research Alliance. These colleagues and programs provided intellectual stimulation, encouragement, financial support, and contact with companies and real industrial problems.

Professional colleagues at universities and industrial companies: Brigham Young University: Prof. Ken Chase; Carnegie–Mellon University: Professors Susan Finger, David Hounshell, and Matthew Mason; Cranfield University: Prof. Tim Baines and Dr. Ip-shing Fan; IPK Berlin: Prof. Dr.-Ing. Frank-Lothar Krause; Lancaster University: Prof. Michael French; l'Université de Franche-Comté: Professors Alain Bourjault and Jean Michel Henrioud; University of Michigan: Professors Walton Hancock, Jack Hu, and Jeffrey Liker; Oxford University: Prof. J. Michael Brady; Stanford University: Professors Mark Cutkosky, Daniel De Bra, and Bernard Roth; Technion: Prof. Dan Braha; USC Information Sciences Institute: Dr. Peter Will; University of Naples Federico II: Professors Francesco Caputo and Salvatore Gerbino; NIST: Dr. Michael Pratt, Dr. Ram Sriram, and Dr. Michael Wozny; University of Pennsylvania: Professors Daniel M. G. Raff and Karl Ulrich; Purdue University: Professors Christoph Hoffmann and Shimon Nof; RPI: Professors Arthur and Susan Sanderson; University of Southern California: Prof. Ari Requicha; University of Tokyo: Professors Takahiro Fujimoto and Fumihiko Kimura; Virginia Polytechnic Institute: Prof. Robert Sturges; WZB Berlin: Dr. Ulrich Jürgens; Adept Technology: Mr. Brian R. Carlisle; Airbus: M. Bernard Vergne, Dr. Benoit Marguet; Analytics: Dr. Anna Thornton; Arvin-Meritor: Mr. John Grace; Boeing: Mr. Tim Copes, Mr. E. L. Helvig, Dr. Stephen Keeler, Dr. Alan K. Jones, Mr. Wencil McClenahan, Mr. Scott P. Muske, Mr. Frederick M. Swanstrom, Dr. Steve Woods; Boothroyd & Dewhurst: Prof. Geoffrey Boothroyd; The Budd Co.: Mr. John M. Vergoz; Cognition: Mr. Michael Cronin; Daimler-Chrysler: Dr. Gustav Olling; Denso Co. Ltd.: Mr. Koichi Fukaya; Eastman Kodak: Mr. Douglass Blanding, Mr. Jon Kriegel, and Dr. Randy Wilson; Fanuc Robotics: Dr. Hadi A. Akeel; Ford Motor Company: Mr. Robert Bonner, Mr. James Darkangelo, Dr. Shuh Liou, Mr. Ting Liu, Dr. Richard Riff, Dr. Agus Sudjianto, and Dr. Nancy Wang; General Motors: Mr. Charles Klein, Mr. Steven Holland; Hitachi, Ltd.: Mr. Toshijiro Ohashi; Lockheed-Martin: Ms. Linda B. Griffin, Mr. Randy Schwemmin; Munro and Associates: Mr. Sandy Munro; M. S. Automation: Dr. Mario Salmon; SDRC: Dr. Albert Klosterman; Telemechanique: Dr. Albert Morelli; Toyota Motor Company: Dr. Christopher Couch; Vought: Mr. Cartie Yzquierdo. These individuals and their companies provided intellectual stimulation, gracious sharing of ideas, and crucial contact with real products and assembly processes to me and my students through summer internships and frequent visits.

Students: Mr. Jeffrey D. Adams, Mr. Jagmeet Singh Arora, Dr. Timothy W. Cunningham, Mr. J. Michael Gray, Dr. Ramakrishna Mantripragada, Mr. Gaurav Shukla, and Mr. Andrew M. Terry. These key students developed much of the theory presented in the first eight chapters of this book. Students whose case studies provided important data and insights include Mrs. Mary Ann Anderson, M. Denis Artzner, Mr. Edward Chung, Mr. Gennadiy Goldenshteyn, Mr. J. Michael Gray, Mr. Brian Landau, Mr. Don Lee, Mr. Craig Moccio, Mr. Guillermo Peschard, Mr. Stephen Rhee, Mr. Tariq Shaukat, and Mr. Jagmeet Singh Arora. Current and former students who wrote important tutorial software include Mr. Michael Hoag, Mr. J. Michael Gray, and Dr. Carol Ann McDevitt. Students whose class projects provided inspiring material of professional quality for the book are named in the chapters where their work appears.

Colleagues and students who read part or all of the book and made valuable comments: Prof. J. T. Black, Prof. Geoffrey Boothroyd, Prof. Christopher L. Magee, Mr. Wesley Margeson, Mr. James L. Nevins, Mr. Stefan von Praun, Mr. Daniel Rinkevich, Mr. Thomas H. Speller, Jr., Prof. Herbert Voelcker, Dr. John Wesner, and Prof. Paul Wright. I also thank several anonymous reviewers for extensive and important comments.

Oxford University Press staff: Peter Gordon, Elyse Dubin, Danielle Christensen, and Brian Kinsey, whose help, forebearance, and enthusiasm are much appreciated.

Copyright holders: Publitec S.r.1, publishers of *Assemblaggio* for many photographs; Sage Publications for many figures reprinted from Gustavson, R., Hennessey, M. J., and Whitney, D. E., "Designing Chamfers," *Robotics Research,* vol. 2, no. 4, pp. 3–18, 1983; ASME International for many figures reprinted from Whitney, D. E., "Quasi-Static Assembly of Compliantly Supported Rigid Parts," *ASME Journal of Dynamic Systems, Measurement and Control,* vol. 104, pp. 65–77, 1982; Whitney, D. E., and Adams, J. D., "Application of Screw Theory to Constraint Analysis of Assemblies Joined by Features," *ASME Journal of Mechanical Design,* vol. 123, no. 1, pp. 26–32, 2001; and Springer-Verlag for many figures reprinted from Whitney, D. E., Gilbert, O., and Jastrzebski, M., "Representation of Geometric Variations Using Matrix Transforms for Statistical Tolerance Analysis in Assemblies,"

Research in Engineering Design, vol. 6, pp. 191–210, 1994; Mantripragada, R., and Whitney, D. E., "The Datum Flow Chain," *Research in Engineering Design,* vol. 10, pp. 150–165, 1998; and Whitney, D. E., Mantripragada, R., Adams, J. D., and Rhee, S. J., "Designing Assemblies," *Research in Engineering Design,* vol. 11, pp. 229–253, 1999; plus many others who are named in connection with the specific items which they permitted to be reproduced.

Funding agencies and respective program managers: U.S. Air Force Wright Laboratory/MTIA, Mr. George Orzel, Program Manager, contracts F33615-94-C-4428 and F33615-94-C-4429; the National Science Foundation grant DMI-9610163, Dr. George Hazelrigg, Program Manager, and Cooperative Agreement No. EEC-9529140, Dr. Fred Betz, Program Manager. Their support and encouragement are gratefully acknowledged.

My family: Dr. Cynthia K. Whitney, Mr. David C. Whitney, and Dr. Karl D. Whitney for love, tolerance, and specific intellectual contributions. This book is dedicated to them.

THE CHAPTER-OPENING QUOTATIONS

Most chapters begin with a quotation that is intended to convey the spirit of the material in the chapter. Every one of these quotations is real and was spoken to me. I have written them down and, in some cases, paraphrased or condensed them before placing them in the book. Where there was no suitable quotation, a chapter does without. Readers are invited to contribute candidate quotes for any chapter and to forward them to me. I will happily collect them and, if appropriate, use them with attribution, should there be a second edition of this book.

MECHANICAL ASSEMBLIES

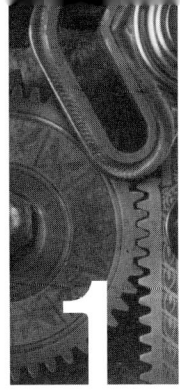

1 WHAT IS ASSEMBLY AND WHY IS IT IMPORTANT?

"Final assembly is the moment of truth."
—Charles H. Fine, MIT

1.A. INTRODUCTION

Assembly is more than putting parts together. Assembly is the capstone process in manufacturing. It brings together all the upstream processes of design, engineering, manufacturing, and logistics to create an object that performs a function. A great deal is known about the unit processes that are required to fabricate and inspect individual parts. Books exist and courses are taught on manufacturing processes and systems for turning and molding, to name a few. Assembly, which actually creates the product, is by comparison much less studied and is by far one of the least understood processes in manufacturing.

Assemblies are the product of the assembly process. But assemblies are also the product of a complex design process. This process involves defining the functions that the item must perform and then defining physical objects (parts and subassemblies) that will work together to deliver those functions. The structure of the item must be defined, including all the interrelationships between the parts. Then each of the parts must be defined and given materials, dimensions, tolerances, surface finishes, and so on. Books exist and courses are taught on how to design machine elements such as gears and shafts that become parts of assemblies. Yet there are no books that tell how to design assemblies, or books that indicate how to tell when an assembly design is good.

This book has several goals:

- To place mechanical assemblies in the context of product development and understand how they mutually affect each other
- To provide representations of assembly requirements, designs, and processes that are under-

standable to design engineers and manufacturing engineers
- To provide a fundamental engineering foundation for designing assemblies
- To connect the design of assemblies with the design of assembly processes and equipment, including technical and economic issues
- To present a systematic approach to understanding assemblies

We are going to address and attempt to answer a number of questions: What does it mean to "design an assembly"? What is a "good" assembly-level design? What must we take into account when designing an assembly? What are the nontechnical, business, and strategic impacts of assembly design decisions? What procedures are available to us to generate good assembly designs? To what degree are the design of the assembly and design of the assembly process separate, and when must they be integrated? How can we represent information about an assembly in a computer? Can we convert the design processes we find necessary or useful into computer-based engineering tools? What information is needed to document design intent for an assembly? Indeed, what is "design intent" for an assembly?

Assembly is different from the traditional unit processes of fabrication like milling and grinding because it is inherently integrative: It brings together parts, for sure, but it also brings (or should bring) together the people and companies who design and make those parts. If people know that the parts they are designing must assemble and work together, they will have a high

1

incentive to work together to ensure that successful integration occurs.

Assembly permits parts to function by working together as a system. Disassembled, they are just a pile of parts. Furthermore, as we shall see again and again in this book, typical assemblies have lots of parts and several functions. There aren't many one-part products.[1] Typical assemblies consist of *many* parts, each with a *few* important geometric features, *all* of which must work together in order to create the product's several functions.

Assembly is different from traditional unit processes in another important way: It is the key link between the unit processes and top-level business processes. For example,

- An appropriate assembly sequence can permit a company to customize a product when it adds the last few parts.
- Properly defined subassemblies permit a company to design them independently or outsource some or all of them from suppliers, as well as to switch between suppliers.
- A well-defined and executed product development process focused on assemblies can make ramp-up to full production faster because problems can be diagnosed faster.
- Properly defined assembly interfaces can allow a company to mix and match parts or subassemblies to create custom products with little or no switching cost.

TABLE 1-1. Assembly Links Unit Manufacturing Processes to Business Processes

Domain	Context	Example Application
Assembly in the large	Business level	• Market size and production volume • Model mix • Upgrade/update • Reuse, carryover • Outsourcing and supply chain
	System level	• Data management and control • Quality management • Subassemblies • Assembly sequences • Involvement of people • Automation • Line layout
Assembly in the small	Technical level	• Individual part quality • Individual part joining • Part logistics, preparation and feeding • Manual vs. automatic • Economics • Ergonomics

In general, assembly is the domain where many business strategies are carried out, all of which depend on careful attention to the strategic aims during product design. Some of these are listed in Table 1-1. In this table, the terms "assembly in the large" and "assembly in the small" are defined in context by means of the items at the far right in the table. They will be discussed in more detail later.

1.B. SOME EXAMPLES

Let us consider some examples to fix our ideas. The first one is a tutorial using a desktop stapler. The second is a panel meter for car dashboards, a product that illustrates how an assembly can embody the business strategy of a company. The third is a portion of the front end of a car. It illustrates the principle that many parts work together to deliver the functional or operating features of a product, and failure to understand how these parts work together can prevent assembly plant workers from understanding and fixing assembly problems. Some examples of poor assembly-related design are described at the end of this section. The stapler, panel meter, and car front end will be used repeatedly throughout the book to illustrate important concepts.

1.B.1. Stapler[2] Tutorial

Even though a desktop stapler may appear simple (see Figure 1-1), it is in fact a precision mechanism that will

---

[1]Crowbars and baseball bats are possible exceptions, as is the diamond engagement ring. The ring is really two parts, of which one is overwhelmingly important and the other is there merely to keep the first one from getting lost. Furthermore, that important one has hundreds of features, all of which are necessary to its function.

[2]A complete analysis of a stapler, together with all part and assembly details, is provided in [Simunovic].

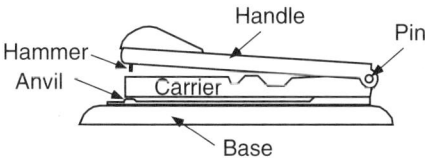

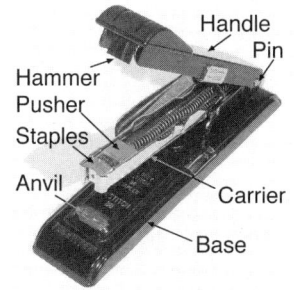

FIGURE 1-1. Desktop Stapler.

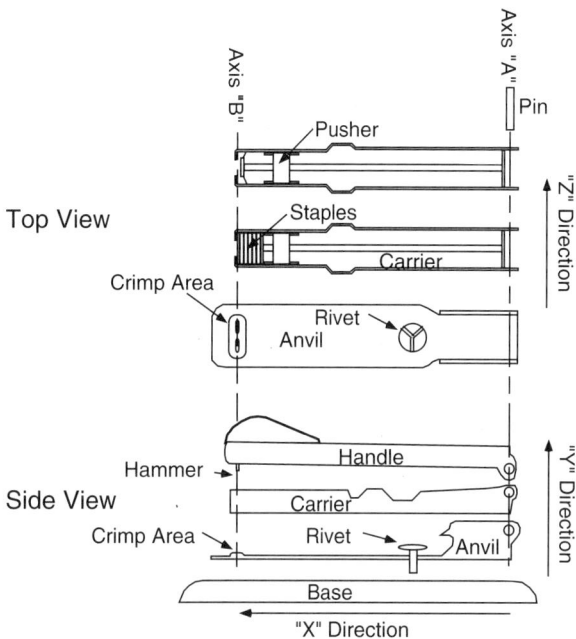

FIGURE 1-2. Stapler Parts. The main parts of the stapler are shown slightly separated from each other in the side view. The top view shows some of these parts plus a few others not visible in the side view. In the top view, the carrier is shown twice, once with staples and once without. The view without staples permits us to see the slot at the left end of the carrier where one staple is pushed out when the user pushes on the handle. The spring that pushes the part called pusher into the staples is not shown. The spring that pops the stapler open after stapling is also not shown.

malfunction badly if its parts are not made to the correct dimensions. A close look at the parts and how they relate to each other reveals why this is true.

The main parts of the stapler, as shown in Figure 1-2, are the base, the anvil (with its crimping area), the carrier (containing the staples and the pusher), and the handle. The anvil, carrier, and handle are tied together along axis "*A*" by the pin. The anvil and the base are tied together by the rivet. Along axis "*B*" we find the slot in the carrier where the last staple will be pushed out, that staple itself, the crimping area of the anvil, and an element of the handle called the hammer, which pushes the staple out of the carrier, through the paper, and onto the anvil, which crimps the staple, completing the stapling operation.

What makes the stapler work? What could cause it not to work? A reader with good mechanical sense can probably figure this out quickly, but products like aircraft and automobiles consist of complex assemblies that are much more difficult to understand. We need help to figure these things out, along with a theory that will help us answer these questions about assemblies that are too complex to be understood just by looking at them.

The way a product is laid out, including which parts perform what functions as well as how the parts are arranged in space, is called its architecture. The architecture of the stapler is relatively simple because it performs only one main function and has so few parts. The architectures of larger products are complex, and the role of architecture extends beyond how the product works into such areas as how it is made, sold, customized, repaired in the field, recycled, and so on.

To simplify this already simple example further, we will consider only one dimension of the stapler, the one called "*X*" in Figure 1-2. The "*Z*" direction in the top view is also important, though not as much, while the direction called "*Y*" in the side view has still less importance. (A thought question at the end of the chapter asks the reader to think more about this.)

In order to understand the stapler, we will use a simple diagram to describe it. This diagram will replace the parts with dots and connections between parts with lines, making a graph called a liaison diagram (Figure 1-3). Using this diagram, we will explain how the stapler works using words and pictures.

Each liaison represents a place where two parts join. Such places are called assembly features in this book.

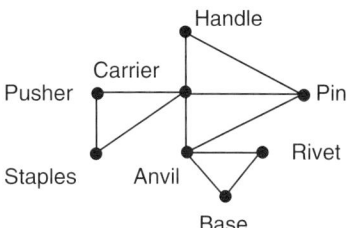

FIGURE 1-3. Liaison Diagram for the Stapler.

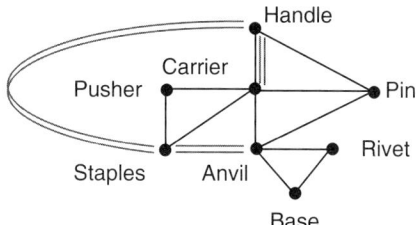

FIGURE 1-4. Liaison Diagram of Stapler with Key Characteristics Indicated by Double Lines.

They serve to position the parts with respect to each other. Some features act to hold a part firmly against another, while other features permit some relative movement between the parts. For example, the liaison between rivet, base, and anvil fixes these parts to each other completely, while the liaison between anvil, pin, and handle permits the handle to rotate about axis A with respect to the anvil.

Using the liaison diagram and the drawing of the stapler, we can make the following statements:

- The rivet connects the anvil to the base.
- The pin connects the anvil, carrier, and handle.
- The carrier connects the pusher and the staples.

In order for the stapler to work properly, the carrier must position the last staple right over the anvil's crimp area in the X direction. In addition, the handle must position its hammer right over the last staple in the X direction so that it strikes it squarely. Also, the hammer must rub against the end of the carrier to gain reinforcement against the buckling force of pushing the staple as well as to guide the end of the hammer against the top of the staple and avoid having the hammer slip off the staple. Finally, the hammer must pass right through the opening in the end of the carrier that the staple passes through, so as to be able to ram the staple firmly against the paper and transfer the necessary staple crimping force through the staple into the crimping area of the anvil. Equivalently, we can say that the operating features (hammer, staple slot, crimp area) must be placed properly inside the parts relative to the assembly features (holes for the pin), and the parts must be positioned relative to each other by the assembly features along axis "A," so that all the operating features align along axis "B."

This long-winded description is captured concisely and unambiguously in Figure 1-4. This figure is the liaison diagram with the addition of some double lines. These lines indicate schematically some important dimensional relationships between the parts at either end of each line pair (in the X direction only). We call these important dimensional relationships *key characteristics* (KC for short). If we get these relationships right, the product will work; if not, then it will not. It is important to understand that the assembly features play the crucial role of positioning the parts properly with respect to each other so that these KCs will be achieved accurately. That is, not only must each part be the correct length in the X direction, but they must assemble to each other properly, repeatably, and firmly.

Note that this diagram is necessarily simplified. In later chapters we will draw such diagrams in more detail so that each of the important operating and assembly features is shown separately. We will also show how to capture actions and relationships along all the axes, the directions of free motion, and so on.

Now, suppose there is some manufacturing variation in the construction of the handle so that on some percentage of the handles the hammer is located a bit too far from axis "A." When staplers are made with these handles, the hammer could strike the end of the carrier instead of sliding smoothly along the inside surface. What if the hammer is located a bit too close to axis "A"? In this case, the hammer might slip off the end of the staple; then it and the staple could become jammed together in the slot. Each of these manufacturing variations leads to an assembly variation in a KC. As another example, suppose a hammer is made too thick; it could jam inside the slot as it pushes the staple out. In either of these last two cases, the user would need pliers or other strong tools to open the stapler and undo the jam. After a few experiences like this, the user will throw the stapler away and buy one from another company.

Are there other ways in which the stapler could malfunction, other than due to mislocation of the hammer in the handle? What if the entire anvil is too long, so that the crimping area is not aligned with axis "B"? What if

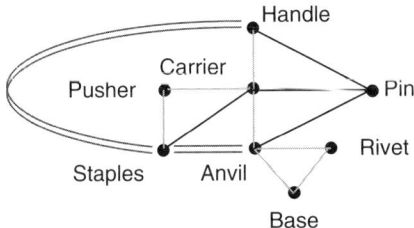

FIGURE 1-5. Liaison Diagram of Stapler with Some Liaisons Grayed Out. The grayed-out liaisons are not involved in delivering the KCs.

the rivet hole is in the wrong place so that the anvil is not where it is supposed to be on the base?

To answer these questions, we need to look more closely at the liaison diagram to determine which parts are really involved in the key characteristics. We will focus on two of the KCs, the one between handle (hammer) and staples and the one between anvil's crimp area and staples. We will assume that the one between handle and carrier is achieved in the same way as the one between handle and staples, using the same parts and assembly features.

With this simplification in mind, consider Figure 1-5. In this diagram, some of the liaisons are shown in gray. We assert that the gray liaisons are not "in the delivery chain for a KC." That is, even if the parts joined by gray liaisons were not installed at all, the key dimensions indicated by the double lines would be achieved anyway. The reader may have to study the stapler in order to be convinced of this. The remaining parts and their black liaisons comprise the parts that are necessary for the KCs to be achieved. The rivet and base are needed to keep the stapler stable on the table during use, and the pusher is needed to force the last staple into position at the end of the carrier, to be sure, but these parts and their relative locations do not affect the KC dimensions.

Not only are some parts not involved in KC delivery, but not all joints in the liaison diagram play the same role in the assembly. Our attention will focus on those joints that link the parts that participate in delivering the KCs. Those joints will be called "mates." Other joints may be important for fastening certain parts to the assembly or for reinforcing the mates. We call these joints "contacts." Knowing which joints are mates and which are contacts is essential in understanding how an assembly works.

The pusher has a *contact* with the staples and pushes them forward firmly against the left end of the carrier. Yet it is the *mate* between the carrier and the staples that positions the last staple properly.

The next step in understanding the stapler is to realize that the two KCs are achieved by different sets of parts, each of which occupy distinct chains. These are shown in Figure 1-6.

From Figure 1-6 we can see that achieving the KC which places the hammer over the last staple requires proper size and relative placement of the staples, via the carrier, the pin, and the handle (which contains the hammer). This chain is shown on the left in the figure. It links the two ends of the double lines that call out the KC. Another way to read this chain is to say that the pin locates the handle (hammer) and the carrier, while the carrier locates the staples. On the right is the chain that places the last staple over the crimping area: The pin locates the anvil (and its crimping area) and it locates the carrier, which locates the staples. Note, too, that both KCs must be achieved in order that the stapler operate properly. Fortunately, different parts are involved in most elements of these chains. This causes the two KCs to be capable of being achieved independently, although we have to be especially careful about the carrier because it is in both chains.

One way to ensure that the stapler achieves the KCs is that both the handle and the anvil be correctly sized and

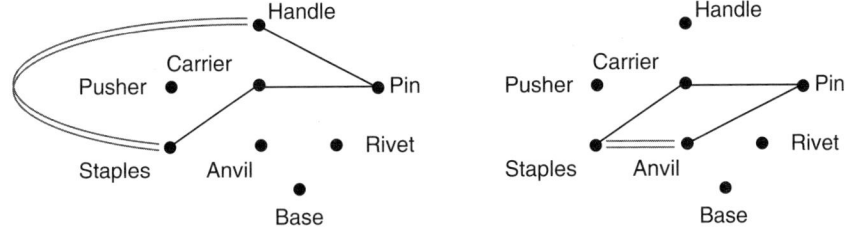

FIGURE 1-6. Delivery Chains for the Two KCs in the Stapler. *Left:* The hammer in the handle lines up with the last staple. *Right:* The last staple lines up with the crimping area of the anvil. Only the liaisons needed to deliver each KC are shown in each drawing. Rivet, base, and pusher are not involved in either KC.

positioned with respect to the carrier. Certainly, the diagram shows that the stapler will not work if this condition is not met. However, we shall see later in the book that it is preferable to be able to make these parts independently of each other, perhaps even to buy them from different suppliers, so that no special selecting, fitting, or measuring is required during assembly. We will also encounter in later chapters many cases where a product has several KCs but the chains that deliver them are coupled. Such situations confront the designer of the assembly with a choice of which KC to favor.

The diagrams in Figure 1-6 can be thought of as the plans for achievement of the KCs. They name the parts in the chain and allow us to identify the assembly features that are involved. These are called *datum flow chains* and are the subject of Chapter 8. When these chains have been designed properly, we can be assured that the assembly has a good chance of working. The chains also tell us where manufacturing or assembly variation could disrupt a KC. If some assemblies do not work, we can refer to the diagram to identify the parts involved, their internal dimensions, and their assembly features, as we search for the cause.

This example has introduced the following concepts and terms. The reader should reread the example if any of these terms are not clearly understood, because they will be used again and again throughout this book:

- Part
- Liaison, and liaison diagram
- Joint, mate, and contact
- Assembly feature
- Operating or functional feature
- Product architecture
- Location of operating feature with respect to assembly feature
- Key characteristic (KC)
- Achievement of a KC
- Chain of parts that achieve a KC
- Manufacturing variation
- Assembly variation
- Customer satisfaction if KCs are achieved; customer dissatisfaction if KCs are not achieved

The topics in the stapler example form the subject matter for Chapters 2 through 8 of this book.

1.B.2. Assembly Implements a Business Strategy[3]

Denso Co. Ltd. is the largest and perhaps the most sophisticated supplier of automotive components in Japan. It designs and manufactures generators, alternators, voltage regulators, fuel injection systems, engine controllers, anti-skid braking systems, and so on, for Toyota and many other automobile builders. Toyota owns 25% of Denso and accounts for almost half its business. Because Toyota manufactures a wide variety of products and wants its components delivered in an arbitrary model mix on a just-in-time (JIT) basis, Toyota puts extreme demands on its suppliers to be responsive and flexible. Over about thirty years, Denso has learned how to use the assembly process to meet Toyota's requirements.[4] Three elements of Denso's strategy are

- The combinatoric method of achieving model-mix production
- In-house development of manufacturing technology
- Jigless assembly methods and minimal changeover time and cost

Denso has applied this strategy to many products over the last thirty years ([Whitney 1993]). One of these is a panel meter for dashboards. This product is mentioned at many points in this book. Here we emphasize Denso's use of the product's architecture to serve the highly variable needs of its main customer, Toyota.

The combinatoric method is the basis of Denso's assembly-driven strategy. A product is divided into generic parts or subassemblies, and necessary varieties are identified. The product is designed so that any combination of varieties of these basic parts will go together physically and comprise a functional product. If there are six basic parts and three varieties of each, for example, then the company could build as many as 3^6 or 729 different versions of the product.

The in-house manufacturing engineering team participates in the design of these parts so that the manufacturing

---

[3]This subsection is adapted from Chapter 3 of [Nevins and Whitney].

[4]The official corporate slogan at Denso is "Conquer Variety," which means do whatever is necessary to accommodate the model mix demands of its customers. In practice, one could say that the slogan is "Never say 'No' to Toyota."

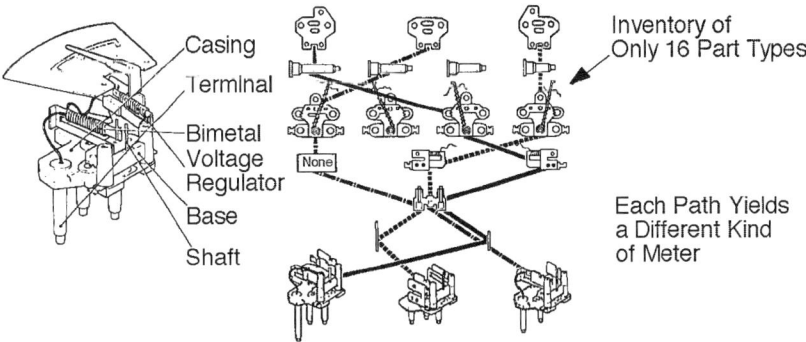

Casing
Terminal
Bimetal
Voltage
Regulator
Base
Shaft

None

Inventory of
Only 16 Part Types

Each Path Yields
a Different Kind
of Meter

288 Different Kinds of Meters Can Be Made
with No Additional Cost or Delay, and
Almost No Change Over Time

FIGURE 1-7. Denso Panel Meters Implement a Business Strategy for Dealing with Fluctuating Model Mix Requirements. This figure shows how Denso can make many different kinds of panel meters from a small group of parts. Three types of casings, four terminals, four bimetals, two voltage regulators, one base, and two shafts can be combined into any of 288 different kinds of meters, of which three possible ones are shown.

system can handle each one. The usual technique is to design common mating features between the parts and common fixturing features to mate to the assembly equipment.

The in-house team also contributes to the flexibility of the assembly process by implementing jigless assembly and quick (or no) changeover activities or equipment. Jigless assembly requires parts that can locate each other or possibly even snap together without the use of fixtures (thus requiring no investment in fixtures or time to change fixtures for different parts). Such parts may "stick" together permanently as is or may require some further fixation such as fasteners, welds, or glue.

Figure 1-7 shows how Denso manages to assemble huge numbers of the panel meters in high variety. Assembly is managed from an inventory of only sixteen part types covering six basic parts. One each of the six basic parts goes into each meter. A fairly ordinary automatic assembly machine assembles these meters one every 0.9 seconds. At the start of each shift, the foreman takes Toyota's orders and dials them into the machine's control panel. The machine then proceeds to make them one model at a time in solid batches. Each meter moves through the machine using its casing as its pallet. Each station adds one of the six basic parts and contains a part feeder for each version of that part. When the last meter in the batch has been launched, a robot at the head end places a dummy casing on the machine that marks the boundary between

batches. As the dummy reaches each assembly station, it strikes a switch that tells the controller that the batch boundary has arrived. The controller tells an air piston to transfer the feed track from one part feeder to another, so that the correct part is available for assembly into meters belonging to the next batch. Since the feeder track never holds more than one part at a time, no extra time or attention is needed to clean out fed but unused parts prior to launching the next batch.[5]

Two points should be noted about this story. First, the strategy is established during the integrated product/ process design phase and does not require sophisticated methods, schedules, or equipment on the factory floor. Second, the operative process for the strategy is assembly. Fabrication schedules are not tightly linked to Toyota's orders. In a typical factory, orders enter the fabrication shops, which generate parts required for the products ordered. These parts are then assembled. In the Denso meter factory, orders enter the assembly shop. All the fabrication shops do is keep the assembly machine's parts feeders full. This can be done using a preliminary schedule with little inventory risk. The risk is that a part would be made but not used, or not used for a long time. But

---

[5]The main cause of feeder problems—namely, multiple parts climbing over each other and jamming the track (called shingling)—is also avoided.

in this method, a part is a member of so many different models (on average 288/16 or 18) that it will not remain an orphan for long.

This example shows that the architecture of the product is a crucial concept in its design, touching the basic approach to making the product when, how, and in what versions the customer wants, all the while aligning to the supply chain and the design of the assembly process and equipment. Denso has made a strategic decision to keep control of all the elements of this process and designs critical assembly equipment in-house.

1.B.3. Many Parts from Many Suppliers Must Work Together

Figure 1-8 shows the assembly tree of the parts of the front end of a Ford Explorer. Reading the tree from bottom to top reveals the assembly sequence. Heavy lines in the figure link a set of connected parts that cooperate to achieve a customer-visible quality KC: that the hood be equally spaced between the two outer fenders and that the gaps between them have equal width along their entire lengths.

The hood and fenders are not assembled directly to each other but instead are joined by a series of other parts. The assembly process is aided by several fixtures. The heavy lines show where these fixtures play their roles. Thus not only must all the intervening parts be made properly but they must also be assembled to each other properly in order for the gaps to be within specification.

Several points can be made about Figure 1-8. First, it took two students over a month to draw it because the assembly is complex and because the parts and necessary tools and fixtures came from so many different companies. The assembly had not been documented this way before, although extensive documentation existed for each part and tool separately. Second, people at the top of the process inside Ford did not know which parts came from where because that responsibility had been given to a capable "full service supplier," the Budd Company, which in turn had outsourced a number of the parts to other smaller firms. Each such organizational boundary is indicated as a dashed line crossing the liaisons in Figure 1-8.

Every time a new car model is launched in an assembly plant, assembly problems are discovered. Rapid

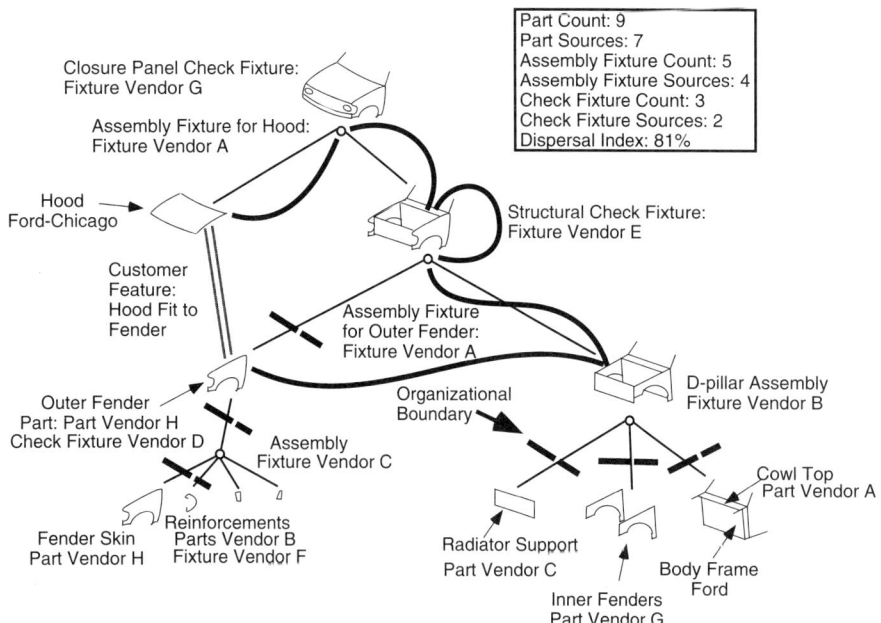

FIGURE 1-8. How the Car Hood and Front Fenders Achieve Quality Fit Requirements. The KC of interest to us is equal hood spacing between the outer fenders. Heavy lines link the main parts, subassemblies, and the necessary assembly fixtures. Also shown are some of the parts that make up the subassemblies. These parts must also be made and assembled properly in order that the KC be delivered. Dashed lines show organizational boundaries between Ford and its suppliers for the parts, subassemblies, and important assembly and checking fixtures. (Drawing based on one by Minho Chang and Narendra Soman.)

launch and ramp-up to full rate production requires diagnosing and solving these problems rapidly. The more complete the information available, the faster the problems can be solved. Information about the individual parts is not enough. In fact, it is typically easier to check if each part has been made "correctly" than to predict if the resulting assembly will work as desired. So, the scheme by which the parts are supposed to assemble and deliver the customer-level requirements must be understood by those doing the diagnosis, who are rarely the ones who did the original design. Even though the problem solving teams comprised cooperating representatives from Ford, Budd, and several other companies, diagnosis was often difficult when an overview like Figure 1-8 was not available.

This example not only makes the point that many parts work together to make an assembly function, but also emphasizes that the supply chain is an integral part of how assemblies are designed, procured assembled, and diagnosed. This is why "supply chain" is listed as one of the highest level of business implications for assemblies in Table 1-1.

1.B.4. Some Examples of Poor Assembly Design

1.B.4.a. Sewing Machine

A sewing machine is the most complex and precise of all consumer home appliances. It contains hundreds of parts, all of which must work together at high speed. Due to the variable nature of cloth and thread, sewing machines need to be adjusted into proper operation. The author and his colleagues visited a sewing machine assembly line some years ago. We found a single line building machines by adding one part at a time. There were no subassemblies and no opportunity to test the functions of the machine individually prior to completion of the assembly. At the end of the line was a hierarchy of adjusters. The first ones applied simple procedures and got perhaps half of the machines to work. The rest were attacked by successively more skilled adjusters. If they all failed, a master mechanic took the machine apart until he found the problem. This is not an economical production method. It could have been

improved by defining functionally testable subassemblies, designing the product to permit those subassemblies to be built and tested efficiently, and designing an assembly system to accommodate that design.

1.B.4.b. Electric Range

Electric ranges are not mechanically complex, but they are electrically complex. The main assembly error involves wires hooked to the wrong terminals. On one line visited by the author and his colleagues, such errors were so hard to find that a range got only two trips through the rework line. If the error was not found by then, the range was set aside and cannibalized for parts. No other option was "economical."

1.B.4.c. Precision Instrument

The instrument in question was an innovative gyroscope that contained some very delicate joints between rotating parts. The entire unit had to be dynamically balanced. This, in turn, required the rotating portion to be disassembled, reassembled in the balancing machine, balanced, disassembled again, and reassembled in the final housing. This is bad enough, since rotating parts should not be disassembled after balancing because it is generally very hard to restore good balance when reassembling. But there is more: These assembly and disassembly steps could not be accomplished without putting force onto the delicate joints. Only the skilled craftsman performing these steps knew this. His direct supervisor did not know; and the chief engineer, who was always warning everyone not to put force on those joints, did not know either. This product was impossible to assemble except by skilled craftsmen, who are in short supply and naturally work slowly and carefully. Thus this product, as originally designed, would have had great difficulty succeeding in its intended market, which required high production rates and low cost. One of the author's colleagues solved the reassembly problem. The excessive force problem was never solved to the author's knowledge because it arose from the layout of the housing, which could not be changed suitably without compromising other important aspects of the design.

1.C. ASSEMBLY IN THE CONTEXT OF PRODUCT DEVELOPMENT

Assemblies are designed in the context of product development. Product development is a process that, according to [Ulrich and Eppinger],

- Identifies a market opportunity and/or customer needs
- Combines these with technological developments

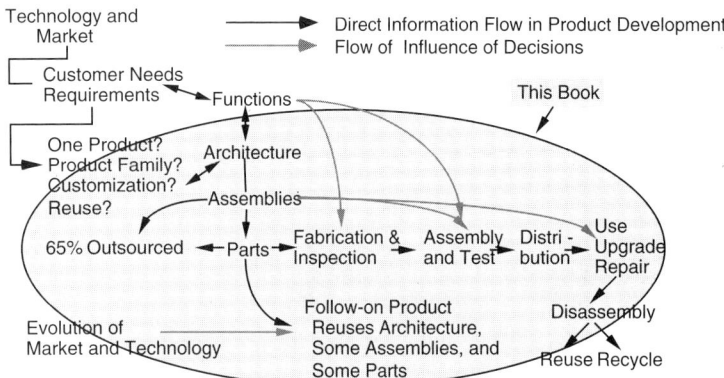

FIGURE 1-9. Diagram of the Life Cycle of a Product Viewed from the Product Development Context.

- Defines requirements and necessary functions of a product
- Generates concepts that could meet the requirements and chooses one, including defining the architecture of the product and, if appropriate, placing the product or the architecture in a product family
- Creates detailed designs that meet the requirements, including related requirements on the subassemblies and parts
- Defines production plans, including part fabrication, assembly, and outsourcing of parts and subassemblies
- Establishes a distribution strategy that may include customizing the product for different customers or sales regions
- Cares for downstream repairs, upgrades, and ultimate reuse or recycling of the product[6]

These interrelated activities are diagrammed in Figure 1-9. This figure places the activities necessary to design assemblies squarely within the larger set of product development activities. It shows that many of the steps, such as conversion of requirements into functions and architectures, directly define assembly-related design tasks. Furthermore, many of the decisions made while designing the assembly strongly influence how (or even if) a number of other strategically important activities will be carried out. These include how or if existing parts or assemblies will be reused in later versions of the product or shared across contemporaneous products, how the product will accommodate field upgrades, and so on.

The life cycle of the product begins with identification of a market need or a technological opportunity. In either case, these must be converted into a detailed list of customer needs and product requirements. A strategic decision must be made whether to make one product or to prepare for a family of similar products that will meet a range of needs or a variety of similar markets. If a family is contemplated, then the range of functions to be provided and the architecture of the product family must be worked out carefully in advance. The architecture describes how the product will be partitioned into subassemblies and parts, which groups of parts will perform which functions, which functions will be in each member of the family, how upgrades or technological advances can be smoothly inserted into the product and its production processes, and so on. Architecture is decided iteratively with consideration of the steps that follow: assembly structure, part design, and manufacturing. Product architecture is the subject of Chapter 14.

Once the architecture of the family has been planned, at least tentatively, the assemblies and parts of each family member can be designed. These must be described with care because typically 65% by cost are outsourced to other companies. Choice of suppliers and the degree to which they are depended upon for key aspects of the product is

---

[6]It is interesting to note that for many products, such as razors, software, jet engines, and copiers, much (if not all) of the profit to the manufacturer is earned after the product is in the customer's hands. These profits are earned through regular maintenance, upgrades, spare parts, or replenishment of consumables. The makers of cars and coffee makers, on the other hand, have not traditionally been strong players in after-markets for fuel, replacement parts, coffee, filters, and so on.

extremely important. A key step in this process is to identify the KCs that the assembly must deliver. Since segments of the chains in delivery of each KC may be made by different suppliers, each supplier must understand the chain and its role in it. KCs are the subject of Chapter 2.

Plans must then be prepared for the fabrication and inspection of each part so that it will be able to perform its function. The same is true of each subassembly and the final product. If important parts or subassemblies are made by other companies or other divisions of the same company, then the final manufacturer must compose very careful specifications for these organizations to follow and must take elaborate steps to ensure that the specifications are met. Design of assemblies to meet functional and physical specifications is the subject of Chapters 4 through 8.

Manufacturing and assembly processes must be prepared. These must not only be able to generate parts that meet the specifications and assemble them properly, but must also be capable of producing at the rate required to meet demand. Different fabrication and assembly processes are appropriate for different production rates. For example, low-volume fabrication might be done by machining while high-volume fabrication could be done by molding or casting. Similarly, low-volume assembly is typically done by people while high-volume assembly is done by machines, but only if the parts and the final assembly are smaller than about 10 cm on a side. These tradeoffs are discussed in Chapters 16 and 17.

Once it is built, the product enters the distribution chain, where dealers, service people, or customers may add parts to customize it or replace ones that need replenishment or repair. When its life is over, it is likely to be recycled, with some parts removed for reuse and the rest separated into like materials and recycled. Thus assembly or disassembly is an ongoing activity throughout the product's life. Typically, it is not possible to perfectly accommodate all of these needs and constraints.

1.D. ASSEMBLING A PRODUCT

The day-to-day process of assembling mechanical products typically involves a long chain or network of activities and actors. A sketch of this process appears in Figure 1-10. The main activities of assembly are

- Marshaling parts in the correct quantity and sequence
- Transporting parts and partially assembled items
- Presenting parts or subassemblies to the assembly workstations
- Mating parts or subassemblies to other assemblies
- Inspecting to confirm correct assembly
- Testing to confirm correct function
- Documentation of the process's operation

Marshaling is a logistic function which may be performed according to one of many strategies. These strategies are based on estimates of work schedules, the planned production of various product types, and lists of the parts needed for each type of assembly. Two types of strategies are generally used, the push type and the pull type.

The push type operates on the basis of a planned production schedule of anticipated final need for finished assemblies. Fabrication or purchase of parts and scheduling of production facilities is initiated on the basis of estimated

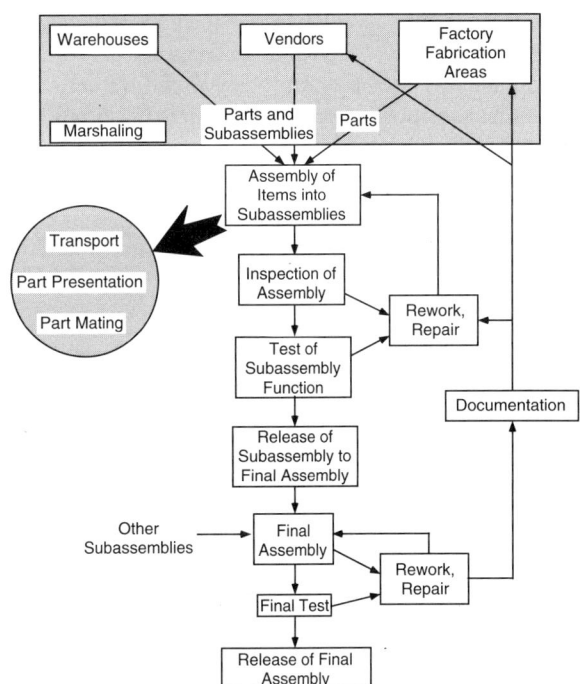

FIGURE 1-10. The Main Processes of Assembly.

lead times (the time between placement of an order and receipt of the item) for producing or obtaining the parts. Since these lead times are typically weeks or months long, push strategies usually cover long time spans. MRP (material requirements planning) is a generic name for many such strategies.

Pull strategies operate in reverse compared to push strategies. The pull method starts with anticipated demand or orders in hand for finished assemblies. As these orders are filled, replacement items are ordered from the next upstream processes. Orders progress upstream in a similar fashion, pulling production downstream as needed. An order is triggered when a rather small safety stock level is reached, often a few hours' or days' work. Due to the much shorter time spans over which such strategies operate, they are called *just-in-time* (JIT) methods[7] ([Monden]).

Transport is the short-term logistic implementation of marshaling. That is, transport accomplishes the actual carrying of parts or assemblies between stations or work areas. The major options for transport are discussed in Chapter 16.

Part presentation takes parts from the transporter and places and orients them so that assembly can occur with only minor adjustments. A person, assembly gripper, tool, or robot may acquire the part either directly from the transporter or from a part feeder (see Chapter 17) and carries it to a point very near where mating takes place.

Part mating is the actual process of fitting parts together. Mechanical mates include peg–hole insertions, interference or force fits (for example, peg larger than hole), insertion of electronic components into sockets or circuit boards, mating of gears, insertion of threaded fasteners, compliant mates like snap fits, and other similar mechanical mates. The physics of such mates is discussed in Chapters 10 and 11. These chapters show how accurately parts must be made and how precisely they must be presented to each other in order that assembly will be successful.

Joining accompanies mating and usually involves fastening in some way. Screws, rivets, adhesive bonding, welding, soldering, crimping, staking, and ultrasonic bonding are examples. Each of these has important implications for assembly, repair, and upgrading, especially decisions regarding use of reversible versus irreversible joining methods.

Inspecting usually involves determining that an assembly operation has been completed correctly. One may check the tightness of a screw or freedom of motion of a shaft in its bearings. This is different from testing, where the issue is to determine that a subassembly functions correctly. The distinction between inspecting and testing is that the latter may be directly related to a functional specification on the assembly or product. Often special test equipment is needed.

In addition to the above direct operations, an important indirect operation is documentation. Assemblers, inspectors, or testers record data such as test results, number of correct assemblies, reasons for failure of tests or assembly equipment, and so on. These documentary data can be crucial to the correct functioning of the factory over the long term, providing the ability to trace problems back to their causes, maintaining control of the processes, and permitting improvement of the factory's performance.

1.E. HISTORY AND PRESENT STATUS OF ASSEMBLY

1.E.1. History

The traditional unit processes of manufacturing have been studied for hundreds of years, and some, like investment casting, are ancient. Today we have well-tested mathematical and computer models for metal cutting, mold filling, sheet metal stamping, and so on. New processes and materials are being introduced constantly, and models of these are naturally less well-developed. Examples are composite parts made of polymer, glass or metal fibers and resins, powder metallurgy, freeform fabrication using polymers, layers, or direct deposition of materials, and a wide range of semiconductor fabrication processes.

Assembly is also an ancient process, and until very recently it was accomplished exclusively by human hands, possibly with some aids to overcome large weight. The evolution of assembly until around the 1940s was largely characterized by increasing efficiency and speed via industrial organization and time studies, usually based on the principle of division of labor. In the nineteenth century, assembly required skilled fitters who adjusted the shapes of parts so that they would assemble. Henry Ford realized that mass production in huge quantities could not be achieved until time-consuming fitting operations were

---

[7]In the Denso panel meter, the parts are made by a push strategy while the assemblies are made by a pull strategy.

eliminated. This he accomplished by increasing the accuracy and repeatability of fabrication machinery. He then organized his assembly workers in teams, each of which built a large subassembly such as a dashboard. This proved too slow, however, because the workers spent too much time getting parts. So he organized the people and parts into an assembly line and brought the work and the parts to the people. At this point, production capacity exploded and the mass production age was born.

Automatic assembly machines were designed early in the twentieth century to perform rapid assembly of simple items. Among the first targets were cigarettes, but today many small products such as ballpoint pens, spray bottle valves, small motors, razor blade cartridges, and other similar items are assembled by the millions on automatic machines. Each workstation on such a machine performs a simple operation like inserting one part or one screw. Basic tests can also be done. This technology is often referred to as "fixed automation" because most such machines are designed to make one product (perhaps in several variants) and are difficult or uneconomical to convert to a different product.

In the 1970s, interest in robot assembly arose, following the success of robot spot welding of car bodies in the late 1960s. Assembly robots were barely out of the university research labs as this effort began. High hopes were placed on robots combined with vision systems, force and touch sensors, powerful computers, and artificial intelligence. For reasons that are discussed below, these hopes were somewhat misplaced. Today there is a place for all assembly technologies: manual, fixed automation, and robots. Chapters 16–18 discuss both the technical and economic features of these technologies, along with criteria for selecting them.

1.E.2. Manual and Automatic Assembly

People cannot scratch chips off a steel casting, so machines for such purposes were needed from the beginning of manufacturing. However, people can always accomplish typical assembly tasks, within limits imposed by weight of parts, required cleanliness in such domains as semiconductors, required dexterity, and so on. So assembly machines were not really needed; they were simply preferable in some cases.

Simple assembly tasks can be done by simple machines, and many products, especially small ones, can be designed to be simple and easy for a machine to assemble.

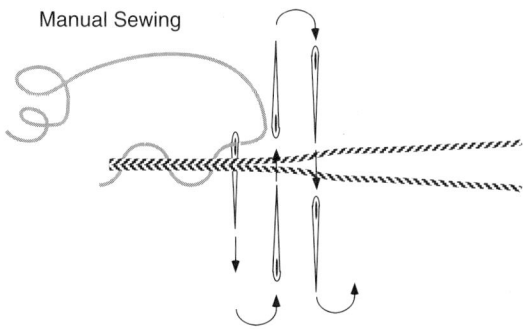

Manual Sewing

FIGURE 1-11. Manual Sewing. ([Whitney 1986]. Reprinted by permission of Harvard Business Review. Copyright © 1986 Harvard Business School Publishing Corp. All rights reserved.)

However, assembly processes can be surprisingly complex, and assembly as an activity often includes many other actions, such as quickly checking that a part is suitable for use. The required dexterity is often greater than one would think just based on watching skilled assembly workers doing their tasks. Such people repeat their tasks thousands or millions of times, and within hours or days they become *very good* at their work.[8] People "just do it," applying a lifetime of experience and practice at using their hands, eyes, and brains. Rarely can they explain how they do it, and rarely are they asked. Assembly thus remains poorly understood. The challenge for the designer of an assembly machine or robot is to accomplish all the things people do (*not* necessarily the same way) or to eliminate the need to do so. An example of this challenge is presented next.

1.E.2.a. Sewing—Contrasting Manual and Mechanized Ways of Doing the Same Thing

Elias Howe invented the sewing machine after watching his mother sew. He studied her method intensely. He then proceeded to invent a new way of sewing that hardly resembled manual sewing at all. He did not attempt to reproduce the manual method.

Figure 1-11 shows manual sewing. In this process, there is one thread that is passed back and forth through the pieces of material by the needle. This process binds the two pieces of material together.

---

[8]Practiced line workers often delight in inviting an unsuspecting visitor to try their task. The visitor is usually embarrassed by his clumsiness and should be a good sport and learn the lesson.

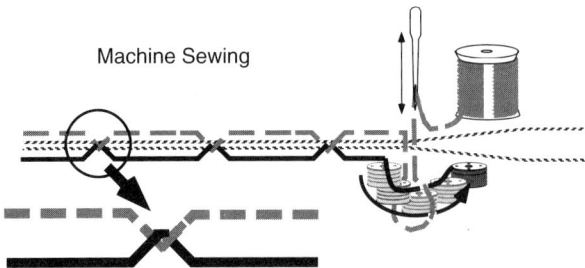

FIGURE 1-12. Machine Sewing. ([Whitney 1986]. Reprinted by permission of Harvard Business Review. Copyright © 1986 Harvard Business School Publishing Corp. All rights reserved.)

TABLE 1-2. Comparison of Manual and Machine Sewing Methods

Characteristic	Manual	Machine
Number of "hands"	Two	One
Number of threads	One	Two
Grasp of needle	Repeated grasp and ungrasp	Never ungrasp
Location of eye	Rear of needle	Tip of needle
Needle movement	Passes through cloth flips 180°	Point penetrates cloth but needle does not pass through
		Needle does not flip
Joining method	One thread passes through cloth repeatedly	Two threads interlock but do not pass through cloth

Figure 1-12 shows the principle of machine sewing. Here, there are two threads which are arranged in a twist that captures the material between them. The needle does not lead the thread through the cloth but instead forms a loop of the upper thread below the material. This loop is caught by a rotating hook that draws it around a bobbin that carries the lower thread. This bobbin normally rests in a cup and has no spindle passing through it. Absence of the spindle makes it possible to pass the loop under the bobbin. Kinematically, what happens is the same as if the bobbin were lifted out of the cup, tossed through the loop, and then returned to the cup. An arm carrying the upper thread then snaps upward, pulling the loop tight. Clearly this process is totally unlike manual sewing (see Table 1-2).

1.E.3. Robotic Assembly

Nowhere has the challenge of automating assembly been as stiff as in the domain of robot assembly. This is partly due to expectations, as noted above. In the 1970s, robots were intended to duplicate a number of human skills deemed necessary for factory automation:

- Behave flexibly, including switching from one task or product to another
- Use their senses to support flexible and dexterous behavior
- React to the unexpected and recover from errors

Quite a few of these skills were deemed "necessary" because people had them, but not because a full economic and technical analysis had shown them to be necessary. Providing them in a robot turned out to be too expensive, either in first cost or in maintenance, in most applications where people were a viable alternative. In addition, some aspects of flexible, sensor-driven adaptability were intended to fix errors, such as lack of repeatability in the parts or reliability of the robots, that should not have occurred in the first place. Careful attention to part and equipment quality eliminates the need in most cases.

In general, machines can't "just do it." They cannot call upon stored skills as people can. They have to be designed explicitly to do each individual action. Furthermore, we probably do not understand, much less appreciate, what people are doing when they assemble parts. Their actions include sensing, moving, and judging in many dimensions at once at high speed. (Sometimes they do the wrong thing, again using to the best of their ability all the above skills.) Until or unless we can dissect and completely describe in mathematical and engineering terms every motion, force, visual analysis, and closed-loop control action performed by a person, we will be unable to recreate them in a robotic machine. Even if it does become possible, it may not be desirable economically and there may be an easier way for a machine to do the same thing.

A sure tipoff that a task is difficult is the case where only one person can do it rapidly or repeatably. If that person is absent, the line slows down or even stops as replacements attempt to take over. As a general rule, if we can't explain a task to another person, we are unlikely to be able to explain it to a machine.

The alternative, also mentioned above, is to alter the design of the product or the assembly tasks themselves, until a simpler machine can accomplish them reliably and repeatably. In many cases, this machine will be simpler and cheaper than a fully capable robot. Sometimes a person can do them well or easily enough that a machine is no longer worth making. Some examples of this process appear in Chapter 15.

Even in such situations, however, a highly ingenious robot solution may be necessary if manual assembly is deemed inappropriate. Notable examples include Sony robots able to assemble Walkmen, camcorders, and Polaroid cameras. Each of these contains a variety of complex tasks, including placing springs, meshing gears, and looping rubber belts around multiple pulleys. Compared to people performing such intricate tasks on precision products, the robots commit fewer errors, damage fewer parts, perform the tasks more repeatably, and thus deliver more consistent product quality. They are impressive to watch.

People who attempt to perform such tasks at high speed may perform inconsistently, damage parts, or develop stress-related injuries.

In summary, we can say that highly dexterous assembly robots are possible and admirable, but not often necessary. The correct attitude, repeated many times in this book, is to understand the technical and economic requirements and *then* to decide what is the most appropriate design for the product and the associated manufacturing and assembly methods. A good rule of thumb is to design for automatic assembly, assume a clumsy sensorless machine, and thereby simplify everything. This will usually improve the ease and success of any kind of assembly, manual or mechanized.

1.E.4. Robotics as a Driver

The effort to develop dexterous assembly robots and to understand their technical requirements, while it did not replace manual assembly, nevertheless raised and confronted a number of key issues that are still important and largely unresolved. They are generic issues in the sense that they will always arise any time technical alternatives are available:

- **Flexibility** versus *generality:* How much effort should be directed toward a solution that can be altered later in response to a possible change in conditions? This choice is difficult because, in general, flexible systems suffer from reduced efficiency, due in part to the time and cost needed for changeover. If the possible change turns out not to be necessary, then the extra cost of the flexible system was wasted, or it must be regarded as if it were an insurance premium.

- **Generality** versus *specificity:* How much effort should be directed toward a solution that can be reused rather than having to design a new solution from scratch each time? The difficulty here is that the general system may be ready sooner than the one designed to suit, but the latter will probably do a better job.

- **Responsiveness** or **adaptation** versus *preplanning* to avoid the need for adaptation. Here the tradeoff is usually cost. It may be easier to adapt to small variations than to design them out.

- **Structure** versus *lack of structure.* "Structure" means preparation, arranging things in neat rows, eliminating the unexpected, and so on, while "lack of structure" means the opposite. People can perform well in new circumstances like emergencies, but they have years of training and ability, some of it developed over millions of years. Some enterprises, such as factories, perform better when they operate repeatably, so relying on adaptable people and machines to make up for lack of structure in the factory represents the wrong mind-set.

In the list above, the words in **boldface** represent the characteristics or enablers of typical machines, whereas those in *italics* represent the presumed and presumably desirable properties of people that robots were intended to duplicate. As researchers and developers attempted to create robots with these properties, the technical and financial difficulties they ran into made the contrasts with the properties of machines interesting. We learned that the properties of people were not well understood, and we further discovered after substituting robots that the true roles of people in many factory operations were barely documented or understood. We also discovered that some of the properties of machines are desirable after all.[9]

---

[9] A story that illuminates this point goes as follows: For years, two venerable ladies inspected fiber emerging from a synthetic fiber spinning machine. An enterprising young engineer decided to replace the ladies with a modern laser inspection system, and the ladies retired. The laser immediately began rejecting about half the fiber produced. Since the fiber-making process had not changed, it became necessary to rehire the ladies and look carefully while they worked. It emerged that they were observing occasional bits of dirt on the fiber, licking their fingers, and gently rubbing the dirt off. So "inspecting" was an inadequate description of what they were doing. They were "cleaning" the fiber as well, no doubt leaving traces of saliva, skin oils, makeup, or whatever else may have been on their fingers at the time.

While the author has betrayed his bias in the case of structure, in general each case must be evaluated anew each time it arises. Researchers and practitioners return to these issues perennially and presumably will continue to do so.

1.E.5. Current Status and Challenges in Assembly

The challenges facing assembly today can be characterized as technical, economic, and managerial. They include product design and manufacture of those products.

From the technical point of view, products are getting more complex. They contain more kinds of technologies, including electronics, mechanics, optics, and, occasionally, chemistry. They are also becoming smaller. Examples include instant cameras, cellular telephones, notebook computers, and small weapons and assistive devices for soldiers. Products are also being made in a wide variety of versions and styles. Variety is driven by globalization and recognition of shades of consumer taste. More customers and more different customer needs are being accommodated. Several of these factors drive the economic and managerial challenges.

Economic challenges arise because consumers want higher quality and choice at lower price. Companies that can deliver in these ways win out over those that cannot. Low-cost assembly is usually found in low-wage regions of the world, so products and product design specifications travel long distances between designer, maker, seller, and buyer. Thus companies choose sources of materials, parts, and labor to reduce their costs and then must manage the logistics and coordination issues that inevitably arise. The entire supply chain must be managed to create balanced flows of work and uniform standards of quality. When customer tastes change, demand for one version may rise while another may fall. These shifts must be communicated quickly down the supply chain or else unwanted items will be made that later must be scrapped or sold at a loss.[10]

The wide variety of technologies and specialized materials force companies to rely on suppliers for design knowledge, proprietary materials, specialized production methods, or even the parts and subassemblies themselves. This fact, aside from the search for low cost, is another reason why supply chains arise.

1.F. ASSEMBLIES ARE SYSTEMS

Assemblies are challenging from an engineering and manufacturing point of view because they are systems. This makes them challenging intellectually as well. Designing them has many features in common with system engineering ([Rechtin]). In both domains:

- It is desirable to work top-down from requirements to realizations, from large systems to smaller subsystems and finally to parts.
- The design process is organized to mimic this so-called "flowdown" by carefully defining element boundaries and interfaces and paying careful attention to interface management.
- The structure and diagram of the system or the assembly is a hierarchical decomposition tree with larger and more composed elements at the top and smaller, simpler elements below.
- A number of elements work together to achieve desired performance.

- Complex patterns of behavior exist in which a symptom may appear "here" but the cause may be "way over there," making diagnosis difficult.[11]

It is not clear that a top-down approach is always the best one, either in system engineering in general or in assemblies. However, it is always helpful to keep the above properties in mind because the system nature of assemblies is unavoidable. Many serious errors in systems, products, and assemblies can be traced to improper, inappropriate, or incomplete definition and follow-up of requirements, as well as to poor definition and management of interfaces.

---

[10] A New York high fashion designer said, "If I manufacture here in New York, I can turn a sketch into dresses on the rack in a week. If I manufacture in Mongolia, it takes six months."

[11] A typical mistake, a consequence of ignoring the system nature of assemblies, is to blame the last part added. One can see the *problem* at this point, but the *cause* may be due to a completely different part, or more likely several parts or previous assembly operations acting together.

1.G. CHAPTER SUMMARY

Assembly is the process with the greatest potential to improve product development methods and manufacturing strategy. Design and production of assemblies is inherently integrative. Most technically and commercially important physical products are assemblies. Assembly considerations link all levels of product development from customer requirements to supply chain design to management of variety and customization. An assembly-driven product realization process can greatly improve a company's prospects for the success of its products. This aspect of assembly is poorly understood and not often exploited.

Assembly is also the least understood of manufacturing processes because people have always done it, people cannot explain how they do it, it is complex at the micro level, it is complex at the macro level, and serious study of it began only recently. To a surprising degree, people are still necessary to manage complex design, manufacturing, and assembly activities, despite ongoing (and occasionally successful) efforts to eliminate them. As technology and products advance and become more complex, people will continue to be necessary, especially where system-level behavior must be understood and managed.

1.H. PROBLEMS AND THOUGHT QUESTIONS

1. Go to a store with lots of brands and models of products. Select a small product, such as a hair dryer, razor, radio, or desktop printer. Examine the different brands or models and note differences that you can see, such as materials choices, fasteners, number of parts, and method of handling wires and switches.

2. If cost is no object, buy a few of these items, take them apart, and continue your comparisons. Further questions below suggest things to think about.

3. For such a product, list a number of design considerations that affect its assembly in the factory or disassembly after sale. How have the product's designers addressed these issues? To answer this question, you need to take note of all the materials and fastening methods used as well as all the different ways the product can be used or misused.

4. For the same product, write down every single action necessary to put it together. Include inspection, assembly, lubrication, cleaning, functional testing, and packaging. Don't forget such support activities as opening boxes that parts arrive in, as well as folding up and recycling the empty boxes.

5. For this same product, identify one or two of the major functions it is designed to perform. List all the parts that participate in delivering each function, noting those parts that participate in more than one function.

6. Look the product up in a magazine such as *Consumer Reports* and see what is said about the pros and cons of the different models tested.

7. Discuss the kinds of operating or quality problems the stapler discussed in Section 1.B.1 would have if errors in part manufacture caused misalignments along the Y axis. What about the Z axis?

8. Using the drawings of the stapler parts in Figure 1-2, sketch the carrier and identify all the important assembly and operating features. Measure these relationships on a real stapler and draw the relevant dimensions on your sketch. Do the same for the anvil.

9. The discussion of the stapler did not mention the pusher. Are there any KCs associated with it? What are they? Make a liaison diagram that includes the pusher and identify the KCs on it.

10. How much do you think a hypothetical robot would be worth if it could really do all the things a person can do? Compare this with what you think it cost to bring you into the world, raise you to adulthood, and pay for your healthcare and schooling, including fairly accounting for the value of your parents' time.

11. Use publicly available information about manufacturing companies, such as company annual reports, to estimate how many hourly employees they have. Assume that most are involved in assembly. Estimate how many people are employed performing assembly in an industry such as automobiles or aircraft. Assume that the final assemblers mainly assemble subassemblies of 5 to 10 parts each and that suppliers build these subassemblies. How many people might be involved in the supply chain for these products?

1.I. FURTHER READING

[Monden] Monden, Y., *Toyota Production System*, 3rd ed., Norcross, GA: Engineering and Management Press, 1998. This is the definitive reference on the TPS.

[Nevins and Whitney] Nevins, J. L., and Whitney, D. E., editors, *Concurrent Design of Products and Processes*, New York: McGraw-Hill, 1989 (out of print, copies available from the editors). This book uses assembly as the vehicle for describing concurrent design, meaning close collaboration of engineers, marketers, finance people, and manufacturing people during product design.

[Rechtin] Rechtin, E., *System Architecting: Creating and Building Complex Systems,* Englewood Cliffs, NJ: Prentice-Hall, 1991. This book provides experience-based advice on the considerations necessary for creating the architecture of a competent system.

[Simunovic] Simunovic, S., "Task Descriptors for Automated Assembly," S.M. thesis, MIT Mechanical Engineering Department, January 1976.

[Ulrich and Eppinger] Ulrich, K. T., and Eppinger, S. D., *Product Design and Development*, New York: McGraw-Hill, 2000. Ulrich and Eppinger present a systematic process for design of new products driven by customer requirements. Equal emphasis is placed on engineering and managerial issues.

[Whitney 1986] Whitney, D. E., "Real Robots Do Need Jigs," *Harvard Business Review,* May–June, pp. 110–116, 1986.

[Whitney 1993] Whitney, D. E., "Nippondenso Co. Ltd.: A Case Study of Strategic Product Design," *Research in Engineering Design,* vol. 5, December, pp. 1–20, 1993.

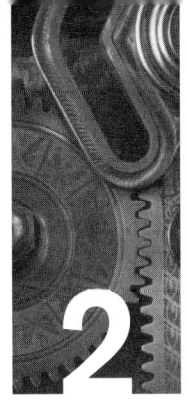

2 ASSEMBLY REQUIREMENTS AND KEY CHARACTERISTICS

"The customer looks at the gap. But it's an empty space and we don't assign anyone to manage empty spaces."

2.A. PROLOG

This chapter begins the first major part of the book. This section, Chapters 2–8, deals with design principles for assemblies. We will approach this in two stages. In the first stage we will deal with products as assemblies of parts, showing how to express top-level requirements in terms of relationships between parts and representing those relationships in a computer. Then we will present a method for creating nominal assembly designs (that is, designs that are assumed free of variation) that meet several rigorous criteria for correctness. Methods for analyzing the effects of variation will be added to the nominal design method in a mathematically consistent way.

Following emerging industry practice, we will refer to critical assembly-level properties of the product as key characteristics (KCs). In different companies, significant characteristics, key product characteristics, functionally important topics, critical-to-function, and critical to quality are all the same thing.

This chapter provides the motivation for KCs and gives a number of examples. As explained in Section 2.E, KCs are intimately related to tolerances and variation. However, before we can deal with tolerances in a systematic way, we need to deal with several other issues. Thus a discussion of tolerances and variation is deferred until Chapters 5 and 6.

Chapter 3 presents the mathematics needed to represent the position and orientation of parts relative to each other in space, as well as how to use the same mathematical representation to express variations in those positions and orientations. Chapter 4 introduces the idea of mechanical constraint and describes how mating features on parts convey constraint from one part to another in order to locate them with respect to each other. Chapters 5 and 6 present the mathematics necessary to describe variation in the position and orientation of these features on parts, as well as to predict the resulting variations in the assembly. Chapter 7 shows how to generate all the feasible assembly sequences for a product whose parts are joined by mating features. Chapter 8 brings all these ideas together to present a design process for assemblies.

2.B. PRODUCT REQUIREMENTS AND TOP-DOWN DESIGN

In both product development and more specifically engineering design, the main task is to develop a product that meets the requirements. Requirements come from many sources, including the following:

- What customers and the market are looking for (functions, appearance, price, and so on)
- What government regulations require (environmental friendliness, absence of toxicity, specific safety rules, and so on)

- What company standards demand (use of particular design methods, materials, off-the-shelf parts, and so on)
- "Given" properties of a high-quality product (reliability, durability, safety, fitness for use, and so on)

In this book, we are principally concerned with performance and quality that arise from, or are "delivered by," mechanical assemblies. For this reason, we will focus on the assembly aspects of the product and look for ways to express desired properties of the assembly that deliver the required performance and quality. These properties are the KCs and are expressible as geometric relationships between the parts in space, usually in the form of positions or angles, occasionally as gaps. These requirements are typically expressed as a range that includes the desired value. Our approach to designing assemblies will be to determine first how to achieve the desired value and then to ensure that it remains within the allowed range. If this is achieved, then the KC will be said to have been "delivered."[1] This process is called top-down design.

Top-down is not the only way to approach design. Some engineers are more comfortable taking a bottom-up approach in which they start with familiar parts and work them together, via some trial and error, into a feasible concept. Several forces contribute to a bottom-up approach.

One is the allure of computer-aided design (CAD) systems, which are much better equipped to support detailed and precise design of individual parts than a top-down decomposition of rough sketches. Before CAD existed, design followed a top-down process in which the most skilled person, a layout man, put down the basic boundaries and centerlines of a concept on blank paper. Gradually a layout emerged. Detail men, the least experienced in the profession, were assigned to design each part, providing detailed geometry, dimensions, and tolerances. A more experienced person took these detail designs and built up an assembly drawing, while a checker looked for errors and interferences by adding up all the dimensions and tolerances. Present-day CAD systems are just beginning to support representation of assemblies. The layout process is hardly supported at all. One of the goals of this book is to provide a mathematical basis on which CAD systems can support a top-down design process for assemblies.

Another factor that leads to a bottom-up approach is the desire of companies to save time and money by reusing existing designs of parts and subassemblies. The savings can be impressive: Designs, tools, equipment, process and test plans, suppliers, along with their specifications and contracts, can all be reused, and the time to redo and verify them all can be saved. But this means that any top-down process has to meet the existing parts coming up from the bottom so that a consistent design results. This design is usually a compromise between novelty, optimal performance, lower cost, and faster time to market.

Third, and most difficult, a top-down process is very challenging intellectually. It requires seeing ahead at each stage of the process, imagining subassemblies and parts before they are known in any detail. Errors in decomposition or in anticipation of how parts and subassemblies will perform together can be quite costly. As the design process advances, several branches of the decomposition tree are being worked on at once, based on common assumptions. If one branch runs into a dead end, it may invalidate assumptions that are important to other branches. These branches must then back up and start over.

To take an example, a top-level requirement for a car may be to have a "smooth ride." Later on, an engineer must set the requirements for a piece of the suspension linkage. Which link design will be the smoothest? One approach is to build several and test them.[2] But this requires design and assembly of many parts, and there may not be time. Often the engineer selects an existing link and modifies it as best he can. Thus top-down morphs into bottom-up.

Nevertheless, we will examine the top-down approach on the assumption that it has the best chance of meeting the performance requirements. The realities of cost and schedule then become constraints that must be included in any design decisions.

2.C. THE CHAIN OF DELIVERY OF QUALITY

Let us return for a moment to a topic first raised in Chapter 1, namely the idea that many parts work together to deliver performance and quality in an assembled product. In Figure 2-1 we show a modified version of Figure 1-8. The liaison diagram contains all the main parts in the front

end and shows the key characteristics for the hood and both front fenders. The assembly fixtures are omitted for simplicity.

This figure is our first step in generating a method for designing assemblies to meet top-level requirements. It

---

[1] In the academic literature on KCs, some authors use the term KC to refer only to the nominal value. In this book, a KC comprises both the nominal and its allowed range.

[2] This strategy is part of an approach called "Set-Based Design." This approach keeps several consistent alternatives in play until late in the process. See [Ward et al.] and [Liker et al.].

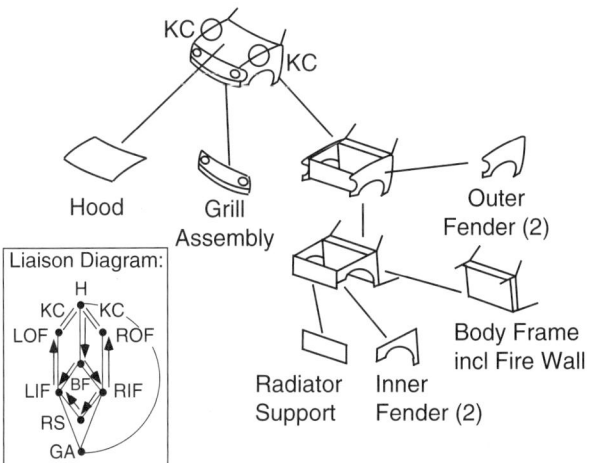

FIGURE 2-1. The "Chain of Delivery of Quality." The drawing at the top shows the assembled parts of the front end of a car. The KCs (marked by circles) require equal size gaps between the car hood (H) and the left and right outer fenders (LOF and ROF). Note that these parts do not assemble directly to each other. The *liaison diagram* ([Bourjault]) depicts the chain of parts leading from one side of the gap to the other. Each part in the liaision diagram is a dot and each part-to-part relationship is a line. The two KCs are represented by parallel lines. The arrows leading from the hood to the two outer fenders pass through the intermediate parts in the chain. At the ends of the chain are the KCs. (Drawn by N. Soman and M. Chang.)

contains a diagramming method and a set of symbols with which we can describe the requirements compactly. Later we will add more symbols and underlying methods that will permit us to capture our strategy, our *design intent,* for achieving the KCs and explaining that strategy to others.

The consequences of failing to meet the requirements can range from field failure to disappointing the customer. Companies routinely pay huge amounts to see that

the product's requirements are met. These costs are incurred in the process of performing calculations and validation experiments, designing equipment, inspecting parts and assemblies, reworking them to remove errors in the factory, recalling them from the field, or even in legal penalties.

In some cases it will be obvious what the KCs should be, while in others it may not be, or may be a subject of debate among the engineers. We take up this topic next.

2.D. KEY CHARACTERISTICS

A useful definition of a KC is provided by [Thornton 1999]:

> Key characteristics are the product, subassembly, part, and process features whose variation from nominal significantly impacts the final cost, performance [including the customer's perception of quality], or safety of a product. Special control should be applied to those KCs if the cost of variation justifies the cost of control.

This definition is similar to those of many companies and was adopted as a reaction to drawings that contained hundreds or thousands of dimensions and tolerances. Certainly, people said, some of these must be more important than others. In an attempt to reduce the number of dimensions and tolerances and, in some cases, to take advantage of the ability of solid model CAD systems to represent complete nominal geometry, KCs were adopted to focus attention on those dimensions that were critical, affected a variation-sensitive characteristic, and were worth controlling.

All elements of the definition must be present in order for a product property to qualify as a KC. If it is not critical, it is not a KC. Similarly, if it does not affect a variation-sensitive characteristic, it should not be a KC. Finally, if the cost to mitigate it is not rewarded, then it should not be controlled. For example, parameters that are deemed to be under control in a statistical sense (as determined by the methods of statistical process control (SPC) described in Chapter 5) need not be KCs. At most they need to be inspected on a sample basis to make sure that they are still in control.

In high-volume production industries such as automobiles, it is customary to use statistical process control because statistically meaningful quantities of measurements can be obtained. During the early days of production of a new product, processes may not be under control and thus several critical parameters may be deemed KCs. As the production process matures, these parameters may come under control and then can be stricken from the KC list.

In low-volume industries such as aircraft, statistically significant quantities of data may never be achieved. In such cases, it may be necessary to keep critical parameters on the KC list indefinitely. Often this implies 100% inspection. The result is a large amount of data gathering and processing.

In both high- and low-volume industries, much of this data gathering is done by suppliers because they make many of the parts and even some of the subassemblies.

Thus KCs can be an important means of communication of quality issues up and down the supply chain.

Although it is clear that KCs, tolerances, and variation are intimately linked, we will defer a mathematical discussion of tolerances and variation propagation until Chapters 5 and 6. Here we will discuss the general strategy of variation risk management. This strategy comprises identification, verification, and mitigation.

2.E. VARIATION RISK MANAGEMENT[3]

Engineering design of an assembly includes three steps:

- Nominal design, that is, the determination of the ideal relative locations and orientations of parts sufficient to deliver the KCs
- Variation design, that is, the determination of how much variation in those locations and orientations can be tolerated and still achieve the KCs
- Process design, that is, the determination of fabrication and assembly processes that will contribute no more than the tolerable variation (possibly including some loosening of the original tolerances because no capable process exists or none that is economical exists)

Nominal design defines mutual positioning constraints by parts while variation design determines how much variation in each constraint relationship can be tolerated. Process design comprises methods for fabricating each part as well as methods for assembling them so that the final assembly varies by no more than the assembly level tolerances. Even for simple assemblies, these steps can be very complex and difficult. Tying them all together to achieve the KCs is called *variation risk management* (VRM).

VRM cannot be approached as a purely technical process because, as implied above, the desired tolerances may not be achievable in an affordable way. Thus every engineering decision about a material, a dimension, a tolerance, a supplier, a process, and so on, must be built on a business case that the cost is justified by the performance

it brings or the risk it avoids. In general, this is extremely complex. The difficulty in deciding how to make the tradeoffs is due to the fact that in most cases there is no engineering model that will tell the engineer how much performance is degraded as variation increases. Among the few examples where this *is* possible are:

- *Hydrodynamic journal bearings.* These support a load that can be calculated from the rotation speed and the properties of the fluid trapped between a journal and a bearing of slightly different radii. We can calculate rather precisely the loss of load-carrying capacity if the radius of cither element is not the required value.
- *Optical systems.* In this case the KC might be sharpness of focus. We can calculate rather precisely how much the beam will diverge if the lenses are not the required shape or separated by the required distance.
- *Pumping efficiency of a compressor.* Here, loss of efficiency can be traced to, among other things, leakage due to running clearances between the rotor and stator. The amount of leakage can be calculated rather precisely given the clearance and the fluid viscosity, so the loss can be related precisely to the amount by which the clearances exceed those desired.

In general, the tradeoffs can be expressed schematically in Figure 2-2.

In this figure, the basic tradeoffs are depicted schematically. Improved performance usually requires tighter tolerances, which in turn usually raises manufacturing and assembly costs: Better equipment or more skilled employees may be needed, more care and time may be needed during each process, additional process steps may be needed, and/or scrap rates may rise. If the improved performance

---

[3]The material in this section is adapted from Section 3 of "Fast and Flexible Information Exchange in the Aerospace Industry, Final Report" ([anon]). That section was written by Anna Thornton, then in the MIT Department of Mechanical Engineering. Thornton coined the term *variation risk management* ([Thornton 2003]).

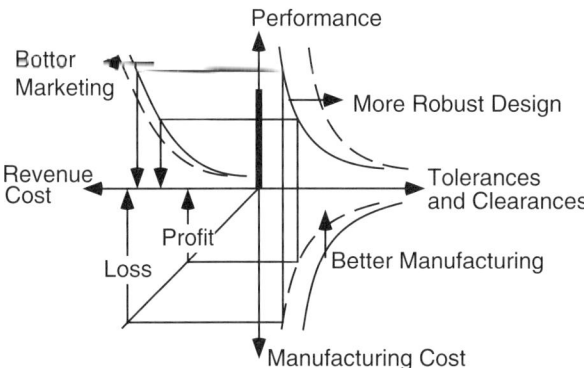

FIGURE 2-2. The Business Case for Variation Risk Management. This figure shows that performance, manufacturing cost, and profit are intimately linked. The outer circuit aspires to high performance but generates a loss. The inner circuit achieves lower performance but generates a profit. The solid bar on the performance axis represents the highest performance that still generates a profit. If the market demands better performance, then several routes to profit could be tried: better marketing, a more robust design, or better manufacturing processes.

can be sold at a higher price, the result may be a profit for the company; if not, a loss may result.

Several remedies are available. A more robust design can deliver the performance with looser tolerances. The methods of [Byrne and Taguchi] and [Phadke] are typical of robust design methods. Better manufacturing methods, including better training and practices by employees, as well as statistical process control, can achieve given variation levels at lower cost.[4] Assembly methods that exploit adjustment or selective assembly can achieve tolerances at the assembly level that are tighter than those achieved at the fabrication level at lower total cost. Better marketing can (sometimes) explain the benefits of increased performance to customers and induce them to pay more, perhaps by demonstrating savings that they will reap. Next, VRM can be used systematically throughout the product development process to align requirements, nominal designs, and tolerances in order to arrive at the most economical tolerances, processes, and process control methods. Failing all these approaches, one must try a different design concept for the product.

A summary of the steps in achieving the desired KC follows:

- Establish the desired nominal value of the KC.
- Establish the permissible variation around the nominal.
- Design the assembly and parts so that the nominal KC value will be achieved if the nominal dimensions on the parts are each achieved on the average (the nominal KC value is the statistical average of all actual KC values generated by the process, and the nominal dimension on each part is the statistical average of all actual part dimensions generated by the process).[5]
- Control variation in part dimensions so that the accumulated variation falls within the limits of the allowed variation of the assembly.

If the last two steps are achieved, then the KC is said to be delivered. The above process corresponds roughly to the steps of the Taguchi design process:

- System design
- Parameter design
- Tolerance design

It should be noted that VRM is still a work in progress and that this book is able to present only a summary of approaches in use as of its writing. This set of definitions and methods provides the baseline against which the current industry practices and research are measured. The method outlined here has been developed through observation of many organizations' processes, and it represents accumulated best practices.

2.E.1. Key Characteristics Flowdown

KC methodologies include tools and processes to identify KCs, to assess their impact on product quality, and to reduce the impact of variation. Central to KC methods is the use of *key characteristics flowdown*. A KC flowdown shows how a top-level customer requirement is decomposed into all the subassemblies, parts, and processes involved.

---

[4] At Xerox, the word *latitude* is used to describe the degree to which a design continues to perform in spite of variations in its operating environment (humidity, operator errors, etc.). This comprises a different aspect of robustness ([Bebb]).

[5] The importance of driving each process so that the process mean is the desired nominal cannot be overemphasized. This point is discussed in detail in Chapter 5.

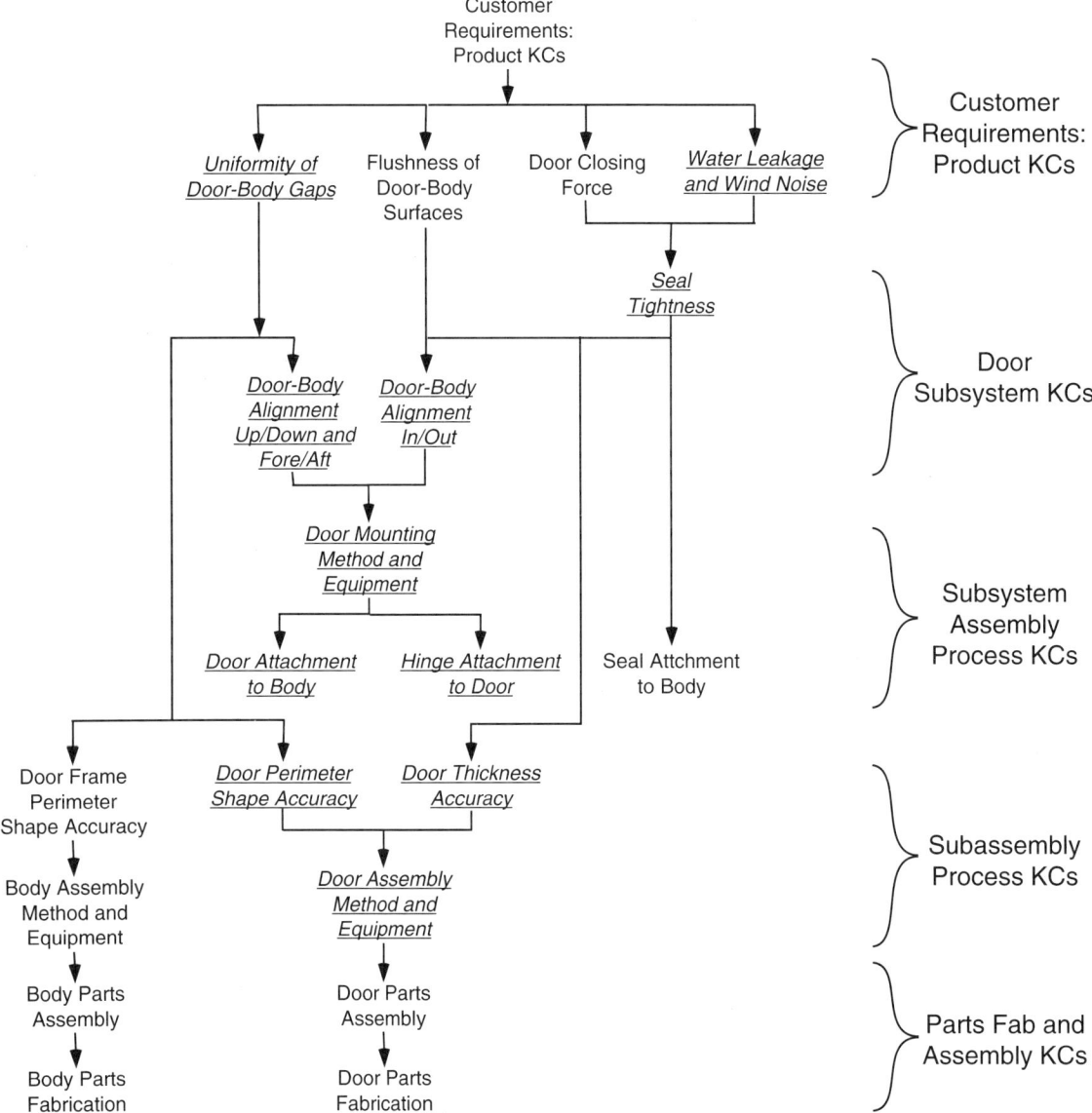

FIGURE 2-3. Key Characteristics Flowdown. The example here is drawn from the auto industry and shows how KCs describing the customer's perception of a door can be flowed down through subassemblies, to features on parts, and finally to manufacturing processes. The ones shown in underlined italic typeface are used later in this chapter. (Figure adapted from [Thornton 1999].)

Figure 2-3 shows a schematic of an example KC flowdown (for customer perception of a car door). The KCs are described in a hierarchical tree structure. At the top of the tree are system KCs—four customer-level requirements. The flowdown process involves identifying additional KCs at the assembly, process, and part level that support the top-level ones. While the top-level KCs are permanent in the sense that they represent permanent customer requirements, the lower-level subassembly, part, feature, and process KCs depend on design or process choices. Alternate subassemblies can be identified, as can alternate assembly sequences, mating features on parts, and so on. These will give rise to different lower-level KCs. These different choices for subassemblies and

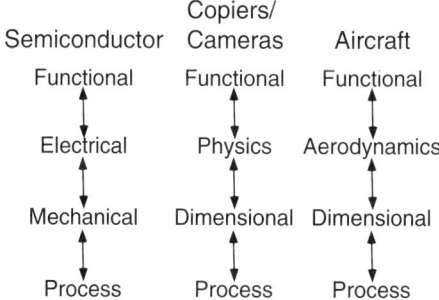

FIGURE 2-4. Typical Flowdown Characteristics for Various Industries.

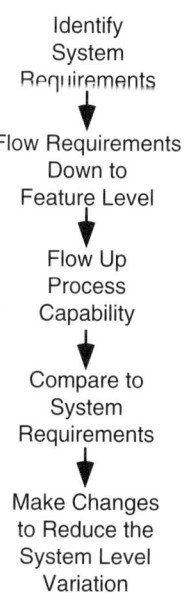

FIGURE 2-5. Ideal Process for Variation Risk Management.

part boundaries represent different architectures for the product.

The KC flowdown is made more complex because subsystem and feature KCs often contribute to more than one system KC. For example, the KCs for "uniformity of door-body gap" and "water leakage and wind noise" both depend on "door-body alignment" in such a way that improving one will degrade the other. We call this *KC conflict* and deal with it in Section 2.G. When KC conflict occurs, we encounter the need to develop priorities. KC prioritization is also dealt with later in this chapter and further in Chapter 8.

Not all KC flowdowns are purely geometric. Different industries have different KC flowdown types. Figure 2-4 shows hierarchies for three industries. For example, semiconductor industries' final system performance is based on its electrical performance, which, in turn, is based on the mechanical properties of the layers in the chip. Similarly, a camera function, such as shutter performance, is based on the physics of springs and actuators. These, in turn, are driven by dimensional and material properties of parts.

2.E.2. Ideal KC Process

Figure 2-5 shows a representation of the ideal variation risk management (VRM) process. First, system requirements that may be affected by process variation are identified. In addition, the maximum amount of variation that can be tolerated by a customer should also be specified. This is referred to as the tolerance.

For aircraft, system requirements can include maximum wing alignment variation, maximum body alignment variation, and maximum steps and gaps between skin panels. Copier requirements can include paper handing and print quality.

The second step involves flowing requirements down to the feature level to generate a KC flowdown such as Figure 2-3. This process is done in parallel with the design process. After the flowdown is completed, manufacturing process capability for the features is collected. The variation in the system requirements is then calculated by *flowing up* process capability. This usually involves deep discussions and negotiations with suppliers. Process capability flow-up can be done using a variety of tools including back-of-the-envelope calculations, such as root sum squared, tolerance stack-ups, or commercial computational methods such as Variation Simulation Analysis (VSA). The predicted performance is then compared to allowable system tolerance.[6]

If unacceptable variation is predicted, variation mitigation strategies are employed. Five strategies are used in industry: design changes, process changes, process improvement, SPC, and inspection. Based on the strategy's cost and effectiveness, the most appropriate method is selected and executed. The process in Figure 2-5 is iterated until the design is complete or the design schedule dictates that it must be released.

---

[6]A little perspective on this process is afforded by an estimate made by a variation risk manager in the auto industry: Approximately 40% of car body sheet metal tolerances can be specified unilaterally by engineering designers and flowed down to manufacturing, while the other 60% are limited by process capability in manufacturing or assembly.

This ideal process is made more complex because early in the design process there is uncertainty about system requirements, product architecture, assembly processes, flowdown, and process capability of fabrication and assembly. The cost of a product is largely determined by decisions made during the early design stages. In addition, robustness improvements are most cheaply made in the early design stages. However, cost is notoriously difficult to predict, and in many cases engineers do not understand processes well enough to navigate through all these tradeoffs successfully. Early in the design stages, when major choices of design concepts, tooling, and assembly are made, there is very little detailed knowledge about variation and its impact on the final product quality. Changes to the design concept may impact performance and/or design schedules. In addition, the effectiveness of the design changes may not be clearly quantifiable.

Figure 2-6 shows a refined approach to VRM that addresses uncertainty in the analysis and data. The first four steps are the same as in Figure 2-5. However, in addition to process capability analysis, uncertainty in predicted values is also quantified. When the fourth step is reached, the designer is faced with three possible cases. In the first case, where the impact of uncertainty is too great, a more detailed model with more accurate data should be used. In the second case, where the system is clearly robust to predicted variation, that system requirement is removed from the current KC list. In the third case, where the system is clearly not robust, changes to the design or manufacturing process must be made.

Negotiation of variation limits between those charged with delivering KCs at the top and those charged with

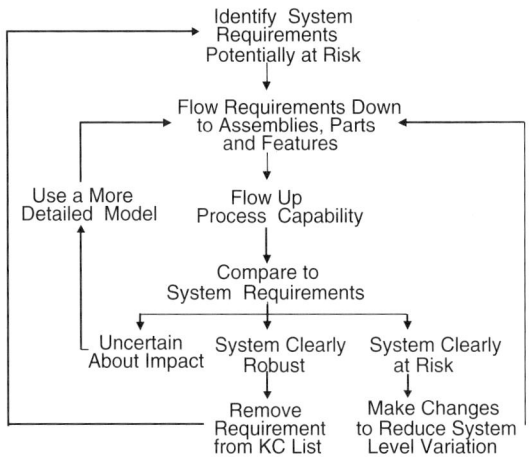

FIGURE 2-6. Level of Analysis Detail as a Function of Time.

fabricating and assembling parts from the bottom can be difficult. Each KC and resulting tolerance needs to be justified, and every effort must be exerted to ensure that the best process combinations of accuracy and cost are employed. If the designers mistrust the shop, they may set the tolerances unnecessarily tight. If the shop mistrusts the designers, they may balk at otherwise reasonable tolerances.[7]

Although KCs were introduced in part to reduce the number of requirements demanding attention to a critical few, there are counter-tendencies that cause KCs to proliferate. These tendencies include an unwillingness to prioritize, plus other organizational or cultural factors. KC proliferation only causes extra work and generation of data that all too often are not analyzed.

2.F. EXAMPLES

In this section, we illustrate the concept of KCs with two examples. The first is an optical disk drive. The other is a car door. Each one illustrates a chain of delivery of important characteristics or dimensions through a chain or chains of parts.

2.F.1. Optical Disk Drive

An optical storage disk stores digital data in the form of pits etched or pressed into its surface. Reading is done optically, as shown in Figure 2-7. The main KC is the cost of storing the data, expressed as cost/megabyte. Additional KCs are size and weight, each important for different

applications such as use in portable CD players or portable computers.

The right side of Figure 2-7 includes arrows that trace the path around the data reading loop. The elements in this

---

[7]A story about gas turbine blades illustrates this situation well. The designers wanted a tolerance to be 0.005" but thought the shop was so unskilled that they put 0.003" on the drawing. "At least we will get a few parts that meet 0.005." The shop tried as hard as it could but, making many blades at 0.004", threw them away because they did not meet the spec of 0.003". A manager inquired about the high scrap rate, causing a lot of confessions on both sides. As a result, the designers agreed to set the tolerance where they really needed it, and the shop promised to institute improved process control so that they could deliver reliably at that tolerance.

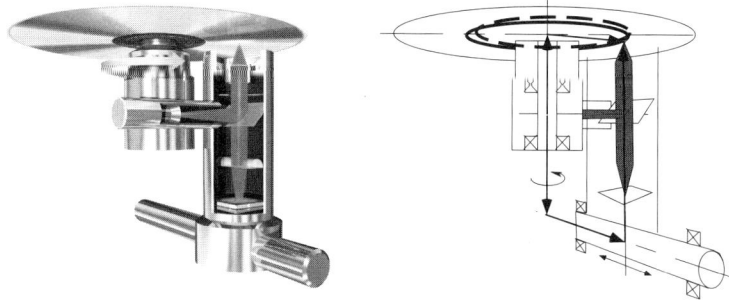

KCs: Size, Weight, $/Megabyte

FIGURE 2-7. Schematic of an Optical Audio or Computer Storage Disk Drive. The data are stored in concentric rings on the disk. One such ring is shown in solid black. The laser beam bounces off the disk and is compared to a reference beam deflected through the half-silvered mirror onto the detector below. Differences in brightness of the beam, caused by the presence or absence of pits on the disk, are interpreted as 1's or 0's. The sequence of points on the disk illuminated by the laser is shown as a dashed-line ring. If the dashed and solid black rings do not coincide, there could be errors reading the data. (Image on the left courtesy of Slim Films. Used by permission.)

loop include a shaft spinning in bearings (shown as boxes with "X" inside), the disk itself with an example data circle, and a linear bearing that positions the laser spot on the disk. Proper performance requires placing the laser spot on the required data circle and keeping it there.

Problems could arise in several ways, all due to errors in fabricating the parts:

- The shaft may not be straight, causing the data circle to waver repeatably in position.
- The disk may not be flat, causing the laser spot to go in and out of focus.
- The bearings may exhibit "runout," which has the same effect as a bent shaft except that it is random.
- The linear bearing may put the spot in the wrong place or not hold it there stably.

The net result of all of these effects is that the spot may read data from more than one data circle, a fatal error. If all efforts to prevent these effects have been exhausted, then further reduction in the risk requires making the circles, as well as the space between them, wider. Then, if it wobbles, moves, or grows out of focus, the spot will still be within the desired data circle. Obviously, making the data circles and spaces wider reduces all of the KCs and makes the product less desirable to the customer.

2.F.2. Car Doors

Car doors must meet a large number of requirements. Several of these are shown in Figure 2-3. Here we look at only

two, based on how the door fits to the car body. One of these is called the appearance KC and is evaluated according to the uniformity of the gaps between the door and other parts of the car, such as the body and adjacent doors. This KC is illustrated in Figure 2-8. The other is called the weather seal KC. It is evaluated according to

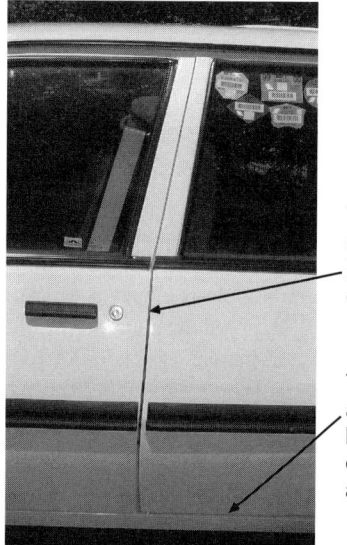

This gap is seen and measured between door outer panels.

This gap is seen and measured between door outer panels and the car body.

FIGURE 2-8. Cosmetic Gaps in Car Doors. The appearance KC is evaluated according to the uniformity of gaps between the door and adjacent regions of the body or other doors. (Photo by the author.)

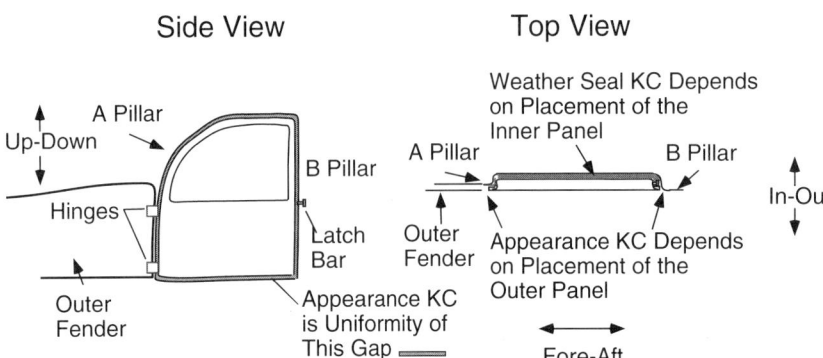

FIGURE 2-9. The KCs for Car Doors. The appearance KC is evaluated according to the uniformity of the gaps while the weather seal KC depends on the tightness of fit between the door and the seal.

how tightly the door fits to seals attached to it and/or to the body. Poor seal fit can cause water leaks or wind noise. These two KCs are explained in Figure 2-9.

Doors are made by joining two stamped steel panels known as the outer panel (the one we see from outside the car) and the inner panel (which is mostly hidden from view by interior trim). The joining process, called hemming, is illustrated in Figure 2-10. For appearance reasons, the inner and outer panels cannot be spot welded together the way other car parts are. Hemming and adhesive are not able to align the parts accurately because the parts mate along flat surfaces that have no definite locating or aligning features on them. Alignment accuracy depends on the hemming tool and can be degraded if the parts shift relative to each other in the tool or while the adhesive is being cured.

After the door parts are joined and the hinges and latch bar attached, they are attached accurately to the car via the hinges. Placement of the hinges on the door and car body is therefore crucial to aligning the door to the car and achieving the KCs. Note in Figure 2-9 that the hinge leaves are open to the 90° position when the door is closed. This is unlike hinges on doors in houses where the leaves of the hinges are closed when the door is closed. With the leaves at 90°, it is possible to adjust their position on the door and the door's position on the car in several different directions independently. Different car makers exploit these freedoms differently.

The car body and doors are then painted, following which, in most car factories, the doors are removed. Cars and doors then go through many assembly steps where interior items like glass, wires, and trim items are installed. "Doors off" assembly also permits the assemblers easier access to the interior of the car and allows doors and cars to be worked on in parallel.

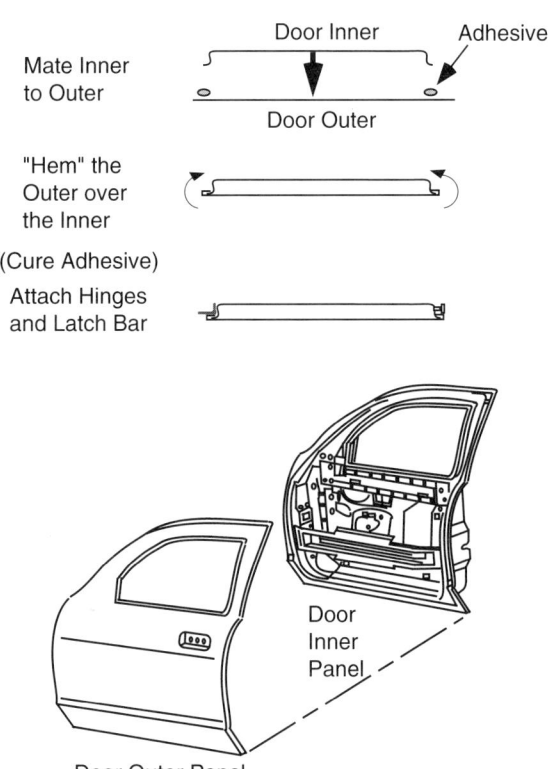

FIGURE 2-10. How Car Doors Are Made. The inner and outer panels are stamped separately and then aligned. Once aligned, they are joined by folding the outer over the inner in an operation called hemming. Often an adhesive is used as well. After the adhesive has cured, the latch bar and the hinges are attached.

Finally the doors must be reinstalled on the car in such a way that the original alignments are reestablished. It is impractical to mount the doors accurately for the first time after painting because any tool or fixture that grasped the door accurately enough to do this properly would scratch the paint. Thus the only way to reattach the doors accurately is to restore the mounting that was in effect before painting. This is always done by again exploiting the hinges, and, again, different firms employ different strategies. These include:

- Using a plastic hinge pin to mount the doors before painting, removing this pin to remove the doors after painting, and replacing the doors later using a steel pin

- Attaching the hinges to the door using locating bolts through close-fitting holes in the hinge, adjusting the door into position on the car using screws through oversize holes, then removing locating bolts from the door to get the door off, and later reinstalling the door repeatably with the locating bolts

- Using hinges that come apart along the pivot axis and reassemble with high repeatability

You can tell which method was used as follows: In the first method, you will see that the pin has what look like arrow tail feathers that prevent it from coming out after installation. In the second method, you will see unpainted bolts fastening the hinge to the door, while the bolts fastening the hinge to the car body are painted. In the third method, you will see an unpainted bolt in the hinge holding it together, while the bolts attaching the hinge to the door and the body are painted.

2.G. KEY CHARACTERISTICS CONFLICT

KC conflict arises when one set of features is involved in the delivery of more than one KC. This event is quite likely to occur because most products must deliver many KCs and there are usually not as many parts as KCs. Each part typically has many features, making each part a participant in many KC delivery chains.[8] This is illustrated for car doors in Figure 2-11. (In Figure 2-3 the participating items are named in italics.) The chain of delivery of both KCs is shown by arrows leading from the body to the door. The hinge is in the chain for both KCs while door inner is in the chain for the weather seal KC and door outer is in the chain for the appearance KC. In addition, door inner and door outer form a subassembly. Once joined, they cannot be adjusted independently into position relative to different locations on the body. Thus, adjusting the door to improve the appearance KC will change and possibly degrade the weather seal KC and vice versa.

Different car firms address this KC conflict differently. Figure 2-12 illustrates two methods used at GM and Ford. The GM method (used on the Grand Am) uses a machine to attach the hinges to the door. This machine mates to the

door using locating features on the door's outer panel. The hinges have oversize holes that permit the machine to adjust their position up–down and in–out. Then the machine attaches a hardened steel locator cone to the free leaf of each hinge, adjusting its position up–down and fore–aft within an oversize hole. In this way, all three directions of the door are addressed. A person, aided by a lifter device, then attaches the door to the car by inserting the

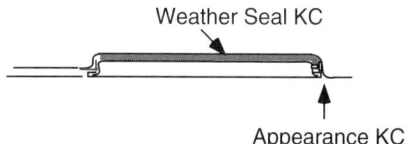

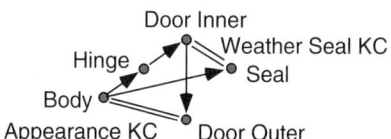

FIGURE 2-11. Chains of Delivery for Two KCs in Car Door Mating to the Car Body. The hinge is in chains that end in both KCs. Door inner is in the weather seal KC chain while door outer is in the appearance KC chain. Door inner and door outer form a subassembly, preventing independent adjustment of each KC by shifting the position of the door's two panels relative to each other. Instead, the whole door is moved by shifting the hinges or by deforming the hinge attachment regions on the door or car body.

---

[8]Manufacturers of six hundred consumer products were surveyed to see what characteristics made them competitive in the marketplace and then were asked what percent of the products' parts participated in delivering those features. In most cases, many parts participated in delivering several distinguishing characteristics ([Ulrich and Ellison]).

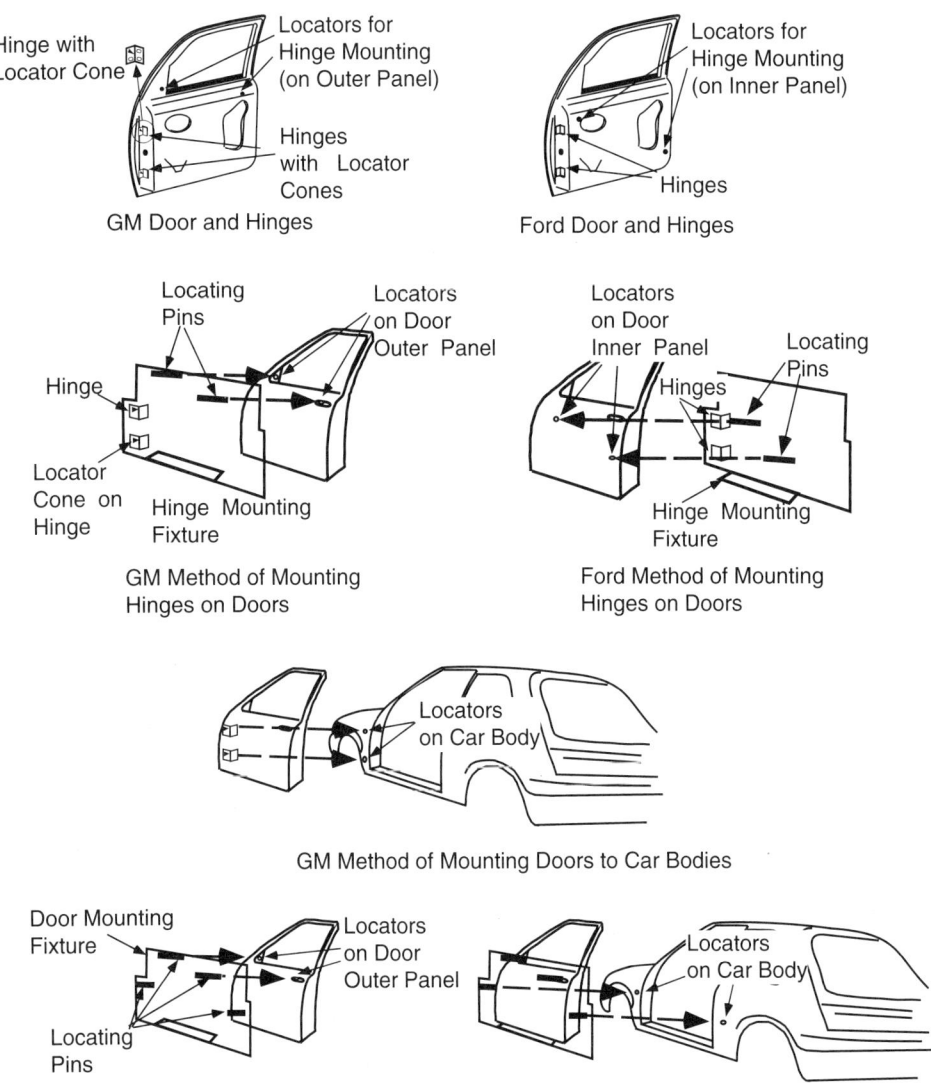

FIGURE 2-12. Two Methods of Attaching Doors to Cars. Each method favors the appearance KC, but each has a different way of doing so. The GM method relies on accurately placing the hinges on the doors as well as accurately placing a cone locator on each hinge leaf. The Ford method, in addition to relying on the accuracy of hinge mounting, also relies on the accuracy of the door mounting fixture that carries the door to the car body because the fixture has locating pins on it that transfer the door outer panel's location to reference holes on the car body.

upper locator cone into a precisely located hole in the car body. The lower hinge's locator cone is inserted into a precisely located vertical slot in the car body.[9] Finally,

the person inserts screws through oversize holes in the hinge leaves. We can see from this figure that the hinge-mounting machine grips the door on the surfaces and edges of the outer panel, thus taking direct control of the appearance KC. The weather seal KC's achievement depends on the accuracy with which the inner

[9]The significance of the hole and slot pattern will be discussed in Chapter 4.

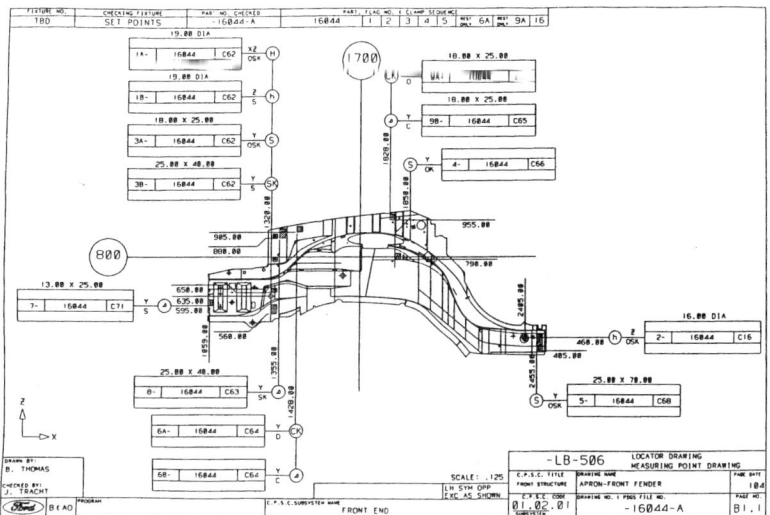

FIGURE 2-13. Ford Locator Drawing. This drawing shows several features on an inner fender that are used to hold the part during fabrication or assembly. Different features have abbreviated names: *H* is a primary locator hole, *S* is a primary locator surface, *h* is a primary locator slot, and HK and SK are permanently transferred locators, used because the original locator has been covered up permanently or in some other way rendered unusable. Small gray squares are reserved locations for clamps or measurements. Only part of this elaborate language is revealed by this drawing. (Courtesy of Ford Motor Company. Used by permission.)

panel is made and attached to the outer panel. The hinges come apart to permit the doors to be removed after painting and repeatedly reinstalled during final assembly.

The Ford method (used on the Mustang and F-150) also uses a machine to attach the hinges to the door. However, it mates to the door using locating features on the inner panel. The hinges are fastened on with locating bolts through oversize holes. A person uses a door mounting fixture to locate the door on the car. This fixture attaches to the door at two locating features on the outer panel: (1) a hole where the handle will later be installed and (2) a slot where the side view mirror will later be installed. The fixture mates to two locators in the car body: (1) a hole just ahead of the door in the inner fender and (2) a slot in a spot behind the door where a sport vent will later be installed. The hinges are joined to the car body with screws through oversize holes. The door stays on after painting. This method similarly favors the appearance KC. Its success depends in part on the accuracy with which the door mounting fixture is made because the chain of KC delivery passes through it. In the GM method, the lifter is not part of any KC chain. The GM method

depends instead on the accuracy of the hinge mounting machine.

Obviously a lot of thought goes into deciding how to achieve the door KCs, even more than indicated here because only two of several KCs have been discussed. No one has figured out how to remove the KC conflict because it is inherent in how sheet metal car doors are made in the first place. A possible remedy would be a polymer outer panel that could be snapped into place after all other assembly and painting was finished, with some adjustments possible. Then the door could be mounted to the car so as to achieve the weather seal KC, and the outer panel could then be mounted to the inner so as to achieve the appearance KC. We will revisit car door mounting in Chapter 8.

To support the achievement of such KCs, car firms use elaborate tooling drawings, such as the one shown in Figure 2-13. Hundreds of these drawings are needed for a complete car. They are created by a team that includes body design engineers, body stamping designers and suppliers, and body assembly and welding equipment designers and suppliers. Figure 2-13 illustrates a piece of the KC flowdown shown schematically in Figure 2-3.

2.H. CHAPTER SUMMARY

This chapter began the first major part of the book, addressing how to design assemblies to achieve top-level customer requirements. These requirements are called key characteristics and are both critical and possibly at risk due to

variation. A variation risk management process was outlined, including flowing each KC down through assemblies, to parts, and finally to features on parts. Examples were given of chains of delivery of KCs in a precision mechanical product (i.e., a CD player) and a sheet metal product (i.e., a car body and door). Finally, the topic of KC conflict was introduced and illustrated with car doors. Each of the topics discussed in this chapter will be revisited in later chapters with increasing mathematical precision.

2.I. PROBLEMS AND THOUGHT QUESTIONS

1. Following the discussion in Chapter 1, take apart an office stapler, identify some KCs, and create KC flowdowns for them analogous to Figure 2-3. Draw the KC delivery diagram for each one analogous to Figure 2-11. Are there any KC conflicts?

2. Look at a refrigerator and sketch the method by which the position of the doors can be adjusted. Identify all close-fitting holes, any slots, and all oversize holes. Figure out which directions of adjustment are permitted by each such hole or slot. Can the door be removed and reinstalled without adjustment, merely by re-achieving the position set at the factory?

3. Look at several cars and determine what method is used to remove and reattach the doors.

4. Identify a product in which there appears to be a KC conflict. Diagram the situation and identify the features or parts that are involved in more than one KC chain. Discuss how the conflict might be resolved. You may conclude that it cannot, in which case you should discuss the risks remaining in the design.

5. Consider the CD drive and assume that the laser spot can be made 1 micron in diameter and perfectly sharp. What is the minimum width and spacing of the data circles? If the useable inner and outer radii of data storage on a disk are 2 cm and 6 cm, respectively, how many bits can be stored on one disk?

6. What if the spot is not perfectly sharp and extends an extra 0.5 micron in diameter at 10% of the center's brightness? What if the bearings run out an amount 0.5 micron? Discuss how each of these errors reduces the storage capacity as well as how *much* the capacity is reduced.

7. Suppose an error-correcting code could be used in conjunction with storing the data on the CD in Problem 5. Such a code would add extra bits but permit software to detect reading errors

and fix them on the fly. More effective codes require adding more bits. What percent of the storage capacity could reasonably be devoted to error-correcting bits?

8. Using Figure 2-11 as a pattern, draw KC delivery diagrams for each of the GM and Ford door-mounting methods shown in Figure 2-12.

9. Consider the car doors and pay attention to the latch bar, which was not discussed in the examples. What role does it play? Which KC delivery chain or chains is it in? Note that the latch bar mates to a latch plate on the car body that is not attached until after painting. Its position can be adjusted. The position of the latch bar on the door usually cannot be adjusted.

10. The Ford Mustang is a sporty two-door car. Each door is therefore rather long and heavy. Suppose Ford chose to switch to a doors-off assembly process. Would Ford use temporary plastic hinge pins? (Such pins are typically about 0.5 cm in diameter.) Explain your answer.

11. What other KCs do car doors have to achieve? Rank them in order of decreasing importance in comparison to the ones discussed in this chapter.

12. A liaison diagram is an example of a network. The simplest form of a liaison diagram is a straight line of nodes. Another more complex form is a hub and spokes arrangement. Most complex is a general network. Let n be the number of parts (nodes). For each of these forms, write an expression using n that gives the maximum number of connections that the diagram can have between its nodes. For example, in a general network with 10 nodes, the maximum number of connections is 45. Do you think that in a typical product with 10 parts there are 45 joints between pairs of parts? If not, why not?

2.J. FURTHER READING

[anon] "Fast and Flexible Communication of Engineering Information in the Aerospace Industry," final report on Air Force Contract number F33615-94-C-4429, available from Wright Laboratory/MTIA, Dayton, OH, May 1998.

[Bebb] Bebb, H. B., "Quality Design Engineering: The Missing Link in US Competitiveness," Keynote address, NSF Design Theory and Methodology Conference, Amherst, MA, June 13, 1989.

[Bourjault] Bourjault, A., "Contribution á une Approche Méthodologique de l'Assemblage Automatisé: Elaboration automatique des séquences Opératoires," Thesis to obtain Grade de Docteur des Sciences Physiques at l'Unitersité de Franche-Comté, November 1984.

[Byrne and Taguchi] Byrne, D. M., and Taguchi, S., "The Taguchi Approach to Parameter Design," 1986 ASQC Quality Congress Transactions, Anaheim, CA.

[Cunningham et al.] Cunningham, T. W., Mantripragada, K., Lee, D. J, Thornton, A., and Whitney, D. E., "Definition, Analysis and Planning of a Flexible Assembly Process," Proceedings of the Japan/USA Symposium on Flexible Automation, Boston, MA, 1996.

[Lee and Thornton] Lee, D., and Thornton, A., "The Identification and Use of Key Characteristics in the Product Development Process," ASME Design Theory and Methodology Conference, Irvine, CA, 1996.

[Liker et al.] Liker, J. K., Sobek, D. K., Ward, A. C., and Christiano, J. J., "Involving Suppliers in Product Development in the United States and Japan: Evidence for Set-Based Concurrent Engineering," *IEEE Transactions on Engineering Management,* vol. 43, no. 2, pp. 165–178, 1996.

[Pahl and Beitz] Pahl, G., and Beitz, W., *Engineering Design,* London: Springer, 1996.

[Phadke] Phadke, M. S., *Quality Engineering Using Robust Design,* Englewood Cliffs, NJ: Prentice-Hall, 1989.

[Thornton 1997] Thornton, A. C., "Using Key Characteristics to Balance Cost and Quality During Product Development," ASME Design Technical Conference, Design Theory and Methodology, Sacramento, 1997.

[Thornton 1999] Thornton, A. C., "A Mathematical Framework for the Key Characteristic Process," *Research in Engineering Design,* vol. 11, pp. 145–157, 1999.

[Thornton 2003] Thornton, A. C., "Variation Risk Management: Focusing Quality Improvements in Product Development and Production," New York: Wiley, 2003.

[Ulrich and Ellison] Ulrich, K. T., and Ellison, D. J., "Customer Requirements and the Design-Select Decision," University of Pennsylvania Wharton School Working Paper 97-07-03, Automotive Engineers Aerospace Technology Conference, 1998.

[Ulrich and Eppinger] Ulrich, K. T., and Eppinger, S. D., *Product Design and Development,* 2nd ed., New York: McGraw-Hill, 2000.

[Ward et al.] Ward, A. C., Liker, J. K., Christiano, J. J., and Sobek, D. K., "The Second Toyota Paradox: How Delaying Decisions Can Make Better Cars Faster," *Sloan Management Review,* vol. 36, no. 3, pp. 43–61, 1995.

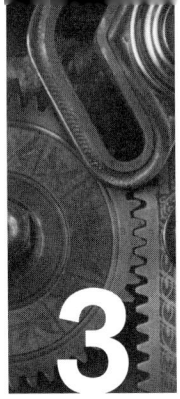

3 MATHEMATICAL AND FEATURE MODELS OF ASSEMBLIES

"Our job is to make holes in parts. If we get the holes right, the parts will go together right."

–Michael Gorden, Ford Motor Company

3.A. INTRODUCTION

This chapter introduces the mathematics necessary to position and orient parts in space with respect to each other. This is the basis of what we call an *assembly model*. It captures mathematically the physical way that parts are located with respect to each other, namely by means of *assembly features,* which are the places on parts that join them to neighboring parts. Our mathematics will permit us to describe features using the same symbols and methods that are used to describe relative positions and orientations of parts in space. Finally, we will see how variations in feature size and location on parts can be analyzed to see their effect on the size and shape of the assembly.

An assembly model should be able to provide the basis for computerized tools for designing assemblies. This means not just permitting the computer to draw the parts on the screen in the correct positions and orientations, but more importantly that the computer's representation should capture the fact that these parts comprise an assembly and that they are assembled to each other in a certain

way. Then it can support assembly design and analysis calculations that use that information.

Most importantly, the assembly model should be able to capture the *design intent* of an assembly. The first level of assembly intent is the KCs. The designer should be able to declare not only the KCs but also how she or he intends to achieve them. Following that are the features, which comprise the plan for constraining the parts in space. Constraint is discussed mathematically in Chapter 4. The model should also seamlessly support analysis of variation, which is the subject of Chapters 5 and 6. CAD systems, as mentioned in Chapter 2, support the capture of design intent for individual parts but lack tools to support assembly intent.

Other aspects of design intent must be captured as well. These include planning the assembly process, designing fixtures and assembly sequences, analyzing variation, understanding the role of suppliers in providing parts and subassemblies, and even showing how the product can be disassembled, repaired, or recycled.

3.B. TYPES OF ASSEMBLIES

Even as there are many aspects of a product in addition to mechanical parts that we will not consider, there are types of assemblies that are out of scope and deserve their own book. Assemblies can be classified as *distributive systems* like pipes, wires, and ducts, *mechanisms* like engines and gearboxes, and *structures* like cars or ships. Each type has its own kinds of KCs and delivers them in its own way. Each therefore requires its own kind of assembly model.

The classification just described can be visualized in Figure 3-1. Each type is discussed next.

3.B.1. Distributive Systems

The basic task of a distributive system is to connect points so that the item being distributed (e.g., fluid, electricity, hydraulic pressure) gets from point to desired point. The

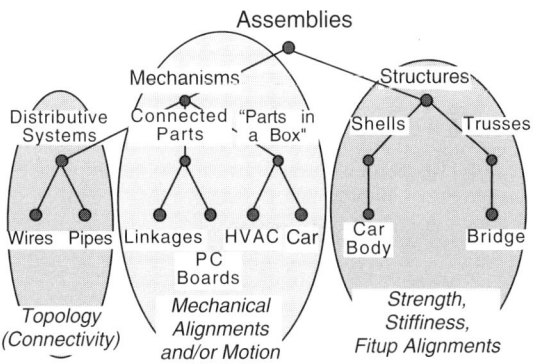

FIGURE 3-1. A Classification of Assemblies.

fundamental nature of a distributive system is its topology or the map of its connectivity. Getting the connections right is the main KC.[1] Usually this can be done without the connecting elements (pipes, wires) being the correct length or size to within tiny fractions of the total length.[2]

Another aspect of distributive systems is that they are often made of combinations of standard items like pipe sections, elbows, valves, joints, pumps, and so on. The difference between instances of these combinations might be how many valves or elbows there are, where they are located on supporting structures, and which ones are connected to which others.

As a result, the main function of a computer model of such assemblies is to account for the connectivity or circuit diagram of the system. A secondary function is to represent multiple copies of the same or almost the same system in an efficient way, which means not simply copying the same valve again into the assembly model every time it is used. The distributive systems of aircraft are especially large and complex. [Callahan and Heisserman] studies models of such systems.

3.B.2. Mechanisms and Structures

Mechanisms and structures differ from distributive systems in that their KCs are more closely tied to exact

geometric shape. Connection errors, however, are not likely to occur. It is easy to see that the microprocessor plugs into the square socket, not the long thin one, or that the connecting rod does not assemble to the exhaust pipe. But the connecting rod must be almost exactly the right size and shape or else it will not assemble to its mating parts and will not function properly.

Similarly, the beams of a structure need to be the right size and shape in order that they can carry the loads required of them. In most structures, shape is much more important than size or mass, because structural stiffness depends on the area moment, whose dimension scales with $(length)\wedge4$, while other geometric quantities scale with lower dimensions of length.

Mechanisms have been further classified in Figure 3-1 as connected parts or "parts in a box." This classification is convenient but not unique. For connected parts, the KCs are often precise dimensions that are critical for balance or efficiency. In many cases a single engineer or a small group can manage these KCs. For parts in a box, the KCs are usually associated with fitting them all within some boundary, a process that is often called *packaging*. Sometimes the relative location of the parts is not critical as long as they are related to each other in some way. Packaging is often a difficult task with many engineers fighting for space for their component. Managing these KCs can be both technically and managerially quite difficult. Chapter 9 of *Car* ([Walton]) vividly describes these issues in a real car design program.

An example is a car heat/vent/air conditioning (HVAC) system. It has to fit into a small space under the instrument panel and contain a fan, heater and cooler coils, and doors that open or close to direct air in different directions. It can be a variety of shapes as long as air flows over the right elements in the right sequence, and there is not too much pressure drop inside. Other designers know this and tend to treat the HVAC as if it were a duffel bag: It can be any shape as long as nothing falls out. The HVAC designer therefore has to work hard to achieve the KCs of his unit.[3]

---

[1] See example in Chapter 1, Section B.4.

[2] An exception might be very high pressure hydraulic pipe. This pipe is quite rigid. If it is not very close to the correct size and shape, it cannot be assembled to previously installed components like valves and pumps without introducing unacceptable stresses or risking leaks.

[3] An experienced car HVAC designer expressed his strategy this way: "I design the biggest, baddest HVAC I can. I push it into the glove compartment, I dig into the radio, I dig into everything. Then I sit back and wait for them to come to me. If I waited until they had designed theirs, there wouldn't be any room for mine." This quote is one of many contained in interviews conducted at Ford by Brian Landau ([Landau]).

3.B.3. Types of Assembly Models

An "assembly model" can mean different things to different people depending on their aspirations for the model. The simplest assembly model is a parts list. It names all the parts in the assembly but says nothing about connectivity or other relationships between the parts.

The next simplest is called a structured bill of materials (BOM). This is a segmented list in which different subassemblies are identified. A subassembly for these purposes is simply a set of parts that can be added as a group to other parts or subassemblies to form a larger subassembly or to form the entire final assembly. If we define the smallest possible subassembly as a single part, then the definition of subassembly is recursive and can be diagramed as shown in Figure 3-2.

More information about connectivity can be obtained from a *liaison diagram* ([Bourjault]). A liaison diagram is a simple graph that uses nodes to represent parts and lines between nodes to represent liaisons or connections between parts. Figure 3-3 shows a simple liaison diagram. Liaison diagrams have a number of interesting uses and properties that are discussed in Chapter 7.

One can add useful information to a liaison diagram by associating information with each liaison, such as what kind of joint it is or what assembly method is used. In addition, one can assign directions to the liaisons to

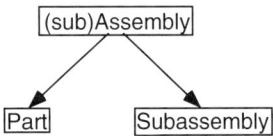

FIGURE 3-2. Recursive Definition of Subassembly. An assembly or subassembly can be decomposed into a part and another subassembly. Since a part is simply a one-component subassembly, this definition is recursive and can be continued until the leaves of the tree consist only of parts.

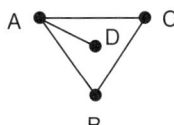

FIGURE 3-3. A Liaison Diagram. This liaison diagram contains four parts. Part A connects to parts B, C, and D. Other connections may be read analogously.

indicate some kind of priority among the parts. This will become important to us when we define constraint and indicate that one part has the responsibility of locating another part. The resulting diagram will be called a *datum flow chain,* the subject of Chapter 8.

For the present, we content ourselves with a liaison diagram, which captures connectivity. Behind this simple diagram lies a lot of information, such as where in space each part is with respect to those with which it has liaisons. We need matrix transformations to capture this information, which leads us to our next topic.

3.C. MATRIX TRANSFORMATIONS

3.C.1. Motivation and Example

This section introduces the mathematics that permits us to add the detail we need to make mathematical models of assemblies. The method used is borrowed from the fields of projective geometry and robotics. It makes use of homogeneous transforms, also called matrix transformations. These transforms permit us to locate (i.e., position and orient) coordinate frames in space with respect to each other. This technique will be used to locate parts, to locate features on parts, and to express both the nominal shape of the assembly and its shape when there is variation in the size or shape of the parts or features.

In a matrix model, each part is assumed to have a base coordinate frame. Mating features on parts each have their own frame as well. A matrix transformation allows us to

say where each feature is on each part with respect to that part's base frame. An assembly is modeled as a chain of these frames. We form the assembly by creating relationships that join the feature frames on mating parts to each other, and the resulting chain of transformations allows us to walk from frame to frame and thus from part to part.

To visualize this, consider the desktop stapler that was presented in Chapter 1. Figure 3-4 shows the stapler in side and top views. Two or more coordinate frames, indicated by orthogonal pairs of arrows, have been added to each part. These frames represent assembly features where parts join to each other, such as the holes where the pin ties the handle, carrier, and anvil together, or they represent important functional features, such as the hammer on the handle, the staple that the hammer must strike, and the crimper that will fold the staple after it passes through the paper.

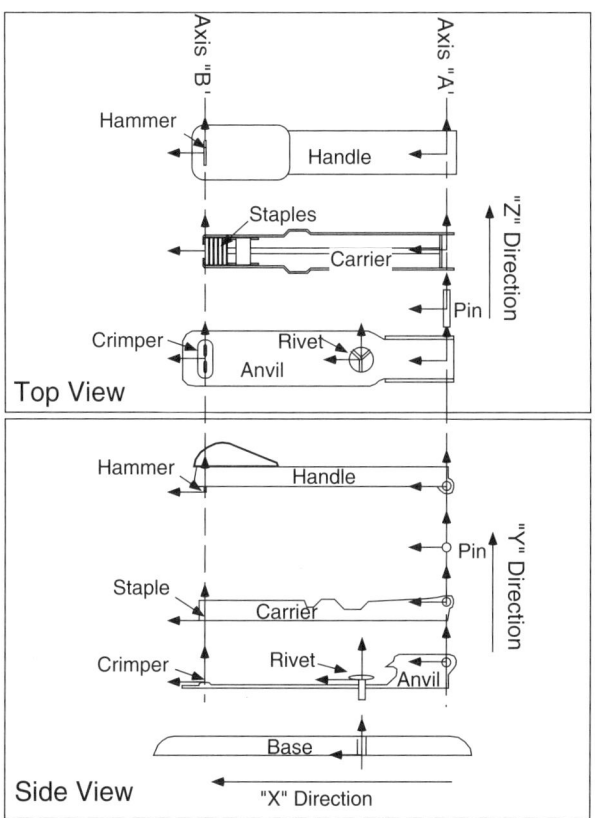

FIGURE 3-4. Diagram of Stapler. Coordinate frames have been added to each part at places where parts join or important functions occur.

Figure 3-5 represents the information in Figure 3-4 in a more abstract way. On the left, the parts have been replaced by shapeless blobs. This is done for clarity as well as to make a point: The exact shapes of the parts are not necessary for much of what we will do in the next few chapters. What is important is what is shown on the left, namely the relationships (called *transforms*) between the frames, shown as arrows. These are of two types. One type of relationship, shown as straight-line arrows, joins two frames inside the same part, such as the one between the pin hole and the hammer on the handle. The transform contains information to allow us to "go" from the hole to the hammer, or equivalently to calculate in XYZ coordinates where the hammer is relative to the hole.

The other type of transform, shown as curved arrows, joins frames on different parts, such as the one between the pin and the hole on the handle into which the pin will

be inserted. To show that the parts will assemble by lining up the three holes on the handle, carrier, and anvil and then putting the pin through this hole, we will make the transforms between these parts identities. An identity transform means "go nowhere" or "be in the same place." To allow us to calculate the relative position in XYZ coordinates of the hammer relative to the end of the staple, we have provided enough frames and transforms to trace a path from the hammer to the handle's hole to the pin to the carrier's hole to the end of the carrier where the staple rests. The KCs are also marked on the right in this figure as double lines.

The mathematical representation of a matrix transform is a 4×4 matrix. This method of modeling spatial relations between objects dates at least as far back as [Denavit and Hartenberg], who used it to represent kinematic linkages. Researchers in assembly and robotics began using it in the late 1960s and early 1970s: [Paul], [Simunovic], [Popplestone], [Wesley, Taylor, and Grossman]. In CAD, researchers used it to represent locations of objects in a computer in the 1960s ([Ahuja and Coons]). In the 1980s, CAD researchers made assembly models of mechanical parts this way ([Lee and Gossard]). The same mathematical model can be used for both chains of links in a linkage and chains of more general parts in an assembly.

In this chapter, we will learn how to work with these frames to build mathematical models of assemblies.

3.C.2. Nominal Location Transforms

3.C.2.a. Basic Properties of Transforms

The 4×4 matrix transformation permits representation of both the relative position of two objects and their relative orientation. It tells us how to express points or vectors in one coordinate frame in terms of another coordinate frame. It is important to understand that the matrix represents the location and orientation of an entire coordinate frame, not simply a point. Figure 3-6 shows the function of the transform schematically. The mathematical form of the transform is

$$T = \begin{bmatrix} R & p \\ 0^T & 1 \end{bmatrix} \qquad (3\text{-}1)$$

In this equation, p is a 3×1 displacement vector indicating the position of the new frame relative to the old one, while R is a 3×3 rotation matrix indicating the orientation of the new frame relative to the old one. (Superscript T indicates a vector or matrix transpose. By convention, all

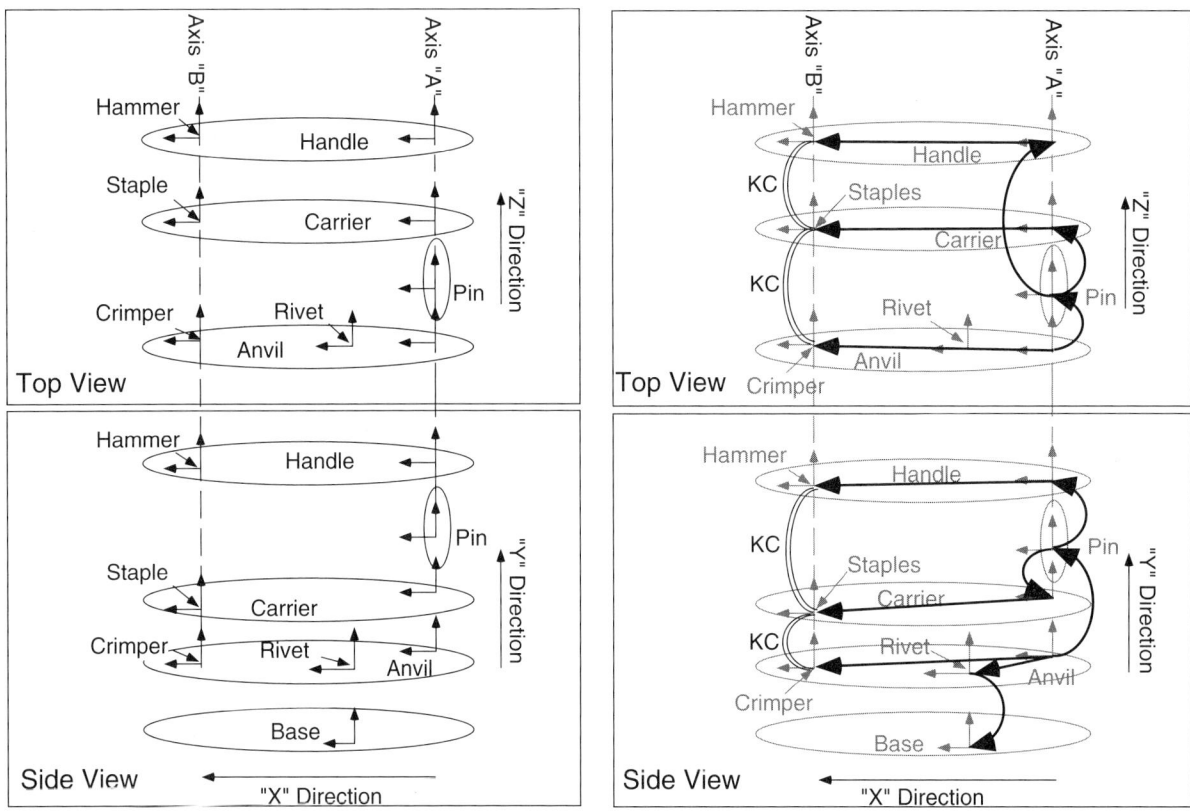

FIGURE 3-5. Schematic Diagram of Matrix Transforms Applied to the Stapler. *Left:* The parts of the stapler have been replaced by blobs. *Right:* Straight-line arrows have been added to relate frames on the same part. Curved arrows have been added linking the coordinate frames of assembly features on different parts to indicate which ones are to be joined in order to assemble the parts. Double curved lines indicate the KCs that were identified in Chapter 1.

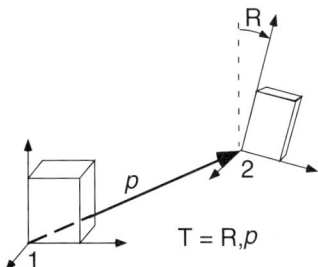

FIGURE 3-6. Schematic Representation of a Transform. The transform T contains a translational part represented by vector p and a rotational part represented by matrix R. Vector p is expressed in the coordinates of frame 1. Matrix R rotates frame 1 into frame 2.

vectors are assumed to be column vectors, so a transposed vector is a row vector.) On a component-by-component

basis, transform T is

$$T = \begin{bmatrix} r_{11} & r_{12} & r_{13} & p_1 \\ r_{21} & r_{22} & r_{23} & p_2 \\ r_{31} & r_{32} & r_{33} & p_3 \\ 0 & 0 & 0 & 1 \end{bmatrix} \quad (3\text{-}2)$$

where vector p is expressed in the coordinates of the original frame and r_{ij} are the direction cosines of axis i in frame 1 to axis j in frame 2.

Transform T can be used to calculate the coordinates of a point in the second coordinate frame in terms of the first coordinate frame. The coordinates of a point are given by

$$p = \begin{bmatrix} x \\ y \\ z \end{bmatrix} \quad (3\text{-}3)$$

Then, in general, if q is a vector in the second frame, its coordinates in the first frame are given by q':

$$q' = \begin{bmatrix} R & p \\ 0^T & 1 \end{bmatrix} \begin{bmatrix} q \\ 1 \end{bmatrix} = Rq + p \qquad (3\text{-}4)$$

This says that q' is obtained by rotating q by R and then adding p.

Suppose a transform T consists only of matrix R, and suppose that we want to find the coordinates of the end of a unit vector along the z axis of the rotated second frame in terms of the unrotated first frame. The calculation is

$$q' = \begin{bmatrix} R & 0 \\ 0^T & 1 \end{bmatrix} \begin{bmatrix} 0 \\ 0 \\ 1 \\ 1 \end{bmatrix} = \begin{bmatrix} r_{13} \\ r_{23} \\ r_{33} \\ 1 \end{bmatrix} \qquad (3\text{-}5)$$

This result shows that the columns of matrix R tell where the coordinate axes have rotated. That is, the first column tells where the x axis went, and so on. The elements of each column are the cosines, respectively, of the x, y, and z components of the new axis expressed in the original frame.

Matrix R can be generated a number of ways. One way is to rotate once about each coordinate axis. This will generate one elemental rotation matrix. Matrix R can then be created by multiplying the elemental matrices into one another. The elemental matrices, as discussed in [Paul], are

$$rot(x, \theta) = \begin{bmatrix} 1 & 0 & 0 & 0 \\ 0 & \cos\theta & -\sin\theta & 0 \\ 0 & \sin\theta & \cos\theta & 0 \\ 0 & 0 & 0 & 1 \end{bmatrix} \qquad (3\text{-}6)$$

$$rot(y, \beta) = \begin{bmatrix} \cos\beta & 0 & \sin\beta & 0 \\ 0 & 1 & 0 & 0 \\ -\sin\beta & 0 & \cos\beta & 0 \\ 0 & 0 & 0 & 1 \end{bmatrix} \qquad (3\text{-}7)$$

$$rot(z, \alpha) = \begin{bmatrix} \cos\alpha & -\sin\alpha & 0 & 0 \\ \sin\alpha & \cos\alpha & 0 & 0 \\ 0 & 0 & 1 & 0 \\ 0 & 0 & 0 & 1 \end{bmatrix} \qquad (3\text{-}8)$$

The order in which T's and R's are multiplied is important, and different sequences will create different results. For example,

$$w = rot(y, 90)rot(z, 90)u \qquad (3\text{-}9)$$

rotates vector u into a new orientation w by first rotating $90°$ about the z axis in the frame in which u is measured,

then $90°$ about y in the same frame. However,

$$w' = rot(z, 90)rot(y, 90)u \qquad (3\text{-}10)$$

rotates vector u into a new orientation w' by first rotating about the y axis and then about the z axis. Equation (3-9) can also be interpreted as saying, Rotate u $90°$ about its original y axis, then $90°$ about its *new* z axis. Similarly, Equation (3-10) can be interpreted as saying: first rotate u $90°$ about its original z axis and then rotate it $90°$ about its *new* y axis.

A transform that simply repositions a frame without reorienting it is

$$T_{trans} = \begin{bmatrix} 1 & 0 & 0 & p_x \\ 0 & 1 & 0 & p_y \\ 0 & 0 & 1 & p_z \\ 0 & 0 & 0 & 1 \end{bmatrix} = trans(p_x, p_y, p_z) \quad (3\text{-}11)$$

A transform T that comprises a translation p_x along x followed by a rotation of $90°$ about the new (translated) z could then be written

$$T = trans(p_x, 0, 0)rot(z, 90) \qquad (3\text{-}12)$$

We can also compute the inverse of a transform. In words, the inverse of T should undo what T did. If

$$\begin{bmatrix} q \\ 1 \end{bmatrix} = T \begin{bmatrix} w \\ 1 \end{bmatrix} \qquad (3\text{-}13)$$

then

$$\begin{bmatrix} w \\ 1 \end{bmatrix} = T^{-1} \begin{bmatrix} q \\ 1 \end{bmatrix} \qquad (3\text{-}14)$$

or, equivalently, if

$$\begin{bmatrix} q \\ 1 \end{bmatrix} = \begin{bmatrix} R & p \\ 0^T & 1 \end{bmatrix} \begin{bmatrix} w \\ 1 \end{bmatrix} \qquad (3\text{-}15)$$

then

$$\begin{bmatrix} w \\ 1 \end{bmatrix} = \begin{bmatrix} R^T & -R^T p \\ 0^T & 1 \end{bmatrix} \begin{bmatrix} q \\ 1 \end{bmatrix} \qquad (3\text{-}16)$$

The transform in Equation (3-16) is the inverse of the transform in Equation (3-15). Embedded in these relationships is the fact that, for rotation matrices,

$$R^{-1} = R^T \qquad (3\text{-}17)$$

3.C.2.b. Examples

Here are some examples that illustrate the rules for using *trans* and *rot*, including the effects of doing so in different sequences.

Equation (3-18) reminds us of the rule regarding sequence of application of a transform. It contains the transforms that we will use in the examples here.

$$\leftarrow \text{use original axes}$$
$$trans(p_x, 0, 0)rot(z, 90) \qquad (3\text{-}18)$$
$$\rightarrow \text{use new axes}$$

We will compare this combined transform with one that contains the same matrices but does something completely different:

$$rot(z, 90)trans(p_x, 0, 0) \qquad (3\text{-}19)$$

We will calculate the effects in both cases, applying the transforms from left to right and from right to left. First, Equation (3-18) is expanded in Equation (3-20). The actions are performed in both sequences in Figure 3-7. It is seen that both sequences result in the same new frame.

$$rot(z, 90) = \begin{bmatrix} 0 & -1 & 0 & 0 \\ 1 & 0 & 0 & 0 \\ 0 & 0 & 1 & 0 \\ 0 & 0 & 0 & 1 \end{bmatrix}$$

$$trans(p_x, 0, 0) = \begin{bmatrix} 1 & 0 & 0 & p_x \\ 0 & 1 & 0 & 0 \\ 0 & 0 & 1 & 0 \\ 0 & 0 & 0 & 1 \end{bmatrix} \qquad (3\text{-}20)$$

$$trans(p_x, 0, 0)rot(z, 90) = \begin{bmatrix} 0 & -1 & 0 & p_x \\ 1 & 0 & 0 & 0 \\ 0 & 0 & 1 & 0 \\ 0 & 0 & 0 & 1 \end{bmatrix}$$

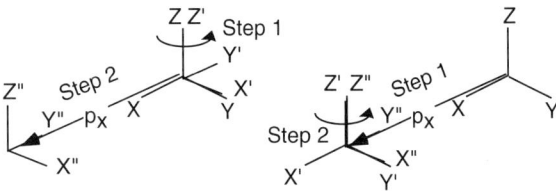

FIGURE 3-7. Illustration of Two Ways of Interpreting Equation (3-20). *Left:* Performing the operations from right to left requires that the original XYZ axes be used throughout the action. Hence, we first rotate 90° about Z and then translate a distance p_x along the original X axis. *Right:* Performing the operations from left to right requires that the new axes be used throughout the action. (For the first operation, new and original have the same orientation.) Hence, we first translate a distance p_x along the original/new X axis and then rotate 90° about the new (translated) frame's Z axis.

Second, we will perform the actions of Equation (3-19) in both sequences. This is illustrated in Figure 3-8. First, Equation (3-19) is expanded in Equation (3-21). Again, we see that the same final frame results. Of course, it is different from the frame that results from the operations in Equation (3-20).

$$rot(z, 90) = \begin{bmatrix} 0 & -1 & 0 & 0 \\ 1 & 0 & 0 & 0 \\ 0 & 0 & 1 & 0 \\ 0 & 0 & 0 & 1 \end{bmatrix}$$

$$trans(p_x, 0, 0) = \begin{bmatrix} 1 & 0 & 0 & p_x \\ 0 & 1 & 0 & 0 \\ 0 & 0 & 1 & 0 \\ 0 & 0 & 0 & 1 \end{bmatrix} \qquad (3\text{-}21)$$

$$rot(z, 90)trans(p_x, 0, 0) = \begin{bmatrix} 0 & -1 & 0 & 0 \\ 1 & 0 & 0 & p_x \\ 0 & 0 & 1 & 0 \\ 0 & 0 & 0 & 1 \end{bmatrix}$$

$$= trans(0, p_x, 0)rot(z, 90)$$

3.C.2.c. Composition of Transforms

The main use of transforms is to permit chaining a series of them together so that we can locate a distant frame by means of several intermediate frames. This is done merely by multiplying one transform by another, as shown in Figure 3-9.

The following forms are equivalent:

$$T_{02} = T_{01}T_{12}$$

$$T_{01} = \begin{bmatrix} R_{01} & p_{01} \\ 0^T & 1 \end{bmatrix}$$

$$T_{12} = \begin{bmatrix} R_{12} & p_{12} \\ 0^T & 1 \end{bmatrix} \qquad (3\text{-}22)$$

$$T_{02} = \begin{bmatrix} R_{01} & p_{01} \\ 0^T & 1 \end{bmatrix} \begin{bmatrix} R_{12} & p_{12} \\ 0^T & 1 \end{bmatrix}$$

$$= \begin{bmatrix} R_{01}R_{12} & R_{01}p_{12} + p_{01} \\ 0^T & 1 \end{bmatrix}$$

The first thing to notice about the matrix in the fifth equation is that it follows the form of the general transform: a rotation matrix in the upper left, a position vector at the right, and a row of three zeroes and a one along the bottom. Thus the composition of two transforms is another transform. This means that we can continue to chain transforms in this way, obtaining another transform each time. The second thing to notice is that we can say

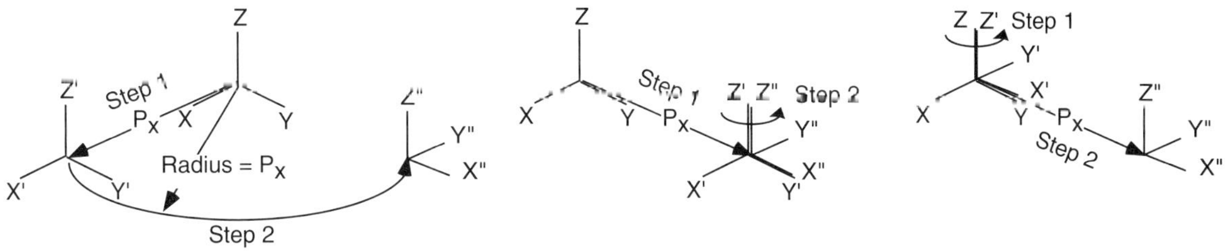

FIGURE 3-8. Illustrating Three Ways to Interpret Equation (3-21). *Left:* Performing the operations (*rotz*, 90) *trans*(p_x, 0, 0) right to left requires using the original axes, including honoring the location of the origin when performing the rotation about (original) *Z*. *Middle:* Performing the operations *trans*(0, p_x, 0)*rot*(*z*, 90) left to right requires using the new axes, again including honoring the location of the origin when performing the rotation about (new) *Z*. *Right:* Performing operations *trans*(0, p_x, 0)*rot*(*z*, 90) right to left requires rotating first 90° about *Z* and then translating a distance p_x along (original) *Y*. These and other interpretations of Equation (3-21) give the same result.

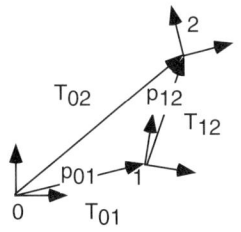

FIGURE 3-9. Illustrating the Composition of Two Transforms. T_{12} locates frame 2 in frame 1 coordinates. T_{01} locates frame 1 in frame 0 coordinates. T_{02} locates frame 2 in frame 0 coordinates.

If frame 2 is rotated in some complex way from frame 0, it may be easier to express the effect (a translation or a rotation) that we want in frame 2 coordinates and then calculate the effect in frame 0 coordinates by writing

$$T_{02} = T_{01}T_{12} \qquad (3-23)$$

The order in which we multiply transforms is important. If T_1 and T_2 are transforms, then

$$T_1T_2 \neq T_2T_1 \qquad (3-24)$$

This fact is used in constructing Equation (3-23), which is the basic equation of matrix transforms, as well as in the examples in Equation (3-20) and Equation (3-21). When we multiply a transform T_{01} from the right by another transform T_{12}, we use T_{01} as the base, effectively adding a coordinate frame T_{12} to a chain of frames that begins at the left end of the chain with a base frame whose transform is I, the identity transform.

Table 3-1 gives some useful MATLAB[4] functions for working with transforms.

If we are careful about how we choose the subscripts of transforms, we can easily read them as a recipe for walking from frame to frame: T_{ij} takes us from frame i to frame j. When we compose two transforms, as in $T_{ij} = T_{ik}T_{kj}$, we can say that subscript k is "used up" when T_{ik} and T_{kj} are chained together to form T_{ij}. This means that frame k no longer needs to be represented explicitly because its effect has been absorbed in T_{ij}. T_{ij} then carries us directly from frame i to frame j. Careful subscripting is very important in debugging complex chains of frames, especially when they are used for variation analysis.

TABLE 3-1. Three Useful MATLAB Functions for Operating on Transforms

Example rotation transform	function Rz = rotz(theta) % creates rotation matrix about axis *Z* % input in radians ct = cos(theta) st = sin(theta) Rz = [ct -st 0 0; st ct 0 0; 0 0 1 0; 0 0 0 1]
Conversion from degrees to radians	function degtorad = dtr(theta) % converts degrees to radians degtorad = theta*pi/180
Translation transform	function Tr = trans(*x*, *y*, *z*) % creates translation matrix Tr = [1 0 0 *x*; 0 1 0 *y*; 0 0 1 *z*; 0 0 0 1]

Note: Function Rz is an example of a rotation operation. Similar functions for rotating about the other axes are easy to write using Equation (3-6) and Equation (3-8).

in words what the composite transform does: It translates along p_{01}, then rotates by R_{01}, then translates along p_{12}, and finally rotates again about R_{12}. The third thing to notice is that the composite transform T_{02} accomplishes in one leap what T_{01} followed by T_{12} do one step at a time.

When we write a transform, say T_{01}, we are able to convert any vector expressed in frame 1 coordinates into frame 0 coordinates. We can also convert any transform expressed in frame 1 coordinates so that its effect appears in frame 0 coordinates. Such a transform might be called T_{12}.

---

[4]MATLAB is a trademark of The MathWorks, Inc.

Examples that use the methods in this section are given in Section 3.E.4.a.

3.C.3. Variation Transforms

We can also express small changes in a transform using a transform. This is highly convenient because it means that we can use the same mathematics to express both the nominal location and the varied location of a frame, and hence of a part or a feature on a part. This is how we will perform variation analyses in Chapter 6.

The kinds of variations that we can express this way are errors in rotation or translation, that is, errors in R or in p. These may be written as follows:

$$Drot(x, \delta\theta_x) = \begin{bmatrix} 1 & 0 & 0 & 0 \\ 0 & \cos\delta\theta_x & -\sin\delta\theta_x & 0 \\ 0 & \sin\delta\theta_x & \cos\delta\theta_x & 0 \\ 0 & 0 & 0 & 1 \end{bmatrix} \quad (3\text{-}25)$$

$$Drot(y, \delta\theta_y) = \begin{bmatrix} \cos\delta\theta_y & 0 & \sin\delta\theta_y & 0 \\ 0 & 1 & 0 & 0 \\ -\sin\delta\theta_y & 0 & \cos\delta\theta_y & 0 \\ 0 & 0 & 0 & 1 \end{bmatrix} \quad (3\text{-}26)$$

$$Drot(z, \delta\theta_z) = \begin{bmatrix} \cos\delta\theta_z & -\sin\delta\theta_z & 0 & 0 \\ \sin\delta\theta_z & \cos\delta\theta_z & 0 & 0 \\ 0 & 0 & 1 & 0 \\ 0 & 0 & 0 & 1 \end{bmatrix} \quad (3\text{-}27)$$

$$Dtrans(dx, dy, dz) = \begin{bmatrix} 1 & 0 & 0 & dx \\ 0 & 1 & 0 & dy \\ 0 & 0 & 1 & dz \\ 0 & 0 & 0 & 1 \end{bmatrix} \quad (3\text{-}28)$$

Multiplying these together creates the error transform DT:

$$DT = \begin{bmatrix} 1 & -\delta\theta_z & \delta\theta_y & dx \\ \delta\theta_z & 1 & -\delta\theta_x & dy \\ -\delta\theta_y & \delta\theta_x & 1 & dz \\ 0 & 0 & 0 & 1 \end{bmatrix} = \begin{bmatrix} I + \delta R & dp \\ 0^T & 1 \end{bmatrix}$$

$$(3\text{-}29)$$

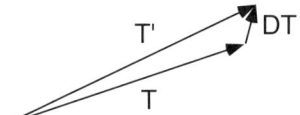

FIGURE 3-10. Properties of the Error Transform. If DT is an error in T, then the erroneous T' is expressed as $T' = T\ DT$.

The upper left 3×3 submatrix δR is a differential rotation matrix. Its elements correspond to a small error in rotation of $\delta\theta_x$ about x, $\delta\theta_y$ about y, and $\delta\theta_z$ about z. The vector dp contains small differential translations dx, dy, and dz. We may write the differential rotation matrix as shown because, if the rotations are small enough, we may consider them to be in the form of a vector like a rotation rate vector, and the order in which they are accomplished does not matter.[5]

The properties of the differential transform are illustrated in Figure 3-10. If there is an error DT in a transform T, then the varied transform is expressed as

$$T' = T\ DT \quad (3\text{-}30)$$

Here, again, the order is important. We accomplish transform T and then we apply the error DT. If the error occurs before transform T is applied, that is, if it occurs in the untransformed frame, then

$$T' = DT\ T \quad (3\text{-}31)$$

For completeness, we introduce the equivalent notation

$$T' = T + dT = T(I + \delta T) \quad (3\text{-}32)$$

where

$$dT = T\ \delta T \quad \text{and} \quad \delta T = \begin{bmatrix} \delta R & dp \\ 0^T & 0 \end{bmatrix}$$

Next, we will show how to use chains of transforms to represent assemblies of parts joined by features.

3.D. ASSEMBLY FEATURES AND FEATURE-BASED DESIGN

This section takes up the topic of features in assembly. First we give some history, then we define manufacturing features and assembly features, and finally we show how to use transforms to locate features on parts and chain parts together via feature frames to create a connective assembly model. This will equip us to use the same mathematical

framework to model assemblies linked by features having either nominal or varied locations.

---

[5]To prove this, form rotation matrices $rot(x, \delta\theta_x)$, $rot(y, \delta\theta_y)$, and $rot(z, \delta\theta_z)$, multiply them together, substitute $\delta\theta$ for $\sin\delta\theta$ and 1 for $\cos\delta\theta$, and eliminate all terms in powers of θ above 1.

3.D.1. History

Feature-based design dates at least to the early 1980s or late 1970s. It was originally an attempt to organize and simplify computer numerical control (CNC) programming of machine tools. Such programs were tedious to write and prone to errors. Many programs were written to cut the same basic shapes, such as drilling and chamfering holes, carving pockets and keyways, and so on. Programming would be easier if one could use a library subroutine capable of carving, say, a general pocket of dimensions L, D, and W, simply by supplying numerical values for L, D, and W, plus coordinates for the position and orientation of the pocket. Pockets, keyways, and so on, soon came to be called *features*. Later, research was done to classify features and provide more comprehensive feature libraries ([Faux and Pratt], [Shah and Rogers]). Finally, it was realized that features provided the opportunity to capture information beyond mere geometry. For example, a keyway feature for holding a key could be given its size based on a calculation of the likely force that the key would encounter. In this way, features rose to become carriers of design knowledge and intent. Below, we call pockets, lightening holes, fillets, and so on, *fabrication features* because they are used to define the shape of the part. Locating holes, keyways, and so on, are called *assembly features* because they are used to define how parts join to each other.

Features were later realized in terms of object-oriented programming. An *object* in this context is a set of data and program code (called a *method*) capable of expressing an item of interest. The data for a keyway feature object might include size parameters of the keyway, while the method could draw a picture of the keyway or contain CNC code for cutting it. Another property of objects, called *inheritance,* is also useful for features. Inheritance means that objects are often subclasses of each other, and the subclasses inherit all of the properties of their superclasses while adding other properties that distinguish them. A *hole* feature is a subclass of the superclass *feature* while a *threaded hole* is a subclass of the hole. Inheritance simplifies programming the objects because only the new elements have to be added when a subclass is defined.

First we will address fabrication features, then assembly features. Finally, we will build assembly models by connecting parts using their assembly features.

3.D.2. Fabrication Features

Fabrication features are the regions of a part that are of importance for the purposes of creating the general shape of the part. Fabrication features present a challenge because their very identity, being for the purpose of defining fabrication instructions, is different depending on the fabrication method used. An example is shown in Figure 3-11. Here we see a series of pockets separated by walls.

The identity of the feature is different depending on whether the pockets are made by removing the metal from the pocket area (say by machining) or by adding metal (say by molding or casting). In the first case, the feature is the pocket, and the rules for creating it are the rules of machining. In the second case, the feature is the wall and the rules for creating it are the rules for molding. In the case of molding, moreover, the walls cannot be parallel as they can be in the case of machining. Instead they must be tapered (sometimes called *drafting* them or giving them a *draft angle*).

The reason why this is a problem is that in many cases the designer does not know, at the time of design, what process will be used to make the part. It depends on the production quantity needed and the economics of the situation. A small number of parts may be machined, but a large number is more likely to be cast or molded. Demand may change over the life of the part. For these reasons, it is worth allowing the designer to use a generic feature to create the shape in the computer independent of fabrication method. Later on, someone will define the fabrication version of the feature, possibly combining or splitting the original generic features. To make this conversion efficient, it is valuable to be able to inspect a CAD design and automatically recognize the shapes that should be combined into features relevant to a specific fabrication process. This is called *feature recognition*. Lacking this, a

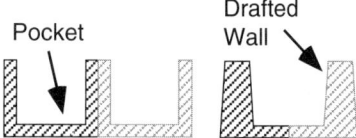

FIGURE 3-11. Pocket and Wall Features. Two pockets are separated by a wall. If the pockets are made by machining, then each pocket is a feature. If the pockets are made by molding or casting, then the walls are the features, and their shape must be tapered ("drafted") so that the part will come out of the mold easily. Note that the pocket feature implicitly defines half a wall, while the wall feature implicitly defines half a pocket.

person must identify the features manually. Feature recognition presents challenges of its own which are beyond the scope of this book. The problem is the subject of ongoing research.

3.D.3. Assembly Features

Assembly features are regions of a part that are important for assembly purposes. Assembly features are made during fabrication, so they are, or correspond to, fabrication features. However, not all fabrication features become assembly features. Furthermore, assembly features carry different design intent and information in their object data and methods. On the stapler, the holes at the right-hand ends of the handle, carrier, and anvil are assembly features, as is the entire outer surface of the pin. The inner rectangular pocket of the carrier is an assembly feature that joins to the outer surface assembly feature of the staples.

Figure 3-12 contains some simple assembly features and the corresponding information that is of interest to us for assembly purposes. Associated with each feature is a transform that holds the location and orientation of the feature's coordinate frame with respect to a coordinate frame on the part. Also noted is the assembly approach and fine motion direction for a compatible mating part. When a

feature has one and only one approach direction, the $-z$ axis of the local frame is used as that direction. This corresponds to a feature hierarchy for robot assembly modeling recommended in [Kim and Wu]. In Chapter 4 we will introduce a library of mating features and demonstrate how they constrain the location of mating parts.

Any geometric shape that is used for assembly can be included in an assembly feature library as long as the required assembly information can be represented. We will learn in Chapter 10, for example, that chamfer width and friction coefficient are important to successful assembly of two parts. The clearance between mating parts is also important, but we cannot calculate that until we know the diameter of the mating feature. This we cannot know until we have a model that permits us to say which parts mate to which other parts using which features on those parts.

Some authors define assembly features as related geometric shapes on mating parts, rather than as single geometric shapes on individual parts as we do here. Each definition has its uses and advantages. In this book we define them as individuals, reserving the flexibility to mate a feature on one part to different possible features on another part. Our definition also permits us to define parts individually with their features if we wish. It also permits us to consider and define variations in feature shape and location individually and later combine their effects.

3.D.4. The Disappearing Fabrication Feature

It is important to realize that many fabrication features are temporary and do not appear on the finished parts. Usually these temporary features serve to hold the parts accurately while other operations, such as machining or grinding, are performed. In fact it has been said that the vast majority of part features do not survive fabrication. Examples include (a) bosses on castings that interface to clamps on machine tool beds and (b) holes or V's punched into sheet metal parts to locate them for later stamping operations. Such features are typically cut or ground off after fabrication is complete.

This fact is important because of its implications for variation analysis. We learned in Chapter 2 that the KCs for subassemblies and parts are subsets of the KCs for the assembly as a whole. In this chapter we are learning that the accuracy of part locations depends on the accuracy with which features are made and placed relative to each other on parts. The accuracy with which we want the assembly features placed on parts can be declared by

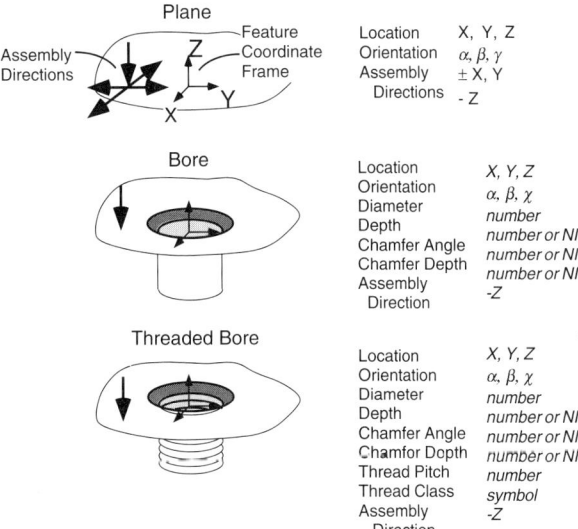

FIGURE 3-12. Three Simple Assembly Features. Each feature is accompanied by information showing where it is and the direction from which a compatible mating part would approach.

imposing tolerances on the transforms that relate them to part center coordinates or to each other.

But those transforms cannot always be generated in their final form. During fabrication, the part must be held and cutting operations, for example, must be performed using transforms that are relative to the features by which the parts are held. Unless an assembly feature can be used to hold a part, the transforms that generate the final feature relationships will pass to and through the fabrication features on their way from one assembly feature to another. The resulting variation in the assembly feature locations thus depends on the variation in many additional transform chain segments that may no longer be on the part once the fabrication features have been removed.

These additional transforms are usually added by process engineers. They have some freedom to choose a set of processes and fixturing features that hopefully will achieve the final desired assembly feature accuracy. In some cases, however, their choices are restricted and achievement of the final tolerances is in doubt. Worse is the situation in which the process engineers do not understand the desired function of the assembly because its KCs have not been declared or made available to them. In such situations, they choose a convenient set of fixturing features sufficient to make the part easily or economically from their point of view regardless of its role in the final assembly. In such cases it is difficult to diagnose assembly problems. When the diagnosis is made, correcting the problem often requires a new fabrication process, fixtures, tools, and measurement plan, a costly consequence. This is one of many reasons why designers of assemblies must keep in close touch with fabrication experts so that they can ensure achievement of the desired final variation limits.

3.E. MATHEMATICAL MODELS OF ASSEMBLIES

We are now in a position to compare a variety of location models of assemblies. These are the world coordinate model, surface-constrained models, and connective models. Each uses 4×4 transforms, but only one, the connective model, permits us to model assemblies as chains, specifically, chains of frames. This is the kind of model we need in order to capture KCs and their delivery paths.

3.E.1. World Coordinate Models

In a world coordinate model, assemblies are placed in a world coordinate frame by expressing each part's coordinate frame and (x, y, z) coordinate location in the world frame. The origin of the world frame of a car or airplane, for example, is normally placed in front of the vehicle a bit beneath the ground plane. This ensures that each part and point in each part has positive coordinates. Each part may be found by estimating its world coordinates and asking for a picture on the computer screen of parts near those coordinates.

Figure 3-13 shows three parts located in a world coordinate frame.

A model like that in Figure 3-13 is often made by drawing each part separately and then carefully placing them in the picture until the desired surfaces touch. A variety of modeling errors could occur. In Figure 3-13b, one such error is shown, namely that part B is in the wrong position.

The result is that it interpenetrates part A, an event called *interference*. CAD systems can detect interferences. However, the same or similar interference could be caused by either part A or part B being the wrong shape even if they are in the correct location, or by part A being in the wrong location. Because this kind of model does not represent the fact that part A should assemble to part B, these kinds of errors cannot be distinguished.

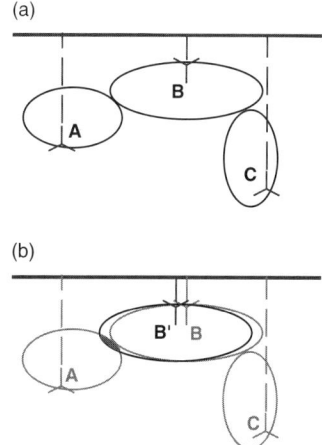

FIGURE 3-13. An Assembly of Three Parts in a World Coordinate Frame. (a) The parts are in their nominal locations. (b) Part B is in the wrong location. It interferes with part A and no longer touches part C.

3.E.2. Surface-Constrained Models

In a surface-constrained assembly model, the user joins items by establishing relationships between different surfaces. Two planes can be made coincident, or two cylinders can be made coaxial, for example. Such operations are often used to build up parts made of elementary surfaces and simpler objects. In some CAD systems, assemblies are built up the same way. The result is that the CAD model cannot distinguish parts and their subparts from assemblies.

3.E.3. Connective Models

In a connective assembly model, the user joins parts by connecting them at their assembly features. This can be done by applying the methods of surface constraint to surfaces on the features. Better, yet, the frames representing the features can be constrained to each other directly. Figure 3-14 shows three parts joined this way. On the left is the nominal situation while on the right a varied situation, caused by an error in placing an assembly feature on part B, is shown. Note that this error can be detected even if the parts are modeled only approximately, as long as the assembly features are modeled and placed on the parts accurately.[6] By contrast, detection of errors in a world coordinate model like that of Figure 3-13 requires that the parts be modeled accurately, since no distinction is made when modeling them between assembly feature surfaces and other surfaces.

A connective assembly model can represent parts, assembly features, and surfaces individually and can tell the difference between them. This makes it possible to model different kinds of variation correctly and to distinguish in the model different sources of error. Consider the situation in Figure 3-15. Part A in this figure is joined to part B by making a surface on one part coincident with a surface on the other. In Figure 3-16, part B mates to an assembly feature "f" on part A. The left sides of each figure show apparently identical nominal situations, while the right sides show apparently identical varied situations. However, in Figure 3-15, we cannot tell the cause of the variation because it does not contain a separate and coordinated group of surfaces called an assembly feature. All sources of error

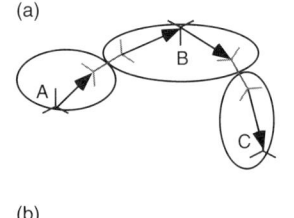

(a)

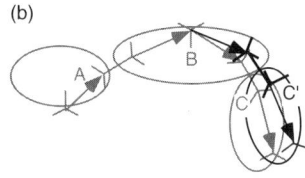

(b)

FIGURE 3-14. Three Parts Joined by a Connective Assembly Model. (a) The nominal situation. (b) A feature on part B is misoriented and mispositioned, causing part C to be in the wrong position.

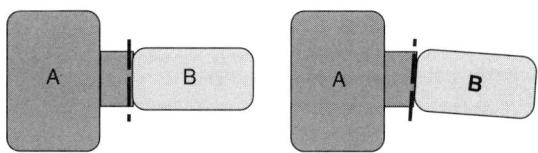

FIGURE 3-15. Two Parts Constrained by Aligning Two Surfaces. The surfaces that are aligned are indicated by the dashed line. *Left:* The nominal situation. *Right:* The situation if the constraint surface on part A is misoriented.

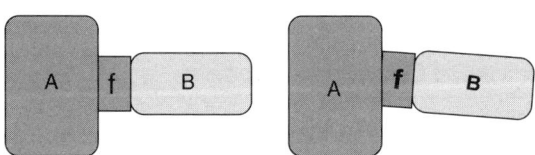

FIGURE 3-16. Two Parts Constrained by Joining Assembly Features. *Left:* The nominal situation. *Right:* The situation if the feature "f" on part A is misoriented.

must therefore be attributed to mislocated surfaces, and all surfaces are treated identically. In fact, in some CAD systems we cannot even tell if the error is on part A or on part B. In Figure 3-16, we can represent the fact that the entire feature on part A is misoriented because we have modeled the feature explicitly. Alternatively, we can represent mismanufacture of the feature leading to its having one misoriented surface. In fact, every kind of error that could occur in practice can be represented individually and unambiguously. This is a huge advantage when analyzing variations.

---

[6]Other errors that could cause interferences between parts can be detected only if the parts' shapes are modeled accurately.

3.E.4. Building a Connective Model of an Assembly by Placing Feature Frames on Parts and Joining Parts Using Features

The connective model of assembly defines a part as having a central coordinate frame plus one or more assembly features, each feature having its own frame. A transform relates each feature's location on the part to the part's central coordinate frame. Features like those in Figure 3-12 consist of a single geometric element. They can be placed on a part by defining a transform from part center coordinates to the feature frame. Alternatively, the transform to the feature frame can be directed from another feature frame.

When two parts join, assembly features on one part are made to coincide with assembly features on the other part. This is done by defining a feature interface transform that relates the frame on one part's assembly feature to the frame on the other part's assembly feature ([Gerbino and Serrano]). If the axes of these two frames are identical, then the interface transform is the identity transform. If not, then typically a reorientation assembly transform must be written to account for the difference between the axes of the two feature frames.

3.E.4.a. General Mathematical Connective Assembly Model

In order to know which parts mate to which other parts and to calculate where the parts are in space as a result, we exploit the fact that each feature has an associated transform which tells where the feature is on the part. "Assembly" of two parts then consists of putting the features' frames together according to some procedure, and then composing several transforms to express the part-to-part relationships. These relationships are illustrated in Figure 3-17. To find or arrive at part B from part A, one starts at the coordinate frame of part A, follows the transform to the coordinate frame of its feature F_A, then goes to the transform of the mating feature F_B on part B, then follows the transform $T_{F_B-B} = T_{B-F_B}^{-1}$ from that feature to part B's coordinate frame. We can express this as

$$T_{AB} = T_{A-F_A} T_{F_A-F_B} T_{B-F_B}^{-1} \qquad (3\text{-}33)$$

The first transform on the right in Equation (3-33) T_{A-F_A} locates part A's feature on part A relative to the part's coordinate frame. The second transform $T_{F_A-F_B}$ is a feature interface transform that captures the relationship

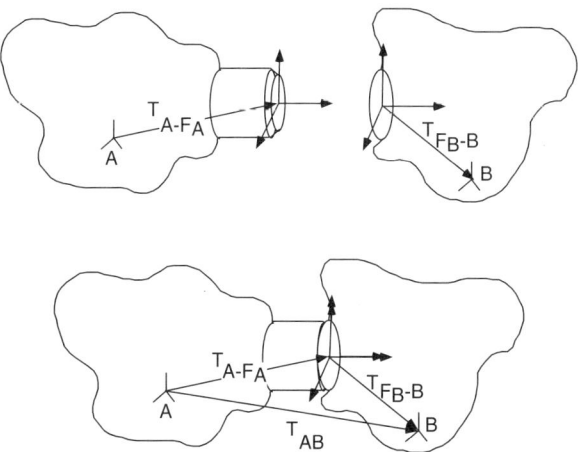

FIGURE 3-17. Mating Two Parts Using Assembly Features. The mathematics of composing transforms may be used to find the location of a mating part relative to another part if we know where the assembly features are on each part. Note that in part B, transform T_{F_B-B} equals $T_{B-F_B}^{-1}$.

between the feature frames on the two parts. The third transform $T_{B-F_B}^{-1}$ is the inverse of the transform T_{B-F_B} that locates part B's feature with respect to part B's coordinate frame. The inverse appears because nominally the transform T_{B-F_B} carries us from part B's origin to part B's feature F_B. The inverse is what we need to carry us from the feature to the part coordinate frame. This step completes the trip from A to B.

The feature interface transform can express any of several constraints between features, such as making their frame origins coincide while making the frames' z axes point toward each other, making the frames' x–y planes coincide while orienting the axes in some specific way, offsets between feature frames, and so on. Using this technique, we can model an assembly as a chain of frames. The model relates the parts in the same way as the physical parts relate to each other: Their assembly features are joined.

In CAD systems it is common to locate parts with respect to each other by constraining certain surfaces to have specific relationships with each other. For example, two planes could be made to coincide, or two cylinders could be made coaxial. This method is less general than the one described here because it fails to capture the fact that the surfaces in question belong to particular parts or features. It also prevents us from using the frame information to calculate relative part locations and variations on them.

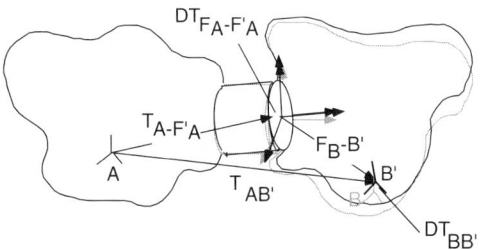

FIGURE 3-18. Varied Location of a Part Based on a Mispositioned Feature on Another Part. The feature on part A is not in the correct orientation, so part B will not be in the correct position with respect to part A. $T_{AB'}$ differs from T_{AB} in a way that can be calculated easily based on knowing $DT_{F_A - F'_A}$.

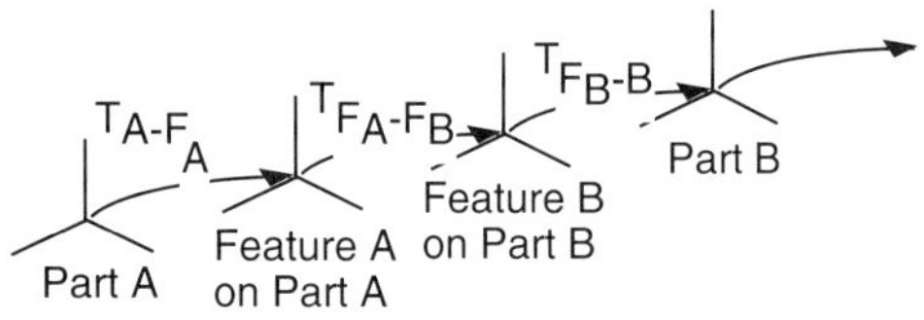

FIGURE 3-19. Sketch of Frame Relationships for General Nominal Connective Assembly Model. This model shows one feature A, located by T_{A-F_A} on part A, one feature B, located by $T_{F_B - B}$ on part B, and an interface relationship $T_{F_A - F_B}$ between them. Presumably, the assembly extends beyond part B in a similar fashion.

If a feature on a part is not placed where it is supposed to be, then we can express the error using Equation (3-30). As shown in Figure 3-18, the feature on part A is mispositioned and/or misoriented. The transform relating it to part A's origin is then

$$T_{A-F'_A} = T_{A-F_A} DT_{F_A-F'_A} \qquad (3\text{-}34)$$

The varied position of part B can then be calculated as

$$T_{AB'} = T_{A-F'_A} T_{F'_A - F_B} T_{F_B - B} \qquad (3\text{-}35)$$

where

$$T_{F'_A - F_B} = T_{F_A - F_B}$$

because there is no feature–feature error

Based on the above examples, we can formulate a general connective model of assemblies. The model will be presented in two forms, the first being the nominal and the second including variations. These are illustrated in Figure 3-19 and Figure 3-20 for the nominal case and the varied case, respectively. The equation for the nominal case is Equation (3-33). A simplified version of the equation for the varied model is Equation (3-34) and Equation (3-35) for the case where there is only variation in the location or shape of the feature. The general equation containing all the variations represented in Figure 3-20 is given in Chapter 6.

The model as presented here appears to assume that only one feature can join part pairs and that assemblies can only consist of single linear chains of parts. In fact, we can represent more complex assemblies, such as those created by joining a part to two previously joined parts by recruiting features on all three parts. We will present assemblies of this type in later chapters. For now we will present only simple chains, but we will show how to join them using single features like pin–hole pairs and compound features made of two or more single features.

3.E.4.b. Examples: Joining Parts Using Single Features

First, we will consider joining two parts using single features, such as those in Figure 3-12. Each such feature is defined by a single frame. In the next subsection we will deal with combinations of such features.

The following examples illustrate basic translation, basic rotation, construction of a part with an assembly feature on it, and construction of an assembly of two parts by placing the frames of their assembly features onto each other. They utilize the MATLAB functions for translation and rotation that appear in Table 3-1. These examples appear in Figure 3-21 through Figure 3-25.

The first example, Figure 3-21, shows how to position a feature whose axes align with part center coordinate axes.

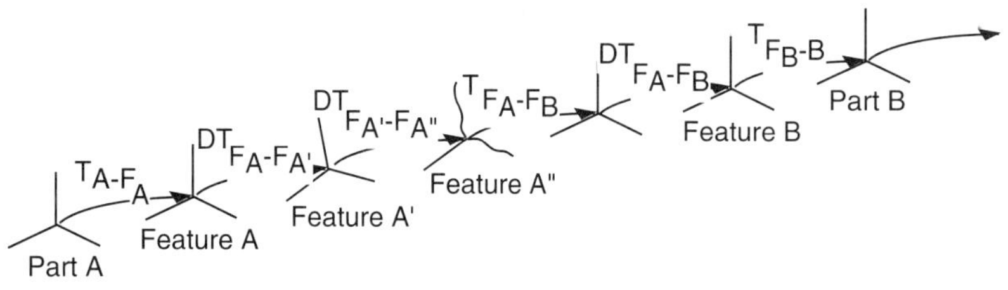

FIGURE 3-20. Sketch of Frame Relationships for General Varied Connective Assembly Model. This model augments the model in Figure 3-19 by the addition of $DT_{F_A - FA'}$ representing mislocation of feature A on part A, $DT_{FA' - FA''}$ representing a misshapen feature A, and DT_{FA-FB} representing variation in the interface relationship.

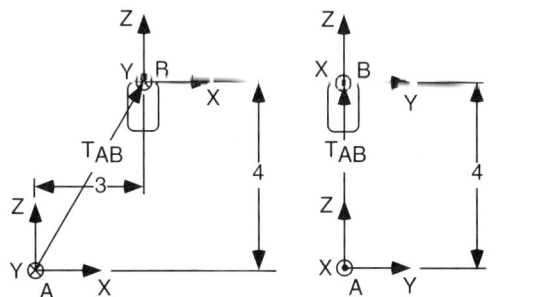

$$>> T_{AB} = trans\,(3,0,4)$$

$$T_{AB} = \begin{bmatrix} 1 & 0 & 0 & 3 \\ 0 & 1 & 0 & 0 \\ 0 & 0 & 1 & 4 \\ 0 & 0 & 0 & 1 \end{bmatrix}$$

FIGURE 3-21. Illustrating How to Write a Transform that Repositions a Feature Without Rotating It. *Left:* View in the *X–Z* plane. *Right:* View in the *Y–Z* plane. The transform equation may be read to say: "To go from frame A to frame B, go 3 units in *X* and 4 units in *Z* (in frame A coordinates)." In this figure, a head-on view of a vector is noted by ⊙ while a tail-on view is noted by ⊗.

This feature is a locating pin. Accordingly, the *Z* axis of the pin's coordinate frame coincides with the pin's centerline.

The next example, Figure 3-22, shows how to position a feature and orient it differently from part center coordinate axes.

The third example, Figure 3-23, shows how to build a part and position and orient a feature on it.

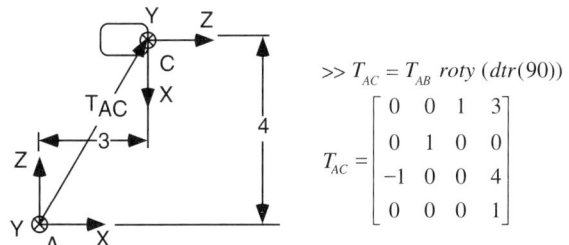

$$>> T_{AC} = T_{AB}\,roty\,(dtr(90))$$

$$T_{AC} = \begin{bmatrix} 0 & 0 & 1 & 3 \\ 0 & 1 & 0 & 0 \\ -1 & 0 & 0 & 4 \\ 0 & 0 & 0 & 1 \end{bmatrix}$$

FIGURE 3-22. Illustrating How to Position and Orient a Feature. This example extends the example in Figure 3-21. The transform equation may be read to say: "To get from frame A to frame C, go 3 units in *X* and 4 units in *Z* (in frame A coordinates) and then rotate 90° about frame A's (relocated) *Y* axis."

Next, Figure 3-24, we build a second part and place a hole on it.

Last, Figure 3-25, we assemble the parts in Figure 3-23 and Figure 3-24. In Chapter 6, we will revisit these examples, inserting variation transforms and determining the varied position of point F.

3.E.4.c. Examples: Joining Parts Using Compound Features

Joining parts becomes a bit more complex if the feature is built up from several elements, such as a hole and a slot. This is called a compound assembly feature (or simply, *compound feature*) and is illustrated with an example in Figure 3-26. Frame $X'Y'Z'$ locates this feature. (Z' is along the hole axis and is not shown in the figure.) Using this compound feature, we could join part A to another part B that had two pins, one that mates with the hole and the other that mates to the slot. Also shown are transforms T_{A1} from the part coordinate center to the hole, T_{A2} from the part center to the slot, and $T_{1'2}$ from the hole to the slot. (The calculations that follow are simplified to the case where the compound feature lies in the XY plane of part A's part center coordinates.)

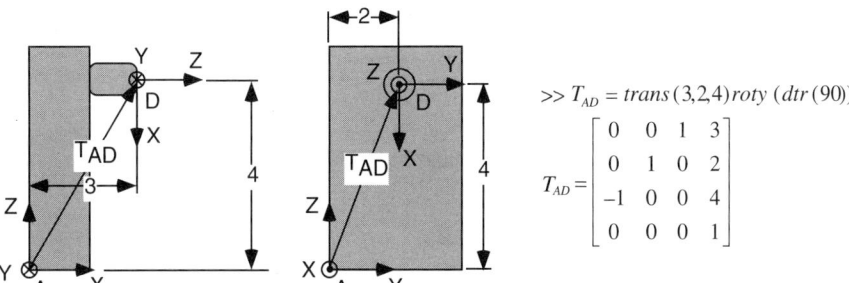

$$>> T_{AD} = trans\,(3,2,4)\,roty\,(dtr\,(90))$$

$$T_{AD} = \begin{bmatrix} 0 & 0 & 1 & 3 \\ 0 & 1 & 0 & 2 \\ -1 & 0 & 0 & 4 \\ 0 & 0 & 0 & 1 \end{bmatrix}$$

FIGURE 3-23. Illustrating How to Build a Part and Place a Feature on It. This example is a slight extension of the one in Figure 3-22. The transform equation may be read to say: "To go from part A's coordinate center at A to the tip of the peg feature at frame D, go 3 units along frame A's *X* axis, 2 units along *Y*, and 4 units along *Z*, and then rotate 90° about frame A's relocated *Y* axis."

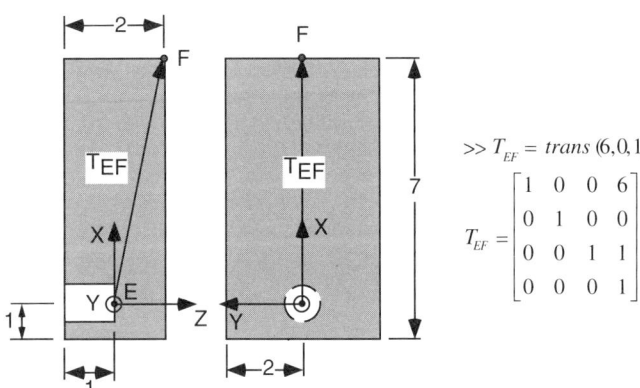

$$\gg T_{EF} = trans\,(6,0,1)$$

$$T_{EF} = \begin{bmatrix} 1 & 0 & 0 & 6 \\ 0 & 1 & 0 & 0 \\ 0 & 0 & 1 & 1 \\ 0 & 0 & 0 & 1 \end{bmatrix}$$

FIGURE 3-24. Illustrating Construction of a Second Part. This part has a hole feature on it as well as a point F that is one end of a KC. The transform equation may be read to say: "To go from the bottom of the hole to point F, go 6 units along frame E's X axis and 1 unit along its Z axis."

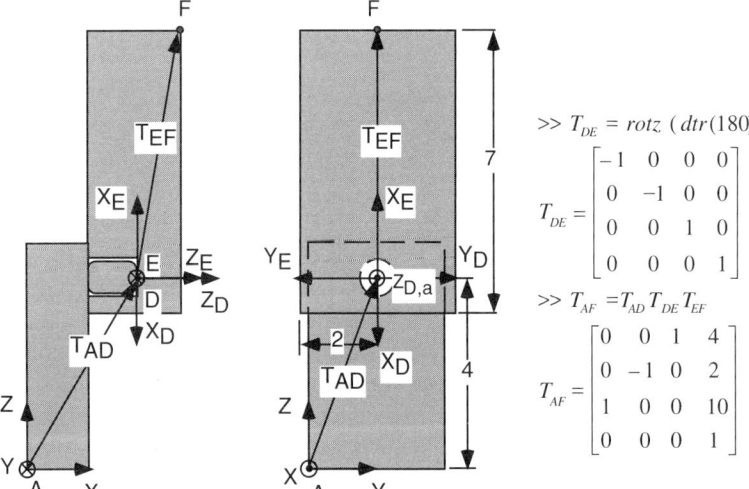

$$\gg T_{DE} = rotz\,(\,dtr(180))$$

$$T_{DE} = \begin{bmatrix} -1 & 0 & 0 & 0 \\ 0 & -1 & 0 & 0 \\ 0 & 0 & 1 & 0 \\ 0 & 0 & 0 & 1 \end{bmatrix}$$

$$\gg T_{AF} = T_{AD}\,T_{DE}\,T_{EF}$$

$$T_{AF} = \begin{bmatrix} 0 & 0 & 1 & 4 \\ 0 & -1 & 0 & 2 \\ 1 & 0 & 0 & 10 \\ 0 & 0 & 0 & 1 \end{bmatrix}$$

FIGURE 3-25. Illustrating Assembling Two Parts and Calculating the Overall Transform T_{AF} from Part A's Coordinate Center to KC Point F. These two parts were built in Figure 3-23 and Figure 3-24. They are "assembled" by placing frame D of the pin on the first part onto frame E of the hole on the second part. An interface transform T_{DE} is needed because the axes of these frames are not aligned in the same way. The equation for assembly transform T_{AF} may be read to say: "To go from frame A to KC point F, follow transform T_{AD} (defined in Figure 3-23), then align frame D and frame E by rotating 180° about frame D's Z axis, then follow transform T_{EF}, which is defined in Figure 3-24."

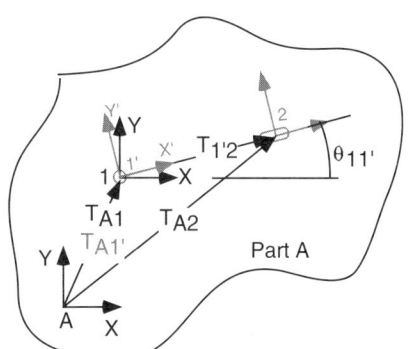

FIGURE 3-26. Compound Feature Consisting of a Hole and a Slot. The feature is made up of a hole and slot. Frame A is at the origin of part A's base coordinate frame. Frame 1 is at the center of the hole. Frame 1′ is the frame of the compound feature. Frame 2 is at the center of the slot. Transform T_{A1} relates hole frame 1 to base frame A, while transform $T_{A1'}$ relates the compound frame 1′ of the hole–slot feature to frame A. The difference between these two frames is the rotation $rot(z, \theta_{11'})$. This rotation can be found by calculating the difference between T_{A1} and T_{A2} as shown in Equations (3-36a)–(3-36j).

The information we want is frame $T_{A1'}$, which describes the position and orientation of the hole–slot feature relative to frame A. This frame is shown in gray in Figure 3-26. We begin by assuming that the locations of the individual hole and slot are known, namely T_{A1} and T_{A2} respectively. The information we need is found from Equations (3-36a)–(3-36j):

$$T_{A1} = \begin{bmatrix} R_{A1} & p_{A1} \\ 0^T & 1 \end{bmatrix} \qquad (3\text{-}36a)$$

$$T_{A2} = \begin{bmatrix} R_{A2} & p_{A2} \\ 0^T & 1 \end{bmatrix} \qquad (3\text{-}36b)$$

$$T_{A1'} = T_{A1} T_{11'} \qquad (3\text{-}36c)$$

where

$$T_{11'} = \begin{bmatrix} R_{12} & p_{11'} \\ 0^T & 1 \end{bmatrix} \qquad \text{where } p_{11'} = 0 \quad (3\text{-}36d)$$

$$T_{A2} = T_{A1} T_{11'} T_{1'2} = T_{A1} T_{12} \qquad (3\text{-}36e)$$

$$T_{12} = T_{A1}^{-1} T_{A2} = \begin{bmatrix} R_{12} & p_{12} \\ 0^T & 1 \end{bmatrix} \qquad (3\text{-}36f)$$

where

$$p_{12} = p_{A2} - p_{A1} = \begin{bmatrix} x_{12} \\ y_{12} \\ 0 \end{bmatrix} \qquad (3\text{-}36g)$$

$$R_{12} = rot(z, \theta_{11'}) \qquad (3\text{-}36h)$$

$$\cos\theta_{11'} = x_{12}/|p_{12}| \quad \text{and} \quad \sin\theta_{11'} = y_{12}/|p_{12}| \quad (3\text{-}36i)$$

where

$$|a| = \text{the length of vector } a \qquad (3\text{-}36j)$$

Now, if we want to place another part onto the first one using the hole and slot, we provide it with two pins, say, one to mate with the hole and one to mate with the slot, and place them together as shown in Figure 3-27. We then equate (or relate) the frame representing the compound hole–slot with the frame representing the compound pin–pin, "assembling" the two frames in the same way as we did when assembling the parts in Figure 3-25. This means that we can assemble parts containing compound features in exactly the same way that we assemble parts with simple features. In Chapter 6, we will see how to do the same thing when the features (and hence their frames) are mislocated or misoriented.

Thought questions at the end of the chapter ask for detailed analyses of these calculations as well as analysis of

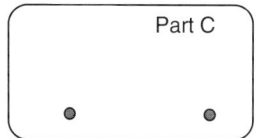

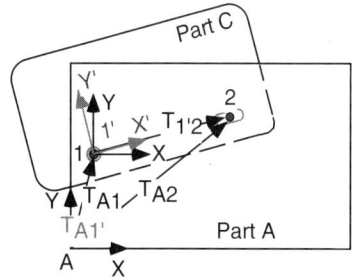

FIGURE 3-27. A Second Part Joined to the First Part Using the Hole and Slot Features. Part C has two pins that engage the hole and slot on part A.

the situation where the hole is on part A but the slot is on another part B which is mated to part A.

In Chapter 6, we show how to find the variation in part C's location with respect to part A if the hole and slot are not positioned correctly.[7]

3.E.5. A Simple Data Model for Assemblies

Knowing only what we know up to now, we can see how rich an assembly model can be. Here we present examples of a part model and a feature model. These are presented in table form without showing any geometry, to emphasize the point that these models are aimed at capturing the *identity* of parts, features, and their *relationships*, rather than the *shapes* of the parts and features or their individual *surfaces*. These relationships form the heart of true assembly models. Once the relationships are known, the details of part shape can be added at any time.

The simple models in Table 3-2, Table 3-3, and Table 3-4 permit us to build up assemblies using the

---

[7]To first order, small errors in the slot's orientation will not have any effect on part C's location with respect to part A as long as the slot's long direction points more or less along vector p_{12}. If the slot should for some reason be oriented perpendicular to this direction, then a condition called overconstraint could occur, causing pathological variations in part C's location. Constraint is discussed in Chapter 4.

TABLE 3-2. A Simple Part Model

PART	NAME_A
COORDINATE_FRAME	X, Y, Z
CONTAINS_FEATURE	FEATURE_1 & TRANSFORM (F1)
CONTAINS_FEATURE	FEATURE_2 & TRANSFORM (F2)
MATES_TO_PART	NAME_X
VIA_FEATURE	FEATURE_1
HOW_TO_FIND(MATES_TO_PART)	USE_TRANSFORM (VIA_FEATURE)

TABLE 3-3. A Simple Feature Entry in a Part Model

FEATURE	ON_PART(NAME)
GENERIC TYPE	FEATURE_TYPE
NAME	FEATURE_NAME
LOCATION	TRANSFORM
TOLERANCE_ON_LOCATION	DIFF_TRANSFORM
FEATURE_INFO	(refer to library)
MATED_TO	FEATURE_ON_PART(NAME)

TABLE 3-4. A Simple Feature Library Entry

FEATURE_TYPE	BEARING_POCKET
MEMBER_OF_CLASS	BLIND_HOLE
COORDINATE_FRAME	X, Y, Z
ASSEMBLY_ESCAPE_DIRECTION	Z
SHAPE_PARAMETERS	DIAM, DEPTH, CHAMFER, RECOMMENDED_CLEARANCE(DIAM) See Bearing Handbook
SHAPE_TOLERANCE	DIFF_PARAMETER_LIST or ANSI-Y-14.5-M callouts
GEOMETRY	

mathematics presented earlier and to navigate among the parts by chaining the transforms together. The entry "HOW_TO_FIND" in Table 3-2 is called a *method* in object-oriented programming. It is intended to represent a program that permits transforms to be multiplied together.

Naturally, a full-strength assembly model would contain much more detail and would probably look considerably different from what has been presented here. However, the basic capabilities are there and the example serves to illustrate the desired capabilities of such a model.

3.F. EXAMPLE ASSEMBLY MODELS

This section presents two example assembly models. One is a complex missile seeker head, while the other is a consumer product, a kitchen juicer.

3.F.1. Seeker Head

A seeker head consists of a sensor (radar, infrared) mounted on a gimbal system that permits the sensor to look up and down or left and right. The seeker locks onto a target and the missile is steered so that the gimbal angles go to zero with respect to coordinates aligned with the missile's body.[8]

The gimbal system consists of two individual gimbals mounted on bearings. Each gimbal is a ring with two shafts called trunnions and a ball bearing on each trunnion. The inner gimbal rides on bearing interfaces to the outer gimbal while the outer gimbal rides on bearing interfaces to the base. The outer gimbal's tilt axis intersects the inner gimbal's tilt axis and is perpendicular to it. Accurate intersection and perpendicularity are crucial KCs for this assembly.

Figure 3-28, Figure 3-29, Table 3-5, and Figure 3-30 show, respectively, an exploded view of the seeker head, the liaison diagram, a table of parts and their constituent features, and an annotated liaison diagram with the feature information on it.[9]

Liaisons 1 and 10 deserve special mention. These liaisons do not exist in the actual seeker head but instead are implemented through the actions of liaisons 2, 3, 4, and 5 (for liaison 1) and 11, 12, 13, and 14 (for liaison 10). Liaisons 1 and 10 had to be added manually to the model to permit assembly sequence analysis to be done. Assembly sequences are nominally sequences of liaisons, and an assembly sequence algorithm tries to find feasible liaison sequences by testing for the existence of approach paths. This is the subject of Chapter 7.

It turns out that no assembly sequence can be generated simply by sequencing liaisons 2, 3, 4, and 5 or 11, 12, 13, and 14. Take the inner gimbal, for example. It must be inserted into the outer gimbal by means of a sort of parallel parking maneuver. See Figure 3-31. If either bearing has already been inserted (from the outside) into a pocket on the outer gimbal, then the inner gimbal cannot be installed. Instead, the inner gimbal must be inserted into an

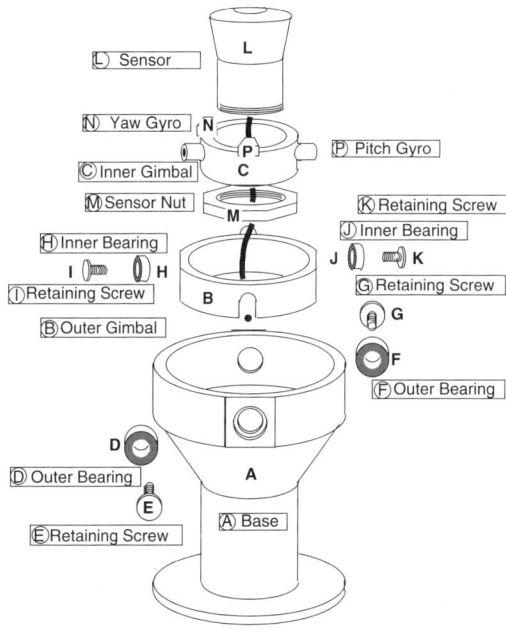

FIGURE 3-28. Exploded View of Seeker Head.

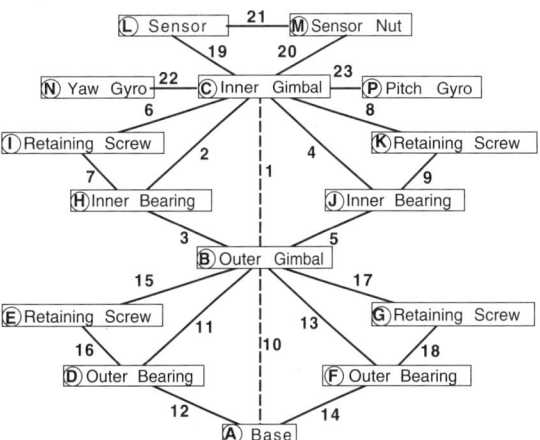

FIGURE 3-29. Liaison Diagram for Seeker Head.

---

[8]Missile steering algorithms are discussed in A. H. Bryson and L.Y.-C. Ho, *Applied Optimal Control,* Ginn/Blaisdell, 1969, pp. 154–155.

[9]These figures and the table were prepared by Alexander Edsall.

TABLE 3-5. Partial List of Parts, Assembly Features, and Assembly Feature Classes in the Seeker Head

Part	Part Name	Feature	Feature Name	Feature Class
A	Base	1	Bearing bore	(Chamfered) bore
		2	Trunnion bore	Bore
		3	Trunnion bore	Bore
		4	Bearing bore	(Chamfered) bore
B	Outer gimbal	1	Bearing bore	(Chamfered) bore
		2	Trunnion bore	Bore
		3	Trunnion bore	Bore
		4	Bearing bore	(Chamfered) bore
		5	Ret. screw hole	Threaded bore
		6	Trunnion	(Chamfered) pin
		7	Trunnion	(Chamfered) pin
		8	Ret. screw hole	Threaded bore
C	Inner gimbal	1	Ret. screw hole	Threaded bore
		2	Trunnion	(Chamfered) pin
		3	Trunnion	(Chamfered) pin
		4	Ret. screw hole	Threaded bore
D	Outer bearing	1	Bore	(Chamfered) bore
		2	Outer diameter	(Chamfered) pin
		3	Inner race face	Plane
E	Retaining screw	1	Thread	Threaded pin
		2	Head	Plane
F	Outer bearing	1	Bore	(Chamfered) bore
		2	Outer diameter	(Chamfered) pin
		3	Inner race face	Plane
G	Retaining screw	1	Thread	Threaded pin
		2	Head	Plane
H	Inner bearing	1	Bore	(Chamfered) bore
		2	Outer diameter	(Chamfered) pin
		3	Inner race face	Plane
I	Retaining screw	1	Thread	Threaded pin
		2	Head	Plane
J	Inner bearing	1	Bore	(Chamfered) bore
		2	Outer diameter	(Chamfered) pin
		3	Inner race face	Plane
K	Retaining screw	1	Thread	Threaded pin
		2	Head	Plane

outer gimbal that contains as yet no bearings. No liaison between parts can be made at this stage, so the inner and outer gimbals must be supported temporarily by a fixture. Liaison 1 represents this temporary situation. It is called a phantom liaison. After this is done, the bearings can be inserted. If liaison 1 were not in the diagram, the assembly sequence algorithm would report that no sequences exist.

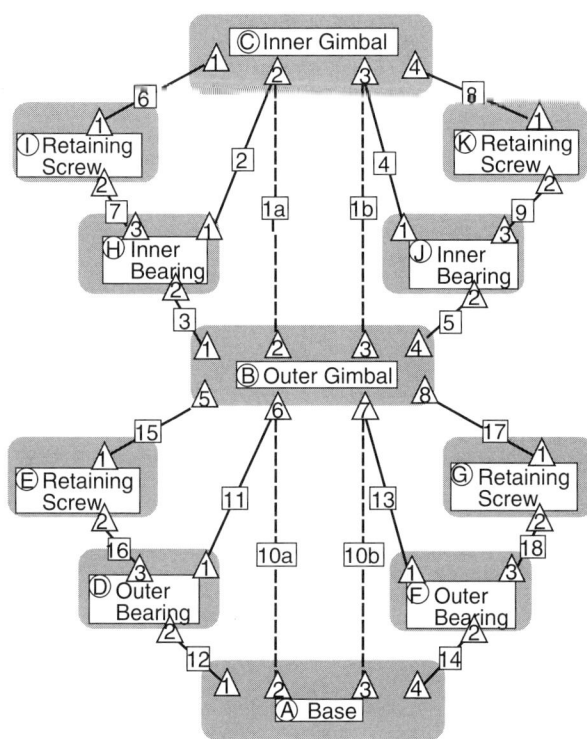

FIGURE 3-30. Annotated Liaison Diagram for Seeker Head. This diagram links all the parts, liaisons, and assembly features.

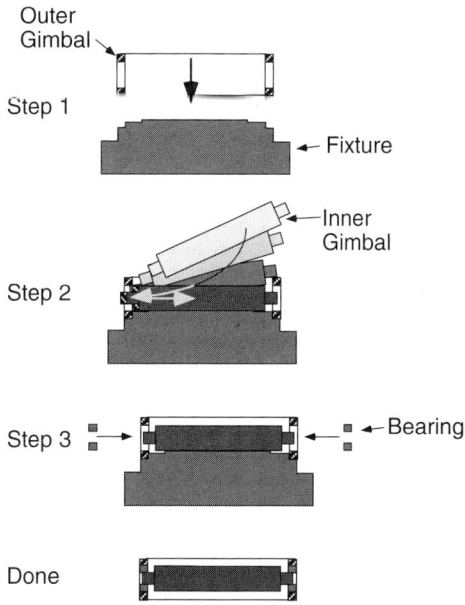

FIGURE 3-31. Illustration of Parallel Parking Maneuver to Install a Gimbal. The bearings install from the outside and must be absent in order to place the inner gimbal in the outer gimbal. Step 1 accomplishes liaison 1. Step 2 accomplishes liaisons 2 and 3 at the same time and accomplishes liaisons 4 and 5 at the same time. A similar process accomplishes liaisons 11–14.

3.F.2. Juicer[10]

The second example product is a home juicer. In this product, manual assembly is possible using the physically defined liaisons but automatic assembly would be very difficult. Addition of phantom liaisons creates new assembly possibilities that open the field for automatic assembly.

The product is illustrated in Figure 3-32. Figure 3-33 is an exploded view with a parts list. Figure 3-34 shows the liaison diagram. On it are shown two phantom liaisons. One links the transmission shaft to the base while the other links it to the container. The difficult part of the assembly involves the transmission shaft, transmission gear, and

base. The shaft mates to the gear via liaison 3, and they trap the base between them. The gear also mates to the base via liaison 2. Without the phantom liaisons, the only way to assemble these parts is for a person to hold the base, push the gear through the hole in it, and hold both while mating the shaft to the gear from the other side. This is not too difficult for a person but would challenge a machine or require intricate fixturing.

If phantom liaison 10 is allowed, the shaft can be placed temporarily in the container while it is upside down. The base can then be placed on the container (liaison 1), following which the gear can be mated to the shaft (liaison 3) and to the base (liaison 2). If phantom liaison 11 is allowed, the shaft can be placed upside down in a fixture, the base can be put on top of it (liaison 11), and then liaison 3 (and liaison 2) can be made.

This example is discussed further in Chapter 7 where we generate the set of allowed assembly sequences with and without the phantom liaisons.

---

[10]The information in this section is drawn from a student project at MIT conducted by Alberto Cividanes, Jocelyn Chen, Clinton Rockwell, Jeffrey Bornheim, Guru Prasanna, Rasheed El-Moslimany, and Victoria Gastelum. Alberto Cividanes prepared the drawings.

FIGURE 3-32. Home Juicer. (Photo by the author. Drawing by Alberto Cividanes.)

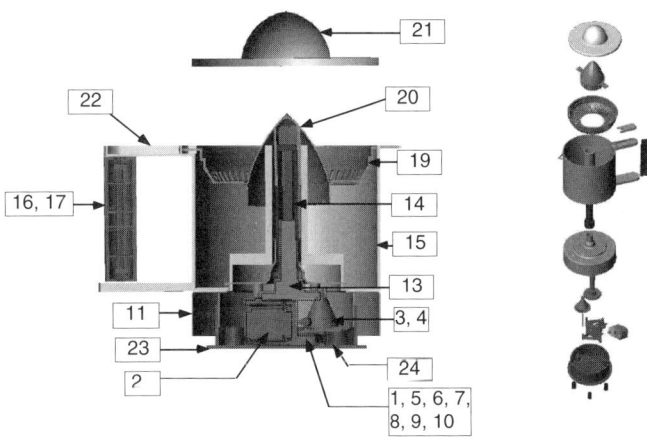

FIGURE 3-33. Cross-Section and Exploded Views of the Juicer with Bill of Materials.

Bill of Materials:

1 – Motor Mount Frame	8 – Conductor	16 – Handle Inside
2 – Motor Assembly	9 – Connector	17 – Handle Outside
3 – Reduction Shaft	10 – Wires	19 – Strainer
4 – Reduction Gear	11 – Base	20 – Squeezer
5 – Power Cord	13 – Transmission Gear	21 – Cover
6 – Spring	14 – Transmission Shaft	22 – Handle Cover
7 – Copper Switch	15 – Container	23 – Bottom Cover
		24 – Rubber Feet

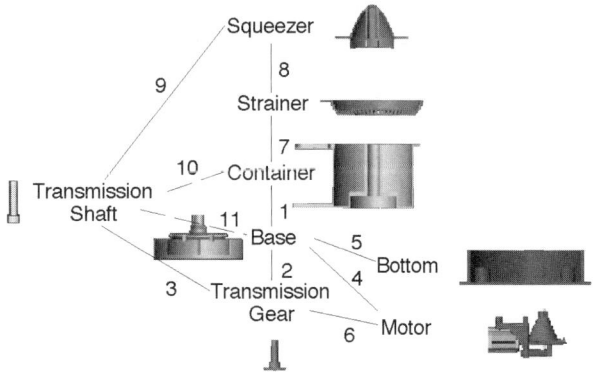

FIGURE 3-34. Liaison Diagram of Juicer Identifying Phantom Liaisons. In the final assembled configuration, the transmission gear mates to the transmission shaft via liaison 3. These parts trap the base between them. Manual assembly is not difficult but machine assembly could be. The possible sequences can be expanded by incorporating phantom liaisons 10 and 11. These permit assembly to be done by either temporarily resting the transmission shaft in the container (via liaison 10) or in the base (via liaison 11). If it is rested in the base, then the gear can mate to it next. If it is rested in the container, then the base must be mated to the container and then the gear can be mated to the shaft. Neither liaison 10 nor liaison 11 exists in the final assembled configuration.

3.G. CHAPTER SUMMARY

This chapter developed the requirements for a connective model of assemblies. This model is based on modeling the connections between parts by analogy with the way physical parts connect to each other, that is, by joining each other at places called assembly features. This kind of model emphasizes the connectivity of the assembly and gives less importance and attention to the detailed shapes of the parts. We will use the connectivity and feature information extensively in later chapters when we analyze constraint, variation, and assembly sequences.

The connective model defines a part as having a central coordinate frame plus one or more assembly features, each feature having its own frame. Transforms relate feature locations on a part to the part's central coordinate frame. When two parts join, assembly features on one part are made to coincide with assembly features on the other part. This is done by defining an interface transform from the frame on one part's assembly feature to the frame on the other part's assembly feature. If the axes of these two frames are identical, then the interface transform is the identity matrix.

3.H. PROBLEMS AND THOUGHT QUESTIONS

1. Take apart a desktop stapler and draw all the parts, or use Figure 3-4. Measure all the distances between the functional and assembly features, such as between the hammer and the hole at the end of the handle. Assign axis names to the frames following the convention for X, Y, and Z shown in Figure 3-4. Using the assembly hole as the origin or base frame for each part, write down the 4×4 transforms that relate each functional feature to its part's base frame. Then, following the example in Figure 3-25, calculate the position of the hammer with respect to the crimper.

2. Find the transform T12 for part A, a rectangular block, shown in Figure 3-35. In the figures, frame 1 is the part's origin frame, while frame 2 is the frame on the assembly feature, a round peg.

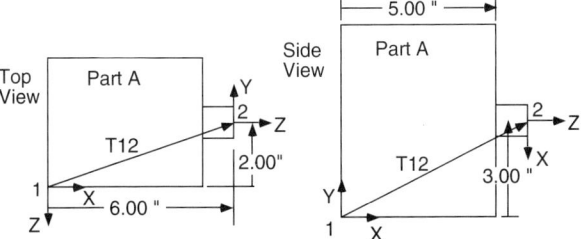

FIGURE 3-35. Figure for Problem 2.

3. Find T34 for part B shown in Figure 3-36, another rectangular block, using the same methods you used in Problem 2. In this case, the assembly feature is a hole at frame 4.

4. Assemble parts A and B by joining the peg to the hole so that their respective coordinate frames align, as shown in Figure 3-37. Find T13 two ways:

   a. From the transform equation $T13 = T12^*T24^*T43$ (Note that you will have to calculate T43, which you can do with knowledge of T34, which is asked for in Problem 3.)

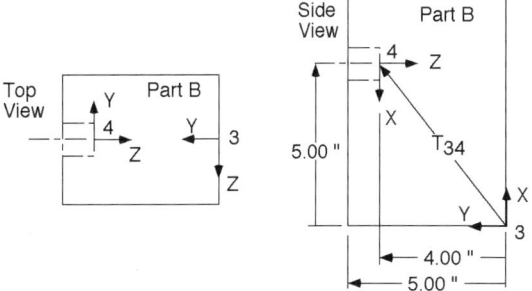

FIGURE 3-36. Figure for Problem 3.

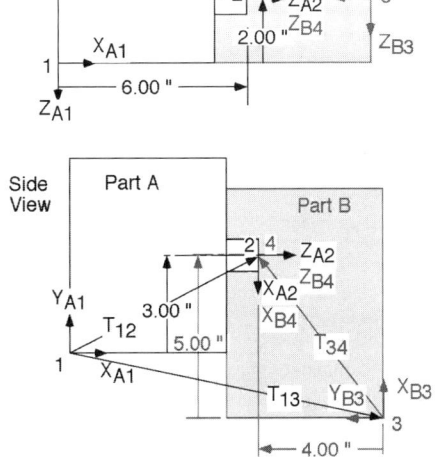

FIGURE 3-37. Figure for Problem 4. For clarity, the coordinate axes have been labeled with the following code: X_{A1} means the X axis of frame 1 on part A. Other axes are labeled consistently with this code.

b. Directly, by inspecting the relationship between frames 1 and 3 and using the methods you used to calculate T12 in Problem 2.

---

5. Find the location in frame 1 coordinates of a point "a" in frame 3 coordinates in part B while joined to part A, such that the point is located 1″ along the $+X$ axis from origin 3. Express the answer as a transform T_{1a}.

---

6. Assume that part A cannot be fabricated by fixturing it at frame 1 but instead that it must be fixtured at frame 0 as defined in Figure 3-38. Find the transform that relates assembly feature frame 2 to fabrication feature frame 0.

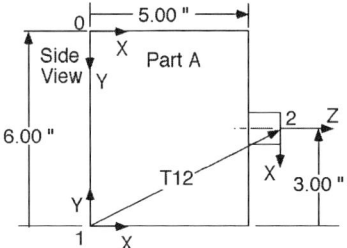

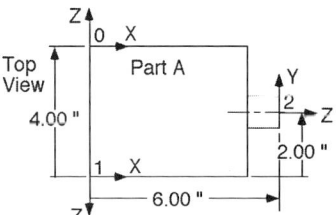

FIGURE 3-38. Figure for Problem 6.

---

7. Figure 3-39 depicts the parts from Problem 4 with the addition of frame 0 on part A.

Consider the two following cases:

- The KC for this assembly is the distance from frame 1 to frame 3.

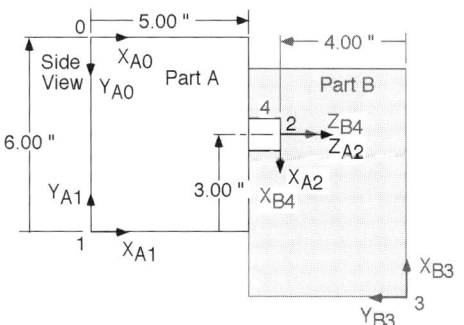

FIGURE 3-39. Figure for Problem 7.

- The KC is the distance from frame 0 to frame 3.

Answer the following questions:

a. Has part A been dimensioned correctly for the case where the KC is the distance from frame 0 to frame 3? Make a drawing to show how you would dimension part A in this case.

b. What surfaces would you use to hold part A while machining the peg on it if the KC is the distance from frame 0 to frame 3? Explain why.

c. What surfaces would you use to hold part A while machining the peg on it if the KC is the distance from frame 1 to frame 3? Explain why.

d. What role does the location of frame 1 play, if any, in each case?

---

8. Your company has just changed suppliers for part B. The new supplier redrew the CAD model, which looks like the diagrams in Figure 3-40.

The assembly of the original part A and the new part B looks like the diagram in Figure 3-41.

Find T13 from the transform equation $T13 = T12^*T24^*T43$. Naturally, you should get the same answer as you did in Problem 4.

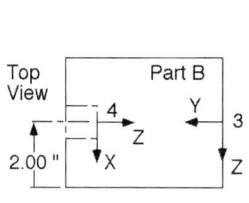 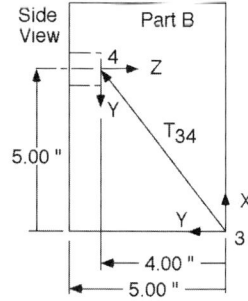

FIGURE 3-40. Figure (a) for Problem 8.

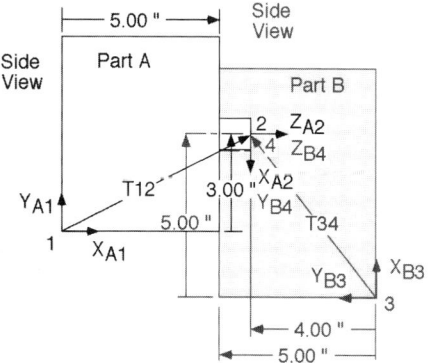

FIGURE 3-41. Figure (b) for Problem 8.

9. Consider the two following transform equations:

$$T_{12a} = A^*B = \begin{bmatrix} 1 & 0 & 0 & 6 \\ 0 & 1 & 0 & 3 \\ 0 & 0 & 1 & -2 \\ 0 & 0 & 0 & 1 \end{bmatrix} \begin{bmatrix} 0 & 0 & 1 & 0 \\ -1 & 0 & 0 & 0 \\ 0 & -1 & 0 & 0 \\ 0 & 0 & 0 & 1 \end{bmatrix}$$

$$T_{12b} = C^*D = \begin{bmatrix} 0 & 0 & 1 & 0 \\ -1 & 0 & 0 & 0 \\ 0 & -1 & 0 & 0 \\ 0 & 0 & 0 & 1 \end{bmatrix} \begin{bmatrix} 1 & 0 & 0 & -3 \\ 0 & 1 & 0 & 2 \\ 0 & 0 & -1 & 6 \\ 0 & 0 & 0 & 1 \end{bmatrix}$$

a. Calculate T_{12a} and T_{12b}.

b. Explain the result.

c. Explain element by element why the last column in matrix A has the values it has.

d. Explain element by element why the last column in matrix D has the values it has.

10. Consider the block below with the pin on it. Frame 1 is the block's base frame while frame 2 is the pin's base frame. Let frame 3 be defined by

$$T_{23} = \begin{bmatrix} R_{23} & p_{23} \\ 0^T & 1 \end{bmatrix} \quad \text{where } p_{23} = \begin{bmatrix} -1 \\ 2 \\ -2 \end{bmatrix} \quad \text{and} \quad R_{23} = I$$

Draw frame 3 in its correct position and orientation on the drawings of the block in all three views (see Figure 3-42).

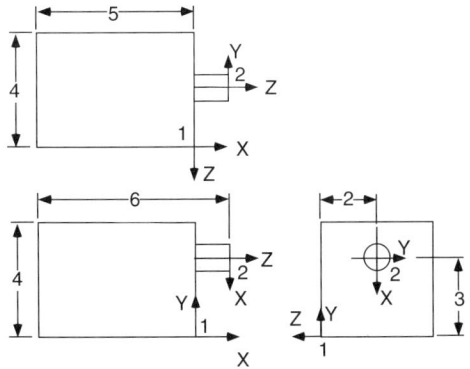

FIGURE 3-42. Figure for Problem 10.

11. Consider the V block shown in Figure 3-43 with the cylinder resting in it.

Find dimensions *A* and *B* algebraically. Explain how you would machine the V block given the dimensions shown.

12. Consider the V block shown in Figure 3-44. Explain how you would machine it given the dimensions shown.

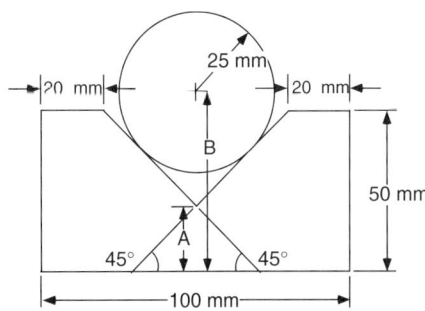

FIGURE 3-43. Figure for Problem 11.

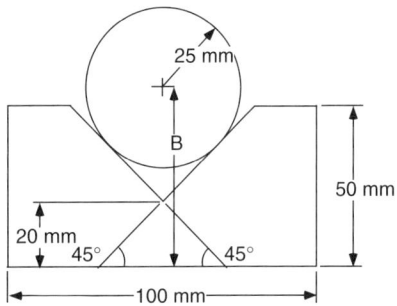

FIGURE 3-44. Figure for Problem 12.

13. Think about and discuss the differences between the dimensioning schemes in Problems 11 and 12. In which case can you say that the V is represented as an assembly feature? In which case can you say that the dimensions explain what the V block is supposed to do? Why is it necessary to solve for dimension A in Problem 11 and not in Problem 12? Is dimension B different in Problem 11 and Problem 12? What is the role of the 50-mm dimension and the 100-mm dimension in each case?

14. Consider the drawing in Figure 3-45, which is an extension of Figure 3-26. Assume that part B is mated to part A using some features that are not shown. You can thus assume that T_{A1}, T_{AB}, and T_{B2} are known. Find expressions for $T_{A1'}$ and $T_{11'}$ along the lines of Equations (3-36a)–(3-36j).

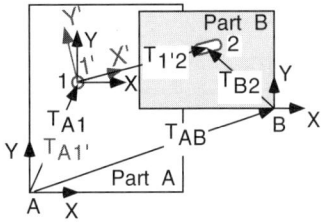

FIGURE 3-45. Figure for Problem 14.

15. Find $T_{A1'}$ in Problem 14 for the case where

$$T_{A1} = \begin{bmatrix} 1 & 0 & 0 & 1 \\ 0 & 1 & 0 & 3 \\ 0 & 0 & 1 & 0 \\ 0 & 0 & 0 & 1 \end{bmatrix}$$

$$T_{AB} = \begin{bmatrix} 1 & 0 & 0 & 6 \\ 0 & 1 & 0 & 2 \\ 0 & 0 & 1 & 0 \\ 0 & 0 & 0 & 1 \end{bmatrix}$$

$$T_{B2} = \begin{bmatrix} 1 & 0 & 0 & -2 \\ -0 & 1 & 0 & 2 \\ 0 & 0 & 1 & 0 \\ 0 & 0 & 0 & 1 \end{bmatrix}$$

16. Derive a subset of the relations expressed in Equation (3-29), as follows:
Define

$$R(z,\theta) = \begin{bmatrix} \cos\theta_z & \sin\theta_z & 0 \\ -\sin\theta_z & \cos\theta_z & 0 \\ 0 & 0 & 1 \end{bmatrix} = \begin{bmatrix} c & s & 0 \\ -s & c & 0 \\ 0 & 0 & 1 \end{bmatrix}$$

$$\Delta\theta_z = \begin{bmatrix} 1 & \gamma & 0 \\ -\gamma & 1 & 0 \\ 0 & 0 & 1 \end{bmatrix} = R(z,\gamma) \quad \text{where } \gamma \text{ is a small angle}$$

$$D\theta_z = \begin{bmatrix} 0 & \gamma & 0 \\ \gamma & 0 & 0 \\ 0 & 0 & 0 \end{bmatrix} = R(z,\gamma) - I$$

$$R' = R(z,\theta+\gamma)$$

$$\delta R = R' - R$$

Then show that

1.

$$\frac{\partial R(z,\theta)}{\partial\theta_z} = \begin{bmatrix} -s & c & 0 \\ -c & -s & 0 \\ 0 & 0 & 0 \end{bmatrix}$$

2.

$$\delta R = R(z,\theta)D\theta_z$$

3.

$$R' = R(z,\theta)\Delta\theta_z = R(z,\theta) + \gamma\frac{\partial R(z,\theta)}{\partial\theta_z}$$

4.

$$\begin{bmatrix} 0 \\ 0 \\ \gamma \end{bmatrix}$$

is the real eigenvector of matrix $\Delta\theta_z$ with eigenvalue equal to 1, that is

$$\Delta\theta_z\begin{bmatrix} 0 \\ 0 \\ \gamma \end{bmatrix} = \lambda\begin{bmatrix} 0 \\ 0 \\ \gamma \end{bmatrix}$$

where $\lambda = 1$.

17. Define

$$\delta R = \begin{bmatrix} 0 & -\delta\theta_z & \delta\theta_y \\ \delta\theta_z & 0 & -\delta\theta_x \\ -\delta\theta_y & \delta\theta_x & 0 \end{bmatrix}$$

Show that

1. $\bar{\delta}$ is the real eigenvector of δR with eigenvalue equal to 1; that is, $[\delta R]\bar{\delta} = \lambda\bar{\delta}$ where $\lambda = 1$ and

$$\bar{\delta} = \begin{bmatrix} \delta\theta_x \\ \delta\theta_y \\ \delta\theta_z \end{bmatrix}$$

2. $[\delta R]p = \bar{\delta} \times p$ (the cross product of $\bar{\delta}$ and p) where

$$p = \begin{bmatrix} p_x \\ p_y \\ p_z \end{bmatrix}$$

is any 3×1 vector

Hint: Use the definition of vector cross product

$$a \times b = \det\begin{bmatrix} i & j & k \\ a_x & a_y & a_z \\ b_x & b_y & b_z \end{bmatrix}$$

3.I. FURTHER READING

[Ahuja and Coons] Ahuja, D. V., and Coons, S. A., "Geometry for Construction and Display," *IBM Systems Journal,* vol. 7, no. 3 & 4, pp. 188–217, 1968.

[Bourjault] Bourjault, A., "Contribution á une Approche Méthodologique de l'Assemblage Automatisé: Elaboration automatique des séquences Opératoires," Thesis to obtain Grade de Docteur des Sciences Physiques at l'Université de Franche-Comté, November 1984.

[Callahan and Heisserman] Callahan, S., and Heisserman, J., "A Product Representation to Support Process Automation," in *Product Modeling for Computer Integrated Design and Manufacture,* Pratt, M., Sriram, R., and Wozny, M., editors, London. Chapman and Hull, 1996.

[De Fazio et al.] De Fazio, T. L., Edsall, A. C., Gustavson, R. E., Hernandez, J. A., Hutchins, P. M., Leung, H.-W., Luby, S. C., Metzinger, R. W., Nevins, J. L., Tung, K. K., and Whitney, D. E., "A Prototype for Feature-Based Design for Assembly," *1990 ASME Design Automation Conference,* vol. DE 23-1, pp. 9–16, Chicago, September 1990, also *ASME Journal of Mechanical Design,* vol. 115, pp. 723–734, 1993.

[Denavit and Hartenberg] Denavit, J., and Hartenberg, R. S., "A Kinematic Notation for Lower Pair Mechanisms Based on Matrices," *Journal of Applied Mechanics,* vol. 22, pp. 215–221, 1955.

[Faux and Pratt] Faux, I. D., and Pratt, M. J., *Computational Geometry for Design and Manufacture,* Chichester: Ellis Horwood Press, 1979.

[Gerbino and Serrano] Gerbino, S., and Serrano, J., "A Feature-Based Tolerancing Model for Functional Analysis in Assemblies of Rigid Parts," Proceedings of the 2nd CIRP ICME 2000 Seminar, Capri, Italy, June 21–23, 2000.

[Kim and Wu] Wu, C. H., and Kim, M. G., "Modeling of Part-Mating Strategies for Automating Assembly Operations for Robots," *IEEE Transactions on Systems, Man, and Cybernetics,* vol. 24, no. 7, pp. 1065–1074, 1994.

[Landau] Landau, B., "Developing the Requirements for an Assembly Advisor," S. M. thesis, MIT Mechanical Engineering Department, February 2000.

[Lee and Gossard] Lee, K., and Gossard, D. C., "A Hierarchical Data Structure for Representing Assemblies, Part 1," *CAD,* vol. 17, no. 1, pp. 15–19, 1985.

[Paul] Paul, R. P., *Robot Manipulators,* Cambridge: MIT Press, 1981. Chapter 1 is a comprehensive treatment of transforms.

[Pieper and Roth] Pieper, D. L., and Roth, B., "The Kinematics of Manipulators Under Computer Control," Proceedings of the 2nd International Conference On Theory of Machines and Mechanisms, Warsaw, September 1969.

[Popplestone] Popplestone, R., "Specifying Manipulation in Terms of Spatial Relationships," Department of Artificial Intelligence, University of Edinburgh, DAI Research Paper 117, 1979.

[Shah and Rogers] Shah, J. J., and Rogers, M., "Assembly Modeling as an Extension of Feature-Based Design," *Research in Engineering Design,* vol. 5, pp. 218–237, 1993.

[Simunovic] Simunovic, S., "Task Descriptors for Automatic Assembly," S. M. thesis, MIT Department of Mechanical Engineering, January 1976.

[Walton] Walton, M., *Car: A Drama of the American Workplace,* New York: Norton, 1997.

[Wesley, Taylor, and Grossman] Wesley, M. A., Taylor, R. H., and Grossman, D. D., "A Geometric Modeling System for Automated Mechanical Assembly," *IBM Journal of Research and Development,* vol. 24, no. 1, pp. 64–74, 1980.

4. CONSTRAINT IN ASSEMBLY

"We like to let the parts fall into place by themselves. The Europeans want to overpower the parts and force them into shape. So we always have to redesign their tooling."

4.A. INTRODUCTION

In Chapter 3 we discussed how to represent assemblies as chains of coordinate frames so that we could capture mathematically the fact that parts connect to each other. Some frames are at the nominal center of parts while others tell the location of assembly features on the parts relative to the part center frame. We noted that each feature frame could be joined to a feature frame on an adjacent part and that there was some mathematical constraint between the adjacent frames. But we did not say how the features *mechanically* operated to secure the location and orientation of one part relative to its neighbor. In this chapter we deal with mechanical constraint between parts and how assembly features impose that constraint.

First, we deal with the basic idea of constraint, that is, how to describe the motions that a part can undergo after some of its degrees of freedom have been constrained. A degree of freedom (dof) is said to be *constrained* when it can have only one value. This is not the same as being *stable,* which means that the dof is held at that value and cannot slip away.[1]

Second, we will define the degree of constraint between two parts and distinguish proper (also known as kinematic or exact) constraint, overconstraint, and underconstraint. Until we establish the necessary mathematical foundations, we will use the following heuristic definitions: Proper constraint means that each part is located in all six degrees of freedom.[2] This is accomplished, speaking roughly, by defining a surface on one part whose

responsibility is to provide a location and value for each degree of freedom for the other part, which it does by mating with a partner surface on the other part. Assembly is accomplished by pushing these surface pairs firmly against each other. In the process, the part loses its degrees of freedom and becomes located with respect to the other parts in the assembly.

Underconstraint means that one or more degrees of freedom are not constrained. That is, for one or more degrees of freedom, there is no mating pair of surfaces capable of defining and locating those degrees of freedom.

Overconstraint means, roughly, that more than one surface on a part seeks to establish the location of a degree of freedom on a mating part. An example is the use of two locating pins normal to the same plane, each of which seeks to locate a part on the plane along the line joining the pins by perfectly mating with two matching holes in the other part. Overconstraint usually causes internal stresses and other problems in the assembly, as will be explained in this chapter.

Third, we describe "kinematic assembly," a method of designing assemblies so there is no overconstraint.

Finally we will define proper constraint, overconstraint, and underconstraint mathematically using Screw Theory, a concept from classical kinematics. We will mathematically define the ability of an arbitrary assembly feature to impose constraint, and we will see how to combine the constraints of several features. At this point we will have a well-defined toolkit for describing a set of parts to a computer so that the location and degree of constraint of each part can be calculated. The computer will then be able to represent the nominal and the varied location

---

[1] This distinction is discussed in Section 4.C.6.b.

[2] Unless it needs one or more degrees of freedom in order to function.

of each part in a way that is completely analogous to how the actual parts mate. The consequences for calculating varied locations will be considered in Chapters 5 and 6.

In the process, we will learn that assemblies can be fit into the following four situations:

- Kinematically constrained

- An attempt at kinematic constraint that, for practical reasons, must be modified by adding small amounts of clearance or small amounts of local overconstraint or interference

- Deliberately overconstrained in order to achieve certain functions

- Constraint mistakes

4.B. THE STAPLER

Let us once again consider the desktop stapler. We can see from Figure 4-1 that it has several unconstrained degrees of freedom. Considering the base to be fixed, we can move the handle and the carrier about the pin, and we can slide the staples and the pusher inside the carrier. A coil spring (not shown) drives the pusher to the left and forces the staples to the left end of the carrier. The carrier thus gives the staples all of their six degrees of freedom and thus provides their constraint; that is, it gives those degrees of freedom their numerical values. The pusher and spring stabilize the staples but do not provide any constraint.

If the pusher were solidly locked to the right end of the carrier and also contacted the staples, then the pusher would be trying to establish the value of the staples' X degree of freedom. Since the left end of the carrier is trying to do the same thing, we would conclude that the staples are overconstrained in the X direction.

Both sides of the carrier appear to be trying to establish the Z degree of freedom of the staples, but in this case there is some clearance that prevents overconstraint. Naturally, if the staples are just a little too wide, they will not fit into the carrier or will not slide freely into position, causing the stapler to stop operating. Overconstraint, or a near-miss, creates delicate situations like this in assemblies.

A lever (also not shown) locks the handle to the carrier, while a friction force induced by interference keeps the carrier from turning freely with respect to the anvil. A second spring (also not shown) pushes the carrier

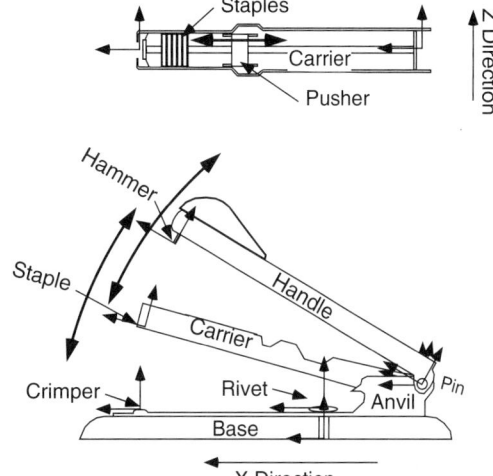

FIGURE 4-1. Degrees of Freedom of the Stapler. Within the stapler there are five parts with one unconstrained degree of freedom each, measured with respect to the base. The carrier, handle, and pin can rotate about the pin's axis, and the pusher and staples can slide inside the carrier.

clockwise with respect to the base so that the user can insert paper between them.

In this chapter we will learn how to describe assemblies more complex than the stapler, determine how many degrees of freedom they have, and decide if they are over- or underconstrained or if all six degrees of freedom are exactly constrained.

4.C. KINEMATIC DESIGN

4.C.1. Principles of Statics

Mechanical assembly is a subset of the classical theory of statics—that is, the description of bodies that may experience external forces and torques but do not accelerate. Statics deals with all the items listed in Figure 3-1: connective systems like pipes, structures like bridges and car bodies, and mechanisms like engines (as long as they are not operating or are moving very slowly).

The theory of statics permits the engineer to calculate the positions of all the parts in the item being analyzed as well as any internal stresses and strains. The analysis requires consideration of the following factors, which are called the principles of statics:

- *Geometric compatibility:* All the parts should have consistent locations with respect to each other; that is, it should be possible to utilize information about the size and shape of the parts, as well as information about contact between them, to calculate the location of any part based on knowledge of the location of any other part.

- *Force–moment equilibrium:* The sum of all forces applied to the parts should be zero, as should the sum of all moments. The same principle applies to each individual part in the assembly with respect to any internal forces and torques that act between the parts.

- *Stress–strain–temperature relations:* Based on the properties of the materials, all deformations caused by the applied or internal forces and torques can be calculated, along with any deformations that arise from changes in temperature. These deformations typically change the size or shape of parts.

In general, these factors interact with each other, giving rise to sets of simultaneous equations.

There is a special case that is of primary interest to us, though not of much interest to most teachers and practitioners of the theory of statics, namely the case where the parts are rigid (or the forces and torques are negligible) so that no deformations arise. Then the internal forces and the locations of the parts can be calculated directly from the applied forces and the geometric compatibility information. Such a situation is called *statically determinate.*

All the assemblies we deal with in this book are statically determinate, and static determinacy is central to the theory of assembly that is the core of the book. The connective model of assembly presented in Chapter 3 is valid only under these circumstances. When we multiply 4×4 matrices together to find the locations of adjacent parts, we are appealing directly to the principle of geometric compatibility and are ignoring the other two principles.

This fact permits us to present the following definition of an assembly, consistent with the theory of statics:

An assembly is a chain of coordinate frames on parts designed to achieve certain dimensional relationships, *called key characteristics, between some of the parts or between features on those parts.*

According to this definition, what makes an assembly an assembly is the chain of frames and its ability to define and deliver a key characteristic (KC). The parts simply provide material from which the assembly features can be fabricated so as to embody the desired constraint actions of the frames. The assembly-as-a-chain-of-frames is defined by the frame mathematics and not by the geometry of the parts.

When all three principles of statics are required to determine the positions of parts, the situation is called *statically indeterminate.* This is the case for most problems analyzed in statics texts and college courses. The positions of the parts can still be calculated, but the calculation is more complex. Merely multiplying the 4×4 matrices together will not give the correct answer.

It is common to call statically determinate assemblies "properly constrained," "fully constrained," or "kinematically constrained." Similarly, statically indeterminate assemblies are sometimes called "improperly constrained" or more commonly "overconstrained." There is *nothing* improper (in the dictionary sense) about statically indeterminate assemblies. We will look at several examples where overconstraint is essential in order for the assembly to deliver its KCs. We will also see that these assemblies must be designed so as to be properly constrained first. After they are assembled into a kinematically constrained condition, external or internal stresses are generated that bring them to their final fully stressed condition. This may be done, for example, by applying external loads or by using shrinkage that results from cooling the parts from elevated temperatures. Another class of "good" overconstrained assemblies contains redundant parts that share large loads.

However, many assemblies contain overconstraint by mistake. Constraint mistakes are so common that some authors feel that something very fundamental is missing from undergraduate engineering education. Correct consideration of constraint is essential in order to design competent assemblies according to the theory presented in this book.[3]

---

[3] Assemblies with intentional overconstraint can also be designed according to the principles in this book. However, their detailed analysis requires consideration of all three principles of statics. This is beyond the scope of this book and is dealt with by many engineering textbooks.

It is crucial to understand that we can decide if an assembly is kinematically constrained or overconstrained by looking at its nominal dimensions. It is not necessary, at first anyway, to examine variations. If an assembly is overconstrained at nominal dimensions, it is overconstrained, period. That is, *constraint is a property of the nominal design,* and the state of constraint of an assembly can be analyzed by inspecting the nominal design.

4.C.2. Degrees of Freedom

The motion of a rigid body can be described by six parameters, three related to linear motion and three related to rotation. Such a body is said to have six degrees of freedom. Usually the motion is referred to three axes at right angles to each other, but that is not necessary as long as each of the six motions can be defined and changed independently of the others.

Figure 4-2 shows a simple cube with three axes attached marked X, Y, and Z, plus α, β, and γ. The first three represent translations along the respective axes while the last three represent rotations about those axes. An object's location (position and orientation) is completely specified with respect to a reference set of axes when these six quantities are known relative to the reference. The object is then said to be *fully constrained.*

For example, suppose the cube is placed on the floor (equivalently, a plane parallel to the X–Y plane shown in Figure 4-2). Then it retains three unspecified degrees of freedom: along X, along Y, and about Z. If we slide the cube along the floor in the Y direction until it comes to rest with its X–Z face flush against a wall, then it has lost two more degrees of freedom (along Y and about Z). If we finally slide it along X until it meets another wall, it

loses its last remaining degree of freedom and becomes fully constrained.

An entirely equivalent way to see what degrees of freedom have been lost is to determine the directions in which we can apply force or torque without the cube moving. If the cube is again placed in the middle of the floor and remains in full contact with the floor, we can push it down in Z or twist it about X and Y as hard as we want and it will not move. Thus those axes are constrained and the others are free.

We may note a few things from this example that are true for any such example, as long as the bodies in question are rigid, there is no friction, and the bodies remain in full contact:

- The sum of the free and constrained degrees of freedom is six.
- In a direction that is constrained, we can push or twist one body against the other as hard as we want and the resulting velocity (linear or angular) will be zero.
- In a direction that is unconstrained, we can push one body relative to the other, while maintaining contact, as fast as we want linearly or angularly and the resulting force or torque will be zero.

These three facts describe an interesting duality between constrained and unconstrained and between force and velocity. These dual properties will be useful to us conceptually when we consider Screw Theory representations of constraint in Section 4.E.

Principles of constraint and degrees of freedom are familiar to technicians and designers who work with jigs and fixtures. The function of a fixture is to immobilize a part, say for the purpose of machining it. The primary task of the fixture is to ensure that all six degrees of freedom of the part are located reliably, repeatably, and solidly. Often, fixture designers speak of "3–2–1" to describe how an object is fixed in space, and they refer to the "3–2–1 principle." What they mean is this: It takes three points to determine the location of a plane, so if one places a part firmly on three sharp points, the part will be located on the imaginary plane that passes through those three points. If one then sets up a new pair of sharp points that do not lie in this plane or on a line normal to it, and push the part until it hits those two new points, the part will lose an additional two degrees of freedom. If one more sharp point is placed so that it does not lie in the first plane or on any line passing through the fourth and fifth points, and the

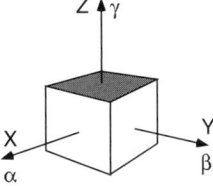

FIGURE 4-2. Degrees of Freedom of a Rigid Body. The three axes represent possibilities for the body's translation and rotation. If values are given for the three positions and angles, then the location (position and orientation) of the body is completely specified, and the body is said to be fully constrained.

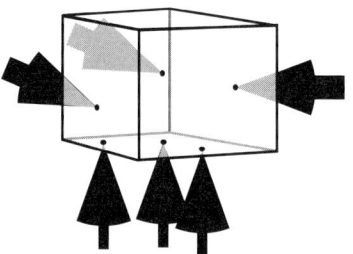

FIGURE 4-3. A Cube Located by the 3–2–1 Principle.

part is pushed up against this point, then the part is fully constrained.

This situation is illustrated in Figure 4-3.

The designer should pick a broad surface for mating to the first three points so that they can be far from each other. This surface must provide orientation stability over two intersecting axes. A long surface should be selected to mate with the next two points so that they, too, can be far from each other to provide stability about the third axis. These choices are discussed further in Chapter 5, where datum surfaces are dealt with.

It is important to understand that sharp points cannot be used in practice. Either they will be crushed or else they will dig into the part. In reality, contacts with finite surface area are used so that they can keep their shape in spite of the stress on them. Such a contact set is often called "semikinematic." The contact points are small regions, and a negligible amount of strain is created in them and in the part being located by them.[4] For most practical purposes, the part may be thought of as kinematically constrained. If the assembly must support large loads, such as those encountered in machine tools, the contacts are usually large plane or cylindrical surfaces. Mathematically, these may be interpreted as creating overconstraint. In practice, they require considerable care in design, fabrication, and assembly.

4.C.3. How Kinematics Addresses Constraint

In classical kinematics, the notions of over- and underconstraint are well known. Kinematicians have defined three kinds of mobility: instantaneous, finite, and full-cycle. Instantaneous mobility is the ability to move an infinitesimal amount; finite mobility is the ability to move an amount that is larger in the calculus sense than infinitesimal; full-cycle mobility allows a mechanism to move continuously without limit. Designers of assemblies other than mechanisms are more interested in proper constraint. That is, the assembly, or certain internal joints, should not be able to move at all, even instantaneously. In practice, instantaneous mobility allows a joint to move a small amount in a "mushy" way instead of being bound in a hard, definite way. Such mushiness is usually undesirable.

Kinematicians have developed mathematical formulas designed to detect the degree of constraint of a mechanism based on counting the links and joints and characterizing the joints' degrees of freedom ([Phillips]). These formulas return a number which can be related to constraint as follows:

- If the number is positive, the mechanism has that many free degrees of freedom and it can move, at least an infinitessimally small amount.

- If the number is zero, then the mechanism cannot move, not even infinitessimally, and it has just enough links to make it immobile.

- If the number is negative, then not only can the mechanism not move but its ability to move is prevented by that many more links than necessary.

For a general mechanism, one uses the Kutzbach criterion to determine this number, called M, the mobility of the mechanism:

$$M = 6(n - g - 1) + \sum f_i \qquad (4\text{-}1)$$

where M is the degree of mobility of the mechanism ($< 0, = 0,$ or > 0), n is the number of parts or links in the mechanism, g is the number of joints in the mechanism and f_i is the number of degrees of freedom available to joint i.

This criterion is customarily used to test for instantaneous mobility. It intended to be applied to general spatial mechanisms. For planar mechanisms, the Grübler criterion is used. This is the same as the Kutzbach criterion except that "6" is replaced by "3" and joints are assumed to have one or two degrees of freedom only.

These criteria usually work as expected, but sometimes they give an obviously wrong answer. Pursuing the reasons why these mistakes occur provides us valuable insight and motivates the quite different method of assessing the state of constraint of an assembly that is presented later in this chapter.

---

[4]If ultra-precision is needed, then these deformations must be taken into account ([Slocum]).

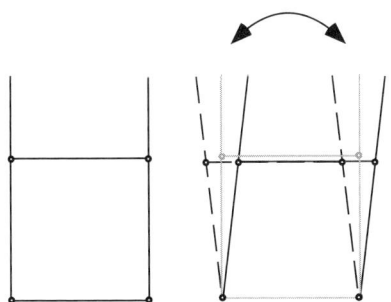

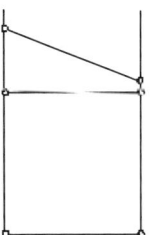

FIGURE 4-5. Four-Bar Linkage with One Additional Link. This mechanism is properly constrained and cannot move.

FIGURE 4-4. Common Planar Four-Bar Linkage. This mechanism has one degree of freedom. It consists of links joined by one degree of freedom pin joints.

Consider first the familiar planar four-bar linkage in Figure 4-4. This linkage has four links and four joints, each of which has one degree of freedom. Employing the Grübler criterion, we obtain

$$M = 3(4 - 4 - 1) + (1 + 1 + 1 + 1) = 1 \qquad (4\text{-}2)$$

That is, the mechanism has one degree of freedom, as we expected.

Now let us add another link, as shown in Figure 4-5. This mechanism has five links and six joints, and application of the Grübler criterion yields

$$M = 3(5 - 6 - 1) + 6 = 0 \qquad (4\text{-}3)$$

That is, the mechanism has zero degrees of freedom and cannot move, as expected. In addition, it has exactly enough links to cause it to be immobile; that is, it is not overconstrained.

However, consider the mechanism in Figure 4-6. It has the same number of links and joints as the one in Figure 4-5, so naturally the Grübler criterion will return the same answer, namely that the mechanism is locked but is not overconstrained. However, it is clear that this mechanism can indeed move. It has one degree of freedom. Moreover, the mechanism is overconstrained and contains, or can contain, locked-in stress. The reason is that the spacing between the upright links is determined by any two of the horizontal crossbars. If we add a third crossbar and vary its length ever so slightly, we can lock a compressive or tensile stress in it and the other two crossbars. However, in the mechanisms in Figure 4-4 and Figure 4-5, it is impossible to lock in stress by changing the length of any link by a small amount. This distinction is one of the main differences between overconstrained assemblies and those that are properly constrained or underconstrained:

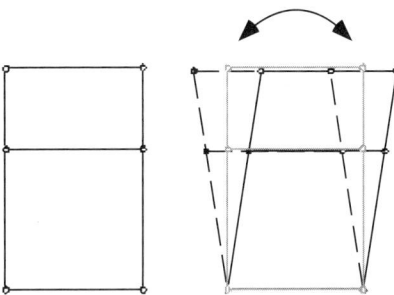

FIGURE 4-6. Mechanism that Is Both Over- and Underconstrained.

Only overconstrained mechanisms can be made to contain locked-in stress merely by infinitessimally changing the length of a link.[5]

So, quite to our surprise, the mechanism in Figure 4-6 is simultaneously over- and underconstrained. The Grübler–Kutzbach criterion cannot report such a condition because it reports only a single number intended to summarize the condition of the entire mechanism.

Why does the Grübler–Kutzbach criterion make this mistake? The reason goes back to the assumptions that underlie the derivation of the criterion itself ([Phillips], vol. 1, p. 14). The basic assumption is that each new link and joint bring with them new constraints that do not duplicate constraints already present in the mechanism. However, if we look at Figure 4-6, we can see that the top crossbar (or whichever one we add last) seeks to establish the distance between the two upright links, whereas this distance is already set by the existing two crossbars. Any degree of freedom that might have existed between these endpoints, which could have been reduced by adding the next link, actually never existed in the first place. Thus, contrary to the assumptions underlying the criteria, we are able in this

---

[5]This is not a necessary condition for overconstraint but is merely illustrative as well as quite common.

case to insert a link that adds constraint without reducing mobility.

Discoveries like this tell us that constraint and mobility are two different things, rather than being different sides of the same coin as assumed by the Grübler–Kutzbach criterion. In order to completely characterize the state of constraint of an assembly, we will need to determine separately its state of mobility and its state of constraint. We discuss a method for doing this beginning at Section 4.F.1. This method does not simply report a number signifying the mobility of the assembly or lack thereof, but instead returns a list of the specific degrees of instantaneous freedom of mobility and/or directions of instantaneous overconstraint of a given part in the assembly.

4.C.4. Kinematic Assemblies

Kinematic or statically determinate assemblies—also known as kinematic design ([Whitehead],[6] [Slocum], [Smith and Chetwynd], [Green]), minimum constraint design or minCD ([Kamm]) and exact constraint design ([Kriegel], [Blanding])—are well known to practitioners of the design of precision instruments and tools. Designers of fixtures and clamps call it "3–2–1" design. The earliest author known to have discussed it explicitly seems to be physicist James Clerk Maxwell. Furthermore, these authors define and give examples of the more practical realizations of statically determinate assemblies, whose joints unavoidably contain either small amounts of locked-in stress or small gaps. "Semikinematic design" ([Whitehead]) and "semi-minCD" ([Kamm]) are common names for this kind of compromise design, consistent with semikinematic fixturing discussed in Section 4.C.2.

The basic engineering idea in a kinematic assembly is that the parts constrain each other in space by virtue of their geometric interfaces. Simply placing them in correct contact, without exerting or locking in any force, completes the act of assembly. This can be done quickly, reliably, repeatably, and with little skill. It is therefore quite well suited to rapid mass production where assembly task times are short and the skill of a trained machinist or technician is not available or required.

---

[6]Whitehead's definition of a precision instrument is basically the same as our definition of an assembly: "An instrument can be regarded as a **chain** of related parts ... any mechanism whose function is directly dependent on the accuracy with which the component parts **achieve their required relationships**." (Emphasis added.)

4.C.5. Constraint Mistakes

Even though the principles of proper and kinematic constraint seem obvious and elementary, the fact is that designers and engineers make constraint mistakes.

If the design is unintentionally *underconstrained,* it neglects to locate a part in some direction. As a result, its location is uncertain and subject to random disturbances. A mathematical check for this condition is given in Section 4.F.1.a. However, an assembly can have underconstrained degree of freedom that support desired functions. A car door has one underconstrained degree of freedom.

If the design is unintentionally *overconstrained,* it contains attempts to locate a part or parts redundantly. For example, the X direction may be located by two different hard locators. The result will or could be locked-in stress along the X axis. The X degree of freedom of the part nevertheless has only one value. According to the definition of constraint at the beginning of the chapter, the part is constrained along X. How can it be both constrained and overconstrained? The answer is that, while a part is indeed constrained when all of its degrees of freedom have single values, the reverse is not true: A part whose degrees of freedom all have single values may still be overconstrained. Just counting how many degrees of freedom have single values overlooks the possibility of overconstraint. A separate check for this condition is necessary. It is given in Section 4.F.1.b. However, an assembly can have overconstrained degrees of freedom that support desired functions. A preloaded pair of ball bearings is overconstrained.

A part or assembly is "properly" or "kinematically" constrained only when it is not underconstrained and not overconstrained. Assemblies that are properly constrained or underconstrained will not contain locked-in stress.

Many CAD systems seek to determine the state of constraint of a part by counting how many degrees of freedom have single values, or by using other methods that take account only of geometry. As a result, these systems fail to detect overconstraints that the methods in this chapter detect.

Three examples of erroneous or questionable overconstraint are given next.

4.C.5.a. Copier
The first example, involving a design error, is shown in Figure 4-7 and involves a desktop copier or printer ([Kriegel]). At the top we see a portion of the original design, consisting of two parallel side panels and a curved

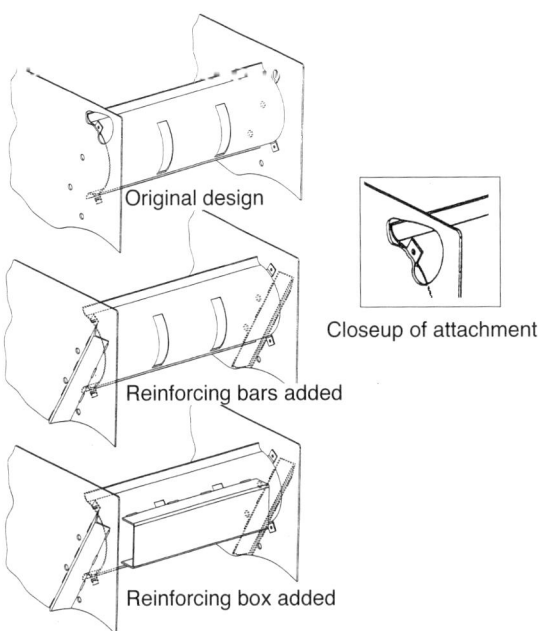

FIGURE 4-7. **An Overconstrained Design.** This figure shows a portion of a desktop printer or copier with an over-constraint mistake ([Kriegel]). The three figures at the left tell a story of efforts by engineers to eliminate the distortion caused by the overconstraint. The figure at the right shows the bent tab by which the curved cross panel attaches to the vertical side panels. The attachment screw is not shown. (Copyright © 1995 ASME. Used by permission.)

panel joining them. The side panels form the sides of the paper path while the curved panel lies in the plane of the paper and serves to turn the paper over. The side panels are mounted firmly in some other portion of the assembly that is not shown. The curved panel joins to the side panels with right-angle tabs and screws. A detail of one such tab is shown in the figure.

An engineer observed that, in some copiers, the side panels were warped inward, causing the paper to jam. To avoid this, he added reinforcing bars, as shown in the middle frame of Figure 4-7. Another engineer later observed that some of the curved panels were warped, and he added a reinforcing box to the curved panel, whereupon the small tabs holding the curved panel to one of the side panels broke.

What we see in this example is a fight between the curved panel and the side panels over which part or parts will set the distance between the side panels. One of the thought questions at the end of the chapter asks you

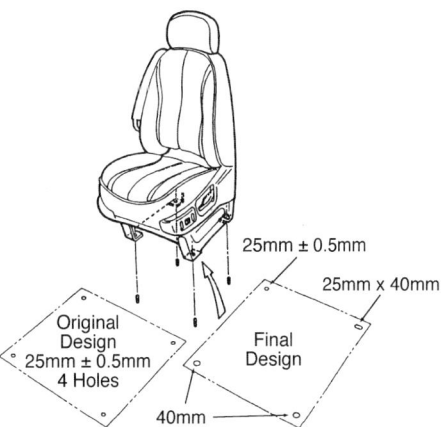

FIGURE 4-8. **Original Overconstrained Car Seat Installation and Final Properly Constrained Installation.** In the original design, there were four 25-mm-diameter studs on the seat that had to be assembled to four 25 ± 0.5-mm-diameter holes. In the final design, the right rear hole has been enlarged into a slot 25 × 40 mm, and the front holes have been enlarged to 40-mm diameter. ([Sweder and Pollack]. Reprinted with permission from SAE Paper 942334. Copyright © 1994 Society of Automotive Engineers.)

to suggest how this problem should have been solved. (Clearly the reinforcing bars were not the correct solution.)

4.C.5.b. Car Seat Mounting

The second example involves over-constraint in a car seat ([Sweder and Pollack]). Figure 4-8 shows how the problem was recognized and solved.

Sweder and Pollack's article is about the use of tolerance analysis software called VSA. It helped them to solve the problem. In the original design, the seat was attached to the car floor by means of four 25-mm-nominal-diameter studs pushed into 25-mm-nominal-diameter holes. Different line workers had different ways of installing the seats,[7] and the seats were inconsistently located with respect to the car doors. VSA detected that the seats were overconstrained. (One of the thought questions at the end of the chapter asks you to explain why the original design is overconstrained.)

The design was modified as follows: The two front holes were enlarged to 40-mm diameter, and the right rear

---

[7]Such an operation is called *operator-dependent,* an undesirable situation. Overconstraint caused this assembly step to be operator-dependent.

hole was enlarged in the side–side direction to create a slot 25 × 40 mm. The fourth hole was left at 25 mm diameter. The line worker was instructed to start by placing the left rear seat stud in this fourth hole. Installation proceeded easily because the right rear stud fell into the slot when the line worker pivoted the seat around the first stud. The two front studs then fell easily into the wide open front holes. Consistent assembly and clearance between the seat and the door were achieved.

4.C.5.c. Aircraft Structure

This example is described in [Hart-Smith]. It involves alternate methods of making aircraft structures containing ribs, skins, and spars, as shown in Figure 4-9. The two spars shown are supposed to provide a structural shape, which is created in turn by spacing the spars apart with a series of ribs. In the past, all ribs were made of separate pieces consisting of a flat panel and separate ends shaped like angle brackets and called shear clips. The spars, which we will assume are rigid, were placed in a fixture to provide the correct fore–aft spacing, and the ribs were built into place to suit the space and maintain it once the parts were removed from the fixture. The spanwise spacing of the ribs was provided by the fixture as well.

Two trends gave rise to a new method. First, design for assembly (DFA) recommended reducing the number of parts. (Chapter 15 discusses DFA in detail, including some of the conflicts and paradoxes raised by this example.) Second, numerical control machining (NC) made it possible to machine complete ribs combining the shear clips and the flat panel. Combining these two trends led to machining the ribs from single large slabs of aluminum. This created apparently efficient ribs in the sense that several assembly steps were eliminated. The new method also dispensed with the fixture and sought to use the ribs to set the distance between the spars. However, it was difficult to make the ribs accurately enough in the fore–aft direction, so such parts often required hand grinding or shimming during assembly, reducing the hoped-for efficiencies. Especially when the ribs were large (a meter or more in some cases), achieving the required tolerances during machining proved uneconomical. If the ribs were not trimmed or shimmed, then compressive or tensile stresses were built up in them, and the spars were warped out of their desired aerodynamic shape. This is undesirable.

[Hart-Smith] recommends making only the two end ribs as single NC-machined pieces, calling them the defining ribs. The remaining intermediate ribs should be made adjustable in the fore–aft direction. Several options are available for doing this, as shown in Figure 4-10 and Figure 4-11. Each method allows enough adjustment in the fore–aft length of the intermediate ribs so that no

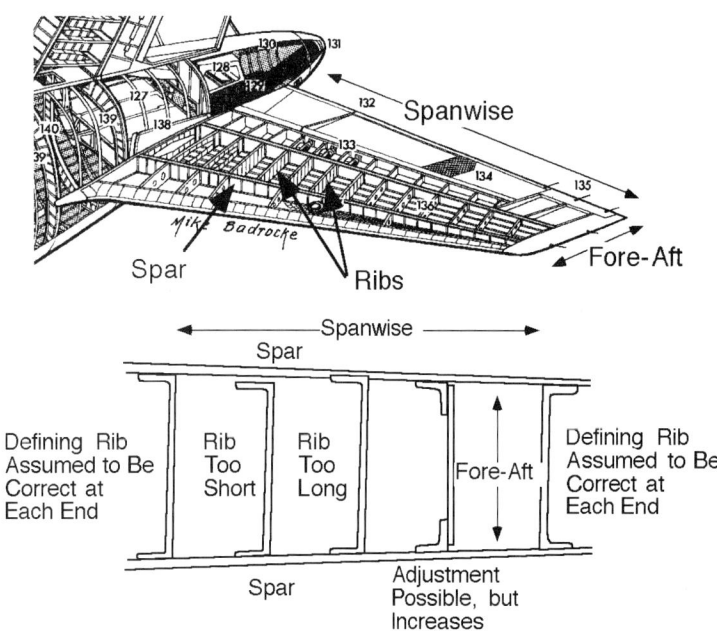

FIGURE 4-9. Example Assembly of Aircraft Structure. *Top:* Cutaway drawing of the horizontal stabilizer near the tail of an airplane. *Bottom:* Detail of ribs and spars, looking down on the stabilizer. The two spars are supposed to be a certain distance apart. The figure compares two methods of doing this. One is based on trying to make all the ribs single pre-machined pieces and relying on them to set the spacing between the spars. The other is based on building up a rib from two angle-shaped shear clips plus a flat piece and relying on an assembly fixture to set the spacing. (Structural cutaway courtesy of *Air International.* Copyright © 1980 *Air International.* Used by permission. Detail of ribs and spars reprinted from [Hart-Smith] with permission. Copyright © 1997 SAE International.)

conflict will occur. Predrilled holes in one spar provide the spanwise location of each rib.

This example makes several points, some of which are made by the previous examples as well. First, if all the

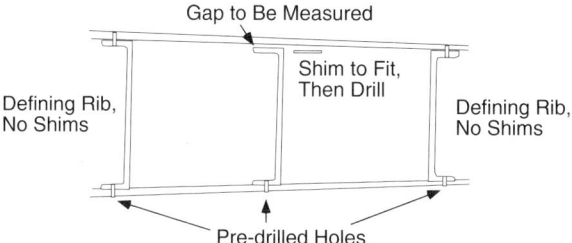

FIGURE 4-10. First Candidate Solution to Spar Spacing Problem. In this solution, one-piece ribs are made, either by NC machining or prefabrication from separate parts. Two defining ribs set the fore–aft distance while each intermediate rib is made a little small and shimmed to fill the gap. The predrilled holes on one spar set the spanwise spacing of the ribs. In practice, aircraft manufacturers usually require that the shim fill the space until the remaining gap is less than some limit, say 0.005". This gap can be drawn up tight by the fastener. The stress thus locked in is called the pull-up stress. Limiting the remaining gap limits the pull-up stress. Pull-up stresses must be subtracted from safety factors in order to determine how much margin there is over anticipated flight loads. Proper management of locked-in stresses in aircraft structure is therefore very important. (Reprinted from [Hart-Smith] with permission. Copyright © 1997 SAE International.)

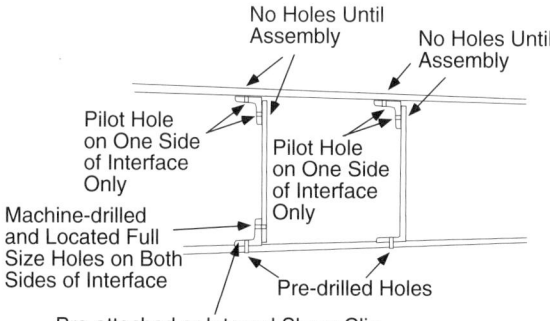

FIGURE 4-11. Second Candidate Solution to Spar Spacing Problem. In this solution, defining ribs (not shown) are again one piece, while multipiece intermediate ribs are used, and some holes are drilled at the time of assembly. On the left, the rib is made entirely from separate pieces, while on the right an NC-made rib has one end premachined to include the shear clip. The predrilled holes on one spar set the spanwise spacing of the ribs. (Reprinted from [Hart-Smith] with permission. Copyright © 1997 SAE International.)

ribs are one piece, then multiple parts will claim a role in setting the fore–aft distance. This is a sure sign of impending trouble. In terms of our previous terminology, the nominal design is overconstrained in the fore–aft direction. The solution proposed is to choose the winners of the conflict in advance and demand that the others be adjusted to suit. Second, there are two important directions, spanwise and fore–aft. Each one is handled separately, using distinct features. In aircraft assembly it is customary to use predrilled holes to set the spanwise spacing of ribs. (Note that these holes appear only on one of the two spars. A thought question asks you to explain why.) Third, several design criteria and recommended practices clash: the desire for assembly accuracy, the desire to reduce the number of parts and assembly steps, and the need for economical machining of parts. The engineer needs to evaluate each of these and decide which to favor, with the goal of achieving the KCs repeatably and economically.

Note that if the spars are not rigid, then either a fixture and adjustable ribs or nonadjustable ribs of sufficient length accuracy will be needed to achieve the desired fore–aft distance.

4.C.6. "Good" Overconstrained Assemblies

To ensure that the reader understands that overconstraint is not always bad, it should be noted that many assemblies achieve their KCs only because there is locked-in stress created by overconstraint. In some cases, this stress is the end goal itself. In other cases, it is required so that the assembly achieves its gross desired shape. In yet others, this stress is used to resist external loads.

Another class of "good" overconstrained assemblies does not contain locked-in stress. Instead the overconstraint arises because there are more locators than needed to provide location. These extra locators can provide symmetry in managing external loads or can provide shape definition for flexible parts like sheet metal or cloth. Several examples of each are given here.

4.C.6.a. Locked-in Stress

The locked-in stress is the end itself in case-hardened armor plate. By means of metallurgical processing and control of the cooling process, the outer surface of the plate is placed in compression. This makes it possible for the plate to resist rupturing caused by crack propagation when it is struck by a projectile.

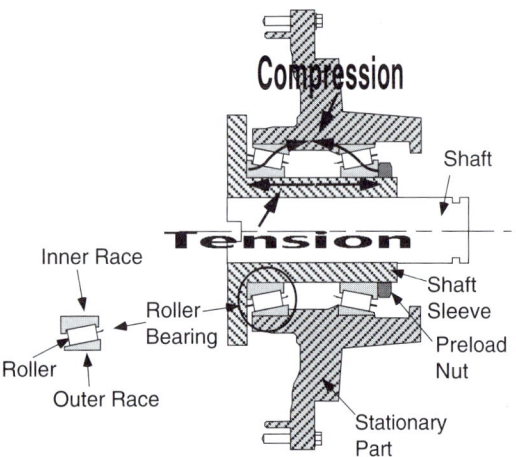

FIGURE 4-12. Preloaded Opposed Pair Bearing Set. The shaft is held by two roller bearings with inclined rollers. When the preload nut at the right is tightened, it forces the inner race of the right bearing into the roller, which is forced into the outer race. The force passes through the stationary part and into the outer race of the left bearing. From there it passes into the left roller, forcing it into the left inner race. The left inner race pushes leftward against the shaft sleeve. The stationary part is thus placed in compression while the rotating shaft sleeve is placed in tension. The rollers and races are in the compressed side of the force loop, placing them in compression and establishing the desired Hertzian stresses.

Locked-in stress creates the gross shape of hot-air balloons, parachutes, suspension bridges, clothing, upholstery, and hang gliders.[8] The stress arises from the force of internal air pressure, the geometry of an internal structure, gravity, and so on.

The preloaded opposed bearing set shown in Figure 4-12 is used in an automatic transmission for trucks. Similar bearing sets are used to support machine tool spindles and other shafts subjected to large loads in arbitrary directions. In any such case, the KC is the ability of the bearing set to hold the shaft aligned in the face of external loads along or normal to the rotation axis. In the transmission, such loads could be caused by the drive shaft, while in a machine tool, cutting forces create them. If the machine tool spindle deflects under such loads, the machine will lose accuracy. Thus the engineer tries to make the bearing system stiff against these loads. This is done by preloading the bearings. The preload sets up Hertzian

stresses in the rollers and races. Due to the geometry of the contact region between races and rollers, the contact area grows rapidly as load builds up, causing the local stiffness to rise rapidly as well. Once the preload is established, any external load will encounter a very stiff system.

Even though each of these examples achieves its KCs by virtue of having locked-in stress due to overconstraint, it is important to realize that the unstressed parts must bear a certain geometric relationship to each other first. Otherwise the locked in stress will not arise or else it will not create the desired shape or stress pattern. That is, behind each of the overconstrained examples is a properly constrained design whose parts must bear certain relationships in the unstressed state in order that the desired overconstrained state will be reached. Thus proper constraint remains an important element of the design even of overconstrained systems.

4.C.6.b. Redundant Locators

Redundant locators exist in many designs where external loads must be reacted and must be distributed uniformly to reduce the load on any one locator or to provide symmetry and avoid an unbalanced situation.

A typical example of this is a planetary gear train (see Figure 4-13). In principle, the relative kinematic locations of the parts are given if only one planet gear is present. However, this configuration will lead to unbalanced loads in the gearbox, leading to a situation where the planet carrier will be tilted off its desired rotation axis. So in

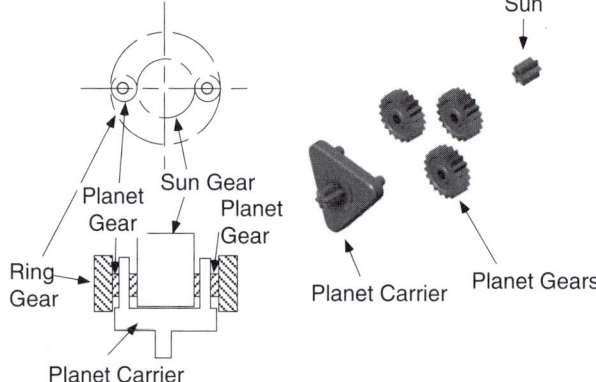

FIGURE 4-13. Planetary Gear Train. *Left:* Schematic drawing showing the pitch circles in the top view and cross sections in the side view. *Right:* Example of a planetary gear train from a cordless screwdriver. (The drawing was prepared by MIT students Jeffrey Dahmus, Yu-Feng Wei, and Meow Yong.)

---

[8]Thanks to Thomas De Fazio for helping create this list.

typical planetary gear trains there are three or four planets.

Naturally, it is impossible to size and tolerance these parts so that there is a line fit among all of them. Either there will be interference, making assembly impossible, or else there will be clearance, making equal distribution of loads impossible. The latter situation is the one used in practice. As the unit is used, the parts with higher loads wear down a little until all the interfaces bear more equal shares of the load. Designers specify the smallest clearances, along with the tightest tolerances on the clearances, that they can, given the price target of the product. Sometimes, selective assembly is used. In this method, discussed in Chapter 6, planets are selected according to a purposeful strategy rather than being taken at random from a bin, so that equal and small clearances are achieved.

4.C.7. Location, Constraint, and Stability

If a part is "located," is it "constrained"? If it is constrained, can it still move? The naked words leave room for confusion. Careful definition can eliminate the confusion.

First, if a part is located, this means that we can definitively calculate its location as long as all its geometric interfaces are in their desired states. Such a part is constrained. A cube on the floor in the corner of a room has a definite location as long as it is snugly against the floor and both walls. It can be moved out of position, however. It needs some kind of retainer or small retaining force to keep it in place, to make it stable. Such a force is regarded

as too small to set up significant locked-in stress. It is called an *effector*.

Each surface that provides a condition of geometric compatibility and eliminates one or more degrees of freedom may be called a *locator*. It is usually considered rigid. Sometimes each of the six degrees of freedom of a body has its own locator, but more often a locator will provide constraint in more than one degree of freedom. Example locators and the degrees of freedom they constrain are the subject of Section 4.E.2.d. In Chapter 8 we will call locators *mates* and effectors *contacts*. Overconstraint occurs when a degree of freedom encounters two or more locators, each of which seeks to rigidly impose location in that direction.

Whitehead says that a part gets its position from locators and keeps it with effectors. The effectors can be small forces or other locators. Several combinations of cases can occur as shown in Table 4-1. This table explores the possibilities using the example of a stool resting on the floor. The examples generally describe a three-dimensional situation in the interests of simplicity.

4.C.8. One-Sided and Two-Sided Constraints—Also Known as Force Closure and Form Closure

Another way to appreciate the difference between locators and effectors is to distinguish one-sided constraint and two-sided constraint ([Blanding]). One-sided constraint exists, for example, when a cube is placed on the floor. A large enough disturbing force directed generally away

TABLE 4-1. Cases of Constraint Defined by Whitehead

	Enough Points of Contact	Too Many Points of Contact
Constraint effected by a "small" force	Pure kinematic design: • Three-legged stool with point-tipped legs Semikinematic design with multiple constraints in the small contact area of each locator: • Three-legged stool with a nonzero contact area tip on each leg, making it impossible to say exactly where the contact point is	Redundant constraint: • Four-legged stool with point-tipped legs (This is really two three-legged stools—your choice which one)
Constraint effected by a large force (called "another locator" by Whitehead)	Semikinematic design: • Three-legged stool with a nonzero contact area tip on each leg and each leg bolted down to the floor	Overconstraint: • Four-legged stool with each leg bolted down to the floor

Note: Whitehead's definitions of kinematic or semikinematic design, redundant constraint, and overconstraint are shown in this table, along with examples that generally cover three degrees of freedom. Underconstraint is not shown.

from the locator surface can dislodge the cube. The floor is the locator while gravity provides the effector force. If the cube is large enough to contact both the floor and the ceiling (assuming floor and ceiling are rigid), then the direction normal to the floor encounters a two-sided constraint and the cube is clearly overconstrained.

The two-sided constraint is also called "form closure" ([Green]). When two parts are joined with a form closure, some surface or surfaces on one part totally surround and intimately contact some mating surfaces on the other part. This means that the part being constrained can be held without calculating how much force is being applied by the effectors. Examples are ideal ball joints or prismatic joints.[9] Ideally, form closures can exist with zero clearance, but this is impossible in practice.

By contrast, one-sided constraint is also called "force closure." In this case, we need to know how much force and friction are available to the effectors and constraint surfaces before we know that a part has been immobilized against a given level of disturbing force. Force closures are used in fixturing parts for the purpose of machining, because they can be arranged to have zero clearance. Thus the parts will be accurately located and will stay located when subjected to large machining forces.

All of the constraint calculations in this book assume zero friction.

Kinematic assembly concepts first emerged in the domain of precision instruments like microscopes where external loads are low or their direction can be predicted from patterns of normal use of the instrument. One-sided constraints are often sufficient in such devices. The constraint surfaces can be oriented to resist expected loads, and relatively low-stiffness effectors, such as screwed-on retaining plates or compression springs, can be used to hold critical parts against their respective constraint surfaces. These effectors will not normally see large loads that could dislodge a part from its location.

Other kinds of machinery can see large or unbalanced loads in more directions or from unexpected directions. These loads must be resisted by substantial load-bearing surfaces. A vertical machine tool shaft cannot run in a V-groove (which provides proper constraint from only one side) but must run in a fully encircling bearing so that it can resist randomly oriented loads and distribute lubrication. The bearing may even operate via hydrodynamic oil pressure rather than direct running contact. The planetary gear train discussed above contains redundant constraints in order to balance the loads, whose direction is known in advance. Front-end loaders might be designed with a single centered support linkage for the shovel. But this would obstruct the operator's view while off-balance loads would twist the linkage. So, instead, these machines have duplicate linkages left and right, creating the potential for overconstraint. Thus the need to support arbitrary loads brings with it the need to introduce opposing surfaces and close running clearances. Along with these conditions comes the chance of overconstraint due to the two-sided constraints that these opposing surfaces generate.

In terms of Table 4-1, we can say that one-sided constraints fall into the classes called pure kinematic constraint or semikinematic constraint. Two-sided constraints fall into the class of redundant constraint if there is a little clearance between the constrained item and its two constraining surfaces.[10] Similarly, two-sided constraints fall into the class of overconstraint if there is interference between the constrained item and its two constraining surfaces.

The case of exactly zero clearance represents pure two-sided constraint and falls into the class of overconstraint. It exists only mathematically and never occurs in practice. However, it is the only consistent way to define form closures that permits us to examine the state of constraint of assemblies that contain them. Zero clearance can be replaced in practical designs with small clearance or small interference as necessary. Essentially, this means that we consider zero clearance to be the fundamental property of form closures, while clearance to avoid overconstraint is an artifact of engineering reality.[11]

Smart engineers will recognize the need for care and the possibility of problems due to overconstraint in situations where two-sided constraints are used to support heavy loads. In Section 4.F.3 we will see that some very common locating methods, such as pins in holes, are actually mathematically overconstrained because they impose two-sided constraint. Nevertheless, these locating methods are very useful and we must learn how to deal with them both mathematically and practically.

---

[9]In kinematics, these are called lower pairs.

[10]Thanks to Stefan von Praun for pointing this out.

[11]This is analogous to Galileo deciding that the fundamentals of mechanics describe cases with zero friction, and that friction is an artifact that can be handled separately as a disturbing force.

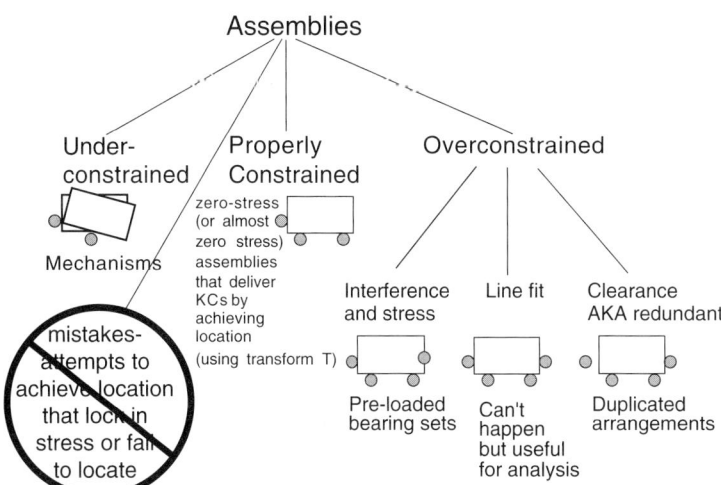

FIGURE 4-14. Summary of Constraint Situations. All three main possibilities, properly constrained, underconstrained, and overconstrained, can occur in well designed mechanical assemblies. Over- and underconstraint can arise for functional or assembly process reasons, or from mistakes. Some of the categories in this figure will be discussed in later chapters, and several elements will be added to this figure at those times.

4.C.9. Force–Motion Ambiguity

If we locate an object using a force closure and a locator, we know exactly where it is, assuming we perfectly know the location of the locator. However, the object cannot resist an arbitrary force. If we try to locate an object using a zero-clearance form closure, it can resist an arbitrary force but the form closure will exert overconstraint. Real form closures contain some clearance. However, this means that we do not know the location of the object any more accurately than the size of the clearance.

Combining these alternatives, we find that we can have perfect knowledge of the location but not of the supportable force, or we can have perfect knowledge of the supportable force but not of the location. This curious fact is similar in spirit to Heisenberg's Uncertainty Principle in physics. Any attempt to decrease location uncertainty in a form closure will require increased care and cost in manufacture and the risk of overconstraint.[12] When the loads on an assembly require form closures, the designer must be prepared to face these costs and risks.

4.C.10. Summary of Constraint Situations

Figure 4-14 summarizes the situations we have considered, plus some that we will consider in later chapters.

Assemblies are either zero stress (or almost zero stress) or else they are nonzero stress. Zero-stress (kinematic or semikinematic) assemblies achieve their KCs by virtue of having their parts located at desired places in space relative to each other. In the absence of stress, geometric compatibility suffices to calculate these locations, which are captured by a series of 4×4 transformations as described in Chapter 3. Zero-stress assemblies can be fully constrained or can have operating degrees of freedom, such as automobile engines do. Underconstraint can arise due to assembly requirements, which will be discussed in later chapters. Fixtures are used to provide the missing constraint. Underconstraint can also arise because of a design mistake. Nonzero-stress assemblies can also occur for two reasons: (a) intentional designs that achieve their KCs by means of locked-in stresses and (b) outright mistakes. Such mistakes are committed surprisingly often. As discussed elsewhere in this chapter, CAD systems as of this writing are of little help in detecting overconstraint mistakes. The theory presented later in this chapter will help in this regard. Finally, remember that deliberately overconstrained designs rest on kinematically constrained unstressed designs. The unstressed kinematic arrangement is set up in such a way that the applied load will generate exactly the desired locked-in stress. This requires skill and forethought.

Remember that 4×4 matrices will not correctly calculate the relative locations of parts if there is locked-in stress. In such a case, all three principles of statics must be brought to bear to perform the calculation. The difference in location between the stressed and unstressed

---

[12] Also, except in some situations where production volumes are very large, the resulting parts will have to be fitted to each other as matched pairs and will not be interchangeable. As discussed in Chapters 5 and 6, interchangeable parts are highly desirable for speeding up assembly and reducing its cost.

conditions may seem small but nevertheless will often be crucial. Whether or not the difference in position is important, the difference in internal forces or stresses will be large and important. Very small changes in the size or shape of parts will raise the locked in stress dramatically, requiring great care in design, fabrication, and assembly. If locked-in stress is not required to achieve the KCs, then overconstraint mistakes can only lead to trouble, including nonrobustness[13] in the state of internal stress.

This is all that we will say about designs that contain deliberate locked-in stress because the focus of this book is on assemblies that deliver their KCs by means of properly placing parts in space at particular locations in the absence of stress.

4.D. FEATURES AS CARRIERS OF CONSTRAINT

The discussion above on constraint is necessary in order to understand the role of assembly features in assembly. Simply put, assembly features are the carriers of constraint between parts. When parts are joined, degrees of freedom are constrained. The shape of the feature determines which degrees of freedom are constrained and which remain free.

Several scenarios for joining parts via features can be envisioned. The simplest one provides a feature that by itself constrains all six degrees of freedom. A square peg in a blind square hole can do this if we consider only the mathematics. From a practical point of view, such a feature may not be the best choice. First, it contains several two-sided constraints, although we can soften their effect if we can tolerate a small amount of locked in stress or a small gap in the joint. Second, if there is a small gap, then repeated alternating loads will lead to a kind of battering wear that will gradually destroy the feature.[14]

A more robust way to join two parts is to use multiple features, each of which provides constraint in some of the six degrees of freedom. The car seat discussed in Section 4.C.5.b provides an example of this. Three degrees of freedom are constrained by the 25-mm pin–hole joint and the surrounding floor (X and Y normal to the hole axis and Z along it), two more by the 25-mm × 40-mm pin–slot joint and the floor around it (θ_z about the hole's axis and θ_y about an axis mutually normal to the hole axis and to the line joining the hole and slot), and the last one (θ_x around the axis joining the hole and the slot) by either of the two front legs resting on the floor pan. (Of course, all four legs are on the floor so in principle the seat is overconstrained in the X–Y plane, but the floor pan is relatively compliant, so the stress built up in it is negligible if the length of the legs is the same within a fraction of a millimeter and the studs can rock slightly in the holes.)

A third way to join two (or more) parts is to use several other parts, each of which contains one or more features that individually constrain some of the third part's degrees of freedom. In order for this method to be definitive in its ability to locate the third part, the first two (or more) must be fully constrained with each other first. That is, they must constitute a fully constrained subassembly. At that point, they are equivalent to a single part with multiple features on it, and the situation is the same as the previous one. The car seat is in fact an example of this since its legs are individual parts that have been joined with other parts to make the seat frame.[15]

As another example, consider the automobile engine assembly situation shown in Figure 4-15. The KC is to center the combustion chambers in the head exactly concentrically with the cylinder bores. Any error in this concentricity will create a thin crescent-shaped shelf on the

---

[13]A design is called robust if its performance is substantially insensitive to variations that might occur, for example, in the size and shape of parts, external loads or operating conditions, or actions by the user of the product. When a design is overconstrained, very small changes in geometry can cause large changes in internal stress. Thanks to Christopher Magee for pointing out the relationship between proper constraint and robustness.

[14]This is a major source of wear in a home breadmaker, for example.

[15]The seat is usually made at a different factory and brought to the final assembly plant for installation in the car. The final assembly process needs to occur in less than a minute. Any constraint mistakes, as noted in [Sweder and Pollack], cause severe problems. The seat therefore not only must be a fully constrained subassembly from the mathematical point of view but must be defined that way from the start of its design so that engineering communication from the car company to the seat factory will be definitive. As discussed in Chapter 1, this issue is one of many that make assembly an issue in supply chain management. At this point in the book, we are starting to develop the necessary vocabulary and supporting mathematics to address this issue systematically.

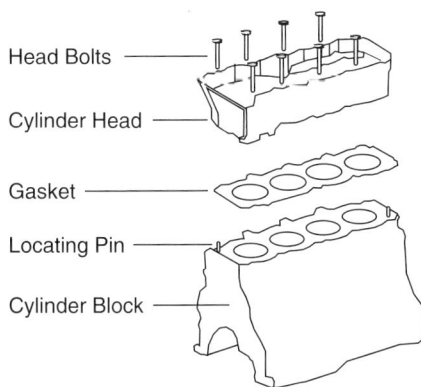

Head Bolts

Cylinder Head

Gasket

Locating Pin

Cylinder Block

FIGURE 4-15. Typical Method of Mounting a Cylinder Head to a Cylinder Block. The KC is to center the combustion chambers on the head over the cylinder bores. Small errors in this location will cause particulates to accumulate on the head-face of the block, leading to failure in an emissions test. Locating pins and holes are the assembly features. The head bolts merely absorb combustion pressure.

top face of the block (called the head-face) where particulates can accumulate, causing the engine eventually to fail an emissions test.

The head is located with respect to the block by means of two locating pins. The head bolts do not locate the head on the block. They merely effect the location, keeping the head from flying off due to the pressure raised by the combustion activity. Thus the pins on the block and their mating holes in the head are the assembly features that create the constraint between head and block in the plane of the head-face. Strictly speaking, the head is overconstrained in this plane, requiring considerable care when each plane is machined, although the gasket can absorb a small amount of this overconstraint. The head is also overconstrained by the two locating pins and holes. The pins are short and the head, made of aluminum, will deform a little around the holes, so the overconstraint can be ignored. If there is concern that this last overconstraint could shift the head in the plane of the head-face, a stress analysis can be performed.

Finally, parts can be constrained by placing them in assembly fixtures. Features on the fixtures can position parts that are not fully constrained by the part features that join them, so that the combined degrees of freedom constrained by the fixtures and the part joints together provide full constraint; the underconstrained part joints can then be fastened, preserving the interpart relationships.

Assembly fixtures are designed for the specific purpose of locating parts during assembly, albeit temporarily. Since no arbitrary or unanticipated loads will be exerted on the fixtured parts, and since easy placement in, and removal from, the fixture are important, assembly fixtures usually provide one-sided constraint for parts. The constraint is effected by clamps that push the parts securely against the constraining (locating) surfaces of the fixture. These surfaces may be either (a) surfaces or edges against which an edge or surface, respectively, of the part is pushed or (b) pins onto which holes or slots in parts are placed. Based on the foregoing, and on what is to come, the reader should note that *placing a part in a fixture is mathematically an act of assembly* just as much as joining one part to another via an assembly feature is. The only difference is a practical one: Parts are always disassembled from fixtures, but not necessarily from each other.

Fixtures often contain overconstraints due to an excessive number of clamps, extra locators, a combination of part–fixture and part–part interfaces, or all of these. We will say more about this in Chapter 8.

4.E. USE OF SCREW THEORY TO REPRESENT AND ANALYZE CONSTRAINT[16]

We are now in a position to give a mathematical basis for constraint. Our approach is called Screw Theory, which is part of classical kinematics. Up to now it has found little use outside of kinematics, except recently when it was applied to establish a theory of mechanical grasping in the field of robotics. All of its underlying ideas have been appreciated intuitively by practical designers of precision machinery, but it is not a common subject in engineering school classes in design. One reason for this is that typical presentations of Screw Theory make it appear impenetrable. The presentation given here goes only deeply enough to permit us to use it conveniently.

4.E.1. History

Application of Screw Theory to kinematic computations began in 1900 with [Ball]. [Waldron] was the first to

---

[16]This section is based on [Adams and Whitney] and [Shukla and Whitney 2001a, 2001b].

apply Screw Theory to the problem of determining the relative degrees of freedom (dof) between any two bodies in a mechanism. [Davies 1981] showed how Kirchhoff's loop and node equations could be applied to mechanism analysis. Using that theory, he developed matrix algebra-based formulae for determining the degrees of mobility and redundancy in planar and spatial mechanisms. He also explored the reciprocal nature of wrenches and screws to determine the rate of work done by wrenches acting on a mechanism, and he showed how stresses can be locked into redundantly constrained mechanisms.

By creating kinematic models of assembly features, researchers have been able to apply kinematic theory to assembly analysis. [Mason and Salisbury] used Screw Theory to characterize the nature of different types of contacts between robot gripper hands and objects. [Ohwovoriole and Roth] extended traditional Screw Theory by deriving two new types of screw systems which they call "repelling" and "contrary" screws. They showed that if one wishes to assemble two parts, the parts can move only along twists that are either reciprocal or repelling. More recently, Konkar created screw system representations of assembly mating features and used the methods of Screw Theory to determine the number of relative degrees of freedom between any two parts in an assembly ([Konkar]). Konkar details the screw system representation of six basic assembly features that correspond to traditional kinematic joints such as prismatic and revolute. He also outlines a computational procedure for implementing the calculation in simple mechanisms. These two elements were lacking in most of the works on kinematics and kinematic feature models described above. Konkar's method was extended to more complex mechanisms in [Shukla and Whitney].

This chapter uses and extends the Konkar–Shukla–Whitney method in two ways: by providing a means of calculating the relative degree to which parts constrain each other, and by defining an extensible toolkit of assembly features which are based on a set of features defined for robot assembly in [Kim and Wu].

4.E.2. Screw Theory Representations of Assembly Features

A *screw* is a way of representing the motions that a rigid body can undergo or of representing the forces and moments exerted on it. Screws representing motion are called *twists* or *twist matrices,* while screws representing forces

are called *wrenches* or *wrench matrices.* A twist or wrench matrix has six columns and one to six rows, one for each degree of freedom being described. Twists and wrenches can be used to describe a wide variety of part-to-part constraints. We will use them to build assembly features from basic surface contacts as well as to construct a toolkit of useful assembly features. We will also define and use algorithms for conducting motion analysis and constraint analysis of assemblies made by joining parts using assembly features.

In all the cases that we treat in this chapter, our goal is to understand what happens when two rigid surfaces come into frictionless contact. The surfaces may be pins in holes, plates on plates, or a variety of other shapes. In every case, the contact can support from zero to five relative motions and from six to one contact forces such that the total number of allowed relative motions and supportable forces is six.

Figure 4-16 shows two surfaces in contact, a plane and a cylinder. Also shown are four directions in which these parts can move relative to each other without losing contact. Each such direction is called a twist. Later we will define such a set of directions as the twist space of the contact. In addition, the figure shows two directions along which the surfaces can exert force or torque on each other. Each such direction is called a wrench. This set of directions will later be defined as the wrench space of the contact. These two spaces do not overlap and have no common directions. Our goal in this chapter is to understand how to determine the twist space and wrench space of contacts that occur inside assembly features and to use that information to determine whether combinations of features provide proper constraint between the parts they join. Screw Theory is our tool for doing this.

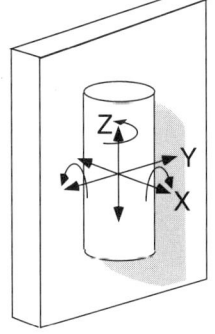

FIGURE 4-16. Two Surfaces in Contact, a Plane and a Cylinder. The coordinate axes are attached to the cylinder, and the Z axis is along the cylinder's axis. While the plane and the cylinder remain in contact, relative motion between the plane and the cylinder can occur along Y and Z and about X **and** Z. They can also exert force on each other along X and about Y.

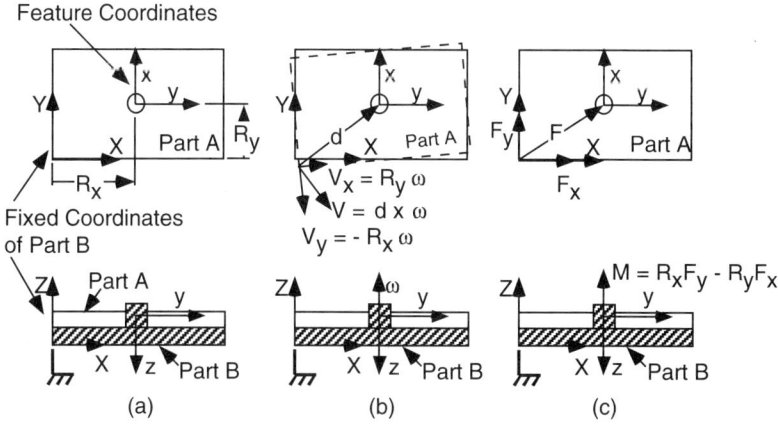

FIGURE 4-17. Two Flat Plates Joined by a Pin–Hole Joint. (a) Definition of coordinates. Part B is considered fixed and contains the reference coordinate frame at the lower left corner. The pin is located a distance R_x in the x direction and R_y in the y direction from the origin of part B's coordinate frame. (b) Part A is free to rotate about the axis of the pin. When it does so at angular rate ω, a point on part A that coincides with the origin of part B's coordinate frame translates at velocity V, which has x and y components V_x and V_y respectively. This is the only motion that part A is capable of. (c) Part A can remain in equilibrium under a variety of applied external forces and moments. The one shown here is a force in the plane of the part directed straight at the center of the pin. It is equivalent to the individual forces F_x and F_y plus the moment $M = R_x F_y - R_y F_x$. Other forces and moments that can be resisted (not shown) are F_z, M_x, and M_y.

4.E.2.a. Example: Two Plates Joined by a Pin–Hole Joint

Before the mathematical definition of a twist or wrench is given, we will motivate the usefulness of Screw Theory for our purposes and then build the twist and wrench matrices of a simple assembly feature using an example. In the discussion that follows, all forces, moments, and velocities will be referred to a common coordinate frame.

Figure 4-17 shows two flat plates that contact each other via two kinds of surface contacts: plate on plate and cylinder inside cylinder. The lower plate, considered fixed, contains a pin that mates by a line fit with a hole in the upper plate. Its coordinate frame is located in its lower left corner, marked by ⌐. The upper plate is capable of one degree of freedom of motion: it can rotate frictionlessly in the X–Y plane around the pin's axis, as shown in the middle of the figure. All other motions are impossible, and attempting to make them happen will give rise to resisting forces and moments. There are in fact five of these resisting forces and moments: F_x, F_y, and F_z, plus M_x and M_y. Thus, as noted above, the number of independent allowed motions plus the number of independent resistable forces and moments adds to six. Furthermore, as noted above, those motions that *can* occur are in directions along or about which forces or moments *cannot* be

resisted; similarly, directions along or about which motion *cannot* occur are precisely those along which forces or moments *can* be resisted. A generalization of this fact is the statement that the wrenches cannot do work along the directions of the twists. Kinematicians say that the twist and the wrench of a surface contact, or of a feature made of several surface contacts, are *reciprocals* of each other. In more complex cases than the one in the following example, we will use an algorithm to calculate the reciprocal for us. This is discussed in Section 4.E.2.d.

Our goal is to determine the twist matrix and wrench matrix of this assembly feature. The twist matrix for this feature is intended to capture the rotation about the pin's axis as well as to say where in the lower plate's coordinate system the axis of rotation is located. This is done by separately writing in the twist matrix the direction about which the rotation (called ω) occurs as well as writing the direction or directions of translation relative to the lower part's coordinate frame caused by the rotation. This translation is captured by defining a point on the upper plate that coincides with the lower plate's coordinate frame before rotation occurs and then noting where that point goes when the rotation occurs. This is shown in Figure 4-18 by the two positions of the small circle at the lower left corner of the upper plate as it rotates.

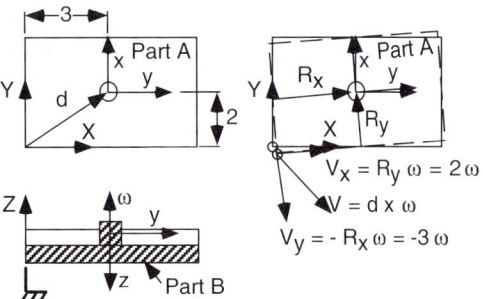

FIGURE 4-18. Example for Calculating a Twist Matrix. When part A rotates in the negative direction about the pin's z axis, the plate also rotates about the positive Z axis of the part B's coordinate frame, denoted by $\overset{\frown}{m}$. Vector d locates the pin in the coordinates of the lower plate. R_x and R_y are the x and y components of vector d, respectively, expressed in part B coordinates. The lower left corner of part A (marked by the small circle) coincides with part B's coordinate frame before rotation occurs. The velocity of this point can be calculated using the vector cross product of d with ω. The formulation shown applies strictly only at the moment when the corners of the two plates coincide.

The general form of a twist is

$$T = [\omega_x \quad \omega_y \quad \omega_z \quad v_x \quad v_y \quad v_z] \qquad (4\text{-}4)$$

A twist matrix contains one or more twists as rows. The first three elements of the twist capture the angular velocity vector ω in part B's coordinates while the last three capture the translational velocity vector v, again in part B's coordinates, of the lower left corner of part A as it swings over part B's coordinate frame origin. For the case in Figure 4-18, there is only one allowed motion, a rotation rate ω_z. This motion, and the velocities v_x and v_y that ω_z imparts to the point on part A at part B's coordinate origin, are captured in the twist matrix as

$$T_{pin\text{-}hole} = [0 \quad 0 \quad \omega_z \quad R_y\omega_z \quad -R_x\omega_z \quad 0] \qquad (4\text{-}5)$$

which can be rewritten as

$$T_{pin\text{-}hole} = [0 \quad 0 \quad 1 \quad R_y \quad -R_x \quad 0] \qquad (4\text{-}6)$$

In Equation (4-6), ω_z is scaled to be unity so that v_x and v_y take the values R_y and $-R_x$, respectively. We have also been careful to convert all translations and rotations from feature coordinates to the fixed coordinates of part B. This will become important when we combine the effects of several features. All these effects must be expressed in a single coordinate frame before they can be combined. In general, one can choose any convenient coordinate frame for this purpose. The same motions, expressed in a different coordinate frame, will give rise to a different (i.e.,

different-looking) twist matrix that nevertheless captures the same information.

Corresponding to the twist matrix is the wrench matrix, which consists of all the forces and moments that the joint can resist. A wrench is defined as

$$W = [f_x \quad f_y \quad f_z \quad m_x \quad m_y \quad m_z] \qquad (4\text{-}7)$$

A wrench matrix contains one or more wrenches as rows. The first three elements of the wrench capture the force component of the wrench while the last three capture its moment. The feature is able to resist this combined force and moment without moving and thus is said to be in static equilibrium with respect to it.

Using coordinates centered at the lower left corner of the lower plate, we can write by inspection the five wrenches corresponding to the situation in Figure 4-17 as follows:

$$
\begin{aligned}
w_1 &= [F_x \quad 0 \quad 0 \quad 0 \quad 0 \quad -R_yF_x] \\
w_2 &= [0 \quad F_y \quad 0 \quad 0 \quad 0 \quad R_xF_y] \\
w_3 &= [0 \quad 0 \quad F_z \quad 0 \quad 0 \quad 0] \qquad (4\text{-}8) \\
w_4 &= [0 \quad 0 \quad 0 \quad M_x \quad 0 \quad 0] \\
w_5 &= [0 \quad 0 \quad 0 \quad 0 \quad M_y \quad 0]
\end{aligned}
$$

The interpretation of Equation (4-8) is as follows:

- First row: Part A will be in static equilibrium under the action of a force along the X axis combined with a suitable moment about the $-Z$ axis. That is, part A will not move when acted on by this wrench.

- Second row: Part A will be in static equilibrium under the action of a force along the Y axis combined with a moment about the $+Z$ axis.

- Taken together, the first and second row are equivalent to a force directed from the origin of part B's coordinates to the axis of the pin. This fact was noted in the discussion surrounding Figure 4-17.

- The other elements of the wrench matrix are self-explanatory.

In Section 4.F.3. entirely algorithmic methods are used to derive such results. Although the answers will be numerical and will apply only to the cases analyzed, these algorithms enable us to analyze much more complex problems than can be done by inspection or symbolically.

If a feature permits only translation, then its twist matrix will contain zeroes in the first three entries and the translation direction(s) in the second three. For example, the square peg sliding in the square slot in Figure 4-19

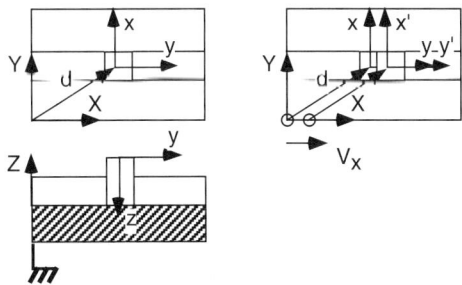

FIGURE 4-19. Square Peg in Square Slot Feature. The plate with the slot is fixed, and a square peg can slide in it in the X direction in fixed plate coordinates.

exhibits only linear motion in the X direction in the fixed part's coordinates. This is captured in its twist matrix as

$$T_{peg-slot} = [0 \quad 0 \quad 0 \quad v_x \quad 0 \quad 0] \tag{4-9}$$

This twist matrix says that a point initially located at the fixed plate's coordinate origin and tied to the moving peg will move in the part coordinate X direction as the peg slides in the slot. Since the peg and slot are square, no rotation can occur, so the entries in the ω section of the twist matrix are zeroes. Since the slot restricts motion to be parallel to one part coordinate direction, there is only one nonzero entry in the v section of the twist matrix. If the slot were inclined in the plate's X–Y plane, then the v section of the twist matrix would contain two entries whose ratio would express the slope of the slot.

If a feature permits more than one independent motion, say a translation and a rotation, then the twist matrix will contain a separate row for each independent motion, and each row will be constructed by procedures similar to those above as if the other rows' motions are frozen.

More generally, the independent rows of a twist matrix span what is called the *twist space* of the feature, describing the independent motions it can support. The rank of the twist matrix tells us how many of these independent motions there are. Correspondingly, the independent rows of a wrench matrix span the feature's *wrench space*. Its rank tells how many independent forces and torques the feature can support.

The ranks of a feature's twist and wrench matrices total six.

If we define the allowed motions of a feature to be along or about certain axes, then we are saying at the same time that the other axes that do not allow motion will support force or torque. In the twist matrices defined for the case shown in Figure 4-17, we said that the pin

did not allow motion along the Z axis. Thus the pin has been given motion resistance capability that shows up in the wrench matrix for this feature as W_3. We could have given Z support capability instead to a separate feature called the upper planar surface of Part B. We are free to define things either way: Ignore the plane and let the pin provide both a pivot and a planar support, or let the pin provide the pivot and let the plane provide support.

4.E.2.b. Formal Definition of Screw, Twist, and Wrench

4.E.2.b.1. Screw. A screw is an ordered six-tuple that may represent either a twist or a wrench. Thus the context has to be given in order to interpret a screw. The first triplet represents a line vector associated with a unique line in space. The second triplet represents a free vector that is not confined to a specific line of action but whose direction only is important. A unique point in space is associated with this second triplet.

The physical interpretation of a screw depends upon whether it is being used to represent a twist or a wrench.

4.E.2.b.2. Twist. A twist is a screw that describes to first order the instantaneous motion of a rigid body: $T = [\omega_x \quad \omega_y \quad \omega_z \quad v_x \quad v_y \quad v_z]$. The first triplet represents the angular velocity of the body with respect to a global reference frame. The second triplet represents the velocity, in the global reference frame, of that point on the body or its extension that is instantaneously located at the origin of the global frame. The line vector represents the rotation vector, if any, of the body and is called the instantaneous spin axis (ISA). The free vector represents the body's translation, whose magnitude may depend on the location of the unique point associated with it. If a body can undergo more than one independent motion, there is a separate twist for each one, and the set of all these independent motions is represented by combining all the twists as a stack of rows called a twist matrix.

The interpretation of the twist is shown in Figure 4-20. This is a generalization of Figure 4-18. The second triplet v is calculated by taking the cross product of the angular velocity ω and the vector r which extends from the origin of the global coordinate frame to a point P on the instantaneous spin axis (ISA). The length of r is the perpendicular distance from the ISA to the origin.

If a twist represents only linear motion, the first triplet entries are zeroes. If a twist represents rotation for which there is no fixed center about which rotation occurs, then

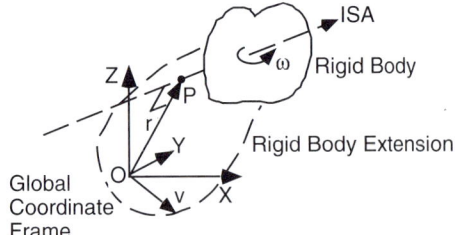

FIGURE 4-20. Graphical Interpretation of Terms Involved in a Twist Matrix. The ISA is the unique line vector ω associated with the twist. The first three elements of the twist give the components of this vector in global coordinates. Vector v is a free vector whose components depend on where it is placed. Here it is placed so that it tracks the velocity of a point on the rigid body (or an imaginary extension of it if necessary) that lies on top of origin O of global coordinates.

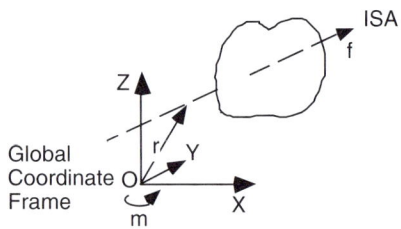

FIGURE 4-21. Graphical Interpretation of Terms Involved in a Wrench Matrix. The ISA is the unique line vector f associated with the wrench. The first three elements of the wrench give the components of this vector in global coordinates. Vector m is a free vector whose components depend on where it is placed. Here it is placed so that it tracks the moment about a point on the rigid body (or an imaginary extension of it if necessary) that lies on top of origin O of global coordinates.

the second triplet entries are zeroes. This occurs, for example, in the case of two flat plates resting one on the other.

4.E.2.b.3. Wrench. A wrench is a screw that describes the resultant force and moment of a force system acting on a rigid body. The first triplet describes the resultant force in a global reference frame. The second triplet represents the resultant moment of the force system about the origin of the global frame. A wrench is also written as a row vector $W = [f_x \quad f_y \quad f_z \quad m_x \quad m_y \quad m_z]$. The first triplet represents independent forces that can be resisted by the wrench, while the second triplet represents moments. If a body is acted on or can resist several independent forces or moments, there is a separate wrench for each one, and the set of all these independent forces and moments is represented by combining all the wrenches as a stack of rows called a wrench matrix. The interpretation of a wrench is shown in Figure 4-21.

Suppose a set of forces and moments is applied to a rigid body as in Figure 4-22. The wrench that describes that system is the resulting force f and moment m about the origin shown in Figure 4-21. Note that the line of action of the resulting force defines the instantaneous screw axis of the wrench. Using the notation of Figure 4-21, the following relations are used to calculate f and m:

$$f = \sum_{i=1}^{n} F_i$$

$$m = \sum_{j=1}^{k} M_j \qquad (4\text{-}10)$$

$$M_i = r_i \times F_i$$

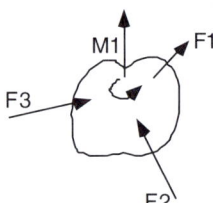

FIGURE 4-22. A Set of Forces and Moments Acting on a Rigid Body.

where vector r_i is the perpendicular distance from the global coordinate frame to the line of action of F_i.

4.E.2.b.4. Reciprocal of a Screw. The reader should note that it is in principle just as easy to write down the twist matrix for a feature as it is to write down its wrench matrix.[17] We know that there is a strong relationship between them and that it should be possible to obtain one from the other. The method for doing so is called finding the *reciprocal* of a screw. A MATLAB routine for doing this is presented in Section 4.E.2.d along with several other useful operations on screws. Here we present a direct demonstration from first principles.

When two rigid bodies interact by contacting without friction, they restrict each others' motions and exert forces and torques on each other. We express the motions as twists and express the forces and torques as wrenches. Under these conditions, the wrench and twist are such that the wrench cannot do any work along the direction of the twist. This condition can be expressed by saying that the vector dot product of the twist and the wrench equals zero.

---

[17]The author prefers to write the twists by inspection and solve for the wrenches.

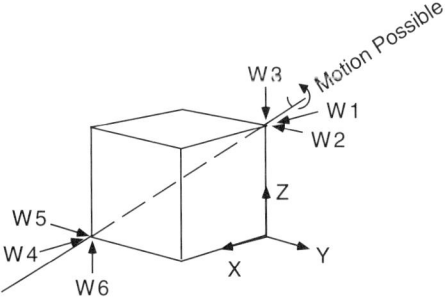

FIGURE 4-23. A Cube Restrained by Six Forces Acting at Two Corners. The cube can spin about an axis that passes through the corners where the forces act ([Roth]).

Because of the sequence in which ω, v, f, and M appear in the twist and wrench respectively, this vector product is defined so as to create the correct component products:

$$T \bullet w = T_1 w_4 + T_2 w_5 + T_3 w_6 + T_4 w_1 + T_5 w_2 + T_6 w_3 \tag{4-11a}$$

or

$$T \bullet w = \omega_x M_x + \omega_y M_y + \omega_z M_z + v_x F_x + v_y F_y + v_z F_z \tag{4-11b}$$

If the product in Equation (4-11) is zero, then T and w are called reciprocals of each other.

As an example, Figure 4-23 shows a cube held by six forces acting three at a time at two opposite corners. If it has any remaining degrees of freedom, they will be along axes along which the existing forces cannot do any work. If we express the existing forces as wrenches as in Equation (4-12), we can solve for the corresponding reciprocal twist using Equation (4-11):

$$
\begin{aligned}
w1 &= [1 \quad 0 \quad 0 \quad 0 \quad 1 \quad 0] \\
w2 &= [0 \quad -1 \quad 0 \quad 1 \quad 0 \quad 0] \\
w3 &= [0 \quad 0 \quad -1 \quad 0 \quad 0 \quad 0] \\
w4 &= [-1 \quad 0 \quad 0 \quad 0 \quad 0 \quad -1] \\
w5 &= [0 \quad 1 \quad 0 \quad 0 \quad 0 \quad 1] \\
w6 &= [0 \quad 0 \quad 1 \quad -1 \quad -1 \quad 0]
\end{aligned}
\tag{4-12}
$$

The result is

$$T = [1 \quad -1 \quad -1 \quad 1 \quad 1 \quad 0] \tag{4-13}$$

This is equivalent to spinning the cube about an axis that runs through the corners where the wrenches act.

4.E.2.c. Formulas for Expressing Feature Twists in Global Coordinates

In this section, we present in mathematical terms the calculations we performed intuitively to construct the twist matrix for the example in Figure 4-18. Assume that a revolute joint, $f1$, with one rotational degree of freedom allowed about its z axis, is placed somewhere in space with respect to a global coordinate frame O. The origin of O and the origin of $f1$ are located on the same rigid part. Let the 4×4 transform that describes the location and orientation of $f1$'s coordinate frame in part coordinates be denoted by A. A is a partitioned matrix of the following form:

$$A = \begin{bmatrix} R & d \\ 0^T & 1 \end{bmatrix} \tag{4-14}$$

where R is a 3×3 rotation matrix, d is a 3×1 displacement vector, and 0^T is a 1×3 row vector of zeros. The unit angular velocity vector that describes the allowable joint rotation in $f1$'s coordinate frame is defined by default as the vector $\omega_z = [0 \quad 0 \quad 1]^T$. That is, rotation occurs about $f1$'s Z axis, which may not correspond to the Z axis of global coordinates. Vector ω_z can be rotated into global coordinates using the relation

$$\omega = (R\omega_z)^T \tag{4-15}$$

The vector r that describes the position of $f1$ with respect to O is also needed to compute the twist matrix. r is contained within A as the displacement vector d. Thus, using this nomenclature, the twist matrix of $f1$ is given by

$$T_{f1} = [\omega \quad v] \tag{4-16}$$

where $\omega = (R\omega_z)^T$, $v = r \times \omega$, and $r = d^T$.

If feature $f2$ describes a translational joint, then it is characterized by a vector k which points along the translation direction in $f2$'s coordinates. This vector must also be transformed to global coordinates using the product Rk. Using this nomenclature, the twist matrix of $f2$ is given by

$$T_{f_2} = [0^T \quad v] \tag{4-17}$$

where $v = (Rk)^T$.

This method of transforming vectors from a local coordinate frame to a part or global frame will be used for all features in this book.

4.E.2.d. Calculating Combinations of Twists and Wrenches

4.E.2.d.1. Definitions. Now that we know how to define a feature and determine its twist matrix, we need a way to combine two or more such features and find the resultant twist matrix that represents the motions permitted by the combination. We do this by calculating the intersection of the twists of each feature. The method for this was developed by Konkar and is given in Section 4.E.2.d.2 in the form of MATLAB routines. For now, we will assume that there are two parts in the assembly, joined by several features. Later we will consider the case where there are more than two parts. First we define some important terms.

Reciprocal of a Screw: The reciprocal of a twist is a wrench and vice versa. If the rank of a twist is n, then the rank of its reciprocal wrench is $6 - n$. The wrench-twist pair that are reciprocals of each other form complementary spaces in the sense discussed in Section 4.E.2.a: If the twist describes directions along which motion is allowed, then the wrench describes directions that can resist forces or moments.[18] The reciprocal of the screw matrix S is calculated in two steps: (1) computing the null space of S and (2) "flipping" the first three elements of the result with the last three. "Flipping" exchanges the columns of the matrix according to the following pattern: $i \rightarrow i + 3 \mod(6)$.[19] Calculating the null space is equivalent to imposing the condition in Equation (4-11). The null space of a matrix is a set of vectors that are transformed to zero by that matrix. The null space is also the orthogonal complement of the space spanned by the vectors in the matrix.

Union of Screws: The union of multiple screws is obtained by concatenating the individual screws into a matrix with each screw occupying a row. Thus if $s_1, s_2, \ldots, s_n$ are the individual screws, the union is given by

$$Union(s_1, s_2, \ldots, s_n) = \begin{bmatrix} s_1 \\ s_2 \\ \cdots \\ s_n \end{bmatrix} \qquad (4\text{-}18)$$

A screw matrix is thus an $n \times 6$ matrix, where n is the number of independent screws represented in the matrix. When the screws are twists or wrenches, the union is called a twist matrix or wrench matrix, respectively.

Intersection of Screws: The intersection of different sets of screws is the set of screws common to all the sets.

The intersection of several twists from several features yields a resultant twist matrix that describes the net motion allowed by the intersected features. To compute the intersection of a set of twist matrices, the reciprocal of each twist matrix is computed to obtain a wrench matrix. The union of these wrench matrices is obtained by gathering the wrenches into one large matrix WU. The reciprocal of this matrix yields the intersection of the original set of twist matrices.

The intersection of several wrenches from several features yields a resultant wrench matrix that describes the forces and torques that the combined features can resist. Equivalently, the resultant wrench describes the directions that are constrained by more than one feature and hence are overconstrained. To compute the intersection of a set of wrench matrices, the reciprocal of each wrench matrix is computed to obtain the corresponding twist matrix. If we obtain a feature from the feature toolkit, its twist matrix is already known. Alternatively, we can analyze a new feature from scratch to determine its twist matrix. In either case, we can skip the step of finding the reciprocal of each wrench matrix and then proceed directly to calculate the union of the twist matrices by collecting them into one large matrix TU. The reciprocal of this matrix yields the intersection of the wrenches acting on the part.

Equation (4-19) expresses the intersection calculation formally. It says that the intersection of a set of screws is the reciprocal of the union of the reciprocals of those screws. The technique applies equally to twists and wrenches.

$$
\begin{aligned}
\text{Intersection}(S_i) &= \bigcap(S_i) \\
&= \text{Reciprocal} \left\{ \bigcup_{i=1}^{n} [\text{Reciprocal } (S_i)] \right\} \\
&= \text{Reciprocal} \left(\begin{bmatrix} \text{Reciprocal } (S_1) \\ \text{Reciprocal } (S_2) \\ \vdots \\ \text{Reciprocal } (S_n) \end{bmatrix} \right)
\end{aligned}
$$

$$(4\text{-}19)$$

4.E.2.d.2. The Twist matrix Intersection Method in MATLAB Form. The process for intersecting several twist matrices comprises four steps.

Step 1: For each feature i on the part, $i = 1$ to n, identify its twist matrix Ti and find the associated wrench

---

[18]In the language of linear algebra, the vector space spanned by the twist matrix is the orthogonal complement of the vector space spanned by the wrench matrix, and vice versa. Together, these matrices span a six degrees of freedom space ([Hoffman and Kunze]).

[19]Strictly speaking, the flip operation is not fundamental to the concept of reciprocal. It is necessary in order for the elements of the resulting wrench to come out in the order $[f \quad M]$.

matrix (Wi) by calculating the reciprocal of (Ti). This is done by

  a. $wi = \mathbf{null}(Ti)$,[20]

  b. $wti = wi'$ where $'$ denotes transpose,

  c. $Wi = \mathbf{flip}(wti)$. The flip operation is done by exchanging columns according to the following pattern: 1 becomes 4, 2 becomes 5, 3 becomes 6. Columns 4, 5, and 6 become 1, 2, and 3, respectively.

Step 1 is accomplished by the MATLAB function called **recip,** which is defined below. From here on, we will simply write recip instead of repeating substeps (a), (b), and (c) above. Thus Step 1 becomes

$$Wi = \mathbf{recip}(Ti)$$

Step 2: Collect all the Wi matrices for these features into a matrix called WU, which is the union of the individual W's.

  a. $WU = [W_1; W_2; \ldots; W_n]$

Step 3: Obtain the resultant twist matrix from the combined action of all part features by applying the recip operation to WU.

  a. $TR = \mathbf{recip}(WU)$

Step 4: (Optional) Obtain the row reduced echelon form of TR for ease of interpretation.

  a. $TRU = \mathbf{rref}(TR)$

Step 4 creates an entirely equivalent version of TR called the *row-reduced echelon form* (rref) that has been scaled to unity in some of its components; **rref** is a library function in MATLAB. This form is much easier to interpret intuitively but is not required for any of the calculations.

4.E.2.d.3. The Wrench Matrix Intersection Method in MATLAB Form

Step 1: For each feature i on the part, find the associated twist matrix (Ti), either by using the twist associated with the feature or by calculating it as follows:

  a. $Ti = \mathbf{recip}(Wi)$

---

[20]**Null** is a library function in MATLAB that finds the null space of a matrix.

Step 2: Collect the individual twists into a union matrix called *TU*.

  a. $TU - [T1; T2; \quad ; Tn]$

Step 3: Obtain the resultant wrench by calculating the reciprocal of *TU*.

  a. $WR = \mathbf{recip}(TU)$

Step 4: (Optional) Obtain the row reduced echelon form of *WR* for ease of interpretation.

  a. $WRU = \mathbf{rref}(WR)$

4.E.2.d.4. MATLAB Functions for Screw Intersection

```
function R = recip(T)
% Takes the reciprocal of a screw
  matrix
p = (null(T))';
[i,j] = size(p);
if i>0
  R = flip(p);
  R = rref(R);
else
  disp('empty matrix')
  R = zeros(0);
end
function W = flip(WU)
% FLIPs columns of WU
% col 1 becomes 4, 2 becomes 5, and
  3 becomes 6
% col 4 becomes 1, 5 becomes 2, and
  6 becomes 3

[i,j] = size(WU);
if j == 6
  for l = 1:i
    for k = 1:3
      W(l,k) = WU(l,k+3);
      W(l,k+3) = WU(l,k);
    end
  end
  W;
else
end
```

4.F. DESIGN AND ANALYSIS OF ASSEMBLY FEATURES USING SCREW THEORY

Screw Theory permits us to represent in a precise mathematical way the interactions between two surfaces in contact. We can represent the ability of those surfaces to translate along or rotate around each other, and we can represent the fact that a surface can resist a contact force directed normally to it. We have now laid a theoretical foundation for representing assembly features using Screw Theory. We will now put this theory to work.

We will use this method to show how we can determine the degrees of freedom, amount of underconstraint, or amount of overconstraint. We will build our way up to models of assembly features in steps, starting with twist matrix models of basic surface contacts. We will see how we can build up familiar engineering features using basic surfaces in combination. In addition, we will see how to build engineering features directly, without starting from basic surface contacts.

4.F.1. Motion and Constraint Analysis[21]

We begin by showing how to use the methods given in Section 4.E.2.d.2 to perform motion and constraint analyses of features. The analysis may result in any of the following situations

- The assembly is kinematically constrained, with no overconstrained degrees of freedom and no underconstrained degrees of freedom.
- Some degrees of freedom can be underconstrained.
- Some degrees of freedom can be overconstrained.
- Some degrees of freedom can be overconstrained while others are underconstrained.

To find out which of these conditions applies to any assembly, we need to use Screw Theory and learn how to interpret the results. The procedure is described next and used repeatedly in the rest of this chapter.

4.F.1.a. Motion Analysis

Motion analysis is used to determine if an assembly has any *underconstrained* degrees of freedom. Suppose that a resultant twist matrix was calculated by intersecting a set of several twist matrices describing the joints connecting

a part to others in an assembly. This resultant, called TR, would contain the intersection of twist matrices describing the joints and would have the form

$$TR = \begin{bmatrix} \omega_{1x} & \omega_{1y} & \omega_{1z} & v_{1x} & v_{1y} & v_{1z} \\ \omega_{2x} & \omega_{2y} & \omega_{2z} & v_{2x} & v_{2y} & v_{2z} \\ \vdots & \vdots & \vdots & \vdots & \vdots & \vdots \end{bmatrix} \quad (4\text{-}20)$$

The number of rows of *TR* is the number of unconstrained degrees of freedom between the part being analyzed and the parts it is connected to. Each row is a twist describing an independent degree of freedom. *If a twist appears in the resultant twist matrix, it means that all of the features connecting the part to others will allow the motion described by that twist.* If the resultant twist matrix is empty, this means that no motion is possible. If it is not empty, then the feature combination contains underconstraint in one or more directions. The first triplet ω gives the axis of an underconstrained angular motion and the second triplet v gives the direction of an underconstrained translation, expressed in the common global coordinate frame used to calculate the intersection.

4.F.1.b. Constraint Analysis

Constraint analysis is used to determine if an assembly has any *overconstrained* degrees of freedom. It does this by finding the intersection of the set of wrenches acting on a part corresponding to the features that connect the part to others in the assembly. Overconstraint may cause assembleability or dimensional control problems when the parts are assembled. Thus the constraint analysis provides a constraint check for the engineer and points out possible problem areas. The constraint analysis results are presented to the engineer as a caution. Often the overconstraint can be removed by redesigning some of the features. The solution to the car seat installation problem in Section 4.C.5.b was to provide more clearance on all but one of the holes in order to remove the over constraint.

Suppose we intersect all the wrenches that are exerted by all the features that join two parts, using the method described in Section 4.E.2.d.3. This produces a resultant wrench matrix *WR*. *WR* has the form

$$WR = \begin{bmatrix} f_{1x} & f_{1y} & f_{1z} & m_{1x} & m_{1y} & m_{1z} \\ f_{2x} & f_{2y} & f_{2z} & m_{2x} & m_{2y} & m_{2z} \\ \vdots & \vdots & \vdots & \vdots & \vdots & \vdots \end{bmatrix} \quad (4\text{-}21)$$

---

[21]The definitions and explanations in this section are adapted in part from [Konkar], [Ball], and [Waldron].

If the resultant wrench matrix is empty, then there is no overconstraint in the combined features. *Any wrench that appears in this matrix can be exerted by all the intersected features.* In other words, each feature is attempting to constrain that particular degree of freedom, which is the definition of overconstraint. Thus, the rows of *WR* describe the overconstrained or redundantly constrained degrees of freedom of the part being analyzed. The first triplet *f* gives the direction of an overconstrained linear motion and the second triplet *m* gives the axis of an overconstrained rotation, expressed in the common global coordinate frame used to calculate the intersection.

A note of caution: The intersection calculation does exactly what it says: It finds the directions that are constrained by all the features involved in the intersection calculation. If *WR* is empty, it simply means that no direction is overconstrained by *those* features. Yet, some subset of those features could create over-constraint. The only way to find out is to intersect all the subsets. A systematic way to do this is illustrated in Section 4.F.6.

4.F.1.c. Analysis Results

Motion analysis tells us if a situation is or is not underconstrained. Constraint analysis tells us if a situation is or is not overconstrained. *If it is not underconstrained and it is not overconstrained, then it is kinematically or properly constrained.* Note that it is possible for a situation to be both over- and underconstrained at the same time. One of the thought questions at the end of the chapter presents such a case.

4.F.2. Basic Surface Contacts and Their Twist Matrices

In this section, we define some basic surfaces, enumerate the ways they can contact each other, and calculate the twist matrices that describe their interactions. This prepares us to build real features from these basic contacts. The general theory of these interactions is very complex, so we will restrict our discussion to some simple cases.

The discussion here extends work from [Herve] and [Charles, Clement, et al.] on identification of fundamental groups of surface interactions and their use in defining tolerancing schemes. This formulation permits us to use the same definitions to describe the degrees of freedom in a feature and to describe displacements that arise when feature surface locations vary. The former permits us to perform constraint analysis on arbitrary feature shapes

TABLE 4-2. Definitions of Some Simple Basic Surfaces and Their Coordinates

Cylinder	Plane	Sphere

while the latter permits us (as discussed in later chapters) to perform variation analysis.

4.F.2.a. Nomenclature for Surface Contacts

The surfaces modeled here are the cylinder, the plane, and the sphere. Each surface contains a coordinate frame. The surface has in principle the ability to move along or about the axes of this frame. Table 4-2 defines these surfaces and frames. The frame definitions are arbitrary. Since the intersection algorithms require identifying a common frame, the ones in the table should be regarded as examples only.

4.F.2.b. Surface Contacts

When two surfaces touch and one is considered fixed in space, the other loses some of its degrees of freedom. The basic surfaces in Table 4-2 can be combined into surface contacting pairs in several combinations, as shown in Table 4-3.

To determine what degrees of freedom remain once two surfaces contact, we can make use of Screw Theory. Take, for example, the cylinder-plane contact illustrated in Figure 4-24. If the cylinder is assumed stationary, then the plane can move in four degrees of freedom as shown. We get the same result if the plane is assumed stationary and the cylinder moves, of course.

Each of these motions can be described by a twist matrix. In this case, the matrix will have four rows, one for each of the possible relative motions shown in Figure 4-24. Figure 4-25 illustrates how to determine one of these rows, the one describing rocking of the plane about the cylinder's *Z* axis. (This is the same as the plane's *Y* axis as shown in Figure 4-24.)

Many other types of contacts are possible between the basic surfaces. For example, we could put the end face of the cylinder onto the plane. However, this is really a plane-plane contact so it is not necessary to account for it separately. Other contacts that we will not deal with, even though they are different from anything in Table 4-3,

TABLE 4-3. Possible Surface Contacts Between the Basic Surfaces in Table 4-2

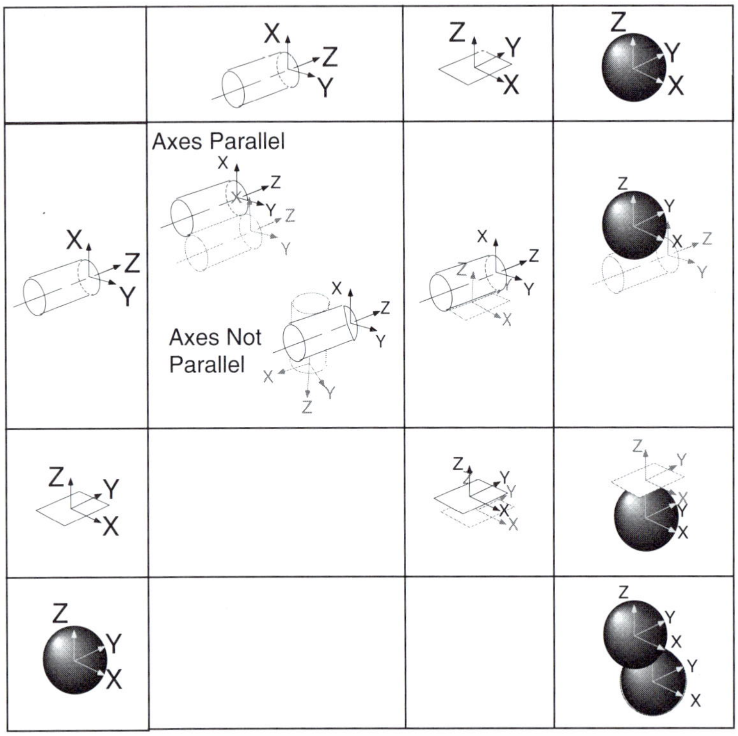
Axes Parallel
Axes Not Parallel

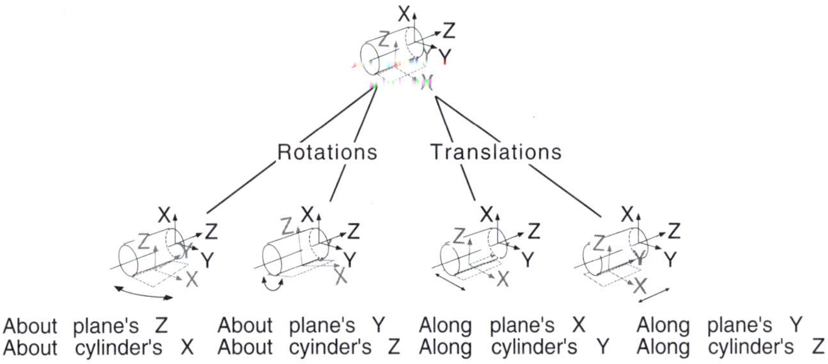
Rotations
Translations
About plane's Z
About cylinder's X
About plane's Y
About cyinder's Z
Along plane's X
Along cylinder's Y
Along plane's Y
Along cylinder's Z

FIGURE 4-24. Four Possible Motions of a Plane in Contact with a Cylinder. The original arrangement is shown at the top, and the four possibilities are classified into rotations at the left and translations at the right. All such motions are considered to be infinitesimal.

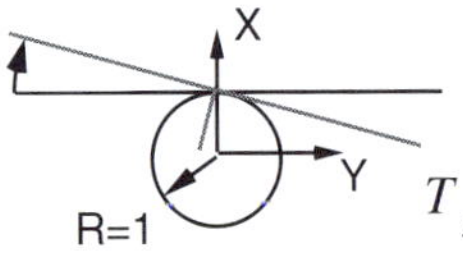
X
Y
R=1

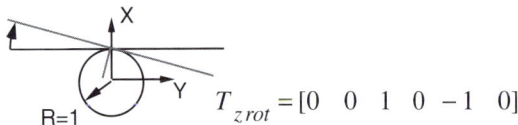

$$T_{zrot} = [0 \quad 0 \quad 1 \quad 0 \quad -1 \quad 0]$$

FIGURE 4-25. Illustrating One of the Possible Motions of the Plane in Contact with the Cylinder, Along with Its Twist Representation. When the plane rocks on the cylinder about the cylinder's Z axis (pointing into the plane of the paper), an imaginary point on the plane that coincides with the origin of the cylinder's frame moves in the cylinder's-Y direction. Thus the twist contains a nonzero entry in the third place representing unit rotation about Z and a nonzero entry in the fifth place representing the resulting translation along $-Y$. The translation entry has unit value because radius R is unity. In general, this twist is $T_{zrot} = [0 \quad 0 \quad \omega_z \quad 0 \quad -R\omega_z \quad 0]$.

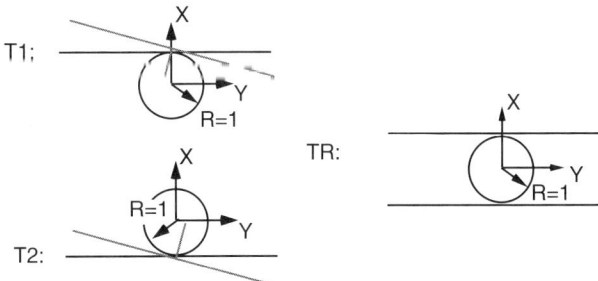

FIGURE 4-26. Two Basic Surface Contacts Made of a Plane and a Cylinder Along with the Pin–Slot Feature Made by Combining Them. *Left:* The two basic contacts. $T1$ and $T2$ are the names of the twist matrices that will be calculated for these contacts. *Right:* The combined feature. TR is the name of the resulting twist of this feature.

include contacting the edge of a plane on the surface of another plane, or the corner of a plane on the surface of a plane, cylinder, or sphere.

4.F.2.c. Construction of Assembly Features Using Surface Contacts

To see how the basic contacts can be used to construct assembly features, we continue with the example of the plane and cylinder, using them to create a pin–slot feature. The slot has two parallel walls made from two planes. These walls are spaced apart exactly the diameter of the pin. Figure 4-26 defines the terms for this process. The common coordinate frame for the necessary calculations is the cylinder's frame.

Table 4-4 shows the development of the individual twist matrices $T1$ and $T2$ that describe the individual basic contacts between each plane and the cylinder. The process of combining the surface contacts involves intersecting their individual twists to find the net motion of both surfaces acting at once on the cylinder. The analysis uses the intersection method in Section 4.E.2.d.2 to calculate the net twist TR of the combination. TR reveals that the pin can translate in two directions and rotate about two others, which is what we expect.

Next, we calculate the internal constraints, if any, that are in this feature by using the constraint analysis. The process appears in Table 4-5. We can see that there are two overconstraints caused by the two-sided nature of this feature, which results from the way it was constructed: Force along X and moment about Y are each imposed by one of the two planes making up the sides of the slot.

We may group the unconstrained degrees of freedom of the pin–slot joint into a set called its twist space, the set of directions along which the pin can move relative to the slot. The twist space of the pin–slot comprises translation along cylinder Y and Z plus rotation about cylinder X and Z. Similarly, we may group the constrained degrees of freedom of the pin–slot joint into a set called its wrench space, the set of directions along which it can support external loads. The wrench space comprises force along cylinder X and torque about cylinder Y. In a similar way, the twist space and the wrench space of any assembly feature may be found. The twist space and wrench space of a feature are reciprocals. The twist and wrench spaces of features will be important to us when we deal with propagation of variation in Chapters 5 and 6. These concepts also will help us distinguish between mates and contacts in Chapter 8.

In this section, we learned how to construct an assembly feature from basic surface contacts. While this is theoretically interesting, it is not always practical or even the simplest way to construct an assembly feature. It is often easier to define its geometry in conventional mechanical terms, such as pin–hole or pin–slot, and determine its twist matrix representation directly. For this reason, we show in the next section how to construct features from typical engineering joints in such a way that they do not contain internal overconstraints but still have twist matrices that express their engineering intent. This section refers you to Section 4.L containing a feature toolkit built using these methods.

A side effect of creating features from basic surface contacts, illustrated by the example above, is that many such features contain inherent overconstraints. This is due to the fact that they are form closures. If we use features constructed this way in an assembly and later analyze this assembly for overconstraints, we will get a constraint report that is cluttered with overconstraints that are actually inside individual features. This clutter will make it harder for us to find overconstraints *between* features, which are the ones of prime interest to us in such an analysis. This is another reason to use the toolkit features in Section 4.L.

4.F.3. Construction of Engineering Features and Their Twist Matrices

Features that are commonly used in assembly fall into two basic classes: those that are involved in the function of the

TABLE 4-4. Creation of the Twist Matrix for the Pin–Slot Feature Constructed from Two Plane-Cylinder Basic Surface Contacts

```
≫T1=[0 0 0 0 0 1;0 0 0 0 1 0;0 0 1 0 -1 0;1 0 0 0 0 0]
T1 =
   0 0 0 0   0 1 translating along cylinder's Z axis
   0 0 0 0   1 0 translating along Y
   0 0 1 0  -1 0 rotating about Z with resulting translation along -Y
   1 0 0 0   0 0 rotating about X
T2 = the same motions as T1, but in a different place in cylinder coordinates
   0 0 0 0 0 1
   0 0 0 0 1 0
   0 0 1 0 1 0
   1 0 0 0 0 0
≫W1=recip(T1)
W1 =
   1 0 0 0 0 0
   0 0 0 0 1 0
≫W2=recip(T2)
W2 =
   1 0 0 0 0 0
   0 0 0 0 1 0
≫WU=[W1;W2]
WU =
   1 0 0 0 0 0
   0 0 0 0 1 0
   1 0 0 0 0 0
   0 0 0 0 1 0
≫TR=recip(WU)
TR =
   1 0 0 0 0 0 rotating about cylinder's X axis
   0 0 1 0 0 0 rotating about Z
   0 0 0 0 1 0 translating along Y
   0 0 0 0 0 1 translating along Z
≫
```

Note: The rows of TR constitute the twist space of the feature.

product and those that are involved in processes for making parts. The former include common joints like cylinder in hole, plate on plate, tongue in groove, and so on. The latter include surface plates, pillow blocks, V-blocks, locating pins and their concave mates (typically holes or slots), V-shaped locators and their concave mates (usually V-shaped notches), and so on. In this section we will construct the twist matrices for a few typical features. Section 4.L contains a toolkit of features complete with their twist matrices. Several of the thought questions at the end of the chapter make use of these features. These

features can be used directly in engineering design. Alternately, the methods described in this chapter can be used to make up new features and find their twists and wrenches.

4.F.3.a. Pin in Hole
We begin by finding the twist and wrench description of a simple pin-on-plane–hole joint. This was done intuitively in Section 4.E.2.a. See Figure 4-27. We will do this two ways: using coordinates attached to the feature, and using global coordinates attached to one of the parts.

TABLE 4-5. Constraint Analysis of a Pin–Slot Feature Built from Basic Surface Contacts

```
T1 =
  0 0 0 0   0 1
  0 0 0 0   1 0
  0 0 1 0  -1 0
  1 0 0 0   0 0
T2 =
  0 0 0 0 0 1
  0 0 0 0 1 0
  0 0 1 0 1 0
  1 0 0 0 0 0
≫TU=[T1;T2]
TU =
  0 0 0 0   0 1
  0 0 0 0   1 0
  0 0 1 0  -1 0
  1 0 0 0   0 0
  0 0 0 0   0 1
  0 0 0 0   1 0
  0 0 1 0   1 0
  1 0 0 0   0 0
≫WR=recip(TU)
WR =
  1 0 0 0 0 0
  0 0 0 0 1 0
≫
```

Note: WR contains two rows, indicating that the two planes confining the cylinder can both support a force in the cylinder's X direction and a moment about the cylinder's Y direction. These two directions are therefore overconstrained.

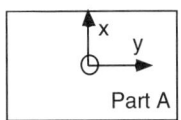

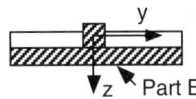

FIGURE 4-27. Example Pin–Hole Feature for Use in Calculating Twists and Wrenches in Feature Coordinates.

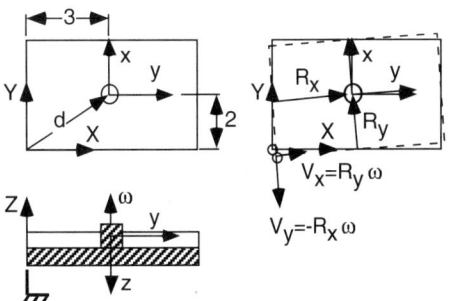

FIGURE 4-28. Example Pin–Hole Feature for Use in Calculating Twists and Wrenches in Global or Part Coordinates.

For the pin–hole joint, a twist representation in *feature* coordinates is given by $T_{\text{pin-hole}}$, as shown in Table 4-6. The entries in the right-hand three positions, corresponding to translation, are zero because all motions are referred to feature coordinates. Since rotation occurs about the origin of feature coordinates, there is no translation at the center of those coordinates. The wrench is given by $W_{\text{pin-hole}}$. Each row represents a force or moment, expressed in feature coordinates, that the feature can support or resist.

To construct the twist matrix for the pin–hole feature using *global* or *part center* coordinates, we refer to Figure 4-28. In this situation, the hole is located two units along the X direction and three units along the Y

direction in fixed (lower) part center coordinates. The upper moving part is shown rotated counterclockwise about the part center coordinate Z axis by a small amount. The twist matrix describes the motion of this origin, marked for clarity by the circle, as it moves from its original position a small amount. Either by inspection or by calculating the required quantities using Equation (4-14) through Equation (4-17), we can write the twist matrix and wrench matrix in part center coordinates as shown in Table 4-7.

for this feature, we begin by identifying the independent motions that the top plate can make relative to the bottom

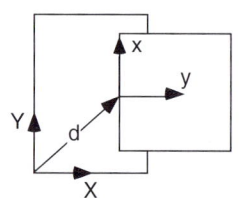

FIGURE 4-29. Feature Comprising a Plate on Another Plate.

4.F.3.b. Plate on Plate

Figure 4-29 shows a feature consisting of two flat plates resting one on the other. To construct the twist matrix

TABLE 4-6. Twist and Wrenches for a Pin–Hole Feature Using Feature Coordinates

≫$T_{Pin-Hole}$=[0 0 1 0 0 0]   (The feature allows rotation about z)

The corresponding wrench matrix is found by taking the reciprocal of $T_{Pin-Hole}$

$W_{Pin-Hole}$ = recip($T_{Pin-Hole}$)

0 0 0 -1 0 0						(The feature can support a moment about x)
0 0 0 0 -1 0						(The feature can support a moment about y)
1 0 0 0 0 0						(The feature can support a force along x)
0 1 0 0 0 0						(The feature can support a force along y)
0 0 1 0 0 0						(The feature can support a force along z)

TABLE 4-7. Twist and Wrenches for a Pin–Hole Feature Using Part-Center Coordinates

≫$T_{Pin-Hole}$=[0 0 1 2 -3 0]   (the feature allows rotation about z, which causes translation of a point aligned with part B's coordinate frame. This is a numerical example of Equation (4-6).)

≫$W_{Pin-Hole}$=recip($T_{Pin-Hole}$)

$W_{Pin-Hole}$ =

1.0000	0	0	0	0	-2.0000	(the feature can support a combination of F_x and $-M_z$)
0	1.0000	0	0	0	3.0000	(the feature can support a combination of F_y and M_z)
0	0	1.0000	0	0	0	(the feature can support F_z)
0	0	0	1.0000	0	0	(the feature can support M_x)
0	0	0	0	1.0000	0.0000	(the feature can support M_y)

Note: This is a numerical example of Equation (4-8).

plate. These are two translations in the plane of contact and one rotation about an axis normal to this plane. The twist matrix will thus have three rows, one for each of these motions.

The first row captures motion along the feature's x axis, which is along the global Y axis. Since we want to express the twist in global coordinates, the result is

$$t_1 = [0 \quad 0 \quad 0 \quad 0 \quad v_y \quad 0] \qquad (4\text{-}22)$$

The second row captures motion along the feature's y axis, which is the global X axis. The result is

$$t_2 = [0 \quad 0 \quad 0 \quad v_x \quad 0 \quad 0] \qquad (4\text{-}23)$$

The third row captures the rotation about the feature's $-z$ axis, which is the global Z axis. There is no fixed axis about which this rotation occurs, so we can arbitrarily place it on the Z axis of global coordinates. The result of doing this is that there is no velocity of the plate generated as a result of this rotation. For this reason, the result is

$$t_3 = [0 \quad 0 \quad \omega_z \quad 0 \quad 0 \quad 0] \qquad (4\text{-}24)$$

The resulting twist matrix for this feature is then

$$T_{Plate-Plate} = \begin{bmatrix} 0 & 0 & 0 & v_x & 0 & 0 \\ 0 & 0 & 0 & 0 & v_y & 0 \\ 0 & 0 & \omega_z & 0 & 0 & 0 \end{bmatrix} \qquad (4\text{-}25)$$

Table 4-8 shows the twist matrix of Equation (4-25) with unit values substituted for the symbolic ones. Since the actual values are arbitrary, unit values are as good as any. The table also shows the corresponding wrench matrix.

4.F.3.c. Pin in Slot

Figure 4-30 shows a pin in a slot in a thick plate, corresponding to the example built from basic surface contacts

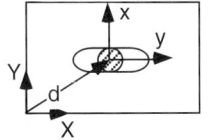
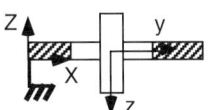

FIGURE 4-30. Pin-in-Slot Feature.

in Section 4.F.2.c. We note that this pin can move relative to the plate in four directions: along feature directions y and z and about feature directions x and z. Each direction will be captured by a row in the twist matrix. The resulting global twist matrix is given by Equation (4-26). Each of these rows should be self-explanatory with the possible exception of the first row. When the pin rotates about feature direction x (global direction Y), an imaginary point extending from the pin to the global coordinate center will move up along the global Z axis. This explains the rightmost entry in the first row. This same logic explains the fourth and fifth entries in the second row.

$$T_{Pin-Slot} = \begin{bmatrix} 0 & \omega_y & 0 & 0 & 0 & d_x\omega_y \\ 0 & 0 & \omega_z & d_y\omega_z & -d_x\omega_z & 0 \\ 0 & 0 & 0 & v_x & 0 & 0 \\ 0 & 0 & 0 & 0 & 0 & v_z \end{bmatrix}$$
$$(4\text{-}26)$$

Assuming that the X and Y components of vector d are $d_x = 3$ and $d_y = 2$, and using $\omega_y = 1$, $v_x = 1$, and $v_z = 1$, we obtain the twist and wrench matrices for this feature shown in Table 4-9.

The reader is encouraged to look at Section 4.L and try making twist matrices for the features found there.

TABLE 4-8. Twist and Wrench Matrices for Flat Plate-on-Plate Feature

```
T =
   0 0 0 1 0 0   (feature allows X translation)
   0 0 0 0 1 0   (feature allows Y translation)
   0 0 1 0 0 0   (feature allows Z rotation)
»W = recip(T)
W =
   0 0 1 0 0 0   (feature can resist Z force)
   0 0 0 1 0 0   (feature can resist X moment)
   0 0 0 0 1 0   (feature can resist Y moment)
```

TABLE 4-9. Twist and Wrench Matrices for Pin in Slot

```
T =
  0 1 0 0  0 3  (feature allows Y rotation)
  0 0 1 2 -3 0  (feature allows Z rotation)
  0 0 0 1  0 0  (feature allows X translation)
  0 0 0 0  0 1  (feature allows Z translation)
»W = recip(T)
W =
  0 1 0 0 0 3  (feature can resist force along Y if accompanied by Z moment)
  0 0 0 1 0 0  (feature can resist moment about X)
```

4.F.4. Use of Screw Theory to Describe Multiple Assembly Features That Join Two Parts

In this section, we use motion and constraint analysis to understand the behavior of several features at once. This is important because parts are often joined by multiple features, such as a bolt circle or several locating pins and holes. A number of these designs are overconstrained, or else they have deliberate clearance in the joints which avoids overconstraint but instead introduces location uncertainty. The method used is exactly the same as that used to determine the degrees of freedom that result from combining two basic surfaces: We form the twist matrix for each feature (or surface contact) individually, and then we use the twist matrix intersection algorithm to find out the net degrees of freedom allowed by the combination.

4.F.4.a. Example

Figure 4-31 shows a feature combination made of one pin–hole feature (toolkit feature 2 from Section 4.L) and one pin–slot feature (toolkit feature 4).

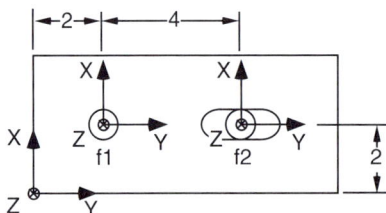

FIGURE 4-31. A Feature Combination Made of One Pin–Hole Toolkit Feature and One Pin–Slot Toolkit Feature.

The motion analysis, using the algorithm in Section 4.E.2.d, is shown in Table 4-10.

TR is the resultant twist matrix for the assembly. It is an empty matrix, meaning that there are no degrees of freedom of motion. Therefore, these parts are unable to move and thus do not have any underconstrained degrees of freedom. To find out if it is properly constrained or overconstrained, we need to carry out a constraint analysis, the next topic.

Constraint analysis is the second step. It requires use of the algorithm in Section 4.E.2.d.3 and is shown in Table 4-11.

Matrix *WR* is not empty. This means that there is overconstraint in the assembly. Considering the second row of *WR* first, it contains a unit moment about global *Y*, indicating that the pin–slot and the pin–hole can both support such a moment.

Refer to Figure 4-32 while we analyze the first row of *WR*. This row contains a unit force along global *Z* and a moment of 6 units about global *X*. These are labeled F_{z1} and M_x in Figure 4-32. We can slide this force to the right 6 units until it is opposite the pin in the slot. It is called F_{z2} in Figure 4-32. This new force also creates a torque about global *X* of 6 units, so the first row of *WR* also describes this configuration. Then we can slide the force along global *X* until it is over the center of the pin, where it is called F_{z3} in Figure 4-32 (and still generates a moment of 6 units about global *X*). This indicates that the pin–slot feature can support a force along the pin's axis. The other pin can also support this force. We know this because any entry in *WR* must be supportable by all the features or else it would not appear in *WR*. In fact, when we defined the twist matrices for the two features, we said that neither could permit motion along global *Z*. Thus each feature is "bidding," so to speak, to be the place where the plate's

TABLE 4-10. Motion Analysis of Pin–Hole and Pin–Slot Joint

```
Analysis of f1:
»T1 = [0 0 1 2 -2 0]
T1=
0 0 1 2 -2 0 (pure rotation about the pin-hole feature's z axis)
Analysis of f2:
»T2 = [0 0 1 6 -2 0;0 0 0 0 1 0;1 0 0 0 0 -6]
T2=
   0 0 1 6 -2  0 (pure rotation about the pin-slot feature's z axis)
   0 0 0 0  1  0 (pure translation about the pin-slot feature's y axis)
   1 0 0 0  0 -6 (rocking of the pin in the slot about the pin-slot feature's x axis)
»W1 = recip(T1)
W1=
   1.0000        0      0       0       0  -2.0000
        0   1.0000      0       0       0   2.0000
        0        0 1.0000       0       0        0
        0        0      0  1.0000       0        0
        0        0      0       0  1.0000  -0.0000
»W2 = recip(T2)
W2=
   1.0000  0       0       0        0 -6.0000
        0  0  1.0000  6.0000        0       0
        0  0       0       0   1.0000       0
»WU = [W1;W2]
WU=
   1.0000        0      0       0       0  -2.0000
        0   1.0000      0       0       0   2.0000
        0        0 1.0000       0       0        0
        0        0      0  1.0000       0        0
        0        0      0       0  1.0000  -0.0000
   1.0000        0      0       0       0  -6.0000
        0        0 1.0000  6.0000       0        0
        0        0      0       0  1.0000        0
»TR = recip(WU)
empty matrix
TR=

[]
```

global Z location will be defined. Thus the upper plate is overconstrained along global Z. Note that sliding the Z force along global X creates a moment about global Y, but the features can support this moment already, as indicated by the second row of WR, so nothing is changed, and the original entries in WR are still sufficient to explain everything.

The feature pair shown in Figure 4-31 therefore contains two elements of overconstraint: along the pin axes and about one axis normal to the pin axes. The first overconstraint cautions us that the upper plate may not lie flat against the lower plate because the lower plate may not be at the same height at each pin location, where height is measured relative to a reference line normal to both pin axes. Equivalently, the left pin may not be perpendicular to the plane of the plate. The second overconstraint cautions us that the two pin axes may not be parallel, or that the hole's axis may not be parallel to the slot's walls,

TABLE 4-11. Constraint Analysis of Pin–Hole and Pin–Slot Joint

```
≫TU = [T1;T2]

TU =
0  0  1  2 -2  0
0  0  1  6 -2  0
0  0  0  0  1  0
1  0  0  0  0 -6
≫WR = recip(TU)

WR =
0  0  1  6  0  0
0  0  0  0  1  0
```

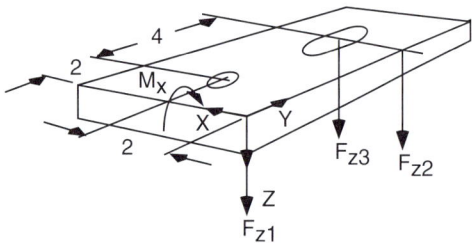

FIGURE 4-32. Illustrating the Equivalence of Three Different Positions for the Z Force in *WRU* in Table 4-11. The arrows for F_{Z1} and M_X represent the original entries in the first row of *WR* and are equivalent to F_{Z2} without M_X and F_{Z3} also without M_X.

measured about the global Y axis. Note that there is no corresponding problem regarding pin axis parallelism about the global X axis because the right pin can rock in the slot about its x axis.

4.F.4.b. Remarks

A constraint analysis of a complex assembly may result in numerous reported overconstraints. Several reasons are possible:

- The engineer made an error.
- The engineer intended that those features be overconstrained.
- The overconstraints are there in a mathematical sense but not in a practical sense.

Let us consider these cases one at a time.

The engineer made an error. In this case, the error may be immediately obvious to the engineer, who can correct it. In a complex assembly, however, this may not be easy, especially in the presence of mathematical overconstraints.

The engineer intended that those features be overconstrained. The engineer does not need to take any action in this case.

The overconstraints are mathematical but not practical. This is the most interesting situation. It arises in particular in the case of two-sided constraints. These, in turn, can arise in several ways:

1. The engineer can create a new two-sided feature by intersecting elementary surface contacts. For example, as we saw in Section 4.F.2.c, a pin–slot feature can be created by intersecting two plane surfaces with a cylinder. The result will contain a two-sided constraint, which will generate an overconstraint report in the analysis. Possibly, the report will be cluttered with overconstraint reports from such features. We avoided this clutter in the examples above by creating a toolkit feature comprising the allowed motions of a pin–slot. The way we described the allowed motions in the twist matrix suppressed the overconstraint that we know is really inside it. This permitted us to focus on achieving desired constraints and detecting errors.

   The engineer may relieve an overconstraint within a feature with two-sided constraint by providing a small amount of clearance in the final design if the resulting location uncertainty, backlash, vibration, or other consequences are tolerable. If the consequences are intolerable, the engineer may provide for a small amount of interference, as long as the resulting compressive stress is tolerable. In the first case, any resulting location uncertainty must be included in the tolerance analysis, while in the other case the resulting stress must be investigated to ensure that it does not cause damage, cracks, fatigue, and so on.

2. The engineer can combine two elementary features that collectively create two-sided or other means of overconstraint. This occurs, for example, in the pin–hole plus pin–slot. The engineer may relieve such overconstraints by providing a little clearance. All the cautions listed above for single features apply here.

Alternatively, he can construct a complex new feature containing the geometries of several elementary features, but defined so that interfeature overconstraints among them are suppressed by definition. This approach should be avoided for the following reason: It is relatively easy to control the dimensions and variations of a single feature, which usually involves creating surfaces that are near each other relative to the size of the feature. Providing the necessary small clearance to avoid overconstraint without compromising accuracy is also relatively easy. However, controlling interfeature dimensions and variations when these features are far from each other relative to the size of each individual feature is much more difficult and prone to errors that can cause overconstraint. It is better to confront this possibility by using simple individual features rather than defining a new complex feature that suppresses the overconstraint and pretending it will not happen. Defining the overconstraint away will simply keep the engineer from finding constraint errors or nonrobust aspects of the design.

It is also up to the engineer to decide which toolkit features best represent the problem at hand, or to design an appropriate feature instead. The overconstraints discussed in the previous subsection will not arise if toolkit features 9, 18, and 19 are used instead of features 2 and 4. One of the thought questions at the end of the chapter asks you to investigate this alternate formulation.

Constraint analysis is useful for finding constraint mistakes. Choosing features like 18 or 19 that optimistically assume away some overconstraint opportunities may result in an optimistic constraint analysis that fails to identify a mistake. A possible design technique is to use the most internally constrained toolkit features like 2 and 4 first, examine the resulting report, and separate the intended constraints from the extraneous ones and the mistakes. Next, eliminate the mistakes. Finally, replace the internally constrained features with similar ones that have a little internal clearance or use some of the less internally constrained features like 18 and 19 and judge whether the intended final constraint arrangement has been achieved.

4.F.5. Graphical Technique for Conducting Twist Matrix Analyses

A simple graphical technique can be used to help keep track of which twist matrices should be collected into unions and which should be intersected ([Shukla and Whitney]). The technique is presented here in a series of increasingly complex examples.

4.F.5.a. A Single Feature with a Single Twist Matrix
Consider the single feature illustrated in Figure 4-33.

To set up the graphical technique, we make a graph that represents parts and the features that join them. A simple graph of this type is shown in Figure 4-34. Then we trace a path or paths in the graph from the moving part to the fixed part, passing through the necessary features and other parts on the way. In this case, part A is the moving part while part B is the fixed part. The diagram is shown in Figure 4-35.

The procedure is:

- Identify every path from the moving part to the fixed part.
- For each path, construct the twist matrix for the moving part for each feature on the path, using the same reference coordinate frame (such as one attached to the fixed part), and form the union of all these twist matrices.

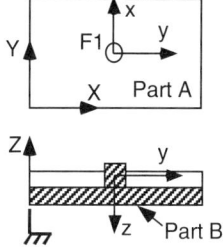

FIGURE 4-33. Single Feature to Illustrate Graphical Technique.

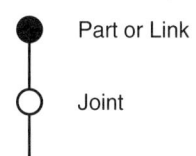

FIGURE 4-34. Definition of Terms for Graph Representation of an Assembly.

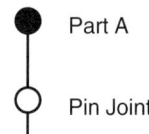

FIGURE 4-35. Diagram for Analyzing the Feature Situation in Figure 4-33. This case is trivial because there is only one path from the moving part (A) to the fixed part (B).

- Form the intersection of all the twist unions using the procedure in 4.E.2.d. A nonempty TR represents underconstraint in the assembly.

In this case, there is only one path and on this path there is only one feature, so the procedure is trivial.

4.F.5.b. Two Parts Joined by Two Features

Next, consider the feature in Figure 4-31. The analysis diagram is shown in Figure 4-36. In this case, there are two paths from the moving part (A) to the fixed part (B). On each path we find one feature. F1 is chosen arbitrarily to be the pin–hole while F2 is chosen to be the pin–slot. The motion analysis matrix *TR* is obtained by intersecting the twist matrices corresponding to F1 and F2, as illustrated in Table 4-10.

4.F.5.c. An Assembly with a Moving Part

Finally, consider the 4-bar linkage shown, together with its analysis diagram, in Figure 4-37. The problem is to determine the degrees of freedom of link L3, considered to be the moving part, with respect to L1, considered to be the fixed part.

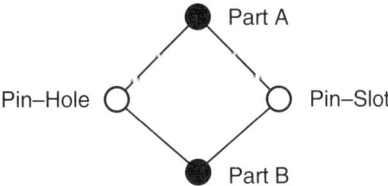

FIGURE 4-36. Diagram for Two Parts Joined by Two Features. In this case there are two paths.

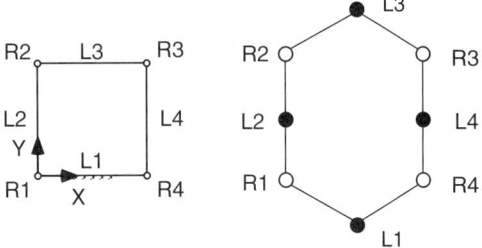

FIGURE 4-37. Four-Bar Linkage and Its Analysis Diagram. The problem is to determine the degrees of freedom of link L3 with respect to link L1. We have two paths with two features on each path. The left path connects L3 to L1 via R2, L2, and R1. The right path connects L3 to L1 via R3, L4, and R4. R1 through R4 are rotary joints, each consisting of a pin–hole feature. Links L1 through L4 are of equal length and all lie nominally in the *X–Y* plane. The *Z* axis points out of the paper.

The motion analysis goes as follows:

- For the left path, we find that there are two features, R2, and R1, between L3 and L1. We need to find how those features generate motion for L3. First, we erase the right path and all its features. Then we form a twist allowing R2 to move L3 while R1 is frozen, and then we form a twist that allows R1 to move L3 while R2 is frozen. Each of these twists is calculated using the same fixed reference associated with L1, say one centered on R1. We then form the union of these two twists to get a representation of the left path.

- For the right path, the process is similar, except that we erase the left path and consider R3 and R4, and we again use a coordinate reference centered on R1.

- Finally, we intersect the left path union and the right path union to find the net motion allowed to L3.

The whole process is shown in Table 4-12.

We have shown that this simple technique permits analysis of single joints made of several features as well as analysis of several parts connected by several joints. If these joints are made of several features, then the user should analyze each joint separately, finding the net twist allowed by all its constituent features by intersecting their individual twists, and then combine the joints using the method shown here.

This method can be used on any assembly or linkage as long as it does not contain cross-coupling. We saw in Figure 4-6 a mechanism that has cross-coupling. The method above will not be able to find the motions of the top horizontal link if the bottom horizontal link is fixed. However, the motion of the top link can be found if the left or right vertical link is considered fixed, and the answer can be rewritten to conform to the situation where the bottom link is fixed.

4.F.6. Graphical Technique for Conducting Constraint Analyses[22]

Systematic constraint analysis begins the same way that motion analysis does, by drawing the graph and enumerating the paths. However, constraint analysis is considerably more tedious because the intersection method has to be applied to all combinations of features

---

[22]See [Shukla and Whitney 2001b].

TABLE 4-12. Motion Analysis of Four-Bar Linkage

Motion analysis:

The first step is to analyze the left path to find L3's motions as if it is connected to the fixed link only by L2. This is done by considering the motion that each remaining feature could give L3 while considering other remaining motions frozen:

Rotate L3 about R2 in R1 coordinates with R1 frozen:

```
»t1 = [0 0 1 1 0 0]
 t1 =
       0 0 1 1 0 0
```

Rotate L3 about R1 in R1 coordinates with R2 frozen:

```
»t2 = [0 0 1 0 0 0]
 t2 =
       0 0 1 0 0 0
```

Form the union of these to get the possible motions of L3 provided by the left path:

```
»tlp = [t1;t2]
 tlp =
       0 0 1 1 0 0
       0 0 1 0 0 0
```

Now analyze the right path to find the motions of L3 as if it is connected to the fixed link only by L4. Again, this is done by considering the motion that each remaining feature could give L3 while considering the other remaining motions frozen:

Rotate L3 about R3:

```
»t3 = [0 0 1 1 -1 0]
 t3 =
       0 0 1 1 -1 0
```

Rotate L3 about R4:

```
»t4 = [0 0 1 0 -1 0]
 t4 =
       0 0 1 0 -1 0
```

Form the union of these to get the motions of L3 provided by the right path:

```
»trp = [t3;t4]
 trp =
       0 0 1 1 -1 0
       0 0 1 0 -1 0
```

Form the intersection of tlp and trp to see what the net allowed motion is

```
»wlp = recip(tlp)
wlp =
0 1.0000    0      0      0      0
    0     0 1.0000    0      0      0
    0     0      0 1.0000    0      0
    0     0      0      0 1.0000  -0.0000
»wrp = recip(trp)
wrp=
0 1.0000    0      0      0 1.0000
    0     0 1.0000    0      0      0
    0     0      0 1.0000    0      0
    0     0      0      0 1.0000 0.0000
```

(continued)

TABLE 4-12. (Continued)

```
≫WU = [wlp;wrp]

WU =

0 1.0000    0       0       0       0
      0     0 1.0000       0       0       0
      0     0     0  1.0000       0       0
      0     0     0       0  1.0000 -0.0000
0 1.0000    0       0       0  1.0000
      0     0 1.0000       0       0       0
      0     0     0  1.0000       0       0
      0     0     0       0  1.0000  0.0000
≫TR = recip(WU)
TR = -0.0000 -0.0000 -0.0000 1.0000 0.0000 0.0000
```

This says that L3 is permitted to move in the X direction in R1 coordinates. This is the answer we expect based on intuition.

The reciprocal of TR, WU, shows what forces and moments can be resisted by the linkage. These are F_y, F_z, M_x, M_y, and M_z. But we do not know if any of these is overconstrained. This question is resolved in Table 4-13.

---

if we want to identify every feature that contributes to overconstraint.

In brief, the process is as follows:

- Choose any path and check to see if its twists, formed into a union, overconstrain the parts.

- Choose a second path and intersect its twist union with the first path's twist union to find a combined feature that allows only those motions common to those paths. Check to see if this combination overconstrains the parts.

- Continue in this way, adding one path's twist union at a time until all have finally been combined and checked for overconstraint.

To see how this works, consider a part A joined to another part B by four features F1, F2, F3, and F4, as shown in Figure 4-38.

On each path there is one twist, so we have four twists, one for each path: T_1, T_2, T_3, and T_4. By using the relationship $W_i = recip(T_i)$, we first find the corresponding

wrenches: W_1, W_2, W_3, and W_4. We then systematically form (the order is arbitrary) T_{12}, T_{123}, and T_{1234}, and W_{12}, W_{123}, and W_{1234}, as follows:

$$T_{12} = \cap(T_1, T_2) = recip(\cup(W_1, W_2))$$
$$= \text{net motions allowed by path 1 and path 2}$$

$$T_{123} = \cap(T_1, T_2, T_3) = recip(\cup(W_1, W_2, W_3))$$
$$= \text{net motions allowed by path 1, path 2,}$$
$$= \text{and path 3}$$

$$T_{1234} = \cap(T_1, T_2, T_3, T_4) = recip(\cup(W_1, W_2, W_3, W_4))$$
$$= \text{net motions allowed by all four paths}$$

$$W_{12} = \cap(W_1, W_2) = recip(\cup(T_1, T_2))$$
$$= \text{overconstraint provided by path 1 and path 2}$$

$$W_{123} = \cap(W_{12}, W_3) = recip(\cup(T_{12}, T_3))$$
$$= \text{overconstraint provided by path 1, path 2,}$$
$$\text{and path 3}$$

$$W_{1234} = \cap(W_{123}, W_4) = recip(\cup(T_{123}, T_4))$$
$$= \text{overconstraint provided by all four paths}$$

Note that

$$W_{1234} \neq \cap(W_1, W_2, W_3, W_4) = recip(\cup(T_1, T_2, T_3, T_4))$$
$$= \text{the force(s) or moment(s) that can be supported}$$
$$\text{by all four paths at once}$$

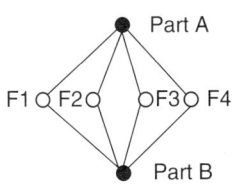

FIGURE 4-38. Path Diagram for Two Parts Joined by Four Features.

TABLE 4-13. Constraint Analysis of Four-Bar Linkage

Constraint analysis:

Intersect the wrenches of the two paths:

```
≫tlrp = [tlp;trp]

tlrp =

    0  0  1  1   0  0
    0  0  1  0   0  0
    0  0  1  1  -1  0
    0  0  1  0  -1  0
≫WR = recip(tlrp)
WR =

    0  0  1.0000      0      0      0
    0  0      0  1.0000      0      0
    0  0      0      0  1.0000 -0.0000
```

This says that L3 is overconstrained via force in the Z direction and moments about X and Y. This is due to the presence of two-sided constraints in the pin joints plus the fact that these joints were defined as able to support Z force.

In Table 4-12 we found that the linkage could support F_y, F_z, M_x, M_y, and M_z. We now know that some of these are overconstrained.

This is important because, in case there is no force or moment that is provided by all four features, W_{1234} will be empty, but we cannot conclude on this basis that there is no overconstraint. One possible reason is that features F1, F2, and F3 share the ability to constrain a particular force or moment that F4 cannot constrain. This possible overconstraint is detected by W_{123}.

For the four-bar linkage, the analysis is shown in Table 4-13.

This process can be looked at two ways. First, it can be used as an existence check for overconstraint. As soon as overconstraint is found, the procedure can be stopped. Second, it can be used to identify overconstrained directions and the features that create them. In this case, the procedure must continue until all feature sets have been combined into the growing intersected set.

Note that choosing the paths in a different sequence will always result in the same number of degrees of freedom, if any, being detected as overconstrained, but the WR matrix reporting the overconstraint may appear different. The reason for this is that, as features are added to the combination, one such set may properly constrain the parts. Any feature added thereafter will necessarily add overconstraint along the direction(s) it is capable of constraining, and these directions will appear in WR. A different sequence of analysis will eventually arrive at proper constraint with a different subset of the features, and the next one added will be different this time than last time. WR will then report this feature's directions as the overconstrained ones rather than another feature's directions. The engineer can use this information to explore the consequences of establishing joints between parts in different sequences, including deciding which features, if any, to redesign in order to remove the overconstraint.

Problem 17 explores this procedure using a simple example. The reader is encouraged to use the procedure even if the examples look simple and the answer is easy to predict by intuition. The method will be welcome indeed when a real industrial strength problem is encountered.

We may look ahead at this point to Chapter 7 on assembly sequences to see that this procedure will apply to sequences of parts as well as sequences of features within two parts. Choosing which sequence to use will depend on, among other things, which one does a better job of delivering the KC.

4.F.7. Why Are the Motion and Constraint Analyses Different?

There appears to be an asymmetry between the motion analysis in Section 4.F.5 and the constraint analysis in Section 4.F.6. Motion analysis requires only intersecting all the twists at once but constraint analysis requires careful accumulation of wrench intersections. The reason is as follows.

If we intersect n twists and find that the intersection is empty, we know that the parts cannot move. We can intersect any subset of these n twists and may indeed find underconstraint, but we do not care because the full set of twists prevents any motion.

On the other hand, if we intersect n wrenches and find an empty intersection, we cannot conclude that there is no overconstraint. We must intersect a series of subsets of these wrenches because one or more of them could cause overconstraint and we want to know if that is the case. Adding more wrenches to the test set will not remove this lurking overconstraint but will just cover it up because the additional wrenches do not share constraining directions with the subset that contains overconstraint.

4.G. ADVANCED CONSTRAINT ANALYSIS TECHNIQUE

On the CD-ROM packaged with this book is an appendix to this chapter, written by J. Michael Gray, explaining a more general method of determining the state of mobility and constraint of an assembly. It is not as simple to apply as the one explained here, but it suffers none of the limitations. It is based on work in [Davies 1981, 1983a, 1983b, and 1983c].

4.H. COMMENT

The reader may have detected that we have dealt extensively with assembly concepts in the last two chapters without talking much about parts in the usual sense. Most books on engineering design deal with parts such as shafts, gears, and bearings. The detailed shape of these parts is important to such studies. We have said virtually nothing about the shape of parts, and deliberately so, for the following reasons.

First, the concepts we need, such as location and orientation in space, and degree of constraint, can be described with mathematical precision and very few symbols (and correspondingly few bytes of memory) using a few numbers in a 4×4 matrix or a twist matrix. To capture the equivalent information, such as the location and orientation of a shaft axis, or the fact that one part can rotate with respect to another, using purely geometric data, would entail thousands or millions of bytes and could possibly be less precise.[23] A major point of the last two chapters is that the main information we need to define a kinematic assembly is not geometric. It amounts to coordinate frames and twist matrices. We can add the geometry later.

Second, we are dealing with only the geometric relationships between parts, not any forces, loads, or deformations that they might experience. We addressed force and deformation only when we showed why pure kinematic constraints consisting of sharp point contacts are not used in practice. A complete engineering design of an assembly must include forces and deformations. Such factors will provide the engineer with most of the information to decide what shape the parts must have. The size and other details of the assembly features will also be influenced by such factors. Nevertheless, the scheme by which the parts will be located in space prior to experiencing loads must be designed with care using the methods described in this book. Whether the loads are considered first and used to influence the locating scheme, or the locating scheme is decided first and then the parts are sized to suit the loads, depends on the engineer's style of working, the materials used, and the degree to which the structure is stressed as a percentage of the yield stresses of its materials.

4.I. CHAPTER SUMMARY

This chapter is one of the most important in this book. It presents a way to design competent assemblies using the principles of kinematic constraint. We distinguish between kinematically constrained assemblies, deliberately overconstrained or underconstrained assemblies, and assemblies that contain constraint errors. Kinematic assemblies are capable of achieving rapid, accurate, and repeatable assembly at reasonable cost. Both Whitehead and Kamm make this point in their books. The car seat example shows this vividly.

The method of Screw Theory permits us to define assembly features as geometric entities capable of establishing constraint relations between the parts they join. Screw Theory also permits us to build up a joint between parts using arbitrary combinations of simpler features and then to examine the state of constraint that is established by that joint.

These concepts and tools permit us to use features and the connective assembly models defined in Chapter 3 to build kinematically constrained assemblies of rigid parts

---

[23] A student once asked the author, "Dr. Whitney, what do you do about the facets?" "What facets?" I asked. "I built a pin-and-hole model in my CAD system, and I sometimes find that my motion algorithm says that the pin cannot turn in the hole because a vertex on the pin interferes with a facet on the hole." Real round pins and holes do not have facets and vertices, of course. Only approximate geometric models of them in CAD systems do. Faceted models are used for approximate interference analysis and to create screen displays. They are appropriate for modeling *assembly drawings* but not for modeling assemblies.

that lie at particular desired places and orientations in space so that they will achieve key characteristics as defined in Chapter 2. These concepts are mathematically precise and consistent, and they capture the most fundamental properties of assemblies. They can be used as the basis for computer models of assemblies, for motion and constraint analyses, and for other analyses, such as variation, that are discussed in later chapters.

4.J. PROBLEMS AND THOUGHT QUESTIONS

1. Prove that three points define a plane using three hemispherical features touching a plate, as shown in Figure 4-39.

Hint: The twist matrix for the plate resting on hemisphere number 1, referred to the lower left corner of the plate, is

$$T_{HS_1} = \begin{bmatrix} 1 & 0 & 0 & 0 & 0 & -1 \\ 0 & 1 & 0 & 0 & 0 & 1 \\ 0 & 0 & 1 & 1 & -1 & 0 \\ 0 & 0 & 0 & 1 & 0 & 0 \\ 0 & 0 & 0 & 0 & 1 & 0 \end{bmatrix}$$

Write down the twist matrices for the plate resting on each of the other two hemispheres, then combine the three twist matrices according to the twist matrix intersection algorithm in Section 4.E.2.d.2. You should get (give or take some minus signs that are not significant)

$$TR_{plane} = \begin{bmatrix} 0 & 0 & 0 & 1 & 0 & 0 \\ 0 & 0 & 0 & 0 & 1 & 0 \\ 0 & 0 & 1 & 0 & 0 & 0 \end{bmatrix}$$

This says, row by row, that the plate can slide in the X direction, it can slide in the Y direction, and it can rotate about Z. This is consistent with the properties of a plane.

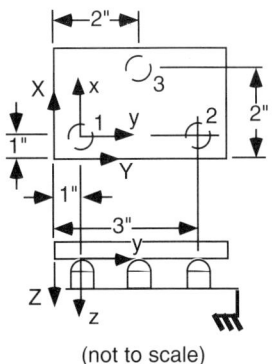

(not to scale)

FIGURE 4-39. Figure for Problem 1.

2. Figure 4-40 shows an arrangement in which part 1 has three hemispherical features under part 2 and two such features at its right. Prove that this configuration leaves part 2 with exactly one unconstrained degree of freedom relative to part 1. Confirm that the result makes sense in terms of the coordinates shown in Figure 4-40.

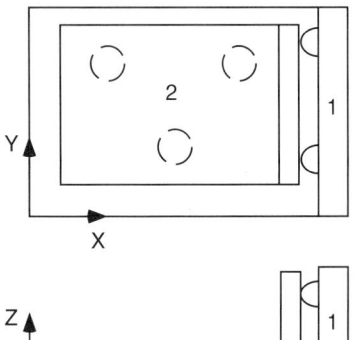

FIGURE 4-40. Figure for Problem 2.

3. Use toolkit features 9, 18, and 19 to analyze the situation shown in Figure 4-41. Follow the methodology in Section 4.F.4.a.

You should be able to show that the upper plate cannot move and it is not overconstrained.

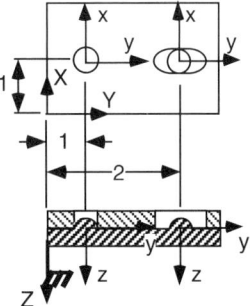

FIGURE 4-41. Figure for Problem 3.

4. Consider the part pair in Figure 4-42, consisting of plate 1 with two pins, mating to plate 2 with one hole and one slot. Analyze the state of motion and the state of constraint using the methods in Section 4.F.4. Compare your answers with those in Table 4-10 and Table 4-11 and explain every similarity or difference between the matrices row by row.

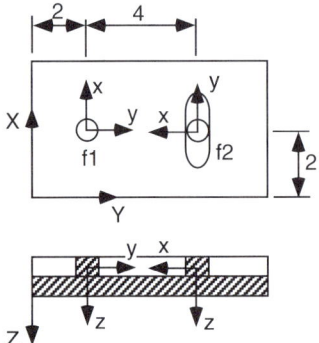

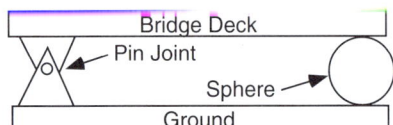

FIGURE 4-42. Figure for Problem 4.

5. Form a joint between two plates using two pin–hole joints (toolkit feature 2). Analyze the state of motion and state of constraint for this joint by forming TR and WR. Explain all the resulting matrices row by row.

6. Figure 4-43 represents, in two dimensions, a common way of supporting the deck of a bridge.

Use motion and constraint analysis to show that this arrangement has one degree of freedom. Now assume that the bridge deck expands due to rising temperature. Show that it is able to do this without encountering overconstraint.

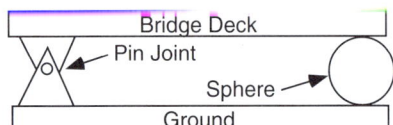

FIGURE 4-43. Figure for Problem 6.

7. Consider the car seat example in Section 4.C.5.b. Reproduce the joint used in the original design by using four instances of toolkit feature 2. Form matrices *TR* and *WR* for this case and explain each resulting matrix row by row.

8. Returning to the car seat example, reproduce the revised design by using the appropriate toolkit features. Explain the resulting matrices row by row and compare them to the results in Problem 4. Discuss any overconstraints that remain.

9. Analyze the state of motion and constraint for the two situations shown in Figure 4-44. In each case, the part to be analyzed contains two slots, through each of which there is a pin. Explain the resulting matrices row by row.

10. An engineer is considering how to join the three parts A, B, and C shown in Figure 4-45. He has decided that he needs a peg and hole to join A and B, and he knows that some kind of

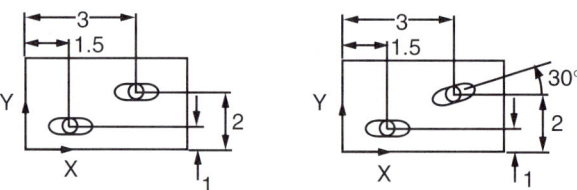

FIGURE 4-44. Figure for Problem 9.

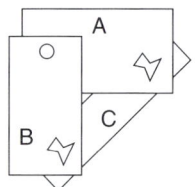

FIGURE 4-45. Figure for Problem 10.

features will be needed between A and C and B and C, respectively. These he has sketched in as irregular polygons for the time being just as reminders. But he has not yet chosen their final shape. Assume that he wants A and B to firmly locate C, and therefore that A and B must be joined first. Is the hole he chose for the A–B mate sufficient, and what are his alternatives for the remaining features? What should he take into consideration when making these choices?

11. Consider the three-bar linkage shown in Figure 4-46. Link A is fixed, while links B and C have pin joints with link A and with each other. Use the twist matrix intersection algorithm to prove that this linkage is rigid and cannot move.

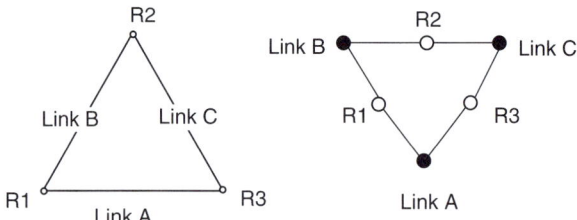

FIGURE 4-46. Figure for Problem 11.

12. Consider the five-bar linkage shown below. Show that the diagram in Figure 4-47 is correct and use it to set up the necessary twist matrices for determining the net motion of L3.

13. Consider the five-bar linkage shown in Figure 4-48. Assume link L2 to be fixed and find the state of motion and constraint of link L1. Repeat this analysis for link L5, again assuming L2 is fixed. Note that the path method as outlined in this chapter cannot

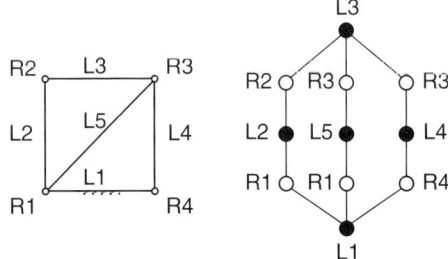

FIGURE 4-47. Figure for Problem 12.

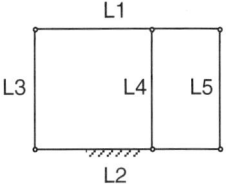

FIGURE 4-48. Figure for Problem 13.

analyze the state of constraint of link L5 if link L3 is considered fixed. What conclusions can you draw concerning the state of constraint of link L5? Does it matter which link is assumed fixed?

14. Consider the aircraft structure example in Section 4.C.5.c and explain why there should not be locating holes for the longitudinal location of the ribs in both the upper and lower spars.

15. Consider the copier example in Section 4.C.5.a. Assume that the two side panels form a rigid unit. Model the joints between the curved panel and the two side panels using toolkit features. Form matrices TR and WR for the curved panel joined to the side panel unit and explain the resulting matrices row by row.

16. Suggest a redesign for the joints in problem 15. Form matrices TR and WR and prove that your design is an improvement.

17. Consider the example given in Figure 4-49, in which two plates are joined by four toolkit hemisphere–slot features. First, decide intuitively whether the plates are underconstrained, fully constrained, or overconstrained. Then find the wrench intersection WR considering all four features at once. Explain the answer, row by row. Then find the intersection of features 1 and 2 and determine how or if they constrain the parts. Then intersect feature 3 with the previously created intersection of features 1 and 2 and determine how or if they constrain the parts. Last, intersect feature 4 with the previously calculated intersection involving features 1, 2, and 3. This last step should reveal the true state of constraint of these parts. Repeat this process using the features in the sequence 2, 3, 4, 1. Explain any differences you observe.

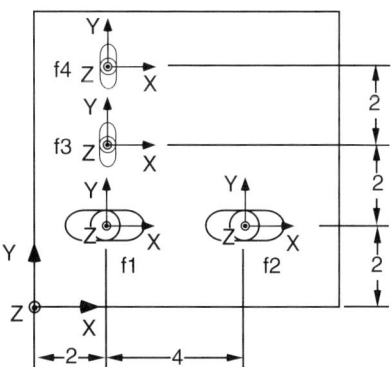

FIGURE 4-49. Figure for Problem 17.

18. Figure 4-50 shows two plates joined by a pin–hole feature. Write the twist matrix for this feature.

Now consider the two situations in Figure 4-51. In each case, a plate is joined to another via a slot in one plate and a tiny pin on the other. Assuming that the pin always stays in contact with the same side of the slot as shown in the figure, prove that a combination of these two situations has the same twist matrix as the pin–hole feature.

Does it matter how big the slots are?

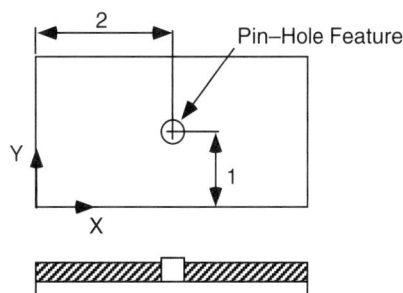

FIGURE 4-50. First Figure for Problem 18.

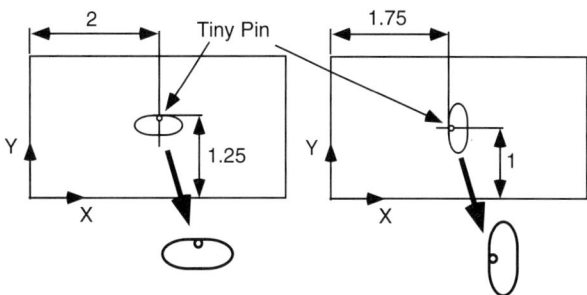

FIGURE 4-51. Second Figure for Problem 18.

4.K. FURTHER READING

[Adams and Whitney] Adams, J. D., and Whitney, D. E., "Application of Screw Theory to Constraint Analysis of Assemblies Joined by Features," *ASME Journal of Mechanical Design,* vol. 123, no. 1, pp. 26–32, 2001.

[Baker 1980a] Baker, J. E., "Screw System Algebra Applied to Special Linkage Configurations," *Mechanism and Machine Theory,* vol. 15, pp. 255–265, 1980.

[Baker 1980b] Baker, J. E., "On Relative Freedom Between Links in Kinematic Chains with Cross-Jointing," *Mechanism and Machine Theory,* vol. 15, pp. 397–413, 1980.

[Baker 1981] Baker, J. E., "On Mobility and Relative Freedoms in Multiloop Linkages and Structures," *Mechanism and Machine Theory,* vol. 16, no. 6, pp. 583–597, 1981.

[Ball] Ball, R. S., *A Treatise on the Theory of Screws,* Cambridge University Press, Cambridge, 1900.

[Blanding] Blanding, D., *Exact Constraint Design,* New York: ASME Press, 1999.

[Charles, Clement, et al.] Charles, B., Clement, A., Desrochers, A., Pelissou, P., and Riviere, A., "Toward a Computer Aided Functional Tolerancing Model," International Conference on CAD/CAM and AMT, CIRP Session on tolerancing for function in a CAD/CAM environment, 1989.

[Davies 1981] Davies, T. H., "Kirchhoff's Circulation Law Applied to Multi-Loop Kinematic Chains," *Mechanism and Machine Theory,* vol. 16, pp. 171–183, 1981.

[Davies 1983a] Davies, T. H., "Mechanical Networks I: Passivity and Redundancy," *Mechanism and Machine Theory,* vol. 18, no. 2, pp. 95–101, 1983.

[Davies 1983b] Davies, T. H., "Mechanical Networks II: Formulae for the Degrees of Mobility and Redundancy," *Mechanism and Machine Theory,* vol. 18, no. 2, pp. 103–106, 1983.

[Davies 1983c] Davies, T. H., "Mechanical Networks III: Wrenches on Circuit Screws," *Mechanism and Machine Theory,* vol. 18, no. 2, pp. 107–112, 1983.

[Green] Green, W. G., *Theory of Machines,* London: Blackie and Son Ltd., 1961.

[Hart-Smith] Hart-Smith, D. J., "Interface Control—The Secret to Making DFMA Succeed," presented at Society of Automotive Engineers, SAE Paper, Reprint #972191, SAE Inc., 1997.

[Herve] Herve, M., "Analyse structurelle des mecanismes par groupe de deplacements," *Mechanism and Machine Theory,* vol. 13, pp. 437–450, 1978.

[Hoffman and Kunze] Hoffman, K., and Kunze, R., *Linear Algebra,* Englewood Cliffs: Prentice-Hall, 1961.

[Hunt] Hunt, K. H., *Kinematic Geometry of Mechanisms,* Oxford: Clarendon Press, 1978.

[Kamm] Kamm, L. J., *Designing Cost-Effective Mechanisms,* New York: McGraw-Hill, 1990.

[Kim and Wu] Kim, M. G., and Wu, C. H., "Modeling of Part-Mating Strategies for Automating Assembly Operations for Robots," *IEEE Transactions on Systems, Man, and Cybernetics,* vol. 24, no. 7, pp. 1065–1074, 1994.

[Konkar] Konkar, R., "Incremental Kinematic Analysis and Symbolic Synthesis of Mechanisms," Ph.D. Dissertation, Stanford University, Stanford, CA 1993.

[Konkar and Cutkosky] Konkar, R., and Cutkosky, M., "Incremental Kinematic Analysis of Mechanisms," *Journal of Mechanical Design,* vol. 117, pp. 589–596, 1995.

[Kriegel] Kriegel, J. M., "Exact Constraint Design," *Mechanical Engineering,* pp. 88–90, May 1995.

[Mason and Salisbury] Mason, M. T., and Salisbury, J. K., *Robot Hands and the Mechanics of Manipulation,* Cambridge, MA: MIT Press, 1985.

[Ohwovoriole and Roth] Ohwovoriole, M. S., and Roth, B., "An Extension of Screw Theory," *Journal of Mechanical Design,* vol. 103, pp. 725–735, 1981.

[Paynter] Paynter, H. M., *Analysis and Design of Engineering Systems,* Cambridge, MA: MIT Press, 1961.

[Phillips] Phillips, J., *Freedom in Machinery* (in two volumes), Cambridge: Cambridge University Press, 1984 (vol. 1) and 1989 (vol. 2).

[Roth] Roth, B., "Screws, Motors, and Wrenches that Cannot Be Bought in a Hardware Store," *Robotics Research: The First Symposium,* Cambridge, MA: MIT Press, pp. 679–735, 1983.

[Shukla and Whitney 2001a] Shukla, G., and Whitney, D. E., "Systematic Evaluation of Constraint Properties of Datum Flow Chain," IEEE ISATP, Fukuoka, Japan, 2001a.

[Shukla and Whitney 2001b] Shukla, G., and Whitney, D. E., "Application of Screw Theory to the Motion and Constraint Analysis of Mechanisms," unpublished manuscript, 2001b.

[Slocum] Slocum, A. H., *Precision Machine Design,* New York: Prentice-Hall, 1991.

[Smith and Chetwynd] Smith, S. T., and Chetwynd, D. G., *Foundations of Ultraprecision Mechanism Design,* Philadelphia: Gordon and Breach, 1992.

[Sweder and Pollack] Sweder, T. A., and Pollock, J., "Full Vehicle Variability Modeling," SAE Paper, Reprint #942334, SAE Inc., 1994.

[Waldron] Waldron, K. J., "The Constraint Analysis of Mechanisms," *Journal of Mechanisms,* vol. 1, pp. 101–114, 1966.

[Whitehead] Whitehead, T. N., *The Design and Use of Instruments and Accurate Mechanism,* New York: Dover Press, 1954.

4.L. APPENDIX: Feature Toolkit

We saw by example how to construct the twist matrix representation of a feature in Section 4.E.2.a as well as how to create features from basic surface contacts in the previous section. Now we can follow that example and create a toolkit of features accompanied by their twist matrices. This appendix presents such a toolkit, but the reader can make up his or her own by following the methods illustrated.

4.L.1. Nomenclature for the Toolkit Features

Each feature below is shown in its nominal mating configuration. Each feature has a coordinate frame whose axes are labeled with lower case letters x, y, and z. This will distinguish feature coordinate axis names from part center coordinate axis names, which are uppercase letters X, Y, and Z. The positive z axis of the feature should always be pointing in the nominal mating direction. By arbitrary choice, the y axis points in the direction of translational freedom for features with only one translational degree of freedom. For features with two translational degrees of freedom, the y axis points in what is considered the primary motion direction. For cases where this does not apply, axis direction assignments are arbitrary but adhered to as convention for each case. Friction is considered negligible in restraining part motion.

In each case, one part in the mating pair is taken to be immobile and is denoted by the attached ground symbol ⊥. The ground symbol shows the location of the part center coordinate frame. For example, in a pin–hole mate, the part with the pin is immobile and the feature's coordinate frame is placed on the cylindrical axis of the pin and centered lengthwise (z direction). It is assumed that all features are at their nominal size and shape and share line-to-line fits unless obvious clearance is shown and discussed. For example in a pin–slot feature, the pin is the same diameter as the width of the slot such that no translation of the pin is allowed along the short axis of the slot and no rotation of the pin is allowed about the long axis.

4.L.2. Toolkit Features

The features appear in Table 4-14.

TABLE 4-14. Toolkit Features

Toolkit Feature Number and Name	Sketch	Twist Matrix	Remarks
1—Prismatic pin in prismatic hole		$T_1 = [0\ 0\ 0\ 0\ 0\ 0]$	In general, twist matrix entries are in part coordinates, but entities like "ω_x" are in feature coordinates.
2—Pin on plate in hole		$T_2 = [\omega\ v]$ where $\omega = (R\omega_z)^T$, $v = r \times \omega$, $r = d^T$	Twist matrix if top plate is very thin: $$T_{2'} = \begin{bmatrix} \omega_1 & v_1 \\ \omega_2 & v_2 \\ \omega_3 & v_3 \end{bmatrix}$$ where $\omega_1 = (R\omega_z)^T$, $\omega_2 = (R\omega_x)^T$, and $\omega_3 = (R\omega_y)^T$ and $v_i = r \times \omega_i$, $i = 1, 2, 3$
3—Prismatic pin and prismatic slot		$T_3 = [0\ v]$ where $v = (Rk_y)^T$ and $0 = (0, 0, 0)$	

(continued)

TABLE 4-14. (Continued)

Toolkit Feature Number and Name	Sketch	Twist Matrix	Remarks
4—Pin on plate in slot		$$T_4 = \begin{bmatrix} \omega_1 & v_1 \\ \omega_2 & v_2 \\ 0 & v_3 \end{bmatrix}$$ where $\omega_1 = (R\omega_x)^T$ and $\omega_2 = (R\omega_z)^T$, $v_i = r \times \omega_i, i = 1, 2,$ and $v_3 = (Rk_y)^T$	One rotation permits the plate to rotate in the XY plane. The other permits the plate to rotate about the X axis of the pin. If the plate is thin, we can add a third rotation: $$T_{4'} = \begin{bmatrix} \omega_1 & v_1 \\ \omega_2 & v_2 \\ 0 & v_3 \\ \omega_4 & v_4 \end{bmatrix}$$ where $\omega_4 = (R\omega_y)^T$
5—Round pin in prismatic slot		Same as feature 4	
6—Round pin in hole		$$T_6 = \begin{bmatrix} \omega & v_1 \\ 0 & v_2 \end{bmatrix}$$ where $\omega = (R\omega_z)^T$ and $v_2 = (Rk_z)^T$	Compared to feature 2, this feature provides a pivot but does not include planar support along the z axis.
7—Threaded joint		$T_7 = [\omega \quad v]$ where $v = r \times \omega + p\omega$	p is the pitch of the threads
8—Elliptical ball and socket		$$T_8 = \begin{bmatrix} \omega_1 & v_1 \\ \omega_2 & v_2 \end{bmatrix}$$	

TABLE 4-14. (Continued)

Toolkit Feature Number and Name	Sketch	Twist Matrix	Remarks
9—Plate–plate lap joint	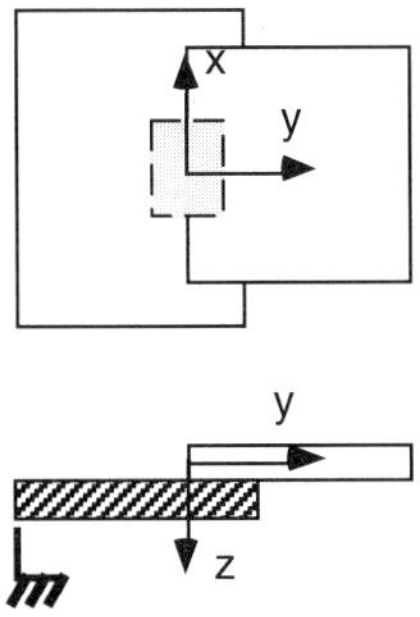	$T_9 = \begin{bmatrix} \omega & 0 \\ 0 & v_1 \\ 0 & v_2 \end{bmatrix}$ where $\omega = (R\omega_z)^T$, $v_1 =$ any vector perpendicular to ω, and v_2 is perpendicular to both v_1 and ω.	The 0 to the right of ω indicates that there is no fixed rotation axis in this case, so there is no definite velocity arising from ω. The lightly shaded area in the sketch represents the allowable location of the coordinate frame for the lapping part.
10—Spherical joint	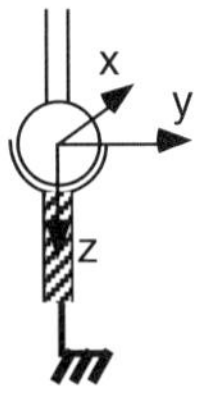	$T_{10} = \begin{bmatrix} \omega_1 & v_1 \\ \omega_2 & v_2 \\ \omega_3 & v_3 \end{bmatrix}$ where $\omega_1 = (R\omega_x)^T$ $\omega_2 = (R\omega_y)^T$ $\omega_3 = (R\omega_z)^T$ $v_1 = r \times \omega_1$ $v_2 = r \times \omega_2$ $v_3 = r \times \omega_3$	
11—Pin in oversize hole	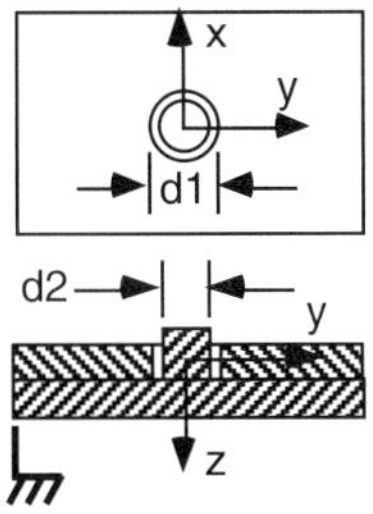	$T_{11} = \begin{bmatrix} \omega & 0 \\ 0 & v_1 \\ 0 & v_2 \end{bmatrix}$ where $\omega = (R\omega_z)^T$, $v_1 =$ any vector perpendicular to ω, and v_2 is perpendicular to both v_1 and ω.	The 0 to the right of ω indicates that there is no fixed rotation axis in this case, so there is no definite velocity arising from ω. We can add two rows to express the fact that the pin can rock in the clearance about the feature's x and y axes. These extra motions are also possible if the upper plate is very thin.
12—Elliptical ball in cylindrical groove	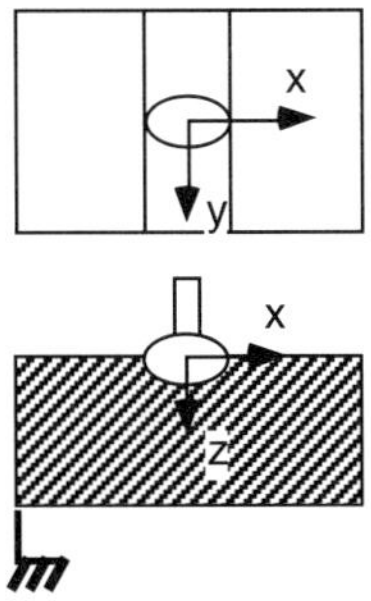	$T_{12} = \begin{bmatrix} \omega_1 & v_1 \\ \omega_2 & v_2 \\ 0 & v_3 \end{bmatrix}$ where v_1 and v_2 are defined as usual, and $v_3 = (Rk)^T$.	

(*continued*)

TABLE 4-14. (Continued)

Toolkit Feature Number and Name	Sketch	Twist Matrix	Remarks
13—Edge on plane		$T_{13} = \begin{bmatrix} \omega_1 & v_1 \\ \omega_2 & 0 \\ 0 & v_2 \\ 0 & v_3 \end{bmatrix}$ where $\omega_1 = (R\omega_x)^T$, $v_2 = (Rk_x)^T$, and $v_3 = (Rk_y)^T$.	
14—Ellipsoid on plane		$T_{14} = \begin{bmatrix} \omega_x & v_1 \\ \omega_y & v_2 \\ \omega_z & v_3 \\ 0 & v_4 \\ 0 & v_5 \end{bmatrix}$ where $v_4 = (Rk_x)^T$ and $v_5 = (Rk_y)^T$.	
15—Sphere in cylindrical trough		$T_{15} = \begin{bmatrix} \omega_1 & v_1 \\ \omega_2 & v_2 \\ \omega_3 & v_3 \\ 0 & v_4 \end{bmatrix}$ where $\omega_1 = (A\omega_x)^T$ $\omega_2 = (A\omega_y)^T$ $\omega_3 = (A\omega_z)^T$ $v_1 = r \times \omega_1$ $v_2 = r \times \omega_2$ $v_3 = r \times \omega_3$ and $v_4 = (Ak_y)^T$.	
16—Pin in slot		$T_{16} = \begin{bmatrix} \omega_1 & v_1 \\ \omega_2 & v_2 \\ 0 & v_3 \\ 0 & v_4 \end{bmatrix}$ where $\omega_1 = (R\omega_z)^T$, $\omega_2 = (R\omega_x)^T$, $v_3 = (Rk_z)^T$, and $v_4 = (Rk_y)^T$.	If we want to capture the case where the upper plate is very thin and the pin can rock about the its y axis, we can add a row to the twist matrix to obtain $T_{16} = \begin{bmatrix} \omega_1 & v_1 \\ \omega_2 & v_2 \\ 0 & v_3 \\ 0 & v_4 \\ \omega_5 & v_5 \end{bmatrix}$ where $\omega_5 = (R\omega_y)^T$.
17—Sphere on plane	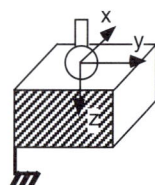	$T_{17} = \begin{bmatrix} \omega_1 & v_1 \\ \omega_2 & v_2 \\ \omega_3 & 0 \\ 0 & v_3 \\ 0 & v_4 \end{bmatrix}$ where the rotations are defined as in feature 15, $v_3 = (Rk_x)^T$, and $v_4 = (Rk_y)^T$.	

TABLE 4-14. (Continued)

Toolkit Feature Number and Name	Sketch	Twist Matrix	Remarks
18—Hemispherical pin in hole		$T_{18} = \begin{bmatrix} \omega_1 & v_1 \\ \omega_2 & v_2 \\ \omega_3 & v_3 \\ 0 & v_4 \end{bmatrix}$	Compared to feature 2, this feature provides a pivot axis about z, permits motion along z, and permits rotation about x and y.
19—Hemispherical pin in slot		$T_{19} = \begin{bmatrix} \omega_1 & v_1 \\ \omega_2 & v_2 \\ \omega_3 & v_3 \\ 0 & v_4 \\ 0 & v_5 \end{bmatrix}$	

5 DIMENSIONING AND TOLERANCING PARTS AND ASSEMBLIES

Customer: "Why are these parts too big?"
Supplier: "It's hotter here than where you are, so we compensated."

5.A. INTRODUCTION

Up to this point, the emphasis in this book has been on the *nominal* design of an assembly. By this we mean the creation of a design for an assembly that will put the parts in certain positions and orientations with respect to each other perfectly, achieving each KC perfectly and ignoring any errors in fabrication of individual parts or errors in assembling them. In this chapter and the next, we consider such errors for the first time.

The word we use when referring to these errors is *variation*. Variation is a physical result of manufacturing processes: Parts and assemblies that are supposed to be identical actually differ from each other and from what we want them to be. Another important word is *tolerance*. Tolerance refers to the amount of variation that we can tolerate in a part or assembly. A third word must be defined for completeness because it is so often confused with tolerance. That word is *clearance*. Clearance is empty space between two parts. One can place a tolerance on a clearance, and clearances can have variation.

Errors in parts and assemblies are inevitable. The actual value of each KC will therefore deviate from the desired value. Variation analysis seeks to ensure that the deviations are acceptable—that is, that each KC lies within the desired range on all, or nearly all, actual assemblies. Two activities are typically involved: *tolerance analysis,* which seeks to determine how the KC will vary given specific variations in parts and fixtures, and *tolerance synthesis,* which seeks to decide values for allowed variations in the parts and fixtures so that desired tolerances at the KC level will not be exceeded.

The parts in an assembly are assembled by connecting a series of features. Presumably these features provide sufficient constraint so that any set of real parts can be placed in repeatably achievable positions and orientations with respect to each other, and that all real parts will achieve these positions and orientations the same way. By "achievable the same way" and "repeatably achievable," we mean that the same surfaces will touch and provide constraint each time, for any set of real parts. This is the same as saying that the assembly is properly constrained. We do not mean that the positions and orientations will have exactly the same values. In fact, the feature locations and orientations have been toleranced in some way, and all acceptable real parts will differ from their nominal designs in some ways within those tolerances. Therefore, one part could have many actual positions and orientations with respect to another part. As a consequence, the KC will not attain its nominal value. Since we defined KCs in Chapter 2 as a nominal value and a range of acceptable variation from that value, we need a way to find out if the KC will be achieved or not, based on knowing or predicting the variation in the parts. This chapter and the next are devoted to this question.

We focus on one kind of variation, namely that which occurs during fabrication of parts and fixtures. These variations cause the assembly or fixturing features to be the wrong shape or be in the wrong position or orientation with respect to some base coordinate frame. The result of such variations is the same in both cases: Some feature of a part will be in the wrong position or orientation with respect to a feature on another part. Such variations will accumulate via chains of frames that pass through parts, and possibly through fixtures. The net result of these variations is that the assembly will be the wrong size or shape, threatening achievement of the KCs.

In terms of the flowdown of KCs presented in Chapter 2, tolerances on parts or feature relationships within parts are the KCs of the parts. Equivalently, these are the manufacturing KCs of the product.

We do not consider variation caused by errors committed during assembly. For example, a part may be placed incorrectly in its assembly fixture and then fastened to its neighbor. Alternatively, two parts may be misplaced on each other and fastened in their incorrect relative positions. We finesse such errors by appealing to kinematic constraint. That is, we assume that our assemblies are kinematically constrained and that all the required constraints are active after each part is added to the assembly or to its fixture. Assembly workers must be trained to place parts firmly against their constraint surfaces, whether those surfaces are on other parts or on fixtures. It is always easier to train them to do this if the parts are kinematically constrained because there is only one obvious right way to do it. Overconstrained assemblies are often "operator-dependent," as discussed in Chapter 4. Nevertheless, assembly errors can happen and can be important.[1]

Our goal in the next two chapters is to learn the strengths and limitations of existing methods for performing tolerance analysis and synthesis. Many methods exist, none completely satisfactory. We describe a few of them and refer the reader to others in the research literature. All of the officially sanctioned national or international standard methods for tolerancing deal exclusively with parts, and none of these deals with assemblies. That is, their focus is exclusively on guaranteeing that randomly selected parts can be mated to each other, either all the time or almost all the time, rather than on determining if a KC is delivered. Tolerancing for function, discussed in Chapter 2, is a third important topic, as is a systematic study of the tradeoffs between better function and higher manufacturing costs usually associated with tighter tolerances. These last topics are beyond the scope of this book.

This chapter will cover the following topics:

- A brief history of efforts to reduce and characterize variation in mechanical parts and assemblies

- Description of geometric dimensioning and tolerancing

- Statistical and worst-case tolerancing

5.B. HISTORY OF DIMENSIONAL ACCURACY IN MANUFACTURING[2]

5.B.1. The Rise of Accuracy and Interchangeability

In the early 1800s at the beginning of the age of manufacturing, each assembly was made of unique parts that were hand-fitted together to make a working product. This required time and skill. The desire to make parts interchangeable created pressure to make them more accurately. As early as 1765, the French army recognized the desirability of making guns from interchangeable parts so that repairs could be made on the field of battle [Hounshell]. The ideal of interchangeable parts comprises the ability to take any randomly selected set of the necessary parts and assemble a working gun from them.

By the late 1810s, it was realized that gages could be used to decide if a part was the correct size and shape. Such gages were made from an example of the final product that was known to function properly. The example product's parts passed the gages, and it was assumed that subsequent parts which passed would not only function but would interchange and still function. The example product therefore stood as the "ideal."

To make this concept work in practice required imposing a lot of discipline on manufacturing activities, including requiring workers to actually use the gages, maintaining a second set of gages to ensure that the workers' gages had not worn out, and maintaining yet a third set of gages as "masters." Additionally, it was realized that if each part had to visit a series of specialized machines, then, to maintain accuracy, the machines would each have to grip the part the same way in the same place. Thus was born the idea of the *jigging surface,* which evolved into the concept of *datum coordination* (discussed in

---

[1] A study was conducted at Ford to see what variations could occur when a sheet metal part is placed in a fixture. Variations as large as 0.5 mm were observed, mostly the result of closing the clamps incorrectly. Since car body assemblers want variations in assemblies to be as small as 2 mm, this part placement variation is significant.

[2] Portions of this section are taken from Chapter 2 of [Nevins and Whitney]. Additional material is adapted from [Voelcker].

Section 5.D.2.a). Not until the mid 1820s was full interchangeability achieved in the manufacture of muskets at one factory. Another ten years were needed to establish two distant factories whose musket parts could be interchanged with each other.

Interchangeability and mechanization were applied after 1850 to commercial products, for which the goal was lower production cost. Although good success was achieved with some products in the period from 1850 to 1900 (such as watches, pistols, and bicycles), great difficulty was experienced with others, notably Singer sewing machines and McCormick reapers. The difficulty was manifested in the need to file the parts to fit, and thus assembly was a time-consuming activity of "fitting" that required large numbers of skilled workers.

By the early 1900s the challenge of manufacturing lay in automobiles. Henry Ford saw the opportunity to create a true mass market entailing production volumes of 2 million or more per year. To achieve such volumes, he knew he could not permit any time-consuming "fitting" during assembly. ("In mass production there are no fitters," he said.) Interchangeability therefore became the route to rapid assembly, while retaining such life-cycle advantages as simplicity of field repair. By 1910 he had achieved sufficient simplicity of design and quality of machines that interchangeability was no longer a problem. His factories were laid out as flow shops. They operated by what is today called just-in-time production with such small inventories that raw iron ore was converted into a car in ten days.

5.B.2. Recent History of Parts Accuracy and Dimensioning and Tolerancing Practices

Even up to the 1920s, the main method of ensuring interchangeability was the use of gages. Until the early 1900s there were no measurement standards, so the master parts and master gages were the standards. To convert to a non-gage method required changing the form of the ideal product. Instead of a physical ideal, a symbolic ideal in the form of a drawing was needed. Drawings could represent the parts and product in a standard way and, with the advent of precision metal gage blocks, could contain dimensions stated in a standard length measure such as inches. Anyone could interpret such dimensions accurately by using the gage blocks to calibrate measuring instruments. The United States established the National Bureau of

Standards in 1901 and the engineering societies set up the American National Standards Institute (ANSI) in 1917. By 1923 Ford had bought the American rights to the famous Johannsen Gage Blocks, which are still widely used.

Drawings with dimensions were common by the late 1800s but drawings with tolerances did not appear until after 1900, when ± dimensions were added to the nominal dimensions to express the acceptable range of a dimension. In the 1940s, the currently used method of "true position tolerancing"—also called geometric dimensioning and tolerancing (GD&T)—was developed in England. It is discussed in the next section of this chapter. It was adopted because prior methods were so ambiguous that parts outsourced to a supply chain could not be relied on to assemble, especially as accuracy requirements increased. While it is the closest to providing unambiguous models of allowed variation, it is challenging to learn, and only a few people become skilled at using it. Efforts to give it a firm mathematical base are ongoing to this day. The existing standard, ANSI Y 14.5-M, applies strictly only to individual parts. There is no internationally accepted standard for dimensioning and tolerancing assemblies. Instead, the standards accepted for parts are used on assemblies. This is not as bad as it might seem, because the methods we describe in this chapter and the next for calculating accumulated variation are essentially the same whether they are applied to single parts or assemblies, as long as the assemblies are kinematically constrained.

Today, a variety of high-precision part fabrication methods exists, ranging from machining to stamping to molding, as shown in Figure 5-1. The precision of some of these methods is remarkable, especially given the fact that they are applied to routine low-cost products like instant cameras, battery-operated screwdrivers, and can openers. Advances in materials, such as glass and nylon-filled polymers, have helped this improvement in part accuracy.

The dominant strategy in use today for making parts in quantity that can be assembled interchangeably and still deliver the KCs is called net build or *build to print*. The assumption is that a drawing or computer model can be given to any competent shop or supplier, or even to multiple suppliers, and with proper care and skill the parts can be counted on to fit. This is an open-loop process that depends on measurement and drawing standards as well

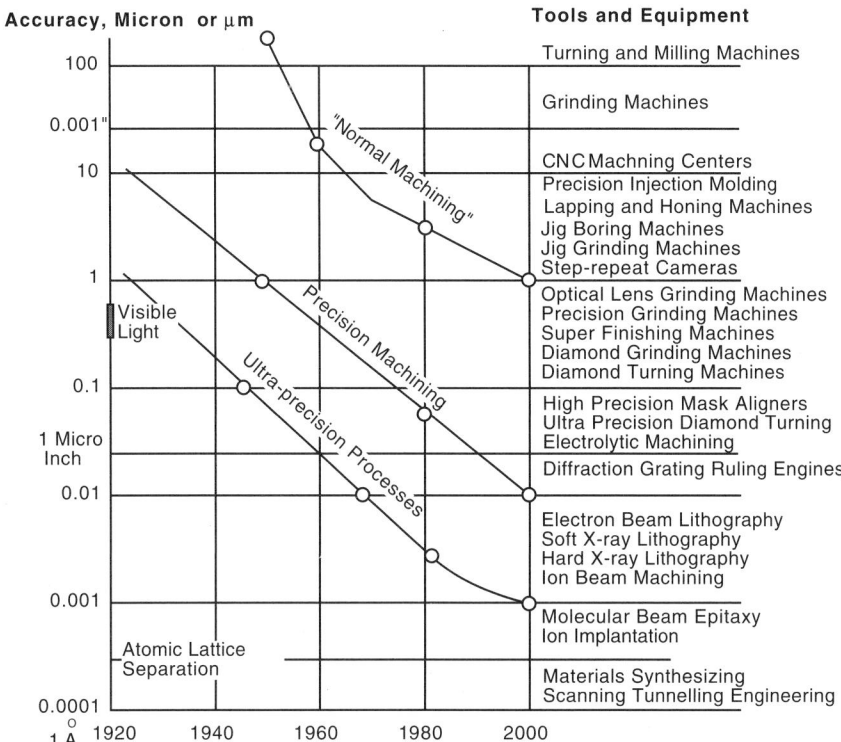

FIGURE 5-1. Accuracy of Several Fabrication Processes. This chart appeared in [Taniguchi] in 1983 and is quite accurate today. In general, it shows a steady rate of improvement in achievable accuracy over time.

as a number of processes that will be discussed below and in the next chapter.

Even though remarkable accuracies can be achieved, the full range from largest dimension to smallest tolerance (generally said to be about 10^{10}) never occurs in a single product. The typical range is about 10^4, with 10^5 or 10^6 in precision products ([Voelcker]). For example:

- The lens of a single-lens reflex camera has tolerances near a wavelength of light ($1\ \mu$m) while the camera's largest dimension is around 10 cm for a total range of 10^5.

- The diameter of the Boeing 777 aircraft fuselage is 22 ft while the tolerance on this dimension is $\pm\ 0.030"$, for a total range of 2.27×10^4 (the coefficient of thermal expansion for aluminum is about 13×10^{-6} per °F; this translates to $0.0034"$ of expansion of a 22-ft diameter for each °F; 10°F tem-

perature change will therefore use up over half the tolerance!).

- The range of fastener diameters from largest to smallest in a given industry is about 3:1 ([Nevins]).[3]

In the last few years, tolerances on car body sheet metal have become so tight that the ideal of interchangeable parts built to print may be unable to meet the tolerances. Some car companies have abandoned the build to print strategy and simply accept parts whose dimensions are close enough, even if they do not fall within the desired

---

[3]These limits on dynamic range of dimensions reflect both technologies and corporate knowledge. When Nevins surveyed manufacturers to determine the range of fastener diameters used, he was told, "If we gave our workers any smaller screws, they would just shear their heads off."

tolerance range, as long as their variation can be kept very small. Then fixtures and tooling are systematically adjusted until satisfactory assembly level KCs and tolerances are achieved repeatably. This is a closed-loop strategy. It is called *functional build* and is discussed in the next chapter.

5.B.3. Remarks

In the 1800s, if the parts of a product fit together, the product probably would work. Thus tolerances were used to generate interchangeability, not just to get the parts to fit, but to gain the benefits of field repair or fast assembly in mass production. Today, products have much higher performance goals and more refined designs. Even if the parts do fit together, the product still may not perform properly. A car door may leak, a gearbox may make noise or wear out too soon, a computer may not run fast enough, or a disk drive may suffer a head crash, even though they all "work" or worked for some period of time.

So the goal of dimensioning and tolerancing is now primarily that of achieving proper performance of the product.

5.C. KCs AND TOLERANCE FLOWDOWN FROM ASSEMBLIES TO PARTS: AN EXAMPLE

A competent assembly achieves its KCs, which means that key dimensions at the assembly level are on their nominals within some specified range called the tolerance. As discussed in Chapter 2, nominal dimensions and tolerances are first established for assemblies and then flowed down to individual parts. Figure 5-2 shows two views of the cross section of an automobile engine. Within the engine is a chain of parts that comprises the combustion and power cycle. This chain connects the crank shaft to the pistons on the one hand and to the valves on the other. The valves open at specific times based on where the pistons are in the cylinders. At various times in the cycle, a valve may open, stroking into the cylinder while the piston is at or near the top of the cylinder. Naturally, we want to avoid a collision between them. We can consider the minimum distance between valves and pistons as a KC for this assembly. Proper operation of the engine depends on achieving this KC, among others. If a piston and valve collide, the engine will be severely damaged.

A diagram of all the parts involved in this KC chain is shown in Figure 5-3. It reveals a series of parts joined by various kinds of features, which are represented by their frames: The crank shaft runs in bearings on the cylinder block, which also contains the cylinders. The cranks join connecting rods which, via wrist pins, join pistons. At one end of the crank shaft is a sprocket on which runs a timing

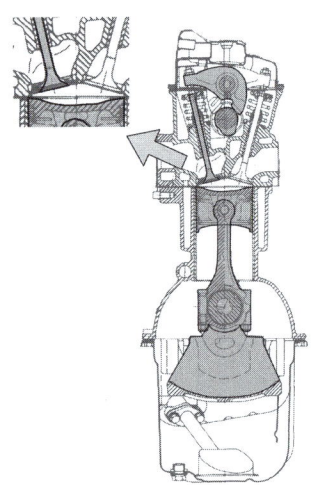
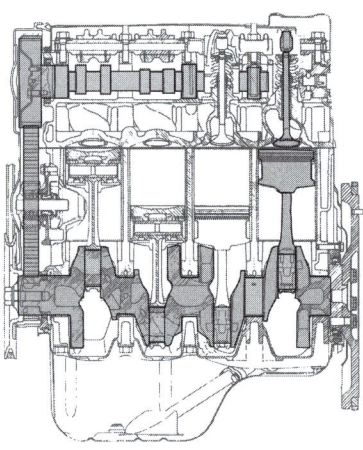

FIGURE 5-2. Automobile Engine Cross Section. Highlighted in gray are the parts that operate together to coordinate the action of the pistons and the valves. The piston at the right is at the top of the cylinder just as the exhaust valve is closing. An important KC is to ensure that the valve stays open as long as possible while the piston is moving up, but that the piston does not collide with it. ([Taylor]. Courtesy of MIT Press. Used by permission.)

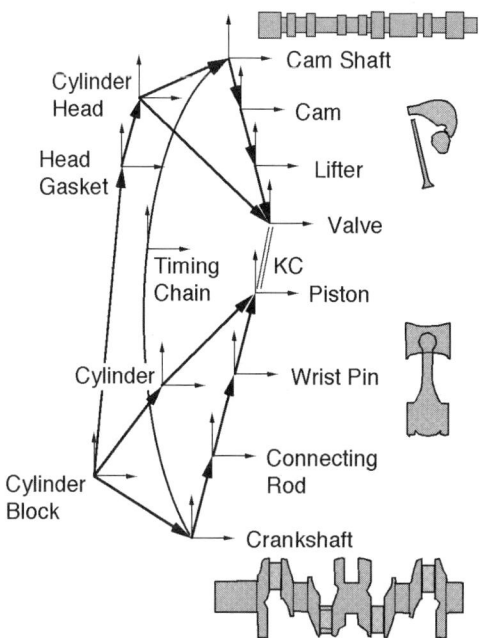

FIGURE 5-3. Chains of Frames in an Engine, Showing Delivery of KC1, the Piston–Valve Clearance. The parts in this chain are shown in Figure 5-2. The KC is the gap between the piston and the valve. This gap is smallest just at the end of the exhaust portion of the cycle when the piston is at the top of the cylinder, the exhaust valve is about to close, and the intake valve is about to open. Thus two valves (four in the case of a four valve/cylinder engine) must each achieve the KC separately.

chain. Another sprocket on the cam shaft also connects to this chain. (The sprockets are not shown in the figure.) The camshaft runs in bearings on the cylinder head, which is bolted to the cylinder block. The head gasket seals the head–block joint. The valve fits in a valve guide in the head. The camshaft contains a cam that contacts the top of the valve stem via a lifter or rocker arm.

To design this chain, the engineer must:

- Define all the parts in the KC chain.
- Define features that will join them to each other.
- Locate those features on the parts, ensure that these features properly constrain the parts (allowing for motions that are needed for function).
- Anticipate or estimate fabrication or fixturing errors that might cause the features to be incorrectly positioned, oriented, or sized, and predict the effect of such errors on achievement of the KC.

At one car company, the first prototype of a new engine suffered a collision between a valve and a piston because one design group assumed that the head–block spacing included the head gasket's thickness while another group did not. If all engineers involved had access to a single connective model of the assembly, this error would not have occurred.

Within the same chain of parts and features is another chain, which lies completely in the cylinder head. It involves contact between the cam or rocker arm and the tip of the valve stem. The parts of a typical design using direct cam-valve actuation are shown in Figure 5-4, and the corresponding chain of frames is shown in Figure 5-5. A part called the lifter is usually placed between the cam

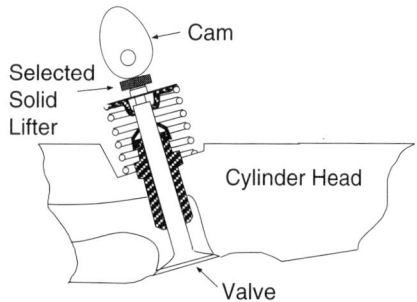

FIGURE 5-4. Engine Valve Actuation Mechanism. This figure shows the use of a solid lifter to just fill the gap between the tip of the valve stem and the cam. Selective assembly is used to find individually the lifter that is the right size for each assembled valve.

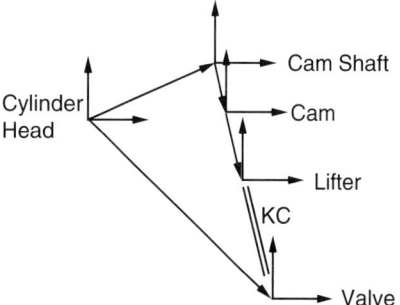

FIGURE 5-5. Chains of Delivery for KC2, the Valve–Cam Clearance. The KC is the gap between the lifter and the valve stem when the valve is closed. If this gap is too big, the engine's timing will be wrong and the engine will be noisy. If it is zero or negative, the valve may not close completely, and over time the stem or the cam will wear, again spoiling the timing. Long before that happens, the driver will notice rough engine performance, and later on the valve and valve seat will burn up.

and the tip of the valve stem. The KC here is that there must be a tiny gap, perhaps only a few microns, between the stem tip and the lifter when the valve is closed. This dimension is so small that it is impractical to achieve it by making the parts in the chain independently to tight tolerances and assembling them interchangeably. The lifter can be a hydraulic type that self-adjusts to fill the gap, or it can be a solid piece.[4] In the latter case, a method called *selective assembly* is used to measure each gap individually and find a lifter that just fills the gap, leaving the required few microns of clearance. Selective assembly is discussed in the next chapter.

5.D. GEOMETRIC DIMENSIONING AND TOLERANCING

In this section, we briefly describe geometric dimensioning and tolerancing (GD&T) and compare it to conventional dimensions on informal drawings. GD&T is a complex topic and is described here mainly in order to show how it reflects basic ideas in kinematic constraint.

5.D.1. Dimensions on Drawings

Double-headed arrows with nominal dimensions and $\pm$ variation limits are the oldest style of dimensioning notation. Anyone who has taken an elementary drafting course has used this method. It is illustrated in Figure 5-6, which shows two views of a cube nominally 1.00" on a side.

There are several problems with this notation. First, it leaves it up to the reader to assume that the desired shape is a cube and thus that the two dimensions shown are representative of all the dimensions of this object. Second, the perpendicularity of the cube's sides is not mentioned and is not affected by the accuracy with which the given dimensions are achieved. In fact, the shape of the object is neither dimensioned nor toleranced. All we know is that there are some lines on the paper that should be 1.00" apart, no more.

In fact, this drawing, together with the statement that it is supposed to be a cube, is really sufficient only to make the drawing shown and tells little about how to make the actual cube or to tell if it meets the requirements for "cubicness." For example, the machinist could fixture the cube along the left side and machine the right side and top, achieving very good perpendicularity between the right and top surfaces. The inspector could place the cube down on the far side and measure its perpendicularity to the bottom. The inspector therefore is not inspecting what the machinist did. Furthermore, neither one knows what the designer wanted. If the part is made by a supplier, another inspector at the customer's shop may choose yet a third way of measuring the part and disagree with the first inspector.

5.D.2. Geometric Dimensioning and Tolerancing[5]

Geometric dimensioning and tolerancing (GD&T), also called true position tolerancing, was developed to deal with solid objects and to avoid the difficulties associated with dimensions that are only good for making drawings.

We can see what GD&T aims to do by considering the alleged cube in Figure 5-6 and asking how many double-headed arrows would be needed to define the relationship between one side of the cube and another side opposite it.

Figure 5-7 shows three sample dimensions, each of which adequately describes a cube that is 1.00" on a side ± 0.02". How do we know if the cube really obeys those tolerances? Have we shown enough such arrows?

In the 1800s, the answer was to place the cube in a gage. In fact, there would have been two gages, called

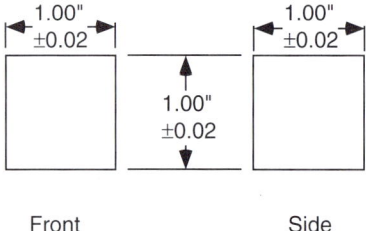

FIGURE 5-6. Example of Double-Headed Arrow Dimensioning.

---

[4] Solid lifters were standard for decades, but selecting them and keeping the gap small as the engine aged was tedious. Hydraulic lifters were an innovation that self-adjusted to fill the gap. But they and the oil inside deform a little under load, slightly spoiling the timing. So solid lifters are making a comeback, especially in high-performance or high-RPM engines, where lifting forces can be high.

[5] Material in this section is based on [Foster] and [Meadows]. The reader is urged to consult books such as these for a complete exposition.

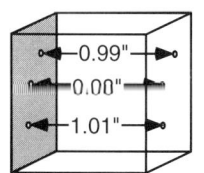

FIGURE 5-7. A Cube with Three Example Dimensions Between Opposite Sides.

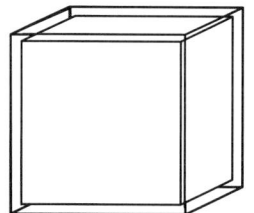

FIGURE 5-8. Two Nested Ideal Cubes. The big cube and the small cube are arranged so that their geometric centers coincide. They represent the maximum and minimum allowable actual cubes. All acceptable actual cubes' outer surfaces lie in the empty space between the big cube and the small cube.

"go" and "no-go." One gage would test if the cube were too big while the other would test if it were too small. If it does not go into the "too small" gage, it is not too small. If it goes in the "too big" gage, it is not too big. If both of these tests are successful, then the trial cube is not too big and not too small, so, by the Goldilocks Principle, it must be good enough, if not just right.

Gages are awkward, they wear and lose their accuracy, and it is not easy to make them, duplicate them for members of the supply chain, or use them for complex parts. Today, the gage idea survives in many of the concepts of GD&T, but measuring machines are often used instead of gages. The GD&T method, as applied to our cube, asserts that each of the cube's surfaces must be within some zone that expresses the tolerance for that surface's location and form with respect to some other surface. The other surface is represented by a datum that is considered to be in the right place by definition. One of the main functions of datums is to assure that the machinist and the inspector(s) use the same surfaces to reference *from* and measure *to* when creating or checking those surfaces. It is up to the designer to choose those datums so that the machinist and inspector do what is intended, and that what is intended contributes to the goal of the assembly.

The "too big" and "too small" gages represent the wish to define an ideal "too big boundary cube" and an ideal "too small boundary cube," as shown in Figure 5-8, and then to say that all acceptable cubes will be smaller than the too big boundary cube and bigger than the too small boundary cube. This means that the outer surfaces of all acceptable cubes must lie in the empty space between the two boundary cubes when the centers of mass of the boundary cubes coincide.[6] The empty space is called the acceptance zone or the tolerance zone. As the inner ideal cube approaches the outer ideal cube, the acceptance zone

becomes smaller, forcing any acceptable real cube to become more "cubic."

It is important to understand that any object, cubical or not, whose outer surfaces lie in the empty space between the boundary cubes is an acceptable "cube" according to this definition. This is basic to how the method works and is not a shortcoming. It reminds us that we have to be careful and thorough if we are going to define a solid object. The double-headed arrow method allows us to be careless, a fact that eludes us until we are confronted with the task of defining a solid object carefully.

In essence, the goal of GD&T is to define each part so that it will assemble interchangeably with any example of its intended mate 100% of the time in spite of unavoidable variations in each part's dimensions, and to provide an unambiguous way of inspecting these parts individually to ensure that this goal will be achieved ([Meadows, p. 5]). GD&T accomplishes this with its more careful specification of three-dimensional shape. By contrast, the goal of an assembly is to deliver its KCs, which means that a sum of several dimensions spanning a chain of parts in the assembly must be within a desired tolerance. These two goals are quite different.

5.D.2.a. Datums and Feature Controls in GD&T

In addition to introducing the idea of the tolerance zone, GD&T also introduced the ideas of the datum and datum hierarchy. These ideas are important to us because they provide a link between GD&T methods for dimensioning and tolerancing *parts* and the coordinate frame method of defining *assemblies of parts* described in Chapter 3. We need this link because GD&T is defined officially only as a method of dimensioning and tolerancing parts, and its approach to assembly is too limited to serve our purposes. The link, as we will see, is accomplished by identifying the

---

[6]Readers who have read Plato will see the connection to Plato's notion of the *ideal* and its contrast with the *real*. There is an ideal cube to which all real cubes aspire but can never be. More precisely, there are two ideal cubes, toward which the real cube may approach from the outside or from the inside, until it lies in the space between them.

TABLE 5-1. List of GD&T Feature Characteristics

Individual Characteristics of a Single Feature	Related Characteristics of More than One Feature
Form: flatness, straightness, circularity, and cylindricity are not related to a datum but instead are related to ideal shapes	Orientation: angle, and its special cases perpendicularity and parallelism, require a datum from which the angle is measured.
Profile of a surface or of a line on a surface is not related to a datum. However, it could be defined as a related characteristic.	Runout requires a datum from which the runout is measured.
	Location: position, symmetry, and concentricity all require a datum from which the characteristic is measured.

datum surfaces with the planes of our coordinate frames and by relating datum hierarchy to the notion of kinematic assembly.

GD&T begins with the notions of the surface and the feature. A feature can be a single surface or a set of related surfaces. A feature needs a location and a tolerance on that location. Some features, like pins and holes, have a size and are called "features of size," while others, such as a plane, have no size. A feature of size, in addition to having a location tolerance, also has a size tolerance. Datums are imaginary perfect geometric shapes that are associated with particular imperfect real surfaces on the part called datum features. The characteristics of features of concern to designers fall into two classes, as shown in Table 5-1. These characteristics differ in the sense that some require a datum while others do not. In general, we will be most concerned with items in the right-hand column of Table 5-1 because the ones in the left column do not contribute much, if any, variation at the assembly level. The ones on the left may be important for some aspects of function, however.

5.D.2.b. The Logic of Datum Assignment[7]

Datum features are real surfaces, while datums are imaginary perfect references like planes, lines, and points. Manufacturing and inspection equipment attempt to simulate these datums with their own real surfaces, which ideally are made to much better tolerances and form than those of the parts they make or measure. Typical gage tolerances are 5% of part tolerances, for example. Datums should be representative of features that are functionally important for the part for the purposes of operation, alignment, or mating to other parts. They should be accessible for fabrication and measurement purposes. Finally, they should be repeatable in the sense that the part

should come to rest on the datums the same way each time as closely as possible. This repeatability comes into play when the part is manufactured, measured, and assembled.

Most of the examples in standard GD&T texts show common circular features sized and positioned relative to plane features. An example would be a bolt circle of four holes or pins placed on a plate with axes nominally perpendicular to the largest plane surface of the plate. The goal of GD&T in such cases is to ensure that the pin pattern on one part mates, with some defined clearance, to the hole pattern on another part. Incorrect bolt circle diameter, incorrect hole or pin position or size, or incorrect angle of the axes all could cause assembly problems. Thus a typical dimensioning and tolerance exercise for such a part begins with the selection of datums and proceeds to stipulating the location and size of the holes and pins.

Datums are assigned in a certain sequence, and that sequence is supposed to be the sequence in which the part will be placed in a machine or measuring apparatus. This sequence can be read from the specification, called a feature control frame, and is often conveniently made alphabetical. Thus the primary datum is often called "A." Datum A is defined by contact between at least three high points on a part's surface and the reference surface of the machine. If the secondary datum B is also a plane, then it is defined by at least two high part points contacting a second reference surface nominally perpendicular to the first, while the tertiary datum C is defined by at least one high part point contact with a third reference surface perpendicular to the first two. It should be clear that the three datums create a kinematic assembly between the part and the machine. It should also be clear that the set A, B, C comprises a fine motion assembly sequence (that is, join A to the previous part, then B, then C) for setting the part in place for the purposes of fabrication, measurement, and final assembly. "Repeatibility" discussed above then means that this fine motion assembly sequence should be used every time.

---

[7]This discussion is based on Chapter 6 and other portions of [Meadows] as well as [Foster].

Suppose angular alignment of the part and of its internal features to another part is important. Then the internal features would be referred to a datum for alignment of their axes or surfaces, and that datum would be relied on to orient the part in the assembly, the fabrication machines, and the inspection equipment. The surface chosen for this task should therefore be a big surface and should have the opportunity to provide three widely spaced points of contact. Thus it is made datum A. If datum B were relied on to align the part, then alignment would be less effective because one of the three points required to establish a plane for alignment purposes would be missing.[8]

Once we know that datum A is for alignment, we can assume that fasteners will pass through it (usually perpendicularly). If the part were intended to be aligned by datum A but the fasteners passed through datum B, then the part would realign itself as the fasteners were tightened until three high points on B were in contact, while datum A would lose contact at one or more of its contact points. Alignment would then be provided by a smaller surface not intended for, or particularly capable of, serving that purpose. Conversely, if fasteners pass through both datum surface A and datum surface B, the part will obviously be overconstrained.

When two parts are supposed to mate, the designer must determine what surfaces need the most physical contact and what surfaces will create the angles at which subsequently related part features will function. According to Meadows, "One must focus on the feature one is defining, thinking only about it and relating it only to features that have been defined prior to it [on that part]. If nothing else has been defined because it is first, then that feature can only be considered for a form control [see Table 5-1]. In this way, one works one's way through a part definition as though through a story, leaving no doubt as to the beginning, the middle, and the end."

Note that only those surfaces which contact others can pass constraint and location to their mating surfaces on other parts. Surfaces that have clearance with their mates generally do not pass constraint or location. They simply succeed in avoiding assembly problems. If the designer wants a surface to pass constraint and location, then the assembly process must be designed so that those surfaces always touch. One-sided constraints with definite

effectors will accomplish this kinematically. If two-sided constraints are used, there are two possibilities: if clearance is allowed, location will be passed only within the uncertainty of the size of the clearance, and, strictly speaking, the assembly will be underconstrained. If a two-sided constraint is designed with interference, then location will be passed and there will be some overconstraint. The exact location will depend on the amount of locked-in stress that results.

5.D.2.c. Dimensions and Feature Control Frames

The dimensions that describe a nominal size or position may be given by ± dimensions or by what are called basic dimensions, which are nominal values without a ± value. Associated with such a dimension is a feature control frame that tells how to verify that dimension, what tolerances it may have, and what datum or datums to use.

The feature control frame contains the basic language and symbols of GD&T. Figure 5-9 and Figure 5-10 show

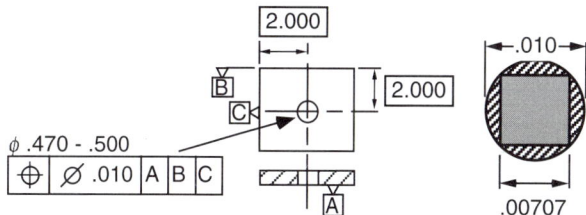

FIGURE 5-9. A Position Tolerance for a Hole or Pin. The control frame is the rectangle with the symbols and numbers in it. On the drawing of the part, the basic dimensions (in boxes) state that the center of the circular feature is nominally 2" from datum surfaces B and C. The diameter of the feature (indicated by ϕ) must lie in the range 0.470" to 0.500". Its position (indicated by the circle with the cross in the control frame) must be inside a cylindrical tolerance zone (indicated by the circle with the diagonal line) whose diameter is 0.010". The orientation of this axis is constrained with respect to the first datum (A), while its position is constrained with respect to the second and third datums (B and C). The square in the circle at the right shows the result of specifying the location of the hole's center by conventional ± dimensions in X and Y separately, while the circle is the acceptance zone for GD&T. No hole center location toleranced by the given ± tolerances would fall outside a region of diameter 0.010". But many holes whose centers are less than 0.010" away from nominal lie inside the circle and outside the square. Thus ± tolerancing would reject them, even though their locations are really just as accurate from an assembly point of view. The circular GD&T zone contains 40% more area and would accept that many more holes if all locations inside the circle were equally likely.

---

[8]That is, datum B could not assert three contact points if datum A already has done so, without causing overconstraint.

some feature control frames together with instructions for how to read them. A short list of GD&T symbols and their meaning appears in Table 5-2.

5.D.2.d. Rule #1: Size Controls Form

Much of the logic behind GD&T reflects the use of gages to determine if parts meet specifications. The size of a cylinder is measured by a gage that fits over its entire length. The hole in this gage is the maximum allowed diameter of the cylinder. If the cylinder is bent then the gage may not function, even though the cylinder's diameter is always within specifications. Thus the cylinder must be straight and round when its diameter is as large as allowed. Similarly for a hole, a plug gage the same depth as the hole is used. The hole must be straight and round when its diameter is as small as allowed. A common term for biggest cylinder and smallest hole is "maximum material condition," abbreviated MMC. Rule #1 states that

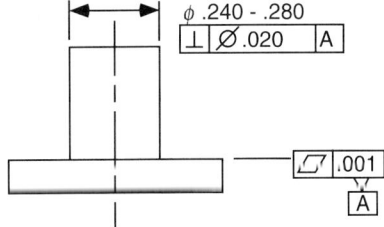

FIGURE 5-10. Orientation Tolerance for a Pin Relative to a Datum Surface with a Flatness Specification. This figure shows two control frames, one for datum surface A and one for the diameter of the pin. Datum surface A must be flat (indicated by the parallelogram). There is no zone symbol inside the control frame next to the flatness tolerance number (0.001) so the tolerance zone consists of two parallel planes spaced apart by 0.001. The pin must have a diameter in the range 0.240–0.280 and its axis must lie in a small cylinder that is perpendicular to datum A and that has a diameter 0.020.

the feature must have perfect form at MMC. This protects the ability of gages to function.

Corresponding to the method for determining size at MMC is the method for determining size at least material condition (LMC). For a cylinder, this would consist of a caliper that would check two opposing points anywhere on the cylinder. There is no requirement for perfect shape at LMC.

These part measuring methods are not entirely satisfactory. For example, calipers are not noted for repeatability. Also, as the cylinder gets longer with respect to its diameter, it must be straighter for the same deviation from perfect diameter, or else the gage will not go on all the way.

Figure 5-11 is an example of GD&T used to specify the height of a block using a zone. D is called the basic dimension or the "true position." It defines the desired location of the upper surface relative to the datum if there is no error. T_s describes the half-height of the tolerance zone in which this surface must lie. While this is a two-dimensional example, it can be extended to cover three dimensions.

Figure 5-12 is a closeup look at the tolerance zone. It shows an example of the actual surface lying inside the zone. The position and angle of the surface are both slightly in error, but the combination of these errors nevertheless leaves the surface inside the zone.

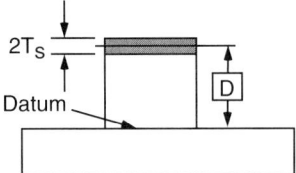

FIGURE 5-11. Example Feature of Size. Dimension D, in the box, is called a basic dimension. It is the ideal value desired by the designer in the absence of variation. The shaded region is the tolerance zone.

TABLE 5-2. Some GD&T Symbols and Corresponding Shape of Tolerance Zones

Symbol	Meaning	Shape of Zone if Diameter Symbol $\varnothing$ Appears	Shape of Zone if No Diameter Symbol Appears
▱	Flat	Does not occur	Two parallel planes
//	Parallel	Cylinder surrounding axis; axis parallel to datum	Two parallel planes parallel to the datum
⊥	Normal	Cylinder surrounding axis; axis is normal to datum	Two parallel planes normal to the datum
◎	Concentric	Cylinder surrounding datum axis	Does not occur
⊕	Position	Cylinder surrounding datum axis	Two parallel planes

Note: The diameter symbol is used when the feature to be located is circular, such as a hole or a bolt circle.

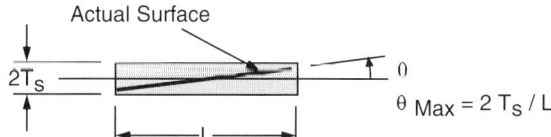

FIGURE 5-12. Acceptable Surface within the Zone. A two-dimensional version of the situation is shown, in which the surface appears as a line. Rule #1 dictates the maximum magnitude of this angle.

As shown in Figure 5-12, the surface can be anywhere in the zone, as long as it lies entirely inside it. This means that size error and angle error are not independent but must follow a relationship like that shown in Figure 5-13. In the full three-dimensional case, we are dealing with a plane that lies inside a tolerance zone shaped like a pizza box and can tilt about either of two axes that lie in a plane parallel to the plane of the box.

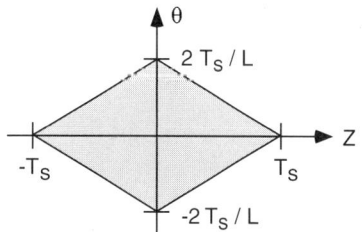

FIGURE 5-13. The Required Relationship Between Size and Angle Error to Obey Rule #1. The height error of the surface relative to the nominal dimension is z, while the angle error is θ. Acceptable top surfaces have height and angle errors that lie in the gray area.

We will use diagrams like that in Figure 5-13, and the underlying mathematical descriptions of them, in the next chapter when we calculate the variations propagated from part to part by errors like those in Figure 5-12.

5.E. STATISTICAL AND WORST-CASE TOLERANCING

This section discusses two common ways to model error accumulation in assemblies, worst-case and statistical. *Worst-case* tolerancing assumes that all parts could be at the extremes of their tolerance zones at the same time, even though this is an unlikely event. Worst-case errors accumulate deterministically, not statistically, and it is necessary to inspect every part to ensure that it does not exceed the worst allowable case. *Statistical tolerancing* assumes that the worst cases are unlikely to occur simultaneously. That is, when one part is a little big, its mate could well be a little small, balancing the errors. An important consequence of this balancing effect is that statistical tolerancing will accept many parts that worst-case tolerancing will reject, saving a lot of money. A statistical attitude is consistent with inspecting a few of the parts, but not all, which saves a lot more money. To ensure that the worst case is unlikely to occur and that sampling inspection is adequate, a method called *statistical process control* is used. Since worst-case tolerancing is a subset of statistical tolerancing, and since statistical process control is necessary for statistical tolerancing, we will discuss these topics in the following sequence: worst-case tolerancing, statistical process control, and statistical tolerancing.

Figure 5-14 and Figure 5-15 compare intuitively worst-case and statistical tolerancing applied to the desktop stapler for the case where the handle is angularly misaligned with respect to the anvil.

Before dealing with statistical and worst-case tolerancing in detail, we need a little philosophy about quality control in general.

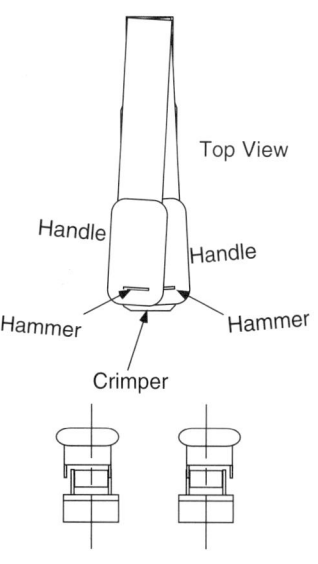

FIGURE 5-14. Top and Front Views of the Stapler with Angular Error Between the Handle and the Anvil. Two extreme errors are shown: handle to the left and handle to the right.

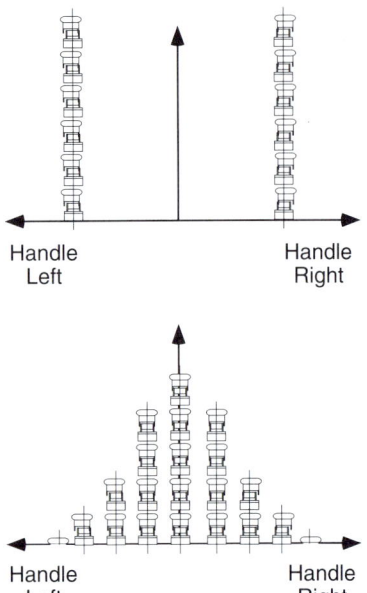

FIGURE 5-15. Intuitive Comparison of Worst-Case and Statistical Tolerancing. Front views of the stapler are shown. *Top:* Worst-case tolerancing. All staplers are assumed to be misaligned to the maximum either leftward or rightward. *Bottom:* Staplers could be misaligned in various ways, mostly not very much but a very few quite a lot. In each case we pile them up in groups with similar misalignments and count how big each pile is. In statistics, such piles are called histograms.

5.E.1. Repeatable and Random Errors, Goalposting, and the Loss Function

Quality control has been studied for nearly a century. The main spokesmen for this activity were W. A. Shewart, J. M. Juran, W. E. Deming, and G. Taguchi. Both technical and organizational approaches have been developed. Here we will deal briefly with the statistical aspects. Readers unfamiliar with the basic properties of distributions of random variables, such as calculating mean and standard deviation, should consult Section 5.J.

Statistical errors can be divided into two categories, called *repeatable cause* and *random or unknown cause*. Statistically, these are measured by the mean and variance, respectively, of a probability density function describing the error. Quality control advocates point out that these two kinds of errors are fundamentally different and are reduced or eliminated using quite different methods.

Repeatable cause errors can usually be traced to a definite and persistent cause, and with some effort they can be substantially reduced or eliminated. Typical causes are design errors in parts or fixtures, or procedural errors by people, such as clamping a fixture too tight. Random cause errors have multiple or intermittent causes, or causes that do not have a fixed effect but vary rapidly. Random errors give an error its spread or deviation about the mean, whereas repeatable cause errors drive the mean, which is fixed or varies quite slowly. Example random errors include temperature fluctuations, variations in material properties, fatigue-induced variation in human performance, and so on. These errors are generally harder to identify and require considerable detective work. They may be reduced but are often impossible to eliminate. It is often suggested that repeatable errors be eliminated first, and then random errors should be addressed.

This is recommended not only because one may be easier to eliminate than the other but also because people confuse the two kinds of errors or do not realize that both are usually present at the same time. Furthermore, Taguchi distinguishes two situations, illustrated at the top in Figure 5-16. This figure shows a tolerance band with the nominal value at zero and a range of about ± 0.0125. Each plot shows the results of measuring many parts and calculating what percent of them exhibit a given measurement, giving rise to a histogram or probability density. The probability density at the top left is highly clustered around -0.01, far from the desired value. It is said to exhibit a *mean shift* error because the mean or average is shifted away from the desired value. The one at the top right is less well clustered but is centered on the desired value. Taguchi says that the one on the right is better because the repeatable cause error has been removed. The random cause error is then visible, and methods suitable to reducing it can be applied. The distribution on the left, while it looks good because it has a narrower spread, is actually consistently wrong and thus less desirable than the one on the right. The distribution at the bottom is the most desirable of the three.

Taguchi says that all points inside the tolerance band are not equally valuable. In fact, the center is the most valuable while value diminishes as the error tends toward the extremes of the band. The idea that all values within the band are equally valuable is often called "goalposting." This is an analogy to goalposts in football or soccer in which all goals have the same value as long as the

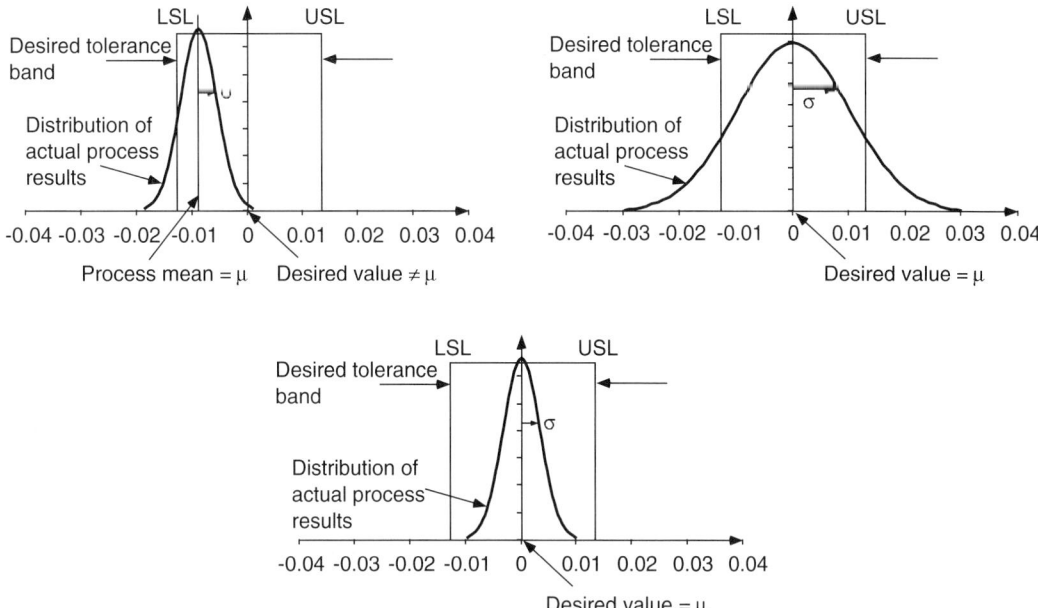

FIGURE 5-16. Three Different Possible Distributions of Size for a Part. *Top Left:* High precision with low accuracy. *Top Right:* High accuracy with low precision. Being consistent, but consistently wrong (left), is less desirable than being right on the average (right), according to Taguchi. *Bottom:* High precision with high accuracy. This is the most desirable result. The symbols in this figure are defined as follows and discussed in detail in this chapter: USL, upper specification limit or tolerance; LSL, lower specification limit or tolerance; μ, the mean of the distribution; σ, the standard deviation of the distribution.

ball passes anywhere between the posts. Taguchi recommends calculating a "loss function" to indicate how value decreases as the error tends away from the desired value. This can often be difficult to do.

In assemblies, goalposting can have serious consequences, and the loss can be calculated readily. Goalposting often creates a mean shift, with the result that all the parts are consistently large or consistently small. Parts made this way create errors that accumulate rapidly in an assembly and can cause the KC to be out of tolerance even though all the parts are "in tolerance." The reason for this apparent paradox is discussed in detail in Section 5.E.4.

5.E.2. Worst-Case Tolerancing

As stated above, worst-case tolerancing assumes that parts could be at the extremes of their tolerance zones all at the same time. This is the only event that this method analyzes. If several parts are placed end to end and their total length must be less than a certain limit, then the tolerance on the length of each part must be such that, when each part is the longest allowed, the total is less than the limit. If the length error on each of n parts is δ_i then the total length error Δ may be found from

$$\Delta = \sum_{i=1}^{n} \delta_i \tag{5-1}$$

If $\delta_i = \delta$ for all i, then the total error will be

$$\Delta = n\delta \tag{5-2}$$

Equation (5-2) tells us that worst-case errors are expected to accumulate in direct proportion to the number of parts in the chain. This is a simple example, but it serves to make the point, which will be contrasted below to the much more gradual way that errors are expected to accumulate if statistical tolerancing is used.

Before we discuss statistical tolerancing, we need to understand statistical process control because it creates and maintains the conditions that make statistical tolerancing valid.

5.E.3. Statistical Process Control[9]

Statistical process control (SPC) is a set of processes designed to bring a process into what is called statistical control. It employs sampling procedures and statistical analyses that avoid the need to inspect every part. Both attributes, such as number of defects, and variables, such as key dimensions, can be treated using SPC. What follows here is SPC applied to variables.

SPC can be applied to measurements on parts or measurements on the processes that make the parts. In many cases, the latter is preferable. For example, to ensure that solder joints on circuit boards are high quality, Motorola determined the required range for solder flux quantity, temperature of the heater, and duration of the heating process. Once the right nominal values and allowed variation for these process variables were determined, it was sufficient to keep them there in order to generate good solder joints. It was not necessary to inspect the joints themselves, which is much more difficult.

SPC calculates statistics called $\bar{X}$ and R on key variables and charts them over time in order to determine if the process is in statistical control. The $\bar{X}$ chart tracks the mean while the R chart tracks the variation. Each chart has a centerline and upper and lower limits whose calculation is described in Section 5.E.3.b. A process is in control if

- Chart data fluctuate in an apparently random fashion.
- The data points cluster near the centerline of the chart.
- Only a few points approach the upper and lower limits.
- Rarely does a point exceed the limits.

Once a process is in control, it is possible to determine if it is capable of delivering the required tolerances repeatably. If it is capable, the charts can be used to detect process drifts that threaten capability before large numbers of out-of-tolerance parts are produced. Process capability is measured by indices called C_p and C_{pk}, which are defined in the next subsection.

SPC is interesting not only because it provides a way to control and improve a process but also because it has important organizational and managerial implications. Once a process is in control and capable, the manufacturer and its customers gain a new level of confidence (a) that the manufacturer knows what it is doing and (b) that when the customers ask for a certain tolerance, they will get it. The customers will state tolerances that they really need and the shop will be able to say with confidence that it can meet those tolerances or can institute SPC to improve the process with the goal of meeting the tolerances. This is essential in flowing KCs down, as discussed in Chapter 2. In the example told there about the turbine blades, the shop instituted SPC as part of the solution to the problem.

5.E.3.a. Distribution of Random Errors

Statistical process control operates under the assumption that parts will have independent randomly distributed errors. While various assumptions can be made about the actual distribution, a Gaussian or normal distribution is often used. This assumption should be made with caution because it is valid only under certain circumstances, which are discussed below.

A normal density function $f(x)$ for a single quantity can be graphed as shown in Figure 5-17. The graph shows the probability that the variable x will lie in the range $[x, x + dx]$. The equation for $f(x)$ is

$$f(x) = \frac{1}{\sigma\sqrt{2\pi}} e^{-\frac{1}{2}\left(\frac{x-\mu}{\sigma}\right)^2} \tag{5-3}$$

where μ is the mean and σ is the standard deviation of the distribution. σ^2 is called the variance. Figure 5-17 is drawn for the case where $\mu = 0$ and $\sigma = 1$. If the data for such a graph are obtained from individual measurements, then they are called a *distribution of individuals*.

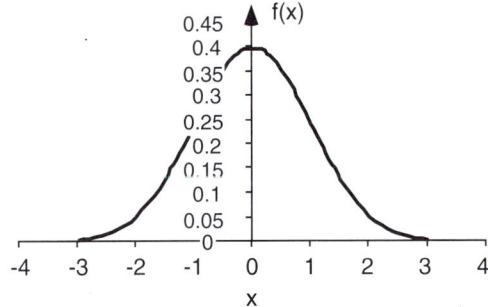

FIGURE 5-17. Normal Density Function for Individuals. In this graph, one standard deviation σ equals one unit on the x axis. The mean μ is zero.

---

[9]This section is based in part on [Swift], [Kolarik], and [DeVor et al.].

TABLE 5-3. Percent of Normal Random Variables That Fall Within and Outside a Range Around the Mean μ, Measured in Units of One Standard Deviation σ

# of σ	% within ±# of σ	% outside	ppm outside
1	68.269%	31.731%	317310.508
2	95.450%	4.550%	45500.264
2.7	99.307%	0.693%	6933.9478
2.8	99.489%	0.511%	5110.2608
2.9	99.627%	0.373%	3731.6268
3	99.730%	0.270%	2699.796
4	99.994%	0.006%	63.3424
5	99.99993%	0.00006%	0.5734

Note: ppm = parts per million.

In a normal distribution, the standard deviation can be used to predict how likely x is to be within certain bounds, as shown in Table 5-3.

So, for example, if (a) the process is kept within $\pm 3\sigma$ of μ, (b) the mean μ is the desired nominal value, and (c) the desired tolerance range is exactly $\pm 3\sigma$ about the nominal, then 99.73% of the parts produced by the process will be within tolerances. The $\pm 3\sigma$ range about the nominal is sometimes called the *natural tolerance range* or *natural tolerance spread* of the process. The upper end of the range is sometimes called the upper natural tolerance limit (UNTL), and the lower end is sometimes called the lower natural tolerance limit (LNTL).

5.E.3.b. Process Control Charts

SPC includes two kinds of process control charts to monitor whether the process is in statistical control: the X-bar ($\bar{X}$) chart and the R chart. The X-bar chart tracks the ability of the process to stay near its mean while the R chart tracks the range or variation of the process. A process that is in statistical control has an X-bar chart that fluctuates randomly near the process mean and an R chart that fluctuates randomly near the mean range. X-bar and R charts are based on small samples of process data taken regularly. If the process is under statistical control, it is not necessary to do 100% sampling.

X-bar and R charts are constructed based on estimates for the process mean μ and process standard deviation σ_x. These are called $\bar{\bar{X}}$ and $\hat{\sigma}$, respectively. Data needed to construct X-bar and R charts consist of n sample values from the process that we will denote as $x_i = \{x_1, x_2, x_3, \ldots, x_n\}$. There are several such samples,

k in number. For each sample, we calculate the sample mean $\bar{x}_i$ and the range R_i as follows:

$$sample\ mean = \bar{x}_i = \frac{\sum_{j=1}^{n} x_j}{n}$$

$$sample\ range = R_i = (x_j \max - x_j \min)_i \tag{5-4}$$

The results from all k samples are averaged to produce a grand process average $\bar{\bar{X}}$ and an average process range $\bar{R}$ as follows

$$\bar{\bar{X}} = \frac{\sum_{i=1}^{k} \bar{x}_i}{k}$$

$$\bar{R} = \frac{\sum_{i=1}^{k} R_i}{k} \tag{5-5}$$

It is known from statistical theory that the standard deviation σ of the process may be estimated from the formula

$$\hat{\sigma} = \frac{\bar{R}}{d_2} \tag{5-6}$$

where d_2 is a factor that depends on the sample size n. Equation (5-6) is valid for $n \leq 10$.

We can also estimate the standard deviation σ_R of range R from

$$\hat{\sigma}_R = d_3 \hat{\sigma} \tag{5-7}$$

The process standard deviation $\sigma_{\bar{x}}$ is estimated using Equation (5-33) in Section 5.J:

$$\hat{\sigma}_{\bar{x}} = \frac{\hat{\sigma}}{\sqrt{n}} \tag{5-8}$$

We can then calculate a number of center values and range limits around them in the following form:

center value = some estimated mean

upper limit = center value + 3 * some estimated standard deviation $\qquad$ (5-9)

lower limit = center value − 3 * some estimated standard deviation

An $\bar{X}$ chart is constructed by drawing a line at the process' estimated mean $\bar{\bar{X}}$ with a line above it called the upper control limit ($UCL_{\bar{x}}$) and a line below called the lower control limit ($LCL_{\bar{x}}$), which are calculated as follows:

$$UCL_{\bar{x}} = \bar{\bar{X}} + 3\hat{\sigma}_{\bar{x}} \quad \text{and} \quad LCL_{\bar{x}} = \bar{\bar{X}} - 3\hat{\sigma}_{\bar{x}} \tag{5-10a}$$

which is equivalent to

$$UCL_{\bar{x}} = \bar{\bar{X}} + 3\frac{\bar{R}}{d_2\sqrt{n}} \quad \text{and} \quad LCL_{\bar{x}} = \bar{\bar{X}} - 3\frac{\bar{R}}{d_2\sqrt{n}}$$

$$(5\text{-}10b)$$

which is abbreviated

$$UCL_{\bar{x}} = \bar{\bar{X}} + A_2\bar{R} \quad \text{and} \quad LCL_{\bar{x}} = \bar{\bar{X}} - A_2\bar{R} \quad (5\text{-}10c)$$

A_2 is a factor that depends on n. Periodically, sample means $\bar{x}_i$ from Equation (5-4) are plotted on this chart.

An R chart is constructed by drawing a line at the mean range $\bar{R}$ with a line above it ($UCL_{\bar{R}}$) and a line below ($LCL_{\bar{R}}$), which are calculated as follows:

$$UCL_{\bar{R}} = \bar{R} + 3\sigma_R \quad \text{and} \quad LCL_{\bar{R}} = \max[0, \bar{R} - 3\sigma_R]$$

$$(5\text{-}11a)$$

which is equivalent to

$$UCL_{\bar{R}} = \bar{R} + \frac{3\bar{R}d_3}{d_2} \quad \text{and} \quad LCL_{\bar{R}} = \max\left[0, \bar{R} - \frac{3\bar{R}d_3}{d_2}\right]$$

$$(5\text{-}11b)$$

which is abbreviated

$$UCL_{\bar{R}} = \bar{R}D_4 \quad \text{and} \quad LCL_{\bar{R}} = \bar{R}D_3 \quad (5\text{-}11c)$$

Factors D_3 and D_4 are functions of n. Periodically, range values R_i from Equation (5-4) are plotted on this chart.

Typical values of the factors needed to calculate X-bar and R chart upper and lower limits are given in Table 5-4.

Various rules exist for interpreting trends in the charts. A good chart has values that vary apparently randomly, stay near the mean, do not show a consistent pattern, and rarely go outside the limits. The process is then said to be in statistical control. Worrisome consistent patterns include an $\bar{X}$ that is steadily rising, falling, cyclic, or consistently above or below $\bar{\bar{X}}$, or an R that is steadily rising, cyclic, or consistently above or below $\bar{R}$.

TABLE 5-4. Values for Coefficients A_2, D_3, and D_4 for X-bar and R Chart Limits

n	A_2	D_3	D_4
5	0.577	0	2.114
10	0.308	0.223	1.777

Note: Values of these factors for other values of n may be found in [Swift].

It should be noted that a sample size of 5 or 10 is not large enough to capture process variations that evolve slowly. For example, a cutting tool may last 300 parts, and, as it wears, the parts drift successively from minimum material to maximum material (or from maximum to minimum depending on whether the tool cuts an interior or an exterior surface). The actual variation over these 300 parts is thus much larger than that revealed by averaging the variation of any 5 in a row. A process control metric called P_{pk} is used to capture these effects. It is similar to C_{pk} but uses much larger samples. P_{pk} may not be useful for immediate feedback to the factory, however, because it takes so long to accumulate the data.

Once we have the process in statistical control, we can keep an eye out for trouble before it causes parts to drift outside the desired tolerance band.

Just because a process is in statistical control does not mean that it is capable of producing parts that meet the tolerances. The tolerances could be too tight, and a situation like that at the top right in Figure 5-16 could occur. Or the process, while staying near its mean, nevertheless has a mean that differs from the desired nominal value of x, giving rise to the situation at the top left in Figure 5-16. To measure the ability of the process to deliver the tolerances, we use process capability indices, which are discussed next.

5.E.3.c. Process Capability Indices

Process capability indices relate the process mean and standard deviation to the tolerance limits, which in SPC books are called the upper specification limit (USL) and the lower specification limit (LSL). The most useful process capability index is called C_{pk}, defined as

$$C_{pk} = \min\left[\frac{(USL - \bar{\bar{X}})}{3\sigma}, \frac{(\bar{\bar{X}} - LSL)}{3\sigma}\right] \quad (5\text{-}12)$$

where $\bar{\bar{X}}$ is defined in Equation (5-5). C_{pk} as defined above compares the process to the $\pm 3\sigma$ range of a normal distribution. Different values of C_{pk} are compared to the actual error distribution in Figure 5-18. If the $\pm 3\sigma$ range lies exactly on the tolerance range, then $C_{pk} = 1$. If the $\pm 3\sigma$ range lies totally inside the tolerance limits, then $C_{pk} > 1$. Larger C_{pk} implies that the process can stay farther from the tolerance limits, so bigger is better. If the process is not centered between USL and LSL, then, for the same process and tolerance limits, C_{pk} will be smaller than if it is centered. If the $\pm 6\sigma$ range lies within the tolerance limits, then $C_{pk} = 2$, implying that only a few parts in a

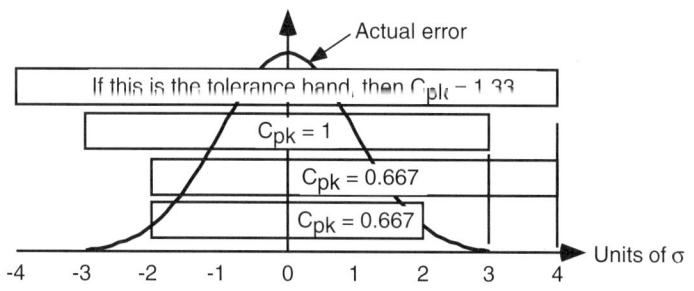

FIGURE 5-18. Examples of Different Values of C_{pk}. *Reading from top to bottom:* If the tolerance is equal to 4σ of typical errors and there is no mean shift, then $C_{pk} = 4/3 = 1.33$; if the tolerance is equal to 3σ of typical errors and there is no mean shift, then $C_{pk} = 1$; but if the tolerance is equal to 3σ of typical errors and there is mean shift equal to 1σ, or if the tolerance is equal to 2σ of typical errors and there is no mean shift, then $C_{pk} = 0.667$.

million will exceed those limits. For typical purposes, a process with $C_{pk} = 1$ is considered to be barely capable.

Occasionally, another capability index, C_p, is used:

$$C_p = \left[\frac{USL - LSL}{6\sigma}\right] \qquad (5\text{-}13)$$

C_p measures the degree to which the process stays within a range of $\pm 3\sigma$, but it does not measure the degree to which the process mean adheres to the desired nominal value. Mean shift could occur and C_p would not detect it. Figure 5-19 shows different situations having the same value of C_p.

[De Vor et al.] point out that C_{pk} is subject to abuse as follows. If a process is improved, or the tolerances are relaxed, C_{pk} becomes quite large. A factory might be able to save money by introducing a mean shift. In a removal process, as discussed in Section 5.E.6, one can stop cutting sooner if a little extra material is left on each piece. This will reduce the value of C_{pk}, but since it was so high, it simply returns to a "good" value and no one notices the mean shift. To encourage factories to deliver parts whose mean dimension is on target, a process capability index

called C_{pmk} is defined ([Kolaric]):

$$C_{pmk} = \frac{C_{pk}}{\sqrt{1 + \left(\dfrac{\mu - T}{\sigma}\right)^2}} \qquad (5\text{-}14)$$

Here, μ and σ are, as before, the process mean and standard deviation, while T is the target value of the dimension. $C_{pmk} = C_{pk}$ when the process mean is on target but is smaller if the process mean is not on target.

Figure 5-20 relates many of the variables we have discussed in the last few sections of this chapter. On the right is an indication of how the process operates, based on measuring every part made (which is almost never done in practice). On the left is an indication of how the sampled statistics $\bar{X}$ and $\sigma_{\bar{x}}$ are distributed. The actual process fluctuates around its mean between its natural limits UNTL and LNTL, while the sample statistic $\bar{X}$ fluctuates around its mean between the control limits UCL and LCL (as long as the process is in statistical control). Note that in this figure the process mean is not centered on the desired mean, indicating that there is a mean shift error. This would be detected by monitoring C_{pk}.

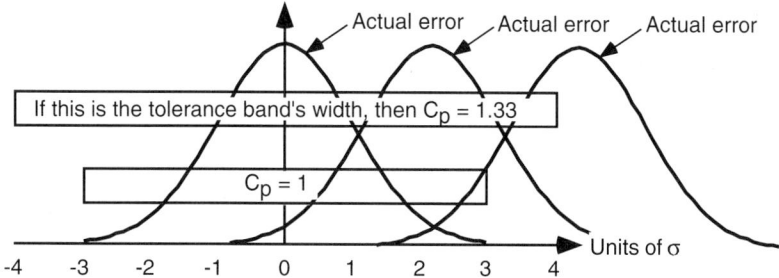

FIGURE 5-19. Illustration of Different Situations Having the Same C_p. C_p is a measure of tolerance relative to the spread of the actual error. Here, three different cases of actual error are shown. One is centered on the desired nominal value while the other two are shifted away from it. All actual errors have the same variance. C_p is the same in all three cases. If the tolerance band is $\pm 4\sigma$ wide, then $C_p = 1.33$. If the tolerance band is $\pm 3\sigma$ wide, then $C_p = 1.0$.

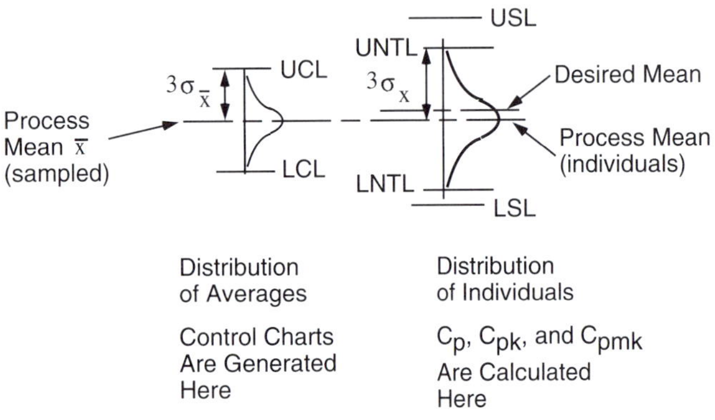

FIGURE 5-20. **Relationships Between Important Variables Associated with SPC.** (Adapted from [Kolarik]. Copyright © 1999 McGraw-Hill. Used by permission.)

5.E.4. Statistical Tolerancing

We now have the tools we need to produce parts that meet stated tolerances, and to do so with zero mean error, or nearly so. We are now in a position to use an economical way of tolerancing the parts called statistical tolerancing. It depends for its validity on the fact that errors in part dimensions have zero mean.

Statistical tolerancing is based on the assumption that the worst case will occur very rarely and that two or more parts to be assembled will not be at the worst extremes of their distributions simultaneously. Sometimes, if that unhappy event occurs, one of the parts can be put back and another chosen, but other times the speed or automation of the process or other circumstances prevents replacement.

However, the case for statistical tolerancing is so strong that the worst eventuality is overlooked. The underlying assumption of statistical tolerancing is that errors will tend to cancel out as parts are added to an assembly. Some parts will be a little large while others will be a little small. This, in turn, depends on part errors having zero mean shift.

The way errors are assumed accumulate for the purposes of statistical tolerancing can be derived as follows, using a simplified case of linear dimensions. Suppose we have k parts, each of which has a length L_i described by

$$L_i = L_{0i} + x_i \quad \text{and} \quad x_i = m_i + \rho_i \quad (5\text{-}15)$$

where the average of ρ_i is given by $\bar{\rho} = 0$ and the variance of ρ_i is σ_i^2.

That is, the desired length of each part is L_{0i} and the error in the length is x_i. x_i has mean m_i and standard deviation σ_i. What is the mean and variance of the length of a row of k such parts?

We can find the mean by using Equation (5-26) in Section 5.J to show that the average total length of k such parts is the average of the sum of the L_i

$$average\left[\sum_{i=1}^{k} L_i\right] = average\left[\sum_{i=1}^{k}(L_{0i} + x_i)\right]$$

$$average\left[\sum_{i=1}^{k} L_{0i}\right] + average\left[\sum_{i=1}^{k} x_i\right] \quad (5\text{-}16)$$

$$= \sum_{i=1}^{k} L_{0i} + km \text{ if all } m_i = m$$

This equation says that errors in the mean of a sum of dimensions accumulate linearly with the number of dimensions being summed. If the mean of each individual part's error is zero, then the mean of the accumulated errors is also zero and the mean of the sum of the dimensions of the k parts is equal to the desired sum. The remaining error is due to ρ, a zero mean error with standard deviation σ. This is the same as saying that some of the remaining errors will be a little large and some a little small, providing some cancellation and less overall error. This is the fundamental assumption underlying statistical tolerancing.

To find out how the remaining errors accumulate under this assumption, we need to know the variance of a sum. Equation (5-29) in Section 5.J says that the variance of a sum of linearly independent[10] random variables is the sum of their individual variances. This means that

---

[10]If x and y are random variables and the mean of xy is the mean of x times the mean of y, then x and y are said to be linearly independent. If x and y have a common assignable cause error, which often happens, then they will not be independent. For example, a batch of sheet metal with more than desired spring back leads to many parts all out of spec in the same direction. (Thanks to Prof. Daniel Frey for emphasizing this point and providing the example.)

variances accumulate linearly with the number of n dimensions being summed. Thus

$$variance\left[\sum_{i=1}^{n} L_i\right] = \sum_{i=1}^{n} \sigma_i^2 = n\sigma^2 \qquad \text{if all } \sigma_i = \sigma$$

$$(5\text{-}17)$$

This in turn means that the standard deviation, the actual variation of the total error, accumulates with the square root of the number of dimensions being summed. Thus the errors around the mean accumulate much more slowly than errors in the mean do, as seen by comparing Equation (5-29) with Equation (5-26). A more general formulation of error accumulation assumes that each part's error x_i contributes to the assembly-level error Y through a sensitivity S_i:

$$Y = \sum_{i=1}^{n} S_i x_i \qquad (5\text{-}18)$$

where

$$S_i = \frac{\partial Y}{\partial x_i} \qquad (5\text{-}19)$$

Then, again assuming that the x_i are zero mean, we have

$$\sigma_Y^2 = \sum_{i=1}^{n} S_i^2 \sigma_{x_i}^2 \qquad (5\text{-}20)$$

For simple linear assemblies, the S_i are usually ± 1. For complex three-dimensional assemblies, the S_i can be almost any number depending on the three dimensional relationships between the features. A full geometric analysis using 4×4 nominal and variational transformation matrices, like that discussed in Chapter 6, is then used. Such an analysis automatically calculates the S_i in the course of the formulation. The formulas and examples in this section of this chapter assume that the $S_i = 1$.

The difference between accumulation with zero and nonzero means is illustrated in Figure 5-21. Here it can be seen that accumulated errors in non-zero mean dimensions rapidly overtake accumulated errors in zero mean dimensions.

Statistical tolerancing assumes that the errors in an assembly can be predicted by summing the variances of the errors in the individual dimensions, taking the square root of the total variance, and calling the resulting standard deviation the error. This is often called *root sum square tolerancing* or RSS. It is stated

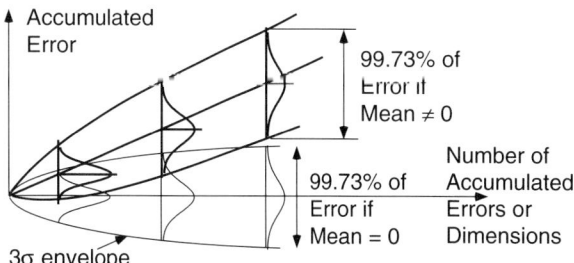

FIGURE 5-21. Comparison of Accumulated Error When the Mean of the Individual Errors is Zero or Nonzero. The percentages shown describe the 3σ limits of a normal distribution. It is clear that if the mean of the errors is not zero, the accumulated error will grow rapidly and soon far exceed the errors when the mean is zero.

mathematically in Equation (5-21).

$$\text{RSS Error in } \sum_{i=1}^{n} L_i = \sqrt{\sum_{i=1}^{n} (\text{Error in } L_i)^2}$$

$$= \sqrt{\sum_{i=1}^{n} \sigma_i^2}$$

$$= \sigma\sqrt{n} \qquad \text{if } \sigma_i = \sigma \quad (5\text{-}21)$$

Then, using Table 5-3, we can say that, for example,

$$\sum_{i=1}^{n} L_i \text{ will lie in the range } \sum_{i=1}^{n} L_{0i} \pm 3\sigma\sqrt{n}$$
99.73% of the time

$$(5\text{-}22)$$

$$\sum_{i=1}^{n} L_i \text{ will lie in the range } \sum_{i=1}^{n} L_{0i} \pm 4\sigma\sqrt{n}$$
99.994% of the time

If the mean error is not zero, then RSS will grossly underestimate the error!

Assume we have a KC chain with n links. If we desire the KC's sum dimension to have a tolerance Tol_{KC}, then we must assign a tolerance to each link, Tol_{link}. This is called *tolerance allocation*. Let us assume that each link is assigned the same tolerance and our goal is that 99.73% of KCs be in tolerance. Then, statistical tolerancing finds Tol_{link} follows:

$$Tol_{KC} = 3\sigma_{KC} \qquad (5\text{-}23a)$$

$$Tol_{link} = 3\sigma_{link} \qquad (5\text{-}23b)$$

$$n\sigma_{link}^2 = \sigma^2 \text{ total for } n \text{ links} = \sigma_{KC}^2 \qquad (5\text{-}23c)$$

$$Tol_{link} = \frac{Tol_{KC}}{\sqrt{n}} \qquad (5\text{-}23d)$$

For worst-case tolerancing, we use Equation (5-2) to write the corresponding formula for Tol_{link}:

$$Tol_{link} = \frac{Tol_{KC}}{n} \qquad (5\text{-}24)$$

Note that there can be several KC links in one part.

An example comparing worst-case and statistical tolerancing is given in Figure 5-22. This figure illustrates an assembly of three parts whose total length must be within ±0.03 of the desired length. Worst-case tolerancing assumes that each part will be at its upper limit all the time, so that the mean part error equals the upper limit.

Equation 5-24 says that each part must then be toleranced to lie within ±0.01 of the nominal, and each one must be inspected to see that it does. Statistical tolerancing, with the process under control as defined in Section 5.E.3, assumes that the mean of the length errors in each part is near zero and that the errors are distributed according to some centrally bunched distribution such as a normal. Equation (5-23) says that each such part should be toleranced to lie within ±0.01732 of the nominal size, so that 99.73% of them meet that tolerance. Then 99.73% of assemblies of three such parts will have overall lengths that lie within ±0.03 ($3\sigma = 0.03$) of the desired value. A 73% larger part tolerance is permitted, and, since C_{pk} is at least 1.0, only a sample of the parts needs to be inspected.

If one wants to tighten the tolerances so that all links lie within ±4σ, meaning $C_{pk} = 1.33$, then "3" in Equation (5-23a) and (5-23b) is replaced by "4." Table 5-3 tells

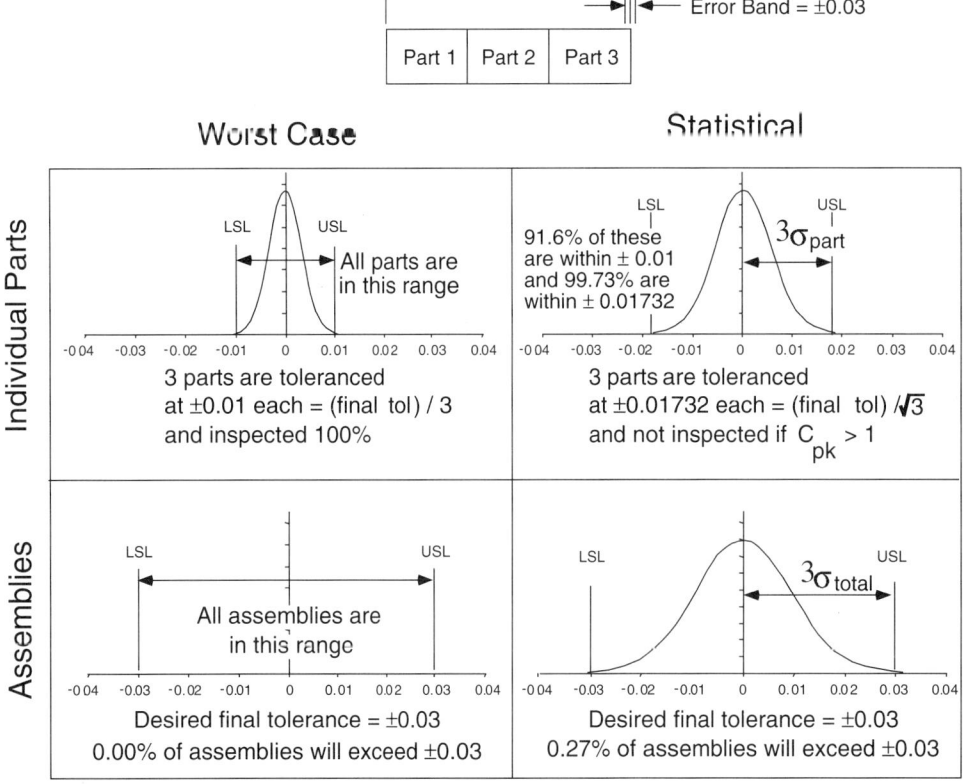

FIGURE 5-22. Comparison of Worst-Case and Statistical Tolerancing for Three Parts Whose Total Length Has a Tolerance of ±0.03. Under statistical tolerancing, each part can be given much larger tolerances than under worst-case tolerancing and still (almost always) meet the assembly level tolerance.

us that 99.994% of links and KCs will be within tolerances. For $\pm 6\sigma$, meaning $C_{pk} = 2$, use "6." However, Equation (5-23d) remains the same in each case. In all such cases, the mean error must be zero or else these error accumulation predictions will be far too optimistic.

If the process is in control so that the mean is near the nominal and if $C_{pk} > 1$, then either the same tolerances on the parts will yield a much better assembly-level tolerance than would the same part tolerance under worst-case assumptions, or much larger part-level tolerances can be accepted for the same assembly-level tolerance as worst-case would yield. In addition, if the process is under control, one need only sample from time to time, whereas under worst-case tolerancing, one must inspect every part. The cost savings in parts not scrapped and inspection time not spent can be so large that there is great incentive to bring the process under control and drive it toward the desired nominal. In Figure 5-22, 8.4% of the statistically toleranced parts are outside the range ± 0.01 and would be scrapped or reworked if subjected to worst-case tolerancing. Yet accepting them causes only 0.27% of assemblies to exceed the range ± 0.03. In both the statistical and worst-case situations, we are talking about the same parts. Thus a major difference between statistical and worst-case tolerancing is the attitude we take toward the parts!

The above analysis applies strictly to the simple case of linear stackup of tolerances in one direction only. A more sophisticated analysis like that in Chapter 6 is needed in the general three-dimensional case. However, the general conclusions remain the same.

Statistical tolerancing requires designers to understand how processes behave and what range of tolerances to expect. In some cases, the considerations can be quite sophisticated. For example, if a circuit designer specifies a 5% resistor, meaning that the resistance will be within $\pm 5\%$ of the specification with a zero mean, then in fact the distribution of resistances will be bimodal. That is, instead of the resistances being bunched around the desired mean, there will be two bunches, one centered perhaps 3% or 4% below the desired mean and the other centered perhaps 3% or 4% above it.

To understand this, we need to know where 5% resistors come from: It is too expensive to make 1% resistors, 5% resistors, and 10% resistors separately. There is too much variation in the resistor-making process. So the manufacturer simply makes resistors as economically as possible, resulting in a spread of about $\pm 20\%$ around the desired value. The manufacturer then removes all the resistors that are within $\pm 1\%$ of the desired value. These may be sold at a much higher price. The distribution of the remainder has a gap around $\pm 1\%$ of the mean. Then, resistors within $\pm 5\%$ are selected and sold for somewhat less, and similarly for 10% resistors. Each of these resistors will have resistance values that are bimodally distributed, reflecting the removal of all the resistors having less variation. That is, 1% and 5% resistors are not made to order but are found by testing. The same applies to microprocessors: 33-MHz 486s were simply 40-MHz 486s that didn't pass the test at 40 MHz but did at 33.

5.E.5. Summary of SPC and Statistical Tolerancing

If the process for making each part in an assembly is in control so that the process mean is kept at or near the desired nominal value, then errors in the assembly-level dimensions will accumulate as the square root of the number of dimensions being added. If the mean is not on the nominal, then errors will accumulate much faster, destroying the assumptions under which statistical tolerancing operates. The strategy under which the part-level tolerances were assigned, based on Equation (5-23), will not be valid, assigned tolerances will be much too large, and the assembly will not achieve its KCs often enough.

Figure 5-23 summarizes the contents of the last few sections. We begin with the desire to deliver the KCs economically. In order to do this, we need to meet the tolerances at the assembly level. Statistical tolerancing is much more economical than worst-case tolerancing, so we need in turn to meet the requirements for statistical tolerancing, namely that the part errors do not exhibit any mean shift. This, in turn, is accomplished by bringing the process under statistical control so we can analyze it and eliminate mean shift and repeatable cause errors. Once the process is under statistical control, we can also monitor the variation. A process that is in statistical control is operating more economically than if it has large variation or assignable cause errors. Finally, to meet the tolerances, we need to bring the process capability up to at least 1.0.

5.E.6. Why Do Mean Shifts and Goalposting Occur?

Why do process means deviate from zero? The main answer is that a tolerance zone is often looked upon as a

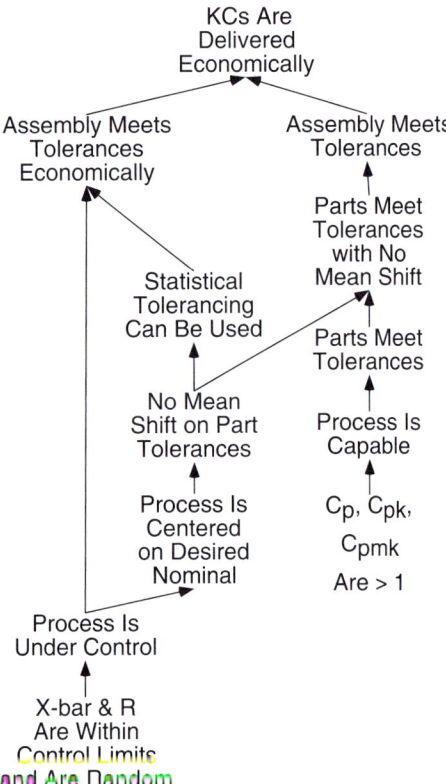

FIGURE 5-23. Summary of Conditions for Economically Delivering KCs. Each item in the figure is enabled by the items whose arrows point up to it.

goal region and that any part whose dimensions lie inside the zone is as good as any other part whose dimensions lie inside the zone. This "goalposting" behavior can lead to parts that are consistently larger or smaller than the nominal so that the mean error is not zero. It is accepted because people working at the part level do not appreciate the implications of nonzero means at the assembly level.

We can see where the tendency to deviate from the mean arises by understanding how actual part dimensions are created on the factory floor. There are two generic kinds of manufacturing processes, those that add material (called additive processes) and those that remove material (called removal processes). Example additive processes include casting, molding, and solid free-form fabrication, also known as rapid prototyping. Example removal processes include milling, turning, and grinding. The process operator is motivated to finish each part rapidly and to

avoid adding too much in an additive process or removing too much in a removal process. "If I leave a little extra material," reasons the milling machine operator, "they can always remove a little more later if they need to, but if I remove too much then the part is scrap. I could easily overshoot while trying to be perfect.[11] Anyway, I don't have time to split hairs inside the tolerance zone."

The resulting situation is illustrated in Figure 5-24. This figure not only illustrates the fact that parts could tend to be made consistently larger or smaller than the desired mean, but also that the resulting distribution of dimensions will be skewed rather than normal and symmetric about the actual mean.

Seeking to place the process mean on the desired nominal value is not the same as seeking perfection in every part. It requires only that the average of each dimension over many parts be close to the desired mean and that there be no consistent error to one side or the other. The X-bar chart is designed to ensure this and to detect problems that may cause the process mean to shift.

5.E.7. Including Mean Shifts in Statistical Tolerancing

If there is a mean shift error, then Table 5-3 does not accurately predict the percentage of parts or assemblies that will exceed the tolerance. The distribution is shifted to one side, and one tail of the distribution will fall outside the upper or lower specification limit, depending on the direction of the shift. The percent of parts or assemblies that exceed the tolerances can be very large, as shown in Table 5-5.

One way to account for mean shift is to make a more pessimistic estimate of the RSS error, as follows:

$$\sigma_Y^+ = M\sigma_Y \qquad (5\text{-}25)$$

where the superscript $+$ indicates that the standard deviation σ_Y of the assembly-level dimension Y has been increased by an inflation factor M. [Chase and Parkinson] discusses several versions of this approach. [Atkinson, Miller, and Scholz] calculates values of M based on

---

[11]This is analogous to the Blackjack player looking at a hand of 17 or 18 and deciding to fold rather than risk taking a hit that carries the score over 21.

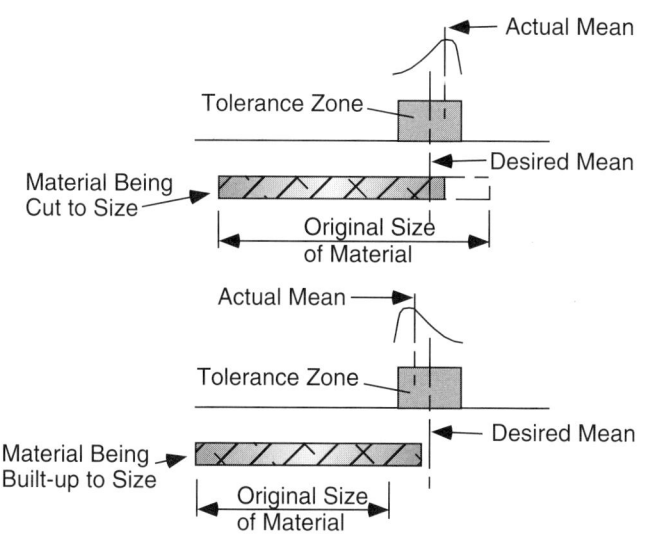

FIGURE 5-24. Comparison of Removal and Additive Processes. *Top:* A removal process can create parts that are on average larger than the desired mean. *Bottom:* An additive process can create parts that are on average smaller than the desired mean. In addition, the distribution of dimensions will be skewed rather than normal and symmetric.

TABLE 5-5. Percent of Parts or Assemblies Exceeding Tolerances in the Presence of Mean Shift

$C_p = \dfrac{USL - LSL}{6\sigma}$	Mean Shift Measured in Units of σ		
	0.0	0.2	0.4
0.5	13.361	20.193	38.556
1.0	0.27	0.836	3.594
1.2	0.0318	0.1363	0.8198
1.4	0.0027	0.0160	0.1350

Note: C_p measures the ability of the process to keep its variation within the specification limits. A capable process has $C_p > 1$. The table compares the percent of parts that will exceed the limits for several values of C_p. The boxed entry 0.27 corresponds to the case where there is no mean shift and $C_p = C_{pk} = 1$. Other entries show that even when C_p is greater than one, mean shift will cause huge numbers of parts to exceed the specification limits.

Source: [Atkinson, Miller, and Scholz].

TABLE 5-6. RSS Modification Factor $M(n)$ to Account for Process Mean Shifts

n	$M(n)$
2	1.104
3	1.136
4	1.162
5	1.186
6	1.207
7	1.227
8	1.245

Source: [Atkinson, Miller, and Scholz].

the number n of dimensions being combined in a linear stack.[12] M is calculated based on several assumptions. The most important of these is that the amount of mean shift in each fabricated feature is proportional to the total tolerance band on that feature from lower to upper limit. The calculation assumes that this shift is 10% of the total band. Normal distributions are also assumed. If each part fabrication process maintains $C_{pk} = 1$, then this mean shift amounts to 10% of $6\sigma_x$ for each

part or $0.6\sigma_x$. The result of this calculation is shown in Table 5-6. Details may be found in [Atkinson, Miller, and Scholz].

For example, if there are three parts contributing to an assembly-level tolerance, then $M(n)$ is 1.136, meaning that the assembly level standard deviation is estimated to be 13.6% larger than if there were no mean shift. In order for the original desired tolerance to be met, each part's tolerances must be decreased accordingly.

5.E.8. What If the Distribution Is Not Normal?

In all likelihood the distribution of part errors will not be normal even if the process mean hews to the desired nominal. As an extreme example, if there is clearance between two parts, then the assembly-level dimension between one

---

[12]Atkinson, Miller, and Scholz work in aircraft fuselage manufacturing at Boeing. In this domain, they estimate that $n = 8$ is typical.

side of one part and the other side of the other part could be any value within the range afforded by the clearance. This is usually modeled as a uniform distribution in which each value has the same probability. What allows us to assume a normal distribution at the assembly level anyway is the fact that, if we add several random numbers from *any* distribution or distributions, the distribution of the sum will tend toward normal anyway. This is called the Central Limit Theorem. Since assembly dimensions are sums of part dimensions, the assembly dimensions will have distributions that are similar to normal. A short discussion of this theorem is in Section 5.I. Here it is shown that as few as four uniformly distributed random variables will have a sum that is nearly normally distributed.

5.E.9. Remarks

Worst-case tolerancing is obviously very pessimistic. It is often used when tolerances are extremely tight and people do not have confidence that the process can deliver. SPC can be used to identify the sources of repeatable errors so that they can be removed. Methods include good shop housekeeping, careful tool change and sharpening methods, and studious repetition of setup procedures. The remaining errors are random. They can often be traced to temperature fluctuations, material property variations, or even differences between individual people. Sometimes additional care or better equipment can reduce these errors as well.

GD&T is essentially a worst-case tolerancing method. It is used to guarantee that the parts will assemble 100% of the time. GD&T does not presently include a way to assign tolerances statistically.

SPC plus a focus on seeking the nominal can be used as a way to maintain independence between suppliers of parts and the final assembler. The supplier hews to the specifications, and the customer need not pay close attention once the supplier's processes are under control. When tolerances are too small, independence must be sacrificed, and various means of coordination are required, such as selective assembly. These are discussed in Chapter 6. Such coordination may be straightforward inside a single shop, but it becomes tedious and costly in a long or complex supply chain.

5.F. CHAPTER SUMMARY

In this chapter we learned that the search for part and assembly accuracy is more than two hundred years old. The ideal has been to achieve parts that will interchange randomly with each other. This has huge advantages in terms of supply chain, assembly and after-sale costs but can increase part fabrication costs. We also learned that guaranteeing 100% interchangeability may not be as economical as allowing a slight possibility that interchanging might fail.

We introduced several ways to model how variation accumulates in assemblies. Worst-case tolerancing is pessimistic but guarantees that parts will mate interchangeably. GD&T essentially adopts worst-case assumptions. Statistical tolerancing and statistical process control were introduced to show how assemblies could achieve their KCs more easily and economically by exploiting the behavior of sums of random variables. The crucial assumption behind statistical tolerancing is that the mean of the fabrication process actually tends toward the desired nominal dimension on each toleranced feature. Statistical process control provides methods for monitoring the process mean as well as monitoring variation around the mean.

The difference between worst-case and statistical tolerancing is the attitude we take toward the parts: Statistical will keep some parts that worst-case will discard. Implementation of statistical tolerancing and statistical process control are much more difficult than doing the underlying calculations because it requires educating the workforce about statistics and random variables, and it requires the workforce to be diligent about process control.

The methods of tolerancing have advanced in recent decades and are better able to describe the true shape of three dimensional parts. However, a satisfying mathematical definition of part variation in the full three dimensional case still eludes us. Internationally recognized ways exist for specifying tolerances on individual parts, mainly for the purpose of supporting interchangeability, but no such standards exist for specifying the full three-dimensional specifications for an assembly, and no single method is widely used.

GD&T contains several principles that are important to our method of modeling assemblies. These include the notion of the datum hierarchy and the kinematic fine motion assembly strategy of mating the surfaces of one part to another in datum hierarchy sequence. The logic required to select datum surfaces is similar to methods we need to plan how parts will locate and constrain each other.

In the next chapter we will bring these ideas together to create ways to model variation statistically at the assembly level.

5.G. PROBLEMS AND THOUGHT QUESTIONS

1. List all the manufacturing errors you can think of that would cause the stapler to not operate properly. There are at least a dozen.

2. In Figure 5-25 are two assembled parts and their KC. Assume that the error in the length of each part is distributed normally with zero mean and $\sigma_p = 0.1$. Use Equation (5-17) to calculate the standard deviation σ_A of the KC. Then assume statistical tolerancing and use Equation (5-22) to predict the range in which the KC will lie 99.73% of the time. Next, assume that $3\sigma_p$ is the worst-case error for the length of each part. Assume worst-case tolerancing and calculate the range in which the KC will lie 100% of the time.

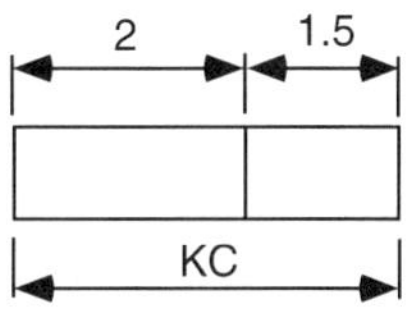

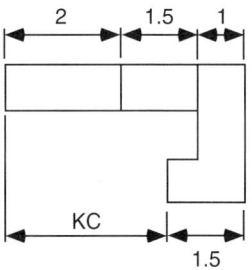

FIGURE 5-25. Two Assembled Parts. The KC is the sum of two dimensions.

3. Repeat Problem 2 for the case of the three parts in Figure 5-26.

FIGURE 5-26. Three Assembled Parts. The KC is the sum of four dimensions.

4. Figure 5-16 compares two situations that can arise in production of a part. In one case there is mean shift but the varia-

tion around the mean is small, while in the other case there is no mean shift but the variation around the mean is large. Some differences between the two situations may be summarized in Table 5-7. Discuss the pro's and con's of a process that delivers parts distributed like those on the left versus those on the right of Figure 5-16.

5. Consider an assembly consisting of a bearing and a shaft.[13] The designer requires a shaft with a nominal diameter of 25 mm and a diametral clearance ranging from 0.020 mm to 0.100 mm. Using worst-case methods, the designer may specify that the range of acceptable diameters for the shaft is 24.980 mm to 25.020 mm. The range of acceptable inside diameters of the bearing is 25.040 mm to 25.080 mm. The smallest shaft mated to the largest bearing will then have a clearance of 25.080 mm − 24.980 mm = 0.100 mm. The largest shaft mated to the smallest bearing will have a clearance of 25.040 mm − 25.020 mm = 0.020 mm. So the designer has achieved his goals.

   a. Let us suppose that after measuring the diameters of 100 shafts we find that the distribution of diameters is Normal and has a mean of 25.000 mm and a standard deviation of 0.010 mm. Calculate C_{pk} for the process of creating shafts. Use Table 5-3 to calculate what percent of shafts will fail inspection. If bearings also have a measured mean of 25.060 mm and a standard deviation of 0.010 mm, what percent of bearings will fail inspection?

   b. Instead of using worst-case methods, let us assume that a statistical approach is used. Calculate the mean value of the clearance. Calculate the variance and then the standard deviation of the clearance by using Equation (5-29). Next, calculate C_{pk} for the clearance. Assuming that this C_{pk} is acceptable, what range of diameters for shafts and

TABLE 5-7.

Probability	Left Side of Figure 5-16	Right Side of Figure 5-16
Probability that the part's dimension will achieve the desired value, or nearly so	Lower	Higher
Probability of a part failing to be in tolerance	Lower	Higher
Probability of an assembly made of parts like this achieving its KCs	Lower	Higher

[13]This example is taken from [Terry], which adapted it from [Cangello].

bearings should be accepted? On this basis, what percent of shaft-bearing sets will fail inspection?

c. On an $X-Y$ plot, lay out the range of the shaft's diameter on the X axis and the upper and lower limits of the bearing's diameter on the Y axis. Put the mean diameters of each at the origin. The range under worst-case tolerancing assumptions is [24.98, 25.02] for the shaft and [25.04, 25.08] for the bearing. Combining these ranges and remembering that each part is selected independently, what shape region in the $X-Y$ plane is occupied by acceptable shafts and

bearings? Using the ranges in the case of statistical tolerancing that you established in part (b), what shape region in the $X-Y$ plane is occupied by acceptable combinations of shafts and bearings?

d. You should note that the same machines, shafts, and bearings are under consideration in (a) and (b) above. The only thing that is different is our attitude toward them. List a few of the factors that would encourage people to take one attitude or the other.

5.H. FURTHER READING

[Atkinson, Miller, and Scholz] Atkinson, R. E., Miller, T. S., and Scholz, F.-W., "Statistical Tolerancing," U.S. Patent 5956251, September 21, 1999.

[Cangello] Cangello, C., "Statistical Methods and Process Control," Corporate Quality Department, United Technologies Carrier, Syracuse, NY, 1991.

[Chase and Parkinson] Chase, K. W., and Parkinson, A. R., "A Survey of Research in the Application of Tolerance Analysis to the Design of Mechanical Assemblies," *Research in Engineering Design,* vol. 3, pp. 23–37, 1991.

[DeVor et al.] DeVor, R. E., Chang, T. H., and Sutherland, J. W., *Statistical Quality Design and Control,* New York: Macmillan, 1992.

[Foster] Foster, L. W., *Geo-metrics-II,* Reading, MA: Addison-Wesley, 1979.

[Hounshell] Hounshell, D., *From the American System to Mass Production, 1800–1932,* Baltimore: Johns Hopkins University Press, 1984.

[Kolarik] Kolarik, W. J., *Creating Quality: Process Design for Results,* New York: WCB McGraw-Hill, 1999.

[Meadows] Meadows, J. D., *Geometric Dimensioning and Tolerancing,* New York: Marcel Dekker, 1995.

[Nevins] Nevins, J. L., "Sensors for Automation," Proceedings of a workshop sponsored by NSF, 1973.

[Nevins and Whitney] Nevins, J. L., and Whitney, D. E., *Concurrent Design of Products and Processes,* New York: McGraw-Hill, 1989.

[Swift] Swift, J. A., *Introduction to Modern Statistical Quality Control and Management,* Delray Beach: St. Lucie Press, 1995.

[Taniguchi] Taniguchi, N., "Current Status in and Future Trends of Ultraprecision Machining and Ultrafine Materials Processing," *Annals of CIRP,* vol. 32, no. 2, pp. 573–582, 1983.

[Taylor] Taylor, C. F., *The Internal Combustion Engine in Theory and Practice,* vol. 2, Cambridge: MIT Press, 1985.

[Terry] Terry, A. M., "Improving Product Manufacturability Through the Integrated Use of Statistics," MIT Master of Science and Master of Business Administration Thesis, June 2000.

[Voelcker] Voelcker, H. B., "The Current State of Affairs in Dimensional Tolerancing: 1997," *Integrated Manufacturing Systems,* vol. 9, no. 4, pp. 205–217, 1998.

5.I. APPENDIX: Central Limit Theorem

Figure 5-27 below compares several discrete probability mass functions representing the sum of one, two, and three random numbers that are each equally likely to have the values -1, 0, and 1. It shows that the sum of as few as three uniformly distributed random numbers has a distribution that is quite similar to a normal distribution. Larger numbers of summed random variables will produce distributions that are even more similar to a normal.

The probability density function of the sum of continuous random variables is calculated by convolving the density functions of the variables being summed. If discretely distributed random variables are used instead, then the distributions are called probability mass functions and the convolution reduces to simple multiplication of polynomials. The multiplication counts up how many ways there are to get a particular sum. Thus if we sum two random variables that each are equally likely to have the values -1, 0, and 1, then there are 9 possible outcomes: one way to get a sum of -2, two ways to get -1, three ways to get zero, and so on. The graphs in Figure 5-27 were generated this way.

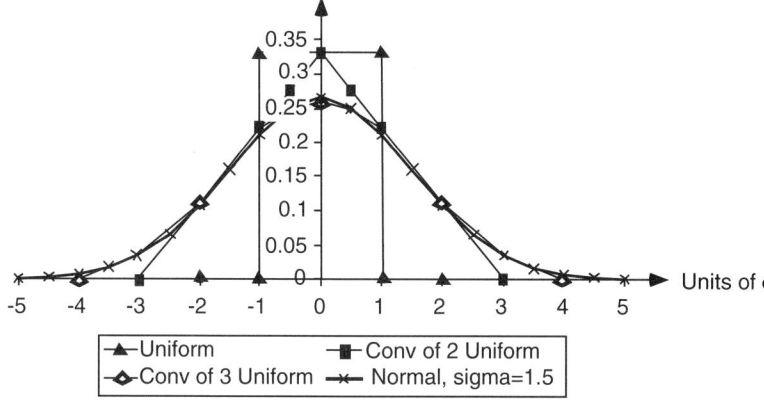

FIGURE 5-27. Comparison of the Probability Density Functions for the Sum of Two or Three Uniformly Distributed Random Numbers. Looking at the graph, it is clear that after only three sums the distribution strongly resembles a normal distribution that has been scaled to have a similar standard deviation.

5.J. APPENDIX: Basic Properties of Distributions of Random Variables

In this section we will briefly prove four useful properties of random variables and their sums: (1) the mean or average of a sum of random variables is the sum of the individual means of these random variables, (2) the variance of the sum of random variables is the sum of their individual variances, (3) the average of a sample of n random variables is the same as their mean, and (4) the variance of the average of a sample of n random variables $1/n$ times their individual variances.

5.J.1. Mean of a Sum

Proof by example:

Let x_1 and x_2 be two random variables. Let $y = x_1 + x_2$. Let $y_i = x_{i1} + x_{i2}$ for $i = 1, 2, 3$ be three examples of y. That is,

$$y_1 = x_{11} + x_{12}$$
$$y_2 = x_{21} + x_{22}$$
$$y_3 = x_{31} + x_{32}$$

Then

$$\text{average}(y) = E(y) = (y_1 + y_2 + y_3)/3$$
$$= (x_{11} + x_{21} + x_{31})/3$$
$$+ (x_{12} + x_{22} + x_{32})/3$$
$$= E(x_1) + E(x_2) \qquad (5\text{-}26)$$

Proof in general:

Just add more x's to the equation for y. That is, the average of a sum of random variables is the sum of their individual averages.

5.J.2. Variance of a Sum

Proof in general:

The definition of variance is

$$\sigma_x^2 = E[(x - E(x))^2] = E[x^2 - 2x E(x) + E^2(x)]$$
$$\sigma_x^2 = E(x^2) - E^2(x) \qquad (5\text{-}27)$$

Let $w = x + y$. Then

$$\sigma_w^2 = \sigma_{x+y}^2 = E[(w - E(w))^2]$$
$$= E[(x - E(x) + y - E(y))^2]$$
$$= E\{[x - E(x)]^2 + [y - E(y)]^2$$
$$+ 2[x - E(x)][y - E(y)]\}$$
$$= \sigma_x^2 + \sigma_y^2 + 2E[xy - yE(x) - xE(y)$$
$$+ E(x)E(y)] \qquad (5\text{-}28)$$

If x and y are linearly independent, then $E(xy) = E(x)E(y)$ so that the last term is zero. Thus

$$\sigma_{x+y}^2 = \sigma_x^2 + \sigma_y^2 \qquad (5\text{-}29)$$

If x and y have the same variance σ^2, then

$$\sigma^2_{sum\,of\,n\,x's} = n\sigma^2_x \qquad (5\text{-}30)$$

That is, the variance of a sum of random variables increases linearly with the number of random variables being summed.

5.J.3. Average of a Sum

Proof in general:

Let $r =$ the sum of n identically distributed independent random variables x where each x has an average $E(x)$ and a variance σ^2_x. Then, from Sections 5.J.1 and 5.J.2 above,

$$E(r) = nE(x)$$
$$\sigma^2_r = n\sigma^2_x \qquad (5\text{-}31)$$

Define $R = r/n = $ an average of n sample sums. Then,

from Equation (5-31), we have

$$E(R) = E(r/n) = E(r)/n = E(x) \qquad (5\text{-}32)$$

That is, the average of n sample sums is the same as the average of the random variables in the sample.

5.J.4. Variance of the Average of a Sum

The following is the proof in the case where the distribution, mean, and variance of the random variables in the sample are known:

$$\sigma^2_R = E(R^2) - E^2(R)$$
$$\sigma^2_r = E(r^2) - E^2(r)$$
$$E(R^2) = [E^2(r)]/n^2 \qquad (5\text{-}33)$$
$$E^2(R) = [E^2(r)]/n^2$$
$$\therefore \sigma^2_R = [\sigma^2_r]/n^2 = [\sigma^2_x]/n$$

That is, the variance in the sample average goes down linearly with the number of members in the sample.

6 MODELING AND MANAGING VARIATION BUILDUP IN ASSEMBLIES

"Those parts passed inspection. I don't understand why they won't assemble."

6.A. INTRODUCTION

Let us recap where we are conceptually: Assemblies achieve their KCs when the parts are in the correct relative positions and orientations, within specified tolerances. Parts get to their desired positions and orientations by joining to each other at places called assembly features. In order that the final assembled state be correct, it is necessary that the features be in the correct positions and orientations with respect to each other on each of the parts, again within some specified tolerances. In Chapter 2, we saw that KC flowdown is the way to identify which parts have a role in delivering each KC. In Chapter 3, we learned that assembly features allow us to build a connective model of an assembly. Chapter 4 showed that each part can do its job locating adjacent parts if its assembly features provide kinematic constraint to the parts it joins. In Chapter 5, we learned that GD&T permits us to define features on parts and provide datums or frames that create kinematic mating of the parts to others. GD&T also focuses on ensuring that the parts will assemble interchangeably 100% of the time. Statistical tolerancing allows us to take advantage of cancellation between positive and negative errors in parts that are assembled, permitting looser tolerances on each part than worst-case tolerancing does. Now we need a way to find out if an assembly will deliver its KCs in the presence of variations in parts and features.

The challenge is that we do not know which parts will be used in each assembly, or we do not want to go to the trouble of individually measuring and selecting compatible parts unless we have no alternative. Thus we have to predict whether the assemblies will achieve the KCs without knowing in advance whether the parts are suitable for this purpose.

Many models of assembly error accumulation are used. For a recent survey, see [Chase and Parkinson]. Each method is based on its own way of deciding if individual parts are acceptable for assembly. If the parts are accepted by a given method, the way that errors accumulate in the assembly must be modeled in a way that is consistent with the statistical distributions of the parts that emerge from that acceptance process. In a few situations, the only way to know if the parts will assemble and deliver the KCs is to actually assemble them.

The approaches used can be grouped into three classes. They vary in the degree of coordination that is used. By coordination, we mean the attention that must be paid to individual parts, pairs of parts, triplets, and so on, in an effort to predict or determine in advance whether an assembly made of those particular parts will achieve the KCs. At one extreme is 100% or deterministic coordination, in which the parts are custom made or chosen for each assembly and become mates for life. No interchangeability is permitted. At the other extreme is zero coordination, in which worst-case tolerancing is used. In this case, the assembly will achieve the KCs no matter what part variations occur within the preset limits. These limits turn out to be uneconomically small in most cases. In between is coordination based on statistical tolerancing, in which a bet is made that the parts will achieve the KCs almost all the time. This is the most economical approach and is used most of the time. Both zero and statistical coordination achieve interchangeability, all or almost all the time, respectively.

In this chapter, we present matrix transform equations for calculating the effects of part variations on assembly-level variation and use matrix transformations to calculate

141

KC variation given part variation. Several examples are given. The same analysis technique can be used to propagate errors of many kinds through fixtures, tools, and assembly equipment. For example, we can predict statistically the net lateral and angular error between the last part already on the assembly and the next part about to be added. As a result, the likelihood of successful assembly can be calculated using Part Mating Theory (see Chapter 10 and Section 6.D.1). Then we show how to express GD&T in terms of 4×4 transforms. Next we treat a variety of methods for managing variation, including tolerance allocation, selective assembly, and functional build. The differences between rigid and sheet metal parts are discussed briefly.

There are several important limitations to our presentation. First, while we will carefully account for the varied location of surfaces on parts and show how this affects the location of one part relative to its mating parts, we will not try to find the varied location of an entire surface on the last part in an assembly under variation relative to the first or other distant part. This is a difficult problem that is the subject of current research. Second, we will model only open-chain assemblies, as we have throughout the book. It is possible to model variation in closed-chain assemblies, such as four-bar linkages, but this is beyond the scope of this book and is also the subject of ongoing research.

6.B. NOMINAL AND VARIED MODELS OF ASSEMBLIES REPRESENTED BY CHAINS OF FRAMES

In this section, we introduce generic equations that represent nominal and varied assemblies as chains of frames. The terminology is shown in Figure 6-1.

Figure 6-1 represents two situations, one being the nominal and the other being the more likely situation in which there is some error. In this case, we have an error related to feature F_{2B} on part B. This feature could be mislocated, misshaped, or both, or there could be a source of variation in the mating relationship between features F_{2B} and F_{1C}. In this section, we write the transform equations for the KC including general error transforms that represent each of these possibilities. We will consider two situations. The first is that in which the assembly feature between two parts is constructed from a single geometric entity like those in the feature toolkit in Chapter 4. The second is that in which the assembly feature is constructed from a combination of such toolkit features.

6.B.1. Calculation of Connective Assembly Model Variation Using Single Features

If each feature comprises a single element that constrains all the intended degrees of freedom, then the calculation of the varied KC is fairly straightforward because each feature is simply described by a frame attached to it. All we have to do is write the frame relationships that position each feature on its part and then chain the frames together, including any error frames that capture misplacement or misorientation of the feature on the part or errors in feature-to-feature relationships. This represents a simple modification of the nominal connective assembly model presented in Chapter 3.

The discussion that follows uses the following naming convention. A nominal transform from a to b is called T_{ab}. If there is variation in this transform, the variation is expressed in a transform called DT_{ab}, and this transform postmultiplies T_{ab}. The necessary equations for the case in Figure 6-1 are as follows. First, we repeat the equations if there are no errors, formulated according to the discussion in Chapter 3 (refer to Figure 6-1):

$$T_{AC} = T_{AB}T_{BC}$$

$$T_{AB} = T_{A-F_{1A}}T_{F_{1A}-F_{1B}}T_{F_{1B}-B} \qquad (6\text{-}1)$$

$$T_{BC} = T_{B-F_{2B}}T_{F_{2B}-F_{1C}}T_{F_{1C}-C}$$

Here, $T_{A-F_{1A}}$ is the transform from part A's coordinate center to feature F_{1A}, $T_{F_{1B}-B}$ is the transform from feature F_{1B} to part B's coordinate center, and so on. $T_{F_{1A}-F_{1B}}$ is

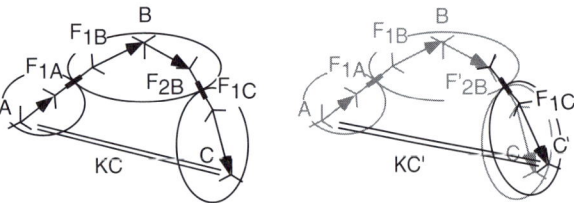

FIGURE 6-1. Parts Joined by a Chain of Frames to Deliver a KC. On the left there are no errors, and the KC is achieved exactly. On the right, there is an error in the construction of part B, and the KC is not achieved exactly. How much the KC deviates from the desired value can be calculated if we know all the frame relations and the error in part B.

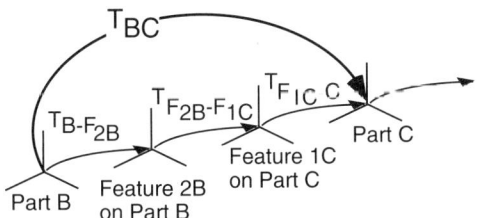

FIGURE 6-2. Illustrating the Nominal Frames in Parts B and C Described in Equation (6-1).

the interface relationship transform between features F_{1A} and F_{1B}. Nominally, this relationship is the identity transform but it can be used to represent a change in coordinate orientation if, for example, the X axis of one feature aligns with the Y axis of the other.

Figure 6-2 illustrates some of the frames in Equation (6-1).

Next, we give the equations representing the three possible sources of variation (refer to Figure 6-3).

First, here are the equations if a feature is placed in error. In this example, feature F_{2B} is misplaced on part B. The placement error, combining location and orientation, is called $DT_{B-F_{2B}}$. This yields an equation for the varied transform from B to C called T'_{BC} that results from this error alone:

$$T'_{BC} = T_{B-F_{2B}} DT_{B-F_{2B}} T_{F_{2B}-F_{1C}} T_{F_{1C}-C} \qquad (6\text{-}2)$$

Second, here are the equations if the feature is misshaped. The shape error due to this error alone is called DT_{F2B}.

$$T'_{BC} = T_{B-F_{2B}} DT_{F2B} T_{F_{2B}-F_{1C}} T_{F_{1C}-C} \qquad (6\text{-}3)$$

Third, here are the equations if there is some variation within the interface relationship between two features. The relationship error is called $DT_{F_{2B}-F_{1C}}$.

$$T'_{BC} = T_{B-F_{2B}} T_{F_{2B}-F_{1C}} DT_{F_{2B}-F_{1C}} T_{F_{1C}-C} \qquad (6\text{-}4)$$

Two possible sources of such variation are clearance and kinematic realignment. Both can be modeled using

methods discussed in Chapter 5. In the case of two plane surfaces, the possible clearance error is motion normal to the planes' surfaces. In the case of a pin–hole feature, clearance provides error motion normal to the pin–hole axis. Kinematic realignment occurs when there is clearance or other under-constraints and the parts must shift position slightly in order to achieve assembly ([Chase et al.]). In the case of plane surfaces, this shift would occur parallel to the planes.

Combining Equation (6-2), Equation (6-3), and Equation (6-4) yields the complete varied transform including all the above error sources:

$$T'_{BC} = T_{B-F_{2b}} \{ [DT_{B-F_{2b}} + DT_{F2B}] T_{F_{2B}-F_{1C}}$$
$$+ T_{F_{2B}-F_{1C}} [DT_{F_{2B}-F_{1C}}] \} T_{F_{1C}-C} \qquad (6\text{-}5)$$

Finally, the transform for the varied KC is

$$T'_{AC} = T_{AB} T'_{BC} \qquad (6\text{-}6)$$

Care must be taken when constructing transform equations that contain variations. As explained in Chapter 3, it is possible in some cases to formulate the error transform in different coordinates. Each different formulation is equivalent but is likely to result in a different transform equation with the error transform having different contents and being inserted into the nominal transform equation in a different place. The general situation is illustrated in Figure 6-4. Equation (6-2) through Equation (6-4) have been written on the assumption that the error in a transform occurs after the nominal transform has been accomplished. Thus all such equation entries read $T_{X-Y} DT_{X-Y}$.

6.B.2. Calculation of Connective Assembly Model Variation Using Compound Features

A feature made up of several elements was called a compound feature in Chapter 3. Calculating the variation it

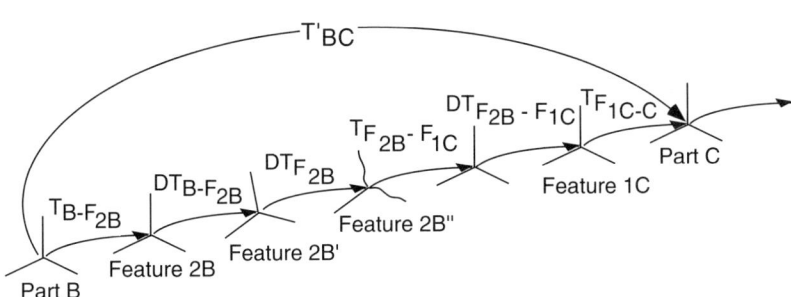

FIGURE 6-3. Illustrating the Varied Frames in Parts B and C Described in Equation (6-5).

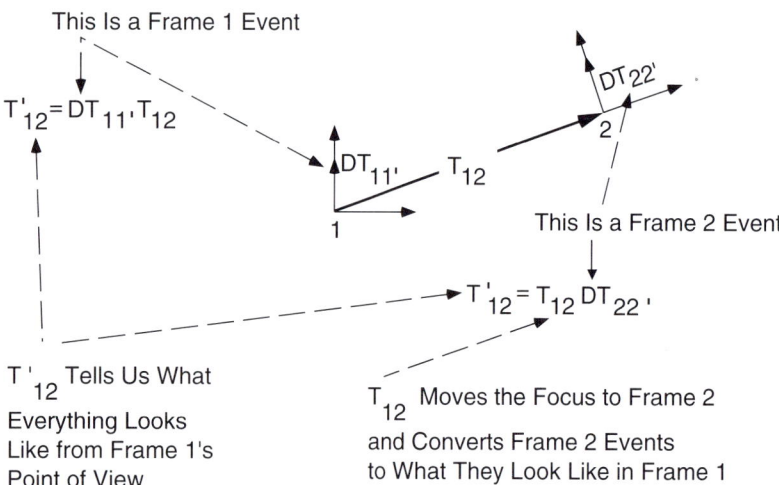

This Is a Frame 1 Event

$T'_{12} = DT_{11'} T_{12}$

T'_{12} Tells Us What
Everything Looks
Like from Frame 1's
Point of View

This Is a Frame 2 Event

$T'_{12} = T_{12} DT_{22'}$

T_{12} Moves the Focus to Frame 2
and Converts Frame 2 Events
to What They Look Like in Frame 1

FIGURE 6-4. Illustrating How Transforms Account for Different Variations and How the Corresponding Transform Equations are Written. If a variation such as $DT_{11'}$ affects frame 1, then it should be placed ahead of transform T_{12} when formulating T'_{12}, because T_{12} shifts the focus to frame 2, which is not where the variation occurs. If a variation such as $DT_{22'}$ affects frame 2, then it should be placed after T_{12} when formulating T'_{12}, so that its effect can be reflected back to frame 1 by T_{12}.

causes is a bit more complex than for a simple feature. The situation is shown in Figure 6-5 for the example of a pin–hole and pin–slot compound feature. We learned in Chapter 3 how to calculate the nominal transform T'_{A1} when there is no error. Here we learn how to calculate T'_{A1}, the varied transform relating part center coordinates to this feature when its elements, the hole and the slot, are not at their nominal positions. The case where the nominal orientation of the slot is not along the line from the hole to the slot can be analyzed using Screw Theory ([Shukla]).

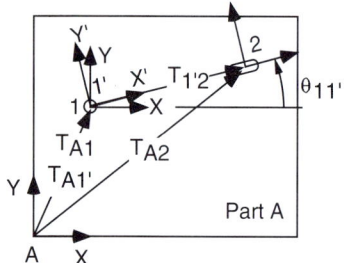

FIGURE 6-5. Compound Feature Without Variation (repeated from Chapter 3). The feature is made up of a hole and slot, each shown at their nominal position and orientation. Frame A is at the origin of part A's base coordinate frame. Frame 1 is at the center of the hole. Frame 1′ is the frame of the compound feature. Frame 2 is at the center of the slot. Transform T_{A1} relates hole frame 1 to base frame A, while transform $T_{A1'}$ relates frame 1′ of the compound hole–slot feature to frame A. The difference between these two frames is the rotation $rot(z, \theta_{11'})$.

In all of the following discussion, we assume that, in spite of all variations, the pin mated to the slot will always lie inside the slot and will never collide with one end or the other. That is, not only is the nominal feature kinematically constrained, but all varied features will be as well. That is, the constraint plan will be robust to variations. As explained in Chapter 8, it is necessary to check explicitly to ensure that this assumption is valid. This check constitutes a kind of variation analysis that is not usually done.

Here are the required equations, simplified for the case where the errors occur in the XY plane of part center coordinates of part A. These equations take as input the nominal locations of the hole and slot plus errors in those locations in x, y coordinates. The output is the varied transform T'_{A1} that locates the varied hole–slot feature with respect to the part coordinate frame. The reader should refer to Figure 6-5 and Figure 6-6 while reading these equations.

First, we repeat in Equations (6-7a)–(6-7j) the equations that describe the nominal transforms T_{A1} and T_{A2} that locate the hole and slot in part A and allow us to calculate the transform $T_{A1'}$ that locates the hole–slot feature:

$$T_{hole\ nominal} = T_{A1} = \begin{bmatrix} R_{A1} & p_{A1} \\ 0^T & 1 \end{bmatrix} \qquad (6\text{-}7a)$$

$$T_{slot\ nominal} = T_{A2} = \begin{bmatrix} R_{A2} & p_{A2} \\ 0^T & 1 \end{bmatrix} \qquad (6\text{-}7b)$$

$$T_{A1'} = T_{A1} T_{11'} \qquad (6\text{-}7c)$$

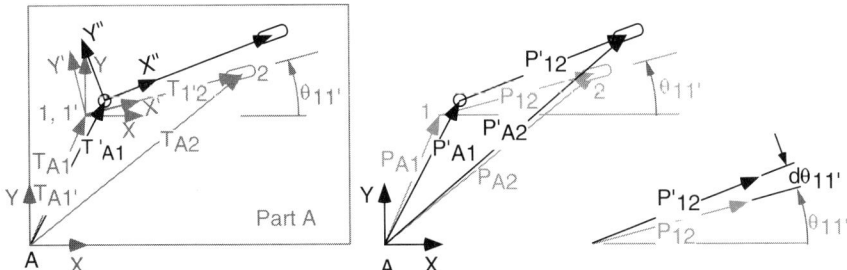

FIGURE 6-6. Compound Feature with Variation. The feature is made up of a hole and a slot, each of which could vary in its location on the part. Variation in the hole's position in either X or Y will affect the position of the mating part. However, only variation in the slot's location normal to the slot's long axis, captured by angle $\delta\theta_{11'}$, will have an effect, and that will be to alter the orientation of the mating part.

where

$$T_{11'} = \begin{bmatrix} R_{11'} & p_{11'} \\ 0^T & 1 \end{bmatrix} \quad \text{where } p_{11'} = 0 \quad (6\text{-}7\text{d})$$

$$T_{A2} = T_{A1}T_{11'}T_{1'2} = T_{A1}T_{12} \quad (6\text{-}7\text{e})$$

$$T_{12} = T_{A1}^{-1}T_{A2} = \begin{bmatrix} R_{12} & p_{12} \\ 0^T & 1 \end{bmatrix} \quad (6\text{-}7\text{f})$$

where

$$p_{12} = p_{A2} - p_{A1} = \begin{bmatrix} x_{12} \\ y_{12} \\ 0 \end{bmatrix} \quad (6\text{-}7\text{g})$$

$$R_{11'} = R_{12} = rot(z, \theta_{11'}) \quad (6\text{-}7\text{h})$$

$$\cos\theta_{11'} = x_{12}/|p_{12}| \quad \text{and} \quad \sin\theta_{11'} = y_{12}/|p_{12}| \quad (6\text{-}7\text{i})$$

where

$$|a| = \text{the length of vector } a \quad (6\text{-}7\text{j})$$

Next, we give equations for the varied locations of the hole and slot, $T_{hole\,varied}$ and $T_{slot\,varied}$, along with other information we need:

$$T_{hole\,varied} = \begin{bmatrix} R_{A1} & p'_{A1} \\ 0^T & 1 \end{bmatrix} \quad \text{and} \quad T_{slot\,varied} = \begin{bmatrix} R_{A2} & p'_{A2} \\ 0^T & 1 \end{bmatrix}$$

$$p'_{12} = p'_{A2} - p'_{A1} = \text{the varied vector from hole}$$
$$\text{to slot}$$

$$\delta\theta_{11} = \cos^{-1}(p'_{12} \bullet p_{12}) \quad \text{where } \bullet \text{ denotes vector}$$
$$\text{inner product}$$

$$dp_{A1} = p'_{A1} - p_{A1} \quad (6\text{-}8)$$

Last, we give the equation for the varied transform T'_{A1} that locates the varied hole–slot feature in part A, based on the nominal transform $T_{A1'}$ and information in Equation (6-8) and Equations (6-7a)–(6-7j):

$$T'_{A1} = T_{A1'}DT_{A1'} \quad (6\text{-}9\text{a})$$

where

$$DT_{A1'} = \begin{bmatrix} 1 & \delta\theta_{11} & 0 & dp_{A1x} \\ -\delta\theta_{11} & 1 & 0 & dp_{A1y} \\ 0 & 0 & 1 & 0 \\ 0 & 0 & 0 & 1 \end{bmatrix} \quad (6\text{-}9\text{b})$$

Error transform $DT_{A1'}$ is analogous to the transform of an error in placing a single feature on a part, such as $DT_{B-F_{2B}}$ in Equation (6-2). Thus error transforms in compound features can be used in varied transform equations the same way error transforms in single features can.

The compound frame in Figure 6-6 could encounter another error in addition to that shown in the figure. The pins that it mates to on the other part could themselves be in the wrong positions on it. These two pins also constitute a compound feature whose frame might be defined as shown on the right in Figure 6-7.

The figures that follow indicate how to calculate the effect of different fabrication errors in part C that misplace the pins, assuming no variation in part A.

The first case is one where the two pins are laterally displaced on part C from their nominal positions. This is shown in Figure 6-8. Figure 6-9 shows the second case where the pin that mates to the slot is displaced in the direction of the slot's long axis. The third case (Figure 6-10) shows the case where the pin that mates to the slot is displaced in a direction normal to the slot's long axis.

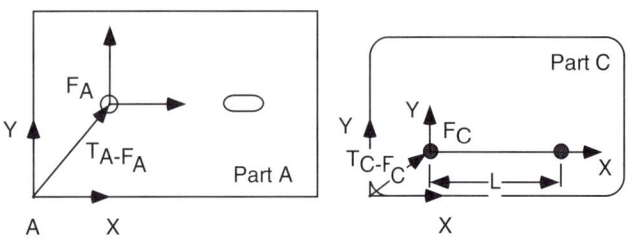

FIGURE 6-7. Part C Has a Compound Feature FC on It Comprising Two Pins That Will Mate to the Hole–Slot Compound Feature FA on Part A.

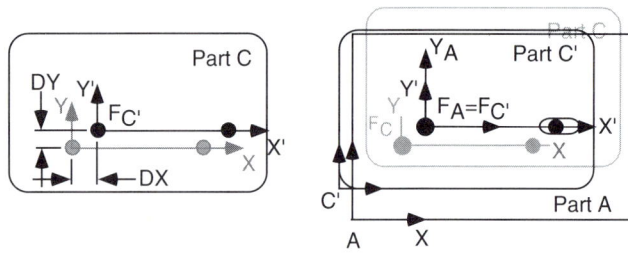

FIGURE 6-8. The Two Pins on Part C Are Displaced by DX and DY but Are Not Rotated in Part C's Coordinate Frame. On the left, the nominal location of the compound feature is shown in gray while the displaced feature is shown in black. On the right, the nominal location of part C with respect to part A is shown in gray while the displaced location C′ is shown in black.

$$T_{AC\,'} = T_{A-FA}T_{FA-FC\,'}T_{FC\,'-FC}T_{FC-C}$$

$$T_{FC'-FC} = trans\,(-DX,-DY\,) = \text{the fabrication error in Part C}$$

$$T_{FA-FC\,'} = I \quad \text{representing assembling the parts by joining FA to FC'}$$

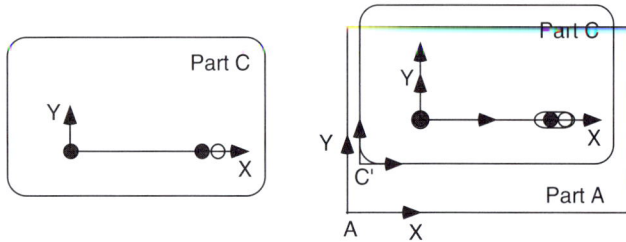

FIGURE 6-9. The Right-Hand Pin Is Displaced Along the Compound Feature's *X* Direction. The displaced pin is shown as an open circle. The feature's displacement direction corresponds to the long axis of the slot. As long as the pin stays within the slot, there is no change in part C's position or orientation with respect to part A.

$$T_{AC\,'} = T_{AC} \quad \text{meaning that there is no error in this case}$$

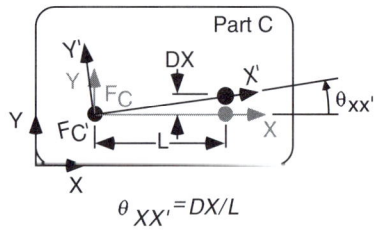

$$\theta_{XX'} = DX/L$$

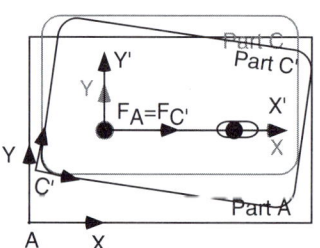

FIGURE 6-10. The Right-Hand Pin Is Displaced Along the Compound Feature's Y Direction, Rotating the Compound Feature Frame Counterclockwise by $\theta_{xx'}$. In this case, part C will be rotated clockwise with respect to part A by $\theta_{xx'}$. On the left, the nominal orientation of the compound feature is shown in gray while the displaced orientation is shown in black. On the right, the nominal location of part C with respect to part A is shown in gray while the displaced location C′ is shown in black.

$$T_{AC'} = T_{A-FA}T_{FA-FC'}T_{FC'-FC}T_{FC-C}$$

$$T_{FA-FC\,'} = I$$

$$T_{FC'-FC} = rot(z,-\theta_{xx'})$$

The analysis is more difficult if the slot's long axis is not oriented along the compound hole–slot feature's X axis. In this case, the angle between these two axes must be taken into consideration. A thought question at the end of the chapter asks you to consider two extreme cases.

If there is clearance between pin and hole or between pin and slot, additional variation can be introduced. This case, among many others, is analyzed in [Bjørke] and is treated in Section 6.B.1 as a random error imposed on the interface transform between features.

6.C. REPRESENTATION OF GD&T PART SPECIFICATIONS AS 4 x 4 TRANSFORMS

In this section we describe a way to convert GD&T or similar part-level tolerances into 4×4 transforms that can be combined mathematically to permit calculation or simulation of assembly level tolerances. The method is based on [Whitney, Gilbert, and Jastrzebski], but similar approaches have been published in [Baartman and Heemskerk], [Lafond and Laperrière], [Pino, Bennis and Fortin], [Mujezinovic, Davidson, and Shah], and [Rivest, Fortin, and Desrochers].

In Section 6.B, we learned how to formulate 4×4 transform models of assemblies built from parts whose assembly features might vary in position and orientation from their nominals. All we said at that time was that error transforms representing variations in feature positions and orientations could be inserted into the calculations, but we delayed showing how to estimate the values of these variations and how to determine the assembly level variation.

The purpose of the following analysis is to model the accumulation of part position and orientation uncertainty as parts are added to an assembly when the parts have been toleranced in accordance with GD&T.

Our challenge is to decide how to formulate the individual features' error transforms. We will approach this by assuming that features or feature surfaces will lie inside tolerance zones defined by the GD&T methodology. Even though GD&T is based on worst-case assumptions, we will look at both worst-case and statistical methods of modeling how errors accumulate. Examples of calculating assembly-level variations are given in Section 6.D. The sections that follow deal with only one GD&T specification, namely position of a surface, but other specifications can be dealt with analogously ([Whitney, Gilbert, and Jastrzebski]).

6.C.1. Representation of Individual Tolerance Zones as 4 x 4 Transforms

The first step is to describe a tolerance zone by allowing small variations only on specific degrees of freedom

associated with the feature, effectively sweeping the complete volume of the zone. Figure 6-11 and Figure 6-12 show an example in two-dimensional form. These figures illustrate the definition of a datum for a feature (whether GD&T requires a datum or not) plus representation of errors as a sequence of coordinate transforms.

Figure 6-13 shows the three-dimensional version of the planar size specification of Figure 6-11.

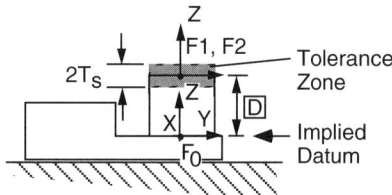

FIGURE 6-11. A Two-Dimensional Feature of Size and Its Tolerance Zone. The top surface of the part is subject to variation. F_0 is the reference frame. F_1 is a frame attached to the center of the tolerance zone. F_2 is a frame on the surface, shown here at nominal position and orientation. This surface may lie anywhere within this zone at any angle but may not extend beyond the zone, guaranteeing correct form at MMC. Thus the allowed angular errors are smaller when the center of the surface is nearer either extreme of the tolerance zone. This fact makes the lateral and angular errors correlated, as discussed in Section 5.D.2.d.

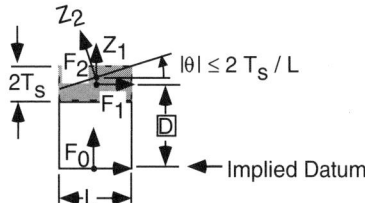

FIGURE 6-12. Description of the Size Tolerance as Three Successive Coordinate Transforms. The shaded volume between the two boundary surfaces represents the tolerance zone within which the actual surface must lie. Cartesian frame F_2 is attached to this surface. Since this surface must lie within the tolerance zone, the maximum value of $|\theta|$ is $2T_s/L$.

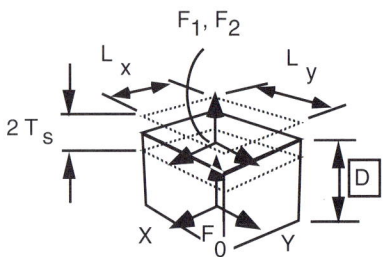

FIGURE 6-13. Defining a Planar Size Tolerance Zone. As in Figure 6-12, the volume between the two boundary surfaces (outlined) represents the tolerance zone within which the actual surface, on which the Cartesian frame F_2 is attached, must lie.

To represent a tolerance zone with a matrix transform, we attach a Cartesian frame, in our case F_1, to the middle of the tolerance zone. This frame also represents the nominal position and orientation of the actual feature with respect to the previous (reference or A datum) frame F_0. To simulate the possibility that the feature could be anywhere in the tolerance zone, we allow specific degrees of freedom defining its position and orientation to vary. The limits on these varying degrees of freedom, centered about the zero value (nominal position and orientation), together constitute a boundary. This is the tolerance zone, outside of which the part would be rejected.

In our case of planar size tolerance, we relax the linear constraint along the Z axis, and the rotational constraints around the X and Y axes. We therefore define three variates on which the individual limits combine to define a maximum boundary. The linear variate Z varies within the range $-T_s$ to $+T_s$, while the rotational variates, θ_x and θ_y, vary respectively within the ranges $\pm 2T_s/L_y$ and $\pm 2T_s/L_x$. These angular limits are obtained using a small angle approximation and are consistent with the requirement that the actual surface must lie entirely within the tolerance zone.

These ranges cannot be considered separately, since if all variates were at their maximum value at the same time, a portion of the surface would be outside of the tolerance zone, violating Rule #1. For this reason, we define the relations between Z, θ_x, and θ_y, resulting in a dipyramid in the three-dimensional $[Z, \theta_x, \theta_y]$ parameter space, as shown in Figure 6-14. This dipyramid is therefore a boundary, resulting from the combined ranges, outside of which the parameters (degrees of freedom) of the surface would result in a violation of the size tolerance specification.

6.C.2. Worst-Case Representation of 4 × 4 Transform Errors

In the worst case, the variations are at the limits of the tolerance zones all the time. An apparently straightforward approach would be simply to substitute these maximum errors into the variation transforms defined in Section 6.B and numerically calculate the assembly-level variation.

There are two problems with this approach. First, each variation could be at either end of the allowed zone, and it is generally impossible in the full six-dimensional case to tell in advance which combination of worst cases at the individual feature level will produce the worst case at the assembly level. Second, situations like those in

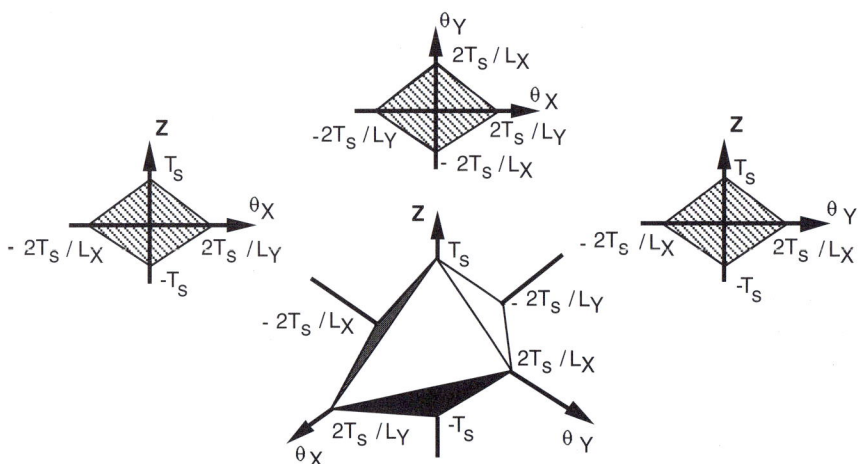

FIGURE 6-14. Region in Dimensional Variation Space Obeying Rule #1 in the Full Three-Dimensional Case of a Planar Feature of Size.

Figure 6-14 show us that the maximum values cannot be used at the same time when Rule #1 is part of the tolerance specification.

To address the first problem, one could try all possible combinations, excluding those that violate Rule #1, but this would be computationally expensive since there could be 2^n possibilities to try, where n is the number of zones involved. Alternatively, one could use linear programming ([Turner and Wozny]) to search for the maximum assembly-level error. To address the second problem, one has little choice except to search over a range of combinations of the related variables by choosing points on the boundary of the diamond-shaped zones illustrated in Figure 6-14. If we build a chain of parts having size tolerances like those in Figure 6-14, the position of a surface on the last part in the chain will lie inside or on the boundary of a shape that is similar to but much more complex than that shown in Figure 6-14. Methods of calculating such shapes are the subject of current research and are beyond the scope of this book. An example of such research is [Mujezinovic, Davidson, and Shah]. In this book we will finesse such complexities by selecting extreme points on the last part and calculating, by worst-case or statistical methods, the possible locations of those points.

A worst-case search is rarely done, although a simple example is shown in Section 6.D.4. Instead some statistical estimating method is used. We consider this approach next.

6.C.3. Statistical Representation of 4 x 4 Transform Errors

The alternative to worst-case analysis is to represent errors like those in Figure 6-13 in statistical terms. To do this we assume, as in Chapter 5, that there is no mean shift. Again, we appeal to the normal distribution. In this case, however, we need to represent several random variables at once. For this purpose we use a zero mean joint normal density function

$$f(z, \theta) = \frac{1}{2\pi\sigma_z\sigma_\theta} e^{-\frac{1}{2}\left(\frac{z^2}{\sigma_z^2} + \frac{\theta^2}{\sigma_\theta^2}\right)} \tag{6-10}$$

for the case of two jointly distributed random variables, and

$$f(z, \theta_x, \theta_y) = \frac{1}{(2\pi)^{3/2}\sigma_z\sigma_{\theta_x}\sigma_{\theta_y}} e^{-\frac{1}{2}\left(\frac{z^2}{\sigma_z^2} + \frac{\theta_x^2}{\sigma_{\theta_x}^2} + \frac{\theta_y^2}{\sigma_{\theta_y}^2}\right)} \tag{6-11}$$

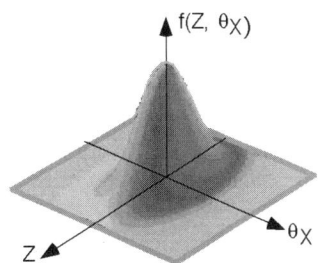

FIGURE 6-15. Joint Normal Density Function Corresponding to Equation (6-10).

for the three-variable case. Equation (6-10) corresponds to the two-dimensional probability density function shown in Figure 6-15. These equations assume *no* correlation between the several variables but we are going to use them anyway, employing an approximation illustrated in the next section to account for the correlation introduced by Rule #1.

Cross sections parallel to the z–θ_x plane in Figure 6-15 are ellipses whose semi-major and semi-minor axes correspond to multiples of the standard deviation in the respective directions. Larger ellipses correspond to larger ranges of possible values of the random variables. In the one-dimensional case shown in Figure 5-17, the $\pm 3\sigma$ range ellipse contains 99.73% of all possible values. In the two-variable case, the same range corresponds to 98.9%, while in the three-variable case, it corresponds to 97.1%. See [Bryson and Ho, p. 311].

6.C.3.a. Plane Feature of Size

All linear methods of adding random variables and calculating statistical tolerance accumulation assume that each random variable is uncorrelated with the others. A two-dimensional plane feature of size can have two errors, a position error and an angle error. Rule #1 obviously correlates these errors, as indicated in Figure 6-14. To model variations in features of size statistically using Equation (6-10) and Equation (6-11), we approximate this correlation by finding the ellipse that best approximates the diamond, as shown in Figure 6-16. This ellipse measures the probability that both independent random variables are inside the ellipse. Since the ellipse and the diamond do not match exactly, the joint probability density model approximates adherence to Rule #1 but cannot represent it exactly. Our reward for accepting this approximation is that we can use normal probability densities and can use relatively simple numerical simulations.

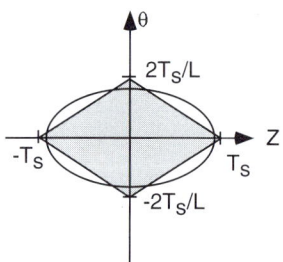

FIGURE 6-16. Cross-Plot of Allowed Lateral and Angular Errors. The diamond shape represents adherence to Rule #1. The ellipse represents approximating adherence using a joint probability distribution of independent variables.

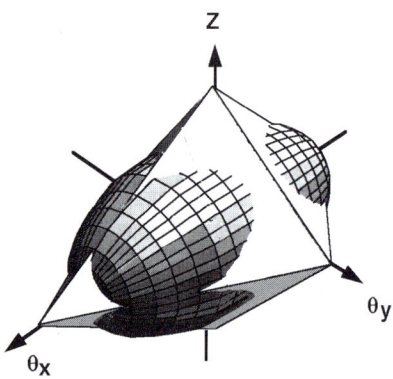

FIGURE 6-17. Dipyramid Showing Allowed Values of Z, θ_x, and θ_y Such that Rule #1 Is Obeyed Along with Its Approximation by the Statistical Ellipsoid.

In the three-variable case corresponding to Figure 6-14 and Equation (6-11), the cross-plot corresponding to Figure 6-16 appears in Figure 6-17.

In order to perform a statistical calculation of accumulated error using a joint normal density function while attempting to represent obedience to Rule #1, we will adjust the parameters in Equation (6-11) to get the best approximation we can between the ellipsoid and the dipyramid. The method for doing this comprises performing a simulation using random numbers and comparing the statistical result to the dipyramid, adjusting the parameters until a good fit is achieved. The details are in [Whitney, Gilbert, and Jastrzebski]. The results are summarized as an adjustment factor f_{opt} that is used to adjust the actual tolerance ranges when inserting them into the theoretical calculations.

Hence, in Figure 6-13, starting from the Cartesian frame attached to the middle of the datum plane F_0, the transformation leading to the frame attached to the middle of the tolerance zone F_1 contains only a translation along Z of the basic size dimension D, without any associated error, as indicated in Equation (6-12):

$$T_{01} = \begin{bmatrix} 1 & 0 & 0 & 0 \\ 0 & 1 & 0 & 0 \\ 0 & 0 & 1 & D \\ 0 & 0 & 0 & 1 \end{bmatrix} \quad (6\text{-}12)$$

Next, the nominal transformation values from the Cartesian frame attached to the middle of the rectangular tolerance zone to the middle of the actual surface F_2 are as shown in Equation (6-13):

$$T_{12}: \quad \begin{aligned} X &= 0, & \theta_x &= 0 \\ Y &= 0, & \theta_y &= 0 \\ Z &= 0, & \theta_z &= 0 \end{aligned} \quad (6\text{-}13)$$

which translates into the identity matrix transform

$$T_{12} = \begin{bmatrix} 1 & 0 & 0 & 0 \\ 0 & 1 & 0 & 0 \\ 0 & 0 & 1 & 0 \\ 0 & 0 & 0 & 1 \end{bmatrix} \quad (6\text{-}14)$$

The elements of the desired error transform DT_{12} are given in Equation (6-15):

$$\begin{aligned} dx &= 0 & \delta\theta_x &= 2 f_{opt} T_s / L_y \\ dy &= 0 & \delta\theta_y &= 2 f_{opt} T_s / L_x \\ dz &= f_{opt} T_s & \delta\theta_z &= 0 \end{aligned} \quad (6\text{-}15)$$

where f_{opt} is equal to 0.95 for this size tolerance case. The dx and $\delta\theta$ values are substituted into Equation (6-16):

$$\begin{aligned} DT_{12} &= \begin{bmatrix} \delta R & dp \\ 0^T & 0 \end{bmatrix} \\ &= \begin{bmatrix} 0 & -\delta\theta_z & \delta\theta_y & dx \\ \delta\theta_z & 0 & -\delta\theta_x & dy \\ -\delta\theta_y & \delta\theta_x & 0 & dz \\ 0 & 0 & 0 & 0 \end{bmatrix} \end{aligned} \quad (6\text{-}16)$$

6.C.3.b. Statistical Combinations of 4 × 4 Transforms

There are two ways that formulations such as those in the previous sections can be combined to predict the variations of an assembly. Both depend on numerical simulation, also called Monte Carlo. MATLAB is a convenient way to perform such calculations. The procedure is as follows:

- Determine the correct form of Equation (6-16) for the error transform DT for each feature in the chain of frames that is subject to variation.

TABLE 6-1. Values of f_{opt} and Entries into 4 × 4 Matrices Associated with Various GD&T Control Frames

TYPES	CHARACTERISTIC	CONTROL FRAME	GEOMETRY	ERRORS
Location	Position @ MMC	⊕ φ Tp Ⓜ		same as below but replace T by $T_{Cp}^2 = s_s^2 + s_p^2 + (T_p + S_{max} + E(S))^2$
	Position	⊕ φ Tp		$DX = f_{opt} T/2$ $D\theta_X = f_{opt} T/L$ $DY = f_{opt} T/2$ $D\theta_Y = f_{opt} T/L$ $DZ = 0$ $D\theta_Z = 0$
	Concentricity	○ φ Tc		
Runout	Circular Runout	↗ TR A		$T = T_p = T_c = T_R$ L = length of cylindrical tolerance zone $f_{opt} = 0.92$
	Total Runout (on surface // to datum axis)	⌖ TR A		
	Total runout (on surface ⊥ to datum axis)	⌖ TR A		$DX = 0$ $D\theta_X = f_{opt} T_R/D$ $DY = 0$ $D\theta_Y = f_{opt} T_R/D$ $DZ = f_{opt} T_s$ $D\theta_Z = 0$ $f_{opt} = 0.95$
Planar Size	Distance			$DX = 0$ $D\theta_X = 2 f_{opt} T_s/L_Y$ $DY = 0$ $D\theta_Y = 2 f_{opt} T_s/L_X$ $DZ = f_{opt} T_s$ $D\theta_Z = 0$ L_X and L_Y are length and width of tolerance zone $f_{opt} = 0.95$
Orientation	Parallelism	∥ Tpa A		$DX = 0$ $D\theta_X = 2 f_{opt} T_o/L_Y$ $DY = 0$ $D\theta_Y = 2 f_{opt} T_o/L_X$ $DZ = f_{opt} T_s$ $D\theta_Z = 0$ $T_o = T_{pa} = T_{pe} = T_a$ L_X and L_y are length and width of tolerance zone $f_{opt} = 0.92$
	Perpendicularity	⊥ Tpe		
	Angularity	∠ Ta A		

Source: [Whitney, Gilbert, and Jastzrebski].

- Assume a probability distribution for each feature variation and calculate the numerical values using Equation (6-15). (A worst-case study can be performed by using only the upper and lower worst-case limits instead of a continuous probability density function.)

- Formulate a model of the chain of frames using Equation (6-5), selecting for analysis the location of apparently critical points on the surfaces of the related parts and placing frames on those points.

- Write a MATLAB model of the chain of frames relating the frames chosen in the previous step, and run it many times with different random values of the feature variations. One may use any distribution for

the individual variations, including uniform distributions with hard limits. Rule #1 can be imposed using program logic, so that $f_{opt} = 1$ in all the equations.

- Alternately, simpler approximations to Rule #1 can be used, in which case one needs to know the values of f_{opt} to use for each kind of tolerance zone. The values to use have been generated by numerical simulation and appear in Table 6-1.

MATLAB code for both approaches is given in Section 6.K. Examples appear in Section 6.D.

The output in either case is a histogram indicating the likelihood of any particular error or combination of errors in several dimensions such as position and angle. These can be compared to the USL and LSL for the points chosen to represent the KC to determine the percent of assemblies that comply with the tolerances. A value of C_{pk} can also be calculated for each individual toleranced dimension.

In addition to Monte Carlo analysis methods, there are analytical techniques that yield closed form solutions if the individual feature errors are modeled with Gaussian distributions ([Veitschegger and Wu], [Jastrzebski]).

6.C.4. Remark: Constraint Inside a Part

We stated in Chapter 4 that the 4×4 matrix method of relating part locations to each other is valid only when the parts are kinematically constrained. This requirement also holds for features: When we use 4×4 matrices to relate features to each other inside a part, those features must be kinematically constrained with respect to each other. This condition can be satisfied if stresses inside a part are too low to cause meaningful deformation. In addition, each feature should be located by a unique and single valued nominal chain of frames. Violations of this latter condition are sometimes called overdimensioning or overtolerancing. Some brands of commercial tolerancing software can detect this error but it is usually an error in the nominal design, not a tolerancing error.[1]

6.D. EXAMPLES

In this section we present four examples. The first illustrates inserting error transforms into nominal transform models of an assembly. The second shows how to calculate assembly process capability. The third introduces fixtures and their errors into the assembly transforms. The last applies worst-case and statistical tolerancing to a three-dimensional analysis of car door assembly.

6.D.1. Addition of Error Transforms to Nominal Transforms

Here are some examples of formulation of varied transform equations, which are extensions of the examples in Chapter 3. These examples also stress the importance of the order in which the transforms are multiplied.

The first example, in Figure 6-18, shows two ways to handle the displacement of a frame (or a feature). One way uses base frame coordinates while the other uses the displaced frame's original coordinates.

The second example involves a rotation error. We begin by misorienting a hole in a part, as shown in Figure 6-19. The figure shows how to calculate the relationship between a point F on the part and the new hole frame E′ if we know T_{EF}, the nominal error free relationship between the original hole frame E and frame F.

The third example shows what happens when the part in Figure 6-19 is assembled to the part from Figure 6-18, as shown in Figure 6-20. The final varied transform T'_{AF} is not shown here. A thought question at the end of the chapter asks the reader to derive this equation and calculate the worst-case values.

6.D.2. Assembly Process Capability

In Chapter 10 we show that successful assembly requires that lateral and angular errors between parts must be kept within certain limits in order to avoid wedging. Figure 6-21 repeats the necessary information from Chapter 10.

In order to know if an assembly is likely to succeed according to these conditions, we need to calculate or estimate the initial lateral error ε_0 and the angular error θ_0. These errors are the result of composing a number of 4×4 transforms comprising a loop that includes the previously assembled parts plus all assembly tooling, fixtures, and equipment, as shown in Figure 6-22.

Figure 6-23 shows schematically the way to calculate the probability that assembly will fail due to wedging or missing the chamfer as a result of accumulated variation in the parts or tooling. If the ellipsoid represents 3σ or 97.1% of all events, then the leftover volume of the ellipsoid outside the parallelogram represents the probability of failure. The process capability C_{pk} may be calculated (in principle, at least) by letting W and c/μ represent the specification limits and the ellipsoid represent the process, and making a calculation analogous to Equation (5-12).

This example illustrates a number of points. First, chains of frames may extend through tools and fixtures as well as parts of an assembly. The relative locations of all of

---

[1] In Chapter 8 we will return to the notion of the "nominal design" and carefully distinguish it from "variation design."

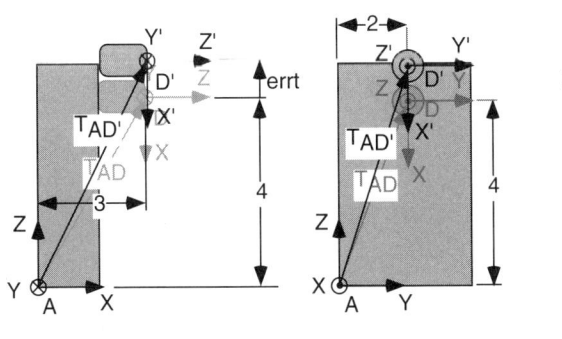

$$\gg T_{AD} = trans\,(3,2,4)\,roty\,(dtr\,(90))$$

$$T_{AD} = \begin{bmatrix} 0 & 0 & 1 & 3 \\ 0 & 1 & 0 & 2 \\ -1 & 0 & 0 & 4 \\ 0 & 0 & 0 & 1 \end{bmatrix}$$

% First method
$$\gg DZ = errt$$
$$\gg DT_{AD1} = trans\,(0,0,DZ\,)$$
$$T_{AD'1} = DT_{AD1}T_{AD}$$
% Second method
$$\gg DX = errt$$
$$\gg DT_{AD2} = trans\,(-DX,0,0)$$
$$\gg T_{AD'2} = T_{AD}DT_{AD2}$$
% $T_{AD'2} = T_{AD}T_{DD'}$

FIGURE 6-18. Illustrating How to Calculate the Varied Transform $T_{AD'}$ from Part A's Coordinate Center to Feature Frame D When Frame D Is Displaced to D' Without Rotation. Two methods are shown at the right in MATLAB notation. (Comment lines in MATLAB begin with "%.") The methods are preceded by the equation for T_{AD} when there is no error, called the nominal equation. Each method of expressing the position of frame D after displacement involves inserting an error transform into the original nominal equation. (A "D" leading off the name of a transform indicates that it is an error transform that will be applied to the transform whose name follows the D.) The first method expresses the displacement (an amount "errt") in frame A. Thus the error transform DT_{AD1} is inserted ahead of the transform T_{AD} that expresses frame D in frame A coordinates. The transform equation can be read to say: "To go from A to D', insert a Z error into A and then go on to D as before." The second method expresses the displacement in frame D coordinates. Thus the error transform DT_{AD2} is inserted after T_{AD} in the second equation. Its equation can be read to say: "To go from A to D', first go to D and then insert a $-X$ error into D."

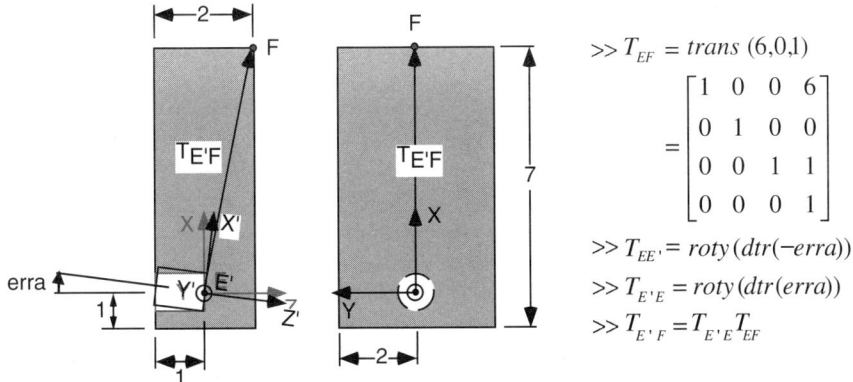

$$\gg T_{EF} = trans\,(6,0,1)$$

$$= \begin{bmatrix} 1 & 0 & 0 & 6 \\ 0 & 1 & 0 & 0 \\ 0 & 0 & 1 & 1 \\ 0 & 0 & 0 & 1 \end{bmatrix}$$

$$\gg T_{EE'} = roty\,(dtr\,(-erra))$$
$$\gg T_{E'E} = roty\,(dtr(erra))$$
$$\gg T_{E'F} = T_{E'E}T_{EF}$$

FIGURE 6-19. This Part Has a Misoriented Hole. Frame E' represents the hole's varied orientation. Transform $T_{EE'}$ (from E to E') indicates that a negative rotation about frame E's Y axis has occurred. Transform $T_{E'E}$ (from E' to E) is a rotation in the opposite direction. Transform $T_{E'F}$ (from E' to F) shows how to get from the misoriented hole to point F. This transform can be read to say: "To go from E' to F, go from E' to E and then from E to F."

these items are described equally well by chains of frames represented as 4×4 transforms. As long as every part is properly constrained with respect to the parts it joins, or properly constrained with respect to the fixture it lies in, then these frames will properly represent the locations. Second, we can analyze both the parts of the assembly and the elements of the tools and fixtures to determine how their respective performance will be affected by variation. Third, we can define a quantity called *assembly process capability* and can calculate it by combining the error propagation methods described here with Part Mating Theory from Chapter 10.

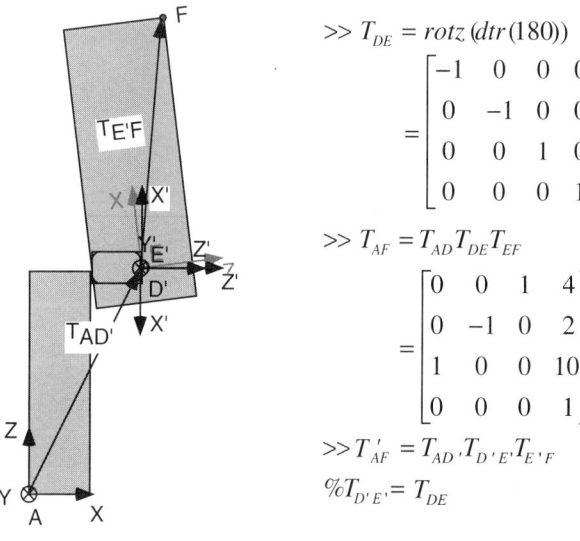

$$\gg T_{DE} = rotz\,(dtr(180))$$

$$=
\begin{bmatrix}
-1 & 0 & 0 & 0 \\
0 & -1 & 0 & 0 \\
0 & 0 & 1 & 0 \\
0 & 0 & 0 & 1
\end{bmatrix}$$

$$\gg T_{AF} = T_{AD}\,T_{DE}\,T_{EF}$$

$$=
\begin{bmatrix}
0 & 0 & 1 & 4 \\
0 & -1 & 0 & 2 \\
1 & 0 & 0 & 10 \\
0 & 0 & 0 & 1
\end{bmatrix}$$

$$\gg T'_{AF} = T_{AD'}\,T_{D'E'}\,T_{E'F}$$

$$\%T_{D'E'} = T_{DE}$$

FIGURE 6-20. The Part with the Mislocated Peg in Figure 6-18 Is Assembled to the Part with the Misoriented Hole in Figure 6-19. Assembly occurs by placing frame D' of part A directly onto frame E' of part B. Transform T'_{AF} tells in frame A coordinates where point F is as a result of including both peg location and hole orientation errors. The last equation can be read to say: "To go from A to F, first go from A to D', then from D' to E', then from E' to F." Because we put D' onto E' when we assembled the parts, the interface transform $T_{D'E'}$ is the same as T_{DE}.

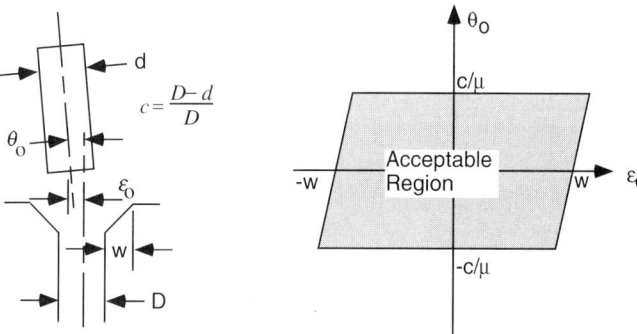

$$c = \frac{D-d}{D}$$

FIGURE 6-21. Wedging Conditions for Assembling Round Pegs and Chamfered Holes. On the left is a simplified model of peg–hole assembly. D and d are hole and peg diameters, respectively. ε_0 and θ_0 are initial lateral and angular error of the peg with respect to the hole. W is the width of the chamfer, μ is the coefficient of friction, and c is the clearance ratio, defined in the figure. On the right is a graph showing values of ε_0 and θ_0 that permit successful assembly, avoiding wedging the parts or a collision outside the chamfer.

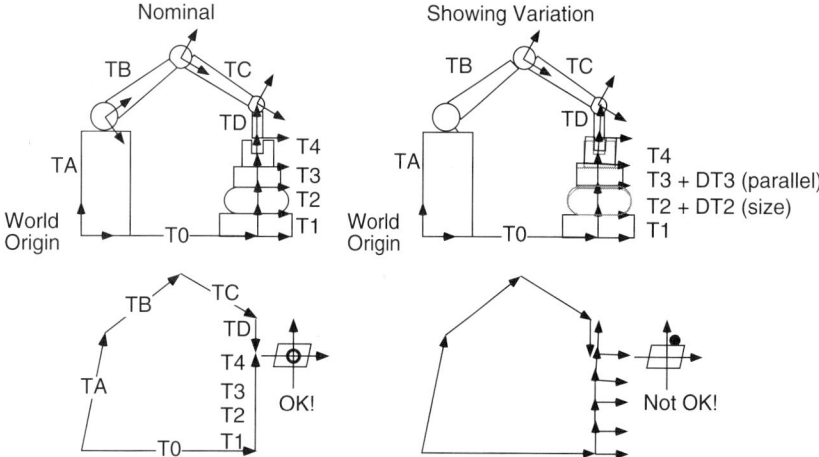

FIGURE 6-22. Illustration of Assembly Process Capability. *Top left:* A robot puts a peg in a hole on a set of assembled parts. The chain of frames at the bottom left TA–TD describes the nominal location of the tip of the next part to be assembled, while the chain of frames T1–T4 describes the nominal location of the receiving part. Transform T0 links these two chains. *Bottom left:* The nominal design is correct, so that the chains meet and the errors in position and angle fall inside the wedging diagram, as indicated by the open circle. On the right there are some errors in the fabrication of the parts so that the chains of frames do not meet exactly. The resulting lateral and angular errors are shown schematically as a black dot just outside the wedging diagram. Not shown, but also possible, are errors in frames TA–TD representing robot errors, along with an error in T0 representing calibration or other errors that misplace the assembly fixture in robot coordinates.

154

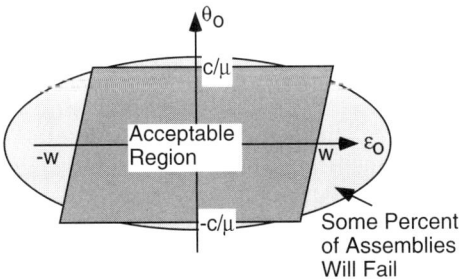

FIGURE 6-23. Combination of Wedging Conditions and Probability Ellipsoid of Position and Angle Error. The area of the ellipse covered by the parallelogram represents the probability that assembly will not fail due to wedging.

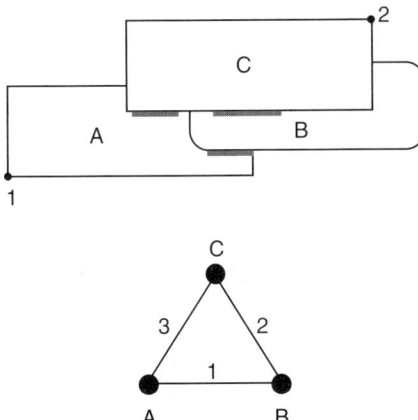

FIGURE 6-24. Three Planar Parts Assembled by Welding, and Their Liaison Diagram. The KC is the relative location of point 1 on part A and point 2 on part C. The thick shaded lines represent welds.

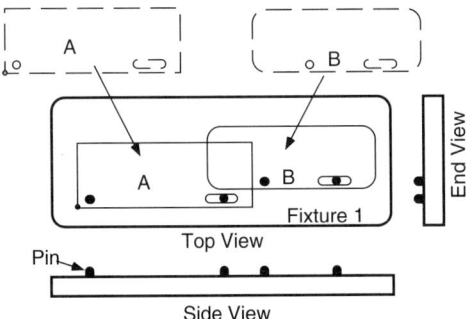

FIGURE 6-25. First Step in the Assembly, Joining Parts A and B Using Fixture 1. Parts A and B are placed in the fixture using pin–hole and pin–slot features. Then they are welded together. The fixture is shown in heavy lines. The state of the parts before they are put on the fixture is shown in dashed lines.

6.D.3. Variation Buildup with Fixtures

In the previous section we looked at error buildup in an assembly and its effect on assembleability of the next part. In this section we look at how errors build up when more than one fixture is used. There are many ways to design an assembly process using fixtures. Some of these are better than others. For example, the fixtures may actually overconstrain the parts, a point that underlies one of the thought questions at the end of the chapter. Another example is studied here, namely different ways that the parts can be fixtured, especially when the assembly consists of several parts, the KC is measured across parts that are not adjacent to each other, and several fixtures are used one after the other to build up the assembly.

Someone has proposed a process for assembling the planar sheet metal parts shown in Figure 6-24. Parts A and B are welded together using fixture 1, and the subassembly of A and B is then moved to fixture 2 in order that part C may be welded on. The KC in question is the relative location of a point on part C with respect to one on part A. The parts in question do not pass constraint or location to each other. Their relative positions and angles are set entirely by the fixtures. We will see as we look at this proposed process that it is not the optimum way to accomplish the assembly. The thought questions at the end of the chapter ask you to consider many alternative fixturing arrangements.

The first step in the proposed assembly process is shown in Figure 6-25, in which parts A and B are joined on fixture 1. The second step is shown in Figure 6-26, in which the subassembly A-B is carried to fixture 2 and joined there to part C. Fixture 2 locates the subassembly using features on part B.

Figure 6-27 uses coordinate frames to show what happens while assembling these parts. Fixture F1 locates parts A and B relative to each other, while fixture F2 locates parts B and C relative to each other.

A coordinate frame representation of the complete assembly and the KC appears in Figure 6-28. It is constructed by placing the two frames labeled "B" in Figure 6-27 on top of each other. The figure shows that, in order to find the relative location of the points on parts A and C that constitute the KC, we need to trace a chain of frames between these points that includes both fixtures. This does not mean that we have to account for the relative location of the fixtures with respect to each other on the factory floor. We can see this because there is no direct chain link

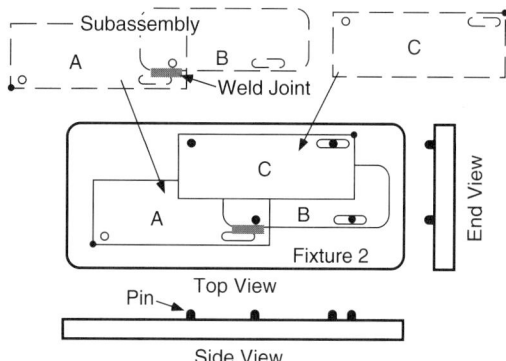

FIGURE 6-26. The Second Step in the Assembly, Adding Part C to the Subassembly of Parts A and B, Using Fixture 2. The weld joint between parts A and B is shown as a thick shaded line. The fixture locates subassembly AB using features on B.

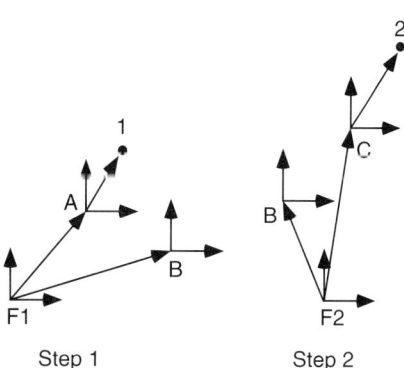

FIGURE 6-27. Coordinate Frame Representation of the Two-Step Assembly of Parts A and B Using Fixtures F1 and F2.

between these two frames in Figure 6-28. What we must do is account for the error that fixture 1 introduces between parts A and B as well as the error that fixture 2 introduces between parts B and C, plus the errors inside each part between the KC points and the features used for fixturing.

Note that this assembly plan locates the first assembly operation by means of features on parts A and B while the second step's operations are done by locating the subassembly using features on parts B and C. In cases like this, we say that a *datum transfer* or *datum shift* has occurred because the second fixture uses different part features than the first fixture does. If fixture 2 located the subassembly using the same part A features that fixture 1 used, then there would be no datum shift and the chain links between fixture 1 and fixture 2 would not appear in Figure 6-28. In fact, neither fixture 1 nor part B would even appear in Figure 6-28! One of the thought questions at the end of the chapter asks for a drawing of the chain under those circumstances.

Consider the instance where the subassembly of A and B is built by a supplier using fixture 1 while C (or a subassembly more complex than just one part) is made by another supplier. Now consider the problem faced by the final assembler who buys these subassemblies and puts them together using fixture 2. If the KC is not achieved, the final assembler must be aware of the entire chain in Figure 6-28 in order to carry out an effective diagnosis of the problem. If the suppliers are far apart, the "length" of this chain could be hundreds or thousands of miles. On the other hand, if step 2 used the features on part A, the final assembler would have an easier diagnosis problem because most of the chain would be contained within his plant. Only that part of the chain representing errors within part A would be outside his plant.

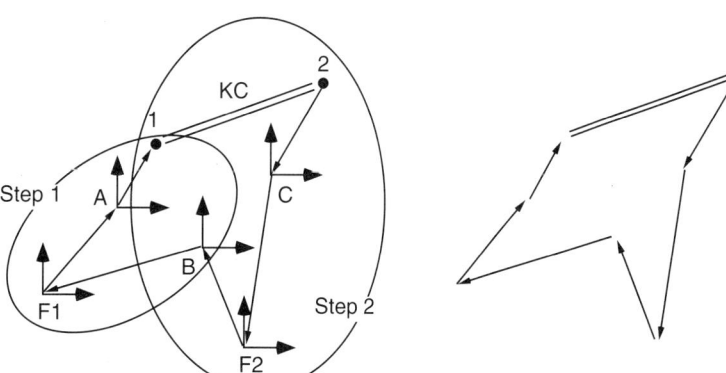

FIGURE 6-28. *Left:* A chain of frames joins the ends of the KC. Steps 1 and 2 are indicated by ellipses. Only frame B is in both ellipses. *Right:* For clarity, the arrows representing the 4 × 4 transforms in the chain are shown separately.

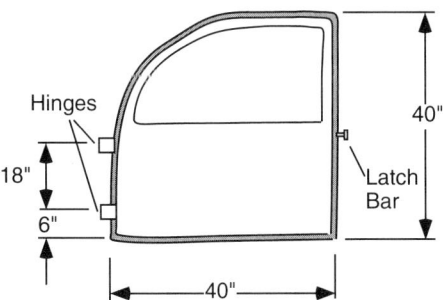

FIGURE 6-29. Car Door Dimensions. These are typical dimensions, taken from the author's car.

6.D.4. Car Doors

In this section we will do some examples that illustrate the following:

- The difference between worst-case and statistical tolerancing assumptions
- The difference between uniform and Gaussian or normal statistics

The MATLAB files that support these examples are on the CD-ROM that is packaged with this book.

Consider the car door sketched in Figure 6-29. We would like to know the effect on the location (position and orientation) of the door in three dimensions of mislocating the hinges on either the door or the car body frame. To do this, we need to define the KC and the relevant dimensions. These are shown in Figure 6-30. The hinges are positioned on the door at coordinate locations shown in this figure but are assumed possibly mislocated in dimensions Y and Z with respect to frame 0, which is the door's base coordinate frame. Errors with respect to X

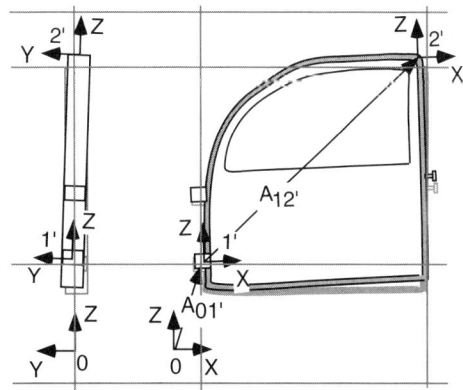

FIGURE 6-31. Example of the Effect on Door Position and Orientation Due to Misplacement of the Hinges. The door is tilted clockwise in the Y–Z plane and counterclockwise in the X–Z plane. It is also lifted along Z. The door's nominal position and orientation are shown in gray while the varied door is shown in black. Some horizontal and vertical grid lines have been added to help make the variation easier to see.

are most likely to occur when the door is mounted to the car body but are modeled below in MATLAB as though they occur when the hinges are mounted to the door.

To perform the analysis, we assume that the two hinges comprise one compound feature as defined in Section 6.B.2. The origin of this feature is the lower hinge whose frame a is nominally located at frame 1, while the other feature component of the compound feature is the upper hinge located at frame b. The tolerance on each hinge's location in X, Y, and Z is assumed to be ±4.5 mm or ±0.1771".

Figure 6-31 shows the door out of position and orientation due to an example set of misplaced hinges.

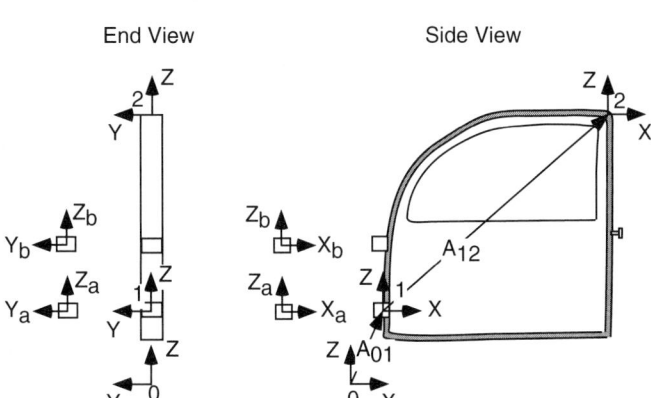

FIGURE 6-30. Coordinates and KC for a Car Door. The KC is the length of the vector joining the origin of the nominal frame 2 and varied origin of frame 2. Frame 0 is the door's base coordinate frame. Frame 1 is the nominal location of the lower hinge, which anchors the compound feature comprising the two hinges. The actual location of the lower hinge is frame a while the actual location of the upper hinge is frame b. For clarity, frames a and b are shown to one side of the two views of the door.

TABLE 6-2. MATLAB Code for Worst-Case Analysis of Door Variation

```
%door_main_worst
%Door Main Program for Worst Case
door_nominal
ERR_MAX=0;
for jj=1:64
    VERRW(jj)=0;
    end
q=0;
for i=0:1;
    for j=0:1;
        for k=0:1;
            for l=0:1;
                for m=0:1;
                    for n=0:1;
                        V=[(-1)^i(-1)^j(-1)^k(-1)^l(-1)^m(-1)^n];
                        q=q+1;
                        door_dev;
                        door_errs;
                        door_act;
                        DT;
                        q;
                        ERR;
                        VERRW(q)=ERR;
                        if ERR>ERR_MAX
                            ERR_MAX=ERR;
                            is=i;
                            js=j;
                            ks=k;
                            ls=l;
                            ms=m;
                            ns=n;
                            qs=q;
                            end
                    end
                end
            end
        end
    end
end
is
js
ks
ls
ms
ns
ERR_MAX
```

TABLE 6-3. Supporting Routines for Worst-Case Door Variation Calculation

```
%door_nominal
%nominal location of door coordinate frames

A1=[1 0 0 5;0 1 0 0;0 0 1 11;0 0 0 1];
A2=[1 0 0 40;0 1 0 0;0 0 1 34;0 0 0 1];

%door_dev
%calculate deviations

E=4.5/25.4;
dx1=E*V(1);
dx2=E*V(2);
dy1=E*V(3);
dy2=E*V(4);
dz1=E*V(5);
dz2=E*V(6);

%door_errs
%door error matrices
dA1=[0 0 0 dx1;0 0 0 dy1;0 0 0 dz1;0 0 0 0];
ddx2=-(dy2-dy1)/18;
ddy2=(dx2-dx1)/18;
ddz2=0;
dA2=[0 -ddz2 ddy2 0;ddz2 0 -ddx2 0;-ddy2 ddx2 0 0;0 0 0 0];

%door_act
%actual door position
T=(eye(4)+dA1)*A1*(eye(4)+dA2)*A2;
DT=T-A1*A2;
ERR=V_L(DT(1:3,4));

%V_L
%calculate length of vector
function vector_length=V_L(x)
vector_length=sqrt(x(1)^2+x(2)^2+x(3)^2);
```

First we will perform a worst-case analysis. For this purpose we assume that each hinge could be at either extreme of its allowed range. With two hinges and three $\pm$ dimensions for each, we have a total of 64 cases to look at. The approach taken is a MATLAB simulation. The main code is in Table 6-2. It is not very imaginative: it simply enumerates all 64 cases.

The supporting routines are given in Table 6-3.

The results are shown in Figure 6-32. All 64 cases appear, of which 8 are identical and worst, with an error of 1.1894".

If a statistical analysis is done instead, we have our choice of probability distributions for the individual errors. If we choose Gaussian, we need to assign a standard deviation. For this purpose, we use $3\sigma = 0.1771$. A sample histogram of individual hinge position errors calculated this way appears in Figure 6-33. Figure 6-34 shows an example uniform distribution of individual hinge location errors analogous to Figure 6-33.

The MATLAB code for conducting the statistical analysis of assembled door variation, given Gaussian or uniformly distributed individual hinge location errors, is given in Table 6-4 and Table 6-5. The results are given in Figure 6-35 for Gaussian and in Figure 6-36 for uniform.

Several things are worth noting about this example. First, it is a full three-dimensional analysis. It is not a

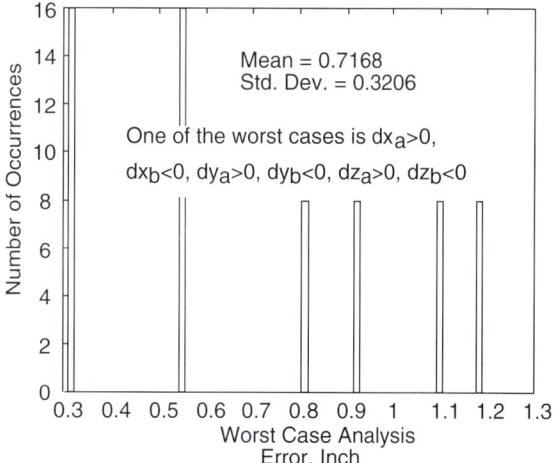

FIGURE 6-32. Histogram of Worst-Case Analysis of Door Variation. All 64 cases appear, and 8 (12.5%) of these have the same worst value of over an inch! One of these 8 is listed in the figure. There, the notation ">0" means that the individual error is at the positive maximum end of the range, namely 0.1771, while the notation "<0" means that the individual error is at the negative maximum end of the range, namely −0.1771.

simple RSS analysis of a linear stack of tolerances. As such, it is difficult to compare worst-case and statistical error accumulation using an analysis like that in Equations (5-23) and (5-24). The best way to make that comparison is to compare Figure 6-32 with Figure 6-35 and Figure 6-36. This comparison indicates that the statistical analysis (whether using Gaussian or uniform distributions of individual feature errors) gives a much smaller range of predicted assembly-level errors and a much smaller maximum value, based on a simulation using 10,000 samples. It is also interesting that the uniform distribution assumption for individual hinge errors, while allowing many more large individual errors, nonetheless gives similar results to the Gaussian analysis, reaffirming our assumption, discussed in Chapter 5, that the sum of several random variables of any distribution tends toward a Gaussian distribution.

Note that a combined vector error of over an inch is huge. It could be argued that typical automobile door fabrication and hinge placement can easily avoid 3σ errors of 0.177". Nevertheless, 0.177" is not large compared to the size of the door. The reason such a small error has such a huge effect is that it enables an angular error which in turn has a lever arm over 40" long in which to operate. Even if we were able to cut the individual feature placement errors by two-thirds to 0.06" corresponding to 1.5 mm (close to the minimum feasible), the resulting error at the

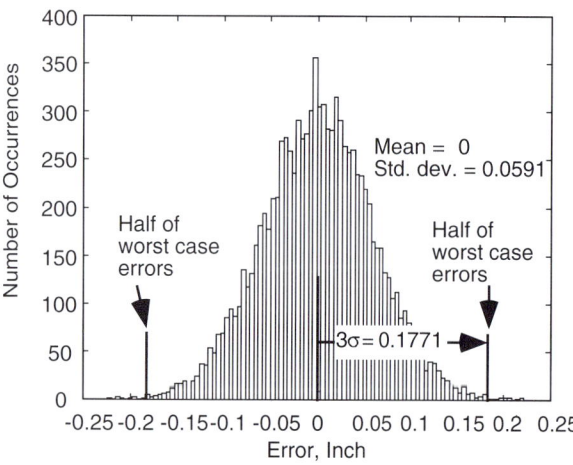

FIGURE 6-33. Sample Histogram of Gaussian Random Individual Hinge Location Errors for Statistical Analysis of Door Variation, Based on 10000 Trials. The figure shows that nearly all the individual errors will fall within the bounds of the worst-case values, which are shown for reference.

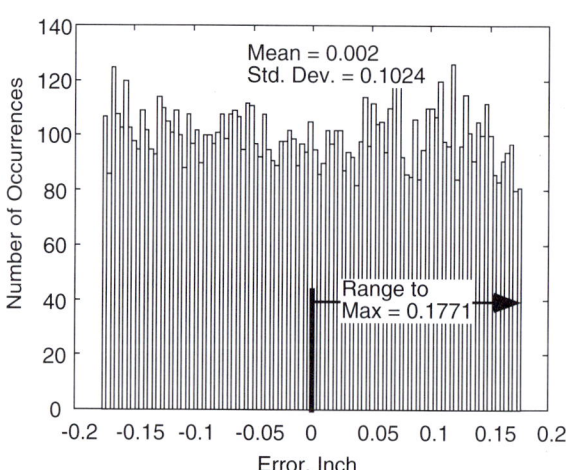

FIGURE 6-34. Sample Histogram of Uniform Random Hinge Location Errors for Statistical Analysis of Door Variation, Based on 10,000 Trials.

opposite corner of the door would still be approximately 0.33", which is unacceptably large. For this reason, door fit accuracy is not achieved by raw control of hinge location errors but instead makes use of a variety of clever fixtures

and hinging and door mounting techniques mentioned in Chapter 2 and discussed in more detail in Chapter 8.[2]

Finally, all these calculations assume no mean shifts. If there are mean shifts, then the errors will be much worse.

TABLE 6-4. MATLAB Code for Statistical Analysis of Door Variations

```
%Door Main Program for Statistical Case
%In line 8, call door_dev_statu for uniform random errors
%In line 8, call door_dev_statg for gaussian random errors
door_nominal
for jj=1:10000
   VERRS(jj)=0;
   end
for kk=1:10000
   door_dev_statu; %door_dev_statg
   door_errs;
   door_act;
   ERR;
   VERRS(kk)=ERR;
end
```

Note: Supporting routines are given in Table 6-3 and Table 6-5.

TABLE 6-5. Supporting MATLAB Routines for Statistical Analysis of Door Variations

```
%door_dev_statu
%statistical door errors, uniform distribution
E=4.5/25.4;
dx1=2*E*(rand-.5);
dx2=2*E*(rand-.5);
dy1=2*E*(rand-.5);
dy2=2*E*(rand-.5);
dz1=2*E*(rand-.5);
dz2=2*E*(rand-.5);

%door_dev_statg
%statistical door errors
E=4.5/(3*25.4);
vrn=E;
dx1=randn*vrn;
dx2=randn*vrn;
dy1=randn*vrn;
dy2=randn*vrn;
dz1=randn*vrn;
dz2=randn*vrn;
```

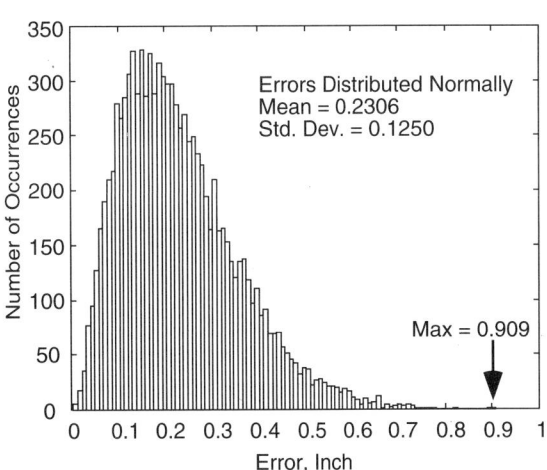

FIGURE 6-35. Histogram of Statistical Analysis of Door Variation Using Gaussian Distribution of Individual Feature Variations Based on 10,000 Trials Using the Code in Table 6-4. This error is the length of the three-dimensional vector from the nominal XYZ position of point 2 on the door (shown in Figure 6-30) to its varied position.

[2]Sometimes, in addition, a person known respectfully as "Big Mike" makes some final adjustments.

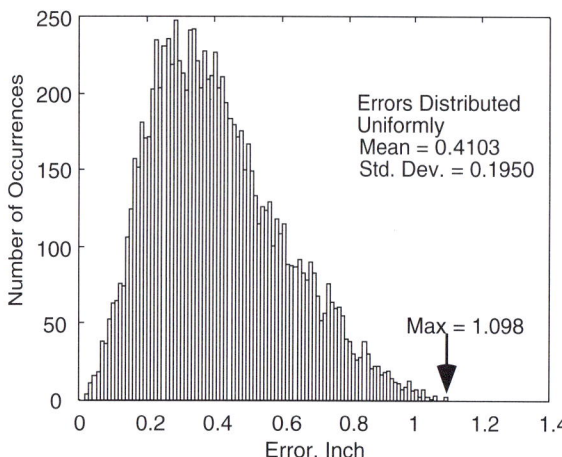

FIGURE 6-36. Histogram of Statistical Analysis of Door Variation Using Uniform Distribution of Individual Feature Variations.

6.E. TOLERANCE ALLOCATION

Tolerances are applied to parts so that the final assembly will achieve its KCs. In most cases, only the sum of part variations over the parts in each assembly matters. For this reason, the amount of variation tolerable from each part is to some degree a decision that the designer can make, as long as the total variation is within limits. This decision is called *tolerance allocation*. The models and examples in Section 6.B are based on assigning the same tolerance to each dimension, but this is neither necessary nor desirable. The typical approach to this problem, other than guessing or giving each part the same tolerance, is to find the minimum cost solution. Another approach, one that is consistent with the theory in this book, is to allocate the tolerances so that the assembly-level dimension achieves some C_{pk} and that the individual part dimensions do so as well. We will briefly discuss each of these approaches in the next two subsections.

6.E.1. Tolerance Allocation to Minimize Fabrication Costs

Tolerance allocation to minimize fabrication costs has been extensively studied by academic researchers. The discussion that follows is based on [Chase, Greenwood, Loosli, and Hauglund], which contains a survey of tolerance cost models.

The basic idea is that cost rises as tolerances get smaller. The reasons include requiring a more expensive machine, requiring more process steps, changing tool bits or measuring more often, paying a more highly skilled operator, taking more time, or scrapping (or reworking) more parts.

Among cost models that have been used are the following:

$$C = A - BT \qquad \text{(Linear)}$$

$$C = A + B/T \qquad \text{(Reciprocal)} \qquad \text{(6-17)}$$

$$C = A + B/T^k \qquad \text{(Reciprocal power)}$$

where A is a fixed cost, B is a tolerance cost factor, and T is the tolerance.

The problem is posed as one of choosing n tolerances T_i, $i = 1, \ldots, n$, for parts involved in delivering a KC so as to minimize the total cost of making the parts while satisfying the constraint that the total variation equal a certain amount. The total variation can be modeled as accumulating according to statistical tolerancing assumptions or worst-case assumptions. The typical way to do this for the statistical accumulation case is the method of Lagrange multipliers:

$$\text{Total cost} = \sum f(\bar{T}), \quad \text{where} \quad \bar{T} = [T_1, T_2, \ldots, T_n] \qquad \text{(6-18a)}$$

$$\text{Total variation} = T_A^2 = \sum_{i=1}^{n} T_i^2 \qquad \text{(6-18b)}$$

Constrained minimization problem:

$$\min_{T_i} C = \sum f(\bar{T}) + \lambda \left(T_A^2 - \sum_{i=1}^{n} T_i^2 \right), \qquad \text{(6-18c)}$$

where λ = the Lagrange multiplier

This is solved by differentiating with respect to T_i and solving for λ. This expression is substituted into the total variation constraint to obtain the individual T_i. The result is

$$\lambda = \frac{\frac{\partial f(\bar{T})}{\partial T_i}}{2T_i} = \frac{\frac{\partial f(\bar{T})}{\partial T_1}}{2T_1} \qquad \text{for all } i \qquad (6\text{-}19)$$

$$T_i = \frac{\frac{\partial f(\bar{T})}{\partial T_i}}{\frac{\partial f(\bar{T})}{\partial T_1}} T_1 \qquad (6\text{-}20)$$

$$T_A^2 = \sum_{i=1}^{n} \left[\frac{\frac{\partial f(\bar{T})}{\partial T_i}}{\frac{\partial f(\bar{T})}{\partial T_1}} \right]^2 T_1^2 \qquad (6\text{-}21)$$

The procedure is to solve Equation (6-21) for T_1 and substitute it into Equation (6-20) for each value of i to obtain the other T_i. This method works for the case where the assembly tolerance is wanted exactly rather than as an upper limit. In addition, the cost functions must be differentiable. Moreover, if different process options are available for different ranges of tolerances, a different method that involves search must be used. Several alternatives are discussed in [Chase, Greenwood, Loosli, and Hauglund].

6.E.2. Tolerance Allocation to Achieve a Given C_{pk} at the Assembly Level and at the Fabrication Level

In this method, developed in [Terry], we can again use either statistical tolerancing or worst-case tolerancing, but statistical tolerancing is assumed in the discussion

that follows. Here, the basic idea is that customer requirements dictate some upper and lower specification limits USL_{KC} and LSL_{KC} for some KC. We assume that we have a model for how the KC will be distributed statistically, so that we can calculate a $C_{pk\text{KC}}$ for it.

We start by assigning a machine to fabricate each feature and determining, from history or experiments, what variance the machine can achieve while fabricating that feature. After obtaining variance data for each machine–feature combination, we must assign a USL and an LSL to each feature (this is the tolerance allocation step) and use the achievable variances to calculate the C_{pkFi} of each feature F_i. Our goal is to have each feature under control and capable as well as to have the KC under control and capable. This typically means achieving $C_{pkFi} = 1.33$ for all the features and $C_{pk\text{KC}} = 1.33$ for the KC. A search algorithm is required to find the appropriate USLs and LSLs.

Several outcomes are possible:

$C_{pk} = 1.33$ or better can be achieved for the KC and each feature with the given machines by assigning the features to machines appropriately (which may require a search of its own).

$C_{pk\text{KC}}$ near 1.33 can be achieved, but 100% inspection might be necessary if one or more processes has $C_{pkFi} = 1.00$ or less.

Some features might have to be assigned to different machines with smaller variances in order to achieve $C_{pk\text{KC}} = 1.33$. In this case, an additional search must be conducted to find the best assignments of features to

TABLE 6-6. Tolerance Allocation Process to Achieve Desired C_{pk} for Each KC and Contributing Feature

Key Tasks and Substeps	Quantitative Evaluation Criteria	Results	Analytical Tools Used, References, and Comments
1 **Define the key characteristic (KC) and its design limits.**		Fully specified KC.	
1.1 Determine the relationship between the KC and product performance.	Efficiency, capacity, other quantified performance measure	Specification limit. For clearances, this is most often the USL.	May require designed experiments, engineering models, etc.
1.2 Determine the failure condition.	Mechanical interference, excessive stress, etc.	Specification limit. For clearances, this is most often the LSL.	Failure modes and effects analysis
2 **Determine mathematical relationship between the KC and its component assembly key characteristics (AKC).**	$Y = f(X_1, X_2, X_3, \ldots, X_m)$ $Y =$ the top-level KC $X_i =$ contributing feature parameters or lower-level KCs	Relationship between the KC and each of its component AKCs.	Engineering analysis
2.1 Determine component AKCs.	Depends on datum structure and feature or dimension location on part	List of contributing features	Engineering analysis
2.2 Calculate sensitivity of KC to each of its AKCs.	$S_i = \frac{\partial Y}{\partial X_i}$	Sensitivity of the KC to each of its AKCs	Engineering analysis

(continued)

TABLE 6-6. (Continued)

Key Tasks and Substeps	Quantitative Evaluation Criteria	Results	Analytical Tools Used, References, and Comments
2.3 If required, modify design to minimize sensitivity to a single characteristic, or to characteristics with large expected variation.			May require different design solution, or adjusting datum structure. This process is not addressed here, but is a crucial part of the robust design process.
3 **Quantify the statistical distribution for each AKC.**		Expected statistical distribution for each AKC	For the purposes of this analysis, distributions are assumed to be normal.
3.1 Obtain statistical summary data for each AKC.	$$\bar{X}_i = \sum_{j=1}^{n} \frac{x_j}{n},$$ $$\sigma_{X_i} = \sqrt{\frac{\sum_{j=1}^{n}(x_j - \bar{X}_i)^2}{n-1}}$$	Mean and standard deviation for each AKC. n = the number of measurements being used. x_j = an individual measurement from the population n.	Data from the actual production process should be used when possible. When it is not available, the mean may be estimated as the nominal dimension and the standard deviation may be estimated from a similar characteristic in a similar production environment. Accuracy of this information determines the quality of the model.
4 **Statistically allocate tolerances to maximize manufacturability.**			
4.1 Statistically combine contributing AKCs to get the expected mean and standard deviation values of the KC.	$$\bar{Y} = \sum_{i=1}^{m}\left(S_i\ \bar{X}_i\right),$$ $$\sigma_Y = \sqrt{\sum_{i=1}^{m}\left(S_i \sigma_{X_i}\right)^2}$$	Fully specified statistical distribution of the KC.	Excel, statistical analysis software
4.2 Evaluate the expected capability of the KC.	$$C_{pkY} = \frac{(USL_Y - abs(\bar{Y} - target))}{3\sigma_Y}$$		USL_Y is determined from the performance requirements in step 1.1.
4.3 Assign preliminary upper specification limits (USL) to each of the AKCs and calculate the standard deviation required to achieve a C_{pkX_i} of 1.33.	$$\sigma_{X_{i*}} = \frac{USL_{X_i} - abs(\bar{X}_i - target)}{4}$$ $$\sigma_{Y*} = \sqrt{\sum_{i=1}^{m}\left(S_i \sigma_{X_{i*}}\right)^2}$$	Starting assumptions for iterative optimization process. These calculated values are distinguished from the measured values by the asterisk.	
4.4 Estimate the C_{pk} of each AKC using the USL_X above.	$$C_{pkY*} = \frac{(USL_Y - abs(\bar{Y} - target))}{3\sigma_{Y*}}$$ $$C_{pkX_i} = \frac{(USL_{X_i} - abs(\bar{X}_i - target))}{3\sigma_{X_i}}$$		The derived standard deviation is used to evaluate Y as a baseline. The empirical standard deviation is used to evaluate X's to allow allocation of tolerance according to capability.
4.5 Using the specification limits on the KC from step 1 as a constraint, iteratively adjust the USL_X for each AKC to maximize the individual C_{pk}s without exceeding a constraint condition.	Adjust each USL_X from 4.4 above to maximize C_{pkX_i} while maintaining C_{pkY*} above 1.33. C_{pkY} will be greater than 1.33 as long as the following relation is true: $\sigma_{Y*} \leq \frac{USL_Y - \bar{Y}}{4}$	An optimized result yields the largest possible tolerances on the individual characteristics while ensuring performance of the design. If the design constraint is related to the LSL, symmetry may be used to obtain USL.	This step can (and should) be automated through the use of an optimizing routine like "Excel Solver" or tolerance optimization software package like "CE Tol."

machines, inside of which another search must be conducted to find the best USL and LSL for each feature.

Better machines might have to be purchased. In this case, the additional search must be over possible variances to find the largest variance (presumably lowering the cost of the machine) that is small enough to achieve the desired C_{pk}. The new machine(s) must be capable of achieving that variance.

In both the least cost and target C_{pk} methods, it is also necessary to have enough machines available to make all the features on all the parts at the desired production rate. This represents an additional design problem that is discussed in the context of assembly in Chapter 16.

Table 6-6 presents Terry's procedure for allocating tolerances to achieve given levels of C_{pk} at the KC and individual feature level simultaneously. Terry implemented his method using the Solver in Excel. The spreadsheet and a detailed description of the technique are on the CD that is packaged with this book.

6.F. VARIATION BUILDUP IN SHEET METAL ASSEMBLIES

Most of the theory and examples in this book deal with rigid parts. However, most of the principles involved also apply to sheet metal and other compliant parts as long as we are careful not to overlook the effects of stress on the shape of the parts. Sheet metal parts differ from typical rigid parts not only because they are more compliant but also because they cannot typically be made as accurately as machined parts or parts molded from relatively rigid polymers. Yet sheet metal assemblies must often meet the same percent tolerances (10^{-3} inches/inch or smaller) as machined ones do. This section therefore deals briefly with some of the technical issues presented by sheet metal parts. More detail may be found in [Hu], [Chang and Gossard], [Ceglarek and Khan], [Ceglarek and Shi], and [Cai, Hu, and Yuan].

6.F.1. Stress–Strain Considerations

Sheet metal parts are formed (stretched and bent) to shape, in contrast to parts that are cut, molded, or literally smashed into shape. Formed parts do not retain the shape of their forming tool or die, but rather spring back because their stress–strain curve contains an elastic segment. This segment stores forming energy that is released when the forming tool or die releases the part. To first order, the important point is that the stamping die cannot be the same shape as the desired part shape since the die must bend the part too far in order that it spring back to the desired shape.

Springback could be predicted if all pieces of "the same" metal had exactly the same material properties, if the dies closed with exactly the same force and speed every time, and if the coefficient of friction between the formed metal and the die were the same all the time. Variations in these quantities result in variations in the shapes of the parts. Calculating what will happen is computationally intensive and prone to error due to the difficulty in modeling some of the plastic deformation phenomena and lack of stability of the other parameters.

Because sheet metal parts are less accurately made than machined parts, and because large ones will sag unless supported, it is customary to assemble them using fixtures. The fixtures support them against gravity and locate them with respect to each other by means of features such as pin–hole, pin–slot, and edge–face. Depending on how part–part joints are designed, the parts may or may not constrain each other during assembly. For this reason, one cannot approach tolerance analysis of sheet metal assemblies the way it is done for machined parts. The latter usually constrain each other and thus pass size errors along to each other. Such errors statistically accumulate in ways that have been discussed earlier in this chapter. Sheet metal parts do not do this necessarily. Thus we have the following admonition: "Erase from your mind the idea that variations accumulate in sheet metal assemblies."[3] As we saw in the previous section, they accumulate partially through the parts and partially through the fixtures. Thus joint design, fixture design, and assembly sequence design are critical in controlling assembly variation in sheet metal assemblies.

When the parts are assembled, it is often necessary to bend them slightly while placing and clamping them into the fixtures in order to make holes line up or edges match. These actions store energy in the parts. The parts are then welded together. (In aircraft assembly, the parts are drilled and riveted together.) Once the parts are joined and the alignment and clamping forces are released, the

---

[3]Walton Hancock, University of Michigan, personal communication.

assembly assumes a new shape that comprises the minimum total stored elastic energy. Calculating this shape is also computationally intensive, even when all the original shapes and material properties are known. Since they are not, such calculations cannot be completely accurate.

The result of all these factors is that sheet metal assemblies contain errors that are difficult to predict. Pretending that they are just like rigid parts does not work. However, some basic strategies can be used to reduce the errors. These include keeping locked-in stresses low, letting the parts align themselves by using part-to-part joints that do not enforce constraint, and manually adjusting the dies and assembly fixtures to get a "best fit" that looks good but may not agree with the original designs.

A simple example of different approaches to sheet metal assembly design concerns the use of butt joints and slip joints. These joints are depicted in Figure 6-37.

Figure 6-38 shows in a simplified way how stamped parts are made. The male and female stamping dies are shown with sharp corners but they are slightly rounded in practice. The corners of the parts are similarly rounded, but the radius is not completely predictable because it depends on spring-back and die friction. As a result, the length of the part from vertical end to end will be

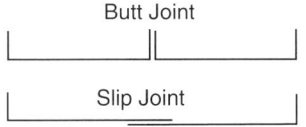

FIGURE 6-37. A Butt Joint and a Slip Joint. *Above:* Cross-section view through two channel-shaped parts. *Below:* Cross-section view through two shallow L-shaped parts. All these parts are stamped from flat sheets.

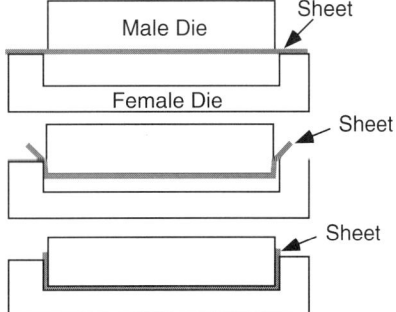

FIGURE 6-38. Simplified Illustration of Stamping one of the Parts in Figure 6-37.

different for each part. Butt joint assemblies will therefore have variable total lengths.

Figure 6-39 shows what could happen if the parts are a bit too long. They may spring back to a flat configuration but they may not because some of their distortion could be captured and retained when the butt joint is welded.

Sometimes, the spring-back is useful for obtaining function and appearance KCs. An example occurs in the design of car hoods where an effect called "overcrown" is designed in. The front of a car hood rests on two posts near the fenders and is held closed by a latch in the center. To keep the hood from rattling when it is closed, it is useful to design the latch so that it positively pulls the hood down into the latched position. One could design a spring into the latch to accomplish this but a clever design makes use of the springiness of the hood itself, as shown in Figure 6-40.

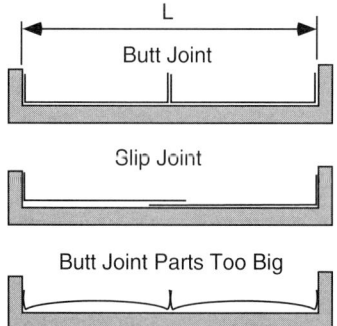

FIGURE 6-39. Behavior of Butt Joints and Slip Joints in a Fixture. The fixture is supposed to create an assembly of desired overall width *L*. If the parts are too long, they will self-adjust in slip-joint configuration but will distort in butt-joint configuration.

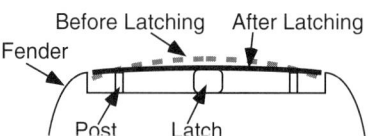

FIGURE 6-40. Use of Overcrown to Fit a Car Hood. The two KCs are as follows: Hood should be held down by a positive spring force; outer edges of hood should be flush with fenders. Before latching, the hood is the wrong shape and is not flush with the fenders, but after latching it is. Thus it must be made "the wrong shape" in order to achieve the KCs when it is latched. (Example provided by Anthony Zambito, Ford Motor Company.)

6.F.2. Assembly Sequence Considerations

If three rigid parts are to be assembled, say by stacking them on top of one another and joining their assembly features, and if each part has some tolerance on its thickness in the vertical direction, the variation of the height of the assembly stack is independent of the sequence in which the parts are assembled. But if the parts are made of sheet metal and are welded to one another one at a time as they are assembled, then the assembly sequence can make an important difference in the assembly-level variation. This is true in most cases where fixtures are used to position the parts relative to each other, whether the parts are flexible or not.

Consider Figure 6-41. It shows a five part assembly: a base plate A, two blocks B and C, and two angle brackets D and E. The KC is the distance between these brackets. Suppose we place the brackets in individual fixtures and weld them to the blocks, and then weld each block to the base plate, as shown in sequence 2. This is unlikely to do as good a job of delivering the KC as sequence 1, in which the KC is directly controlled by fixture F2. In this simple example, it is not hard to see what is the best thing to do, but in complex assemblies with joints that face in general three-dimensional directions, it can be difficult. More subtle effects may arise due to the heating caused by welding, and different sequences can cause different heat-related distortion effects.

6.F.3. Adjustment Considerations

Slip joints are often used as opportunities to adjust the parts before fastening them together. Fastening can be done by welding, drilling and riveting, or adhesive bonding. In Chapter 4 we made clear our preference for kinematically constrained assemblies. Clearly, a slip joint is underconstrained. We can look on this as a curse or a blessing. In the case of sheet metal parts whose variability usually exceeds those of rigid parts, it is a blessing. Adjustment can be active, based on measuring the parts and actively moving them into the correct configuration. Alternately, adjustment can be passive: placing the parts in a fixture and moving them firmly into kinematically constrained assembly with the fixture while the slip joint slips. This accomplishes the measurement and the adjustment at the same time.

Complex assemblies are often deliberately designed so that close tolerance KCs can be achieved by means of adjustments when no other way is practical. In sheet metal assemblies, this often involves placing a slip joint somewhere and taking care to fasten it last, after the parts are in the correct configuration. This slip joint is usually placed where it is invisible. Practitioners call this "washing uncertainty to someplace where it doesn't matter."

We will return to these issues with more examples in Chapter 8.

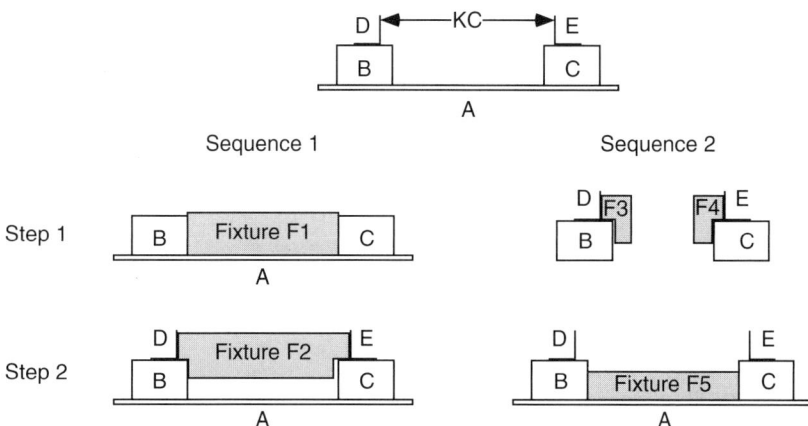

FIGURE 6-41. Two Candidate Assembly Sequences for a Five-Part Assembly. The KC is the distance between the two angle brackets D and E. Sequence 1 directly controls this KC with fixture F2 while sequence 2 controls it only indirectly via a chain of fixtures.

6.G. VARIATION REDUCTION STRATEGIES

Let us review where we are. We have learned that real parts always have some variation in their shape as well as variation the size, shape, and location of features on them, and that we can model those variations in order to predict how an assembly's KCs will vary. Several strategies are used to manage variation at the part level and/or at the assembly level in order to minimize the assembly-level impact:

1. We can place hard bounds on the limits of part-level variation and inspect each part to be sure that it is within the limits. This utilizes worst-case tolerancing. It is expensive and often unnecessary if one takes a statistical view.

2. We can take a statistical view and carefully distinguish between mean shift and variation around the mean. Then we have several alternatives:

   a. We can drive out the mean shift and then reduce the variation as much as possible, and we can utilize statistical tolerancing.

   b. We can identify those tolerances on which the C_{pk} is the highest and use our control of the process to tighten those tolerances and allow larger tolerances elsewhere in the assembly. This is an example of tolerance allocation.

   c. We can measure each part and choose one that is the right size to fit. This is called selective assembly and is discussed below.

3. In case we are unable to drive out the mean shift, we can still try to reduce the variation as much as possible and use other means to accommodate the mean shift. The consistency afforded by reduced variation gives us these alternatives:

   a. We can adjust the parts into the correct configuration. Consistency in the parts allows us to institute a systematic adjustment process.

   b. We can just live with the mean shift as long as we do not insist on making the parts to print. This is called "functional build" and is also discussed below. It, too, depends on the consistency that results from reducing variation.

Several of the above strategies involve what economists call coordination. It means that the parts are treated as individuals rather than statistically identical members of an ensemble. This is usually expensive but it is used when the alternative, namely making the parts accurately enough for interchangeability, is even more expensive.

Below we discuss a few of these strategies.

6.G.1. Selective Assembly

Selective assembly is used when process variability is too large for the required tolerances and it is not economical to reduce the variability. In Chapter 5 we discussed the valve train of an automobile engine and drew its tolerance vector diagram. Figure 6-42 reviews the situation. On the left is the overhead cam valve actuation system shown in Figure 5-4. The clearance between the cam and the end of the valve stem must be less than a few microns. No amount of statistical process control can generate parts that can be selected at random and assembled to meet such a tolerance economically. So a large empty space, perhaps 3 mm, is left between the cam and the valve, and this space

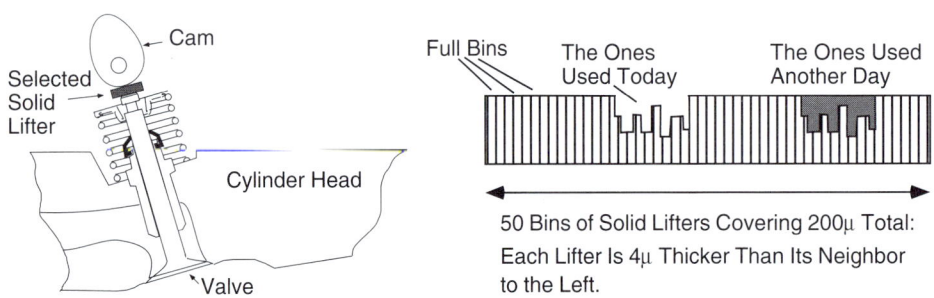

FIGURE 6-42. *Left:* **An Engine Overhead Cam Valve Mechanism with a Selected Solid Lifter.** *Right:* **Lifters of Different Thicknesses Stacked in Bins.** Only a few lifters have been used, indicated by the fact that their bins are not full. The rest of the bins are full, indicating that no lifters of those sizes have been used.

is measured individually. A lifter of the correct thickness is selected from a bin containing premeasured lifters and is installed. This is usually done automatically.

No attempt is made to manufacture lifters of a certain thickness. Instead, they are merely machined, ground, and measured. Then, like the 486 microprocessors discussed in Chapter 5, they are placed in the bin reserved for lifters of that thickness.

At one factory visited by the author, there were fifty such bins, as indicated on the right in Figure 6-42. The difference in thickness from the thinnest to the thickest lifter was 200 μm, indicating that the builders of this engine cared about differences in cam-valve spacing of 4 μm and would use the next larger lifter to keep the spacing from exceeding that amount. Furthermore, on the day of his visit, the author observed that the bins were all full except for a few bunched together as shown in Figure 6-42. Since the factory had been running for several hours at that moment, it is clear that overall the assemblies being made were highly consistent.

When two parts must be selected together to meet a sum or difference dimension between them, it is important that the range of sizes of each part is sufficient to make it easy to find a mate for each part. Suppose part A is a shaft that goes inside bearing B. The desired situation is shown in Figure 6-43. If part A is overrepresented by ones that are too small while part B is overrepresented by ones that are too big, then big A's and small B's will be quickly used up and the process will stop with a lot of small A's and big B's unable to find mates. This issue is illustrated in Figure 6-44.

6.G.2. Functional Build and Build to Print

Functional build is a pragmatic strategy used by some automobile manufacturers to shorten the time needed to create car body sheet metal assemblies that fit together. It involves (a) accepting an existing mean shift as long as the variation is small and (b) adjusting shims in fixtures or other parts of the assembly process so that the parts can be assembled. Toyota, a company famous for its ability to drive variation out of processes, uses this method, as does Honda. Other firms, such as Ford, deliberately avoid it. Instead, Ford enforces high C_{pk} on its manufacturing operations and suppliers. This is called *build to print* or net build. Its goal is to (a) design the parts so that they will assemble interchangeably and (b) build them to conform to the designs.

MacDuffie and Helper provide the following interesting quotes from Tower Automotive, a supplier of sheet metal items for Ford and Honda:

> Ford has focused on [quality] systems. They believe that if you have good quality control systems, you'll have good parts. After the systems are in place, they leave you alone as long as you're performing.... Honda cares about making the part fit the car, while Ford cares about making the part fit the blueprint. During product launch, Honda takes parts as soon as they are made and runs back to try them on the car. Then they tell us to change this, change that. Ford usually isn't here during our trials. They just want to be sure that we are meeting the spec. If there is a problem, they eventually issue an engineering change. But at

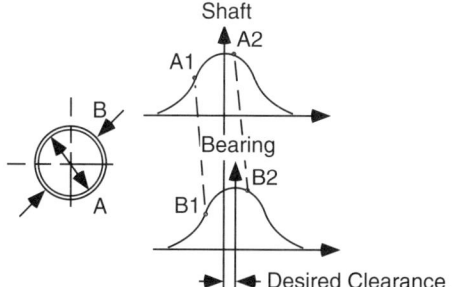

FIGURE 6-43. Illustrating Selective Assembly of a Shaft and a Bearing. The procedure is to measure shaft diameter A and pick a bearing B whose diameter is larger by the desired clearance. If the size distributions of A and B are similar, then each shaft A will find a suitable mate B among the available bearings.

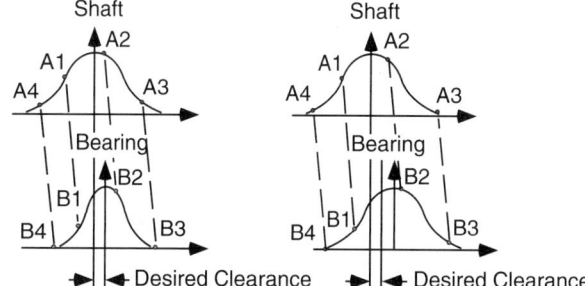

FIGURE 6-44. Illustrating Two Cases Where Selective Assembly Has Difficulty. On the left, the distribution of bearing diameters is narrower than that of shafts, so there are orphan shafts whose diameters are bigger than A_3 and smaller than A_4. On the right, the distribution of bearing diameters is similar to that of shaft diameters, but the mean bearing diameter is too large. Shafts with diameters smaller than A_4 and bearings with diameters larger than B_3 will be orphans.

Honda, things happen in a matter of days. At first we thought they were nuts. But... you get what you want, a part that works on the vehicle, right away. Everything else, like whether the blueprint is up to date, is secondary. ([MacDuffie and Helper], pp. 167–168)

This quote discloses all the plusses and minuses of functional build: It requires a lot of close attention and communication but it saves time. It pragmatically produces parts that fit, but engineering documentation of them is late or nonexistent because many of the changes are made by hand-grinding the stamping dies or experimentally adjusting shims, activities that are difficult to document. If there is no documented nominal, it can become difficult later on to trace the reason for a deviation from "good" parts. If the original design engineers are left out of this adjustment process, then they will fail to learn about any design mistakes they may have made.

Build to print seeks the ideal of interchangeable parts built to specifications that are passed down the supply chain. It requires more up-front communication between the customer and the supplier during product design to be sure that the supplier can deliver the required C_{pk}, but then the supplier can operate open loop. Conventional SPC methods can be used to monitor the supplier's performance. In this kind of arrangement, suppliers must demonstrate their ability to use SPC to gain and maintain control of their processes before they will be awarded contracts. This process works for traditional rigid part assemblies where there usually is good enough process control to meet the assembly-level tolerances. In assemblies like car body sheet metal, it may not work when assembly level tolerances are as small as ±1 mm. Functional build may be the only workable method.

A technical example of the functional build process for sheet metal stamping dies is provided in [Glenn], who compares it to build to print from the point of view of development time and cost. Consider the problem of making dies for two parts that are then welded together. In the build to print process, each die is built to print and test parts are made. If the parts are each within tolerances, the dies are accepted. If either part is out of specification, its die is reworked until it meets the specification. Each die is considered independently of the other die. In the functional build process, the test parts are considered together and their total error is calculated according to formulas discussed below. If the total error is in a certain band, then no rework is needed. The analysis below shows conditions under which functional build requires significantly less

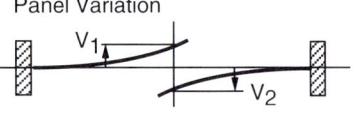

Panel Variation

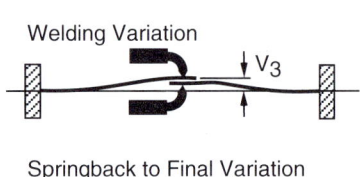

Welding Variation

Springback to Final Variation

FIGURE 6-45. Illustration of Springback After Spot Welding. ([Liu and Hu]. Copyright © Elsevier Science. Used by permission.)

rework of the dies, even if each part is considerably out of specification.

Figure 6-45 shows two simple parts being spot welded in a slip joint configuration. Suppose each part is made by a different stamping die, and due to die errors, the parts have errors v_1 and v_2 as shown in the figure. The assembly will have springback error v_a, which is given by

$$v_a = 0.125v_1 + 0.125v_2 \qquad (6\text{-}22)$$

To ensure quality, v_a must satisfy

$$|v_a| < 0.25 \qquad (6\text{-}23)$$

If a build to print strategy is used, then each die and the parts it makes are considered in isolation. Each die must then be reworked until v_1 and v_2 are both within their individual limits. In the worst case, these limits are

$$|v_1| < 1 \quad \text{and} \quad |v_2| < 1 \quad \text{independently} \qquad (6\text{-}24)$$

On the other hand, if a functional build strategy is used, then the dies are considered together and there are many combinations of v_1 and v_2 that do not require either die to be reworked. The assembly will have small enough v_a if v_1 and v_2 are related as shown in Figure 6-46. That is, v_1 and v_2 do not both need to be small as long as their sum is small enough. Some combinations of v_1 and v_2 require that only one die be reworked. The various combinations are shown in Figure 6-47. Here it is seen that functional build is much more tolerant.

If we know the C_{pk} and probability distributions of v_1 and v_2 then, for functional and build to print, we can calculate the probabilities of having to rework one or both

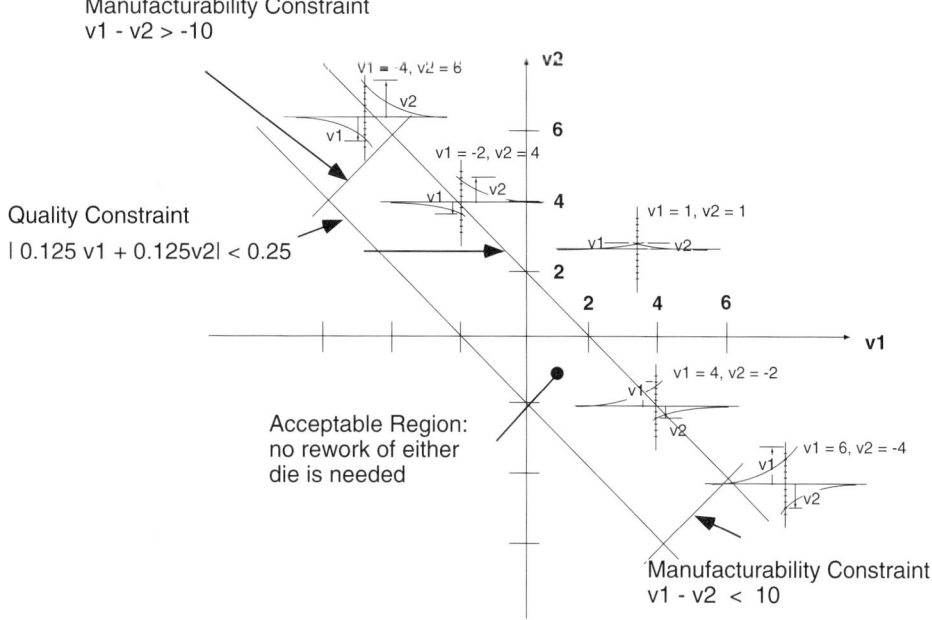

Manufacturability Constraint
v1 - v2 > -10

Quality Constraint
| 0.125 v1 + 0.125v2| < 0.25

Acceptable Region:
no rework of either
die is needed

Manufacturability Constraint
v1 - v2 < 10

FIGURE 6-46. Feasible Combinations of Part Variation in a Welded Slip Joint. The amount of assembly error as a function of part errors is calculated by considering the parts' elasticity. ([Liu and Hu], [Glenn]. Copyright © David Glenn. Used by permission.)

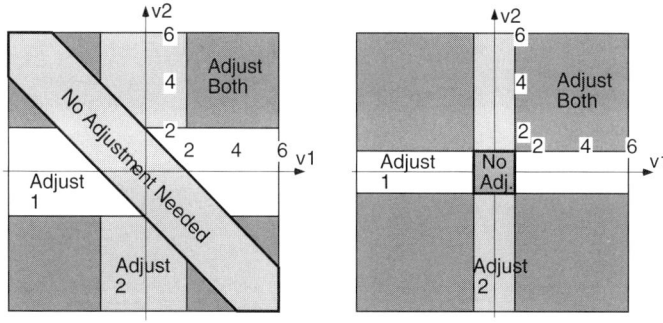

FIGURE 6-47. Comparison of Build to Print and Functional Build. *Left:* Functional build accepts both dies without modification if part variation lies in the large diagonal region, and requires adjusting only one die in each large horizontal or vertical band. *Right:* Build to print accepts both dies without modification only in the small 1 × 1 region in the center, and accepts either one in only the small horizontal and vertical bands. The region marked "adjust both" is much larger on the right than on the left. ([Glenn]. Copyright © David Glenn. Used by permission.)

dies. It should be clear from Figure 6-47 that rework will be less likely under functional build. However, all its disadvantages must be weighed in deciding whether to adopt it or not.

6.H. CHAPTER SUMMARY

This chapter showed how to build models of varied open chain assemblies using 4 × 4 matrix transforms. Methods were developed to represent single features, compound features, and features toleranced using GD&T. Several numerical examples were done.

Approaches were developed for both rigid and compliant parts. These examples span the range from assembly work cells to sheet metal parts of automobiles and aircraft.

We saw that, in spite of advances in computer modeling of parts and tolerances, some assemblies cannot

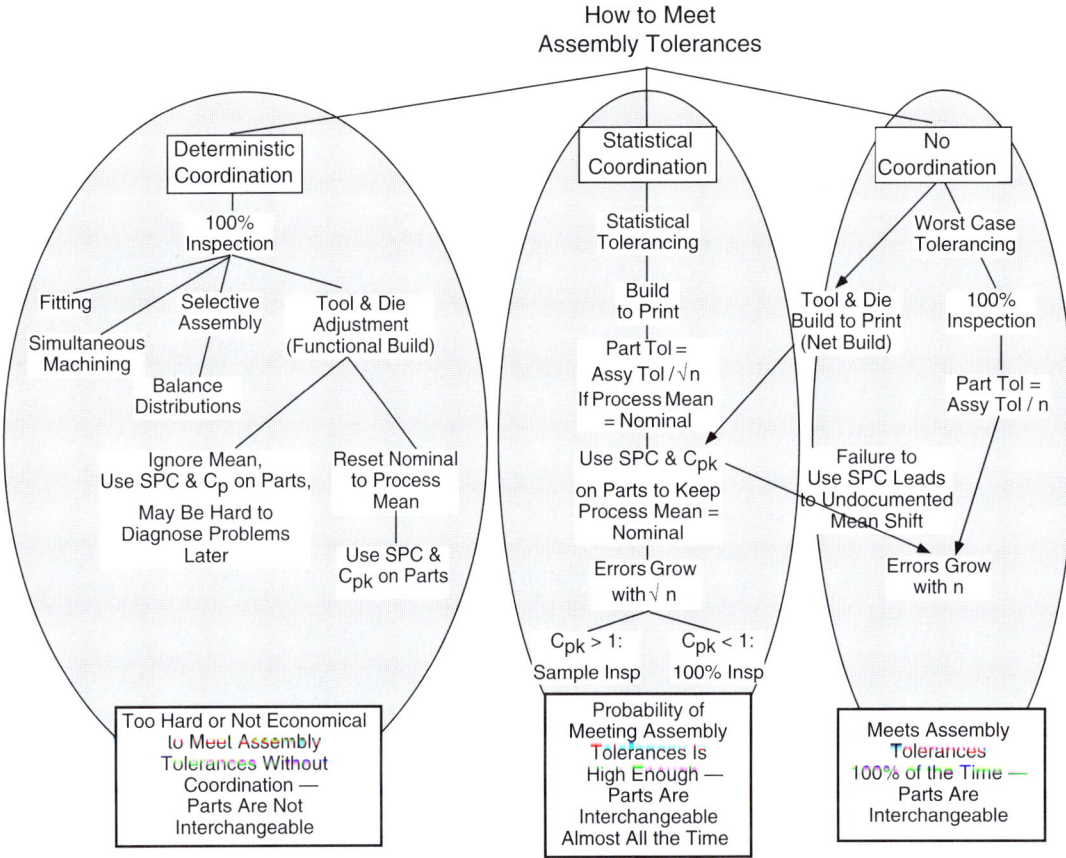

FIGURE 6-48. Logic Tree of Tolerancing Parts to Meet Assembly Tolerances. All these methods achieve the assembly KCs, but only the center and right regions describe methods that include interchangeability of parts and minimal or no coordination.

be made via interchangeable parts and still achieve their KCs. In such cases, various strategies are employed. Most of these require coordination, which involves premeasurement and sorting of parts, meetings between customers and suppliers, and close communication. Such issues go beyond the mathematics and draw us more deeply into the issues of assembly in the large. In general, we can say that tolerancing and variation management are not simply mathematical but also intensely people-oriented.

Figure 6-48 summarizes the factors that we discussed in this chapter. It arranges the possible situations from left to right according to how much coordination is required. *Deterministic coordination* implies 100% inspection and requires that each part be considered and dealt with as a partner with an intended "mate for life." Interchangeability is abandoned. Deterministic coordination is an economi-

cal way to deal with tools and dies that have mating male and female parts because they are one of a kind. The effort and cost can be spread out over all the parts they will make. Interchangeability of the die halves is not needed anyway. *Statistical coordination* makes a bet that the benefits of deterministic coordination can be had without the expense and effort. The bet is that two randomly selected parts will be able to mate for life with high enough probability that each can be dealt with individually until the moment of assembly. This constitutes the most economical route to interchangeable parts and successful assemblies. A number of processes, such as SPC, must be put in place in order that this bet will be successful. *No coordination* is the case where there is little or no confidence in such a bet and no desire to abandon interchangeable parts. Which of these strategies is the correct one must be evaluated on a case by case basis.

6.I. PROBLEMS AND THOUGHT QUESTIONS

1. Complete the examples in Figure 6-18 through Figure 6-20 by finding the final transforms of the parts and the assembly assuming a maximum position error of ± 0.1 unit and a maximum angular error of $\pm 0.1°$. Report all four combinations of $\pm$ errors.

---

2. Consider the parts joined by peg–hole features from Section 3.H. A third part has been added, and the peg on part A, which now has a *square* cross section, has been lengthened, as shown in Figure 6-49.

  a. Based on the nominal dimensions (i.e., ignoring any $\pm$ dimensions) given in this drawing, state in words which features and surfaces determine the location of part B with respect to part C.

  b. Now consider the variations shown in two of the dimensions. One of these has to do with the location of the square peg feature on part A while the other has to do with the perpendicularity of the square hole feature on part B, measured in the plane of the paper as shown. Assume that the + dimension (such as +0.003) is three standard deviations of a normal distribution in each case. Approximately what percent of assemblies will experience problems, and where will these problems occur? How much of the error is attributable to each of the two sources? Show all your work, matrices, computer code, spreadsheet formulas, and so on.

One way to approach this problem is to set up the 4×4 matrices that locate the points of interest. Most of the work for this is in problem 3 at the end of Chapter 3. You can use MATLAB to multiply the matrices out for you and you can insert random variables with the correct standard deviation, loop a lot of times, and plot a histogram. The following MATLAB code does useful things you may need:

  a. y = randn returns a normal random number y with zero mean and unit standard deviation (SD). To get mean m, add m to y. To get SD $= s$, multiply y by s.

  b. The following code makes normal random numbers and stores them in a vector z, and then makes a histogram of z with 200 bins:

```
»for i = 1:10000
z(i) = randn;
end
»hist(z,200)
»
```

Note that the semicolon after $z(i)$ keeps MATLAB from printing out every intermediate z while it is working.

  c. The following code makes a 2×2 matrix y in which the 1, 2 element is normal random:

```
»y = [1 randn;2 3]
```

  d. The following code calculates an error transform dx, multiplies it into a fixed transform *transx* 10,000 times, saves the randomized values of the (1, 4) coordinate of the resulting transform (the X coordinate of the location vector), and makes a histogram of them with 200 bins:

```
»transx = [1 0 0 3;0 -1 0 0;0 0 1 0;0 0 0 1]
transx =
   1  0  0  3
   0 -1  0  0
   0  0  1  0
   0  0  0  1
»for i = 1:10000
dx = [1 0 0 randn/10;0 1 0 0;0 0 1 0; 0 0 0 1];
z = transx*dx;
m(i) = z(1,4);
end
»hist(m,200)
```

The resulting histogram appears in Figure 6-50. Play with this code until you can make it work, then apply the ideas to the problem.

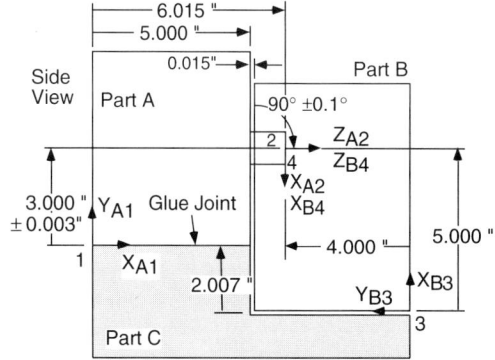

FIGURE 6-49. First Figure for Problem 2.

---

3. Repeat Problem 2 for the case shown in Figure 6-51. Be sure to express your MATLAB output in frame 3 coordinates. How much of the error is attributable to each of the two sources?

---

4. Consider the part pair shown in Figure 6-52, consisting of plate 1 with two pins, mating to plate 2 with one hole and one slot.

How much will the angle of plate 2, moving about the Z axis, change if the location of pin f2 changes by ± 0.003 in either the X or Y directions in part 1 home coordinates (at the lower left)? Answer separately for X and Y. Provide numerical answers.

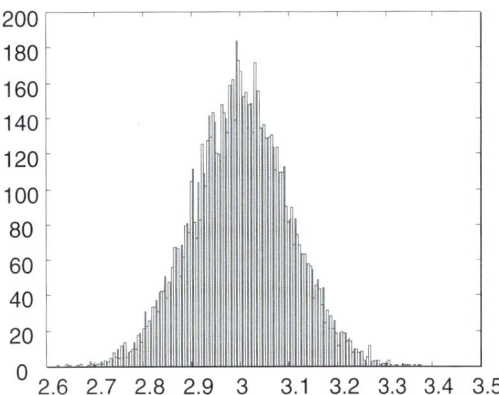

FIGURE 6-50. Second Figure for Problem 2.

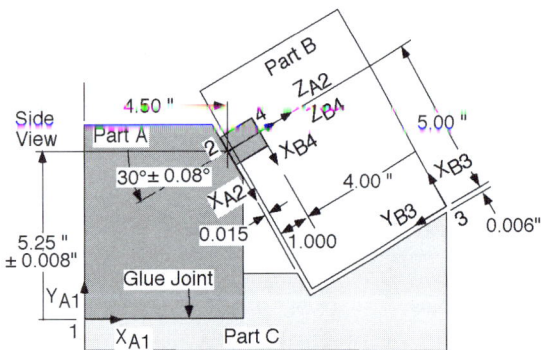

FIGURE 6-51. Figure for Problem 3.

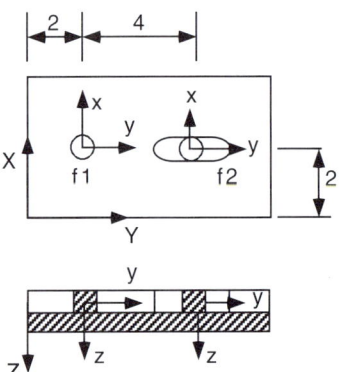

FIGURE 6-52. Figure for Problem 4.

5. Answer the same question as in problem 4 but instead consider the features as shown in Figure 6-53. Again, express your answer in terms of part home coordinates (at the lower left). Under what circumstances is it possible to assemble the two parts, given that the pins are in their varied positions?

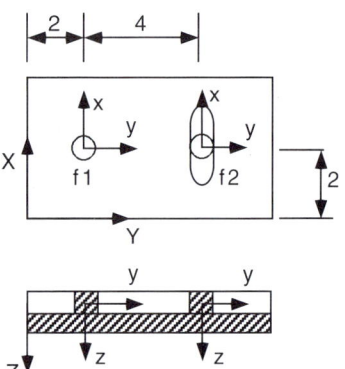

FIGURE 6-53. Figure for Problem 5.

6. Figure 6-54 corresponds to one in a thought question at the end of Chapter 3 where we found how to calculate the frame of a compound feature when one element of that feature lay in one part while the other element lay in a different part. Here we are interested in what happens to the compound feature when the second part has a varied position and orientation with respect to the first part, so that the compound feature is mislocated or misoriented. Write the necessary equations to find $T_{A1'}$, the varied frame that locates the varied compound feature with respect to part A's coordinate center.

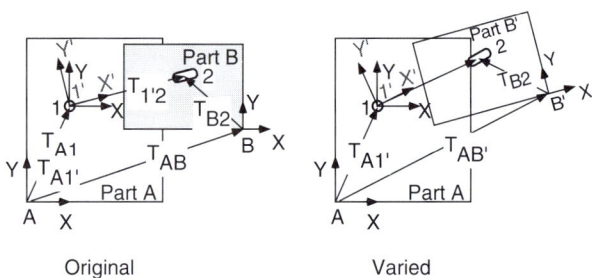

FIGURE 6-54. Figure for Problem 6.

7. Show how to combine the effects of errors in compound features based on misplacement or misorientation of feature elements on both parts. That is, show how to combine the errors described in Figure 6-6 with those shown in Figure 6-7 through Figure 6-10.

8. Consider the situation shown in Figure 6-55. The drawing shows a plate with a hole and a slot that could have any position a distance R from the center of the hole. This plate is to be placed over another plate having two pins of diameter D a distance R apart, so that one pin goes in the hole and the other pin goes in the slot.

Write an equation that allows you to calculate the amount of angular rotation permitted of one plate with respect to the other using $R\theta/\varepsilon$ as a parameter. What is the largest value of $R\theta/\varepsilon$ that you would recommend as a design guideline?

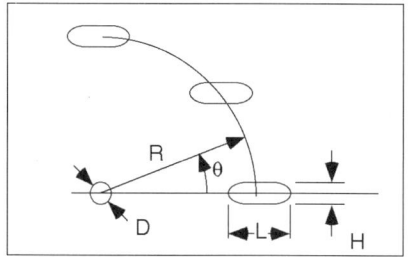

$$H = D + \varepsilon$$

FIGURE 6-55. Figure for Problem 8.

9. Consider the assembly fixture problem shown in Figure 6-56.

This corresponds to the problem described in Figure 6-26 except that the second step (shown above) is accomplished using different fixturing features on the parts. Draw the vector chain diagram for this step corresponding to the chain in Figure 6-26 as well as the full KC delivery chain corresponding to Figure 6-27. Explain in words what the difference(s) is (are) between the two fixturing strategies.

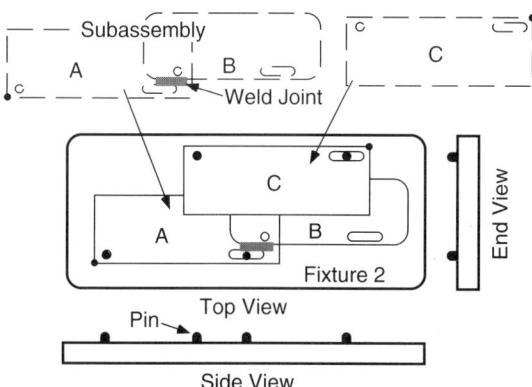

FIGURE 6-56. Figure for Problem 9.

10. In Chapter 8, we define product-level KCs called PKCs and distinguish them from assembly process KCs, calling them AKCs. In Figure 6-26, the PKCs were defined as point 1 on part A and point 2 on part C. Identify the AKCs in Figure 6-26 as well as in Problem 9 above, noting which are the same and which are different.

11. Repeat Problem 9 for the situation in Figure 6-57 and compare it to Problem 4 as well as to Figure 6-26. Here, part B has two alternate sets of fixturing features, an upper pin–slot combination and a lower pin–slot combination. Analyze two cases: (a) parts A and B are joined using the lower pin and slot combination on B; (b) parts A and B are joined using the upper pin and slot combination on B. Identify the AKCs in each case.

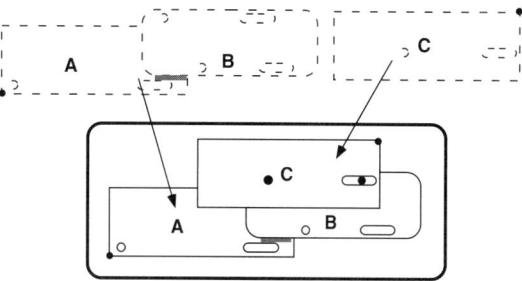

FIGURE 6-57. Figure for Problem 11.

12. Repeat problem 9 for the case shown in Figure 6-58: Assume that parts A and B are joined using the lower pin and slot feature pair on both A and B. Identify the AKCs in each case.

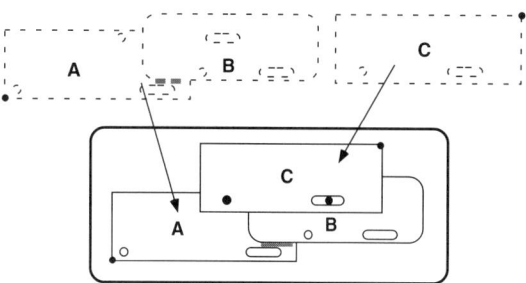

FIGURE 6-58. Figure for Problem 12.

13. Repeat Problem 9 for the case shown in Figure 6-59.

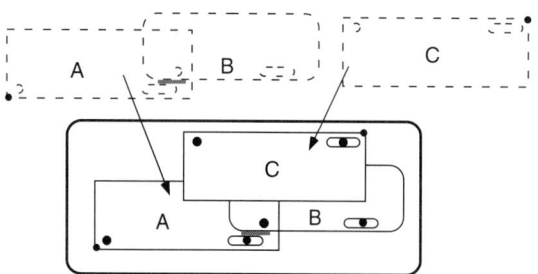

FIGURE 6-59. Figure for Problem 13.

14. Consider Figure 6-47. Assume that errors $v1$ and $v2$ are independent and normal with mean of zero and $3\sigma = 2$. What is the probability that dies built according to the build to print strategy will need no adjustment? Answer the same question regarding dies built according to the functional build strategy. A good way to approach this is to make a simple MATLAB simulation.

6.J. FURTHER READING

[Baartman and Heemskerk] Baartman, J. P., and Heemskerk, C. J. M., "On Process Planning with Spatial Uncertainties in Assembly Environments," *Manufacturing Systems,* vol. 20, no. 2, pp. 143–152, 1991.

[Bjørke] Bjørke, O., *Computer-Aided Tolerancing,* 2nd ed., New York: ASME Press, 1989.

[Bryson and Ho] Bryson, A. E., and Ho, Y.-C., *Applied Optimal Control,* Waltham: Blaisdell, 1969.

[Byrne and Taguchi] Byrne, D. M., and Taguchi, S., "The Taguchi Approach to Parameter Design," ASQC Quality Congress, Anaheim, CA, 1986.

[Cai, Hu, and Yuan] Cai, W., Hu, S. J., and Yuan, J. X., "Deformable Sheet Metal Fixturing: Principles, Algorithms, and Simulations," *ASME Journal of Manufacturing Science and Engineering,* vol. 118, pp. 318–324, 1996.

[Ceglarek and Khan] Ceglarek, D., and Khan, A., "Optimal Fault Diagnosis in Multi-Fixture Assembly Systems with Distributed Sensing," *Transactions of ASME, Journal of Manufacturing Science and Engineering,* vol. 122, no. 1, pp. 215–226, 2000.

[Ceglarek and Shi] Ceglarek, D., and Shi, J., "Dimensional Variation Reduction for Automotive Body Assembly," *Manufacturing Review,* vol. 8, no. 2, pp. 139–154, 1995.

[Chang and Gossard] Chang, M., and Gossard, D. C., "Modeling the Assembly of Compliant, Non-ideal Parts," *Computer-Aided Design,* vol. 29, no. 10, 1997, pp. 701–708.

[Chase, Greenwood, Loosli, and Hauglund] Chase, K. W., Greenwood, W. H., Loosli, B. G., and Hauglund, L. F., "Least Cost Tolerance Allocation for Mechanical Assemblies with Automated Process Selection," *Manufacturing Review,* vol. 3, no. 1, pp. 49–59, 1990.

[Chase et al.] Chase, K. W., Magleby, S. P., and Glancy, G., "A Comprehensive System for Computer-Aided Tolerance Analysis of 2-D and 3-D Mechanical Assemblies," in *Geometric Design Tolerancing: Theories, Standards, and Applications,* ElMaraghy, H. A., editor, London: Chapman and Hall, pp. 294–307, 1995.

[Chase and Parkinson] Chase, K. W., and Parkinson, A. R., "A Survey of Research in the Application of Tolerance Analysis to the Design of Mechanical Assemblies," *Research in Engineering Design,* vol. 3, pp. 23–37, 1991.

[Glenn] Glenn, D. W., "Modeling Supplier Coordination in Manufacturing Process Validation," Ph.D. thesis, University of Michigan IOE Department, 2000.

[Hu] Hu, S. J., "Stream of Variation Theory for Automotive Body Assemblies," *Annals of CIRP,* vol. 46, no. 1, pp. 1–6, 1997.

[Jastrzebski] Jastrzebski, M. J., "Software for Analysis of 3D Statistical Tolerance Propagation in Assemblies Using Closed Form Matrix Transforms," S.M. thesis, MIT Department of Mechanical Engineering, June 1991.

[Lafond and Laperrière] Lafond, P., and Laperrière, L., "Jacobian-based Modeling of Dispersions Affecting Predefined Functional Requirements of Mechanical Assemblies, Parts 1 and 2," 1999 IEEE International Symposium on Assembly and Task Planning, Porto, August 1999.

[Lapierrière and Lafond] Lapierrière, L., and Lafond, P., "Modeling Tolerances and Dispersions of Mechanical Assemblies Using Virtual Joints," 1999 ASME Design Engineering Technical Conference, Las Vegas, paper no. DETC99/DAC-8702, September 1999.

[Liu and Hu] Liu, S. C., and Hu, S. J., "A Parametric Study of Joint Performance in Sheet Metal Assembly," *International Journal of Machine Tools Manufacturing,* vol. 37, no. 6, pp. 873–884, 1997.

[MacDuffie and Helper] MacDuffie, J.-P., and Helper, S., "Creating Lean Suppliers: Diffusing Lean Production Throughout the Supply Chain," in *Remade in America: Transforming and Transplanting Japanese Management Systems,* Adler, P., Fruin, M., and Liker, J., editors, New York: Oxford University Press, 1999.

[Mujezinovic, Davidson, and Shah] Mujezinovic, A., Davidson, J. K., and Shah, J. J., "A New Mathematical Model for Geometric Tolerances as Applied to Rectangular Faces," Proceedings of DETC 01, paper no. DETC2001/DAC-21046, ASME Design Engineering Technical Conferences, Pittsburgh, September 2001.

[Pino, Bennis, and Fortin] Pino, L., Bennis, F., and Fortin, C., "The Use of a Kinematic Model to Analyze Positional Tolerances in Assembly," IEEE International Conference on Robotics and Automation, Detroit, May 1999.

[Rivest, Fortin, and Desrochers] Rivest, L., Fortin, C., and Desrochers, A., "Tolerance Modeling for 3D Analysis," 3rd CIRP International Working Seminar on Computer-Aided Tolerancing, Paris, April 1993.

[Shukla] Shukla, G., "Augmenting Datum Flow Chain Method to Support the Top-Down Design Process for Mechanical Assemblies," S.M. Thesis, MIT Department of Mechanical Engineering, June 2001.

[Terry] Terry, Andrew M., "Improving Product Manufacturability Through the Integrated Use of Statistics," MIT Master of Science and Master of Business Administration Thesis, June 2000.

[Turner and Wozny] Turner, J. U., and Wozny, M. J., "A Framework for Tolerances Using Solid Models," 3rd International Conference on Computer-Aided Production Engineering, Ann Arbor, MI, June 1988.

[Veitschegger and Wu] Veitschegger, W. K., and Wu, C.-H., "Robot Accuracy Analysis Based on Kinematics," *IEEE Journal of Robotics and Automation,* vol. RA-2, no. 3, pp. 171–179, 1986.

[Whitney, Gilbert, and Jastrzebski] Whitney, D. E., Gilbert, O. L., and Jastrzebski, M. "Representation of Geometric Variations Using Matrix Transforms for Statistical Tolerance Analysis in Assemblies," *Research in Engineering Design,* vol. 6, pp. 191–210, 1994.

6.K. APPENDIX: MATLAB Routines for Obeying and Approximating Rule #1

If we want to perform a Monte Carlo simulation of variation accumulation in an assembly and want to simulate GD&T tolerance specifications, we need a way to impose or approximate Rule #1. One way to do this is to pass every random feature variation through a Rule #1 filter that imitates the process of inspecting each part and rejecting those that fail to meet the part level tolerances as defined by GD&T.

Table 6-7 is MATLAB code that imposes Rule #1 on a plane feature of size like that shown in Figure 6-13.

TABLE 6-7. MATLAB Code that Imposes Rule #1 on Random Variations for a Plane Feature of Size

```
function [z,thx,thy,qq]=Rule1g3D(LX,LY,TS)
%calculates gaussian variations thetax, thetay, and z of 3D size %variation to
  obey Rule #1

a1=randn*2*TS/(3*LY);
a2=randn*2*TS/(3*LX);
b=randn*TS/3;
c=LY*abs(a1)/2+LX*abs(a2)/2+abs(b);
if c<TS
        thx=a1;
        thy=a2;
        z=b;
        qq=1;
else
        thx=a1;
        thy=a2;
        z=b;
        qq=0;
end
%[z,tx,ty,q]=Rule1(1,2,.005)
%a(i)=z
%b(i)=tx
%c(i)=ty
%d(i)=q
%plot(a,b,'gx') crossplots two variates
%plot(b,d,'rx') shows range of variate b with and without %imposing Rule #1
```

Note: The comments at the end of the code illustrate how to use it.

Figure 6-60 shows the result, based on a hypothetical feature in which $LX = 1$, $LY = 2$, and $TS = 0.005$. Figure 6-61 gives an idea of where the points are that are rejected when Rule #1 is imposed. This is a sample two-dimensional view of a complex three-dimensional space, so some points appear to be inside the acceptance zone when in fact they are outside in the direction normal to the plane of the image.

Table 6-8 gives MATLAB code for approximating imposition of Rule #1 based on use of Gaussian random variables and precalculated adjustment factors f_{opt}. The appropriate value for a plane feature of size is used.

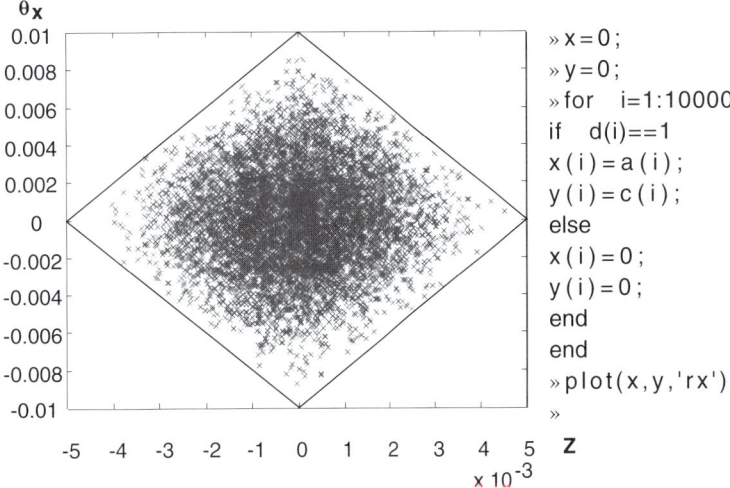

FIGURE 6-60. Sample Distribution of Errors in a Three-Dimensional Feature of Size with Rule #1 Imposed. The diamond-shaped region is a cross section through the diamond shown in Figure 6-14 that shows the boundaries of the acceptance region for Rule #1 for a plane feature of size measuring 1×2 with a tolerance zone ±0.005. The MATLAB code at the right shows how the plot was generated, based on use of vectors $a(i)$ and $c(i)$ generated by the code in Table 6-7.

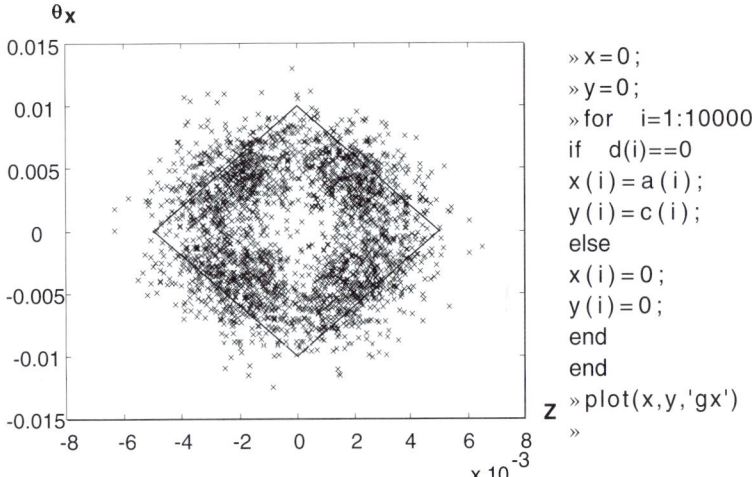

FIGURE 6-61. Sample Distribution of the Points Rejected from Figure 6-60 by Rule #1. The diamond-shaped region is a cross section through the diamond shown in Figure 6-14 indicating where Rule #1 is imposed. The MATLAB code at the right shows how the plot was generated, based on use of vectors $a(i)$ and $c(i)$ generated by the code in Table 6-7.

TABLE 6-8. MATLAB Code for Approximating Imposition of Rule #1 for a Plane Feature of Size

```
function [z,thx,thy]=noRule1g3D(LX,LY,TS)
%calculates gaussian variations thetax, thetay, and z of 3D size %variation to
   approximate Rule #1 using Olivier Gilbert's calculations

a1=.95*randn*2*TS/3*LY;
a2=.95*randn*2*TS/3*LX;
b=.95*randn*TS/3;
thx=a1;
thy=a2;
z=b;


%[z,tx,ty]=Rule1(1,2,.005)
%a(i)=z
%b(i)=tx
%c(i)=ty
%plot(a,b,'gx')
```

Note: The comments at the end of the code illustrate how to use it.

7 ASSEMBLY SEQUENCE ANALYSIS

"We flip it over to be sure that any bonus parts fall out."

7.A. INTRODUCTION

This chapter addresses the question of generating a good assembly sequence for a product.[1] The mathematics of assembly sequence analysis and the data models needed to support it take their form from the feature models of Chapter 3 and the constraint concepts of Chapter 4. Traditionally, choice of assembly sequence was the province of industrial or manufacturing engineers, and the choice was made after the product was designed, based on criteria that are relevant to factory operations. On this basis, the reader might expect to see this chapter grouped with others related to manufacture of assemblies. However, assembly sequence affects many aspects of product design and production, and is relevant to many life cycle issues of the product, so assembly sequence analysis should be part of early product design. In fact, assembly sequence choice focuses attention on so many strategic and tactical aspects of the product that this issue can serve as a natural launch pad for integrative product design.

Imagine a hypothetical product of six parts. We can build it *many* ways, among them bottom up, top down, or from three subassemblies of two parts each. What makes any of these ways better than the others?

There are construction reasons, such as space for tools that address fasteners or lubrication points. Similar considerations apply to ease of assembly, since some sequences may include some tricky part mates or awkward maneuvers whose success may be doubtful, whose failure might damage some parts, or whose action might injure or fatigue the assemblers.

There are quality control reasons, such as (a) the ability to test the function of a subassembly or (b) the avoidance of a sequence that installs fragile parts early in the process. Some sequences might not offer the opportunity to test some function until it was buried beneath many other parts, making rework expensive.

There are process reasons. Some sequences may not allow a part to be jigged or gripped from an accurately made surface, making assembly success doubtful. Some sequences may require many unproductive moves, such as fixture or tool changes or the need to flip a subassembly over. Flipovers (more generally, reorientations) may be unavoidable, but some sequences may require reorienting before the subassembly is fully fastened together, risking the possibility that it will disassemble spontaneously unless extra (costly) fixtures are provided. Additionally, reorientations may be easy for people but difficult, awkward, or costly for machines due to the extra axes and controls needed. Thus a sequence without reorientations may be sought if automatic assembly is a goal. Product redesign may be necessary to permit such a sequence.

Finally, there are production strategy reasons. These include being able to make some subassemblies to stock, since they are common to many models, so that final assembly to order can be done quickly by adding only the remaining parts. Similarly, some products are designed to support a strategy called "delayed commitment" or "plain vanilla box" ([Lee], [Swaminathan and Tayur]). In this approach, the product is customized for each buyer or class of buyers by adding a few parts specific to that buyer. It is often convenient to add these parts at or near the end of the sequence. Perhaps the distributor or even the buyer

---

[1] Portions of this chapter are taken from Chapters 8 and 9 of [Nevins and Whitney 1989].

will add these parts.[2] Parts with long lead times are also conveniently placed at the end of the assembly sequence to maximize the time available to procure or make them ([Mather]). However, this technique is of limited value because assembly takes such a short time relative to the time to make or buy and ship something. Production strategy impacts of assembly sequence choice are discussed in more detail in Chapter 14.

Some of these reasons clearly can have a major impact on how the product is designed, and bringing them up will spur discussion of the many topics just discussed. If assembly sequence analysis is delayed until the product's design is "finished," then some other way must be found to expose the detailed, architectural, and strategic issues that assembly sequence analysis brings up. Failing that,

the design will be frozen, and changes to implement any of the above considerations will be very costly.

Since assembly sequence choice is both important and difficult, it is fortunate that computer-based algorithms exist to address it. This chapter presents a general approach, explains one algorithm in detail, and gives several examples that illuminate the way assembly sequence analysis links design and manufacture of assemblies. The methods discussed here, like those in the general literature, address gross motion planning only.[3] If a sequence says "join parts A and B," it assumes that the required fine motions are possible and can be planned later using other methods. Naturally, such assumptions have to be checked. Even very small changes in part size or shape can invalidate them.

7.B. HISTORY OF ASSEMBLY SEQUENCE ANALYSIS

Traditionally, assembly sequence analysis was done by industrial or manufacturing engineers to improve the efficiency of a manual assembly line. As we will see later in this chapter, a product that has at least one assembly sequence will typically have hundreds or thousands, so the industrial engineer does not lack for choices. The goal is to balance the line. [Scholl] presents a thorough treatment of assembly line balancing. Line balancing involves choosing a feasible assembly sequence and assigning the different steps to the people so that each person has a quantity of work that takes approximately the same time to accomplish. Different assembly or test activities take different amounts of time, and different people, due to skill level or handedness, will take different amounts of time to do equivalent tasks. Different models or styles of the product may contain different parts or different numbers of parts, requiring partial redesign of the assembly process when production shifts from one model to another during the

day or week. New workers will arrive as others leave, and on different days certain people will be sick or on vacation. The engineer must know the tasks and the people well in order to do this job.

One of the first algorithms to assist line balancing was developed in [Prenting and Battaglin]. This algorithm took as input a diagram called a precedence graph, which indicates the order in which assembly tasks may be performed. This graph in principle contains all feasible assembly sequences in the form of a network. At that time, there was no algorithm capable of generating this network, so it had to be created by hand. Later in this chapter, we will discuss algorithmic methods for creating networks of feasible sequences. The Prenting and Battaglin algorithm looked for sequences in this network that had the best balance. In most cases, such sequences take the shortest time to accomplish.

Little additional research was done on assembly sequence analysis until the advent of robot assembly in the 1970s. As mentioned in Chapter 1, robotics spurred interest in many basic assembly issues that had been conveniently ignored when people did nearly all assembly. Attempts to have machines perform assembly revealed many knowledge gaps. In the case of assembly sequences, machine assembly provided additional constraints and

---

[2]Hewlett-Packard used delayed commitment for power supplies in printers. Power requirements are different in different countries, and it proved impossible to predict how many printers in which power configurations would be sold in a given time period. Power supplies are cheap relative to printers. So HP decided to ship printers without power supplies and provide large quantities of different power supplies separately. Distributors took orders and installed the correct power supplies before shipping the product. This strategy requires that the printers be designed so that the power supplies can be installed easily—that is, that the assembly sequence support adding them last. Their installation must also be foolproof since HP cannot train every distributor's personnel in this task.

---

[3]Gross and fine motions are discussed in detail in Chapter 9. For our purposes here, gross motions carry parts from place to place while fine motions are the final maneuvers of assembly after parts touch each other.

opportunities in addition to line balancing that required a new approach to the problem. Among these new approaches were several heuristics.

One heuristic simply says to "start with the base part." This sounds reasonable except that (1) it may not be obvious which part this is, and (2) this heuristic may cause good sequences to be ignored. Later in this chapter we will discuss some counterexamples to this heuristic.

Another heuristic starts with the observation that fasteners provide a kind of punctuation to assembly processes ([Tseng and Li], [Akagi, Osaki, and Kikuchi]). That is, assembly processes follow a pattern in which several parts are added, and then a fastener or set of fasteners binds them all together. This pattern is then repeated until the product's assembly is finished. Fasteners provide closure to a phase of the assembly and stabilize the parts so that they can be reoriented or passed to another station. Between fastening operations there may be many choices for sub-sequences, many of which do not differ from each other significantly. Thus assembly sequence identification and choice become fastening sequence identification and choice. There are many fewer fasteners than parts,[4] so this approach reduces the size of the problem and simplifies it greatly.

Heuristic methods have the advantage that they usually work fast. However, they do not guarantee results. They may miss feasible sequences or generate sequences that are incorrect. Algorithms, in contrast to heuristics, promise *correctness and completeness,* but they tend to operate slowly if there are many parts in the assembly. Assembly sequence identification is a combinatorial problem and in principle grows extremely rapidly as part count increases. Thus the design of the algorithm is crucial if it is not to bog down and become unsuitable for normal use on industrially realistic problems.

The first algorithm that generated all feasible assembly sequences was published in [Bourjault]. This method is described in detail later in this chapter. *Feasible* means that the sequence can be finished and no parts will be left over. Like successful algorithms, feasible sequences are correct and complete. Bourjault's method utilized the liaison diagram and expressed sequences in terms of the sequence of liaisons to be established. Like most subsequently developed methods, Bourjault's method

consists of testing which liaisons can or cannot be accomplished at a given stage of assembly. It then combines this information to formulate the feasible sequences. Testing which liaisons can be accomplished involves a combination of queries to a person and/or algorithmic and geometric analyses by the computer based on the liaison diagram and answers to previously asked questions.

De Fazio and Whitney and their students built on Bourjault's method, increasing the size of problem it could solve efficiently and linking it to CAD or other data that describe how the parts are connected to each other ([Baldwin et al.], [De Fazio et al.], [Whipple]). These methods paid careful attention to which parts might possibly be added at a given stage, reducing the number of queries that the engineer had to answer.

Other methods developed since Bourjault include those based on exploded view heuristics ([Gustavson], [Rivero and Kroll]), methods that address additional concerns like formation of suitable subassemblies or stability during assembly ([Lee and Shin]), and methods that use robot motion planning techniques from artificial intelligence to decide whether parts can be added ([Halperin, Latombe, and Wilson], [Homem de Mello and Sanderson]). The exploded view approach exploits common architectural themes in products first observed in [Kondoleon], namely that assembly typically involves adding a series of parts all in one direction. Often, following the principles of design for assembly (discussed in Chapter 15), there is one dominant assembly direction. Sequence identification thus begins by identifying these directions, choosing a sequence of directions to investigate, and choosing part sequences along each direction. Within one direction, Gustavson sequences the parts in the order in which their centers of mass, or the centers of mass of their bounding boxes, appear. Often, only minor corrections have to be made to sequences generated this way. Several approaches to assembly sequence planning are explored in [Nof, Wilhelm, and Warnecke].

Today, assembly sequence analysis is relatively mature, and research methods are beginning to appear in commercial software. The methods available solve different problems, and it is important to distinguish between them. The alternatives are as follows:

- Find *all* feasible sequences. This is the most ambitious goal, and it gives the engineer the greatest scope for choice. "All" means all, including many that are

---

[4]This statement is based on counting all fasteners put in at the same time or one right after the other as one fastener.

functionally redundant. For example, one sequence will build subassembly A, then subassembly B, while another sequence will build B before A. A third sequence will turn its attention from A to B, adding one or more parts to A, then to B, in one of many possible combinations. Most algorithms contain heuristic ways of eliminating these redundant sequences. [Amblard] discusses this problem.

- Find *all linear* sequences. Linear sequences do not build any intermediate subassemblies but instead build the product in one process by adding one part at a time to a single growing set of parts. There are vastly fewer linear sequences than general sequences, so the problem is easier to solve. However, subassemblies are useful for a variety of reasons. They permit parallel operations, reducing the time from start to finish. They can reduce space requirements but increase transportation requirements. They allow asynchronous scheduling of different portions of the assembly and break the assembly process into independent segments. This reduces the vulnerability of the system to breakdowns in one region. They also permit functional in-process testing as well as the outsourcing of subassemblies.

- Find *one* feasible sequence or *one feasible linear* sequence. In some circumstances, it is enough to have one sequence. For example, to repair a broken machine, build a space station, or perform surgery, one sequence may suffice or even be mandated.

One-sequence methods are often favored by robotics researchers. They fill research objectives, which often focus on planning gross and fine motions or testing planning methods. All-sequence methods are favored by product designers. Their challenge is *sequence design*. In general, design is a matter of generating alternatives, generating criteria, and then choosing an alternative to satisfy the criteria. In industry, an assembly sequence may be used for years. For this reason, its design must be carried out very carefully. The first step in such a process, as discussed below, is to have access to all the alternatives.

All-sequence methods are also useful for generating disassembly sequences suitable for repair and maintenance. Regardless of the amount of damage, if any, that it has suffered, a broken machine may require a different disassembly sequence than the reverse of the original assembly sequence. There are many reasons for this, and one of the thought questions at the end of the chapter asks the reader to think about this issue.

7.C. THE ASSEMBLY SEQUENCE DESIGN PROCESS

This section presents a general approach to assembly sequence design. We discuss the criteria that come into play and show how to include assembly model and assembly feature data.

7.C.1. Summary of the Method

The method presented here generates all feasible sequences. It contains two main phases. In the first phase, all infeasible (i.e., truly impossible) sequences are eliminated. Impossible means that the sequence cannot be completed and/or there will be parts left over. For example, suppose we have cookies, a cookie jar, and a lid for the jar. The sequence that begins "put the lid on the jar" is not feasible. While we will discuss in detail only Bourjault's method for doing this phase, many algorithms for doing so are available. Most operate by composing questions for the computer or the engineer to address. These questions

are designed to determine if a proposed assembly step is possible or not.

In most methods, the answer to each question can be formulated by considering only the step at hand. No lookahead is required to see if the choice at this step could lead to trouble later. This fact makes it much easier to answer the questions.

Once the infeasible sequences have been eliminated, what remain are the good, the bad, and the ugly. The second phase requires the engineer to generate and apply criteria that will reveal the good, of which there may be many. For example, the sequence that begins "put the lid upside down on a table and carefully pile the cookies on top of it" is a feasible sequence but not a good one. As we will see below, the criteria typically emphasize different things, making the choice of a sequence interesting.

There is no algorithm for accomplishing the second phase, and there is unlikely ever to be one. Eliminating infeasible sequences is a mechanistic task that evaluates unambiguous choices. If a part (the jar's lid) blocks access by another part (cookies), then it blocks access, period. This can be detected definitively by a variety of methods, including intervention by the engineer. Ranking and selecting among conflicting criteria, even generating criteria in the first place, requires judgement and evaluation. Values are contingent and open to dispute. Only people can address such situations.

The next two sections discuss these two phases in general, giving rules that apply to every method for accomplishing them. Following this, a detailed engineering process for accomplishing them is given.

7.C.2. Methods for Finding Feasible Sequences

Most methods for generating the set of feasible sequences begin with the liaison diagram. Bourjault's method treats this diagram as a graph and uses a variety of graph theory and circuit analysis methods to determine a set of questions for the computer or the engineer to answer. Bourjault's questions take one of two forms: "Is it possible to add this set of parts if that set of parts has already been assembled?" or "Is it possible to add this set of parts if that set of parts has not already been assembled?" Equivalent to Bourjault's method is one in which the questions are generated by systematic textual manipulation of a string of liaisons and application of several rules that govern liaison diagrams.[5] More recent methods ([Homem de Mello and Sanderson], [Baldwin]) also consider the liaison diagram as a graph and enumerate all the cuts that can be made through it. A cut is a line drawn through the diagram that divides it into two parts. Since this cut severs one or more liaisons, it is equivalent to separating the parts joined by them. Each cut is converted into a question for the engineer or the computer of the form "Can this set of parts be added to that set of parts?" A class of methods called "onion skin methods" ([Whipple]) considers first those cuts that involve parts that are physically on the outside of the assembly. This reduces the number of questions.

In many algorithms, the questions are posed in terms of removing parts rather than installing them. The question reads, "Can this set of parts be removed from that set of parts?" This is conceptually easier for some people, especially if the algorithm is presented to the assembly engineer in conjunction with drawings or sketches of the parts. Nevertheless, the question addresses the feasibility of assembly, not disassembly.

Not only can the questions be posed in terms of disassembly actions, but the algorithm itself can operate in reverse assembly order, asking about parts on the outside first and working its way in. Algorithms that operate from the outside of the product inward generate *disassembly* sequences and have the advantage that these will never dead-end. Algorithms that operate from the inside out generate *assembly* sequences. Unless an assembly sequence algorithm contains look-ahead features, it will generate some sequences that dead-end. These are not hard to detect and remove, but this is an additional step that is not really necessary. Under the rules to be stated below, the reverse of a feasible disassembly sequence is a feasible assembly sequence, so the order in which the algorithm addresses the parts is not important.

The Bourjault method and ones like it generate liaison sequences rather than part sequences. In liaison sequences there is a first liaison rather than a first part. This has advantages and disadvantages. On the favorable side, it focuses on the operations to be performed rather than on the parts. A bonus is that many nonassembly operations can be included by calling them liaisons. These include lubrication, inspection, temporary removal and later reinstallation of a part, and reorientation of the assembly. These may be added to the liaison diagram and handled in the same way as part liaisons when questions are generated. An example of this kind of liaison is given in Section 7.H. For example, "Can this test be performed when these parts have been assembled?" is a question that the engineer can answer, with the result that all feasible sequences will put the test at the right place(s) in the assembly process. The engineer will then have the opportunity to choose the sequence that puts the test at the earliest feasible place, if that is important.

On the other hand, liaison sequences do not directly address fixturing issues.[6] The first part is usually added to a fixture rather than to another part. Some parts are easier to fixture or are more appropriate as first parts than others. A focus on part-to-part liaisons makes it more difficult to bring such issues into assembly sequence design. A simple workaround consists of considering the fixtures to be

---

[5]The text analysis method is presented in Section 7.D. Its rules are listed in Section 7.L.

[6]The phantom liaison discussed in Chapter 3 is an example.

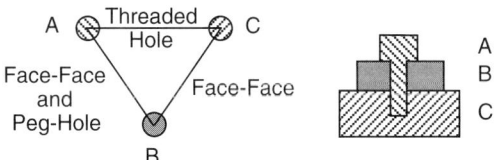

FIGURE 7-1. Example Liaison Diagram. Part A is a screw that fastens part B to part C.

parts and identifying liaisons between them and the parts. We will have many occasions to think this way, especially in the next chapter, which deals with the datum flow chain and ties many of the ideas from Chapters 2–8 together.

7.C.2.a. Rules Governing Liaison Diagrams and Sequence Generation

The success of assembly sequence algorithms—that is, their ability to be correct and complete—depends on being systematic about the data that are provided to them. The liaison diagram is the most basic input data. See Figure 7-1. This section discusses several important properties of liaison diagrams.

In a liaison diagram, each part is a node and each link is a liaison. Liaisons indicate the fact that two parts join. While the diagram can be augmented to include information about the joint, as is done in Figure 7-1, the sequence generation algorithms currently in use do not take account of many details of the joint. For example, an assembly feature may consist of a pin–hole joint and a pin–slot joint, but the liaison diagram represents this as a single line. For most algorithms, this is sufficient, especially if the escape direction of a compound feature like a pin–hole and pin–slot is represented by its twist matrix. Fine motions needed by a particular feature pair are not addressed by current assembly sequence algorithms.

7.C.2.a.1. The Loop Closure Rule. An important property of liaison diagrams, first demonstrated by Bourjault and essential to the efficient operation of his and other algorithms, is the *loop closure rule*. This rule applies to any loop in a liaison diagram and reads as follows: If at some point in an assembly process, a loop of n liaisons stands with $n - 2$ liaisons already made, then the next step applied to that loop will close *both* of the remaining open liaisons. In other words, it is impossible for a partial assembly to exist in which there is a loop with only one undone liaison. This can be illustrated by referring to Figure 7-2. Here, parts B and C have been joined and A is

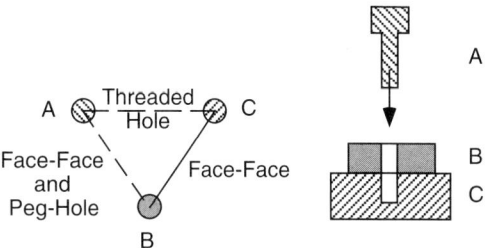

FIGURE 7-2. The Liaison Diagram and Assembly from Figure 7-1 Before Part A is Added. Installing A correctly will make both of the remaining liaisons.

about to be added. As long as we understand that installing A means making all its liaisons, it should be clear that both of the undone liaisons in the diagram will be made at the same time[7] and that there is no way that one liaison could remain undone if A is installed correctly.

In addition to the loop closure rule, but related to it, are two additional rules:

- The parts are rigid.
- The liaisons are also "rigid" in the sense that a liaison stays made once it is made.

Why are these rules necessary? They clarify the relationships between parts and eliminate many ambiguities that would baffle an algorithm. For example, referring to Figure 7-2, we might stretch part B so that its hole enlarges. This would permit sliding part A through B without touching it, or sliding B over A after A had been installed. Or, we could squeeze B so far to one side that A had access to the hole in C directly. If these actions were possible, then we would not know how to answer a question like "Can we make the liaison between A and C?"

7.C.2.b. Rules Governing Posing and Answering Questions

Elastic parts seem like unusual situations, to be sure, but it turns out that forbidding them is important to the validity of two additional rules called the subset rule and the superset rule, also enunciated by Bourjault. These rules, in turn, greatly improve the efficiency of the algorithms by easily eliminating some questions based on the answers to previously answered questions.

---

[7]"At the same time" means "during the process of completely installing the part." It does not mean literally "at the same instant of time."

7.C.2.b.1. The Subset Rule. The subset rule says, "If it is possible to add part X to the set of parts $\{Y\}$, then it is possible to add X to any subset of $\{Y\}$ that permits the necessary liaisons." The proof is easy: We know that $\{Y\}$ does not contain any parts that block the addition of X. Therefore, no subset of $\{Y\}$ can contain any X-blockers either. Since parts and liaisons are rigid, there is no way to bend parts in the subset or alter their liaisons in order to convert them into blockers.

This rule is also true if we consider adding a set of parts $\{X\}$ to set $\{Y\}$.

7.C.2.b.2. The Superset Rule. The superset rule says, "If it is impossible to add part X to parts $\{Y\}$, then it is impossible to add X to any superset of $\{Y\}$." This proof is also easy. Since $\{Y\}$ contains parts that block access for X, adding more parts to $\{Y\}$ will not eliminate those blockers unless we violate the rule that says parts and liaisons are rigid.

This rule is also true if we consider adding a set of parts $\{X\}$ to set $\{Y\}$.

7.C.2.b.3. Violations of the Subset and Superset Rules. When violations of these rules occur, they can be pretty interesting. For example, elastic parts can be used, and these can be bent aside to permit the addition of parts that would otherwise be blocked.[8] More interesting is a case observed by the author in Japan in 1986 in which some liaisons were "elastic." The product was a Sony video recorder tape deck containing about one hundred parts. This unit contained an electric motor that was used to reconfigure the deck between forward and reverse direction, as well as to permit the cassette to be inserted and removed. The assembly line in question comprised twenty-five robot stations and two manual stations. It was in fact the first large-scale application of Sony robots.

At around robot station 20, after the motor had been installed, a robot picked up a tool with electric contacts on it and applied them to the contacts on the motor. The motor operated and reconfigured the assembly so that parts

previously obscured were now accessible, permitting more parts to be added to them. This product had powered degrees of freedom and thus could take an active role in its own assembly.

Powered fixtures can serve the same purpose, reconfiguring a product during assembly to permit other operations to occur that would not otherwise be possible.

Clever features like this are beyond the capabilities of typical assembly sequence algorithms. Within their scope, however, these algorithms are extremely useful.

7.C.3. Methods of Finding Good Sequences from the Feasible Sequences

If the algorithm generates all feasible sequences, it is the engineer's job to find good ones or throw away bad ones. This is called sequence editing. The algorithm and supporting software described in [Baldwin et al.] operates this way. It also permits the user to impose additional constraints on sequences as well as to directly delete elements of the feasible set in order to narrow it down. The ways it supports these activities are discussed in Section 7.G.1.

If the algorithm used to find feasible sequences generates only one sequence, then the engineer must decide if it is a good one or not, without having others with which to compare it. The algorithm and supporting software described in [Kaufman et al.] operates this way. It allows the user to request another sequence that obeys certain rules that the user can select. This algorithm is discussed in Section 7.G.2.

7.C.4. An Engineering-Based Process for Assembly Sequence Design

This section describes in detail the two-phase method for generating assembly sequences and choosing one or a few. In the first phase, all the feasible sequences are generated. In the second phase, good sequences are culled from all the feasible ones. The description emphasizes the engineering aspects and assumes that some kind of algorithm is available to manage the computational burden. If the engineer is not too concerned about having absolutely every feasible sequence available for consideration, he or she can generate a reasonable set of candidate sequences manually and follow the process described here to guide the engineering decisions.

---

[8] Elastic parts can cause confusion if the algorithm poses disassembly questions. A door with a spring-loaded latch can be closed without the doorknob being installed but cannot be opened unless the doorknob is present. The user has to remember that the algorithm is asking about assembly (closing the door) when he is asked, "Can the door be opened if the doorknob is not installed?" The correct answer is YES.

7.C.4.a. Generating the Feasible Sequences

The procedure for generating feasible sequences is diagrammed as a simple flowchart in Figure 7-3. The steps in this process will now be discussed in turn.

The parts list and assembly drawing needed to begin assembly sequence analysis may be quite preliminary. Sketches or cross-section drawings are enough to get started. The point is to begin assessing the assembly issues as early in product design as possible. The design will be quite fluid at this point and changes are to be expected. The design engineers are often unaware of the assembly implications of their designs except in the most general way, and they will not be sensitive to the degree to which trivial changes to the design will require the assembly analysts to scrap their previous work and start over. The assembly analysts should never seek to slow the design engineers down but instead should be glad they have a fast computer-based method that helps them keep up as the design evolves.[9] Of course, the analysts could just wait until the design is finished, but this would sacrifice the opportunity to critique and improve many assembly-driven attributes of the product.

Once the liaison diagram is available, the analysis can begin generating and addressing the precedence questions. The algorithm from [Baldwin et al.] will ask the questions and answer many of them behind the scenes by using the subset and superset rules (and other methods described below), asking the user only when necessary. The goal of these questions is to find out what moves are forbidden. As will become clear in the example below, this process usually requires several questions to be generated and answered for each blockage situation that is detected. If X cannot be added to (or removed from, depending on how the question is posed) set $\{Y\}$, then follow-up questions will be needed to find out which parts among $\{Y\}$ are the actual blockers.

Two types of blockage must be checked. These are called *local constraints* and *global constraints*. Local constraints are usually easy for the computer to detect, while the engineer has an easier time than the computer finding global constraints. In a typical local constraint, a part is

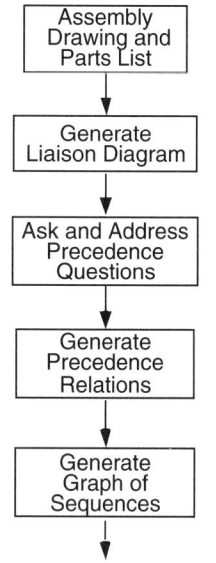

FIGURE 7-3. Flowchart for Generating the Feasible Assembly Sequences.

literally trapped by its neighbors due to the combination of local assembly or escape directions that they impose on it. These escape directions are properties of the assembly features and the relative locations of the parts in space, so as soon as the assembly model (Chapter 3) and the liaison diagram are available, the computer can determine these directions and be ready when it is necessary to evaluate local constraints. A part is locally unblocked when all its extant liaisons share at least one common escape direction. Otherwise the part is trapped and cannot be removed. The parts' detailed shapes do not need to be known in order to detect local blockage.

Even if it is not locally blocked, it may still be globally blocked. This can happen because, like cookies in the jar with the lid on, the way out is blocked. Or it can happen because the path is too narrow at some point and the part will interfere with one or more others, even if it does not have liaisons with them. It can be difficult and computationally intensive to determine if there are or are not any global blockages. In general, this is a path-finding problem, often called the piano mover's problem in Artificial Intelligence. The difficulty lies in part in the fact that one does not know if all possible paths have been tried. Furthermore, a feasible path may be contorted or may require that the part in question squirm or perform the equivalent of acrobatic twists and turns in order to escape. Lots of computation is required to find such contorted paths because so many degrees of freedom must be tested in

---

[9]Some companies aspire to have their design engineers be capable of including assembly issues in their designs, but this has proven to be elusive. Too much knowledge of assembly, both in the large and in the small, is needed to identify all the constituents and their priorities and to choose wisely among them.

so many combinations. In many cases, if the assembly engineer is presented with a reasonably accurate sketch or drawing, he or she can quickly determine whether a path exists. Alternatively, the engineer can use one of a growing array of assembly simulation software to try out a path using CAD models of the parts. In any case, some knowledge of the three-dimensional shape of the parts is needed to detect global blockage.

While many assembly sequence algorithms in the research literature seek to answer all the questions automatically without involving the engineer, it may not be worth the effort. No algorithm has so far been able to guarantee finding all the global blockages in general, while people can usually eyeball a path fairly easily. More is to be gained by confronting the engineer with the design and involving him or her in its details.

On the other hand, if a design requires a part to perform acrobatic flips in order to be assembled, then the product probably should be redesigned, thus removing the worst challenges to an algorithm that seeks to answer the questions by itself. In fact, the parts in most well-designed practical products have a lot of mobility, for good engineering reasons. Most parts have only a few liaisons with their neighbors, with the result that locally each part imposes assembly constraint on only a few others. This makes sequences plentiful and easy to find.[10] Parts tend to cluster into functionally related groups, making identification of subassemblies fairly easy. Finally, most parts can be installed using simple paths. Thus algorithms that use exploded view heuristics or check for global blockages by searching only in the direction of local escape are often successful or close to it.

When the algorithm from [Baldwin et al.] is finished asking questions and processing the answers, it reports its results in two ways. One is a textual statement of precedence relations, expressed as ordering constraints on liaisons. The other is a tree, diagram, or network that displays or contains the sequences (see Figure 7-4). For example, a precedence relation for the cookies and jar problem

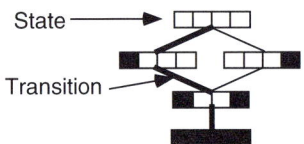

State
Transition

FIGURE 7-4. Example Liaison Sequence Diagram. Each row contains one or more state elements containing empty or filled-in cells. Each state corresponds to a feasible subassembly or as many as two feasible subassemblies. Each cell in a state corresponds to a liaison. Empty cells indicate liaisons that have not been done, while filled-in cells indicate completed liaisons. Each line between states is a transition, during which one or more liaisons are done. A path from the top state (no liaisons done) to the bottom state (all liaisons done) is a feasible liaison sequence. This diagram expresses two feasible sequences.

TABLE 7-1. Feasible Sequences for the Liaison Sequence Diagram in Figure 7-4

	Liaisons for Sequence 1	Liaisons for Sequence 2
Step 1	1	4
Step 2	4	1
Step 3	2 and 3 at once	2 and 3 at once

would read "cookies-to-jar > lid-to-jar." The symbol ">" means "before." The list of precedence relations can be analyzed to reveal which liaisons are unprecedented—that is, which liaisons do not appear on the right-hand side of any precedence relation. These unprecedented liaisons can be first. Once these have been done, they can be erased wherever they appear on the left-hand side in the precedence relations, permitting identification of those relations that are satisfied and of liaisons that can be done next. Proceeding in this way generates a list of liaisons that can be done at each succeeding stage. This information is used to create the diagram that contains all the feasible sequences.

The algorithms developed in [Homem de Mello and Sanderson] and [Halperin, Latombe, and Wilson] do not generate precedence relations. Instead, they generate a graph called an AND/OR tree that implicitly contains all the feasible sequences. This tree can then be used to build any desired sequence. An advantage of having precedence relations, illustrated in Section 7.H.4.c and Chapter 8, is that sequences can be selected or eliminated by writing additional precedence relations and imposing them on the initial set of sequences.

---

[10]Fewer liaisons per part is a characteristic that leads to large numbers of assembly sequences in a product. That is, if two assemblies have the same number of parts, the assembly with fewer liaisons per part is likely to have more assembly sequences. The minimum number of liaisons per part for an assembly of n parts is $(n-1)/n$ while the maximum is $(n-1)/2$. A small survey of assemblies shows that the average number of liaisons per part is less than 2. Section 7.M discusses this interesting topic in more detail.

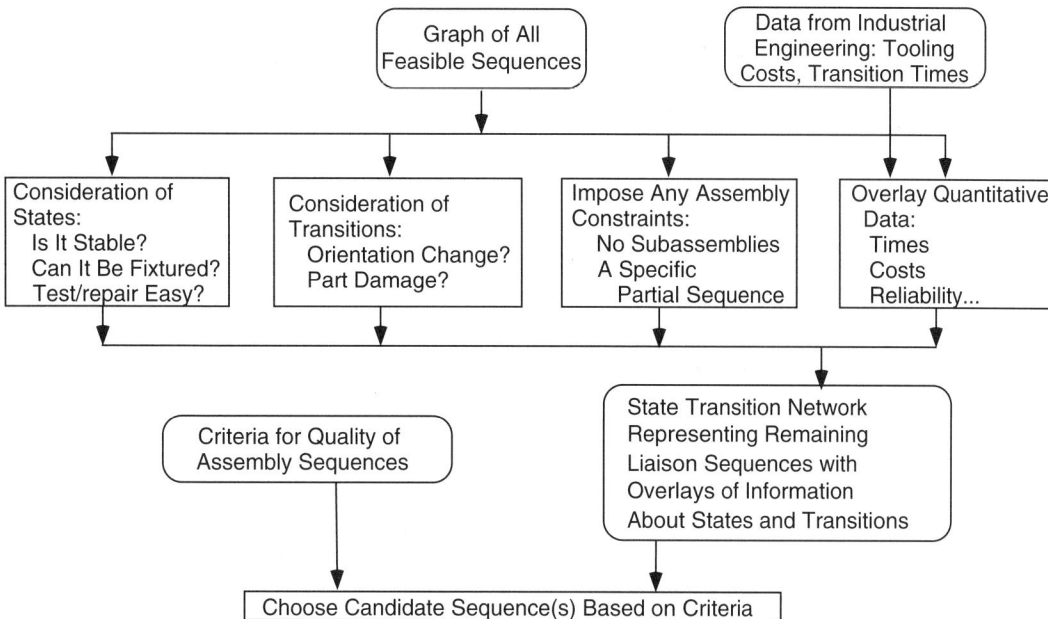

FIGURE 7-5. Flowchart of Sequence Selection Process. This process operates by finding undesirable states and transitions and deleting them.

A simplified version of a liaison sequence diagram is shown in Figure 7-4. It contains states representing feasible intermediate subassemblies and feasible transitions between states that represent accomplishment of one or more liaisons. At the top is a blank state indicating that no liaisons have been done. At the bottom is a full state indicating that all liaisons have been done. Each path from top to bottom is a feasible liaison sequence. The feasible sequences are listed in Table 7-1.

7.C.4.b. Selecting Good Sequences

Once the diagram (or other representation) of sequences is available, it can be used as an index into the feasible sequences, states, and transitions as an aid in selecting a good sequence. The method described here consists of examining states and transitions and deleting undesirable ones. This is a matter of judgement, and the engineer must balance many competing criteria. A flowchart of the process is in Figure 7-5.

In typical problems there are more sequences than transitions and more transitions than states. For this reason, it is efficient to delete states and transitions because this will eliminate many sequences at once. There is no need in general to examine every sequence. If a state has no predecessors or no successors as a result of a deletion, then the stranded state can be automatically purged.

Examining Figure 7-5 one box at a time, we can identify the different issues that the engineer takes into account when judging candidate sequences. Some are directly visible from the parts themselves, such as whether a state or transition contains some undesirable factor or is especially attractive for some reason.[11] One state may be easier for a person or machine to reach into, or it may be more stable or easier to fixture. It may place the parts in such a way that datum surfaces can be used to achieve accurate location. Transitions can similarly be examined for their suitability, ease for a person or machine to execute them, risk of part damage, and so on. Software supporting this phase of the process can present a picture of the parts involved in any state or an animation of a transition in order to aid the engineer in visualizing the situation.

---

[11]Note that a state may be arrived at by several different transitions. In calling it a state, we are appealing to the formal meaning of that word, which asserts that the arrival path does not matter, that the state will be the same state regardless of which path may have been used. Thus the transitions can be ignored while judging the states. However, each possible transition into a state must be judged on its own.

It may happen that considerations like these will leave the engineer with no desirable states or transitions at all, in which case there will be no feasible sequences. Early discovery of such situations is especially valuable because the engineer can seek remedies quickly and relatively inexpensively.

The assembly engineer can also impose restrictions on the feasible sequences, such as requiring the sequence to be linear, or requiring that a particular liaison be made as soon after another as possible, and so on. For example, it would be desirable to put a lid on a container of liquid right after pouring in the liquid to avoid spillage or contamination. Since reorientations are not represented as distinct operations in current algorithms, it is up to the engineer to impose the precedence constraint that liaisons requiring reorientation cannot be imposed on a container of liquid until after its lid has been installed.

Sequences can also be compared based on time, cost, number of reorientations, number of fixtures, and so on. The force of such criteria depends on each case and requires access to data on the cost of people and machines, the loss incurred if a product unit is damaged, and so on. For example, typical large fixtures can cost $3,000 each. Consider a moving assembly line that supports a takt time or cycle time of one minute and extends over a distance of 200 m in order to accommodate a series of assembly steps. The entire 200 m must be populated with fixtures and so must the pipeline that returns empty fixtures to the beginning of the line. If three hundred fixtures are needed, the investment will be nearly a million dollars. If assembly requires two fixtures, then the investment will be nearly two million dollars. A sequence that saves a fixture can thus be a significant advantage.

The procedure for making assembly cost comparisons is nontrivial. Many analyses of assembly cost assume that each transition can be assigned a cost. If this were true, then the lowest cost sequence would be the one whose path length, measured as the sum of all the transition costs along it, was the smallest. Efficient shortest-path algorithms could then be used to find the least-cost sequence. Unfortunately, each transition cannot be assigned a cost. As will be shown in Chapter 18, assembly cost is a function of the entire sequence, and each sequence, even if it shares some transitions with other sequences, will have its own cost. Among the reasons for this is the fact that several assembly steps might feasibly be done at a single workstation, depending on the sequence chosen. The cost of a workstation would therefore depend on which operations were done there. If several operations used the same tool, then that tool could be purchased once. If a different sequence separated those operations onto several stations, then several copies of that tool would have to be purchased, increasing the cost of that sequence. [Klein] found that different manual assembly sequences for automobile subassemblies that required assist devices and tools could differ in total cost by up to 20%, which is considered a huge difference. [Milner, Graves, and Whitney] discusses a way to search a diagram of all feasible sequences in conjunction with a path cost calculator described in Chapter 16 to find the least-cost sequence.

Different sequences can create finished assemblies of different quality. We have dealt with this issue in previous chapters and will do so again later. In general, assemblies joined by fully constrained assembly features will have the same assembly-level variation regardless of assembly sequence, whereas, for a variety of reasons, assemblies with some underconstrained assembly features will have sequence-dependent variation at the assembly level. The next chapter deals with this issue.

7.D. THE BOURJAULT METHOD OF GENERATING ALL FEASIBLE SEQUENCES

Since the purpose of this book is to present a consistent approach to assembly rather than to explain every algorithm in detail, we will give the reader only the flavor of how an assembly sequence algorithm works by illustrating the Bourjault method on a simple example. More efficient algorithms exist, but they are more complex than necessary to give the flavor. The rules applied in this example implement a textual analysis rather than the graph theory methods originally used by Bourjault. The textual rules are stated in complete form in Section 7.L.

The method will be applied to the simple planar two-dimensional four-part assembly shown in Figure 7-6. These are polygonal parts whose relative assembly/escape directions are shown in the figure. The figure also shows

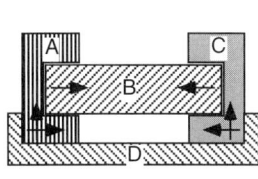
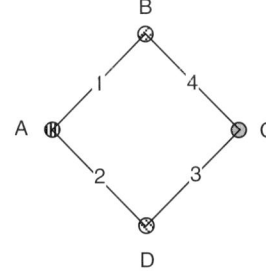

FIGURE 7-6. Simple Example Assembly and Its Liaison Diagram. Arrows on the assembly drawing indicate escape directions for part B relative to parts A and C, and for parts A and C relative to part D.

the liaison diagram. The questions are in the form $R(i; B)$, where i is a liaison number and B is a set of liaisons. This is interpreted to mean "Can liaison i be done when liaisons B have already been done?"[12] If the answer is NO, then the reason must be pursued by finding out which members of B are the cause. A question of this form must be asked for every liaison i in the assembly. In this example, we will therefore be faced with four questions. The sequence in which they come up is arbitrary, as are the numbers assigned to the liaisons.

7.D.1. First Question: $R(1;2,3,4)$

This question asks if liaison 1 can be done when liaisons 2, 3, and 4 have already been done. This question cannot be addressed as asked because liaisons 2, 3, and 4, if already done, comprise a loop in the liaison diagram with only one undone arc. According to the loop closure rule, no such situation can exist. We must therefore decompose this question, preserving its intent, by creating legal open loops against which to test liaison 1. We do this by successively opening liaisons 2, 3, and 4 and restating the question three times, once for each opened liaison.

For example, let us open liaison 2, creating the question $R(1;3,4)$. That is, can we do liaison 1 when liaisons 3 and 4 have already been done? Referring to Figure 7-6, we can see that this is the same as asking if we can mate part

A to part B when a subassembly of parts B, C, and D has already been made. The answer is obviously NO, but we are not done because we do not know which of the liaisons 3 and/or 4 are the reason for the NO. To find this out, we must further decompose the question and generate two subquestions: $R(1;3)$ and $R(1;4)$. The first asks if we can mate A and B when C and D have been mated, to which the answer is YES. Similarly, the second subquestion asks if we can mate A and B when B and C have been mated, to which the answer is again YES. Therefore, the only reason why the original question generates NO is because of the combination of liaisons 3 and 4. Therefore, we conclude that we cannot do liaison 1 if liaisons 3 and 4 have already been done. Since 1 cannot be *after* the combination of 3 and 4, then 1 must occur *before or at the same time as* 3 and 4. We write this as 1 >= 3,4. So we have our first precedence relation.

Having opened liaison 2, let us now close it and open liaison 3, creating the question $R(1;2,4)$. The answer to this question is NO, so we must again decompose it into two subquestions $R(1;2)$ and $R(1;4)$. The latter was answered YES already, and the former yields the answer YES as well. We conclude therefore that it is the combination of 2 and 4 that prevents 1 from being done, resulting in the second precedence relation: 1 >= 2,4.

Having opened liaison 3, let us now close it and open liaison 4, creating the question $R(1;2,3)$. The answer to this question is NO, so again we must decompose it. When we do, we find that it yields questions that we have already answered YES, so we end up with our third precedence relation 1 >= 2,3.

7.D.2. Second Question: $R(2;1,3,4)$

This question is approached in exactly the same way, and yields two additional precedence relations: 2 >= 1,3 and 2 >= 3,4.

7.D.3. Third Question: $R(3;1,2,4)$

This question yields two additional precedence relations: 3 >= 1,2 and 3 >= 2,4.

7.D.4. Fourth Question: $R(4;1,2,3)$

This question yields three additional precedence relations: 4 >= 1,2, 4 >= 1,3, and 4 >= 2,3.

---

[12] In Bourjault's original method there was a second question $S(j, B)$, which means "Can liaison j be done when liaisons B have not been done?" Experience showed that unless there were flexible parts, question S never resulted in any new precedence relations beyond those generated by question R.

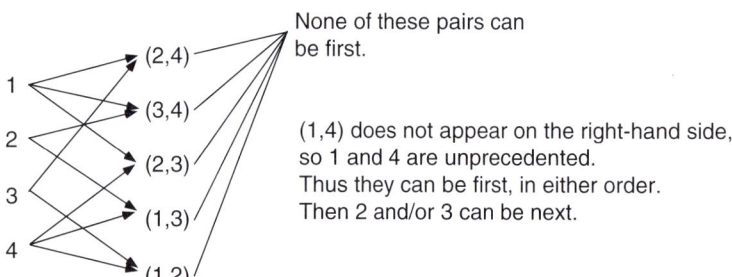

FIGURE 7-7. **Diagrammatic Summary of Precedence Relations for an Example Assembly.** To read the diagram, pick an arrow and read it to say: "The liaison at the tail must be done before the liaison(s) at the head." Liaisons 1 and 4 do not appear on the right-hand side, so they are unprecedented. Either one can be first, followed by the other.

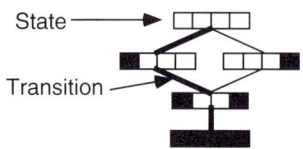

FIGURE 7-8. **Liaison Sequence Diagram for the Example Assembly in Figure 7-6.** There are two feasible liaison sequences. Both involve making a subassembly of parts A, B, and C, and then placing this subassembly in part D. The subassembly can be made in one of two subsequences.

7.D.5. Reconciliation of the Answers

Even though each individual precedence relation contains the relation ">=" (such as $1 >= 3,4$), it is not possible to perform liaisons 1, 3, and 4 simultaneously. Doing so would violate the loop closure rule. An attempt to remedy this violation by closing the loop would require doing 1, 2, 3, and 4 simultaneously, which violates several other precedence relations. The result is that the "=" must

be dropped from the relations, so that they all contain only ">."[13]

7.D.6. Precedence Question Results

A diagram summarizing the results of answering the questions appears in Figure 7-7. This diagram shows that all of the precedence relations can be summarized as $1,4 > 2,3$. This means that liaisons 1 and 4 can be first in either order and must both be done before 2 and 3 (in either order) can be done. The loop closure rule requires 2 and 3 to be done simultaneously, however. Only local constraints had to be investigated in order to find all NO answers, of which there are 10, corresponding to the 10 precedence relations. Global constraints had to be investigated to verify all YES answers, of which there are 6 unique ones. Many of these were looked at multiple times but were answered trivially easily because they had been answered already.

The liaison sequence diagram is shown in Figure 7-8. It shows that there are two feasible liaison sequences. This concludes the example.

7.E. THE CUTSET METHOD

Here we briefly illustrate the cutset method of generating precedence relations ([Baldwin], [Whipple]). The method is illustrated in Figure 7-9. Eight cuts are possible in the liaison diagram, of which two are shown. Each cut shown removes one part at a time. There are four such cuts. In addition, there are two cuts that address removing two parts

at a time. Questions concerning removing two parts at a time do not reveal any new precedence relations.

---

[13]It should be noted that the distinction between ">" and ">=" cannot be made with the conventional precedence diagram used for line balancing.

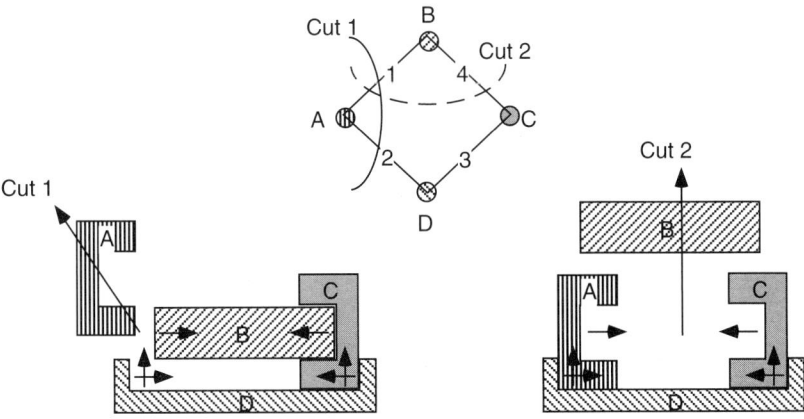

All Questions Except the Last Are Answered by Inspecting Local Freedom.

Cut 1: R(1,2;3,4)? No: 1,2 >= 3,4

Cut 2: R(1,4;2,3)? No: 1,4 >= 2,3

R(3,4;1,2)? No: 3,4 >= 1,2

R(2,3;1,4)? Yes.

1,4 are unprecedented, so they can be first.

FIGURE 7-9. Example of the Cutset Method Applied to the Example Assembly in Figure 7-6.

7.F. CHECKING THE STABILITY OF SUBASSEMBLIES

A classic problem in assembly planning is to determine if a subassembly is stable. Stability can be defined in a number of ways. Our definition is as follows: A subassembly is stable if it does not spontaneously disassemble under the effects of gravity or the accelerations imposed by material handling or assembly activities. Sufficient conditions for stability include the presence of fasteners, press fits, or sufficient friction to resist external forces. If one or more of these conditions apply to each liaison, then it is easy to prove that an assembly is stable. However, detecting necessary conditions can be difficult. Here is a simplified approach to the problem.

Any subassembly is defined by its liaisons and the motion freedom they permit. Liaisons are simply abbreviations for assembly features and their constraints, as discussed in Chapter 4. In many cases, complex combinations of liaisons can restrain a part from being able to move even when no liaison by itself is capable of completely preventing the part from moving.

With this preamble, the method can be stated as follows:

1. Start with the part that is in the fixture and mark it "stable."

2. Inspect each liaison that this part has with other parts. If that liaison can stabilize the part it links to (by means of a fastener, press fit, adhesive, etc.) or if the escape direction of the linked part has a negative dot product with gravity and the set of all disturbing forces (allowing for friction, if any), then mark that part "stable."

3. Move on to another part that is marked "stable" and repeat step 2 for that part, considering one at a time all liaisons it has with unmarked parts.

4. When all parts have been considered by step 2, then if all are marked "stable," then the subassembly is stable. Otherwise go to step 5.

5. If a part is not marked "stable," then consider all the liaisons it has with stable parts and see if combinations of liaisons to the unmarked part can stabilize it. This can be ascertained by taking dot products of their escape directions to see if a common escape

direction exists. If no common escape direction exists, or if the common direction has negative dot product with gravity and the set of all disturbing forces (allowing for friction, if any), then the part is stable.

6. If it is possible by repeated application of step 5 to mark all the parts "stable," then the subassembly is stable. Otherwise, it is not.

7.G. SOFTWARE FOR DERIVING ASSEMBLY SEQUENCES

This section briefly describes two research-based software systems for assisting an assembly engineer to find assembly sequences.

7.G.1. Draper Laboratory/MIT Liaison Sequence Method

The Draper/MIT system is based on methods described in Section 7.C and [Baldwin et al.]. It generates all feasible sequences for which no intermediate state consists of more than two subassemblies.[14] Such sequences are called two-handed because only two hands are required to mate all the items present in any given state. It consists of several linked modules:

- Input of the liaison diagram and mutual escape directions for parts
- Question–answer method of eliciting precedence relations
- Generation of the diagram of all feasible sequences
- Editing of the diagram by the engineer by eliminating states and transitions and imposing additional precedence relations and other conditions

A sample window from this software appears in Figure 7-10.

7.G.2. Sandia Laboratory Archimedes System

The Archimedes system is described in [Kaufman et al.], [Jones, Wilson, and Calton], [Halperin, Latombe, and Wilson], and [Latombe]. This method makes use of robot motion planning methods that assume the parts are polyhedral. In this case, it can be shown that the algorithms run in polynomial time. That is, the time it takes to automatically generate the assembly sequences for a product containing

K parts is proportional to K^n. The method makes use of many of the rules cited above, such as the subset rule and the superset rule, but in general it seeks to avoid asking the user any questions. Instead, it makes use of a technique called a nondirectional blocking graph to determine local blockages as well as a limited set of global blockages. Blockages are detected by moving CAD models of the parts in different directions a small amount and testing for interferences. The global blockages found are those that can be detected by moving a part infinitely far in the same direction as the local escape direction(s). With additional computational burdens, the method can investigate global escape paths that consist of two straight line paths, or even more, though the computational burden will grow rapidly. This method is therefore best suited to products whose parts primarily assemble locally and globally along the same single axis.

After the system has determined the local and one- (or two-) direction global constraints, it generates an AND/OR tree ([Homem de Mello and Sanderson]) representation of the feasible assembly sequences and selects one at random to present to the assembly engineer. The assembly engineer can alter this sequence using filters, of which a partial list appears in Table 7-2.

TABLE 7-2. Example Filters Available in Archimedes

Filter Name	Meaning
REQ_TOOL	The step must use this tool
REQ_VERTICAL	Motion must be vertical
REQ_FASTENER	The next step must install this fastener
REQ_LINEAR	The assembly process must be linear, with no subassemblies
REQ_SUBSEQUENCE	A particular set of steps must be done in the given sequence
PHB_STATE	This state is prohibited
MIN_REORIENT	Pick a sequence with minimum reorientations

Source: [Jones, Wilson, and Calton].

---

[14] As always in this book, a single part qualifies as a subassembly for counting purposes.

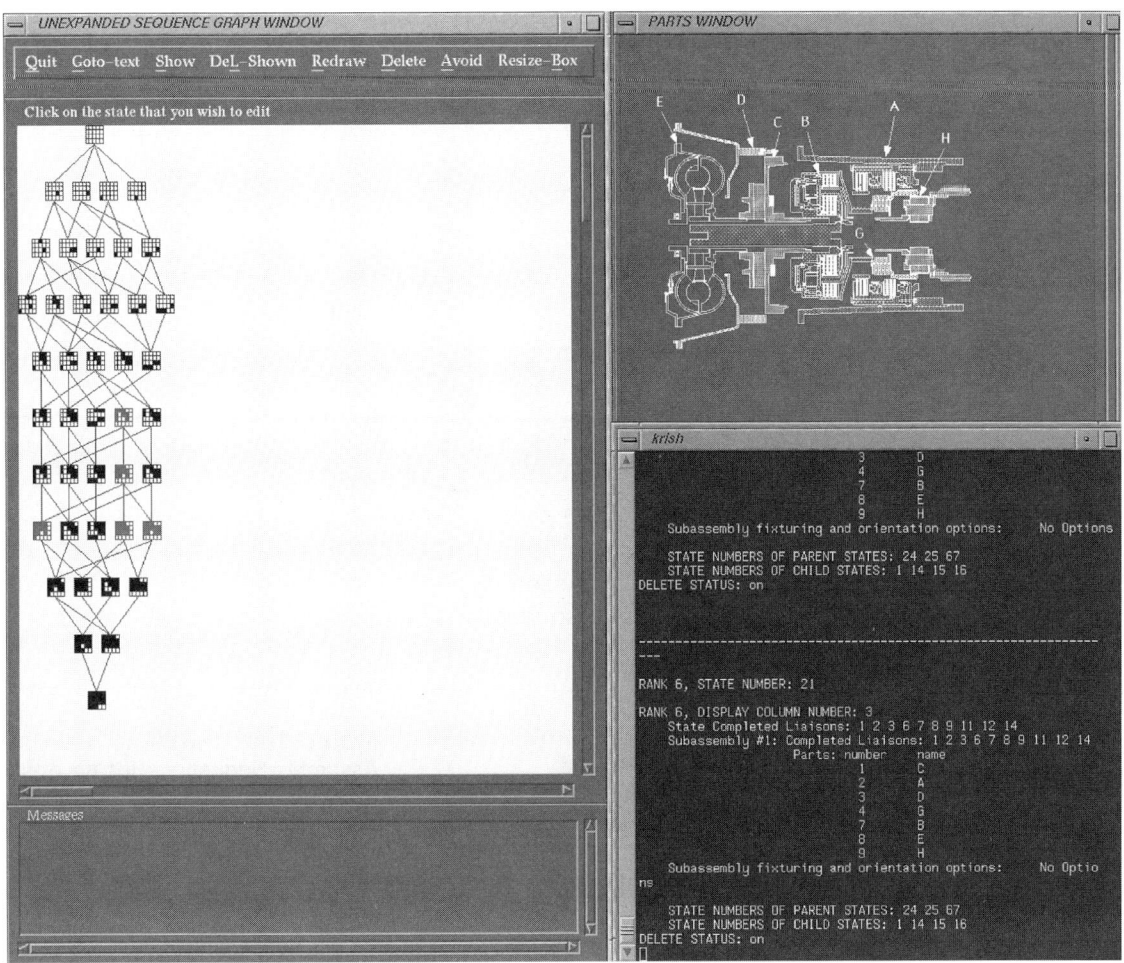

FIGURE 7-10. Sample Window from Draper/MIT Assembly Sequence Software. This window shows a phase of editing the sequences of a large truck automatic transmission. The state illustrated in the parts window at the upper right is the one that is in the 7th row, 4th position from the left, in the sequence diagram at the left. The lower right window allows commands to be typed in. A row of command buttons is available at the top left. These permit the user to delete states or transitions or to request that the diagram be redrawn.

7.H. EXAMPLES

This section presents several examples of assembly sequence derivation and/or selection. These are (1) an automobile alternator sequence suitable for robot assembly, (2) redesign of a sequence to improve the assembly process capability and prevent assembly errors, (3) assembly of a consumer product illustrating the use of phantom liaisons to increase the number of feasible sequences, and (4) a detailed analysis of an industrial product.

7.H.1. Automobile Alternator

In 1977–1978, the author and his colleagues at the Charles Stark Draper Laboratory, Inc., built a pioneering robot assembly system that assembled automobile alternators ([Nevins and Whitney, 1978]). Figure 7-11 and Figure 7-12 show several possible sequences for the alternator. We used sequence 3 in Figure 7-12. It is one of

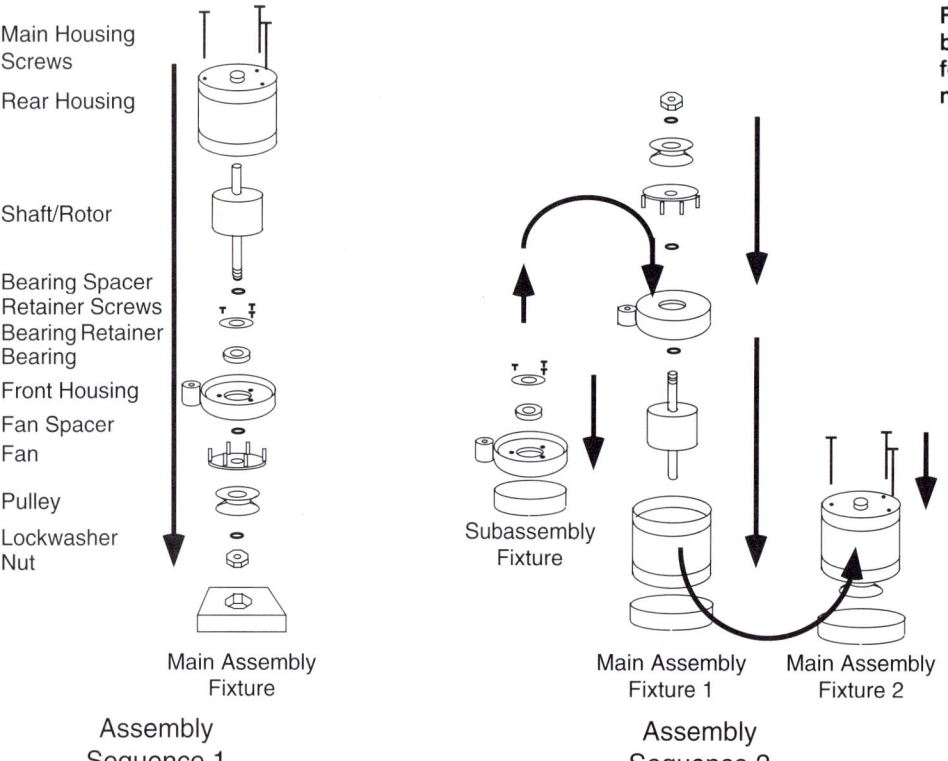

Main Housing
Screws

Rear Housing

Shaft/Rotor

Bearing Spacer
Retainer Screws
Bearing Retainer
Bearing
Front Housing
Fan Spacer
Fan

Pulley

Lockwasher
Nut

Main Assembly
Fixture

Assembly
Sequence 1

Subassembly
Fixture

Main Assembly
Fixture 1

Main Assembly
Fixture 2

Assembly
Sequence 2

FIGURE 7-11. Two Feasible Assembly Sequences for an Automobile Alternator.

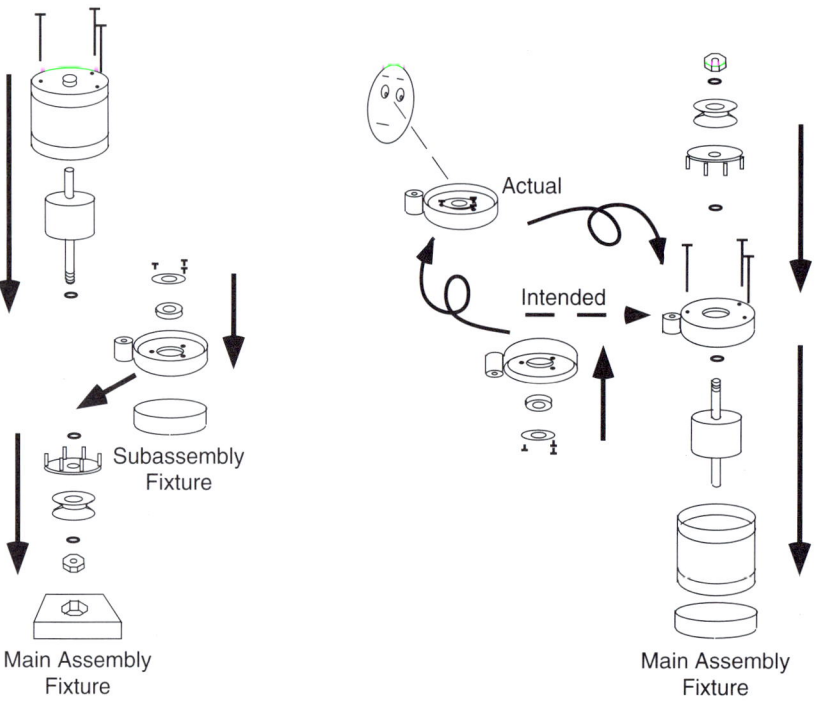

Subassembly
Fixture

Main Assembly
Fixture

Assembly
Sequence 3

Actual

Intended

Main Assembly
Fixture

Assembly
Sequence 4

FIGURE 7-12. Two More Feasible Assembly Sequences for an Automobile Alternator.

196

many that were generated manually by holding the parts in our hands or waking up in the middle of the night after another one popped into our heads. The sequence we chose avoids reorientations of the product and permitted us to use a robot with only four degrees of freedom.

Sequence 1 in Figure 7-11 looks attractive because it can be done in one unbroken sequence of approaches from a single direction. No reorientations are needed. The difficulty with this sequence is that it is hard to access the front housing in the presence of the fan in such a way that the bearing retainer can be kept stationary while its screws are being driven. If the retainer is not held, it will twist out of position when the first screw is tightened.

Sequence 2 uses the "base part" heuristic and starts the assembly sequence by placing the rear housing in a fixture and adding parts to it. The front housing is built as a subassembly, permitting use of a fixture that can grasp the retainer from below and hold it while the screws are inserted. However, this sequence requires two reorientations, one for the front housing and the other for the whole assembly so that the main housing screws can be inserted and tightened. The last reorientation therefore involves reorienting an unfastened subassembly, something that is not desirable.

Sequence 3, like sequence 1, would have been skipped by the "base part" heuristic. It differs from sequence 1 in that the front housing is built on a separate fixture like sequence 2 uses. A separate fixture represents additional cost. However, no reorientations are needed.

Sequence 4 was observed by the author at a large Japanese automobile components manufacturer. It conforms to the "base part" heuristic, although it is not known if this heuristic played a role in generating the sequence. This sequence also builds the front housing as a separate subassembly but it is clear that the manufacturing engineers intended to avoid reorientations. Thus the subassembly was built in the orientation shown, with the bearing, retainer, and screws somewhat precariously installed upside down from below. Yet the author observed a person picking up each housing as it exited the subassembly station and inspecting it to be sure that all the parts were present. If the engineers had been willing to pay for a person to reorient the subassembly, they could just as well have built it right side up in the first place and taken advantage of easier assembly from above. The subassembly station they built must have proven to be less than

completely reliable, requiring a person to check each unit.[15]

This example shows that the same criteria applied to the same product can result in many candidate sequences surviving the winnowing process. It also shows that different engineers can select different sequences and that sometimes their best intentions for how the assembly system will operate are not borne out.

7.H.2. Pump Impeller System

Figure 7-13 shows the parts of a simple pump impeller system. Manual assembly consists of screwing the bottom washer to the shaft, then sliding on the impeller, then screwing the top washer onto the shaft. The finished impeller system is installed into a housing that is not shown. This is a precision item and part-part clearances are small. It was assembled manually for years, but the manufacturer wanted to switch to a robot or other automatic method. The obvious assembly sequence is shown in the figure. It places the bottom washer in a fixture that interfaces to the washer on a nonfunctional surface. To avoid overconstraint, the small hole in the fixture is large enough that the end of the shaft will never touch it.

A variation analysis of the type described in Chapter 6, Section D.2, showed that approximately 4% of assemblies would fail due to wedging[16] or because the hole in the impeller and/or the top washer would miss the end of the shaft. The cause of this failure is the relatively poor tolerance on the orientation of the nonfunctional fixturing surface of the bottom washer. The operative surface, the one that interfaces to the impeller, is carefully toleranced to be perpendicular to the axis of the hole through it so that it will address the impeller uniformly. The nonfunctional surface has no role in the operation of the product and also plays no role in manual assembly. The operator simply feels around when mating the impeller to the shaft. No one anticipated

---

[15]Two points to note: First, in sequences 2 and 4, the nut is tightened by rapidly twisting it against the inertia of the rotor. Second, in sequence 4, the alternator was redesigned so that the main housing screws are installed from the front. This redesign is relatively easy to accomplish. Applying it to sequence 2 would save one reorientation.

[16]Wedging is explained in Chapter 10. When the clearance is small and the coefficient of friction is high, the parts can become locked in a tilted position and are said to be wedged.

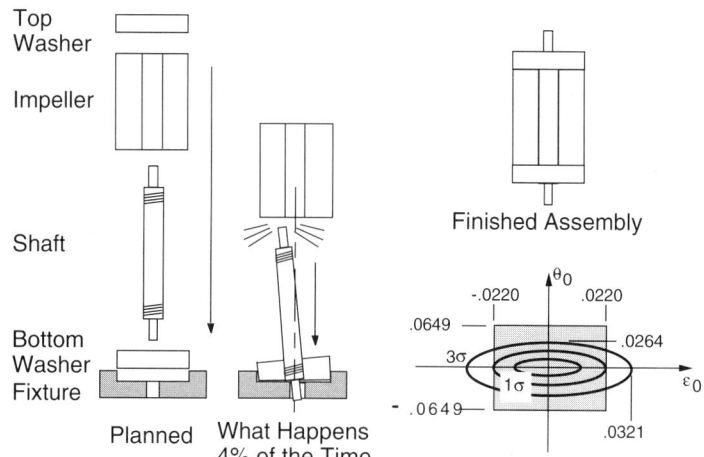

FIGURE 7-13. First Candidate Assembly Sequence for the Pump. The bottom surface of the bottom washer is used as a fixturing surface. This surface is nonfunctional and was not toleranced to be close to perpendicular to the hole into which the shaft assembles. The result is that the shaft could be tilted in the fixture, making assembly of the impeller impossible. The graph at the lower right shows the predicted 3σ limits of lateral and angular error of the top end of the shaft relative to an impeller centered over the fixture, based on the given tolerances. The shaded region is the combination of lateral and angular errors that will permit assembly, according to Part Mating Theory as explained in Chapter 10. The 2σ contour is inside the permissible region but the 3σ contour is partly outside. The extra area inside the 3σ contour and outside the permissible region represents the percentage of assembly attempts that could cause the impeller's hole to miss the shaft.

automatic assembly or thought that this nonfunctional surface would need to be held to a tight tolerance perpendicular to the hole for any other reason. Since the orientation of this surface was not toleranced tightly, the shaft ended up varying widely in orientation relative to the fixture, and, more critically, the location of its upper end varied so much that the impeller could not be mated to it with 100% certainty by a machine that carries the impeller to a nominal mating point on the fixture's vertical axis. Due to large investments in tooling and validation of the design, no redesigns of any kind were permitted. Instead, a different assembly sequence and fixturing strategy was adopted.

Figure 7-14 shows the sequence that was adopted. The sequence is based on the observation that the ends of the shaft must assemble to the outer housing of the pump very accurately. Thus they are carefully toleranced in terms of their diameter as well as their concentricity with the axis of the shaft. Although this region was small, it was the only surface available that had the ability to place the top end of the shaft repeatably in position to receive the remaining parts. The first fixture interfaces to the washer and does not present enough angular variation to threaten the thread

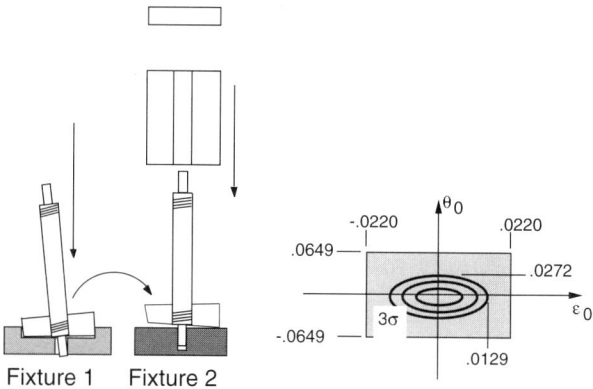

Fixture 1   Fixture 2

FIGURE 7-14. Second Assembly Sequence for the Pump. This sequence transfers the assembly from the poorly toleranced bottom washer to the well-toleranced end of the shaft in order to permit successful assembly of the impeller and top washer. The graph at the right shows the same wedging-avoidance region as in Figure 7-13 but indicates that virtually all assemblies will succeed since the 3σ curve of position and angle error falls well inside this region.

mating required when the shaft is inserted. (The relative forgiveness of threaded joints to angular error is discussed in Chapter 10.) This fixture carefully avoids touching the end of the shaft. The second fixture interfaces to the end of the shaft and carefully avoids touching the washer except at the minimal point required to constrain the shaft–washer subassembly in the vertical direction.

This example shows that small design features in a product can have fatal but hard to predict effects on certain assembly strategies. These effects often are invisible if manual assembly is used but intervene forcefully in mechanized assembly. Their influence will not be understood until too late if assembly analysis is delayed until after the parts are designed and toleranced for function. The example also illustrates the difference between assembly features and functional features. If a feature is intended for functional use, the design engineer will devote some care to its tolerances, but there is no guarantee that these tolerances will suit any assembly applications that the assembly engineer may want to use the feature for. In this example, the assembly engineers were lucky to find a functional feature that they could recruit as an assembly feature.

A feature that is intended from the start to be an assembly feature can be designed as such, starting from the idea that it will provide constraint at least temporarily, and concluding by giving it the tolerances it needs for its assembly role. These may or may not be more stringent than the tolerances needed for a permanent functional role. In order to take advantage of this opportunity or to avoid problems later during assembly, the design of assembly sequences should begin during concept design of the product. This was not done in the case of the pump, and no harm came of it until the switch from manual to machine assembly was attempted.

7.H.3. Consumer Product Example[17]

This product, a home juicer, was introduced in Chapter 3, where the idea of the phantom liaison was introduced. Here we will see the difference between assembly sequences possible with and without these phantom liaisons. The liaison diagram appears in Figure 7-15. The Draper/MIT software was used to develop the sequences.

---

[17]This example was prepared from materials developed by MIT students Alberto Cividanes, Jocelyn Chen, Clinton Rockwell, Jeffrey Bornheim, Guru Prasanna, Rasheed El-Moslimany, and Victoria Gastelum. Alberto Cividanes prepared the part drawings.

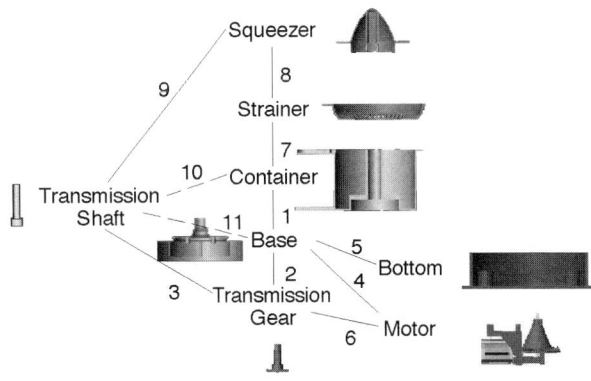

FIGURE 7-15. Liaison Diagram for the Juicer, Showing Phantom Liaisons 10 and 11.

If we do not allow the phantom liaisons, the precedence relations are given by

$$
\begin{aligned}
&7 >= 8 \,\&\, 9 \\
&2 \,\&\, 6 >= 4 \\
&3 >= 1 \,\&\, 2 \\
&4 \,\&\, 6 >= 2 \,\&\, 5 \\
&7 \,\&\, 8 >= 1 \,\&\, 2 \,\&\, 3 \,\&\, 9 \\
&8 >= 7 \,\&\, 9 \\
&2 >= 3 \\
&2 \,\&\, 4 >= 5 \,\&\, 6 \\
&9 >= 1 \,\&\, 7 \,\&\, 8
\end{aligned}
\tag{7-1}
$$

In this case, the software asked 46 questions which took the author about 25 minutes to answer. The resulting liaison sequence diagram appears in Figure 7-16. Sequences selected from this diagram are suitable for manual or automatic assembly using one hand and a fixture to hold the base upside down.

If the phantom liaisons are allowed (Figure 7-17), the precedence relations are given by

$$
\begin{aligned}
&10 >= 9 \\
&2 \,\&\, 6 >= 4 \\
&10 \,\&\, 11 >= 1 \\
&4 \,\&\, 6 >= 2 \,\&\, 5 \\
&7 \,\&\, 8 >= 9 \,\&\, 10 \\
&2 \,\&\, 11 >= 3 \\
&2 \,\&\, 4 >= 5 \,\&\, 6
\end{aligned}
\tag{7-2}
$$

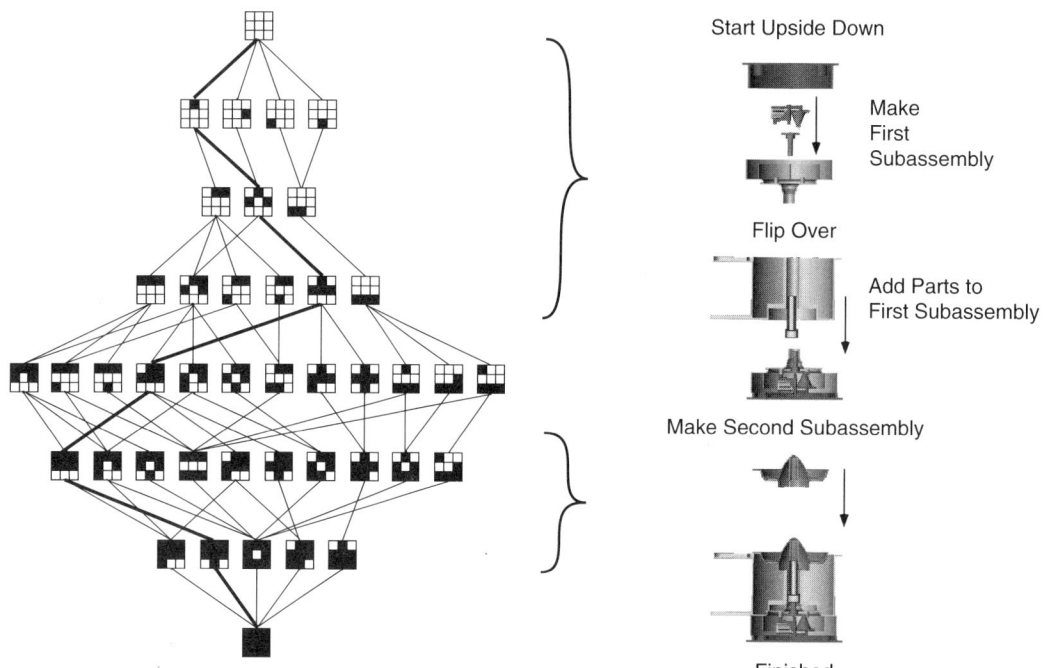

FIGURE 7-16. Liaison Sequence Diagram for Juicer without Phantom Liaisons. One feasible assembly sequence, shown by heavy lines in the liaison sequence diagram, is illustrated on the right. This sequence is suitable for automatic assembly because every part is placed so that a functional liaison is made. Each such liaison properly constrains the parts and holds them in controlled locations, which is important for automatic assembly.

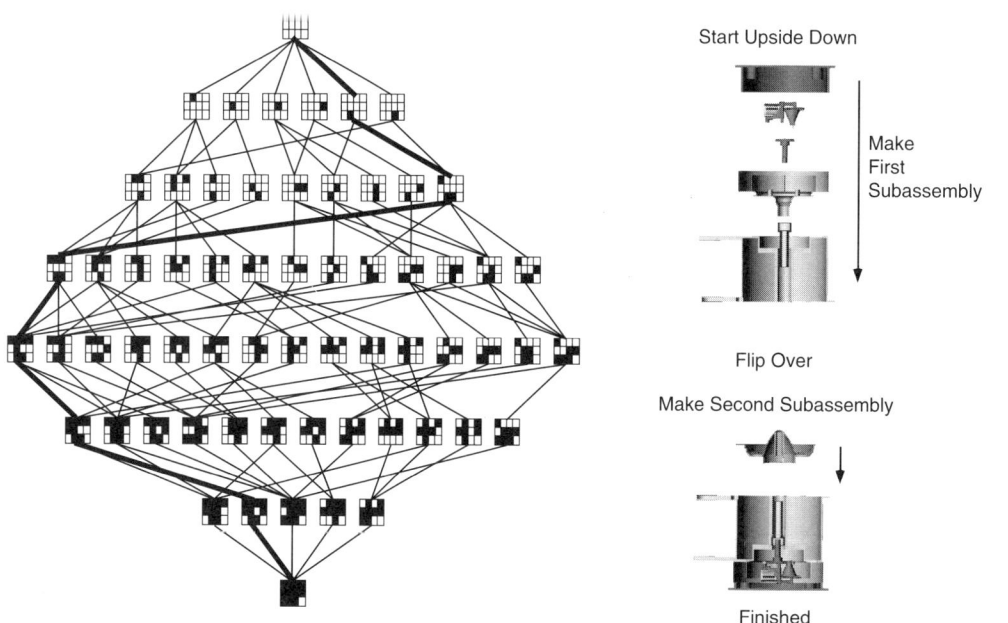

FIGURE 7-17. Liaison Sequence Diagram for Juicer if Phantom Liaisons Are Allowed. One feasible assembly sequence is shown. It is relatively easy for manual assembly but not so for automatic assembly since the transmission shaft is not tightly constrained when it is first placed in the container. A person can easily stabilize it by holding container and shaft in one hand while putting the gear in with the other hand and pressing the gear and shaft together.

200

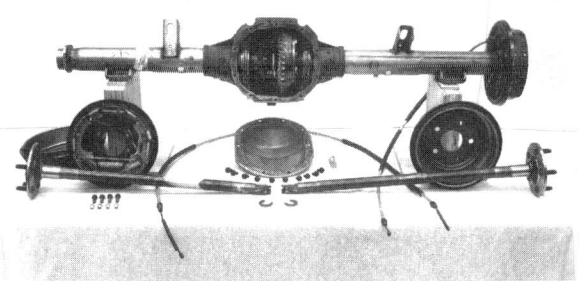

FIGURE 7-18. Photo of Ford Rear Axle Assembly. At the rear of the photo is the axle carrier subassembly. It comprises the differential housing, into which the axle tubes have been pressed. Inside the housing are the differential gears. This subassembly arrives at the final assembly line, where the rest of the parts are added. One of these parts, a brake shoe assembly (called a backing plate) has already been installed on the end of the right hand axle tube. In the center foreground at the end of each axle are the C-washers that are used to retain the axle shafts in the assembly.

7.H.4. Industrial Assembly Sequence Example

The author and his Draper Laboratory colleagues planned a robot assembly system for Ford that addressed heavy rear axle assemblies. Figure 7-18 is a photo of this assembly and its parts. [Figure 7-21 contains, among other things, a parts list and drawing.] It was (and still is) assembled manually, and robot assembly would have been a challenge. Part of our work included a thorough assembly sequence analysis.

7.H.4.a. Discussion of the Assembly

Several aspects of the axle's assembly are unusual. The one that departs most from the assumptions of the analysis methods described in this chapter is the need to partially disassemble the axle during its assembly. The axle arrives at the final assembly line partially assembled. The assembled parts include the carrier assembly (differential housing and axle tubes) and the differential gears. The gears are held together by a pinion shaft. This shaft must be partially removed in order that the axles can be installed and locked in place with parts called C-washers. See Figure 7-19 and Figure 7-20. The procedure to install the C-washers is as follows:

1. Loosen and withdraw the pinion shaft bolt.
2. Withdraw the pinion shaft until it clears the central region of the differential.

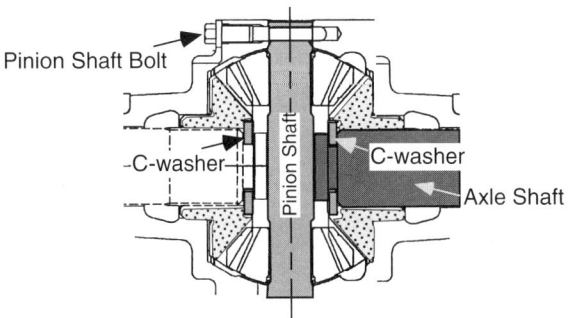

FIGURE 7-19. Drawing of Differential Showing Axle Shaft, Pinion Shaft, and C-Washer in Final Assembled Positions.

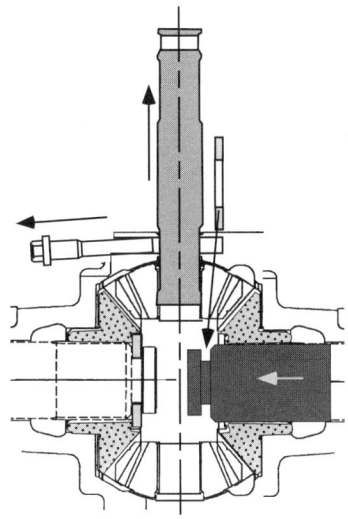

FIGURE 7-20. Drawing of Differential Showing Pinion Shaft Removed and the Right Axle Shaft Pushed in and Ready for Installation of the C-Washer.

3. Insert an axle until its end protrudes into the central region of the differential.
4. Insert a C-washer in the groove in the end of the axle.
5. Pull the axle back out as far as it will go.
6. Repeat for the other axle.
7. Replace the pinion shaft and the bolt.

7.H.4.b. Precedence Relations and Liaison Sequences

To simplify the analysis of this assembly, certain steps have been consolidated and called liaisons. (See Figure 7-21, which shows the liaison diagram.) One of these

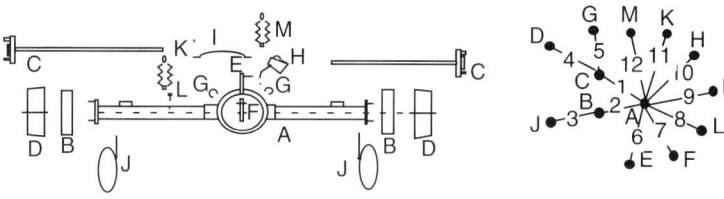

Parts of Rear Axle

Liaison Diagram

FIGURE 7-21. Parts List, Liaison Diagram, and Precedence Relations for the Rear Axle.

Parts	Liaisons	Precedence Relations
A = Carrier Assy	1 = C to A	2 > 1
B = Backing Plate	2 = B to A	5 > 4
C = Shaft	3 = J to B	1 & 2 & 6 > 5
D = Brake Drum and Nut	4 = D to C	5 > 7
E = Pinion Shaft & Bolt	5 = G to C	11 > 8
Withdrawn	6 = E to A	10 > 9
F = Pinion Shaft & Bolt	7 = F to A	12 > 10
Re-inserted	8 = L to A	12 > 11
G = Push In Shaft & Insert C-	9 = I to A	3 > 1 & 4 & 5
Washer & Push Shaft Out	10 = H to A	7 > 10
H = Oil	11 = K to A	9 > 11
I = Cover	12 = M to A	
J = Brake Cable, Coiled		
K = Final Press Test		
L = Air Test Plug		
M = First Pressure Test		

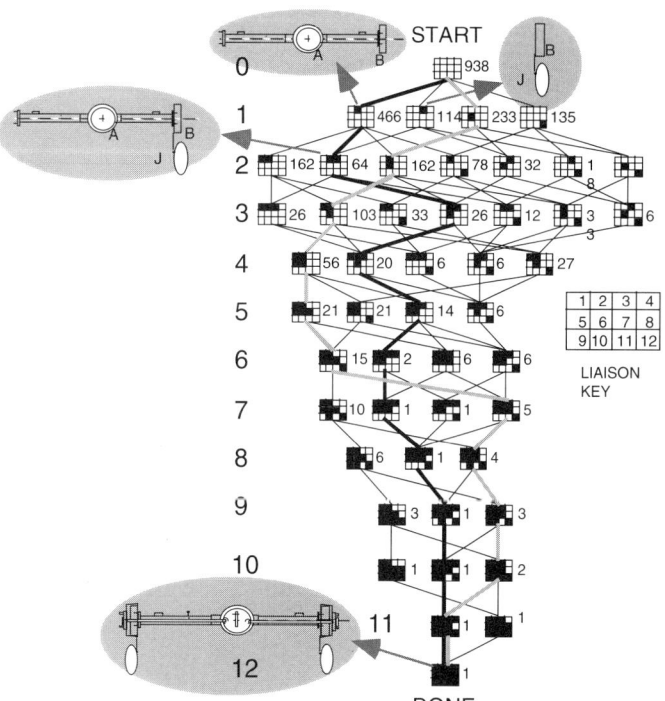

FIGURE 7-22. Liaison Sequence Diagram for the Ford Rear Axle. This figure shows all the feasible assembly sequences for the axle. A few are highlighted in thick lines. Each state corresponds to a feasible subassembly, of which a few are shown. There are 938 sequences altogether. This total is determined by starting at the bottom of the diagram and marking each state according to the following rules. Mark the bottom state with a 1. Go up one row and mark each state by adding up the numbers in the states below that link to it by transitions. Continue up row by row until the top state is reached. The number assigned to this state is the total number of sequences. The number at each intermediate state is the number of sequences possible starting from that state.

is "loosen pinion shaft bolt and withdraw pinion shaft." This is liaison E. Reversing this set of steps is called liaison F. "Push an axle shaft into the empty space left when the pinion shaft is withdrawn, install a C-washer, and pull the shaft out as far as it will go" is called liaison G. While this operation must be done twice, once for each axle, it is called out only once since the precedence relations for each are the same and thus they must be done one right after the other.

Figure 7-21 also shows the results of the precedence relation analysis. This analysis was done using the method from [De Fazio and Whitney]. Figure 7-22 shows the liaison sequence diagram. A few of the feasible states are illustrated.

Let us consider a few of the precedence relations in Figure 7-21. The first one states "2 > 1." Liaison 1 is C to A, meaning "put the axle shaft into the carrier assembly." Liaison 2 is B to A, meaning "mate the backing plate to the carrier." The precedence relation says that the backing plate must be installed before the axle shaft can be put in. If the axle is put in first, then the large hub on the end of the axle will make it impossible to put the backing plate on the carrier and tighten the fasteners. If we look at Figure 7-22, we can see that anywhere liaison 1 has been done, liaison 2 has been done at a previous state.

Look at the first row of states in Figure 7-22. It says that liaisons 2, 3, 6, and 12 can be done first. In the existing manual sequence, liaison 2 is done first. But liaison 12 is an interesting alternative. It is an air pressure test to see that there are no leaks. It was determined that most leaks are the result of split axle tubes due to the press operation that inserts them into the differential housing. Since these splits are there right at the beginning, it makes sense to detect them before adding a lot of other parts, none of which can repair the leak. To accomplish this test, it would be necessary to plug up all the openings in the carrier assembly. Currently the test is done at the end of assembly, when these openings are filled with parts and differential lubricant. While artificially plugging the holes would be inconvenient, disassembling a leaking axle assembly currently involves dumping out the lubricant (put in as liaison H) and handling oily parts, which is a messy job.

Another appealing alternative first liaison is 3, which joins the brake cable to the backing plate. In the current assembly, this is done after the backing plate is mated to the carrier. The nominal orientation of the carrier at this point places the opening of the differential case up so that the C-washers can be inserted. But this places the installation point for the brake cables inconveniently underneath. To make brake cable insertion possible from above, which is easier, the carrier must be manually rotated 180°. This is fatiguing for the operators.

If the cable were to be attached to the backing plate first as a separate subassembly, then the backing plate could be oriented at the operator's convenience independently of the bulky and heavy carrier. Ford chose, however, to put the cable on after the backing plate was installed on the carrier, and after the C-washers had been installed. The reason was that if a C-washer fell into the differential case, then it would fall out when the carrier was rotated to permit the cable to be installed.

7.H.4.c. Winnowing the Sequences to a Few

Once we have all the feasible sequences, we can impose some commonsense orders to state transitions or assembly moves:

1. The axles shafts should be secured with the C-washers immediately following axle-shaft insertion. This constraint avoids certain unstable or conditionally stable states in favor of fully stable states. Symbolically, this means that whenever liaison 1 appears, it should be immediately followed by liaison 5.

2. The differential-carrier cover should be placed immediately after filling the carrier with oil. This is a straightforward precaution against contamination either by spillage outward or foreign objects in the carrier and lube. Liaison 9 thus should immediately follow liaison 10.

3. Seal the axle by inserting the test–hole plug immediately after a successful leak-check, a simple precaution against leakage. Liaison 8 should thus immediately follow liaison 11.

4. The initial air test (liaison 12) should be done first.

5. The carrier should be inverted and the brake cable installed only after the C-washers have been installed. This adds a new precedence relation 7 > 3.

Imposing these five liaison sub-sequences prunes the diagram to eight sequences as shown in Figure 7-23. Choosing among eight sequences may be fairly easy. There are three branches to consider. Looking at the top region (A), one may prefer taking the left-hand branch, representing adding backing plates (liaison 2), before

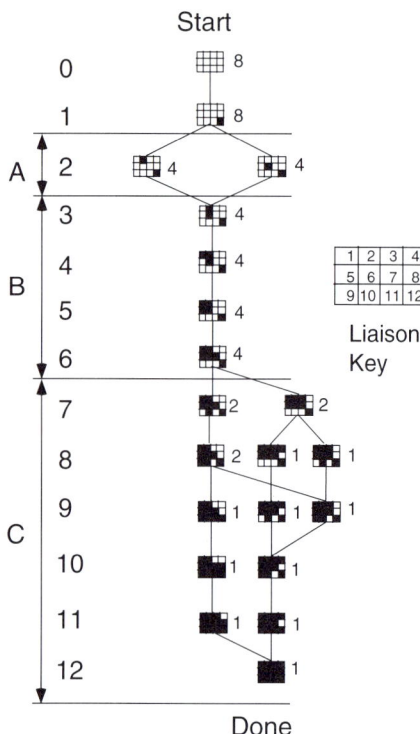

FIGURE 7-23. Remaining Rear Axle Assembly Sequences After Imposing Five "Commonsense" Constraints. Regions A and B comprise work inside the differential case. Region C represents the rest.

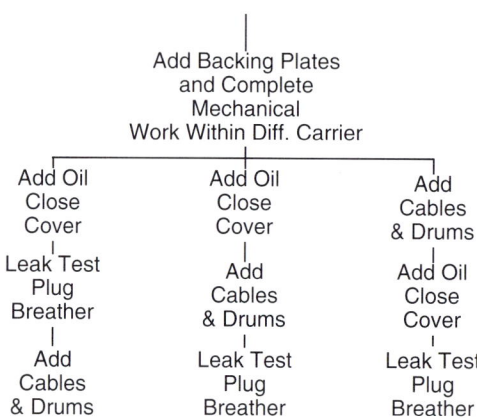

FIGURE 7-24. Representation of Region C of Figure 7-23.

The top stem of the tree, portions A and B, implies firstly finishing all mechanical work *within* the differential carrier, followed by any of the three sequence choices for region C represented in Figure 7-24.

Again, choices of sequence are best made in the context of layout and assembly technology choice for the assembly line. Even ignoring these issues, some observations influencing sequence choice can be made. First, all the operations are fairly easily automated, except those involving assembling and securing the brake cables, which are quite difficult to automate. If automation is to be considered, it is generally convenient to separate manual and automated stations to the extent possible. This consideration suggests the left branch of Figure 7-24 because a manual station to add the cables can be outside the automated region of the line. Second, the leak test is an operation that is followed by an implied but not shown branch allowing any failed units to be reconsidered and reworked. Where rework is a possibility, one may wish to branch as early in the assembly as possible, that is, at a state of minimum value added. This again is consistent with choice of the left-hand branch of Figure 7-24.

The above discussion does not by any means exhaust the issues raised by assembly sequence generation and choice. It should be noted that only a full economic analysis, simulation, risk analysis, and consideration of the proper roles of people on the line will tell which assembly sequence is the best one. This point is underscored by the complex issues related to the safety of people who are near assembly robots and the influence of these issues on redesign options and sequence choices.

Not at all represented in the winnowing methods displayed here are quantitative network techniques, namely,

the right-hand branch, withdrawing the pinion shaft bolt and pinion shaft (liaison 6), on the basis that it represents avoiding one unstable or conditionally stable state in which the pinion shaft could fall into the gearbox.

The choice implied by the first branch perhaps should be made in the light of ideas about the layout and technology of choice for assembling the rear axle. For example, pulling out the pinion shaft bolt and pinion shaft is much easier manually than mechanically unless the parts are redesigned. People could do this operation at the head of the line, keeping them away from robots or other mechanization within the line. If suitable redesign is possible, then replacing the pinion shaft can be made easier as well, so possibly an entire stretch of an automated line could be free of people, an important safety consideration. However, if the necessary redesign cannot be accomplished, people will be needed, eliminating any advantage from putting liaison 6 first. These people will be flanked on both sides by automatic assembly equipment, and safety needs may then rule out the equipment.

assigning data about times and costs to appropriate states and state transitions and using shortest-path or other network optimization routines for choosing the assembly sequence candidate.

Where quantitative network estimation or optimization techniques are to be used, it is important that the network data or estimates be real, dimensioned quantities (time in seconds, money in dollars, for example). The temptation, if any, to characterize the ease or difficulty of a state transition (assembly move) on an arbitrary nondimensioned scale, from one to ten, say, should be avoided. Questions of judgment should be treated as such rather than trying to quantify the not necessarily quantifiable. The engineer may face the judgmental issues in a binary way: Good states and transitions are kept, bad ones are cut.

7.I. CHAPTER SUMMARY

This chapter takes the position that assembly sequences can be designed based on technical and business criteria, once the set of feasible alternatives is available. A wide variety of criteria can be brought to bear on the choice inasmuch as everything from worker fatigue to enterprise strategy is influenced by the choice. One must therefore take great care in making this choice. The analytical methods available today vary in the degree to which they depend on the assembly engineer and the degree to which they can generate every feasible sequence, the two degrees moving in opposite directions. Regardless, they all can use information about the relationships between the parts that is available from the basic connective assembly model and twist matrix representations of the assembly features.

The kinds of filters that eliminate feasible sequences include those that protect parts against damage and people against injury, avoid costs associated with wasteful motion or extra fixtures, or impose a desired style on the process. All of these are basically geometric and usually leave the assembly engineer with thousands of sequences to choose from. In the next chapter we will discover a new class of filters related to constraint as discussed in Chapter 4. These constraint-driven filters eliminate the vast majority of sequences.

7.J. PROBLEMS AND THOUGHT QUESTIONS

1. Why might a product need a disassembly sequence that is different from the reverse of its original factory assembly sequence? Give some examples.

2. Estimate the C_{pk} of the step in the assembly sequence in Figure 7-13 where the impeller is put onto the shaft.

3. Consider the example of the cookies and jar. Using the liaison diagram shown in Figure 7-25, do the complete assembly sequence analysis using the Bourjault method, following the example in Section 7.D.

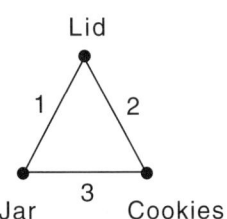

FIGURE 7-25. Figure for Problem 3.

4. At the beginning of the chapter it was claimed that assembly sequence analysis can ignore fine motion planning issues. Examine the validity of this claim and see if you can find a counterexample where two ways of doing the same fine motion for a pair of parts might require a different sequence for the assembly of some of the other parts.

5. Identify as many feasible assembly sequences for the stapler as you can. You may need an actual stapler in order to do this. Don't forget that "unattractive" and "impossible" are completely different evaluations.

6. Draw an exploded view of the alternator. Does it reveal all of the interesting assembly sequence options?

7. Shown in Figure 7-26 are two diagrams called assembly trees. They represent the two ways of putting together the four parts shown in Figure 7-6.

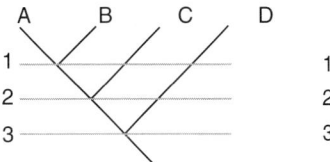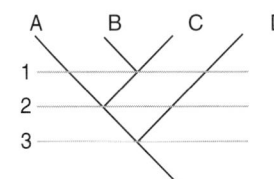

FIGURE 7-26. Figure for Problem 7.

These diagrams are interpreted as follows: The parts are listed across the top. Down the side are rows indicating which action in the sequence is happening. When lines meet, the parts whose names are on the ends are joined. For example, on the left, at the first stage, parts A and B are joined, while at the second level, part C is added to AB. An assembly tree can contain more than one sequence because in general more than one part pair could be joined at a given level. With this information, draw as many unique assembly trees for the alternator (Section 7.H.1) as you can.

8. Compare the assembly tree method with a method like Bourjault's that generates every sequence explicitly. Do they contain the same information?

9. Consider assembling pistons, wrist pins, connecting rods, crank shafts, and cylinder blocks in car engines. Draw as many assembly trees as you can and discuss their pro's and con's.

10. Divide up rear axle assembly into phases bounded by the installation of fasteners. Assume that the C-washers are fasteners, along with the screws that attach the backing plates and the differential cover. What can you learn about the axle's assembly sequence options by doing this?

11. Choose a simple consumer product like a can opener, draw its liaison diagram, and generate all the feasible assembly sequences, using any available method.

12. Compare some of the sequences you generated for the product in Problem 11 and explain their pro's and con's.

13. Discuss the relative merits of using the end of the shaft in Figure 7-14 as an assembly feature compared to improving the tolerances of the nonfunctional side of the end washer.

14. Determine the number of assembly sequences for the juicer under the assumption that (a) the phantom liaisons are not allowed (Figure 7-16) and (b) they are allowed (Figure 7-17).

15. Derive a version of Equation (7-5) and Table 7-4 in Section 7.M under the assumption that the product has one degree of freedom rather than zero. Discuss the results in relation to the data in Figure 7-29.

7.K. FURTHER READING

[Akagi, Osaki, and Kikuchi] Akagi, F., Osaki, H., and Kikuchi, S., "The Method of Analysis of Assembly Work Based on the Fastener Method," *Bulletin of the JSME,* vol. 23, no. 184, pp. 1670–1675, 1980.

[Amblard] Amblard, G. P., "Rationale for the use of Subassemblies in Production Systems," S.M. thesis, MIT Sloan School of Management and Operations Research Center, May 1988.

[Baldwin] Baldwin, D. F., "Algorithmic Methods and Software Tools for the Generation of Mechanical Assembly Sequences," S.M. thesis, Mechanical Engineering Department, MIT, February 1990.

[Baldwin et al.] Baldwin, D. F., Abell, T. E., Lui, M. M.-C., De Fazio, T. L., and Whitney, D. E., "An Integrated Computer Aid for Generating and Evaluating Assembly Sequences for Mechanical Products," *IEEE Transactions on Robotics and Automation,* vol. RA-7, no. 1, pp. 78–94, 1991.

[Bourjault] Bourjault, A., "Contribution á une Approche Méthodologique de l'Assemblage Automatisé: Elaboration automatique des séquences Opératoires," Thesis to obtain Grade de Docteur des Sciences Physiques at l'Université de Franche-Comté, November 1984.

[De Fazio and Whitney] De Fazio, T. L., and Whitney, D. E., "Simplified Generation of All Mechanical Assembly Sequences," *IEEE Journal of Robotics and Automation,* vol. RA-3, no. 6, pp. 640–658, 1987.

[Gustavson] Gustavson, R. E., "SPM: A Connection from Product to Assembly System," C. S. Draper Laboratory internal memo #BTI-120, April 27, 1990.

[Halperin, Latombe, and Wilson] Halperin, D., Latombe, J.-C., and Wilson, R. H., "A General Framework for Assembly Planning: The Motion Space Approach," *Algorithmica,* vol. 26, no. 3–4, pp. 577–601, 2000.

[Homem de Mello and Sanderson] Homem de Mello, L. S., and Sanderson, A. C., "A Correct and Complete Algorithm for the Generation of Mechanical Assembly Sequences," *IEEE Transactions on Robotics and Automation,* vol. 7, no. 2, pp. 228–240, 1991.

[Huang and Lee, C.] Huang, Y. F., and Lee, C. S. G., "Precedence Knowledge in Feature Mating Operation Assembly Planning," Purdue University Engineering Research Center Report #TR-ERC 88-32, October 1988.

[Jones, Wilson, and Calton] Jones, R. E., Wilson, R. H., and Calton, T. L., "On Constraints in Assembly Planning," *IEEE Transactions on Robotics and Automation,* vol. 14, no. 6, pp. 849–863, 1998.

[Kaufman et al.] Kaufman, S. G., Wilson, R. H., Jones, R. E., Calton, T. L., and Ames, A. L., "The Archimedes 2 Mechanical Assembly Planning System," Proceeding of the IEEE Conference on Robotics and Automation, 1996.

[Klein] Klein, C. J., "Generation and Evaluation of Assembly Sequence Alternatives," S.M. thesis, MIT Department of Mechanical Engineering, November 1986.

[Kondoleon] Kondoleon, A. S., "Application of Technology-Economic Model of Assembly Techniques to Programmable Assembly Machine Configuration," S.M. thesis, MIT Department of Mechanical Engineering, May 1976.

[Latombe] Latombe, J.-C., *Robot Motion Planning*, Norwell, MA: Kluwer Academic Publishers, 1991.

[Lee] Lee, H., "Effective Inventory and Service Management Through Product and Process Redesign," *Operations Research,* vol. 44, no. 1, pp. 151–159, 1996.

[Lee and Shin] Lee, S., and Shin, Y. G., "Assembly Coplanner: Cooperative Assembly Planner Based on Subassembly Extraction," Chapter 13 in *Computer-Aided Mechanical Assembly Planning,* Homem de Mello, L. S., and Lee, S., editors, Boston: Kluwer Academic Publishers, 1991.

[Milner, Graves, and Whitney] Milner, J., Graves, S., and Whitney, D. E., "Using Simulated Annealing to Select Least-Cost Assembly Sequences," Proceedings of the IEEE Robotics and Automation Conference, 1994.

[Nevins and Whitney 1978] Nevins, J. L., and Whitney, D. E., "Computer-Controlled Assembly," *Scientific American,* vol. 238, no. 2, pp. 62–74, 1978.

[Nevins and Whitney 1989] Nevins, J. L., and Whitney, D. E., *Concurrent Design of Products and Processes*, New York: McGraw-Hill, 1989.

[Nof, Wilhelm, and Warnecke] Nof, S. Y., Wilhelm, W. E., and Warnecke, H.-J., *Industrial Assembly*, New York: Chapman and Hall, 1997.

[Prenting and Battaglin] Prenting, T. O., and Battaglin, R. M., "The Precedence Diagram: A Tool for Analysis in Assembly Line Balancing," *Journal of Industrial Engineering,* vol. XV, no. 4, pp. 208–213, 1964.

[Rivero and Kroll] Rivero, A., and Kroll, E., "Derivation of Multiple Assembly Sequences from Exploded Views," ASME Advances in Design Automation Conference, Minneapolis, DE-Vol 69-2, pp. 101–106, 1994.

[Scholl] Scholl, A., *Balancing and Sequencing of Assembly Lines,* Heidelberg: Physica–Verlag, 1995.

[Swaminathan and Tayur] Swaminathan, J. M., and Tayur, S. R., "Managing Broader Product Lines Through Delayed Differentiation Using Vanilla Boxes," *Management Science,* vol. 44, no. 12, part 2 of 2, pp. S161–S172, 1998.

[Tseng and Li] Tseng, H.-E., and Li, R.-K., "A Novel Means of Generating Assembly Sequences Using the Connector Concept," *Journal of Intelligent Manufacturing,* vol. 10, pp. 423–435, 1999.

[Whipple] Whipple, R. W., "Assembly Sequence Generation Using the 'Onion Skin' Method and Sequence Editing Using Stability Analysis," S.M. thesis, MIT Mechanical Engineering Department, June 1990.

7.L. APPENDIX: Statement of the Rules of the Bourjault Method

$R(i; B)$ is the abbreviation for a question that means "can liaison i be done when liaisons B are done?" The following are the basic rules:

1. $\{B, i\}$ must form a connected liaison diagram. If they do not, then the portion of $\{B, i\}$ that is connected is retained and called B'. Ask $R(i; B' - i)$.

1a. As a corollary we can conclude that $R(i,$ the disconnected part) = YES.

2. $R(i; \phi)$ is always YES, where ϕ is the empty set; that is, it contains no liaisons.

3. If there is a loop with N arcs in the liaison diagram, then, when $N - 2$ arcs have been done, the last 2 must be done simultaneously, as long as all the parts are rigid. This is the loop closure rule. Such consequential loop closings can close other loops, and the process may cascade.

4. If B contains any loops with only one undone arc, then such loops must be opened further by successive removal, one at a time, of each of the other arcs in each such loop in all combinations before any questions can be asked. This is necessary in order that B be definable in view of Rule 3.

5. If $R(i; B)$ is YES, then $R(i; B-)$ is also YES, where $B-$ is any subset of B. This is the subset rule.

6. If $R(i; B)$ is NO, then $R(i; B+)$ is also NO, where $B+$ is any superset of B (that does not contain i). This is the superset rule.

7. If we find that $R(i; j)$ and $R(j; i)$ are both NO, then i and j must be done simultaneously, or else the assembly is impossible.

8. If $R(i; B)$ is analyzed and if i closes loops and thereby forces liaisons C to be done at the same time as i, then the answer to $R(c_i; B)$ for each c_i in C is the same as the answer to $R(i; B)$. This is especially powerful in automatically answering questions in problems with many loops.

9. If $R(i;B) = $ NO and $R(i;B-) = $ YES for each subset $B-$ of B having one fewer liaisons in it than B has (including $B- = \phi$), then $R(i;B)$ represents a precedence relation of the form $i >= B$. That is, if $R(i;B)$ is NO, then B cannot be done before i. The reverse must therefore hold for every legal sequence. The reverse of "not before" is "after or simultaneously with," so we write, "B after or si-

multaneously with i." Equivalently, $i >= B$, where the $=$ is relevant only in cases where there are loops in the liaison diagram or where rule 7 holds.

Rules 5, 6, and 8 contain the information to prove the validity of rule 9, including the fact that $>=$ is the correct relation if there are loops in the liaison diagram.

7.M. APPENDIX: Statistics on Number of Feasible Assembly Sequences a Product Can Have and Its Relation to Liaisons Per Part for Several Products

This appendix deals briefly with the question of what factors influence the number of feasible assembly sequences that an assembly can have. There is no way to generalize about the number of feasible sequences based only on the number of parts and the number of liaisons. However, a few trends can be observed by looking at real products and a few reasonable guesses can be made about other trends. Clearly, if there are many parts and few precedence relations, then there will be many possible sequences. The minimum number of sequences is clearly one while the maximum for an assembly of n parts and no precedence relations is close to $n!$ ([De Fazio and Whitney]). Liaisons are the only way that we can express part-to-part constraints without knowing details about the parts' shapes, so the discussion and data that follow focus on what we can learn by counting liaisons.

The more parts there are in an assembly, the more feasible sequences there will be, other things being equal. On the other hand, the more liaisons there are, the more opportunities there are for precedence relations, and the more precedence relations there are, the more restrictions there are on feasibility of sequences. Thus we should expect the number of parts and the number of sequences to be roughly directly correlated if the number of liaisons is roughly the same, while the number of liaisons and the number of sequences should be roughly inversely correlated if the number of parts is roughly the same.

The number of liaisons per part (the same as the network complexity factor in graph theory) allows us to normalize the amount of constraint among parts in many assemblies. The network complexity factor is defined as the ratio of the number of arcs in a network to the number of nodes. In our case, nodes are parts and arcs are liaisons. If the network has n nodes, then the maximum number of pairwise connections possible is $n(n-1)/2$ whereas the

minimum is $n - 1$. The maximum ratio of connections to nodes is therefore $(n-1)/2$ and the minimum is $(n-1)/n$. This factor is used to make the comparisons that follow.

The data in Table 7-3, plotted in Figure 7-27, bear out the expectations listed above. The only assembly in this dataset with a small number of assembly sequences and a large number of parts is the one with an unusually large number of liaisons per part, namely the Chinese puzzle (see Figure 7-28). The extra constraints imposed by so many liaisons is obviously what makes the puzzle a puzzle: to find the one and only assembly sequence.

We have more data on liaisons per part than we have about number of feasible sequences. These data are shown in Figure 7-29. They show the number of liaisons per part for seventeen different assemblies along with the theoretical minimum and maximum number of liaisons per part possible for the number of parts that each assembly has. What is interesting here is that for typical engineered

TABLE 7-3. Data on Assembly Sequences, Parts, and Liaisons/Part for Several Assemblies

	Number of Parts	Number of Liaisons	Liaisons/Part	Number of Sequences
Throttlebody	5	7	1.4	10
Ballpoint pen	6	5	0.83333333	12
Juicer	8	9	1.125	71
Rear axle	13	12	0.92307692	938
Transaxle	9	15	1.66666667	2450
Six-speed transmission	11	18	1.63636364	3318
Chinese puzzle	14	84	6	1

Note: For most products in this table, the number of liaisons per part is less than 2. For these products, more parts and liaisons are associated with more feasible assembly sequences. The outlier in this table is the Chinese puzzle, which has a relatively unusual and large six liaisons per part and only one assembly sequence.

products (as contrasted with puzzles) the number of liaisons per part hovers around the theoretical minimum, and none exceeds two.

We can speculate about the reason for this. Many engineering factors are involved. More liaisons mean more toleranced interfaces, more complexity, more cost, and more places where failure could occur. If the engineer wants parts to interface with many others, the result will likely be that almost all the interior space of the product will be filled with material comprising one part or another. The Chinese

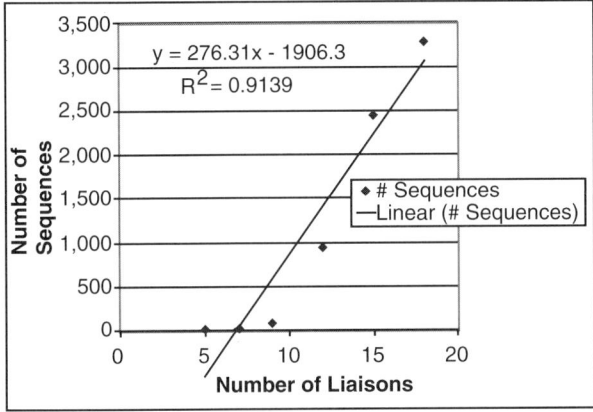

FIGURE 7-27. Correlation Between Number of Assembly Sequences and Number of Liaisons for Typical Products. This chart omits the Chinese puzzle.

FIGURE 7-28. Chinese Puzzle. This is the puzzle documented in Table 7-3.

puzzle is solid inside, for example. If there is to be empty space inside the outer shell of an assembly, say to allow for cooling air, insertion of other parts, service, vibration without collision, and so on, then there must be relatively few connections between the parts themselves. Assembly is also obviously easier if inserting a part requires paying attention to only a few joints with other parts.

From a theoretical point of view, we can show that more liaisons per part increases the likelihood of overconstraint. To see why, we begin with the Grübler criterion for planar mechanisms introduced in Chapter 4 for determining the numerical value of the number of degrees of freedom in a mechanism:

$$M = 3(n - g - 1) + \sum \text{joint freedoms } f_i \qquad (7\text{-}3)$$

where n is the number of parts, g is the number of joints, and f_i is the degrees of freedom of joint i.

If we define α to be the number of joints (liaisons) per part and define the average number of degrees of freedom per joint in the mechanism as β, then we have

$$g = \alpha n$$
$$\sum f_i = g\beta = \alpha\beta n \qquad (7\text{-}4)$$
$$M = 3(n - \alpha n - 1) + \alpha\beta n$$

If the mechanism is to be exactly constrained, then $M = 0$ and Equation (7-4) can be solved for α to yield

$$\alpha = \frac{3 - 3n}{n(\beta - 3)} \rightarrow \frac{3}{3 - \beta} \quad \text{as } n \text{ gets large} \qquad (7\text{-}5)$$

This expression is based on assuming that the mechanism is planar. If it is spatial, then "3" is replaced by "6" but everything else stays the same. Table 7-4 evaluates Equation (7-5) for both planar and spatial mechanisms.

Table 7-4 shows that α, the number of liaisons per part, cannot be very large or else the mechanism will be overconstrained. If a planar mechanism has several two degree-

TABLE 7-4. Relationship Between Number of Liaisons Per Part and Number of Joint Freedoms for Exactly Constrained Mechanisms

β	α Planar	α Spatial
0	1	1
1	1.5	1.2
2	3	1.5

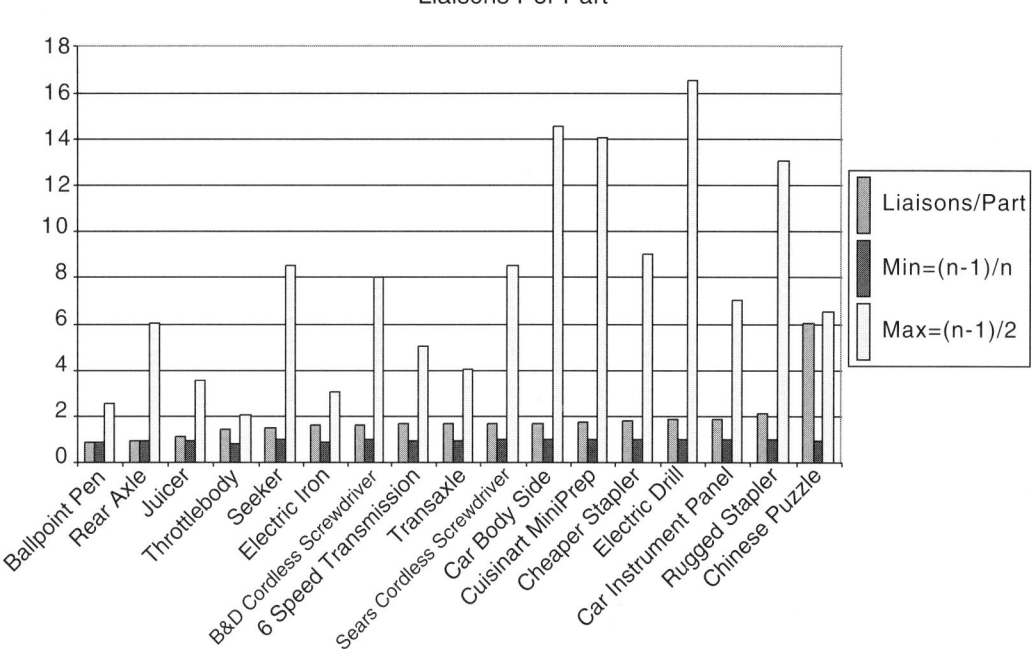

FIGURE 7-29. Statistics on Liaisons Per Part. Data are from the author's observations, work performed at Charles Stark Draper Laboratory, and projects by MIT students. The data cover products with as few as six parts and as many as twenty-nine. (The number of parts n for these products can be calculated from the data on maximum number of liaisons/part and the formula max $liaisons/part = (n - 1)/2$.) It is clear that, except for the Chinese puzzle, the number of liaisons per part is close to the theoretical minimum for all of these assemblies. The ballpoint pen has a roughly linear shape liaison diagram while the rear axle's liaison diagram has a roughly hub and spokes shape. Both of these products have liaisons/part less than one. The other products have liaison diagrams that are general networks and their liaisons/part are without exception larger than one but rarely more than 2, except for the Chinese puzzle. The data confirm the theoretical prediction in Equation (7-5).

of-freedom joints (pin–slot, for example), then a relatively large number of liaisons per part can be tolerated. Otherwise, the numbers in this table confirm the data in Figure 7-29. Most assemblies are exactly constrained or have one operating degree of freedom. Thus $\beta = 0$ or $\beta = 1$, yielding small values for α, consistent with our data.

The liaison diagrams for many of these assemblies appear in this book. In Chapter 3, it was said that liaison diagrams can be classified as being linear in shape, or hub and spokes shaped, or general networks. It should be clear

that to attain the minimum number of liaisons per part, the liaison diagram must be linear. Hub and spokes shapes generate a number of liaisons per part near the minimum and probably less than one. General networks will likely have a number larger than one. The fact that assemblies as complex as the seeker head and the two vehicle transmissions have a number that is less than two and so far short of their respective maximums is remarkable. The reader should find the various liaison diagrams and compare their shapes to the data in Figure 7-29.

8 THE DATUM FLOW CHAIN

"We don't design assemblies. We design parts and try to assemble them."

8.A. INTRODUCTION

This chapter addresses the questions "What does it mean to design an assembly?" and "How do we know when we have a good design?" To answer these questions, we will bring together all the items from the previous six chapters. Our aim is to be able to present a unified way to lay out, analyze, outsource, assemble, and debug complex assemblies. To accomplish this, we need ways to capture their fundamental structure in a top-down design process that shows how the assembly is supposed to go together and deliver its key characteristics (KCs). This process should

- Represent the top-level goals of the assembly
- Link these goals to engineering requirements on the assembly and its parts in the form of KCs
- Show how the parts will be constrained, and what features will be used to establish constraint, so that the parts will acquire their desired spatial relationships that achieve the KCs
- Show where the parts will be in space relative to each other both under nominal conditions and under variation
- Show how each part should be designed, dimensioned, and toleranced to support the plan
- Assure that the plan is robust

A clear statement of these elements for a given assembly is called the *design intent* for that assembly.

This chapter describes a concept called the datum flow chain (DFC) to capture assembly design intent.[1] A DFC

is a delivery chain for a KC, defining the chain of parts and features that link one end of a KC to the other. It provides a method, together with a vocabulary and a set of symbols, for documenting a location strategy for the parts and for relating that strategy explicitly to achievement of the product's key characteristics. It helps the designer choose mating features on the parts and provides the information needed for assembly sequence and variation analyses.

Many assembly problems occur due to a lack of such a plan or lack of access to it, especially when trying to diagnose assembly problems on the factory floor. Assembly problems occur in most industries, and the character of the assembly process in the factory (smooth, confused, etc.) is a strong indicator of the quality of all the upstream design and fabrication processes.[2]

In this chapter, we will define the DFC and show how to use it to represent common assembly situations. We will show that assemblies can be classified into two types: *Type 1* assemblies are properly constrained. The assembly process for Type 1s puts their parts together at their prefabricated mating features. *Type 2* assemblies are underconstrained. The assembly process for Type 2s involves fixtures and can incorporate in-process adjustments to redistribute variation. We will see below that the DFC for a Type 1 assembly directly defines the assembly itself. However, the DFC for a Type 2 assembly directly defines

---

[1]Portions of this chapter are adapted from [Mantripragada and Whitney] and [Whitney, Mantripragada, Adams, and Rhee]. Additional material comes from class projects conducted by MIT students.

[2]Much of the research presented in this book was motivated by observing the difficulty of diagnosing assembly problems in factories. See [Cunningham] for a comparison of corrective action methods used in assembly factories in the automobile and aircraft industries, along with examples of how the DFC would help diagnose those problems.

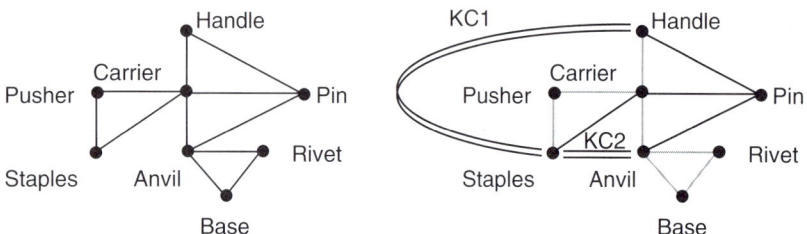

FIGURE 8-1. The Stapler, Its Liaison Diagram (*left*), and Two Key Characteristics (*right*). Several irrelevant liaisons have been colored gray because they play no role in positioning parts to deliver either KC.

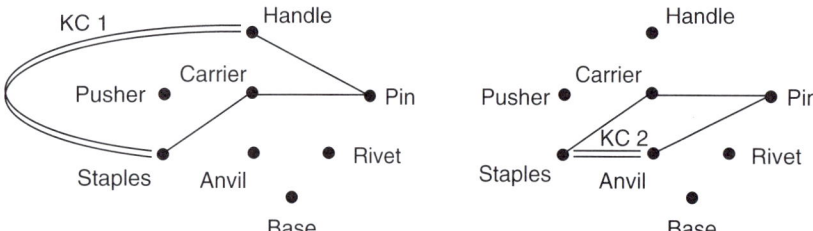

FIGURE 8-2. The KCs of the Stapler Shown Separately with the Liaisons That Deliver Them. Irrelevant liaisons are not shown.

the process for creating it and thus only indirectly defines the assembly.

We will also define two types of assembly joints, called *mates* and *contacts:* Mates pass dimensional constraint from part to part, while contacts merely provide support, reinforcement, or partial constraint along axes that do not involve delivery of a KC. Some joints act as mates along some degrees of freedom and as contacts along others. Symbols for each of these types of joints will be introduced. We will then present the scope of the DFC in assembly planning using several examples.

Finally, we will see that the DFC contains all the information needed to carry out a variation analysis of the KC it delivers. This fact links the scheme by which the parts are located in space to the sources of variation in their locations.

To visualize the ideas to be presented in this chapter, we again turn to the desktop stapler. In this chapter, we will learn how to characterize the liaisons of an assembly as delivery chains for key characteristics. This is illustrated in Figure 8-1, where some of the liaisons are shown in gray to denote that they play no role in KC delivery. It is further emphasized in Figure 8-2, where each KC chain is shown separately and the irrelevant liaisons are omitted altogether. The stapler also illustrates

the difference between mates and contacts. The difference is illustrated in Figure 8-3. All these concepts will be made concrete in this chapter and related to their underlying mathematical representations introduced in earlier chapters.

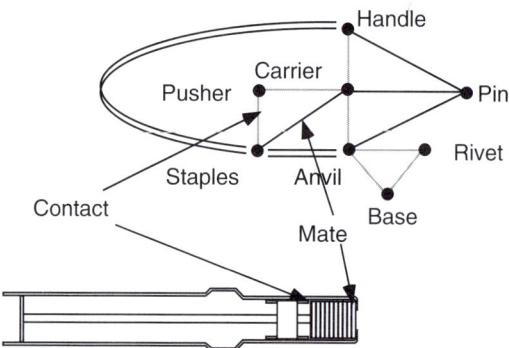

FIGURE 8-3. Illustrating the Difference Between a Mate and a Contact. The mate provides constraint for the staples by establishing their position relative to the end of the carrier. The pusher and staples share a contact, which reinforces or stabilizes the stapler-carrier mate. In the vocabulary of Chapter 4, the staples are properly constrained along the axis of the carrier. Note that the contact is colored gray, indicating that it does not participate in KC delivery.

8.B. HISTORY AND RELATED WORK

Assemblies have been modeled systematically by [Lee and Gossard], [Sodhi and Turner], [Srikanth and Turner], and [Roy et al.] and others. Such methods are intended to capture relative part location and function, and they enable linkage of design to functional analysis methods like kinematics, dynamics, and, in some cases, tolerances. Almost all of them need detailed descriptions of parts to start with, in order to apply their techniques. [Gui and Mäntylä] applies a function-oriented structure model to visualize assemblies and represents them in varying levels of detail. In this book, we have not attempted to model assemblies functionally. Our work begins at the point where the functional requirements have been established and there is at least a concept sketch.

Top-down design of assemblies emphasizes the shift in focus from managing design of individual parts to managing the design of the entire assembly in terms of mechanical "interfaces" between parts. We saw in Chapter 4 that [Smith] proposes eliminating or at least minimizing critical interfaces, rather than part-count reduction, in the structural assembly of aircraft as a means of reducing costs. He emphasizes that, at every location in the assembly structure, there should only be one controlling element that defines location, and everything else should be designed to "drape to fit." In our terms, the controlling element is a mate and the joints that drape to fit are contacts. [Muske] describes the application of dimensional management techniques on 747 fuselage sections. He describes a top-down design methodology to systematically translate key characteristics to critical features on parts and then to choose consistent assembly and fabrication methods. These and other papers by practitioners indicate that several of the ideas to be presented here are already in use in some form but that there is a need for a theoretical foundation for top-down design of assemblies.

Academic researchers have generated portions of this foundation. [Shah and Rogers] proposes an attributed graph model to interactively allocate tolerances, perform tolerance analysis, and validate dimensioning and tolerancing schemes at the part level. This model defines chains of dimensional relationships between different features on a part and can be used to detect over- and underdimensioning (analogous to over- and underconstraint) of parts. [Wang and Ozsoy] provides a method for automatically generating tolerance chains based on assembly features in one dimensional assemblies. [Shalon et al.] shows how to analyze complex assemblies, including detecting inconsistent tolerancing datums, by adding coordinate frames to assembly features and propagating the tolerances by means of 4×4 matrices. [Zhang and Porchet] presents the oriented functional relationship graph, which is similar to the DFC, including the idea of a root node, propagation of location, checking of constraints, and propagation of tolerances. A similar approach is reported in [Tsai and Cutkosky] and [Söderberg and Johannesson]. The DFC is an extension of these ideas, emphasizing the concept of designing assemblies by designing the DFC first, then defining the interfaces between parts at an abstract level, and finally providing detailed part geometry.

CAD today bountifully supports design of individual parts. It thus tends to encourage premature definition of part geometry, allowing designers to skip systematic consideration of part–part relationships. Most textbooks on engineering design also concentrate on design of machine elements (i.e., parts) rather than assemblies.

Current CAD systems provide only rudimentary assembly modeling capabilities once part geometry exists, but these capabilities basically simulate an assembly drawing. Most often the dimensional relations that are explicitly defined to build an assembly model in CAD are those most convenient to construct the CAD model and are not necessarily the ones that need to be controlled for proper functioning of the assembly. What is missing is a way to represent and display the designer's strategy for locating the parts with respect to each other, which amounts to the underlying structure of dimensional references and mutual constraint between parts. The DFC is intended to capture this logic and to give designers a way to think clearly about that logic and how to implement it.

8.C. SUMMARY OF THE METHOD FOR DESIGNING ASSEMBLIES

Ideally, the design of a complex assembly starts by a general description of the top-level requirements in the form of KCs for the whole assembly. These requirements are then systematically formalized and flowed down to subassemblies and finally down to individual parts. The assembly designer's task is to create a plan for delivering each KC. To do this, he or she defines a DFC for each KC, showing how the parts in each DFC will be given their desired nominal locations in space. This is equivalent to properly constraining each part. During these early stages of design, the designer has to do the following:

- Systematically relate the identified KCs to important datums on subassemblies, parts, and fixtures at the various assembly levels from parts to subassemblies to the final assembly.
- Design consistent dimensional and tolerance relationships or locating schemes among elements of the assembly so as to deliver these KC relationships.
- Identify assembly procedures that best deliver the KCs repeatedly without driving the costs too high.

These major elements of the assembly design process are implemented by establishing three basic kinds of information about an assembly:

- "Location responsibility": Which parts or fixtures locate which other parts.
- Constraint: Which degrees of freedom of a part are constrained by which surfaces on which features on which other parts or fixtures, including checking for inappropriate over- or underconstraint.
- Variation: How much uncertainty there is in the location of each of the parts relative to some base part or fixture which represents the reference dimension.

The design process comprises two steps: *nominal design* and *variation design*. The nominal design phase creates the constraint structure described above, by using the concepts in Chapter 4, and assuming that the parts and their features are rigid and have nominal size, shape, and location. The variation design phase comprises making the DFC robust against variations away from nominal dimensions, plus checking each DFC using traditional tolerance analysis, as described in Chapters 5 and 6, to determine if each KC can be delivered. A KC, as described in Chapter 2, is said to be "delivered" when the required geometric relationship is achieved within some specified tolerance an acceptable percent of the time.

The DFC provides a way to define a competent nominal assembly. *Nominal* means that the assembly has all its dimensions at their ideal values and that there is no variation. *Competent* means that the assembly is capable of properly constraining all its parts, that all its KCs have been identified, and that a way to deliver each KC has been provided.

We will see below that these elements of "competency" are all related to each other and that they are really different ways of saying the same thing. Furthermore, they can be addressed using the nominal dimensions. Once we are sure that the nominal design is competent, we can examine it for its vulnerability to variation. Portions of this step are included in conventional tolerance analysis, but it will become clear that we mean much more than that.

The method is capable of describing assemblies that are built simply by joining parts as well as those that are built using fixtures. In either case, the participating elements (parts and fixtures) are linked by the DFC and its underlying constraint scheme. A typical assembly sequence builds the DFC beginning at its root or datum reference and working its way out to the KCs. Sequences that "build the DFC" are a very small subset of the feasible sequences found by methods described in Chapter 7. When DFCs are found to be deficient during the design process, it often emerges that a different assembly sequence is associated with an alternate DFC design. This fact links assembly sequence analysis to assembly design, variation buildup, and assembly process planning.

The method also provides guidance in the surprisingly common situation in which there are more KCs than the degrees of freedom of the assembly can deliver independently. This situation is called *KC conflict*. We will see that KC conflict can be detected using the methods of constraint evaluation presented in Chapter 4.

In this method, parts[3] are merely frameworks that hold assembly features, while assembly features are the links that establish the desired state of constraint among adjacent parts, leading to the achievement of the assembly-level geometric relationships. The DFC is an abstract version of this framework, providing a kind of skeleton for the assembly.

The mathematical foundation of the method is the 4×4 transform and Screw Theory, which are used to describe the three-dimensional locations of parts and features, to determine the degrees of freedom constrained by individual features, and to check for proper constraint when parts are joined by sets of features. These elements of the method were presented in Chapters 3 and 4.

---

[3] Here, we mean parts considered only from the point of view of their membership in the assembly, not as, for example, carriers of load or liquids, barriers against heat flow, and so on. These factors comprise significant requirements on parts that must be considered as part of their design.

An important conclusion from this method is that most of the information required to support it can be stored as text. Very little detailed geometry is needed, and its use is isolated to a few steps in the process and a few places on the parts. This is important because it reflects the fact that the most important steps in designing an assembly comprise establishing connectivity and constraint, not defining geometry. This, in turn, is important because it provides a route to representing assembly information more abstractly, richly, and compactly than is permitted by geometry alone. This, in turn, provides a language and other constructs for capturing this information as a natural part of the design process, avoiding the need to discover it by analyzing geometry, as many CAD systems do today.

A corollary is that the method describes steps that demand the careful definition of a data and decision record that constitutes declaration of the consistent design intent for the assembly. This record can be used to judge the adequacy of the design as well as to manage its realization up and down the supply chain and debug that realization on the factory floor and in the field.

8.D. DEFINITION OF A DFC

8.D.1. The DFC Is a Graph of Constraint Relationships

A datum flow chain is a directed acyclic graphical representation of an assembly with nodes representing the parts and arcs representing mates between them. "Directed" means that there are arrows on the arcs. "Acyclic" means that there are no cycles in the graph; that is, there are no paths in the graph that follow the arrows and return to the start of the path. Loops or cycles in a DFC would mean that a part locates itself once the entire cycle is traversed, and hence are not permitted. Every node represents a part or a fixture, and every arc transfers dimensional constraint along one or more degrees of freedom from the node at the tail to that at the head. Each arc has an associated 4×4 transformation matrix that represents mathematically where the part at the head of the arc is located with respect to the part at the tail of the arc. A DFC has only one root node that has no arcs directed toward it, which represents the item from which the locating scheme begins. This could be either a carefully chosen base part or a fixture. A DFC can be a single chain of nodes or it can branch and converge. For example, if two assembled parts together constrain a third part, the DFC branches in order to enter each of the first two parts and converges again on the third part.

Figure 8-4 shows a simple liaison diagram and associated DFC. In this DFC, part A is the root. It completely locates parts B and C. Parts A and C together locate part D. A thought question at the end of the chapter asks the reader to define some assembly features that are able to accomplish this locating scheme.

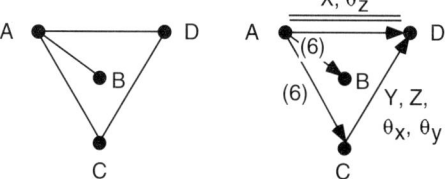

FIGURE 8-4. A Simple Liaison Diagram and Datum Flow Chain. The liaison diagram (*left*) shows which parts are connected to each other. The DFC (*right*) shows how they are connected and constrained. Each arc is labeled with the degrees of freedom it constrains or the names of those degrees of freedom in any convenient coordinate system. This DFC is intended to deliver a KC between parts A and D. The KC is indicated by the double line next to the arrow. No information is given regarding which degrees of freedom are of interest in this KC.

Every arc in a DFC is labeled to show which degrees of freedom it constrains, which depends on the type of mating conditions it represents. The sum of the unique degrees of freedom constrained by all the incoming[4] arcs to a node in a DFC should be equal to six (less if there are some kinematic properties in the assembly or designed mating conditions such as bearings or slip joints which can accommodate some amount of predetermined motion; more if locked-in stress is necessary such as in preloaded bearings). This is equivalent to saying that each part should be properly constrained, except for cases where over- or underconstraint is necessary for a desired function.

---

[4]Arcs that are "incoming" to a node are defined as arcs whose arrows point toward the node.

A DFC is similar in many ways to an electric circuit diagram. A circuit diagram defines a connection structure or network that has many properties of its own, independent of the resistors, capacitors, and other individual circuit elements. It has a unique ground or reference voltage. Many operating characteristics of the circuit can be calculated from its graphical properties, such as spanning trees and independent loops. Both the nominal operating behavior and the sensitivity to component variations can be calculated from the circuit. We will see that many of these properties of electric circuits are shared by DFCs, including their ability to set the agenda for design and analysis.

8.D.2. Nominal Design and Variation Design

The DFC represents the designer's intent concerning how the parts will obtain their locations in space in all six degrees of freedom. Each KC will have its own DFC, and thus each DFC is responsible for delivering its KC. If the parts are perfect, then the KC will be delivered perfectly. If they are not, then a variation analysis like those in Chapter 6 must be undertaken. Variation in parts passes from part to part along the DFC and accumulates to determine the variation in the KC. Thus the DFC acts as a tolerance chain that guides the designer in finding all the variations that contribute to each KC. It is not necessary to perform a separate analysis to find the tolerance chain in order to carry out the variation analysis of a KC.

8.D.3. Assumptions for the DFC Method

The following assumptions are made to model the assembly process using a DFC:

1. All parts in the assembly are assumed rigid. Hence each part is completely located once its position and orientation in three dimensional space are determined.

2. Each assembly operation completely locates the part being assembled with respect to previously assembled parts or an assembly fixture. Only after the part is completely located is it fastened to the remaining parts in the assembly.

Assumption 1 states that each part is considered to be fully constrained once three translations and three rotations are established. If an assembly, such as a preloaded pair of ball bearings, must contain locked-in stress in order to deliver its KCs, the parts should still be sensibly constrained and located kinematically first, and then a plan should be included for imposing the overconstraint in the desired way, starting from the unstressed state. If flexible parts are included in an assembly, they should be assumed rigid first, and a sensible locating plan should be designed for them on that basis. Modifications to this plan may be necessary to support them against sagging under gravity or other effects of flexibility that might cause some of their features to deviate from their desired locations in the assembly.

Assumption 2 is included in order to rationalize the assembly process and to make incomplete DFCs make sense. An incomplete DFC represents a partially completed assembly. If the parts in a partially completed assembly are not completely constrained by each other or by fixtures, it is not reasonable to expect that they will be in a proper condition for receipt of subsequent parts, in-process measurements, transport, or other actions that may require an incomplete assembly to be dimensionally coherent and robust. This assumption enables us to critique alternate assembly sequences, as explained in Section 8.K.

8.D.4. The Role of Assembly Features in a DFC

The DFC comprises design intent for the purpose of locating the parts but it does not say how the parts will be located. Providing location means providing constraint. We know from the foregoing chapters that assembly features are the vehicles we use to apply constraint between parts. Thus the next step after defining the DFC is to choose features to provide the constraint. Once features have been declared, we can calculate the nominal locations of all the parts by chaining their 4×4 transforms together, and we can check for over- or underconstraint, using methods that are by now familiar.

In order to be precise about our locating scheme, however, we need to distinguish two kinds of feature joints: mates and contacts. These are the subject of the next section.

8.E. MATES AND CONTACTS

A typical part in an assembly has multiple joints with other parts in the assembly. Not all of these joints transfer locational and dimensional constraint, and it is essential to distinguish the ones that do from the ones that are redundant location-wise and merely provide support or strength. We define the joints that establish constraint and dimensional relationships between parts as *mates,* while joints that merely support and fasten the part once it is located are called *contacts.* Hence mates are directly associated with the KCs for the assembly because they define the resulting spatial assembly relationships and dimensions. The DFC therefore defines a chain of mates between the parts. If we recall that the liaison diagram includes all the joints between the parts, then it is clear that the DFC is a subset of the liaison diagram. The process of assembly is not just of fastening parts together but should be thought of as a process that first defines the location of parts using the mates and then reinforces their location, if necessary, using contacts.

8.E.1. Examples of DFCs

This section uses some simple examples to illustrate how to draw a DFC starting from the KC(s). The first example is assembly of an automobile wheel to an axle. The second is assembly of three simple sheet metal parts. Both examples illustrate the difference between mates and contacts.

8.E.1.a. Wheel and Axle

Consider Figure 8-5, a simplified automobile axle and wheel. The axle hub includes a rim plus four studs. The wheel contains a round opening in the center, plus four holes, larger than the studs, centered around this opening. When the wheel is mounted to the hub, the opening fits snugly over the rim and the studs protrude through the holes, ready for the nuts to be installed.

The designer's goal for this design is to achieve dynamic balance and a smooth ride. The KCs he has chosen to achieve this goal are as follows:

- Make the wheel concentric with the axle shaft's axis.
- Make the plane of the wheel perpendicular to this axis.

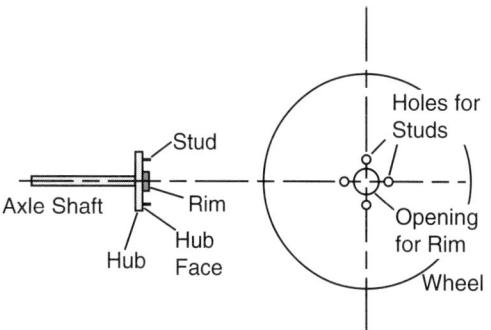

FIGURE 8-5. A Wheel and Axle Illustrating the Difference Between Mates and Contacts. The dimensional and constraint relationships between the wheel and axle are established by the mate between the wheel's opening and the axle's rim, as well as by the mate between the planar face of the wheel and the planar face of the hub. All other interfaces between these parts provide no constraint and are contacts.

To deliver these KCs, the designer has chosen two features on the axle, the face of the hub and the rim. The hub face must be perpendicular to the axle's axis and the rim must be concentric with this axis. Similarly, he has chosen two features on the wheel, namely, the plane of the wheel and the opening in the center. The plane must be in the coordinate frame in which the wheel's inertia matrix is diagonal, and the opening must be centered on this frame.[5] In our terms, the hub face and rim constitute mate features, as do the wheel plane and opening. The studs and their holes constitute contacts. They play no role in achieving the KCs. They merely keep the wheel from falling off. Of course, this is important and we could have called it a KC, but achieving it does not depend on how the parts involved are geometrically located. The important constraint relationships between the axle and wheel are completely determined by the mate features already defined.

A DFC for the wheel and hub is shown in Figure 8-6. It represents mates as graph arcs with arrows on them as well as a number indicating how many degrees of freedom are located by the mate. Contacts are shown as dashed lines. All the important features are defined, and their roles in establishing constraint relationships and KCs are shown.

---

[5]Small errors in the wheel features are inevitable due to the unpredictability of the mass distribution of the rubber tire. These are removed by dynamically balancing the wheel using small lead weights.

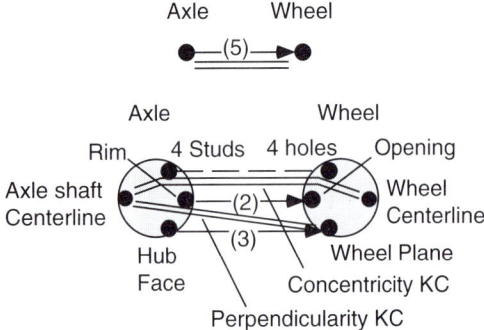

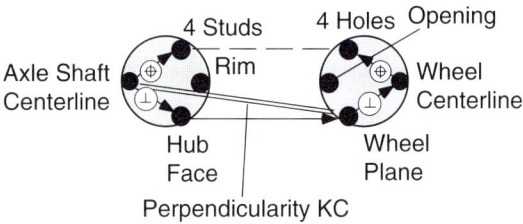

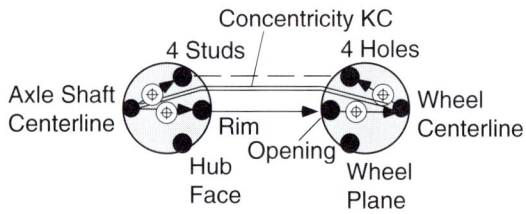

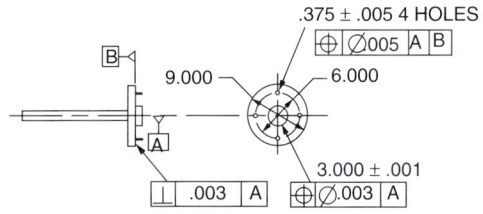

FIGURE 8-6. DFC and KCs for the Wheel and Axle in Figure 8-5. *Top:* The simplest representation of the DFC for this assembly consists of two nodes representing the parts, a set of parallel lines representing a KC, and one arrow with the number 5 on it, indicating that the axle has a mate with the wheel that defines 5 of its degrees of freedom. *Bottom:* A little more detail (adapted from [Zhang and Porchet]) reveals that the KC can be decomposed into two separate KCs and that different features on the parts are involved in delivering them. The features on the axle and wheel are related in different ways. The hub and rim on the axle each have mates with the opening and plane on the wheel, respectively. Together, these features define 5 of the wheel's six degrees of freedom and all the KCs. The joint between the studs and holes is a dashed line, indicating that it is a contact. When the nuts are tightened onto the studs, the sixth degree of freedom is fastened, but its exact value is not of interest to us. There is no KC on this dimension. The studs fit easily into oversize holes, and any orientation of the wheel within the stud–hole clearance is acceptable.

Note that one of these datum features is the axle's centerline. This is not a piece of geometry itself. Calling it a feature is, however, perfectly consistent with GD&T.

Figure 8-7 expands the DFC for the assembly to show all the necessary features on each part and their relative location requirements. The symbolic blobs in Figure 8-6 representing the two parts, with their four black dots representing the important features, have been expanded to show the perpendicularity and concentricity relationships between the features. Also shown is a possible simplified statement of these requirements for the axle using the symbols of GD&T as discussed in Chapter 5. Figure 8-5, Figure 8-6, and Figure 8-7 present together a simple example of definition of assembly requirements, their capture as KCs, the definition of DFCs to deliver these KCs, the identification of feature-to-feature relationships between the parts that create the necessary mates, and finally definition of the resulting requirements on mutual feature relationships inside one of the parts of this assembly. It

FIGURE 8-7. DFC with Features and Their Required Mutual Locations Inside the Parts. Above is an expanded view of the assembly in symbolic form. It shows all the interpart relationships between features. These features play essential roles in delivering the axle-wheel assembly's KCs. Below is a possible simplified rendition of a GD&T specification for realizing the necessary feature-to-feature relationships inside one of the parts. The interpart relationships express the requirements that the hub must be perpendicular to the axle shaft's centerline and that the rim must be located with respect to the centerline, both within some tolerances. The circle on which the studs lie must also be located with respect to the shaft centerline, but a larger tolerance is allowed. The root of the DFC in the axle's centerline is also the A datum for the axle.

should be clear from these figures that the DFC represents a continuous chain not only between parts but inside them as well. The only difference between the arcs of a DFC between parts and the arcs inside a part is that only mate relationships exist inside parts. Contact relationships exist only between parts.

An alternate design for joining these parts is commonly used. It dispenses with the rim and its mating opening and uses five studs and holes instead. The nuts have generous chamfers on them where they engage chamfered holes in the wheel. A thought question at the end of the chapter asks the reader to compare this alternate design with the one described here.

TABLE 8-1. Distinguishing Mates and Contacts

Function	Mate?	Contact?	Example
Full six dof constrained	Yes	No	Square peg in square hole
No dof constrained	No	Yes	Nuts attaching wheel to axle hub
Some dof constrained along a KC	Yes along KC directions	Yes along non-KC directions	Rim on axle hub; slip joint in sheet metal

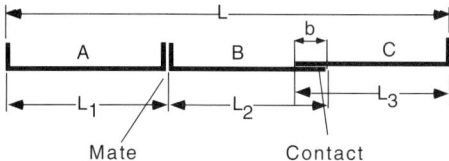

FIGURE 8-8. An Assembly with a Mate and a Contact. The KC is the overall length L of the assembly. In the direction of the KC, the A–B joint provides location and constraint, but the B–C joint does not. It simply joins B and C and will do so as long as overlap dimension b is large enough.

8.E.1.b. Sheet Metal Parts

Figure 8-8 above shows three simplified sheet metal automobile body parts. Between them they have two joints, namely, one butt joint called a mate and one slip joint called a contact.[6] The KC is the overall length L of the assembly. The slip joint can be adjusted in the direction of the KC.

If we consider this to be a full three-dimensional assembly, then it is obviously underconstrained, and neither of the joints would then be called a mate. However, if we consider the KC, which specifies one dimension only, then we could argue that the joint between A and B is a mate because it constrains the part-to-part relationship in a direction that contributes to delivery of the KC. Similarly, we could argue that the joint between B and C is a contact because it does not provide such constraint.

However, the B–C joint clearly does provide constraint in the direction normal to the planes of the parts. Why then call it a contact? The reason is that there is no KC specified in that direction to which this joint makes a contribution. This leads us to a rule, namely that every assembly must be properly constrained (up to the limit where function may require some unconstrained degrees of freedom) but not every joint that provides constraint in some direction(s)

has to be a mate. Underconstrained assemblies need help to achieve proper constraint beyond what the joints themselves can provide. As we will see below, fixtures are usually used to provide the missing constraint. Typically, the parts will have joints with the fixtures at these points and the DFC will pass through these part-fixture joints, causing us to call them mates.

Table 8-1 combines these definitions. Later in this chapter we will use the name "hybrid mate-contact" to refer to joints that provide incomplete constraint and which act as mates along the directions they constrain. In terms of the definitions used in Chapter 4, joints that provide full six degree of freedom (dof) constraint play the role of "locators" while joints that provide no constraint play the role of "effectors."

8.E.2. Formal Definition of Mate and Contact

Generalizing on Table 8-1, we can categorize all joints between parts as shown in Figure 8-9. This figure makes use of the concepts of wrench space and twist space introduced in Chapter 4. It permits us to examine a joint systematically, surface contact by surface contact, to determine the function of each surface contact in the assembly.

The categorization in Figure 8-9 can be applied to joints or to fundamental surface-to-surface contacts as discussed in Chapter 4. For example, Figure 8-10 reviews the cylinder-plane contact and shows its twist space and wrench space. Constraint and variation occur only along the directions in the wrench space.

8.E.3. Discussion

Explicit identification and definition of the mates in an assembly is an integral part of assembly design and is a prerequisite to assembly process planning and variation analysis. The choice of which joints will be mates and which ones will be contacts is made by the designer at the conceptual design stage.

---

[6]Butt joints and slip joints were introduced in Chapter 6. In the auto industry, the butt joints are called coach joints.

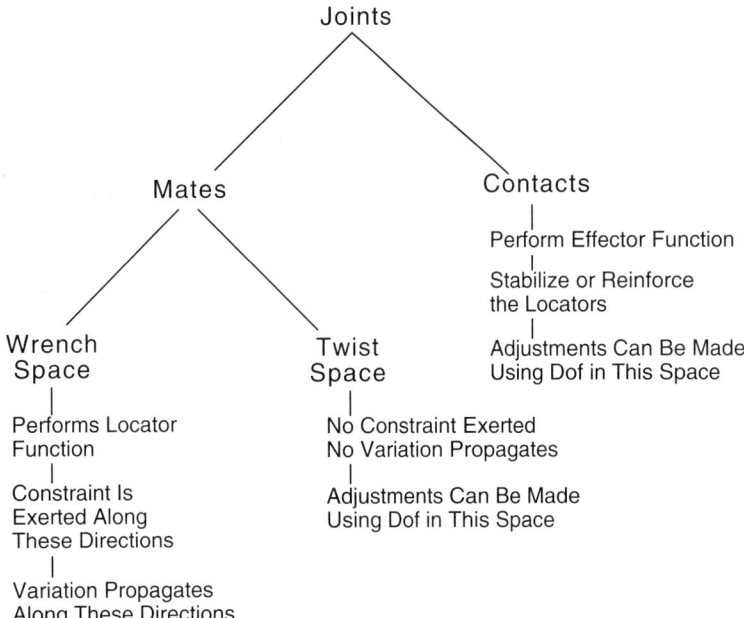

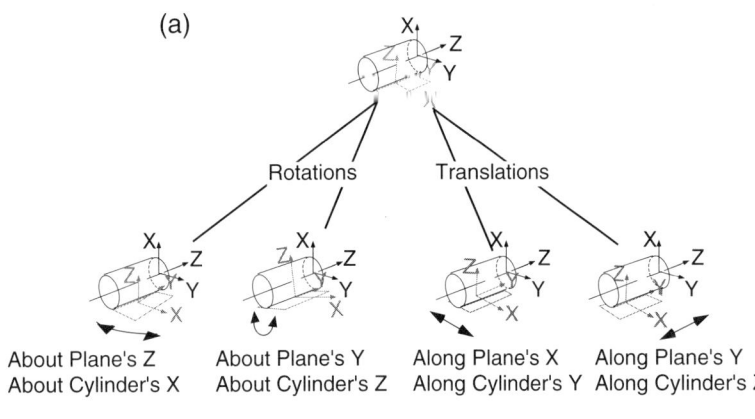

FIGURE 8-9. **Categories of Joints Between Parts.** Some joints are mates while others are contacts. Within each mate is a twist space and a wrench space. Constraint behavior characteristic of a mate occurs in its wrench space. Adjustment behavior (typically associated with contacts) can occur in its twist space. Joints where this occurs are called hybrid mate-contacts.

FIGURE 8-10. Twist Space (a) and Wrench Space (b) for the Cylinder–Plane Surface Contact.

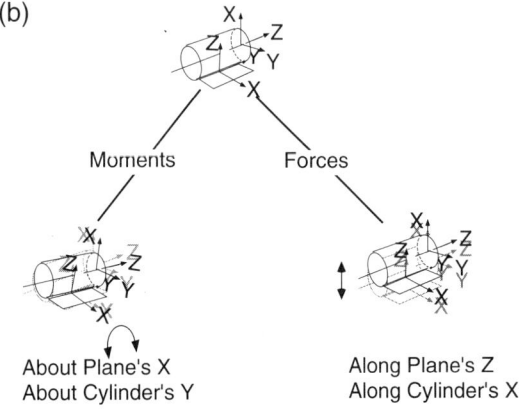

When defining the DFC, the designer must define explicitly the surfaces or reference axes on mating features which are intended to carry dimensional constraint to the mating part. This approach makes it unnecessary, even counterproductive, to construct algorithms that "identify" tolerance chains or loops, since the DFC equips the designer to define them purposefully as a main objective of assembly design. On the other hand, defining the DFC and its implementing features prepares the designer to carry out the steps of GD&T or some other systematic tolerancing scheme for each part, as illustrated by the example in Figure 8-5 through Figure 8-7.

We turn next to the distinction between two types of assemblies, called Type 1 and Type 2. The DFCs for these, and the strategies used to achieve their KCs, are quite different.

8.F. TYPE 1 AND TYPE 2 ASSEMBLIES EXAMPLE

To clarify our approach to designing assemblies, we need to distinguish between two kinds of assemblies, which we call Type 1 and Type 2. Type 1 assemblies are constrained completely by feature relations between their parts. Type 2 assemblies are underconstrained by their features and need fixtures or measurements to add the missing constraint. We will illustrate the difference with an example from the automobile industry.

Figure 8-11 shows a simplified car floor pan.[7] This assembly consists of three stamped sheet metal parts. The KC is the overall width of the car, which is nominally of dimension L. The design shown in the figure consists of parts with flanges that are spot-welded together to form butt or coach joints. On the right in Figure 8-11 is the liaison diagram for this assembly, showing the KC as a double line joining parts A and C. Parts A and C contain the features that must be a distance L apart in order to deliver the KC.

The way this assembly has been designed, each part locates the adjacent part in the left–right direction by means of a flange, a short piece of metal that is intended to be perpendicular to the plane of the part. This flange is formed by stamping the part from flat stock. The flanges are typically spot welded together. As discussed in Chapter 6, when such a part is stamped, there is some uncertainty in the bend radii at each end. The result of this is that the overall width of the part from flange to flange is uncertain.

Figure 8-12 shows a DFC for this assembly. Because each part locates the adjacent part, we say that it has a mate with that part. We indicate this with arrows between the parts in the DFC. Figure 8-12 can be read to say: "Part A locates part B and part B locates part C. The KC is a geometric relationship between part A and part C." Note that we can trace a chain of mates from one end of the KC to the other. Note, too, that the flange joints completely constrain the adjacent parts along this chain. On this basis, we say that this assembly is a Type 1. The direction of the chain, as well as the designation of part A as the root, is arbitrary. A feasible assembly sequence for this assembly is

1. Mate parts A and B;

2. Mate parts B and C.

All of the foregoing, together with the DFC, comprise the documentation of the design intent for this simple assembly.

Figure 8-13 shows an alternate design for this assembly. It differs from that shown in Figure 8-11 in that there is a contact between part B and part C. The designer has proposed this design because he predicts that the sizes of the parts measured between the flanges will not be accurate enough to ensure delivery of the KC. He knows that only the overall width L matters, so he has shown parts B and C joined by a slip joint. This joint can be adjusted so that width L will be achieved.

However, this design differs fundamentally from the original. A candidate DFC appears in Figure 8-14. This DFC does not contain a chain of mates from one end of the KC to the other. In fact, we can see that part B and part C do not constrain each other in the direction of the KC. These two facts tell us that this is a Type 2 assembly and that we need a fixture or measurement to provide the missing constraint.

Figure 8-15 shows a candidate fixture designed to remove the under-constraint from this assembly, while Figure 8-16 shows the DFC that applies to the assembly when this fixture is used. A number of points are worth noticing. First, it is now possible to trace a chain of mates through the DFC from one end of the KC to the other, although this

---

[7]This example was provided by Robert Bonner and James D'Arkangelo of Ford Motor Company.

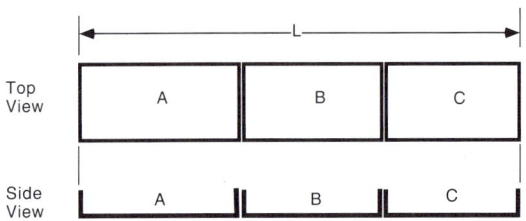 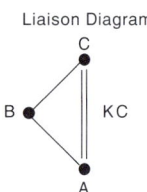

FIGURE 8-11. Example Simplified Car Floor Pan. *Left:* Top and side views of a three-part sheet metal car floor pan. These are U- or channel-shaped parts stamped from flat stock. The KC for this assembly is its overall width *L*. *Right:* The liaison diagram and the KC.

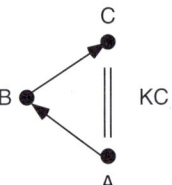

FIGURE 8-12. Datum Flow Chain for the Assembly in Figure 8-11. Part A locates part B while part B locates part C. The KC is a geometric relationship between A and C.

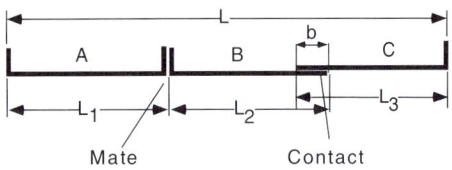

FIGURE 8-13. Alternate Design for Car Floor Pan. The KC is the same as in Figure 8-11, but in this design there is a slip joint contact between part B and part C.

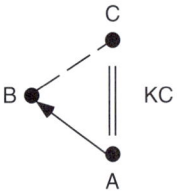

FIGURE 8-14. Proposed DFC for the Assembly in Figure 8-13. The mate is shown by an arrow as in Figure 8-12, while the contact is shown as a dashed line. In this DFC it is not possible to trace a chain of mates from one end of the KC to the other. This DFC does not completely constrain the parts. It is therefore not capable of delivering the KC.

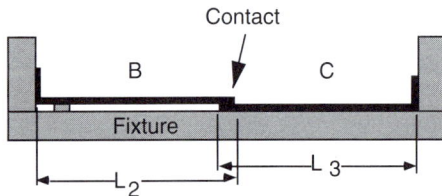

FIGURE 8-15. Fixture for Providing Constraint for Parts B and C.

chain, unlike that in Figure 8-12, not only passes through parts but also passes through the fixture. Indeed, whereas parts B and C have a contact with each other, they have mates with the fixture. The fixture provides the missing constraint in the direction of the KC via these mates. We can read Figure 8-16 to say: "The fixture locates parts B and C, while part B locates part A." All the methods we learned in Chapter 4 about assessing the adequacy of constraint can be used on feature mates between parts and fixtures, just as they can on feature mates between parts. In this case, such an analysis will reveal that the fixture is free to constrain parts B and C because the contact between these parts applies no constraint of its own in the direction of interest.[8] If this contact were a mate, then there would be overconstraint in this fixture design.

The assembly process implied by Figure 8-13, Figure 8-15, and Figure 8-16 is as follows:

1. Place parts B and C in the fixture and weld them together.

2. Weld part A to part B, completing the assembly.

No other assembly sequence is possible using the fixture in Figure 8-15.

It may appear that we are finished, but in fact we are not. There is an alternate way to remove the under-constraint from this assembly. It is shown in Figure 8-17. The corresponding DFC is shown in Figure 8-18. We can read Figure 8-18 to say: "The fixture locates parts A and C, while part A locates part B." Again, we can trace a chain of mates in this DFC from one end of the KC to the other, and again it passes through the fixture. The assembly

---

[8]Remember that this is a one-dimensional example, so only the left–right dimension matters. The KC is measured in this direction, and the mates between parts, and between parts and fixtures, apply constraint only in this direction.

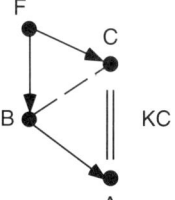

FIGURE 8-16. Improved DFC for the Assembly in Figure 8-13 Using the Fixture in Figure 8-15. In this DFC, it is possible to trace a chain of mates from one end of the KC to the other. However, this chain passes through the fixture.

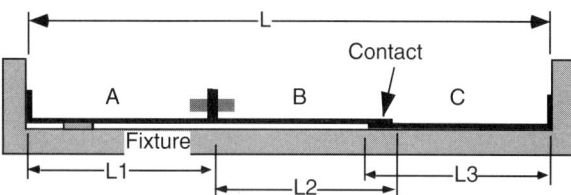

FIGURE 8-17. Alternate Fixture Design for the Assembly in Figure 8-13. If this fixture is used, then part A is welded to part B first using the mate. The weld is indicated as the fat gray line. Then the subassembly of A and B is placed in the fixture on top of part C. The parts are pushed firmly against the ends of the fixture to create the A–F and C–F mates, and finally the contact is fastened. Alternately, all the parts can be placed in the fixture at once as long as the A–B, B–C fastening sequence is followed.

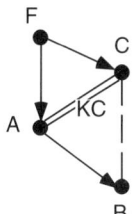

FIGURE 8-18. DFC for the Assembly in Figure 8-13 Using the Fixture in Figure 8-17.

process implied by Figure 8-13, Figure 8-17, and Figure 8-18 is as follows:

1. Mate part A and part B.
2. Put the A–B subassembly in the fixture and join part C to part B.

Alternately, do the following:

1. Place all the parts in the fixture.
2. Weld part A to part B, then weld parts B and C together.

No other sequences are possible using this fixture, and only one joining sequence for the parts has a chance of delivering the KC.

Are the two assembly strategies, fixtures, and DFCs for this Type 2 assembly shown in Figure 8-16 and Figure 8-18 equivalent? Let us recall the reason why the designer chose to investigate a Type 2 in the first place. He wanted the ability to adjust one of the joints (he chose B–C) so as to improve the likelihood of delivering the KC. We are not done comparing the design alternatives, including the Type 1 in Figure 8-12, until we examine, at least in principle, the variation that could result from each so that we can compare their ability to deliver the KC.

First, examine the DFC in Figure 8-12. The variation in the KC arises from the combination of the individual part variations. Since we know that stamped flanges could contain variation affecting the size of the part, we know that this design is vulnerable to this kind of error. Second, examine the design in Figure 8-16. Without performing a detailed variation analysis, we can see that it could suffer from the same difficulty as the design in Figure 8-12 because variation from the stamping of Part A could still be a factor. In addition, there is going to be some variation due to the construction of the fixture. Finally, examine the design in Figure 8-18. Here, the KC is completely under the control of the fixture, and fixture variation will be the only contributor. While we will not conduct a full variation analysis, it is a good bet that the third design will have the least variation. A thought question at the end of the chapter asks the reader to analyze this situation quantitatively.

The designs in Figure 8-12 and Figure 8-16 both suffer from error due to stamping the flanges, although the total variation is larger in Figure 8-12. Only the design in Figure 8-18 really eliminates error due to flange stamping. The reason it can while that in Figure 8-16 cannot is a fundamental one that we will make into a rule. This rule states that in an assembly where a part is connected to others, or to fixtures, by both mates and contacts, the incoming mates should be fully fastened before any of the contacts are fastened.[9] The reason is that the incoming mates define the location of the part. If a contact is fastened before all the incoming mates are fastened, then the part will be positioned at least in part by the one to which it has a contact. The contact has thus been given a role it does not have the capability to handle, namely to provide location for another part. When the remaining mate(s) is/are fastened, there are two possibilities. First, the variation

---

[9]In Figure 8-16, part C's contact with part B is made before its mate with the fixture. In Figure 8-18, this sequence is reversed, and the B–C contact is made after C acquires its mate with the fixture.

in the KC will be larger than it would have been if all the mates were fastened first. Second, an overconstrained situation could result.

We can make another observation from this example that will hold for others: In a Type 1 assembly, the overall variation depends directly on the variation in the individual parts. Furthermore, the assembly sequence does not matter; every assembly sequence will give the same final variation in the assembly. In a Type 2 assembly, the overall variation depends not only on the parts, but also on fixtures or, more generally, on the assembly process. Equivalently, we can say that different Type 2 assemblies are in fact different assembly processes, with different fixtures, different assembly sequences, and different final variation in the assembly, even though they assemble the same parts. Type 1 assemblies may thus be called part-driven, while Type 2 assemblies are called assembly process-driven.

8.G. KC CONFLICT AND ITS RELATION TO ASSEMBLY SEQUENCE AND KC PRIORITIES

As mentioned in Chapter 2, a single assembly can often have several KCs associated with it. Because each assembly has a limited number of mates and assembly steps, it is possible that achievement of the KCs cannot be guaranteed independently. Multiple KCs in the same assembly can be classified as follows:

- *Independent:* The delivery chains of the KCs share no degrees of freedom of any mates. The variation in each KC is completely independent of the variation in every other KC. For example, in Figure 8-6, the concentricity KC and the perpendicularity KC arise from different degrees of freedom and feature surfaces, and follow separate DFC delivery chains.

- *Correlated:* The delivery chains share some degrees of freedom. Variation in these degrees of freedom will affect all KCs that share them. However, there is still some opportunity to improve the variation of each KC without degrading the variation of the others. In Figure 8-2, the variation of each KC in the stapler is lower-bounded by the carrier's variation. However, the probability of each KC achieving its tolerance also depends on variation in other parts that are not shared. The correlation may in some cases be broken by such means as providing an adjustment or resorting to selective assembly in one or more of the legs of the DFC that are not shared ([Goldenshteyn]). But these are serious redesigns and are often unavailable.

- *Conflicting:* The KC delivery chains share so many degrees of freedom that attempts to improve one KC will always degrade another; or the probability of achieving one KCs tolerance requirement will always be lower than the probability of achieving another.

KC conflict can arise in two ways. In one situation, there is no remedy short of drastic redesign of the parts, while in the other, the conflict can be resolved by choosing another assembly sequence.

1. The DFCs for different KCs share so many arcs that any adjustments or statistical error accumulations will be identical or additive, preventing independent achievement of the KCs. This is illustrated in Figure 8-19. There is no possibility in such situations of relieving the problem by choosing a different assembly sequence. Instead, one must choose a priority for the KCs, and the one that is finished first in the assembly sequence is the one that will have the higher probability of being achieved, or is the one that may be given tighter tolerances. This situation may occur in Type 1 or Type 2 assemblies. Here, too, redesign of the parts to permit adjustments or selective assembly may relieve the situation.

2. Due to the requirement that each subassembly be completely constrained, some KCs may be impossible to adjust into achievement because the available degrees of freedom were "used up" during prior assembly steps. Some assembly sequences permit independent achievement of the KCs, while others do not. This is illustrated in Figure 8-20 and is discussed in connection with aircraft assembly in Section 8.I.3. This situation occurs only in Type 2 assemblies.

An indicator that KC conflict could arise is the case where more than one KC chain is completed at the same assembly step ([Arora]). This is illustrated for car doors later in this chapter in Figure 8-46.

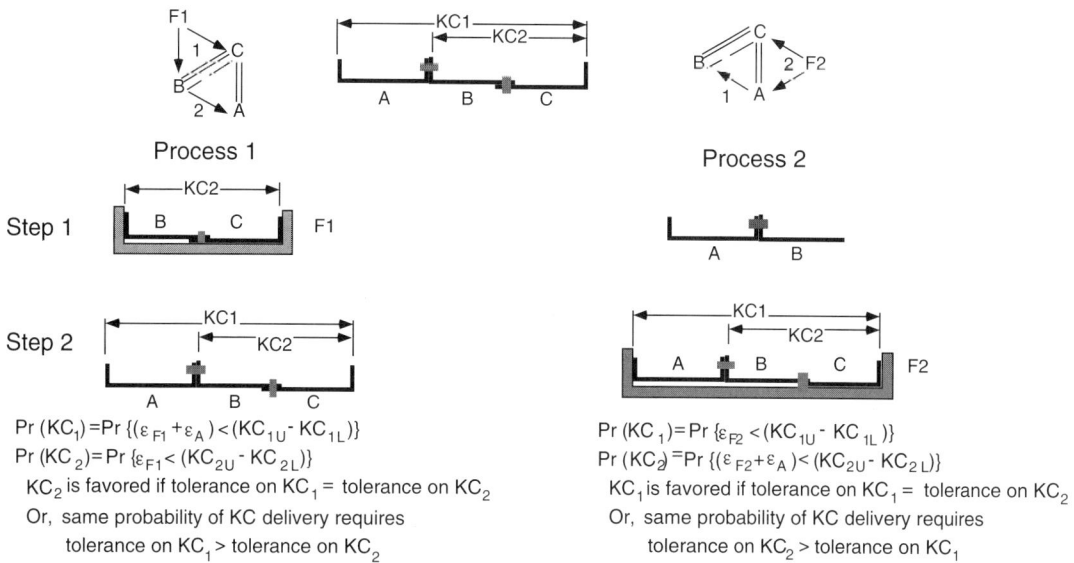

$$Pr(KC_1) = Pr\{(\varepsilon_{F1} + \varepsilon_A) < (KC_{1U} - KC_{1L})\}$$
$$Pr(KC_2) = Pr\{\varepsilon_{F1} < (KC_{2U} - KC_{2L})\}$$
KC_2 is favored if tolerance on KC_1 = tolerance on KC_2
Or, same probability of KC delivery requires
tolerance on KC_1 > tolerance on KC_2

$$Pr(KC_1) = Pr\{\varepsilon_{F2} < (KC_{1U} - KC_{1L})\}$$
$$Pr(KC_2) = Pr\{(\varepsilon_{F2} + \varepsilon_A) < (KC_{2U} - KC_{2L})\}$$
KC_1 is favored if tolerance on KC_1 = tolerance on KC_2
Or, same probability of KC delivery requires
tolerance on KC_2 > tolerance on KC_1

FIGURE 8-19. Example of KC Conflict. This example is similar to that in Figure 8-13 with the addition of a second KC. (Pr (KC_1) means probability of achieving KC_1. ε_{F1} means error in fixture F1. KC_{1U} means upper specification limit on KC_1. Other notation is to be interpreted similarly.) There is a chain of mates from one side of each KC to the other side, but these chains contain arcs that are part of both chains. Since there is only one contact by which to adjust two KCs into compliance, one is bound to be achieved with lower probability than the other, or else one must be given looser tolerances than the other. This problem exists in both of the assembly sequences shown here.

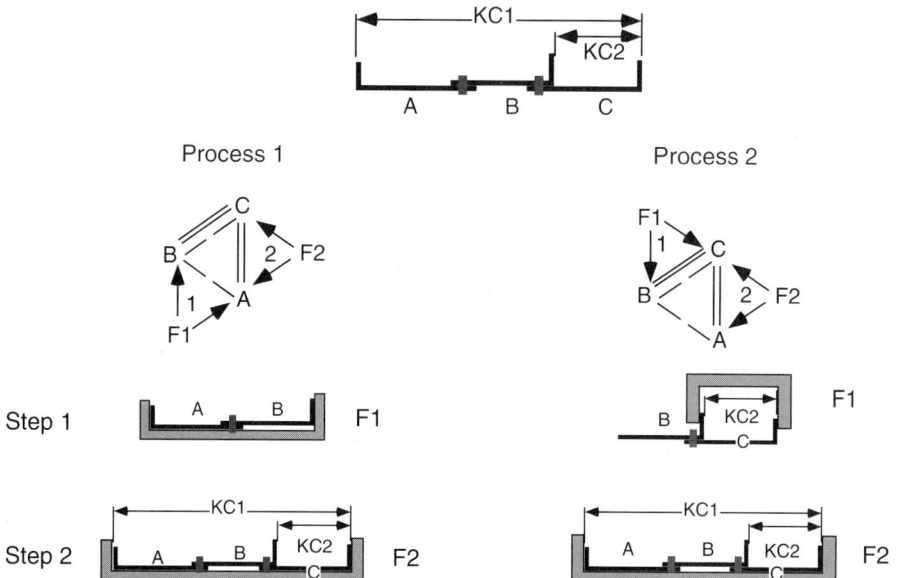

FIGURE 8-20. Example of KC Sensitivity to Assembly Sequence. The parts in Figure 8-19 have been rearranged so that there are two contacts in the assembly and no mates. Now there are in principle enough degrees of freedom to adjust both KCs into compliance but, if the wrong assembly sequence is used (process 1), one of these degrees of freedom will be used up before any adjustment can be made, rendering the situation similar to that in Figure 8-19. In process 1 the chains of mates connecting the ends of the KCs share some arcs, whereas in process 2 the chains are independent.

[Hu] points out that errors in car body assembly can be traced to their sources by observing whether the errors are correlated at the final assembly level, subassembly levels, or not at any level. A correlation occurs when measurements relating an entire group of parts to a reference location all show errors in the same direction or errors of similar magnitude and direction. For example, correlations at the assembly level imply that an entire subassembly was built correctly but was installed in the final assembly incorrectly. In this case, there may be no need to seek error sources at the subassembly level or below.

The situation in Figure 8-19 occurs in practice, as illustrated in Section 8.I.1, which discusses assembly of car doors.

8.H. EXAMPLE TYPE 1 ASSEMBLIES

In this section we will look at some Type 1 assemblies and learn a few more things about using the DFC: a fan motor, a front wheel drive automobile transmission, a Cuisinart food processor, the pump impeller discussed in Chapter 7, and a machined part assembly for an automobile called a throttle body.

8.H.1. Fan Motor[10]

The fan motor is part of a low-cost table fan. It is shown in Figure 8-21.

The motor consists of four main parts, plus fasteners: the stator, the rotor, and front and rear end housings, as shown in Figure 8-22. Four long screws hold the assembly together. The rotor shaft runs in solid oil-impregnated self-aligning bronze bearings mounted in the housings. Self-aligning means that the bearings can wobble slightly about two axes normal to the bearing axis. They can do this because their outer shape is spherical and they mount in spherical pockets pressed into the housings. The axial location of the rotor with respect to the stator is adjusted by selecting the right number of spacers and putting them on the shaft before assembling it to the end housings.

Figure 8-23 shows the DFC for the fan motor. The KC relates the rotor and the stator. Actually, two dimensions must be controlled, namely, the axial and radial relationships between rotor and stator that are discussed in the caption of Figure 8-22. These are called out explicitly in Figure 8-24, which identifies at least schematically the features inside each part that play roles in delivering each KC or controlling each degree of freedom in the assembly. Figure 8-27 is a similarly detailed DFC for the rotor.

Important features on the end housings and rotor convey dimensional relationships between these parts. Details about how they are constructed are in Figure 8-25. In spite of the apparently casual way these features are formed, they are able to provide the necessary accuracy.

Several points about the fan motor are worth mentioning. First, the self-aligning bearings in the end housings provide both position and angular location for the rotor. The design is symmetric, so we could have chosen either housing and its bearing as the root of the DFC. Once we pick one, we say that its bearing provides X, Y, and Z location. Without the other housing in place, however, the shaft can wobble about θ_x and θ_y. For this reason, we note on the DFC that the other housing provides angular alignment about X and Y.

Second, the cast raised bevel features used to align the housings to the stator strictly speaking create an overconstrained situation unless a small amount of clearance is provided. Cast-in features are not very accurate, however, so interference could occur some of the time. The designer probably felt that any excess material on the bevels would be crushed when the screws were tightened and that the variation, if any, would be smaller than the tolerance sought on radial centering.

Third, the self-aligning feature of the bearings prevents overconstraint from developing between the housings, the stator, and the rotor.

Finally, we can easily see from the detailed DFCs in Figure 8-24 and Figure 8-27 how to choose datum features and the dimensioning scheme for locating the features of each part so that each part will be able to carry its branch of the DFC. For example, we must carefully make the rotor so that the core is the right diameter and that the outer-diameter surface is concentric with the shaft centerline. Similarly, we must center the bearings in the housings with respect to the raised bevels. These features are important for delivering the axial and radial alignment KCs necessary for the efficient operation of the motor. A thought question at the end of the chapter asks the reader to consider the rear housing in detail.

---

[10]This section makes use of report material prepared by MIT students Cesar Bocanegra, Winston Fan, Sascha Haffner, Yogesh Joshi, Tsz-Sin Siu, and Carlos Tapia.

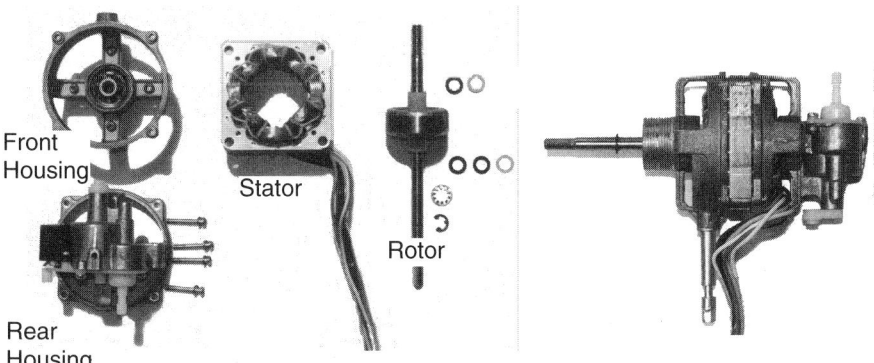

FIGURE 8-21. Small Fan Motor. *Left:* The parts—stator and windings, rotor, front and rear housings, plus screws, washers, and spacers. *Right:* The motor assembled. (Photos by the author.)

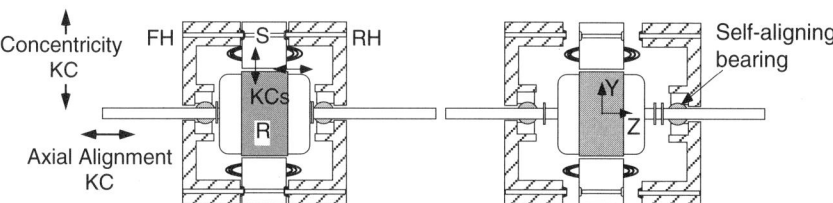

FIGURE 8-22. Schematic of Fan Motor. (S, stator; FH, front housing; RH, rear housing; R, rotor.) *Left:* Assembled. *Right:* Showing front housing and rear housing slid away from stator along shaft. The partially obscured spheres on the shaft represent self-aligning bronze bearings. The KCs are the correct axial (Z) and radial (X and Y) positions of the rotor with respect to the stator. These are important for the efficiency of the motor. To achieve these KCs, the edges of the rotor core must be opposite the edges of the stator, and the outside diameter of the rotor core must be centered radially with respect to the inside diameter of the stator. The screws that join the housings and the stator are not shown. Thin thrust washers lie on the rotor shaft between the rotor and each housing. The correct number of these is selected to just barely fill the axial (Z) gap and center the rotor axially.

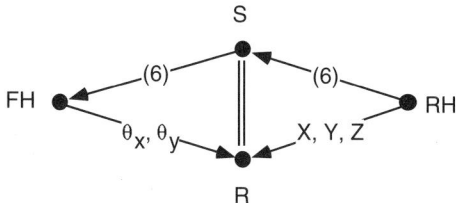

FIGURE 8-23. DFC for Fan Motor. This DFC delivers the KC that requires the rotor to be centered, radially and axially, inside the stator. The rear housing aligns the front housing radially and axially via the stator. It aligns the rotor radially by means of a spherical self-aligning bearing. The front housing has a similar bearing. Together, the front and rear housings locate the rotor. Cast-in raised bevels on the rear and front housings mate to holes in the stator. These bevels and holes are visible in Figure 8-22 and details of their construction are shown in Figure 8-25. The bevels can also be seen in Figure 8-26.

The discussion about the rotor and housings can be generalized to an important rule: The chain of feature constraints within a part is, or should be, a little DFC of its own, obeying all the rules of a DFC.[11] It should be a subset of the whole assembly's DFC for that KC. This rule implements the top-down nature of this approach to design of assemblies and creates the starting point for detailed design, dimensioning, and tolerancing of individual parts so that they will play their desired role in delivery of the KCs.

8.H.2. Automobile Transmission

Automobile transmissions are complex assemblies comprising a die-cast case, a number of planetary gear sets, shafts, and subassemblies called clutches. The general

---

[11]Recall that the features inside a rigid part are always properly constrained with respect to each other. For this reason, all feature relations within a part will be mates, never contacts.

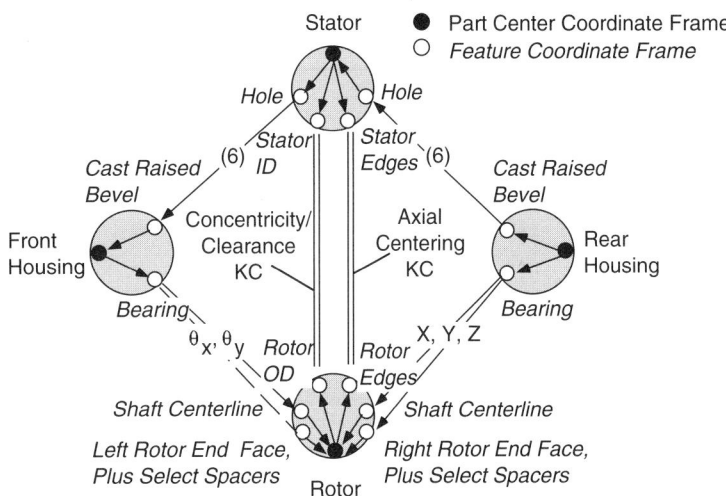

FIGURE 8-24. Detailed DFC for the Fan Motor. This DFC defines two separate KCs and includes details of the individual features in each part that are involved in locating the parts with respect to each other and delivering each KC. A separate DFC can be drawn for each KC. A thought question at the end of the chapter asks the reader to do this.

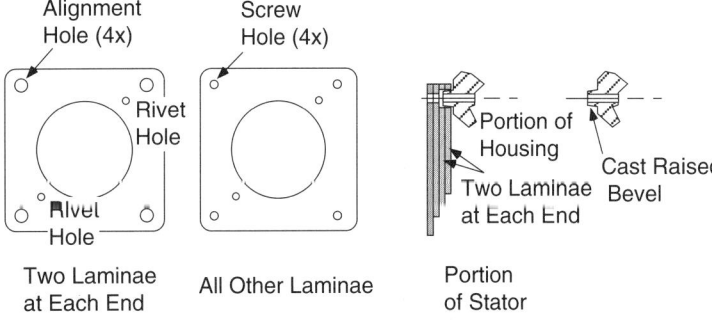

FIGURE 8-25. Detail of Stator Construction. The stators are built by stacking laminae over pins through the rivet holes. The first two and last two laminae have extra large holes at the four corners. These enlarged holes are the locating features on the stator that accept cast raised bevels on the front and rear end housings for the purpose of aligning the housings and the stator. The screws that fasten these three parts together play no role in locating them because they are smaller in diameter than the holes they pass through. The cast raised bevels can be seen in Figure 8-26.

FIGURE 8-26. Detail of Cast Raised Bevel on Motor Housing. (Photo by the author.)

layout of a front wheel drive transmission consists of two parallel shafts, one concentric with the engine's crankshaft and the other offset to one side that carries the output power to the differential and the wheels. The clutches are used to immobilize rings, planets, or suns of different planetary gear sets, thereby causing the transmission to have a different gear ratio. The clutches in turn are activated by pistons powered by oil pressure provided by an oil pump at one end of the transmission. A transfer chain carries power from the input side to the output side. These parts and their relationships are shown in Figure 8-28 and Figure 8-29.

The internal moving parts of the transmission, consisting of the rotating clutches and transfer chain hub, make up a stack that must fit between the bottom, formed by the oil pump, and the top, formed by the bell housing. Elements of this stack include layers of clutch plates made of metal with friction material bound to them. Since the thickness of the friction material is difficult to control, the height of this stack is quite uncertain. To allow for this uncertainty, the opening between the oil pump and the bell housing is made deliberately large, and the space is filled by a select thrust washer. The assembly process, here greatly simplified, involves joining the oil pump to the case, inserting a thrust washer for the rotating clutches to thrust against, inserting the rotating clutches and transfer chain hub, and measuring the empty space between the top of the case and

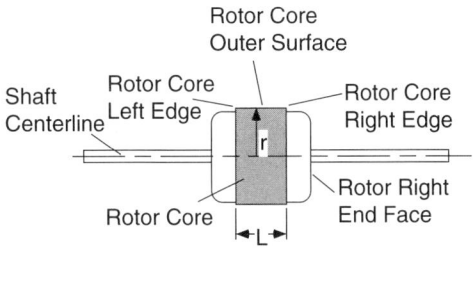

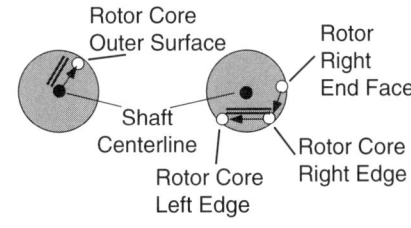

Radial KC
Inside Rotor

Axial KC
Inside Rotor

FIGURE 8-27. Details of Rotor Construction. *Top:* Two key dimensions of the rotor are the length *L* of the core and the core's radius *r*. In addition, the outer surface of the core must be concentric with the shaft. These two requirements can be expressed as local DFCs inside the rotor. Two such DFCs are shown at the bottom in this figure together with the KCs that they deliver.

the top of the hub. Another thrust washer of the correct thickness is selected and placed on top of the hub, and the bell housing is installed on top, closing the case.

A problem with this assembly sequence is that the case also contains a band clutch that must be installed in the case from the oil pump end before the oil pump is attached to the case. This is a wide sheet of spring steel with friction material on the inside. It wraps around the outside of the rotating clutches and can stop them from rotating if it is pulled tight around their outer diameter. This action provides an additional gear ratio. The assembly problem arises because this band is not perfectly circular when it is installed. It could protrude into the region where the rotating clutches are to be inserted. If it does, then the rotating clutches could collide with it during assembly, stripping off the friction material or doing worse damage.

The author and his Draper colleagues attempted to avoid this problem by choosing a different assembly sequence. This sequence builds the transmission upside down from the one described above. It starts with the bell housing, then places a thrust washer on it, then places the transfer chain hub and rotating clutches on the washer. The empty space is measured between the bottom of the rotating clutches and the case at the oil pump end. A washer of the correct thickness is selected and inserted, the band clutch is inserted, and the case is closed up by inserting the oil pump. Since the band clutch is inserted after the rotating clutches and in full view of the operator, damage is avoided.

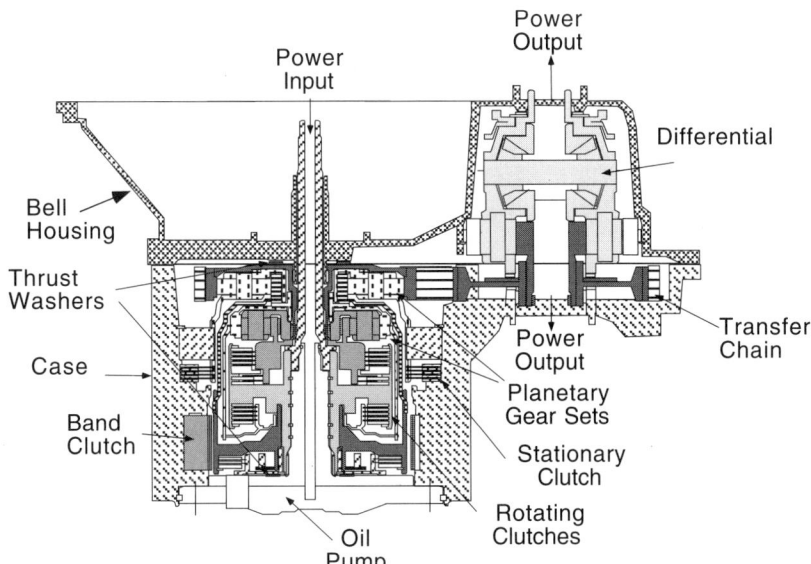

FIGURE 8-28. Cross-Sectional View of a Typical Front Wheel Drive Transmission. Input power comes in from the engine to the central shaft. It passes through the gears and rotating clutches, then to the transfer chain, and finally through the differential and out to the wheels. The oil pump provides hydraulic power to activate the pistons that operate the clutches to change gears.

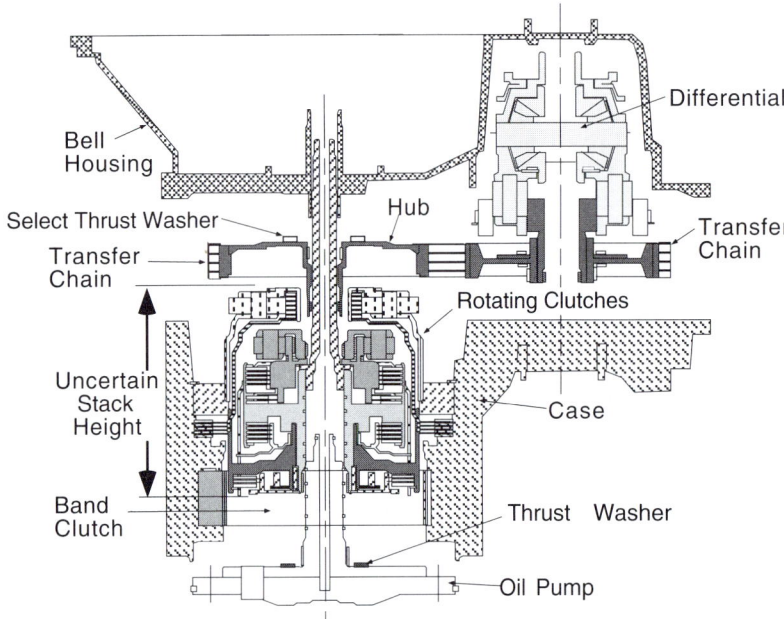

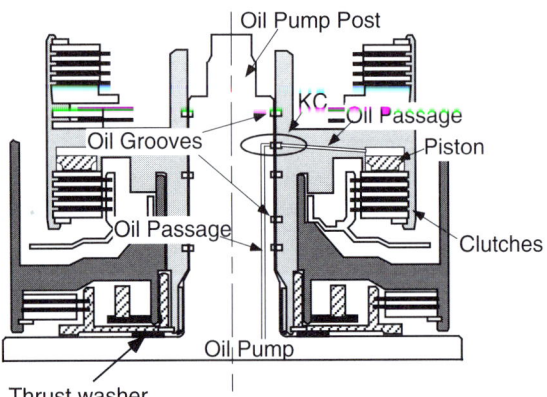

FIGURE 8-30. Detail of Oil Pump and Rotating Clutches Showing Alignment of Oil Grooves. These grooves guide high-pressure oil from the oil pump to the pistons in the rotating clutches. Alignment of the oil grooves on the oil pump and on the rotating clutches is the KC.

Unfortunately, this alternate sequence cannot be used. The reason is that the two thrust washers are not equivalent. This can be seen by examining Figure 8-29 or Figure 8-30 in detail. The oil pump feeds oil to the individual rotating clutches through circumferential grooves in its central post. Each of these grooves must line up axially (vertically in the figures) with a corresponding groove on the inner diameter of the rotating clutches. If the grooves are

misaligned axially, high-pressure oil will be fed to more than one piston at the same time. It is then possible that the transmission will shift into the wrong gear or that it will try to be in two gears at once. This could cause rough shifting or even serious damage to the transmission.

We can describe this situation using our vocabulary and symbols as follows. The alignment of the grooves is obviously an important KC for this assembly. We can create a DFC for this KC by tracing a path from the face of the oil pump through intermediate parts and features to each of the grooves, as shown in Figure 8-30 and Figure 8-31. This DFC clearly passes through the thrust washer at the oil pump end. If we selected this washer based on the height of the rotating clutch stack, we would have no ability to control its size for the purpose of achieving the KC. A tolerance analysis of this DFC would reveal unacceptable variation in oil groove alignment.

The select thrust washer at the bell housing end is not involved in delivering a KC. Its job is merely to fill empty space. It does not control the location of any part or feature. It is appropriate to say that it is involved in a contact. However, the thrust washer at the oil pump end must be involved in a mate because it is in the chain that controls the location of the rotating clutches with respect to the oil pump. For this reason, it is part of a DFC.

In terms of constraint and degree of freedom analysis, we can say that the rotating clutches have only one

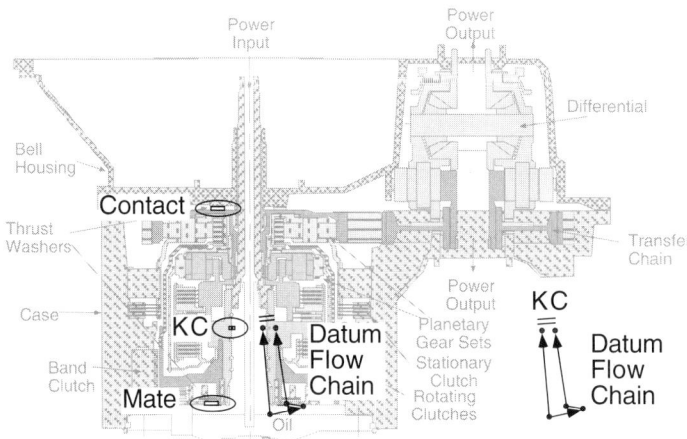

FIGURE 8-31. DFC to Align Oil Grooves in the Oil Pump Hub and the Rotating Clutches. This DFC starts at the face of the oil pump that mates to the case and follows two paths. One path leads to the oil grooves on the oil pump post while the other path leads to the oil grooves on the rotating clutches. On the way, the second path passes through the thrust washer. The lower thrust washer participates in a mate between the rotating clutches and the oil pump, while the upper thrust washer participates in a contact between the transfer chain hub and the bell housing.

axial degree of freedom, and if we locate it using a select washer, we no longer have that degree of freedom available to us to ensure that the oil grooves line up. Our only alternative would be to provide some means to adjust the oil pump axially with respect to the case or adjust the hub axially inside the oil pump, but either would probably be too expensive and prone to oil leaks.

Finally, once we have identified the washer at the bell housing end as a contact and the washer at the oil pump end as a mate, then we can invoke the rule that says "make the mates before the contacts" to give us a clue that the bell housing end washer must be the last part installed in the internal stack prior to closing the case.

This example shows, among other things, that we can use the DFC to describe situations that involve selective assembly. It also shows that we can seek alternate assembly sequences to solve assembly problems, but it may occur that the alternate sequence is unavailable. We will encounter this problem again and again, reinforcing the idea that assembly sequence analysis is an essential element in design of the delivery strategy for the KCs.

8.H.3. Cuisinart[12]

Figure 8-32 shows a Cuisinart food processor. Figure 8-33 shows an exploded view and names the parts while Figure 8-34 shows the DFCs of interest. The rotating blade is

driven by an electric motor through a planetary gear train. The sun gear is on the motor shaft, while the ring gear is held stationary by the top frame. The three planet gears drive the shaft that turns the knife.

The DFC of interest to us here is the one that aligns the sun gear on the motor shaft to the center of the combined pitch circles of the three planet gears. Misalignment means a noisy unit that will wear out rapidly. One could imagine delivering this KC either of two ways. One way would prescribe a mate between planets and sun and a contact between the motor and the top frame. Assembly would consist of carefully establishing the mate and then fastening the contact. The other possibility is to establish a mate between the motor and the top frame and a contact

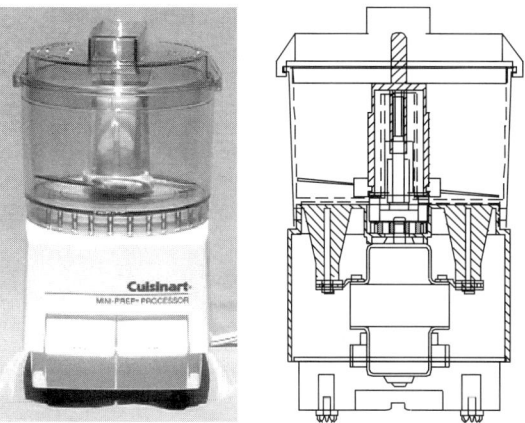

FIGURE 8-32. A Cuisinart Food Processor. (Photo by the author. Drawing by the students.)

---

[12]This section makes use of report material prepared by MIT students Chris Anthony, Cristen Baca, Eric Cahill, Gennadiy Goldenshteyn, and Amy Rabatin.

between the motor gear and the planets. Even though the second chain is longer and subject to several uncertainties due to the presence of the motor mount gaskets, this is the way the designer intended the assembly to work. The gaskets fit tightly over the ends of the posts on the top frame and into slots in the motor brackets. There is no way to adjust the motor's position relative to the gears. There is a little running clearance between sun and planets and lots of grease.

In fact, the design as intended follows recommended practice for setting up gear trains: One designs the case to hold the shafts in the correct relative positions in order that the pitch circles of mating gears are as nearly tangent as possible. In this case, the product is intended only for home use. The gaskets introduce location uncertainty but they dampen noise. It is unlikely that the product will be used so heavily that the gears will wear out rapidly from a user's point of view.

This example shows that certain recommended design practices can be captured in DFCs and employed over and over in different situations.

8.H.4. Pump Impeller

The pump impeller assembly was discussed in Chapter 7. There we saw that one assembly sequence presented a high probability of assembly problems due to loose tolerances on the features used for fixturing. A different sequence, fixtures, and fixturing features had to be used. We can use the DFC to represent the different sequences and at the same time we can learn a little more about the DFC method.

Figure 8-35 shows the two processes. Figure 8-36 shows DFC representations for these two processes. These diagrams show unambiguously how the two processes differ. A thought question at the end of the chapter asks the reader to think about this assembly, especially the design of fixture 2.

This example shows that we can use the DFC to analyze assembly processes as well as assemblies, and it also shows that we can describe part mating criteria as well as assembly quality with KCs.

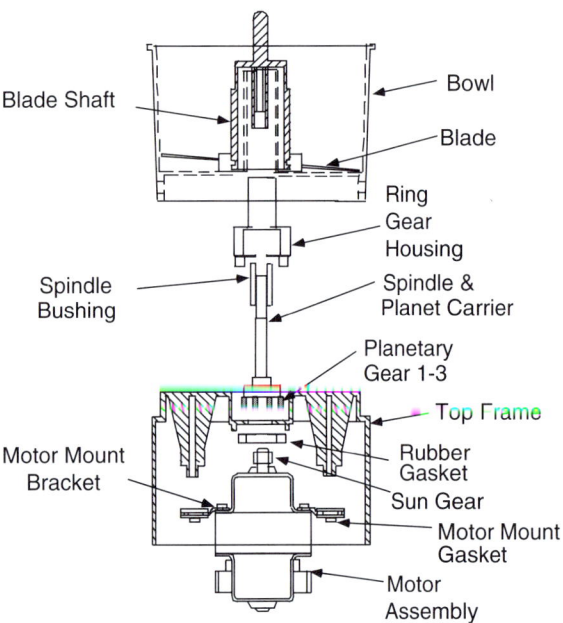

FIGURE 8-33. Exploded View of the Cuisinart.

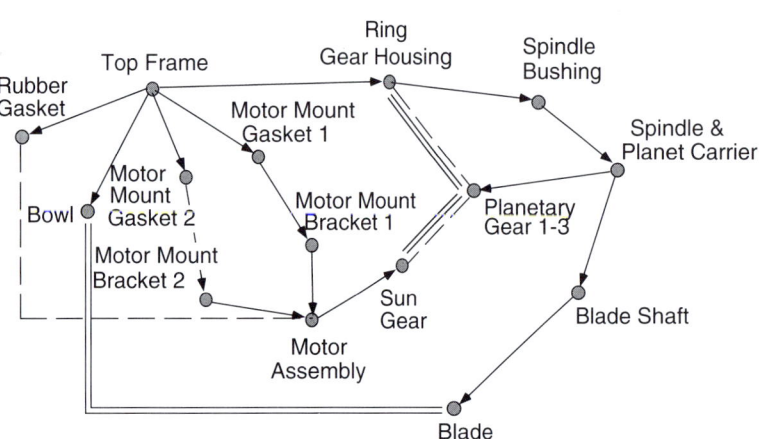

FIGURE 8-34. DFCs for the Cuisinart. Two DFCs are shown: One provides correct clearance between the blade and the bowl, and the other provides proper alignment of the planetary gear system. Especially important is the relation between the motor (sun) gear and the planetary gears.

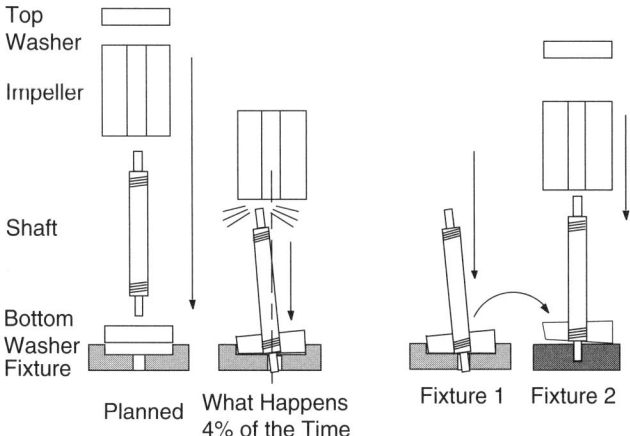

FIGURE 8-35. Comparison of Two Assembly Processes for the Pump Impeller. *Left:* The original process. *Right:* The improved process.

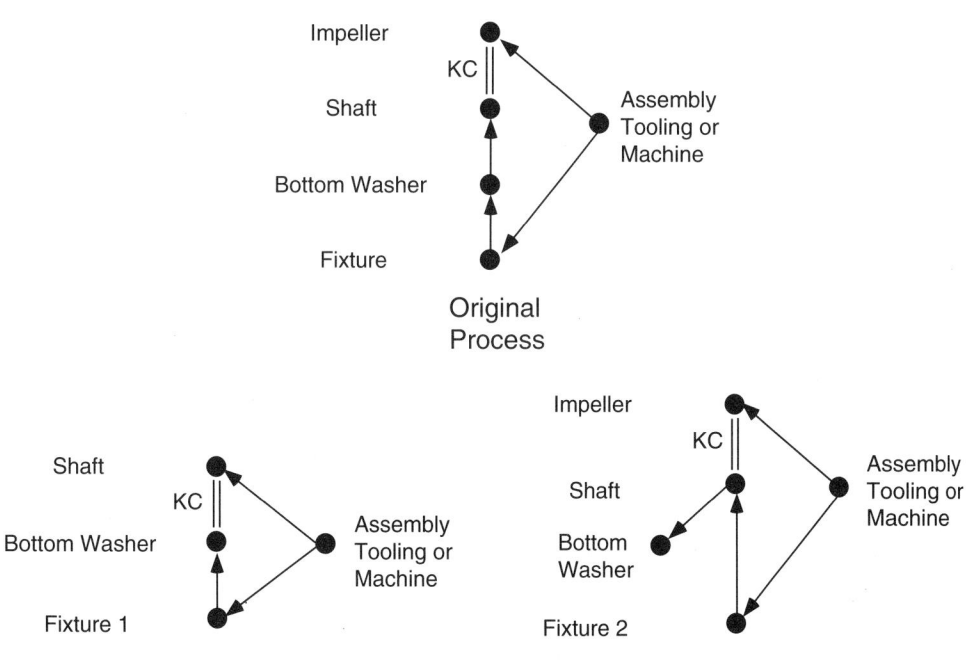

FIGURE 8-36. DFCs for the Two Assembly Processes Shown in Figure 8-35. *Top:* DFC for the original process, drawn to describe the assembly of the impeller to the shaft. This DFC is intended deliver the KC shown, which describes the part mating criteria that avoid wedging and collision with the part chamfer. This DFC is unable to deliver the KC a high enough percent of the time. *Bottom:* DFC(s) for the improved process. At the left is the first phase, which joins the shaft to the bottom washer. The KC for this step describes the conditions for successful shaft–washer assembly. This process is relatively easy because thread mating operations are relatively tolerant of angular error. At the right is the second phase, which joins the impeller to the shaft. The KC for this step describes the conditions for shaft–impeller assembly. This operation is by far the most difficult since the clearance is so small. The DFC shows that the bottom washer plays no role in this process.

8.H.5. Throttle Body

A throttle body mounts to the intake manifold of an auto-mobile engine and controls air flow to the engine. A photo of a typical throttle body is in Figure 8-37, while drawings of the main parts appear in Figure 8-38. At the left end of the shaft is a cam and lever to which is attached the cable from the accelerator pedal. Also at this end is a return spring that closes off the air flow when the pedal is released. At the right end of the shaft, mounted to the bore, is a potentiometer called the throttle position sensor that reports shaft angle to the engine control computer. Midway along the shaft is mounted a disk that serves as the air flow control device.

The main KC for this device is that the disk close tightly within the bore, while subsidiary KCs are that the disk not stick in the open or closed positions. See

Figure 8-39 for details about how the disk fits to the shaft, and see Figure 8-40 for details of how the disk fits in the bore.

We will consider two ways of designing the throttle body to deliver these KCs. They appear in Figure 8-41. The bore locates the shaft in five degrees of freedom, with the remaining degree of freedom being rotation about the X axis to provide the operating motion of opening and closing the air flow passage. The bore and the shaft share in locating the disk. In the DFC on the left, the disk is fastened to the shaft by means of screws that pass through clearance holes in the disk. The screws have a contact with the disk. In the DFC on the right, accurate location of the disk inside the bore is sought by means of locating pins that mate the disk to the shaft. A thought question at the end of the chapter asks the reader to compare these two designs.

FIGURE 8-37. Photograph of Throttle Body. (Photo by the author.)

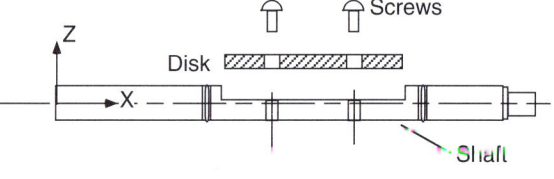

FIGURE 8-39. Detail of Shaft and Disk. The shaft has a recess into which the disk fits with a little clearance in the X direction. The screws go through clearance holes in the disk and into threaded holes in the shaft.

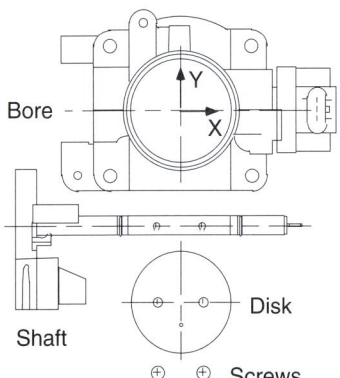

FIGURE 8-38. Throttle Body Parts and Assembled. (Drawing prepared by Stephen Rhee.)

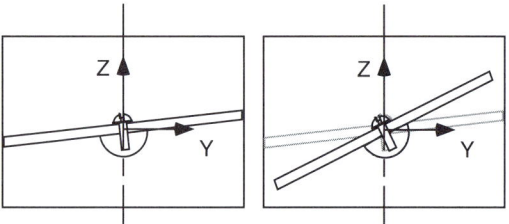

FIGURE 8-40. Detail of Disk in Throttle Body Bore. This view is along the X or shaft axis. On the left, the disk is in the closed position. On the right, the disk is open, and the closed position is shown in light gray. Note that the disk in this view is not a rectangle but is a parallelogram so that its skewed edges conform to the inside diameter of the bore when the disk is closed at an angle that is not perpendicular to the bore's axis. An important KC of this assembly is that the disk fit tightly inside the bore.

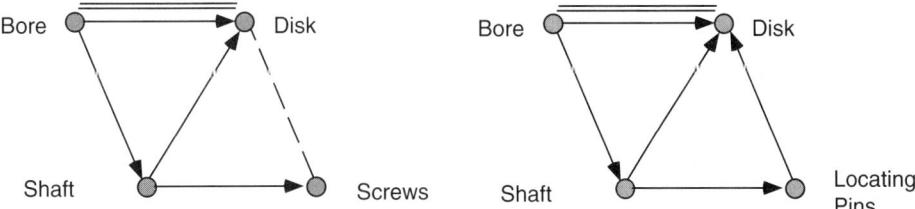

FIGURE 8-41. Two Possible DFCs for the Throttle Body. *Left:* The DFC for the design shown in Figure 8-38. *Right:* An alternate design. All the KCs regarding how the disk fits to the bore without sticking are condensed into one KC symbol.

8.I. EXAMPLE TYPE 2 ASSEMBLIES

In this section we will look at some Type 2 assemblies. These assemblies cannot be built merely by joining their mating features because some of them provide insufficient constraint. There are several possible reasons for this. Most revolve around the fact that it may be impossible or uneconomical to make the parts with sufficient accuracy or repeatability to deliver their KCs as Type 1s. Sheet metal assemblies are commonly of this type, but a number of machined parts assemblies fall into this class as well. Car doors and aircraft assemblies are both made of flexible parts, and assembly using fixtures is common. Below we consider one example of each.

8.I.1. Car Doors

We considered car doors in Chapter 2, where we noted that they typically have two conflicting KCs. Figure 8-42 shows a typical car door assembly process, while Figure 8-43 repeats a figure from that chapter, showing the two KCs and a diagram that we now recognize as a DFC. This DFC assumes that there are features on the inner panel that permit it to completely locate the outer panel, as well as features on the car body that permit it to completely locate the door via complete location of the hinges. If only it were so! In fact, no one tries to make car doors this way because, as we showed in Chapter 6, the tolerances on gaps and flushness at the assembly level are too small, on the order of ±2 mm or less, while tolerances on the parts are nearly as large (±1.5 mm or so).

In fact, fixtures are needed to support the process that is used to build the subassembly, place the hinges properly, and install the door onto the car body. Figure 8-44 and Figure 8-45 show two possible DFCs for this process that include fixtures. One of them appears to achieve both

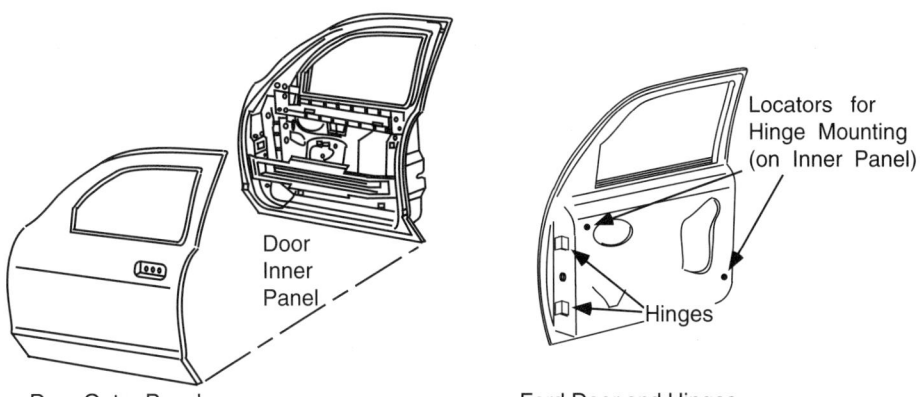

FIGURE 8-42. Typical Car Door Assembly Process. *Left:* The door is made by joining an outer panel and an inner panel. *Right:* The subassembly of door inner and door outer plus hinges and latch bar is ready to be attached to the car body. At the subassembly level, the hinges are used to adjust the door in the in/out and up/down positions. At the final assembly level, the hinges are used to adjust the fore/aft (and possibly the up/down) position. Other strategies are possible.

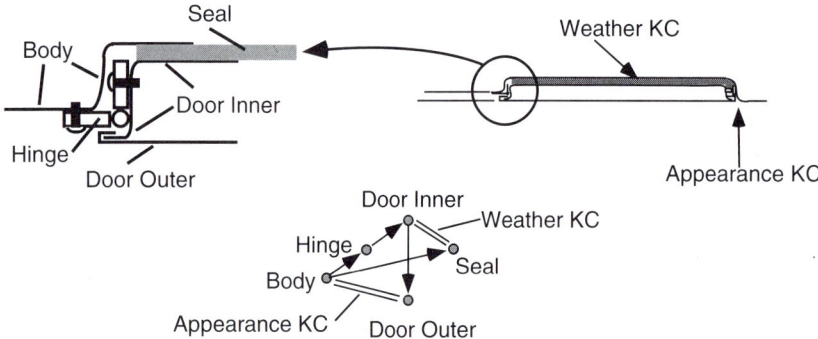

FIGURE 8-43. Car Door, Its Two KCs, and a DFC. This DFC imagines making a door and installing it as a Type 1 assembly. Also shown is a detail of how the hinge interfaces to the door inner panel and the car body.

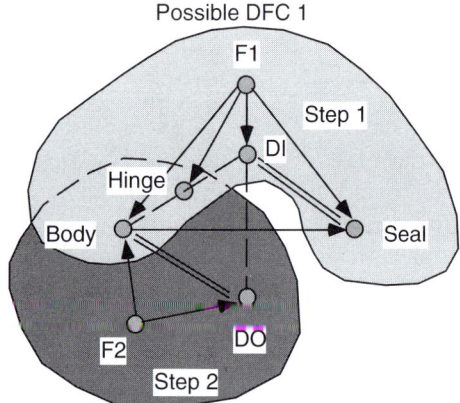

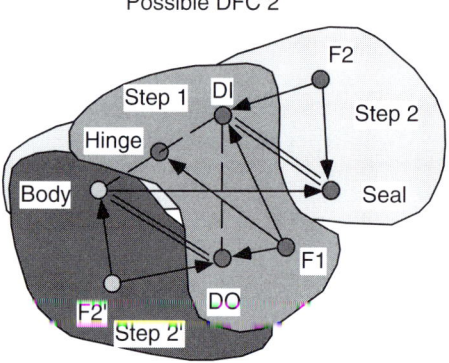

FIGURE 8-44. First Candidate DFC for Car Doors. This DFC starts by installing the door inner panel (DI) to the car body, using the hinges to achieve the weather seal KC. Fixture F1 is used for this step. Then fixture F2 is used to assemble the door outer panel (DO) to the door inner panel in such a way as to achieve the appearance KC. Unfortunately, this assembly sequence is impossible using today's door construction methods.

KCs independently but is in fact impossible by today's methods. The other suffers from KC conflict but is used anyway for lack of a better alternative.

In each of these door assembly processes, we can see that the fixtures are unconstrained by the joints between the parts, which are contacts or at least have unconstrained degrees of freedom in the directions controlled by the fixtures. A thought question at the end of the chapter asks the reader to label the arcs of these DFCs with explicit degree of freedom notations. Also, in Figure 8-45, there are not enough degrees of freedom using fixture F2 or F2′ alone to achieve both KCs independently. A thought question

FIGURE 8-45. Second Candidate DFC for Car Doors. This DFC first uses fixture F1 to make a subassembly of door inner and door outer plus hinges. There are then two possibilities for step 2. Either fixture F2 achieves the weather seal KC or fixture F2′ achieves the appearance KC. In either case, the KC that is not directly controlled is achieved with larger tolerances or lower probability.

at the end of the chapter asks the reader to use the twist matrix intersection algorithm to prove this.

Figure 8-46 uses the notation of Chapter 7 to describe the alternate assembly sequences for the car doors. Note that several apparently feasible sequences are in fact unavailable once we take the KCs and constraints into account.

8.I.2. Ford and GM Door Methods

In this section we consider in some detail methods of attaching doors to cars used on some models of cars at GM and Ford, respectively. These are examples of widely differing methods used by different car manufacturers. They

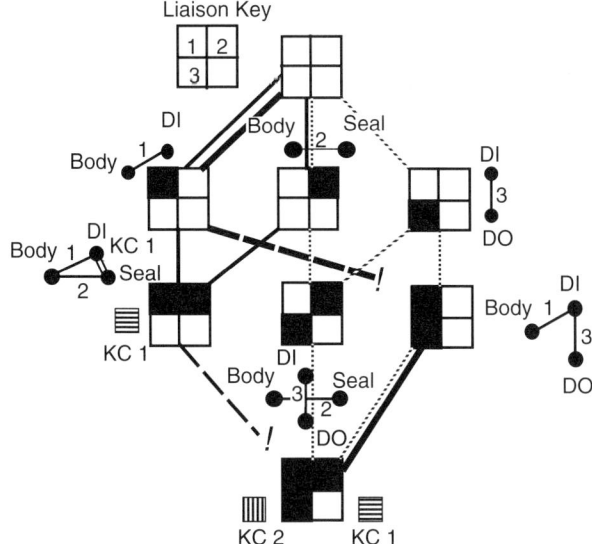

FIGURE 8-46. Alternate Liaison Sequences for Car Doors. The liaison diagram is at the lower right. The achievement of each KC is indicated by a shaded square next to a state in the liaison sequence diagram. Sequences that travel down the left side of the diagram appear able to achieve the KCs one at a time. These sequences, unfortunately, are impossible. The surviving sequences that travel down the right side of the diagram necessarily achieve the KCs simultaneously at the last step.

help us illustrate additional properties of DFCs and help us understand the behavior of the hybrid mate-contact type of joint. We do not know the variations of all the fixtures and parts in these processes, so we will not make any judgment concerning which one is better. The discussion that follows has the flavor of analysis because we are reverse engineering existing processes. If we were designing new processes, then everything we are depicting would be determined by the designers of the parts and fixtures as they sought to determine the design intent of the assembly and deliver the KCs. No reverse engineering would or should be necessary.

Figure 8-47 shows schematically the two methods we will discuss. (The reader may also refer to Figures 2-9 through 2-12 for additional views of these processes.) Each method comprises two stages. In the first stage, hinges are attached to a door subassembly consisting of door inner and door outer. In the second phase, this subassembly is attached to the car. In both cases the KCs are as discussed above. Both KCs are affected by how the door's position with respect to the body varies in all three directions: up/down, in/out, and fore/aft. However, each company's method delivers these KCs in different ways. Furthermore, we need to look carefully at each of the three directions in order to see all the differences, check for constraint violations, and calculate accumulated variation.

Note that each KC could have different tolerances in each direction, making these distinctions important. This

point is true in general: any KC may be defined and toleranced in one or a few directions and be undefined or untoleranced in others.

The GM method applies the hinges to the door while gripping the door on the outer panel. Not only are the hinges attached in this setup, but a locator cone is attached to each hinge. The hinge mounting machine positions and fastens each hinge in the in/out and up/down directions, and then it places and fastens the locator cone on each free hinge flap carefully in the up/down and fore/aft directions. The door is then attached to the car by mating the locator cones in a hole and slot compound feature set vertically on the body just forward of the door opening. Screws fasten the hinges to the frame. These hinges come apart at the pivot, permitting the door to be removed after painting so that the door and car final assembly processes can occur independently. The doors are rehung by reconnecting the hinges.

In the Ford process, the hinge mounting machine mates to the door's inner panel. It locates the hinges in the in/out and up/down directions, as does the GM hinge mounting machine. A moveable fixture then is used to pick up the door by mating to the door's outer panel using a hole and slot feature set shown in Figure 8-48. An operator carries the door to the car using this fixture and mates the fixture's two locator pins to a hole and slot compound feature on the body. The hole is in the body just ahead of the door opening while the slot is in the body just behind the door opening. Screws fasten the hinges to the body.

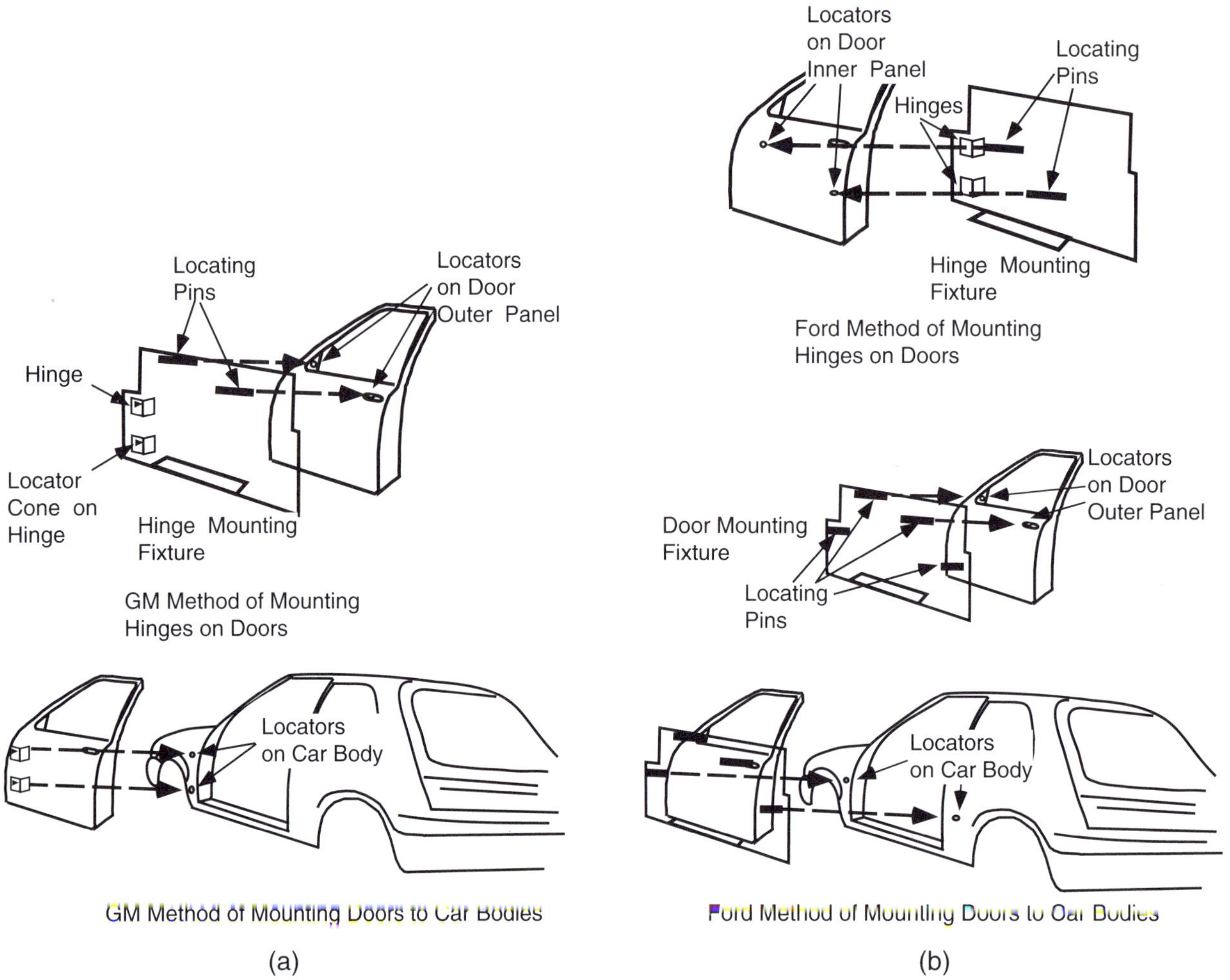

Locators
on Door
Inner Panel

Locating
Pins

Hinges

Locators
on Door
Outer Panel

Locating
Pins

Hinge

Locator
Cone on
Hinge

Hinge Mounting
Fixture

Hinge Mounting
Fixture

Ford Method of Mounting
Hinges on Doors

GM Method of Mounting
Hinges on Doors

Door Mounting
Fixture

Locators
on Door
Outer Panel

Locating
Pins

Locators
on Car Body

Locators
on Car Body

GM Method of Mounting Doors to Car Bodies

Ford Method of Mounting Doors to Car Bodies

(a)

(b)

FIGURE 8-47. Comparison of Two Door Attachment Methods. (a) The GM method allows the doors to be removed and remounted accurately later because the hinge flaps can be separated from each other and rejoined repeatibly. Hinges contain all the features needed to locate the door to the car body. The hinge mounting fixture is responsible for placing the hinges accurately on the door as well as attaching a cone locator accurately to each hinge. The door is aligned to the body by mating the cone locators to features on the body. (b) The Ford method leaves the doors on once they are attached to the body. Hinges contain the features necessary to locate the door in the in/out direction. However, the door mounting fixture contains the features necessary to position the door in the up/down and fore/aft directions. As the screws are tightened, the door slides in/out along the pins on the mounting fixture.

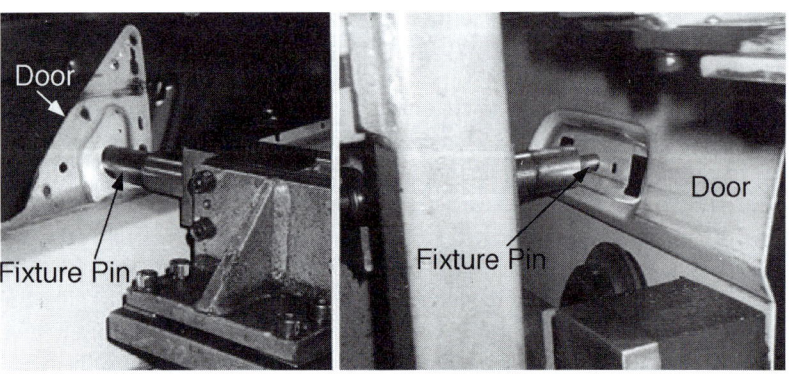

Door

Fixture Pin

Fixture Pin

Door

FIGURE 8-48. Photo of How Ford Door Mounting Fixture Mates to Door. *Left:* A pin in the fixture mates to a hole at the front of the door where the side-view mirror will be attached later. *Right:* Another pin mates to a horizontal slot at the rear of the door where the handle will be attached later. (Photo by the author. Used by permission of Ford Motor Company.)

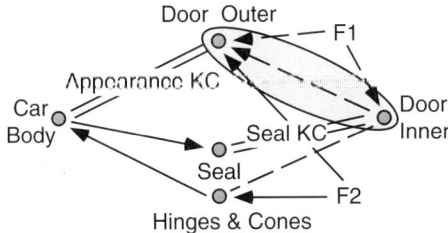

FIGURE 8-49. DFC for GM Door Process. Fixture F1 joins door inner to door outer. Hinge mounting machine F2 locates the hinges and cone locators with respect to the door outer. Each dashed line arrow indicates a hybrid mate-contact relationship in which some directions are constrained and act as mates while others are contacts and provide no constraint. Door inner locates door outer in the in/out direction, while fixture F1 locates these parts with respect to each other in the up/down and fore/aft directions. Variations in these directions will be different because the fixture is present along some directions but not others.

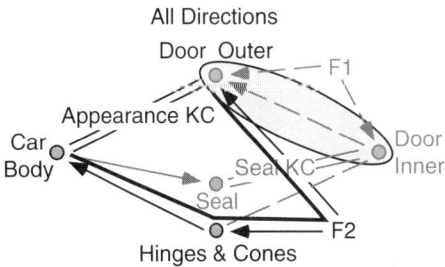

FIGURE 8-50. Chains for Determining Constraint and Assembly-Level Variation for the Appearance KC in the GM Method. The chain (traced by the heavy line) comprises properly constrained items that link one end of the KC to the other end. All three directions that could contribute variation follow the same chain. Fixture F1 and the door inner to door outer joint play no role in this KC and thus are shown in gray. Because all directions are portrayed in this figure, these joints are shown in their original hybrid form without distinguishing which directions play a role in which KC.

These hinges do not come apart, and the doors stay attached to the car from that point on.

Let us consider each of these processes in detail, first the GM process, then the Ford process. The DFC for the GM process is shown in Figure 8-49.[13] It is drawn, like all other DFCs, as a chain of mates from one end of a KC to the other. It shows that two fixtures are involved, one for joining door inner to door outer, and the other for attaching the hinges (and the locator cones which are not shown separately) to the door. No fixture is shown for attaching the door to the body because the hinges and locator cones provide all the location constraints. Note in this figure that, in addition to the arrow indicating a mate and a dashed line indicating a contact, there is an arrow with a dashed line. This indicates a hybrid mate-contact. This symbol is necessary in order to describe carefully how the door subassembly is made, and it will play an additional role when we consider the Ford process.

The shape of door inner constrains the location of door outer with respect to door inner in the in/out direction, but fixture F1 constrains door inner with respect to door outer

in the up/down and fore/aft directions. Thus each relationship is a mate in one direction and a contact in another. As long as neither tries to constrain a direction that is constrained by the other, there is no problem. What is important to us is that variation in the in/out direction is governed by the process capability of stamping while variation in the other two directions includes stamping variation plus that contributed by fixture F1. In order to determine which directions are mates and which are contacts in these hybrid joints, we need to consider the different directions separately and carefully.

The question before us is to determine how each direction is constrained and how variation will accumulate along the chain of mates that joins each end of the KC and delivers it. To accomplish this, we need to identify each feature in the chain for each direction and the internal surfaces of each feature that affect the chain. Figure 8-50 and Figure 8-51 do this for the GM process. Figure 8-50 shows that all directions of the appearance KC are delivered in the same way. Figure 8-51 shows that the weather seal KC is delivered differently in the different directions. Each hybrid joint is shown as a mate for the direction or directions it constrains and to which it contributes variation, and as a contact for the directions in which it plays no role.

Figure 8-52 shows the DFC for the Ford door process. This process differs from the GM process in several ways, as discussed above. The hinge mounting machine F2 interfaces with the door inner panel, and a door mounting tool F3 aligns the door in two directions with respect to

---

[13]In this figure, there is an arrow from the hinge on the door to the body of the car, implying that the door or the hinge locates the body. Obviously, the body is big and heavy while the door is relatively small and light, so in a physical sense the door cannot change the position of the body. Thus it may not make sense physically to have the arrow point from the hinge to the body. The reader may reverse this arrow if it makes him or her feel better, but the nature of the chain as a whole will not change.

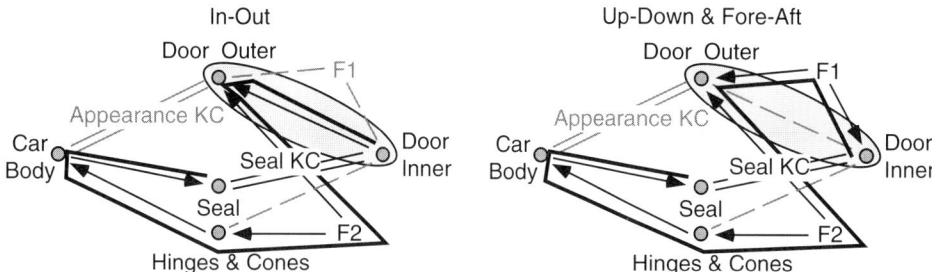

FIGURE 8-51. Chains for Determining Constraint and Assembly-Level Variation for the Weather Seal KC in the GM Method. *Left:* The in/out direction's chain follows the hinge mounting process and includes the stamping variation in the door inner panel's relationship to the outer panel. Fixture F1 plays no role in this direction so it appears gray with contact relationships to these panels. *Right:* The up/down and fore/aft directions' chain includes both F1 and F2. The door inner to door outer relationship plays no role and is shown as a contact.

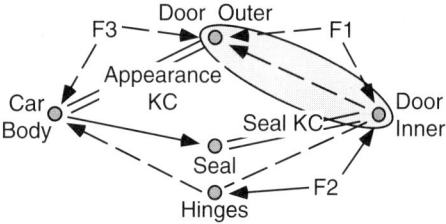

FIGURE 8-52. DFC for the Ford Door Process. Fixture F1 joins door inner to door outer. Hinge mounting fixture F2 locates the hinges with respect to door inner. Door mounting fixture F3 locates the door to the body with respect to door outer. In this process, the hinge is located with respect to the door inner panel, and a separate fixture aligns the door to the body. The hinge to body, door inner to door outer, F3 to door and body, and F1 to door relationships are shown as hybrids because they act as mates in some directions and contacts in others.

the body. Several joints are shown as hybrid mate-contacts because they are mates in some directions and contacts in others. A detailed direction-by-direction drawing is necessary to define which direction is which.

Figure 8-53 and Figure 8-54 show how each KC is delivered in each direction in the Ford process.

This example has shown how to decompose a multi-direction KC into its components in a convenient coordinate frame and then to draw the DFC and identify constrained directions separately for each coordinate direction. Once this is done, the variation contributed by each feature in each chain in each direction can be computed and combined into a unified 4 × 4 matrix model of the entire KC delivery process. We can identify arcs in individual directions that are claimed by more than one KC chain, such as (a) F2-hinge-door-inner in the in/out direc-

tion of the Ford process and (b) F2-hinge-door-outer in all directions of the GM process. Then a rational discussion may be conducted to determine the effect of giving one or another of these coupled KCs priority. Different joint schemes, assembly sequences, and fixture designs can be considered as part of the KC delivery design process.

8.I.3. Aircraft Final Body Join

Most large aircraft fuselages are assembled using fixtures that are even larger. These fixtures are made of aluminum, as are the aircraft themselves, to equalize temperature-induced expansion or contraction. The tolerances sought on such aircraft are challenging. As a result, the fixtures cost a great deal to design, build, and keep exactly the right shape. The description that follows is generic but is similar to that used by major airframe manufacturers.

A simplified version of this process is shown in Figure 8-55. It creates a full 360° fuselage tube, ready to be joined to another one. Typical individual sections of large aircraft are about 40 feet long and 12 to 24 feet in diameter. Figure 8-56 shows the DFC for controlling the diameter and circumference, including contributions by suppliers. Figure 8-57 shows the joining process for two of these tubes. Typical aircraft have between two and four such joints, depending on the length of the aircraft.

The DFC(s) required to achieve the diameter and circumference KCs are shown in Figure 8-56.

Figure 8-57 shows all the KCs that are sought during final body join. It is easy to see that there are more than can be individually adjusted or given independent tolerances, inasmuch as the two sections are practically rigid. The most important KC is structural, requiring minimum edge

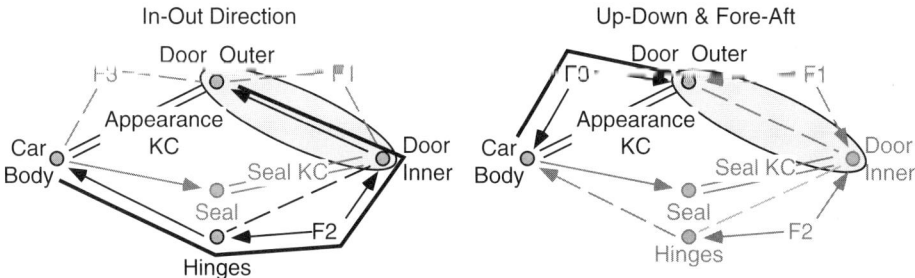

FIGURE 8-53. Chains for Determining Constraint and Assembly-Level Variation for the Appearance KC in the Ford Method. *Left:* In the in/out direction, fixture F2 is responsible for locating the hinge with respect to the door, using door inner as a reference. The hinge transfers this in/out location when it mounts to the side of the body onto a plane that is normal to this direction. The hinge's relation to the body is shown as a solid arrow to indicate that in this direction it acts as a mate. Door inner transfers in/out location to door outer, completing the chain. *Right:* In the up/down and fore/aft directions, F1, F2, and the hinge exert no constraint and play no role. F3 is shown as a mate between door outer and the body to deliver this KC.

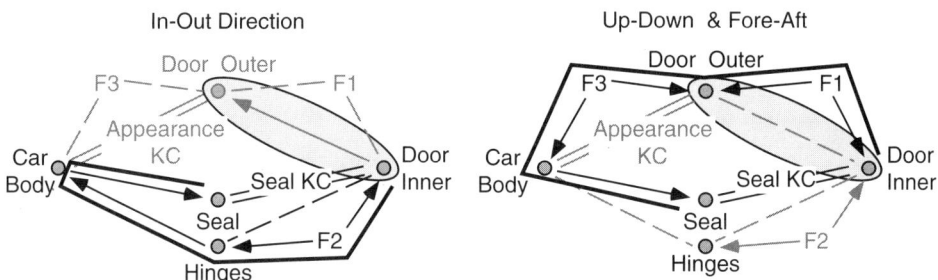

FIGURE 8-54. Chains for Determining Constraint and Assembly-Level Variation for the Weather Seal KC in the Ford Door Method. *Left:* In the in/out direction, fixture F2 is responsible for locating the hinge with respect to the door. The hinge transfers this in/out location when it mounts to the side of the body onto a plane that is normal to the in/out direction. The hinge's relation to the body is shown as a solid arrow to indicate that in this direction it acts as a mate. The in/out relationship between door inner and door outer is shown as a solid arrow for the same reason. However, it plays no role in delivering this KC. *Right:* In the up/down and fore/aft directions, fixtures F1 and F3 are responsible for delivering this KC. F1 relates door inner to door outer in these directions, while F3 positions the door on the body using door outer and body features that provide location in these directions. The hinge and the door inner-door outer joint play no role in these directions, so they are shown as gray or dashed lines.

distance for rivet holes where keel sections join. Next most important are KCs associated with the passenger floor and the seat track. These KCs permit seats to be mounted anywhere, including straddling a joint. Equally important is the skin gap around the circumference, which is controlled to achieve aerodynamic drag requirements. Least important are the horizontal skin alignments. These are purely cosmetic. Each of these KCs has its advocates who all watch final body join carefully. Obviously not all of them can be completely satisfied.

One can imagine a different assembly sequence for this product. Acknowledging that some KCs are overwhelmingly important, we could start by guaranteeing that they

would be achieved. A way to do this might be to build the entire length of the aircraft keel, cargo floor, lower side panels, and passenger floor with seat tracks as one subassembly upside down on one long fixture. This long subassembly would then be turned right side up, and the upper side panels would be added along the entire length, using some spacers or fixtures to preserve the diameter until the crown panels can be added. The crown panels would simply be put on top and allowed to settle into place, whereupon they would be fastened in. This sequence would create a strong keel and a straight passenger floor and deliver all the seat track KCs. The fuselage body would be straight, and all major circumferences would

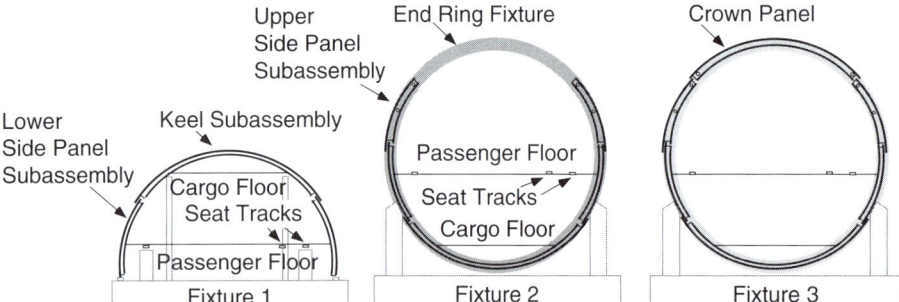

FIGURE 8-55. Aircraft Fuselage Assembly. The individual subassemblies are assembled from the bottom of the aircraft to the top. Fixtures are in gray. A photograph of a typical fixture appears in Chapter 19, which is on the CD-ROM packaged with this book. The keel subassembly and lower side panels are assembled to the passenger floor upside down, using the passenger floor and its seat tracks as the main interfaces to the fixture. Then this subassembly is inverted and the upper side panels are added at a second workstation. The crown panel is added at a third workstation. The required circumference is achieved partly by precision locating holes on the parts but mostly by the thick gray end ring fixtures at the second and third assembly stations. At one major aircraft manufacturer, each of the subassemblies shown is outsourced, except for the passenger floor.

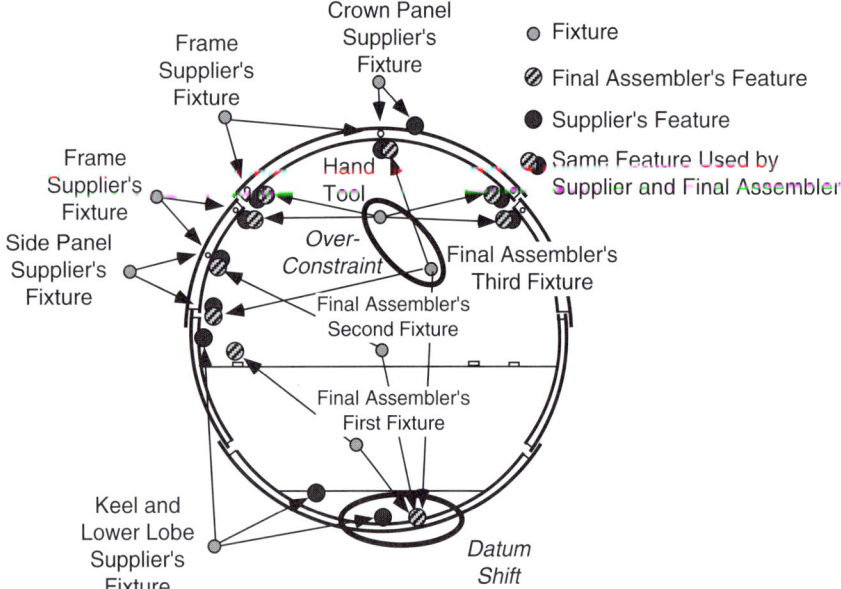

FIGURE 8-56. DFCs for Aircraft Body Section Assembly to Achieve Diameter and Circumference KCs. Three companies are involved in this complex DFC. The figure greatly oversimplifies the situation but captures the essence. The assembly comprises several skin panel subassemblies made by assembling skins to circumferential stiffening ribs called frames. The skin panels are shown slightly separated from each other for clarity. The first supplier makes the frames and drills several precision holes in each frame. The second supplier makes the skins, attaches longitudinal stiffeners called stringers to them (stringers are not shown), and then attaches the frames, using the precision holes to align and join the frames to the stringers and the skins. This process creates the keel panel and cargo floor, lower lobes, side panels, and crown panel. The final assembler adds the passenger floor and joins these panels into a full ring. In most cases, the suppliers and the final assembler share certain features so that the chains join all around the circumference. However, there appears to be one datum shift in the keel panel between the second supplier and the final assembler. That is, the supplier and the final assembler do not use the same feature at the keel for the same purpose. Also, the process for adding the crown panel appears to involve some overconstraint because both the third fixture and the hand tool seek to establish the chord spacing across the lower edges of the crown.

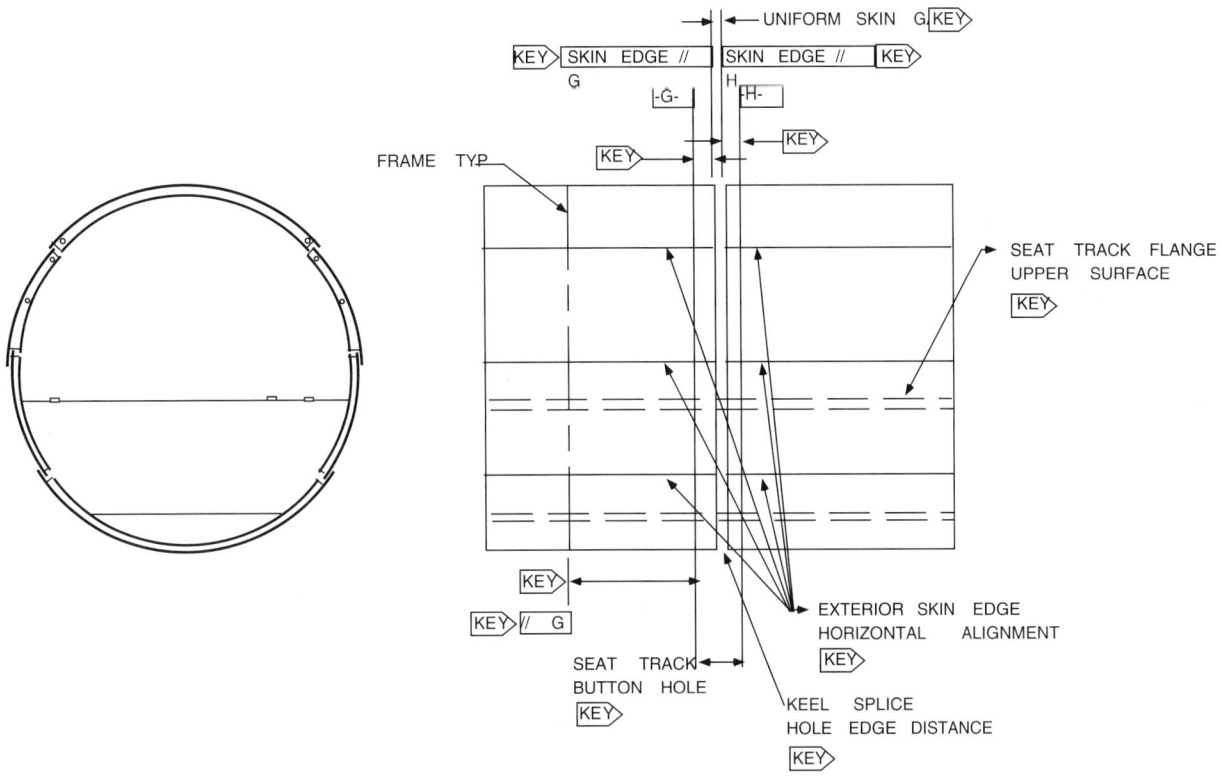

FIGURE 8-57. KCs for Final Body Join of Aircraft Fuselage. *Left:* End view of fuselage. *Right:* Two fuselage tubes about to be joined. All the KCs are noted.

be continuous at the places where current final body joins occur. There would be fewer overconstraints, especially circumferentially, because the important KCs would be achieved one at a time instead of simultaneously as is done in the current process. However, the alternate process would not be able to guarantee the horizontal skin alignment KCs. These are the least important from a structural point of view. The foregoing discussion is an example of KC prioritizing, as defined in Chapter 2.

The above assembly process is clearly impractical for a large aircraft but not for small ones.

An experienced aircraft assembly process engineer once described the process in Figure 8-57 as follows: "Wash the uncertainty from the nose and tail toward the middle where it has no place to go."

8.J. SUMMARY OF ASSEMBLY SITUATIONS THAT ARE ADDRESSED BY THE DFC METHOD

8.J.1. Conventional Assembly Fitup Analysis

A DFC can be drawn for any situation typically called assembly fitup, that is, situations where the KC represents the need for a stack of parts to fit within a boundary with a minimum and maximum dimension on the gap.

8.J.2. Assembly Capability Analysis

A DFC can be drawn to include a partially built assembly, the fixture that holds it, a part about to be added to it, and any tooling or assembly equipment that is guiding that part toward its destination on the assembly. The KC represents satisfaction of the part mating conditions that permit the chamfers to meet and avoid wedging.

8.J.3. Assemblies Involving Fixtures or Adjustments

A DFC can be drawn for situations where the parts do not completely constrain each other and require an outside aid to provide both the missing constraint and a critical dimension. The KC is that dimension. The outside aid may be a fixture or a measurement and adjustment apparatus. In either case, the DFC passes through the parts and such apparatus or fixture.

8.J.4. Selective Assembly

A DFC can be drawn around a set of parts whose KC is achieved by making a measurement and selecting a suitable last part that is the correct size given the measurement. This case is a combination of cases 1 and 3 above.

8.K. ASSEMBLY PRECEDENCE CONSTRAINTS

We noted in Sections 8.E.1.b and 8.F that there were two rules to follow in the design of DFCs. The first one states that all mates to a part should be made before any contacts. The second states that all subassemblies during assembly should be completely constrained. Each of these can be used to generate additional precedence constraints to add to the constraints generated by typical assembly sequence algorithms. These two rules eliminate sequences that do not contribute toward achievement of KCs. Any assembly sequence that survives application of these two rules permits us to monitor the growth of each DFC as assembly proceeds. The logic behind these rules is as follows:

The design of a DFC involves the conscious decision of designing mates and contacts. As mentioned earlier, contacts do not define any dimensional relationships between parts and have to be established only after the mates that define the dimensional relationships are made. Using this argument, the following rule is imposed by the DFC:

Contact rule: *Only connected subgraphs of a DFC can form permissible subassemblies.*

Subassemblies with only "contacts" between any two parts are not permitted because contacts do not contribute to a KC. This rule will thus generate additional assembly precedence constraints that eliminate subassemblies whose parts do not establish part of a DFC.

If the location of a part is defined by more than one part in the assembly, all the defining parts should be present in the subassembly before the part can be assembled. This argument is captured in the following rule:

Constraint rule: *Subassemblies with incompletely located (underconstrained) parts are not permitted.*

The constraint rule imposes the condition that the unique degrees of freedom on DFC arcs coming into all but one of the nodes in a subassembly must add up to six minus any operating degrees of freedom. The one exceptional node could represent either a base part or a fixture, and has no incoming arcs. This rule ensures that every subassembly has fully constrained parts.

An algorithm for generating the extra constraints is in Section 8.R.

It is usually possible to define alternate DFCs for the same assembly. Different DFCs will generate different extra assembly sequence precedence constraints. Combining these with the basic precedence rules obtained via the methods of Chapter 7 will result in a generally different set of allowed sequences. Each such set of sequences, associated with a different DFC, is called a family of sequences. If the assembly is Type 1, then only one member of the family needs to be analyzed to see if its final assembly tolerances are satisfactory. If it is Type 2, then each sequence has to be examined individually.

If a DFC contains hybrid mate-contacts, then the directions containing the mates take precedence.

It is entirely possible that application of all these rules will result in a deadlock in which no feasible assembly sequence is found. This is likely to be caused by KC conflict. In such cases, KC priorities and the design of parts, features, constraint schemes, and fixtures must be reexamined and revised.

8.L. DFCs, TOLERANCES, AND CONSTRAINT

We learned in Chapter 4 that assemblies could be properly constrained or overconstrained. We showed that we could tell the difference using only nominal dimensions and did not need to consider tolerances. The distinction between properly and overconstrained assemblies is crucial for understanding the role of the DFC. *In essence, there is no such thing as a DFC as we have defined it if the assembly is overconstrained.* In such a case, we cannot trace a chain of mates that determine the relative locations of the parts merely by knowing geometric information about the position and shape of assembly features. We must instead engage in a stress and deformation analysis in order to find out where the parts are relative to each other. This is neither bad nor impossible, but it is different from using the DFC and a chain of 4 × 4 matrices.

Knowing that an assembly is properly constrained when the dimensions take on their nominal values is not enough, however. Variation will occur, and we must ensure that the DFC(s) that we designed apply under the conditions imposed by all allowed variations. For example, if the clearance we designed into a contact goes to zero or tries to become negative under some allowed combination of variations, then that contact essentially will try to exert dimensional constraint and usurp the role of some mate elsewhere in the assembly, creating an overconstraint. The result is that we no longer know where the chain of mates is. In effect, there is no unique and permanent DFC for that assembly that applies under all allowed variations. Such a thing could occur, for example, in the wheel-axle example in Section 8.E.1.a. If the circles defining the locations of

the studs on the axle or the stud holes on the wheel are allowed to vary too much, or if the diameters of the studs or stud holes were allowed to vary too much, there could be a conflict between the hole–rim mate and the stud–stud hole joint in some percentage of the parts. This must not be allowed to happen.

To protect us against this eventuality, we need to conduct a kind of variation analysis that is unlike conventional variation analysis. It ensures that the DFC is robust and will function the same way when subjected to allowed variations as it does when variations are zero. That is, all designated mates will remain mates and all designated contacts will remain contacts, regardless of allowed variations.

Such an analysis may have been conducted in the case of the car seat discussed in Chapter 4, where the two forward mounting holes were enlarged from 25 mm to 40 mm. This creates a huge diametral clearance between the hole and the 25-mm diameter stud and virtually ensures that the stud will never touch the edge of the hole. On this basis, the rear hole and slot will always determine the relation between the seat and the car floor. These rear features comprise the mate between the seat and the floor and contribute to delivering the KC that defines the clearance between the seat and the door.

If the assembly is overconstrained at nominal dimensions or becomes overconstrained for some allowed combination of variations, the design is fundamentally flawed and must be corrected. There is no sense proceeding to conventional tolerance analysis until the design error is corrected.

8.M. A DESIGN PROCEDURE FOR ASSEMBLIES

The preceding sections of this chapter can be summarized into a procedure for designing assemblies. This procedure consists of two phases, the nominal design phase and the variation design phase. Within the nominal design phase is a constraint analysis phase. This is illustrated in Figure 8-58 and described in words below.

8.M.1. Nominal Design Phase

- Identify the key characteristics (KCs) that the assembly must deliver.
- Sketch the parts and draw a liaison diagram. Mark

each KC on the liaison diagram by adding a specially marked arc between the parts related by the KC.

- Tentatively classify the assembly as either Type 1 or Type 2.
- Establish a tentative DFC for each KC, identifying possible constraint requirements between parts (and fixtures if necessary). Mark which liaison diagram arcs would be mates and which would be contacts.
- Identify places where fixtures or measurements will be needed by noting the existence of KCs between parts that are not joined by a chain of mates.
- Define a tentative set of features that can carry the

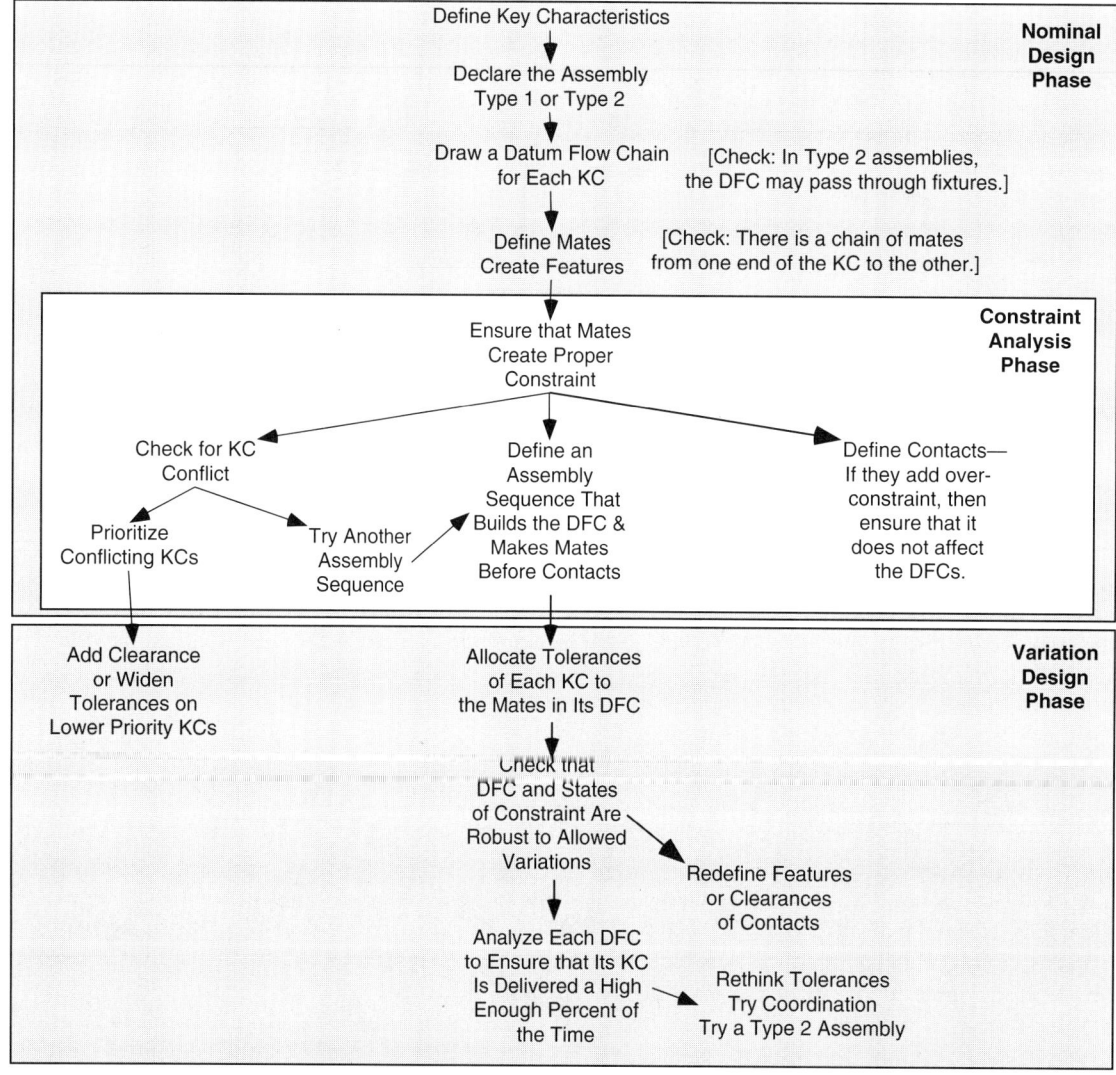

FIGURE 8-58. Diagram of Assembly Design Process.

desired constraint, consistent with functional requirements on the features.

- Examine these feature sets for over- or underconstraint, making necessary corrections.

- If the assembly contains several KCs, examine it for the possibility that there are not enough degrees of freedom to adjust them independently or to achieve them within tolerances with statistical independence. This occurs because more than one KC lays claim to the same degrees of freedom on the same arc of a DFC. It sometimes occurs because the chosen assembly sequence achieves the KCs all at once. Possibly another assembly sequence can achieve them one at a time, relieving the conflict. Otherwise, either the conflict must be accepted by prioritizing the KCs, the KCs must be redefined, or major changes to the features must be considered.

- Identify geometrically feasible assembly sequences, utilizing local constraint knowledge deduced from the features. If fixtures are part of the assembly process, identify only sub-sequences that utilize a single fixture, and string together such sub-sequences into a final sequence.

- Restrict the assembly sequences to those that build fully constrained subassemblies and which make all the mates on a part before any of its contacts.

8.M.2. Variation Design Phase

- Allocate tolerances from each KC to the mates in its DFC.
- Examine each arc in each DFC to determine if variation in the size and location of a feature, mate, or contact could alter the DFC. Improve the design, tolerances, or clearances related to these items until the DFC is robust against such variations.

- Analyze the ability of the candidate DFC, feature set, fixtures, and sequence to deliver the KC(s) by performing a three-dimensional variation analysis of each DFC. Extend the analysis over chains of fixtures if necessary, being careful to include any datum transfers that occur between fixtures. If the KC cannot be delivered with the required accuracy or frequency, then some portion of the design must be repaired, starting with the assembly sequence and fixtures, if any, and retreating to different DFCs and features if nothing else works. Possibly the assembly cannot be made as a Type 1 and will have to be redesignated as a Type 2. Then the whole process begins again.

8.N. SUMMARY OF KINEMATIC ASSEMBLY

Chapters 2–8 present a theory of assembly applicable to assemblies that are statically determinate. As such, it provides a formalism and a set of mathematical models and algorithms to support what good designers have done for a long time. It rests on the mathematical basis of statics and kinematics, especially the 4×4 matrix for creating chains of coordinate frames and Screw Theory for examining states of mutual constraint. This theory is summarized in Figure 8-59.

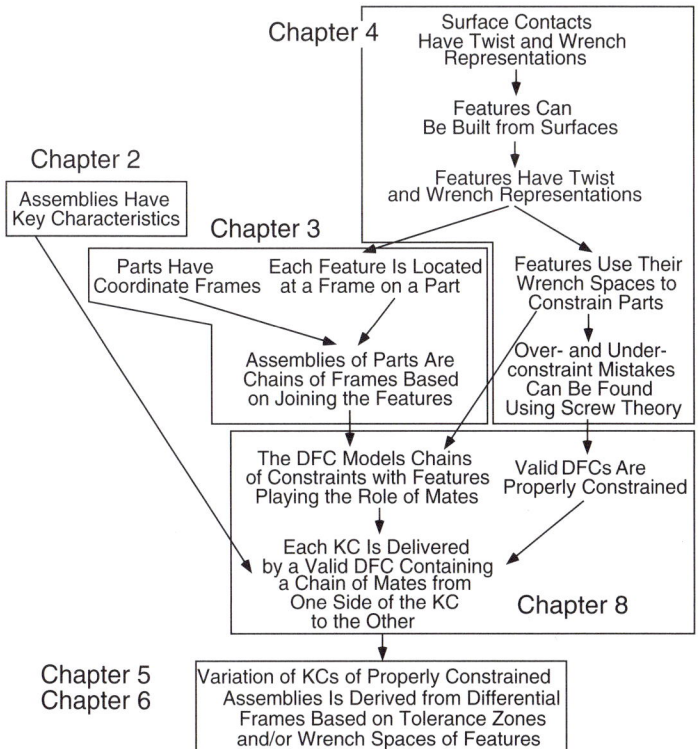

FIGURE 8-59. Summary of Kinematic Assembly.

8.O. CHAPTER SUMMARY

This chapter concludes the part of the book devoted to designing assemblies that began in Chapter 2. A systematic process was described for accomplishing top-down design of assemblies. The datum flow chain was presented as a way of describing the constraint structure of an assembly, capable of documenting the way that each KC will be delivered. Competent assemblies are properly constrained and have a distinct and well-defined DFC for each KC.

Features join parts and provide the means by which constraint is passed from part to part. Variation propagates along the DFC and is exerted by the wrench space of surface contacts within features. No separate tolerance chain needs to be defined.

From here on, the chapters deal first with design of assembly processes and then with the role of assembly in product development.

8.P. PROBLEMS AND THOUGHT QUESTIONS

1. Figure 8-4 shows in symbolic form a locating scheme by which two parts labeled A and C work together to locate a part labeled D. The degrees of freedom that each of A and C constrain on D are named, but no features are declared. Sketch two different solutions containing features that could implement the indicated scheme. Best results are obtained when you draw a perspective sketch that shows all the parts at once and contains all three coordinate axes clearly marked. For extra credit, use motion and constraint analysis from Chapter 4 to verify your design.

2. In Section 8.E, we discussed a method for mounting an automobile wheel to its axle. This method uses a set of features and creates a kinematically constrained assembly that is capable of delivering the KCs. Another method is commonly used. There is no rim on the hub and no opening in the wheel. Instead there are 4 or 5 studs that pass through clearance holes in the wheel. Onto these studs go nuts that have generous chamfers on them that engage chamfers on the wheel. Describe the state of constraint of this design and explain why the instructions for mounting this kind of wheel tell us to hand tighten all the nuts first and then wrench-tighten them in opposite pairs, gradually bringing all the nuts to equal tightness. The studs in the alternate design are likely to have larger diameters than those in the rim design. Why?

3. Figure 8-12, Figure 8-14, Figure 8-16, and Figure 8-18 present alternate DFCs for realizing a car floor pan design to deliver the car's width as the KC. Write down the 4×4 matrices that represent each of these DFCs. Assume that the distance between two flanges on one sheet metal piece has a zero mean 3σ error of 0.2 mm and that the distance between the ends of a fixture has a zero mean 3σ error of 0.1 mm. Calculate a worst-case **and** a 3σ statistical estimate of the error that each DFC suffers.

4. Draw individual DFCs for each of the two KCs of the fan motor. Write all the constrained degrees of freedom on each arc. Identify on its DFC the degrees of freedom associated with each KC. Draw only the necessary arcs. Show on each arc only those degrees of freedom that this particular DFC needs in order to deliver its KC.

5. Carefully draw the rear housing of the fan motor and identify the important relationships between its features that are needed so that this part can play its role in the DFCs. Follow the examples in Figure 8-7 and Figure 8-27. (There is no need to attempt a GD&T representation, however.)

6. Draw a DFC for eliminating end-shake between the bell housing and the rotating clutches of the transmission in Figure 8-29 by using a select washer at the bell housing end of the automatic transmission.

7. Assume that the thrust washer at the bell housing end of the transmission is declared a mate and that a washer of the same size is used there in every transmission. Assume further that end-shake is removed by using a select washer at the oil pump end. Draw a DFC or DFCs that support the two KCs, namely a gap small enough to eliminate end shake between the oil pump and the rotating clutches at the oil pump end, and alignment of the oil grooves on the oil pump and the corresponding grooves on the rotating clutches. Analyze the degrees of freedom involved in each KC and explain why the KCs are in conflict.

8. In Figure 8-35, the bottom washer appears to touch fixture 2. Does it in fact do so? Was it intended to do so? Justify your answers.

9. Figure 8-51 shows DFCs for the GM door process. Write the necessary 4×4 matrices that describe the DFCs for each KC. First, identify each constrained direction at each interface between parts or between parts and fixtures. Then define a coordinate frame for each part and locate each feature on the part in that frame. Next, identify each direction at each interface that could contribute variation, including both parts and fixtures, and compose an error matrix that expresses each error. Finally, write an equation containing these matrices that predicts the error in each KC. Each equation should be valid for one KC and all directions.

10. Two candidate DFCs for the throttle body are shown in Figure 8-41. Add to each of these DFCs a note on each arc stating which degrees of freedom are constrained by each mate. Comment on the suitability of each DFC to deliver the KCs. Describe a feasible assembly sequence for each candidate DFC. Take explicit account of the rule that says, "Join the mates before joining the contacts."

11. Consider the assembly in Figure 8-60 below. It consists of three parts, A, B, and C. This is a Type 2 assembly according to the theory of the DFC. When these parts are assembled one at a time, the slots may be used to make small adjustments in relative part position, after which they are fastened, perhaps by welding or riveting. After each joint is fastened, the next part is added, adjusted, and fastened. "Adjustment" may be done by active measuring and repositioning or by placing the parts in a fixture and letting them self-adjust. According to the theory of the DFC, the parts have mates with the fixture and contacts with each other.

Use twist matrices and twist matrix intersection to show the following:

a. That part C is adjustable in the X and Y directions with respect to part A while part B is adjustable in the Y direction with respect to part A.

b. That a KC requiring X axis adjustment is compatible with these freedoms, while one requiring adjustment in Z rotation is incompatible.

c. That if there are two KCs between part A and part C, one that requires X axis adjustment and another that requires Y axis adjustment, but the assembly process fastens Parts A and B before the adjustment can be done, then the KCs cannot be achieved independently.

Compare Figure 8-60 and Figure 8-43 and decide which of the parts {door inner, door outer, and car body} correspond to parts A, B, and C.

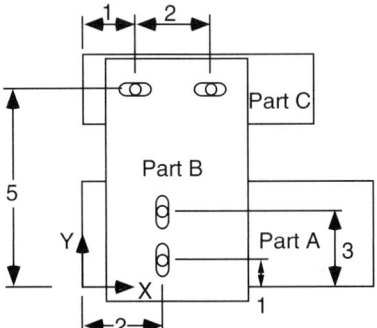

FIGURE 8-60. Parts A, B, and C for Consideration in Problem 11.

12. Two sheet metal parts are shown in Figure 8-61. They are to be spot-welded together along the flanges. (The welds are shown

as black blobs. Spot welding squeezes the flanges together.) When the parts are on the fixture, the flanges are supposed to touch gently. The KCs are that the large flat planes of the parts be coplanar and that their edges at one end be colinear. Someone has designed a fixture to hold these parts while they are being welded. Each part has a hole–slot feature, and there are four pins on the fixture. The shop floor people notice that it is often hard to get the parts out of the fixture after they have been welded together. The manufacturing engineer tells them that it would cost too much to modify the fixture to include a power ejector to force the parts out. So the shop floor people take a hand grinder and grind down the diameters of the pins. Draw a DFC for this assembly. Then explain (a) why the parts are sometimes hard to remove from the fixture; (b) whether it is also sometimes hard to put the second part into the fixture; (c) what these facts tell you about the fixturing scheme; (d) whether it makes any difference how much the shop floor people reduce the pin diameters; (e) what a better feature set or fixture design might be for these parts, given that the parts must still mate via the flange. Draw a careful sketch of your new design. Explain how each degree of freedom of each part is constrained. Explain how each KC is delivered by drawing a DFC for it.

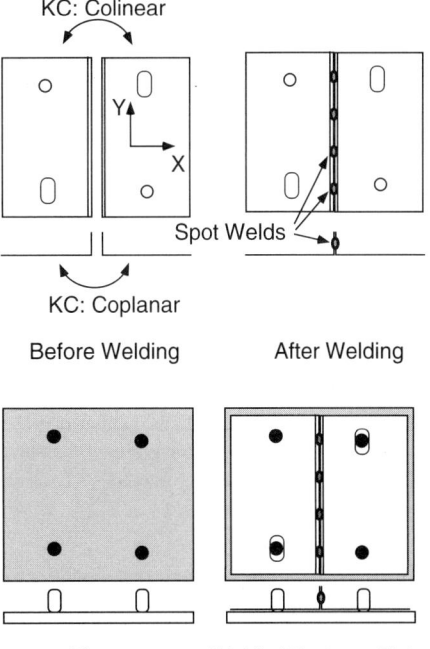

FIGURE 8-61. Parts and Fixtures for Consideration in Problem 12.

13. Someone has suggested the redesign in Figure 8-62 for the parts in Problem 12.

At the top are the parts, which have the same flange joint f_{AB} as those in Problem 12. However, different features are shown on the parts that interface to the pins on the fixture. At the bottom is a path diagram for conducting a constraint analysis of this situation according to the methods of Chapter 4. Use this diagram to examine the state of motion and constraint for this design.

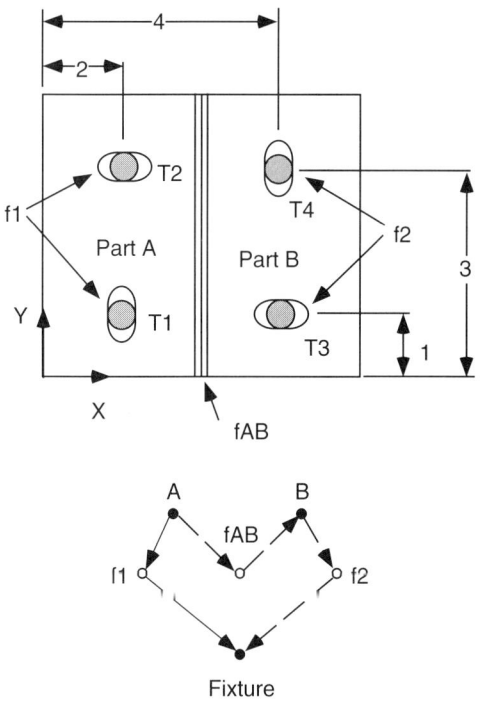

FIGURE 8-62. Figure for Problem 13.

14. Draw DFCs for the automobile engine KCs in Chapter 5 (valve–piston clearance and valve–cam clearance). Use a coordinate system whose Z axis is along the crankshaft and whose X axis is vertical (pointing from the crankshaft to the cylinder head). Identify each feature in each DFC and indicate which degrees of freedom it constrains. Identify places where these DFCs share arcs. Does KC conflict occur between these two KCs? Explain your answer carefully.

15. Which one of the examples in this chapter illustrates a violation of the rule that all of a part's incoming mates should be made before its contacts are made?

16. Consider the two feature mate designs in Figure 8-63. As shown in Questions 3 and 4 at the end of Chapter 6, these two arrangements differ greatly in their robustness to variations in the Y direction spacing of the two pins. Taguchi says that adjusting a parameter can often improve the robustness of a design. What parameter would you adjust in the design on the top in order to accomplish this?

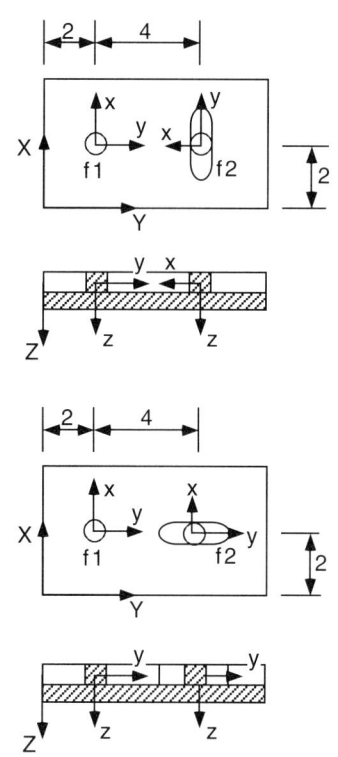

FIGURE 8-63. Figure for Problem 16.

8.Q. FURTHER READING

[Cunningham] Cunningham, T. C., "Migratable Methods and Tools for Performing Corrective Actions in Automotive and Aircraft Assembly," S.M. thesis, MIT Mechanical Engineering Department, February 1996.

[Goldenshteyn] Goldenshteyn, G., "Design of Assemblies with Compliant Parts: Application to Datum Flow Chain," S.M. thesis, MIT Mechanical Engineering Department, May 2002.

[Gui and Mäntylä] Gui, J., and Mäntylä, M., "Functional Understanding of Assembly Modeling," *CAD*, vol. 26, pp. 435–451, 1994.

[Hu] Hu, S. J., "Stream-of-Variation Theory for Automotive Body Assembly," *Annals of the CIRP*, vol. 46, no. 1, pp. 1–6, 1997.

[Lee and Gossard] Lee, K., and Gossard, D. C., "A Hierarchical

Data Structure for Representing Assemblies, Part 1," *CAD,* vol. 17, no. 1, pp. 15–19, 1985.

[Mantripragada and Whitney] Mantripragada, R., and Whitney, D. E., "The Datum Flow Chain," *Research in Engineering Design,* vol. 10, pp. 150–165, 1998.

[Muske] Muske, S., "Application of Dimensional Management on 747 Fuselage," SAE Paper 975605.

[Roy et al.] Roy, U., Bannerjee, P., and Liu, C. R., "Design of an Automated Assembly Environment," *CAD,* vol. 21, pp. 561–569, 1989.

[Shah and Rogers] Shah, J. J., and Rogers, M., "Assembly Modeling as an Extension of Feature-Based Design," *Research in Engineering Design,* vol. 5, pp. 218–237, 1993.

[Shalon et al.] Shalon, D., Gossard, D., Ulrich, K., and Fitzpatrick, D., "Representing Geometric Variations in Complex Structural Assemblies on CAD Systems," *ASME Advances in Design Automation,* DE-Vol 44-2, pp. 121–132, 1992.

[Söderberg and Johannesson] Söderberg, R., and Johannesson, H., "Spatial Incompatibility—Part Integration and Tolerance Allocation in Configuration Design," ASME Paper DETC98/DTM-5643, Atlanta, September 1998.

[Sodhi and Turner] Sodhi, R., and Turner, J., "Towards a Unified Framework for Assembly Modeling in Product Design," Rensselaer Polytechnic Institute, Troy, NY, Technical report 92014, 1992.

[Srikanth and Turner] Srikanth, S., and Turner, J., "Toward a Unified Representation of Mechanical Assemblies," *Engineering with Computers,* vol. 6, pp. 103–112, 1990.

[Tsai and Cutkosky] Tsai, J.-C., and Cutkosky, M. R., "Representation and Reasoning of Geometric Tolerances in Design," *AIEDAM,* vol. 11, pp. 325–341, 1997.

[Wang and Ozsoy] Wang, N., and Ozsoy, T. M., "Automatic Generation of Tolerance Chains from Mating Relations Represented in Assembly Models," Proceedings of the ASME Advances in Design Automation Conference, Chicago, September 1990.

[Whitney, Mantripragada, Adams, and Rhee] Whitney, D. E., Mantripragada, R., Adams, J. D., and Rhee, S. J., "Designing Assemblies," *Research in Engineering Design,* vol. 11, pp. 229–253, 1999.

[Zhang and Porchet] Zhang, G., and Porchet, M., "Some New Developments in Tolerance Design in CAD," *ASME Advances in Design Automation,* DE-Vol 65-2, pp. 175–185, 1993.

8.R. APPENDIX: Generating Assembly Sequence Constraints That Obey the Contact Rule and the Constraint Rule

The following algorithm generates assembly sequence constraints that obey the contact rule and the constraint rule. The liaison diagram and the DFC for the assembly are represented using their incidence matrices. In these matrices rows represent the nodes in the graphs and columns indicate the liaisons (mates in the case of the DFC). Table 8-2 shows these matrices for the aircraft wing subassembly in Figure 8-64. The liaison diagram and DFC for this assembly are shown in Figure 8-65.[14]

A computer program reads these matrices as inputs and applies the contact and constraint rules as follows:

Contact rule: To eliminate the possibility of subassemblies with only contacts between parts, the incidence matrices for the liaison diagram and the DFC are compared to determine which liaisons are contacts. Then, for each contact, a precedence relation is generated stating that all mates in the DFC pointing to the parts the contact connects must be completed before the contact can be completed.

For example, in Figure 8-65, liaison 3 joining Plus-chord and splice stringer-3 is a contact. Incoming mates to Plus-chord and splice stringer-3 include liaisons 2 and 4. Thus, liaisons 2 and 4 must be completed prior to or simultaneously with liaison 3, yielding the precedence constraint $2 \& 4 >= 3$. This type of precedence relation will ensure that subassemblies with only contacts between parts will not be allowed. Subassemblies involving only plus-chord and stringers are not permitted by this rule, as there are no designed mating features between these parts.

Constraint rule: To ensure that subassemblies with incompletely constrained parts are not allowed, each row in the DFC matrix is examined one at a time. If a part (row) has more than one incoming mate (element with value '−1'), then all incoming mates must be simultaneously completed to ensure that the part be fully constrained when assembled. For example, looking at the first row of the DFC matrix in Table 8-2, the Plus-chord has two incoming mates, liaisons 2 and 4. Thus, liaisons 2 and 4 must be completed simultaneously ($2 >= 4$ and $4 >= 2$). The Constraint rule prevents subassemblies such as (Fwd-skin, Plus-chord, Aft-skin) subassembly since it has incompletely constrained parts.

---

[14]This subassembly is the subject of a chapter-length example; see Chapter 19 on the CD-ROM that is packaged with this book.

TABLE 8-2. Incidence Matrices for the Liaison Diagram and DFC Shown in Figure 8-65

	Liaison Diagram										Datum Flow Chain					
Arcs ⇒	1	2	3	4	5	6	7	8	9		2	4	6	7	8	9
Parts ⇓																
Plus-Chord	1	1	1	1	1	0	0	0	0		−1	−1	0	0	0	0
Str 1–2	1	0	0	0	0	1	0	0	0		0	0	−1	0	0	0
Aft-Skin	0	1	0	0	0	1	1	0	0		1	0	1	1	0	0
Splice-Str3	0	0	1	0	0	0	1	1	0		0	0	0	−1	1	0
Fwd-Skin	0	0	0	1	0	0	0	1	1		0	1	0	0	−1	1
Str 4–11	0	0	0	0	1	0	0	0	1		0	0	0	0	0	−1

Note: In the liaison diagram incidence matrix at the left, a 1 in a row means that a liaison connects to the part in that row. Similarly, in the DFC incidence matrix at the right, a 1 indicates that an arc leaves that part, while a −1 indicates that an arc comes into that part.

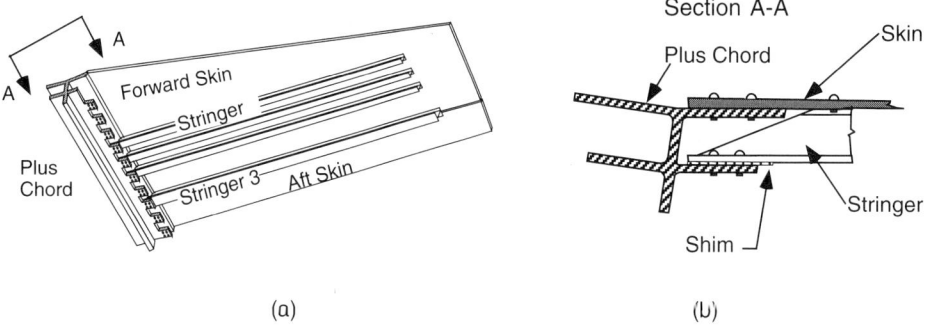

FIGURE 8-64. Aircraft Wing Subassembly. The wing skin has a forward part and an aft part which are spliced together by stringer 3. Additional stringers 1 and 2 are on the aft skin while additional stringers 4–11 are on the forward skin.

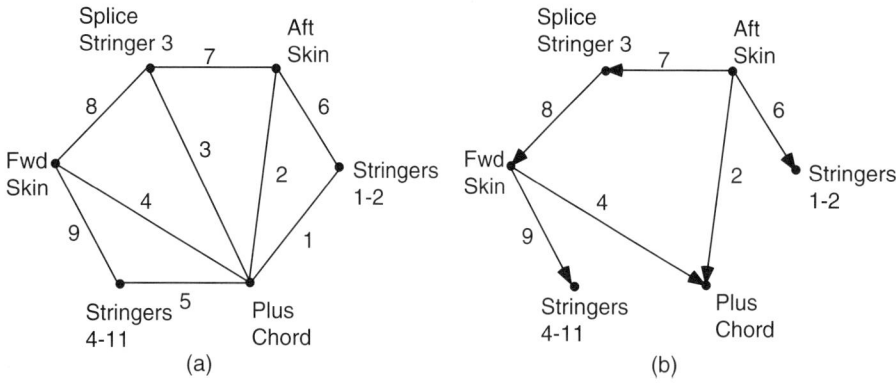

FIGURE 8-65. Liaison Diagram (a) and DFC (b) for the Wing Skin Subassembly in Figure 8-64.

This procedure assumes that all parts in the finished assembly are properly constrained by their neighbors. That is, a constraint analysis should have been performed on the assembly before assembly sequence analysis is undertaken. If it has not, then all feasible subassemblies generated by assembly sequence analysis need to be tested individually using Screw Theory to determine their state of constraint. The goal is to guarantee that at each stage of assembly (that is, in each rank of the assembly sequence network) there is at least one properly constrained subassembly. If this test fails, then fixtures need to be added or the design needs to be repaired.

9 ASSEMBLY GROSS AND FINE MOTIONS

9.A. PROLOG

This chapter begins the second major part of the book, which deals with basic assembly processes. The chapters in this part deal with basic motions and forces between parts (Chapter 9), the physics of part mating for rigid parts (Chapter 10), and compliant parts (Chapter 11). The theory in these chapters was developed in the early 1970s to help define the requirements for robot assembly. However, it is also useful for identifying assembly processes that might be difficult for people.

Following this part of the book, the remainder will explore assembly in the large more deeply by investigating concurrent engineering of products and processes, product architecture, design for assembly, design and economic analysis of assembly systems and workstations, and a complete case study.[1]

Assembly in the small is governed by several phenomena and conditions:

- Motion in space with parts in contact and not in contact
- Geometry of parts
- Compliance of parts or the tools, hands, or fixtures that hold and maneuver them
- Friction between parts in contact

This chapter deals with motion as well as the forces that arise when parts contact each other. It presents some methods for enabling a machine to respond to these forces. It also reinforces the three- and six-dimensionality of assembly and use of matrices to represent assembly phenomena that characterized the first part of the book.

9.B. KINDS OF ASSEMBLY MOTIONS

Assembly motions can be classified into two types: *gross motion* and *fine motion*. We take these up in turn.

9.B.1. Gross Motions

Gross motions generally carry parts from place to place over distances that are large compared to the size of the parts. During these motions, parts generally do not touch each other. As long as parts do not touch, gross motions can be fast and do not need high accuracy, except possibly at the end of the move, where a transition to fine motion occurs. This transition is the first critical

phase in assembly. A variety of conditions must be met in order for this transition to be successful. These will be the subject of Chapter 10.

All assembly activity involves gross motions, which can be accomplished by people, robots, or single-purpose fixed automation. Approximately half of all assembly time is consumed by gross motions.

9.B.2. Fine Motions

Fine motions are small compared to the size of a part and occur when parts are touching during actual part mating. These motions are likely to be slower than gross motions, and they usually require high accuracy. Fine motions must

---

[1] The case study is on the CD-ROM that is packaged with this book.

also meet a variety of conditions in order that assembly can start and finish successfully. These conditions are also discussed in Chapter 10, along with mathematical models that can be used for error prevention or to design apparatus that will aid assembly.

Parts contact each other in a variety of ways during fine motion, and in a sense this phase of assembly can be considered to be a series of controlled collisions. When we assemble parts with our bare hands, we are usually unaware of these collisions, but we semiconsciously sense them and adjust the paths of the parts to cause assembly to proceed.

9.B.3. Gross and Fine Motions Compared

The dimensions on which we will compare gross and fine motions are those of errors and preplanning—that is, what kinds of errors can occur during each kind of motion and what, if anything, can we do about them. One goal of making this comparison is to enforce the idea that these two kinds of motions are different, while another is to introduce a mode of thinking about problems of this type, which will do us well later in the book. The basis of this mode of thinking is to understand the different nature of people, robots, and machines and to learn how to take advantage of the different capabilities of each without suffering from inaccurate expectations.

In gross motion, errors are generally in the positions of stationary parts or the trajectories of moving ones. Because gross motion speeds are high, such errors can cause catastrophic collisions. These errors can be seen by suitable sensors, but if they are first detected by touch, then it is too late and the damage will happen before any economically feasible or technically effective response can be mounted.

People generally avoid gross motion errors using vision or, less often, touch or hearing. Currently, machines cannot use vision in this way because, while it is very easy to make a machine move rapidly, it is difficult for it to process visual data fast enough to muster a response to an unexpected obstacle. For this reason, people designing machine systems in factories for gross motion tend to arrange the environment and trajectories so that the chance of a collision is extremely low. It is usually quite easy and inexpensive to take this approach. The savings are more than recovered if the number of collisions is thereby reduced even from one to zero because collisions cost so much. An alternative is to make gross motions very

slowly, so that the collisions will not cause much damage. This is uneconomical because time is money, as we shall see in Chapter 18. Since contact between parts is not necessary for gross motion, it is cheaper to avoid contact altogether.

In Chapter 1 we introduced the idea of *structure,* meaning a deliberate plan to arrange an environment and to design assembly processes within that environment that ensure success by virtue of planning ahead. By its nature, such an approach does not require sensing and feedback, so it is often characterized as *open loop.* In general, if reasonable and usually affordable precautions are taken, the history and outcome of gross motion are rarely in doubt.

In fine motion, many of the characteristics of errors and feasible responses are reversed from the gross motion case. Here, tiny errors the same size as clearances between parts can cause assembly to be disrupted, slow down, or fail. In many cases, the errors are too small to see with reasonable effort, but they can easily be felt, due to the forces and moments they generate. These errors are the result of intimate short-range interactions. Vision is thus not able to deliver its most valuable capability, its long range. As the size of errors diminishes, the cost of eliminating them by prearranging part locations and feature sizes rises dramatically. On the other hand, due to slow speeds, the collisions are not catastrophic and, as hinted above, are usually necessary to guide the parts together. Thus the expense of avoiding errors during fine motion by preplanning every move and location down to the last thousandth of a millimeter is usually not rewarded.

In fact, the forces that arise when parts touch each other during assembly actually comprise signals that can be used in a *closed-loop* approach, in which forces are sensed and motions are commanded in response to the felt forces. We take up this topic in Section 9.C.

The contrasts between gross and fine motion are of additional interest because they suggest that different apparatus might be usefully employed to accomplish them. In the next section it is shown that the speed with which small adjustments must be made during fine motion exceeds the bandwidth of most gross motion devices. On the other hand, fine motion devices can be smaller and have more limited motion range than gross motion devices. As a result, they are likely to have higher bandwidth. The human arm and hand are constructed this way, as are most robots that perform assembly that requires small motions of the parts.

9.C. FORCE FEEDBACK IN FINE MOTIONS

9.C.1. The Role of Force in Assembly Motions

The forces that arise in fine motions are usually due to glancing blows rather than head-on collisions, and some motion usually continues along a deflected path. Figure 9-1 shows two common kinds of contacts between parts and the kinds of forces and moments that arise.

On the left in Figure 9-1 we see a part approaching another from above. The moving part's axis is parallel to the fixed part's axis but is displaced to the left. This is called *lateral error*. When they touch, a force to the right is generated at the tip of the moving part. A proper response is that the moving part should translate to the right without rotating (its angle of approach is correct already).

On the right in this figure we see a part approaching with a counterclockwise angle error, even though the trajectories are such that the tip of the moving part will enter the mouth of the stationary one. This is called *angular error*. When they touch, a clockwise torque is generated at the tip of the moving part. A proper response is that the moving part should rotate clockwise about its tip without translating (its lateral position at the tip is correct already).

This discussion contains most of the elements of what is called a force feedback strategy for removing lateral and angular errors in fine motion. We know what we want to happen when each kind of error occurs. How do we design a strategy that will do so? We can generalize the approach as follows:

- Make an engineering model of the assembly situation, enumerating the possible error states in position and angle, and calculating what forces are generated when collisions occur in each of the identified error states.

- Rearrange the model so that the forces generated are indicative of what error state could have generated them.

- Create a motion strategy that will convert the sensed forces into motions that are consistent with the error state and that tend to reduce the sensed forces, thereby reducing or eliminating the error.

- Design a controller that will close the loop by making the recommended motion, sensing any forces that are generated immediately afterward, deducing the error state that caused the new forces, and generating new corrective motions.

While this method sounds good in principle, there are several pitfalls. First, while analyzing an error state to see what forces and torques it generates is relatively straightforward, inverting this analysis so that sensed forces can be used to identify the error state can be quite difficult. Force measurements can be noisy, and different error states can generate similar force patterns. Second, friction will introduce a kind of noise that can make it difficult or impossible to determine what the applied forces in fact are, irrespective of the accuracy of the sensors. Finally, the controller may be subject to instability, inasmuch as it generates motions that in turn generate more force signals, to which it can overreact.

We leave until Chapter 10 the question of how to model what forces arise from what error states and how friction can interfere. In this chapter we concentrate on how to represent the forces and motions and how to keep the controller stable.

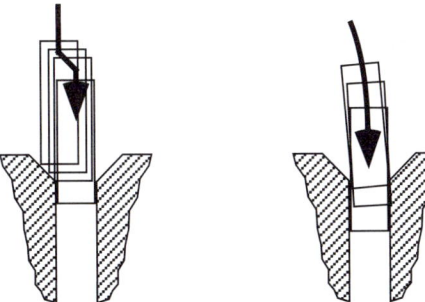

FIGURE 9-1. Force–Motion Events in Response to Typical Part–Part Errors. *Left:* Two parts are aligned parallel to each other but the approach trajectory is offset to one side. A force on the moving part is needed to bring the trajectory of the peg into coincidence with the axis of the hole. *Right:* Two parts are angularly misaligned. A torque on the moving part is needed to align the parts. In each case, it is assumed that the moving part is held in some way that is responsive to the applied force or torque and moves accordingly.

9.C.2. Modeling Fine Motions, Applied Forces, and Moments

In order to model fine motions and the forces that arise when parts contact during fine motion, we need to define some coordinate frames and terminology. Figure 9-2

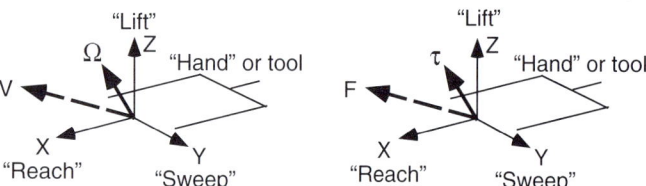

FIGURE 9-2. Velocity and Force Vectors of a Hand or Tool Expressed in Hand Coordinates.

shows a generic hand or assembly tool with a coordinate frame located at the grip point. In the left-hand part of the figure, the coordinate frame contains two vectors marked V and Ω.[2] These represent linear and angular velocity, respectively. Together they describe the six degrees of freedom of motion of which the hand is capable, each vector having x, y, and z components in hand coordinates.

In the right-hand part of Figure 9-2, the coordinate frame contains two other vectors marked F and τ. These represent force and torque, respectively. It is assumed that when the hand or something it is holding contacts another object, such forces and torques will arise, causing changes in the motions of the hand. These concepts are used in the next section where we derive and analyze a simple force feedback algorithm for steering a hand in response to sensed forces.

9.C.3. The Accommodation Force Feedback Algorithm

The basis of a force feedback algorithm for an assembly machine or robot is to (a) use the sensed forces to indicate what position or angular errors exist between parts that are touching each other and (b) use that information to make corrective motions that remove the errors and help the assembly process to advance. The algorithms we will look at are particularly simple and are presented mainly to stimulate thought about what happens when parts are assembled as well as to exercise the required mathematical models. In particular, these algorithms ignore the effects of friction. In Chapter 10, friction will be taken into account and shown to be a crucial element of the problem.

In Figure 9-2 we distinguished between linear and angular velocities as well as between "linear" and "angular" forces (i.e., forces and torques). In Figure 9-1 we indicated that two kinds of part-to-part misalignments could occur, one arising from lateral error and the other arising from angular error. The basic idea of the force feedback algorithm is to associate the lateral misalignments with sensed forces and to associate the angular misalignments with sensed torques. The strategy of the algorithm is to translate in the direction of felt forces and to rotate in the direction of sensed torques. Usually this can be expected to cause the sensed forces and torques to decrease, which indicates that the error is decreasing. This logic follows the pattern indicated in Section 9.C.1 for generating a force feedback algorithm.

A flowchart of such an algorithm is shown in Figure 9-3. At the left the original fine motions are shown. These could be thought of as the nominal program of an assembly machine or robot. If there are misalignments (and there almost always are due to small errors in part locations or motion trajectories) the parts will touch, contact forces will arise, and the sensor will sense them. These forces are shown at the bottom of the figure. The figure shows the sensor decomposing the contact forces into separate force

---

[2]It is important to remember that it is possible to write a rotation rate Ω as a vector, but it is not possible to write large rotations this way. Instead, as we saw in Chapter 3, large rotations must be represented by matrices, and the order in which those rotations occur about the individual axes is important and noncommutative. For very small rotations and for rotation rates, the order does not matter and so they can be gathered together into a vector.

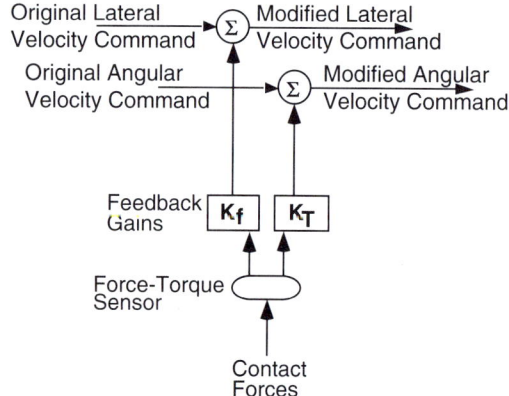

FIGURE 9-3. Diagram of a Force–Motion Strategy.

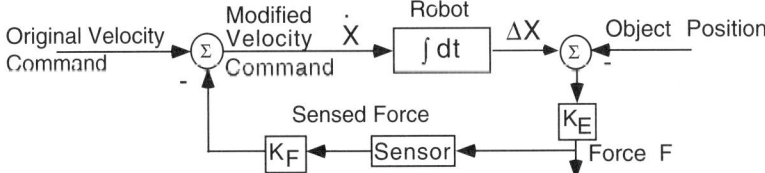

FIGURE 9-4. Closed-Loop Force–Motion System. When the robot strikes an object in the environment, a force F is generated via stiffness matrix K_E. This force is sensed by the sensor and is converted into a modified motion command via feedback gain K_F, which represents the separate gains K_f and K_T in Figure 9-3.

and torque vectors.[3] A controller interprets each vector, multiplying it by a feedback gain K_f for the forces and K_T for the torques, thereby converting it in effect into a velocity. The dimensions of the gains are thus velocity units/force unit and angular rate units/torque unit, respectively. These gains essentially describe the urgency with which a reaction to a unit force or torque should be mustered. Low gain means sluggish response, while high gain means quick response. The velocities thus generated are used to modify the original commanded velocities, generating a change in the path of the hand.

This algorithm is called the *accommodation method* (see [Whitney 1977]). It essentially mechanizes a damping style approach to force feedback because it responds like a damper: A force generates a velocity. [Whitney 1987] contains a survey of force feedback algorithms that follow similar patterns. Some generate a position change in response to a sensed force and thus behave like springs. Others generate a combination of position, velocity, and acceleration and thus respond like more general dynamic systems.

It would be nice if that were all to the problem of correcting the errors in a trajectory during fine motion. However, we need to remember that each motion command generates a new force, which in turn will cause the force feedback algorithm to generate a new corrective motion. The situation is shown in the feedback loop diagram in Figure 9-4. In this figure, the logic of Figure 9-3 is combined with the physics of contact between the machine and the outside world. Here, closed-loop nature of the situation is made clear. In particular, the motions generate forces through the mechanism of a stiffness K_E, which represents the combined stiffness of the hand, the sensor,

the part held in the hand, and the outside world in contact with the hand.

We need to take account of the closed-loop nature of the situation because the wrong choice of feedback gains could cause the system to go unstable. A great deal of attention has been devoted to this problem by the research and industrial communities, and we can see some of the broad conclusions of these studies using the qualitative analysis presented here. More detail may be found in [Brady et al.].

The basis of the analysis is to assume that the loop in Figure 9-4 operates at discrete intervals of time ΔT. At each discrete point in time, the sensor is read and a correction to the commanded velocities is computed and added to the original commands. The new velocity is held constant until the next time ΔT later when the process repeats. Then the criterion for keeping this loop stable is

$$K_F K_E \Delta T < 1 \qquad (9\text{-}1)$$

This criterion says that if any of the three quantities on the left becomes too large, the system can become unstable, meaning that any force generated during one time interval will generate a larger force at the next interval. Note that while we can specify K_F and ΔT, we cannot necessarily specify K_E. This means that if the environment is stiff, the algorithm can get into stability trouble. Note, too, that while we can prescribe a smaller ΔT, the dynamics of the robot or tooling will limit the speed at which it can respond. Smaller ΔT will simply be ignored. Finally, we could try making K_F smaller, but that will mean sluggish response. This can be interpreted to mean that the force does not cause a meaningful motion response, a consequence of which is that large forces will build up.

The accommodation algorithm is mechanized by creating particular feedback gains in the form of matrices. These matrices have dimension 6×6 because they convert three forces and three torques into three linear and three

---

[3]Commercial sensors exist that sense and report the complete vector of sensed forces and torques.

angular velocities. If this matrix is diagonal, then each sensed force component will generate a corresponding velocity component: Force in the X direction generates velocity in the X direction, torque about the X axis generates angular velocity about the X axis, and so on.

For example, suppose we choose the feedback matrix to be[4]

$$K_F = \begin{bmatrix} 1 & 0 & 0 & 0 & 0 & 0 \\ 0 & 100 & 0 & 0 & 0 & 0 \\ 0 & 0 & 100 & 0 & 0 & 0 \\ 0 & 0 & 0 & 100 & 0 & 0 \\ 0 & 0 & 0 & 0 & 100 & 0 \\ 0 & 0 & 0 & 0 & 0 & 100 \end{bmatrix} \quad (9\text{-}2)$$

This matrix says that the hand should barely notice forces it feels in the X direction while it should respond fairly urgently to forces and torques felt along the other five axes. Assuming that X is the direction of insertion of one part into another, say a peg and hole as shown in Figure 9-1, the matrix in Equation (9-2) assigns low gain to X forces and high gains to all the others. In this way, the peg, commanded to move in the X direction, will continue to do so until it feels a strong force at the bottom of the hole. However, if it feels forces or torques along or around the Y and Z axes, indicating lateral or angular misalignment, it will move in such a way as to reduce those forces and become better aligned.

We can get creative with these matrices. For example, the matrix

$$K_F = \begin{bmatrix} 0 & 200 & 0 & 0 & 0 & 0 \\ 0 & 100 & 0 & 0 & 0 & 0 \\ 0 & 0 & 0 & 0 & 0 & 0 \\ 0 & 0 & 0 & 0 & 0 & 0 \\ 0 & 0 & 0 & 0 & 0 & 0 \\ 0 & 0 & 0 & 0 & 0 & 0 \end{bmatrix} \quad (9\text{-}3)$$

will respond with motion in the X direction to forces felt in the Y direction. If the hand is commanded to move in the negative Y direction, then when it strikes a surface oriented along X, it will accommodate the Y direction contact force (via the gain 100) and begin moving laterally along the surface (via gain 200), maintaining

---

[4]The sign convention adopted for these example matrices is consistent with that of Figure 9-4. However, in an actual sensor whose axes align with hand velocities, motion in the $+X$ direction that causes contact with an obstacle will generate a negative F_x. If this fact is taken into account, then all the signs in the example K_F matrices must be reversed.

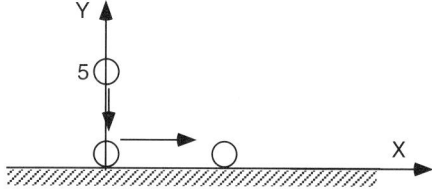

FIGURE 9-5. Coordinates for Edge-Following Algorithm. The circle represents a peg held by a hand or gripper that is not shown. The peg begins at coordinates (0, 5) and moves in the negative Y direction. When it reaches $Y = 0$, it will stop moving along Y and begin moving along X.

contact with the surface because the original Y direction motion command is still in effect. This action is illustrated in Figure 9-5. This behavior is quite unlike that generated by Equation (9-2) because it is gyroscopic: A motion is generated at right angles to the sensed force. Thus it stops moving in the direction where it detects resistance (Y) and begins moving where it encounters none (X).

A slight modification to this algorithm will cause the hand to follow a curved contour.

Finally, we can create a matrix that will help a hand or robot put a disk with a hole over a pin. The matrix is

$$K_F = \begin{bmatrix} 0 & 0 & 0 & 0 & V_R & 0 \\ 0 & 0 & 0 & -V_R & 0 & 0 \\ 0 & 0 & 100 & 0 & 0 & 0 \\ 0 & 0 & 0 & 0 & 0 & 0 \\ 0 & 0 & 0 & 0 & 0 & 0 \\ 0 & 0 & 0 & 0 & 0 & 0 \end{bmatrix} \quad (9\text{-}4)$$

This matrix detects torques created when the disk moves down in the $-Z$ direction and touches the pin off center. These torques are proportional to the Z contact force between the disk and the pin and to the distance in X and Y from the center of the disk (assumed to be at the center of the hand coordinate system) and the pin. The ratio of the two torques is the arctangent of the azimuth angle from the center of the disk to the pin. The matrix then generates velocity commands in X and Y proportional to the distance from the disk center to the pin in the same ratio as the arctangent of the angle, thus sending the disk toward the pin.

9.C.4. Mason's Compliant Motion Algorithm

A different approach to force feedback is taken in [Mason], which defines two complementary *natural constraints* and *artificial constraints* that govern force and motion of objects in contact. The natural constraints consist of those positions and velocities that keep a fixed and moving object in contact, plus forces and torques along directions that are not governed by constrained positions or velocities. The artificial constraints are calculated so that a desired motion can occur. Along any direction where a force (or velocity) is defined, the corresponding velocity (or force) cannot be defined, but it can be determined by the physics of the situation. The result is that the subspaces of natural and artificial constraints are orthogonal complements of each other.

For example, if a round peg is sliding along a straight surface, as shown in Figure 9-5, the natural constraints after the peg contacts the surface are that $V_y = 0$ and (if there is no friction) $F_x = 0$. The artificial constraints are $V_x = anything$ and $F_y = anything$. These four vectors comprise all the conditions on the peg's motion, and each set is orthogonal to the other in the sense that if a component in one set is held to a given value like zero, then the corresponding component in the other set is not held to any value.

Mason's method involves using geometry to establish the natural constraints by prescribing an "ideal trajectory" for the moving part, and using orthogonal complementarity to solve for the artificial constraints. A controller is then designed to command the gripper to move along the artificial motion directions or exert force along the artificial force directions while accepting the natural constraints. Although his method applies strictly only to situations where the parts are completely surrounded by the natural constraints, it can be extended to situations like that in Figure 9-5 where the natural constraint applies only to motion in the $-Y$ direction. This is done by pretending

that the peg rides between two parallel surfaces and making it stay on the single surface by adding a bias force in the $-Y$ direction to the solution.

Comparing the accommodation method with Mason's method reveals that accommodation makes no attempt to solve explicitly for the complementary constraints. Instead, it approximates them by making some of the feedback gains very high (where natural force or velocity constraints can exert their effect) and making others very low (so that artificial force or velocity constraints can exert their influence).

Other hybrids of force and velocity control have been proposed. Interested readers can find out more in [Brady et al.] and [Whitney 1987]. All of them present interesting stability challenges. A more complete discussion of these issues may be found in [Asada and Slotine].

9.C.5. Bandwidth of Fine Motions

How fast does a peg move laterally during chamfer crossing, such as that illustrated in Figure 9-1? This is important because, as suggested by the discussion following Equation (9-1), a force feedback algorithm can respond only so fast, and failure to respond will result in excessive forces building up.

We can answer this question by modeling the chamfer surface as a segment of a periodic waveform and applying Fourier theory to determine its basic frequency content. The force feedback control system, including the robot, will have to respond at the highest frequency of interest in this waveform in order to perform adequately. Figure 9-6 shows how to convert the chamfer crossing event into a time-based waveform whose shape depends on the

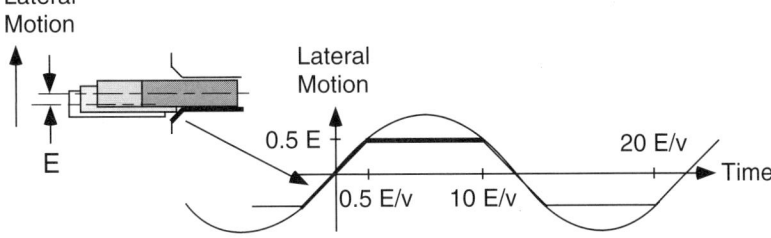

FIGURE 9-6. Conversion of Chamfer Crossing into a Periodic Waveform. The heavy black line in the graph represents the chamfer. The peg's traverse of the chamfer is expanded into a periodic trapezoidal waveform in time and a sine wave is fitted to it. The speed with which the peg moves laterally to absorb a lateral error E is proportional to the approach velocity V and inversely proportional to E. The waveform shown assumes that the insertion is finished when the peg is a distance $9.5E/V$ into the hole.

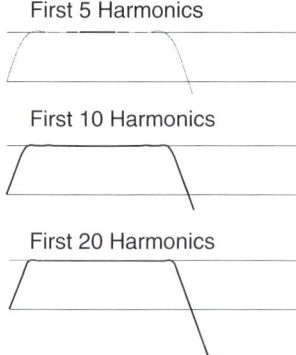

First 5 Harmonics

First 10 Harmonics

First 20 Harmonics

FIGURE 9-7. Fourier Approximation to a Trapezoidal Periodic Waveform. The figure shows how many harmonics of the Fourier series representation of this waveform are needed in order to get a good approximation to the waveform in Figure 9-6. Clearly, at least the first five harmonics are needed.

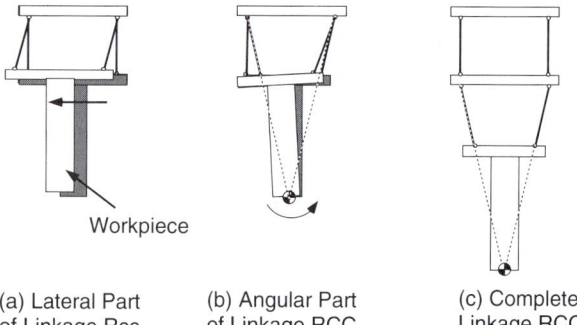

Workpiece

(a) Lateral Part of Linkage Rcc

(b) Angular Part of Linkage RCC

(c) Complete Linkage RCC

FIGURE 9-8. Schematic of the Linkage Version of the Remote Center Compliance. The linkage RCC consists of two parts, one (a) that responds to lateral forces by moving laterally without rotating, and another (b) that responds to torques about a distant point by rotating about that point without translating. Combining the two parts (c) into one yields a device that can support a workpiece in such a way that lateral and angular errors can be removed independently.

amount of lateral error E that must be absorbed when the peg is approaching at velocity V. The period of the waveform shown is $T = 20E/V$, so the base frequency of the wave is $f = V/20E$. If $V = 200$ mm/sec and $E = 1$ mm, then $f = 10$ Hz. Figure 9-7 shows that at least five harmonics will be needed to approximate this waveform to reasonable accuracy, implying that the bandwidth of the lateral error absorption maneuver shown is at least 50 Hz. This is extremely high for any known robot's main joints. Only a specially designed hand with small mass and strong high-speed actuators would be able to attain such a bandwidth.

9.C.6. The Remote Center Compliance

The previous subsections described active force feedback methods for accomplishing assembly fine motions that

remove lateral and angular error. The implication from Section 9.C.5 is that active force feedback will have difficulty accomplishing these maneuvers at an economically attractive speed. In this section we describe a passive device that can accomplish them fast enough. A theoretical explanation of why it works is deferred to Chapter 10, but an intuitive explanation can be given here.

The device in question is called a remote center compliance (RCC). In its original form, it is a linkage device similar to the sketches shown in Figure 9-8. Commercial versions are similar to the devices shown in Figure 9-9. Both versions create a so-called remote center of rotation, indicated by the target mark ✪, about which the workpiece can rotate in response to externally applied torques at its tip. Typically, an RCC is integrated into the wrist or gripper of assembly tooling or a robot gripper.

The RCC is successful because it is simple, low cost, and fast. It implements a compliance-based response

3 Shear Pads

6 Shear Pads

Shear Pad

FIGURE 9-9. Commercial RCCs. These devices mechanize the actions of the linkage RCC by means of elastic elements called shear pads. These units are made of alternating layers of metal and soft polymer, and they exhibit small axial compliance and relatively large lateral or shear compliance.

to applied forces and torques that arise due to lateral and angular error, whereas the accommodation algorithm described in Section 9.C.3 implements a damping-based response. The forces that arise disappear in the case of the accommodation algorithm whereas they remain locked

into the RCC until the gripper releases the workpiece. This disadvantage is more than compensated by the fact that the accommodation algorithm is active and has limited bandwidth whereas the RCC is passive and comprises a small mass and can thus attain much higher bandwidths.

9.D. CHAPTER SUMMARY

This chapter introduced basic concepts of force and motion that are necessary for understanding assembly in the small. Gross motions and fine motions were compared and found to differ in many ways, the most important being the degree to which it makes sense to remove errors or uncertainty prior to launching the motion: for gross motions it does while for fine motions it does not, below a certain limit that depends on a variety of factors. The next few chapters investigate some of these factors. They include

the clearance between parts being assembled, the friction forces between the parts, and the force–motion or compliance characteristics of the tooling, grippers, and the parts themselves.

Several methods of removing assembly errors during fine motion were introduced, including active and passive means. Both operate by separating errors into lateral and angular components and by arranging to remove them independently.

9.E. PROBLEMS AND THOUGHT QUESTIONS

1. The stability criterion for the accommodation algorithm is given as

$$K_F K_E \Delta T < 1 \qquad (9\text{-}5)$$

This criterion may be interpreted in words to say "At a given time step, not all of the accumulated contact force can be removed during the next time step. Prove this interpretation. *Hint:* You should begin with

$$F_k = K_E(X_k - X_e)$$
$$V_k = V_0 - K_F F_k \qquad (9\text{-}6)$$
$$X_{k+1} = X_k + V_k \Delta T$$

and note that (ignoring the constants V_0 and X_e)

$$\Delta X_{k+1} = V_k \Delta T$$
$$\Delta F_{k+1} = K_E \Delta X_{k+1} \qquad (9\text{-}7)$$
$$V_k = -K_F F_k$$

Then multiply both sides of Equation (9-5) by V_k and solve for ΔF_{k+1}.

2. Program the force feedback simulation in Section 9.G and try different values of time step, feedback gains, and stiffness matrices. You should discover combinations of values for which the simulation goes unstable.

3. Imagine that you are blindfolded and walk through an unknown environment. Would you reach forward with your hand loosely probing and your elbow bent and then walk slowly? Or would you close your fist, straighten your arm, and charge ahead? Explain your choice in terms of Equation (9-5).

4. The accommodation algorithm imitates the behavior of a damper. That is, it responds with a velocity when it feels a force. Modify the program in Section 9.G so that it responds like a spring.

9.F. FURTHER READING

[Asada and Slotine] Asada, H., and Slotine, J.-J., *Robot Analysis and Control*, New York: Wiley Interscience, 1986.

[Brady et al.] Brady, M., Hollerbach, J. M., Johnson, T. L., Lozano-Perez, T., and Mason, M. T., editors, *Robot Motion, Planning, and Control*, Cambridge: MIT Press, 1982. Chapter 5, "Compliant Motion" (pp. 305–471), contains several papers that illustrate different formulations of force control of robots and manipulators.

[Mason] Mason, M. T., "Compliance and Force Control for Computer Controlled Manipulators," in *Robot Motion,*

Planning, and Control, Brady, M., Hollerbach, J. M., Johnson, T. L., Lozano-Perez, T., and Mason, M. T., editors, Cambridge: MIT Press, pp. 373–404, 1982.

[Whitney 1977] Whitney, D. E., "Force Feedback Control of Manipulator Fine Motions," *ASME Journal of Dynamic Systems, Measurement and Control,* June, pp. 91–97, 1977.

[Whitney 1987] Whitney, D. E., "Historical Perspective and State of the Art in Robot Force Feedback," *International Journal of Robotics Research,* vol. 6, no. 1, pp. 3–14, 1987.

9.G. APPENDIX

This appendix contains the text of a simple TRUE BASIC program that simulates the behavior of the matrix in Equation (9-2). The coordinates are defined in Figure 9-5. If your version of BASIC does not support "mat" expressions, then you will have to rewrite the code so that each matrix multiplication is handled on a component-by-component basis.

TABLE 9-1. TRUE BASIC Code for Simple Force Feedback Demonstration

```
dim newx(2,1), x(2,1), v0(2,1),
   xtemp1(2,1),xtemp2(2,1)
dim KF(2,2), KE(2,2)
set window -1,2,-5,5
plot lines: -1,0;1,0
plot lines: 0,2;0,-2
plot lines: -.01,-.5;.01,-.5
plot lines: -.01,-1;.01,-1
plot lines: .5,-.05;.5,.05
plot lines: 1,-.05;1,.05
input prompt "V0 ": v0(1,1), v0(2,1)
input prompt "time step ": dt
let newx(1,1) = 0
let newx(2,1) = 5
let KF(1,1) = 0
let KF(2,1) = 0
let KF(1,2) = 1
let KF(2,2) = 7
let KE(1,1) = 1
let KE(2,2) = 1
let t = 0
do while t < 3
let t = t + dt
mat x = newx
mat xtemp1 = dt*v0
mat newx = x + xtemp1
mat xtemp2 = KE*x
mat xtemp2 = KF*xtemp2
mat xtemp2 = dt*xtemp2
if x(2,1) < 0 then mat newx = newx - xtemp2
plot lines: newx(1,1),newx(2,1);
loop
print "yss ", v0(2,1)/(KF(2,2)*KE(2,2))
end
```

The equations represented in the program are

$$F_k = K_E(X_k - X_e)$$
$$V_k = V_0 - K_F F_k \qquad (9\text{-}8)$$
$$X_{k+1} = X_k + V_k \Delta T$$

where

$$X_k = \begin{bmatrix} x \\ y \end{bmatrix}_k$$

$$V_k = \begin{bmatrix} \dot{x} \\ \dot{y} \end{bmatrix}_k$$

$$K_F = \begin{bmatrix} k_{fxx} & k_{fxy} \\ k_{fyx} & k_{fyy} \end{bmatrix} \qquad (9\text{-}9)$$

$$K_E = \begin{bmatrix} k_{exx} & k_{exy} \\ k_{eyx} & k_{eyy} \end{bmatrix}$$

$$V_0 = initial\_velocity$$

The program is set up so that

$$V_0 = \begin{bmatrix} 0 \\ -5 \end{bmatrix}$$

$$K_F = \begin{bmatrix} 0 & 1 \\ 0 & 7 \end{bmatrix}$$

$$K_E = \begin{bmatrix} 1 & 0 \\ 0 & 1 \end{bmatrix} \qquad (9\text{-}10)$$

$$X_0 = \begin{bmatrix} 0 \\ 5 \end{bmatrix}$$

Sample output from the program in Table 9-1, using time step $dt = 0.05$, is shown in Figure 9-10.

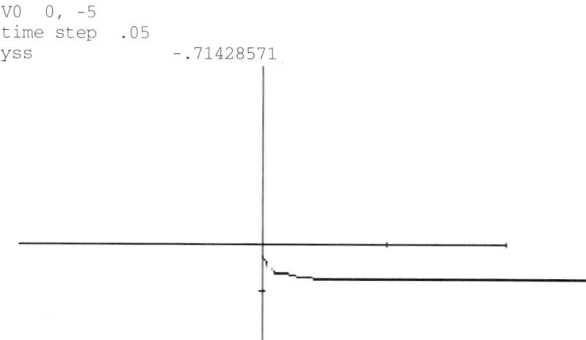

FIGURE 9-10. Sample Output from Program in Table 9-1.

10 ASSEMBLY OF COMPLIANTLY SUPPORTED RIGID PARTS

10.A. INTRODUCTION

The next two chapters present the basic processes of assembly. This requires looking closely at the details of the fine motions. The engineering results give a designer the conditions (part size, shape, and part-to-part error) for successful assembly. This information influences both the design of parts and design of fabrication and assembly equipment.

The most common type of assembly involves *rigid parts,* which generally do not change shape during assembly. They are the focus of this chapter. Their assembly is easier to model than that of *compliant parts,* covered in Chapter 11, which deform in expected and acceptable ways during assembly. Typical examples include snap fits, assembly of electrical connectors, stretching springs or elastic belts over pins or pulleys, and so on.

This chapter describes the requirements for successful assembly of compliantly supported rigid parts.[1] "Compli-

antly supported" means that, by design or accident, the grippers, tooling, and/or assembly apparatus can deform as necessary in order to accommodate small errors in the trajectory of the parts as they begin to mate. "Rigid" means that the parts deform negligibly in comparison to what supports them. We will see below that it is possible for an assembly action to fail even when the parts are dimensioned and made properly so as to permit a final assembled state to exist.

An important property of these compliant supports is that they are passive. That is, they contain no sensors, motors, or other active elements. When they move to accommodate relative errors in part positions or orientations, it is because their compliant elements deform under the action of forces and torques that arise between the parts due to these errors.

10.B. TYPES OF RIGID PARTS AND MATING CONDITIONS

Rigid parts and their mating conditions may be classified by the shapes of the parts and the clearance between them. See Figure 10-1 through Figure 10-5. The most common shape is round, although rectangular and tongue–groove shapes are also used. These types usually mate via motion in one direction only. Two simultaneous coordinated motion directions are required to mate threaded parts. Most commonly, there is clearance between the parts, although the clearance can sometimes be extremely small. Some parts are made with deliberate interference, effectively

negative clearance. Such parts are assembled by using extra force (*force* or *interference fits*) or by cooling one part and/or heating the other to temporarily create positive clearance (*shrink fits*). One may also classify mates by noting whether one or both parts have chamfers. If neither part has a chamfer, the mate is called *chamferless.*

Three types of mates will be considered here: pegs and holes, screw threads, and gears. The theory that predicts required mating force and gives error conditions is much more highly developed for round pegs and holes than for any other geometry. Extension to rectangular shapes results in extremely complex equations, which may be found in the references ([Sturges]). There is a small amount of

---

[1]This chapter is based on Chapter 5 of [Nevins and Whitney].

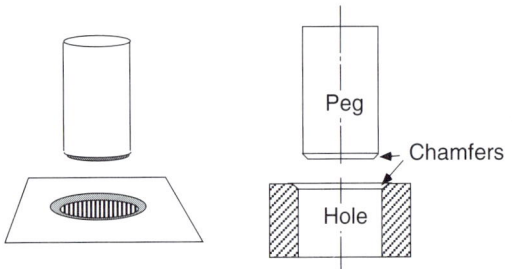

FIGURE 10-1. Schematic Illustration of Mating of Round Pegs and Holes with Chamfers on Both Peg and Hole.

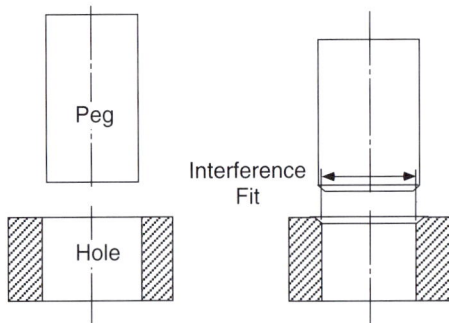

FIGURE 10-2. *Left:* Schematic Illustration of Mating Two Parts Without Chamfers. *Right:* Schematic Illustration of Interference Mating of Two Parts.

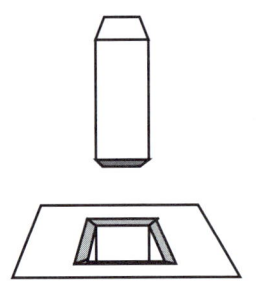

Rectangular Peg–Hole Mate
with Chamfers

FIGURE 10-3. Schematic Illustration of Rectangular Peg–Hole Mate.

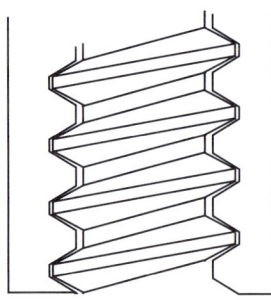

FIGURE 10-4. Schematic Illustration of Screw Thread Mating.

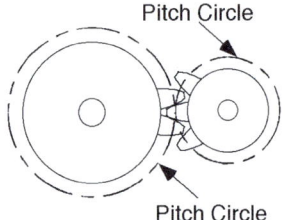

FIGURE 10-5. Schematic Illustration of Gear Mating. The pitch circles of the gears are tangent when the gears are properly mated.

theory concerning threaded mates and some intuitive information about gears, which are summarized in this chapter with details in the references. Other types of mates, such as push-twists associated with bayonet-base light bulbs, exist but are not discussed here.

The other main class of part mates, discussed in Chapter 11, is compliant part mates, in which the parts deform during assembly. Similar but more complex equations may be derived to describe assembly forces and the influence of errors for such parts. A particularly important feature of such analyses is the ability to determine how to redesign the shapes of the mating surfaces or adjust their compliance in order to enhance mating, prevent unmating, deal with alignment errors, and so on. These opportunities also apply to rigid parts, but the discussion of them is concentrated in Chapter 11.

10.C. PART MATING THEORY FOR ROUND PARTS WITH CLEARANCE AND CHAMFERS

Figure 10-6 schematically represents assembly of a round peg and hole in two dimensions, although we should remember that assembly is three-dimensional in general. The various models of assembly derived in this book are mostly two dimensional and are accurate enough for most purposes ([Whitney], [Simunovic], [Arai and Kinoshita]). The figure defines five typical phases of assembly: approach, chamfer crossing, one-point contact, two-point contact, and line contact. Not every assembly event contains all of these phases, but most do. Throughout

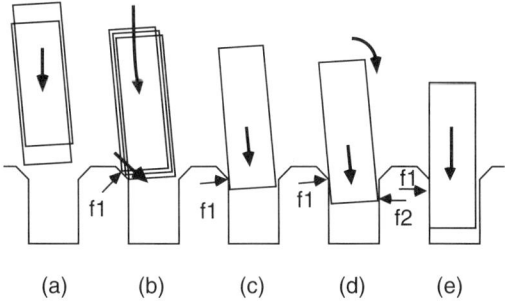

FIGURE 10-6. Phases of Mating Pegs and Holes: (a) Approach, (b) Chamfer Crossing, (c) One-Point Contact, (d) Two-Point Contact, (e) Line Contact. The contact forces are f_1 and f_2. The figure is drawn for the case where initial lateral error is to the left and initial angular error is counterclockwise. Three other cases are possible.

this chapter and the next, we will assume that the peg approaches the hole along a path that is parallel to the hole's axis.

Parts typically begin mating with some relative lateral and angular error, so the first contact occurs on the chamfers. During chamfer crossing, the contact point moves down the chamfer toward the rim of the hole as the parts try to move laterally to remove the lateral error. The part is pushed laterally by the force acting on it at the contact point. Once the contact point reaches the rim, it remains there,[2] acting as the "one point" of one-point contact. As the peg advances farther into the hole it finally strikes the opposite side, establishing a second contact point. During the two-point contact phase, the parts try to rotate with respect to each other to remove angular errors. The part is turned angularly by the torque created by the forces acting at the two contact points. In some cases, two-point contact may be followed by line contact, in which the parts are in contact all along one wall of the hole and their axes are parallel.

These moves constitute the fine motions of a typical simple assembly. Other assemblies, such as push–twist, snap actions, and thread mating, include other fine motions.

---

[2]This is what happens if the peg is shifted to the left and tilted counterclockwise, as shown in Figure 10-6. If it is tilted clockwise, the first contact point will travel with the peg's tip down the left inside wall of the hole until the second contact occurs on the right side of the rim.

10.C.1. Conditions for Successful Assembly

The mechanics of part mating are governed by the geometry of the parts, the compliance of the parts and supports, the friction between parts as they move past each other during assembly, and the amount of lateral and angular error between the parts as mating begins. The interplay of these factors determines whether assembly will be successful and how large will be the forces exerted on the parts by the tooling and each other.

The success or failure of a peg–hole assembly depends on whether or not the parts pass successfully through two potential danger zones. First, the lateral or angular errors before assembly could be so large that the parts fail to meet within the bounds of the chamfers (or part diameters if there are no chamfers). Second, there are two forms of failure associated with two-point contact during the fine motion phase; these are called *wedging* and *jamming*. While the names sound similar, the events are different and have different causes and cures. Later sections of this chapter will put firm mathematical formulations behind these ideas. For now, we concentrate on an intuitive understanding.

Wedging is an event in which the contact forces between peg and hole can set up compressive forces inside the peg, effectively trapping it part way in the hole. Figure 10-7 is a schematic illustration. To avoid wedging, one must keep the angular error between peg and hole at the moment of first two-point contact small enough. The equations describing successful assembly, developed in Section 10.C.4, show that there is a relation between avoiding wedging and ensuring that the chamfers meet.

Jamming is an event in which the peg cannot advance into the hole because the insertion force vector points too far off the axis of the hole. Figure 10-8 is a schematic illustration. To avoid jamming, one must support the peg so that the reaction forces set up by the two contact points

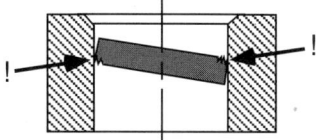

FIGURE 10-7. Schematic Illustration of Wedging. The part is elastically or plastically deformed by the opposed contact forces. The remedy is either to push harder and risk damage or pull the part out and try again.

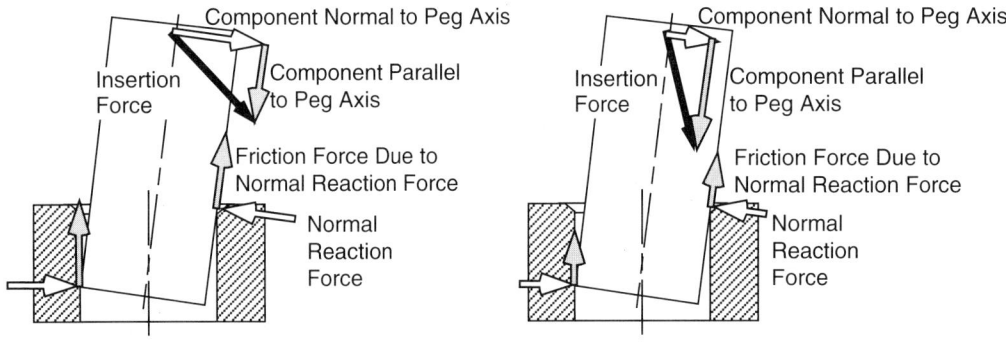

FIGURE 10-8. Schematic Illustration of Jamming. The insertion force and the reaction forces it generates are shown, separated into components along and normal to the peg's axis. *Left:* The peg is jammed because the insertion force does not have a large enough component along the peg's axis to overcome the friction forces raised by reactions to the component normal to the peg's axis. *Right:* The peg is not jammed because the insertion force component along the insertion direction is large enough to overcome friction.

are able to turn the peg parallel to the hole's axis. Such supports comprise the grippers or tooling. These supports are also important in chamfer crossing and avoidance of wedging.

Wedging and jamming are both failure events that could occur during two-point contact. Wedging can be avoided if two-point contact is delayed until the peg is deeper in the hole than a certain critical depth. Jamming can be avoided if the forces and moments acting on the peg during two-point contact are properly managed. This will be accomplished by properly designing the compliant support that holds the peg, the topic of the next section. In all of the following discussions, the mathematical results are stated, but the derivations are given in Section 10.J.

10.C.2. A Model for Compliant Support of Mating Parts

During assembly, parts are supported by jigs, fixtures, hands, robots, grippers, and so on. These supports have some compliance, either by accident or by design. Compliance is the inverse of stiffness: Stiff things are not very compliant, and vice versa. Since the parts also have some compliance in general, we define rigid parts as those whose compliance is small compared to the compliance of the supports. Correct design of these supports is a crucial issue in successful assembly, along with control of alignment errors between the parts.

Therefore we call a deliberately designed assembly tool compliance an *engineered compliance* to distinguish

it from the *undocumented compliance* that always exists in tooling and parts.[3] A properly engineered compliance can guarantee successful assembly and low contact force in spite of lateral and angular errors between the parts prior to assembly, whereas an undocumented compliance cannot make such a guarantee. In the presence of undocumented compliance, the assembly task may appear to "work" successfully but the reasons why may not be understood. A small change in conditions, such as temperature or part geometry, will cause it to fail. A number of early research experiments in robot assembly depended on undocumented compliance.

The geometry of the peg and hole is defined in Figure 10-9. It shows an idealized peg and hole with some initial relative lateral error ε_0 and angular error θ_0. Because of these initial errors, the peg must both rotate and translate in order to mate with the hole. The support must therefore be compliant both laterally and angularly to permit these motions to occur. Thus it is important to be able to model the compliance of the support.

A one-dimensional spring is the simplest example of a compliance. In assembly, forces and torques can act on a part from any direction, so it is necessary to think of multi-axis compliances in order to understand how the parts will move in response to these forces and torques. The force F acting on a part is in general a 6-vector (3 forces,

---

[3]"Compliance" is used in this book both as a noun and as an adjective. The adjective means the property of being compliant. The noun means an object that is compliant.

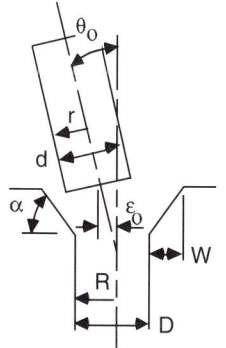

FIGURE 10-9. A Peg–Hole Mate Modeled as a Two-Dimensional Tab–Slot Mate. The peg approaches the hole along a path parallel to the hole's centerline, and is shown with an initial angular error θ_0 and an initial lateral error ε_0. The diameter of the hole is D, radius R. The diameter of the peg is d, radius r. The width of the chamfer is W, and its slope angle is α.

rotation. This point is called the *compliance center*. The part behaves as if it were supported at this point by three independent lateral springs in the X, Y, and Z directions, plus three independent angular springs about those axes. The stiffness or compliance of the part when acted on at this point is then simply described by only six numbers, namely the three XYZ lateral stiffnesses and the three angular stiffnesses about the XYZ axes. The origin of these axes is at the compliance center.

In the case of a planar model such as in Figure 10-9, the compliance at the compliance center consists of just one lateral spring and one angular spring, as illustrated in Figure 10-10. The point marked ◕ in Figure 10-10 is the compliance center of the support. It might be called the mathematically equivalent support point for the peg because, in general, it is not the point at which the peg is physically supported. Instead, the physical support has been replaced mathematically and equivalently by one lateral spring of stiffness K_x and one angular spring of stiffness K_θ located a distance L_g from the tip of the peg.

Figure 10-11 shows how the compliance center's location affects how the part moves when a lateral force is applied. If the force passes through the compliance center, the part translates but does not rotate.

A part mating event can then be represented by the path of the supported part (constrained by its shape and the shape of the part it mates to), the path of the support

3 torques), and the resulting motion δ is also a 6-vector (3 translations, 3 rotations). These two vectors are related by a 6×6 matrix C called the compliance matrix of the part and its support:

$$\delta = CF \qquad (10\text{-}1)$$

Since the part will move differently depending on the point at which it is pushed, each such point has associated with it its own, usually different, compliance matrix. For a three dimensional object, the matrix contains 36 generally nonzero entries. In many cases there is a special point on the supported part where the matrix is diagonal (only the six diagonal entries are nonzero). That is, pushing at this point with a pure force causes only lateral motion, and applying a pure torque about this point causes only

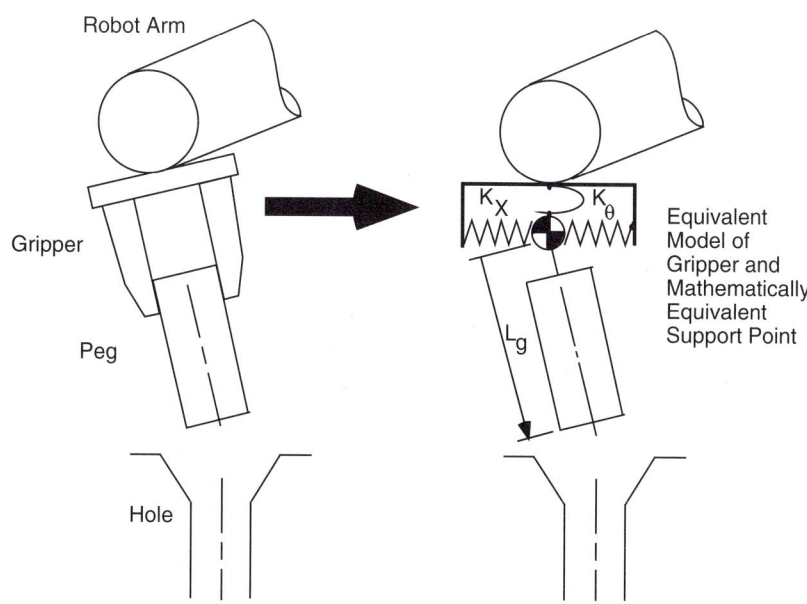

Robot Arm

Gripper

Peg

Hole

K_X K_θ

L_g

Equivalent Model of Gripper and Mathematically Equivalent Support Point

FIGURE 10-10. Illustrating the Mathematical Support Point of the Peg at the Compliance Center. The physical support shown at the left is a typical robot gripper, but any support could be represented this way. The equivalent mathematical support at the right consists of one lateral and one angular spring that characterize the compliance of the physical support. The compliance center ◕ is located a distance L_g from the tip of the peg.

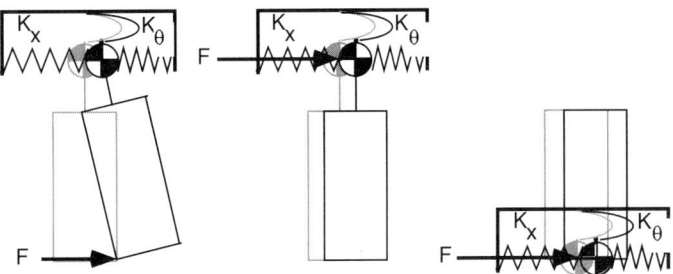

FIGURE 10-11. Deflections Arising from Applied Force at or not at the Compliance Center. *Left:* The compliance center is far from the tip of the part, where force *F* is applied. The peg translates and rotates. *Center:* The force is applied at the compliance center. The part translates but does not rotate. *Right:* The compliance center is at the tip of the peg, where force *F* acts. The peg translates but does not rotate. The forces in the left and right illustrations are typical of reaction forces when a peg encounters a chamfer at the start of assembly.

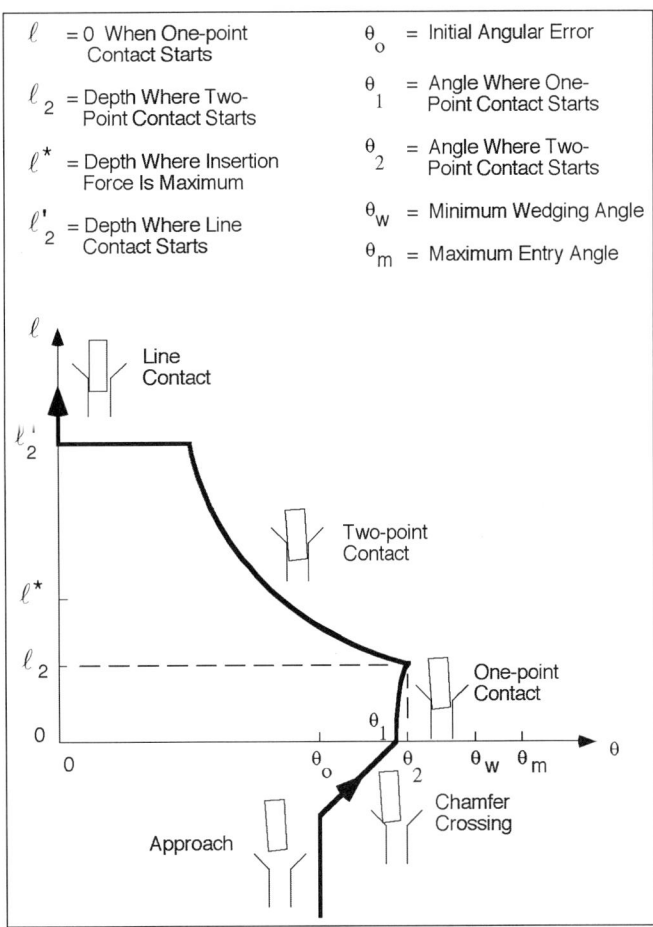

FIGURE 10-12. The Life Cycle of an Assembly. This plot traces the history of insertion depth and angle between peg and hole as the peg passes through the five phases of assembly defined in Figure 10-6. During a typical part mating event, as ℓ increases, θ remains constant until the peg hits the chamfer. During chamfer crossing, θ generally (but not always) increases, as it generally does during one-point contact. During two-point contact, θ decreases, and it may go to zero if line contact occurs. $\ell = 0$ when chamfer crossing ends and one-point contact begins. $\theta = 0$ when peg and hole axes are parallel.

(constrained by the machine doing the assembly), the forces and moments applied to the part by the compliances of the support as these paths deviate and the compliances stretch and compress, and the forces applied to the parts by the contact and friction forces during assembly. Figure 10-12 sketches the five phases of assembly again in terms of the angle θ between peg and hole and the depth of insertion ℓ, while Figure 10-13 shows schematically how the springs deform during assembly if the compliance center is far from

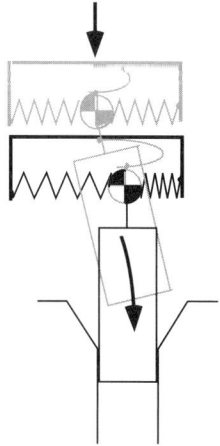

FIGURE 10-13. Schematic Illustration of the Deformation of the Lateral and Angular Springs as Assembly Proceeds. The deformation of these springs gives rise to contact forces between the tip of the peg and the walls of the hole.

the tip of the peg. Compare this to Figure 10-23, which shows the deflections when $L_g \approx 0$.

10.C.3. Kinematic Description of Part Motions During Assembly

The behavior of pegs and holes during assembly is strongly affected by the location of the compliance center. To illustrate this, let us consider the two situations depicted in Figure 10-14. In (a) the compliance center is located far from the tip of the peg. If there is some lateral error, the peg will encounter a lateral force acting on its tip arising from the contact between the tip of the peg and the chamfer. In response, it will both translate and rotate. This happens because the contact force exerts a lateral force on the lateral spring and a moment on the angular spring, causing both springs to deflect. The rotation will combine with and

possibly increase any initial angular error and may cause the parts to have a two-point contact. Since two-point contact is a prime danger zone where wedging or jamming could occur, we would like to prevent this type of contact or delay it until the peg is far into the hole and the risk of wedging and jamming is low. Clearly, the smaller the angular error and deflection, the farther into the hole two-point contact will occur.

In (b) the compliance center is approximately at the tip of the peg and the case of pure lateral error is shown. In this case, chamfer crossing removes the lateral error without introducing any angular error, clearly a desirable situation. This happens because the peg reponds to the lateral contact force at its tip by moving only laterally. This happens, in turn, because the part is effectively supported at its tip by a lateral spring. It is also supported there by an angular spring but there is no moment about this spring, so it does not deflect and the peg thus does not rotate.

In (c) there is both lateral and angular error. Chamfer crossing removes the lateral error while two-point contact removes the angular error. We will show later that (c) represents the safest two-point contact situation if there is angular error, and that the contact forces between the parts are as small as possible.

To proceed further, we need to consider the geometry of two-point contact in more detail. Figure 10-15 shows a peg part way into a hole. It is easy to show that insertion depth ℓ and wobble angle θ are approximately related by

$$\ell\theta = cD \qquad (10\text{-}2)$$

where

$$c = (D - d)/D \qquad (10\text{-}3)$$

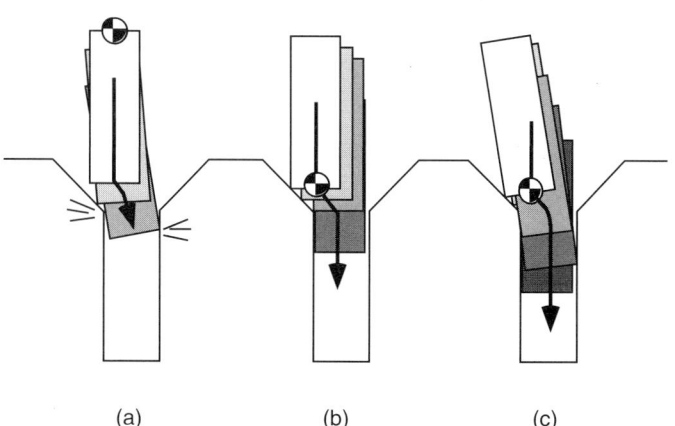

(a)   (b)   (c)

FIGURE 10-14. Comparison of Part Mating Behavior for Two Different Locations for the Compliance Center. In (a), the compliance center is at the rear of the peg. Initially the peg has some lateral error but no angular error. The lateral error becomes angular error as the peg passes over the chamfer and one-point contact begins. Wedging or jamming could occur. In (b) and (c), the compliance center is at the tip of the peg. In (b), there is initially only lateral error, all of which is removed during chamfer crossing without introducing any angular error. In (c), there is initially both lateral and angular error. Again, the lateral error is removed during chamfer crossing, while the angular error is removed during two-point contact.

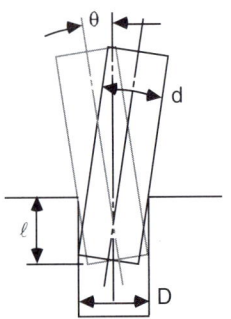

FIGURE 10-15. Geometry of a Two-Point Contact.

The variable c is called the clearance ratio. It is the dimensionless clearance between peg and hole. Figure 10-16 shows that the clearance ratio describes different kinds of parts rather well. That is, knowing the name of the part and its approximate size, one can predict the clearance ratio with good accuracy. The data in this figure are derived from industry recommended practices and ASME standard fit classes ([Baumeister and Marks]).

Equation (10-2) shows that as the peg goes deeper into the hole, angle θ gets smaller and the peg becomes more parallel to the axis of the hole. This fact is reflected in the long curved portion of Figure 10-12.

Figure 10-17 plots the exact version of Equation (10-2) for different values of clearance ratio c. Note particularly the very small values of θ that apply to parts with small values of c. Intuitively we know that small θ implies difficult assembly. Combining Figure 10-17 with data such as that in Figure 10-16 permits us to predict which kinds of parts might present assembly difficulties.

The dashed line in Figure 10-17 represents the fact that there is a maximum value for θ above which the peg cannot even enter the hole. This value is given by

$$\theta_{\max} = \sqrt{2c} \qquad (10\text{-}4)$$

It turns out in practice that the condition in Equation (10-4) is very easy to satisfy and that in fact a smaller maximum value for θ usually governs. This is called the wedging angle θ_w. Wedging and jamming are discussed next.

10.C.4. Wedging and Jamming

Wedging and jamming are conditions that arise from the interplay of forces between the parts. To unify the discussion, we use the definitions in Figure 10-9, Figure 10-10, and Figure 10-18. The forces applied to the peg by the compliances are represented by F_x, F_z, and M at or about the tip of the peg. The forces applied to the peg by its contact with the hole are represented by f_1, f_2, and the friction forces normal to the contacted surfaces. The coefficient of friction is μ. (In the case of one-point contact, there is only one contact force and its associated friction force.) The analyses that follow assume that these forces are in approximate static equilibrium. This means in practice that there is always some contact—either one point or two—and that accelerations are negligible. The analyses also assume that the support for the peg can be described as having a compliance center.

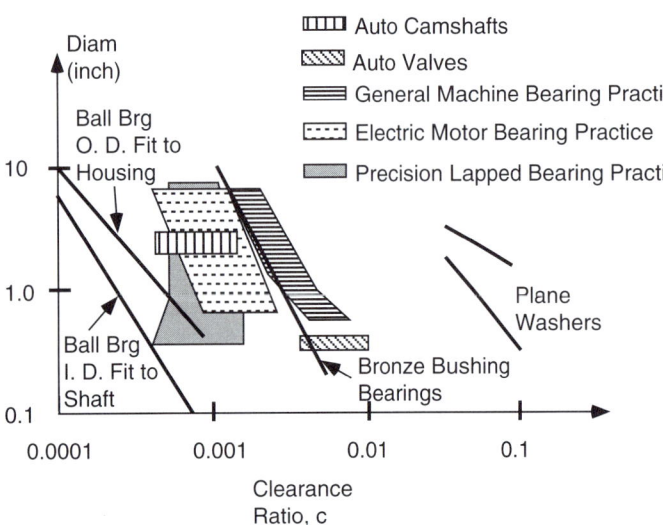

FIGURE 10-16. Survey of Dimensioning Practice for Rigid Parts. This figure shows that for a given type of part and a two-decade range in diameters, the clearance ratio varies by a decade or less, indicating that the clearance ratio can be well estimated simply by knowing the name of the part.

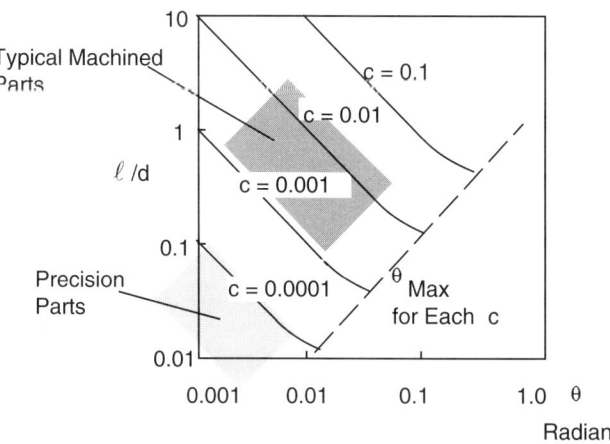

FIGURE 10-17. Wobble Angle Versus Dimensionless Insertion Depth. Parts with smaller clearance ratio are limited to very small wobble angles during two-point contact, even for small insertion depths. Since successful assembly requires alignment errors between peg and hole axes to be less than the wobble angle, and since smaller errors imply more difficult assembly, it is clear that assembly difficulty increases as clearance ratio (rather than clearance itself) decreases.

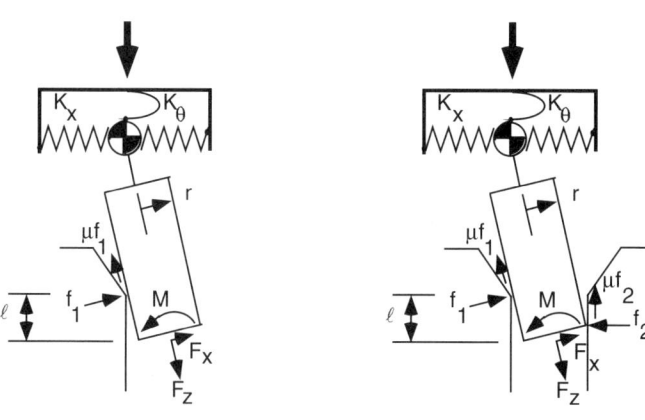

FIGURE 10-18. Forces and Moments on a Peg Supported by a Lateral Stiffness and an Angular Stiffness. *Left:* The peg is in one-point contact in the hole. *Right:* The peg is in two-point contact.

10.C.4.a. Wedging

Wedging can occur if two-point contact occurs when the peg is not very far into the hole. A wedged peg and hole are shown in Figure 10-19. The contact forces f_1 and f_2 are pointing directly toward the opposite contact point and thus directly at each other, creating a compressive force inside the peg. The largest value of insertion depth ℓ and angle θ for which this can occur are given by

$$\ell = \mu d \qquad (10\text{-}5)$$

and

$$\theta_w = c/\mu \qquad (10\text{-}6)$$

respectively. These formulas are valid for $\theta \ll \tan^{-1}(\mu)$.

A force–moment equilibrium analysis of the peg in one-point contact shows that the angle of the peg with respect to the hole's axis is given by

$$\theta = \theta_0 + S\varepsilon_0 \qquad (10\text{-}7)$$

where

$$S = \frac{K_x L_g}{K_x L_g^2 + K_\theta}$$

ε_0 and θ_0, the initial lateral and angular error between peg and hole, are defined in Figure 10-9, while L_g, the distance from the tip of the peg to the mathematical support point, is defined in Figure 10-10.

We can now state the geometric conditions for stage 1, the successful entry of the peg into the hole and the avoidance of wedging, in terms of the initial lateral and angular errors. To cross the chamfer and enter the hole, we need

$$|\varepsilon_0| < W \qquad (10\text{-}8)$$

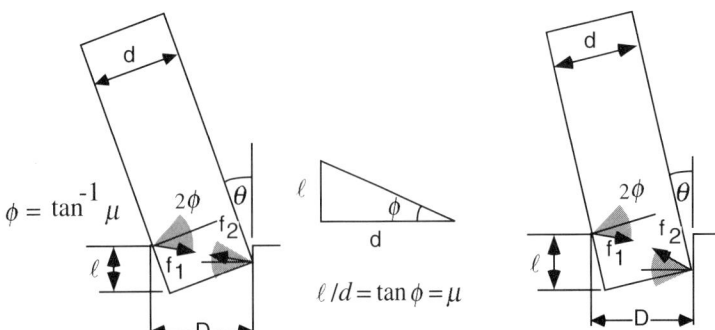

FIGURE 10-19. Geometry of Wedging Condition. *Left:* The peg is shown with the smallest θ and largest ℓ for which wedging can occur, namely $\ell = \mu d$. The shaded regions, enclosing angle 2ϕ, are the friction cones for the two contact forces. The contact force can be anywhere inside this cone. The two contact forces are able to point directly toward the opposite contact point and thus directly at each other. This creates a compressive force inside the peg and sets up the wedge. This can happen only if each friction cone contains the opposite contact point. *Right:* Once $\ell > \mu d$, this can no longer happen. Contact force f_1 is at the lower limit of its friction cone while f_2 is at the upper limit of its cone, so that they cannot point right at each other.

where W is the sum of chamfer widths on the peg and hole, and

$$|\theta_0 + S\varepsilon_0| < \pm c/\mu \qquad (10\text{-}9)$$

If parts become wedged, there is generally no cure (if we wish to avoid potentially damaging the parts) except to withdraw the peg and try again. It is best to avoid wedging in the first place. The conditions for achieving this, Equation (10-8) and Equation (10-9), can be plotted together as in Figure 10-20. This figure shows that avoiding wedging is related to success in initial entry and that both are governed by control of the initial lateral and angular errors. We can see from the figure that the amount of permitted lateral error depends on the amount of angular error and vice versa. For example, we can tolerate more angular error to the right when there is lateral error to the left because this combination tends to reduce the angular error during chamfer crossing. Since we cannot plan to have such optimistic combinations occur, however, the extra tolerance does us no good, and in fact we must plan for the more pessimistic case. This forces us to consider the smallest error window.

Note particularly what happens if $L_g = 0$. In this case the parallelogram in Figure 10-20 becomes a rectangle and all interaction between lateral and angular errors disappears. The reason for this is discussed above in connection with Figure 10-14. This makes planning of an assembly the easiest and makes the error window the largest.

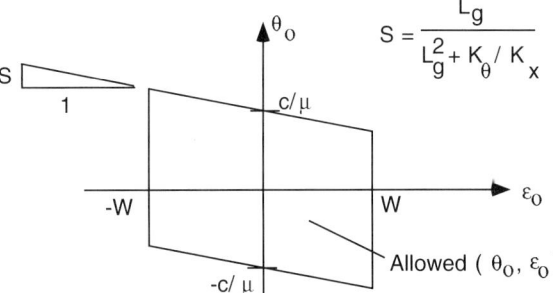

FIGURE 10-20. Geometry Constraints on Allowed Lateral and Angular Error To Permit Chamfer Crossing and Avoid Wedging. Bigger W, c, and ε, and smaller μ make the parallelogram bigger, making wedging easier to avoid. Not only must the error angle between peg and hole be less than the allowed wobble angle, as shown in Figure 10-17, but the maximum angular error is also governed by the coefficient of friction if wedging is to be avoided. If L_g is not zero, then if there is also some initial lateral error, this error could be converted to angular error after chamfer crossing. So, avoiding wedging places conditions on both initial lateral error and initial angular error. The interaction between these conditions disappears if $L_g = 0$. This fact is shown intuitively in Figure 10-14.

10.C.4.b. Jamming

Jamming can occur because the wrong combination of applied forces is acting on the peg. Figure 10-21 states that any combinations of the applied forces F_x, F_z, and M which lie inside the parallelogram guarantee avoidance

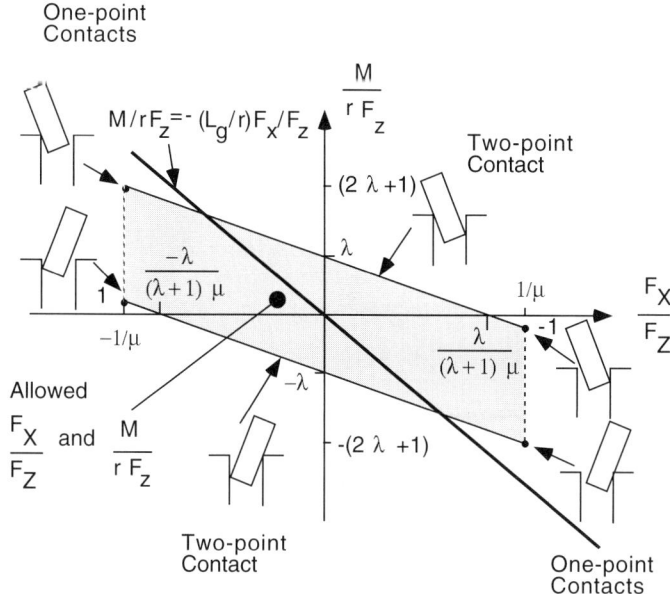

FIGURE 10-21. The Jamming Diagram. This diagram shows what combinations of applied forces and moments on the peg F_x/F_z and M/rF_z will permit assembly without jamming. These combinations are represented by points that lie inside or on the boundary of the parallelogram. λ is the dimensionless insertion depth given in Equation (10-10). When λ is small, insertion is just beginning, and the parallelogram is very small, making jamming hard to avoid. As insertion proceeds and λ gets bigger, the parallelogram expands as its upper left corner moves vertically upward and its lower right corner moves vertically downward. As the parallelogram expands, jamming becomes easier to avoid.

of jamming. The equations that underlie this figure are derived in Section 10.J.4. To understand this figure, it is important to see the effect of the variable λ. This variable is the dimensionless insertion depth and is given by

$$\lambda = \ell/\mu D \qquad (10\text{-}10)$$

As insertion proceeds, both ℓ and λ get bigger. This in turn makes the parallelogram in Figure 10-21 get taller, expanding the region of successful assembly. The region is smallest when λ is smallest, near the beginning of assembly. We may conclude that jamming is most likely when the region is smallest. (Since the vertical sides of the region are governed by the coefficient of friction μ, the parallelogram does not change width during insertion as long as μ is constant.)

If we analyze the forces shown on the right side of Figure 10-18 to determine what F_x, F_z, and M are for the case where K_θ is small, we find that

$$F_x = -F \text{ arising from deformation of } K_x$$
$$M = L_g F = -L_g F_x \qquad (10\text{-}11a)$$

Dividing both sides by rF_z yields

$$\frac{M}{rF_z} = -\left(\frac{L_g}{r}\right)\left(\frac{F_x}{F_z}\right) \qquad (10\text{-}11b)$$

which says that the combined forces and moments on the peg F_x/F_z and M/rF_z must lie on a line of slope $-(L_g/r)$ passing through the origin in Figure 10-21. If L_g/r is big, this line will be steep and the chances of F_x/F_z and M/rF_z falling inside the parallelogram will be small. Similarly, if M/rF_z and F_x/F_z are large, the combination of these two quantities will define a point on the line that is far from the origin and thus likely to lie outside the parallelogram.

On the other hand, if L_g/r is small so that the line is about parallel to the sloping sides of the parallelogram when λ is small, then the chance of the applied forces falling inside the parallelogram will be as large as possible and will only increase as λ increases. Similarly if M/rF_z and F_x/F_z are small, they will define a point on the line that is close to the origin and thus be likely to lie inside the parallelogram. When λ is small and jamming is most likely, the slope of sides of the parallelogram is approximately μ. Thus, if L_g/r is approximately equal to μ, then the line, and thus applied forces and moments, have the best chance to lie inside the parallelogram. Since μ is typically 0.1 to 0.3, we see that the compliance center should be quite near, but just inside, the end of the peg to avoid jamming.

Instead of considering a single lateral spring supporting the peg at the compliance center, let us imagine that we have attached a string to the peg at this point.

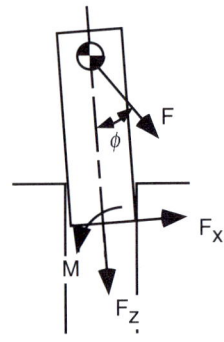

FIGURE 10-22. Peg in Two-Point Contact Pulled by Vector F. This models pulling the peg from the compliance center by means of a string.

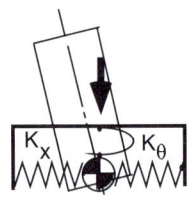

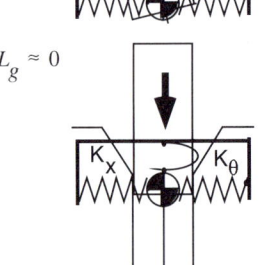

FIGURE 10-23. When L_g is Almost Zero, the Lateral Support Spring Hardly Deforms Under Angular Error. Compare the deformation of the springs with that in Figure 10-13, which shows the case where $L_g \gg 0$.

See Figure 10-22. This again represents a pure force F acting on the peg. In this case, F can be separated into components along F_x and F_z to yield

$$
\begin{aligned}
F_x &= F \sin \phi \\
F_z &= F \cos \phi \\
M &= -F_x L_g
\end{aligned}
\tag{10-12}
$$

so that

$$
\frac{M}{r F_z} = -\frac{F_x L_g}{r F_z} = -\frac{L_g}{r} \tan \phi
\tag{10-13}
$$

which is similar to Equation (10-11). In this case, we can aim the string anywhere we want but we cannot independently set F_x and F_z. But, by aiming the force, which means choosing ϕ, we can make F_x as small as we want, forcing the peg into the hole. As $L_g \to 0$, we can aim ϕ increasingly away from the axis of the hole and still make M and F_x both very small.

In Chapter 9, a particular type of compliant support called a Remote Center Compliance, or RCC, is described which succeeds in placing a compliance center outside itself. The compliance center is far enough away that there is space to put a gripper and workpiece between the RCC and the compliance center, allowing the compliance center to be at or near the tip of the peg. Thus $L_g \to 0$ if an RCC is used.

Figure 10-23 shows the configuration of the peg, the hole, and the supporting stiffnesses when $L_g \cong 0$. In this case, K_x hardly deforms at all. This removes the source of a large lateral force on the peg that would have acted at distance L_g from the tip of the peg, exerting a considerable moment and giving rise to large contact forces during two-point contact. The product of these contact

forces with friction coefficient μ is the main source of insertion force. Drastically reducing these contact forces consequently drastically reduces the insertion force for a given lateral and angular error. Section 10.J derives all these forces and presents a short computer program that permits study of different part mating conditions by calculating insertion forces and deflections as functions of insertion depth. The next section shows example experimental data and compares them with these equations.

10.C.5. Typical Insertion Force Histories

We can get an idea of the meaning of the above relations by looking at a few insertion force histories. These were obtained by mounting a peg and hole on a milling machine and lowering the quill to insert the peg into the hole. A 6-axis force–torque sensor recorded the forces. The peg was held by an RCC. The experimental conditions are given in Table 10-1.

TABLE 10-1. Experimental Conditions for Part Mating Experiments

Support: Draper Laboratory Remote Center Compliance
Lateral stiffness = K_X = 7 N/mm (40 lb/in.)
Angular stiffness = K_Θ = 53,000 N-mm/rad (470 in.-lb/rad)
Peg and hole: Steel, hardened and ground
Hole diameter = 12.705 mm (0.5002 in.)
Peg diameter = 12.672 mm (0.4989 in.)
Clearance ratio = 0.0026
Coefficient of friction $\cong$ 0.1 (determined empirically from one-point contact data)

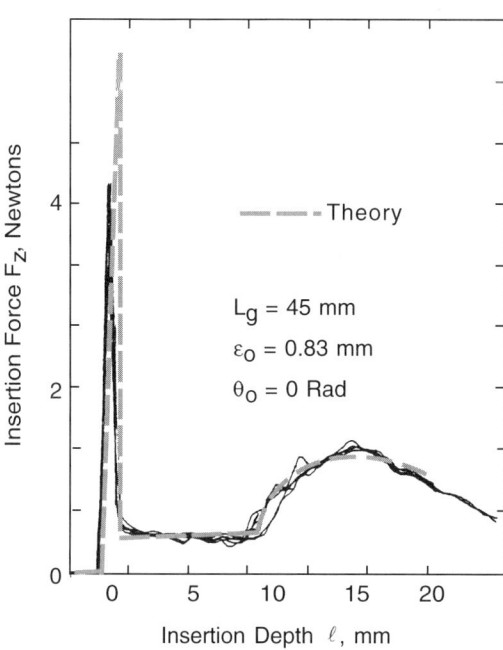

FIGURE 10-24. Insertion Force History. The compliance center is 4r back inside the peg from the tip. There is lateral error only, no angular error. As expected, two-point contact occurs, giving rise to the peak in the insertion force at a depth of about 18 mm. The peak at around 0 mm is due to chamfer crossing. Also shown on the plot is a theoretical estimate of insertion force based on equations given in the Section 10.J. A computer program in Section 10.J was used to create the theoretical plot.

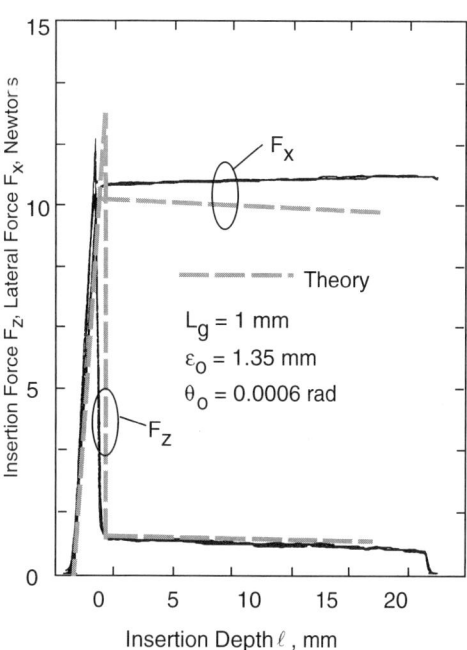

FIGURE 10-25. Insertion and Lateral Force History. The peg, hole, and compliant support are the same as in Figure 10-24, but L_g is essentially zero. As predicted, two-point contact does not occur, even though there is initially more lateral error than in Figure 10-24. This additional lateral error also is responsible for the larger chamfer crossing force (the large spike at $\ell = 0$) in this case compared to Figure 10-24.

Figure 10-24 shows a typical history of F_z for a case where there is only lateral error and the compliance center is about 4r away from the tip of the peg. The first peak in the force indicates chamfer crossing. Between $\ell = 1$ mm and $\ell = 9$ mm is one-point contact, following which two-point contact occurs. The maximum force occurs at about $\ell = 18$ mm or about twice the depth at which two-point contact began. For many cases, we can prove that the peak force will occur at this depth. A sketch of the proof is in Section 10.J.

Figure 10-25 shows the insertion force for the case where the lateral error is larger than that in Figure 10-24, but L_g is almost zero. Here, there is essentially no two-point contact, as predicted intuitively by Figure 10-14 and Figure 10-23. Also shown is the lateral force F_x. These results show the merit of placing the compliance center near the tip of the peg.

Figure 10-26 summarizes the conditions for successful chamfered compliantly supported rigid peg–hole mating.

10.C.6. Comment on Chamfers

Chamfers play a central role in part mating. Clearly, wider chamfers make assembly easier since they lessen the restrictions on the permissible lateral error. Chapter 17 discusses the relationships among the various sources of error in an assembly workstation and describes how to calculate the width of chamfers needed.

While all of the figures in this chapter show chamfers on the hole, the same conclusions can be drawn if the chamfer is on the peg. If both peg and hole have chamfers, then W in Equation (10-7) and Figure 10-20 is the sum of the widths of these chamfers.

Also, it is significant that if a properly designed compliant support is used, with its compliance center at the tip

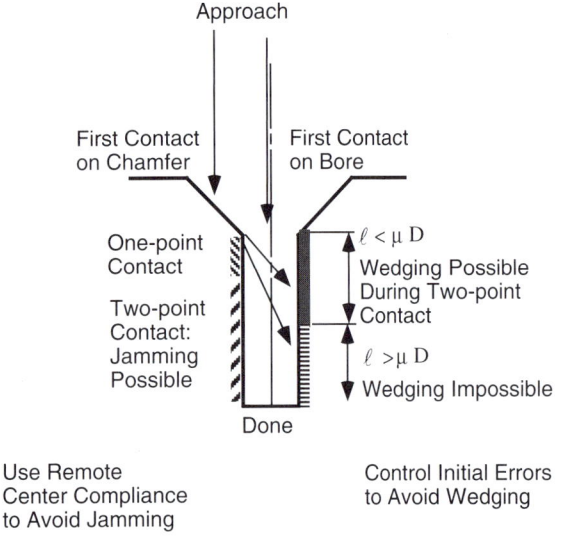

FIGURE 10-26. Pictorial Summary of Conditions for Successful Assembly of Round Pegs and Holes with Chamfers.

of the peg, there will be little insertion force except that generated by chamfer crossing. As Chapter 11 shows, the magnitude of this force depends heavily on the slope and shape of the chamfers.

While most chamfers are flat 45-degree bevels, some solutions to rigid part mating problems have been based on chamfers of other shapes. Figure 10-27 shows two examples of designs for the ends of plug gauges. Plug gauges are measuring tools used to determine if a hole is the correct diameter. To make this determination accurately requires that the clearance between hole and gauge be very small, making it difficult and time-consuming to insert and remove the gauge, and to avoid wedging it in the hole. The designs in Figure 10-27 specifically prevent wedging by making the ends of the gauges spheres whose radii are equal to the peg's diameter. The small undercut in the second design also helps to avoid damaging the rim of the hole.

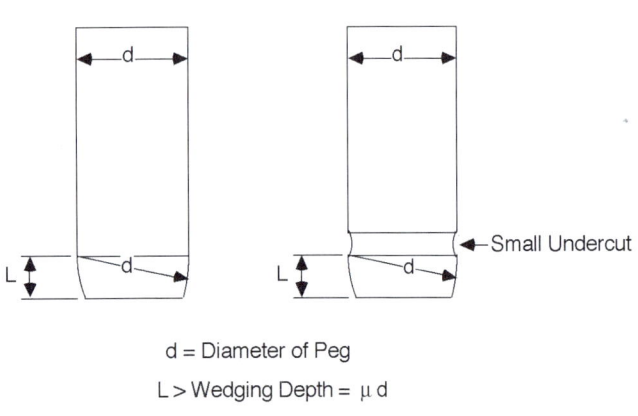

d = Diameter of Peg

L > Wedging Depth = μ d

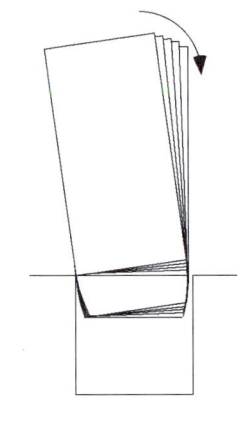

FIGURE 10-27. Two Designs of Chamfer That Prevent Wedging. Note that the radius of the arc forming the nose of the peg is equal in length to the diameter of the peg. In order to avoid wedging, it is necessary to pivot the peg about the point where the nose becomes tangent to the straight side, as shown at the right.

10.D. CHAMFERLESS ASSEMBLY

Chamferless assembly is a rare event compared to chamfered insertion because only a few parts have to be made without chamfers. Many of these are parts of hydraulic valves, whose sharp edges are essential for obtaining the correct fluid flow patterns inside the valves. In other cases, chamfers must be very small due to lack of space; a chamfer always adds length to a part, and sometimes there is a severe length constraint, either on a part or on the whole product. Chamferless assemblies are, of course, more difficult than chamfered ones because W in Equation (10-8)

is essentially zero. An attempt to assemble such parts by directly controlling the lateral error to be less than the clearance is almost certain to fail. This is especially true of hydraulic valve parts, whose clearances are only 10 or 20 μm (0.0004" to 0.0008").

In spite of their relative rarity, chamferless assemblies have attracted much research interest and some solutions that require active control, such as that in Figure 10-28. This is a multiphase method in which the peg is lowered until it strikes the surface well to one side of the hole. The

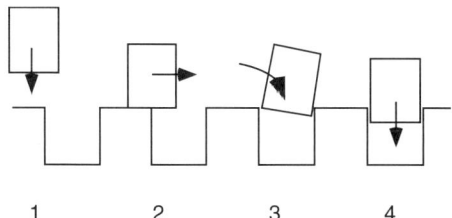

FIGURE 10-28. A Chamferless Assembly Strategy: (1) Approach, (2) Slide laterally, (3) Catch the Rim of the Hole and Tilt, (4) Lower Peg into Hole.

lateral error may not be known exactly but the direction toward the hole is known well enough for the method to proceed. The peg is then slid sideways toward the hole. It is held compliantly near the top so that when it passes over the edge of the hole its tip catches the rim of the hole and it starts to tip over. A sensor detects this tilt and lateral motion is stopped and reversed slightly. Hopefully this allows the tip to fall slightly into the hole. The peg is then lowered carefully. Rocking and lowering are repeated until the peg is in.

An elaboration of this strategy is employed by the Hi-Ti Hand ([Goto et al.]), a motorized fine motion device invented by Hitachi, Ltd. In this method, if the peg meets resistance during the lowering phase, it is gently rocked side to side in two perpendicular planes. The limits of this rocking are detected by sensors, and the top of the peg is then positioned midway between the limits. The peg is then pushed down some more or until resistance is again detected. This push and rock procedure is repeated as necessary until the peg is all the way in. In the case of the Hi-Ti Hand, mating time is typically 3 to 5 seconds. This method is good if the parts are delicate because it specifically limits the insertion force. For parts that can stand a little contact force, however, it is far too slow. Typical assembly times for chamfered parts held by an RCC are of the order of 0.2 seconds.

Figure 10-29 shows an entirely passive chamferless assembly method ([Gustavson, Selvage, and Whitney]). "Passive" means that it contains no sensors or motors. Figure 10-30 is a schematic of the apparatus itself. It has several novel features, including two centers of compliance which operate one after the other. The operation begins with the peg deliberately tilted into an angular error and as little lateral error as possible. (Note that this is the opposite of the initial conditions for the Hi-Ti Hand, where initial angular error is zero and there is deliberate lateral error.) When the peg is tilted, one side of the peg

Chamferless Inserter
U.S. Patent 4,324,032

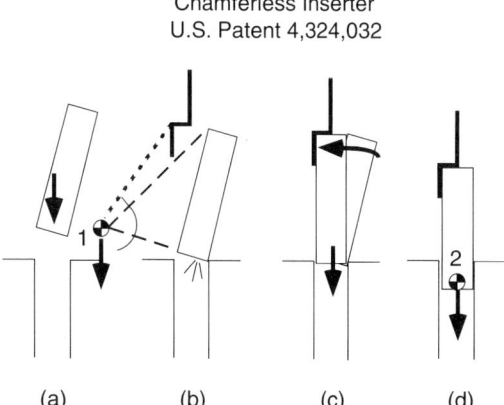

(a) (b) (c) (d)

FIGURE 10-29. Passive Chamferless Assembly Strategy. The inserter works by first permitting the peg to approach the hole tilted and then to turn up to an upright orientation with one edge slightly in the mouth of the hole. Insertion proceeds from that point with the aid of a conventional RCC. The details of how this is accomplished are shown in Figure 10-30.

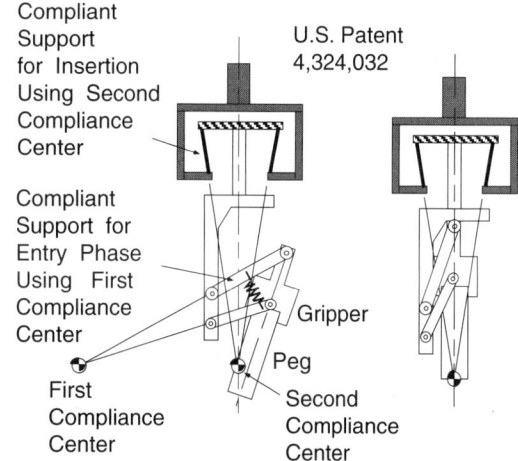

FIGURE 10-30. Schematic of Passive Chamferless Inserter. *Left:* Arrangement of the device while the peg is approaching the hole. The first compliance center is active and the part can rotate around it because of the sprung linkage attached to the gripper. The linkage is designed so that the tip of the peg does not move laterally very much while the peg is rotating up to vertical. What little tip motion there is will be in a direction away from the first compliance center so as to keep the tip pressed firmly against the rim of the hole. By this means the peg is most likely to remain in the mouth of the hole. *Right:* The part has engaged the mouth of the hole and is now locked into the vertical position. Insertion proceeds from here the same as if there had been chamfers and chamfer crossing were complete.

effectively acts as a chamfer, and it is almost certain that the tip of the peg and mouth of the hole will meet. Once they meet, the gripper continues moving down while the peg tilts up to approximately vertical under the influence of the linkage which creates the first compliance center. Upon reaching vertical, the peg locks into the gripper and comes under the influence of the compliant support above it, having the second center of compliance at the tip of the peg. The peg's tip stays in the mouth of the hole while rotating up to vertical. Insertion then proceeds as if the parts had chamfers, starting from the point where chamfer crossing is complete.

Examples of the apparatus in Figure 10-30 are in use installing valves into automobile engine cylinder heads.

10.E. SCREW THREAD MATING

Figure 10-4 showed normally mated screws. Assembling screws involves a chamfer mate similar to peg–hole mating followed by thread engagement. The screw (or nut) is then turned several turns until it starts to tighten. The last stage comprises tightening a specified amount.

Aside from missing the mouth of the hole, screw mating can fail in two possible ways. One is a mismatch of threads caused by angular error normal to the insertion direction. The other is a mismatch caused by having the peaks of the screw miss the valleys of the hole due to angular error along the insertion direction. Both of these are interchangeably called "cross-threading."

In order for the threads to mismate angularly normal to the insertion direction, the angular error must be greater than the angle α between successive peaks or valleys, defined in Figure 10-31.

If we define the angle between peaks as α, the diameter of the screw as d, and the thread pitch as p threads per unit length, then

$$\alpha = p/d \qquad (10\text{-}14)$$

Values for α for different standard screw thread sizes are shown in Figure 10-32. They indicate that for very small screws, an angular error of 1.14 mrad or 0.8 degree is enough to cause a tilt mismatch. Angular control at this level is comparable to that required to mate precision pegs and holes, as indicated in Figure 10-17. For larger screws, the angles become comfortably large, indicating what is

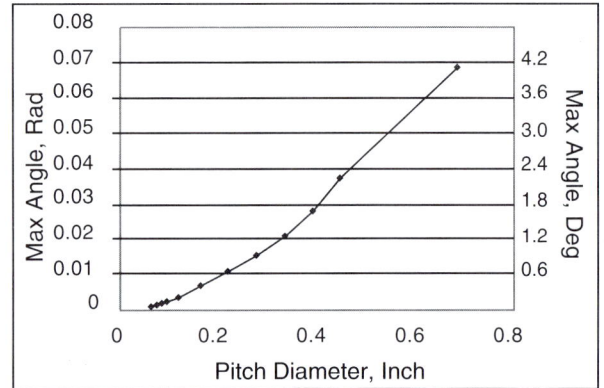

FIGURE 10-32. Maximum Permissible Angular Error Versus Screw Size for UNC Threads to Prevent Tilt Mismatch Between Threads. Since angular errors are relatively easy to keep below a few tenths of a degree, angular cross-threading is fairly easy to avoid for all but the smallest screws.

found in practice, namely that this kind of error does not happen very often since angular control as good as a degree or so is easy to obtain, even from simple tools and fixtures.

The other kind of screw mating error is illustrated in Figure 10-33. Here, the error is also angular, but the angle in question is about the insertion axis in the twist direction. That is, the thread helices are out of phase. Unless the materials of either the screw or the hole are soft, this kind of error is also difficult to create.

Some study of this problem may be found in Russian papers. Figure 10-34 and Figure 10-35 are from [Romanov]. The screw has a taper or chamfer of angle α while the hole thread has a taper of angle γ. The analysis in this paper is entirely geometric, with no consideration of friction. The conclusion is that α should be greater than γ (see Figure 10-36). This is an interesting conclusion because the Russian standards at the time the paper was written were $\alpha = 45$ degrees, $\gamma = 60$.

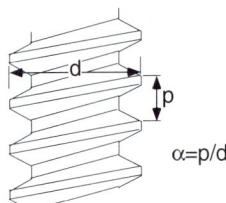

FIGURE 10-31. Schematic of Screw Thread Defining _p_ and _d_. In order for threads to mismate due to tilt angle error, the tilt must be greater than α.

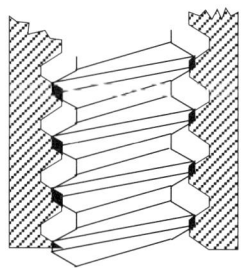

FIGURE 10-33. Mismated Screws Due to Helical Phase Error. The helices of the screw's threads and the hole's threads are out of phase and have interfered plastically with each other.

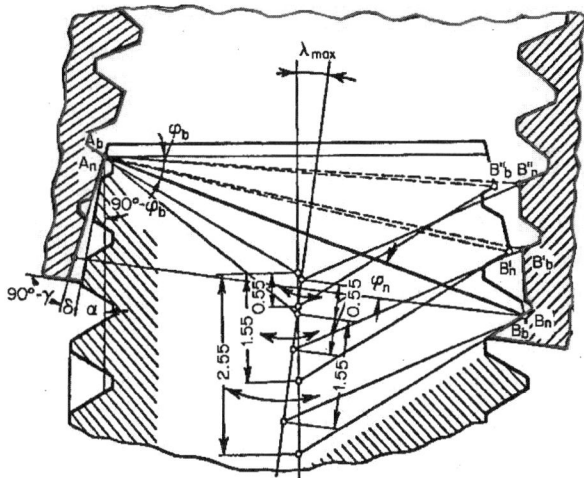

FIGURE 10-34. Variables Involved in Predicting Screw Cross-Threading. ([Romanov])

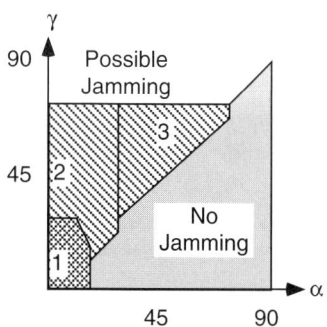

Region 1: Adjacent Threads Crossed

Region 2: Screw Tilted ~ p/d

Region 3: Screw Tilted ~ 2p/d

Note: The graph is drawn for p/d = 0.156, but graphs for other p/d are similar.

FIGURE 10-35. Sample Diagram of Good and Bad Values of α and γ. ([Romanov])

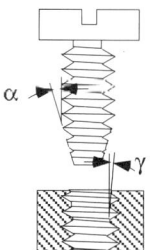

FIGURE 10-36. Screw and Threaded Hole with Screw Chamfer Steeper than Hole Chamfer.

Another method of aiding the starting of screws is to drastically change the shape of the tip. Two examples are shown in Figure 10-37. These are called "dog point" and "cone point" screws. Each has two disadvantages—extra cost and extra length—but the advantages are valuable. The dog point is a short cylinder that assures that the screw is centered in the hole and parallel to it. The cone point provides the largest possible chamfer, making it easier to put the screw in a poorly toleranced or uncertainly located hole, such as in sheet metal.

The above methods of assembling screws all depend on the helices mating with the correct phase without doing anything explicit to ensure that correct phase is achieved. A method that searches for the correct phase is the "turn backwards first" method, known to work well with lids of peanut butter jars. Usually this method requires sensing. To utilize it, one places screw and hole mouth-to-mouth and turns the screw backwards until one senses that the it has advanced suddenly. The magnitude of this advance is approximately one thread pitch. At this point, the threads are in a dangerous configuration, with chamfered peaks almost exactly facing each other. So it is necessary to turn an additional amount back, perhaps 45 degrees. Then it is safe to begin turning forwards. If a full turn is made without an advance being detected, successful mating will not be possible, and the parts should be separated. This method is slow and, as stated, requires sensing, but it works well and may be necessary in the case of unusually large diameters and small thread pitches, where even small angular errors can cause mismating.

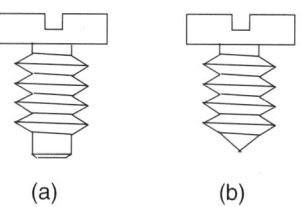

(a)   (b)

FIGURE 10-37. (a) Dog Point and (b) Cone Point Screws.

The last phase of screw mating is the tightening phase. Screw tightening must be done with care in order to obtain a properly and safely secured joint without risking stripping the threads. A commonly used but unreliable method is to measure the torque required to tighten the screw. The unreliability is based on the fact that the felt torque is a combination of tightening torque and friction torque between the head of the screw and the hole face. Because of the extra friction torque, one typically feels more torque than is actually being exerted on the threads. Errors of 50% or more are not unusual.

A more reliable method measures both turn angle and torque and seeks to set a certain amount of elongation into the screw rather than to achieve a certain amount of torque. To achieve this, it is necessary to sense torque versus turn angle and try to determine the inflection point of the curve. This point is related to the point at which the screw starts to deform plastically, at which it has achieved its maximum safe stretch. For many screws, the entire tightening event occurs within 1 to 10 degrees of rotation, as indicated in Figure 10-38. Since screws are typically turned rapidly

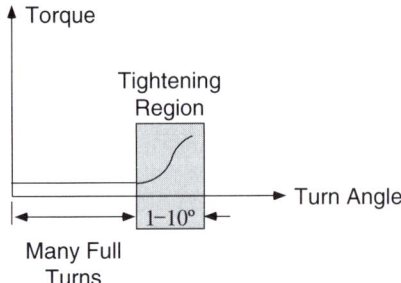

FIGURE 10-38. Schematic of Screw Tightening Torque Versus Screw Turn Angle. The torque rises very quickly after many turns with little or no torque. Torque is applied until the inflection point on the curve is reached. If significant torque is detected after only one turn or less, then some kind of mismating has probably occurred.

by automatic screwdrivers, the measuring apparatus and brakes on the screwdriver must act quickly. Commercial devices are available that operate on this principle. A study of torque-angle-controlled tightening of precision threads by automatic control is given in [Dunne].

10.F. GEAR MATING

The last topic in this chapter is the assembly of gears. This is a complex topic on which only a little research has been done. We will assume that one gear has already been installed, and it is necessary to install and mate another or others to it. There are several cases to consider. In each case the common element is that gear mating requires two separate alignments to occur. One is to bring the pitch circles into tangency, and the other is to fit the teeth together. These two steps can be done in either order, depending on the circumstances. Pitch circles are illustrated in Figure 10-5.

The first case analyzed is the easiest. There is plenty of space near the insertion point so the arriving gear may be brought down to one side of its mate as shown in Figure 10-39. Once it is near, the tool rotates the gear about its spin axis while bringing it laterally toward its mate. The mating direction is perpendicular to the spin axis of the gears. Eventually the teeth mesh and assembly can continue. So this method mates the teeth first and then the pitch circles.

If the arriving gear is on a shaft that must be inserted into a bearing, the above method works if the teeth can be mated before shaft and bearing. If shaft and bearing

must mate first, then the best method is to spin the shaft and gear while inserting along the spin axis, in the hope of mating the teeth. The same problem arises if two gears that are linked together must mate simultaneously with a third gear, as shown in Figure 10-40. Thus this method approximately mates the pitch circles first and then mates the teeth.

However, an approach along the spin axis may not succeed as easily as one perpendicular to it. Gears are

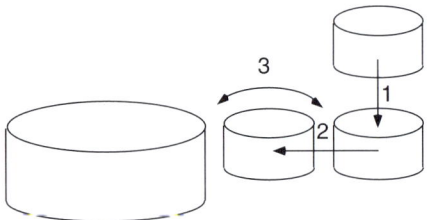

FIGURE 10-39. The Side-Approach Method of Mating Gears. In step 1, the gear is placed next to the mating gear. In steps 2 and 3, the gear is moved toward its mate and is simultaneously rotated in one direction or in oscillation, until the teeth mate.

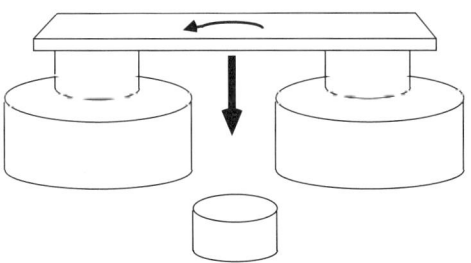

FIGURE 10-40. The Spin-Axis-Approach Method of Mating Gears. This method is often needed with planetary gear trains.

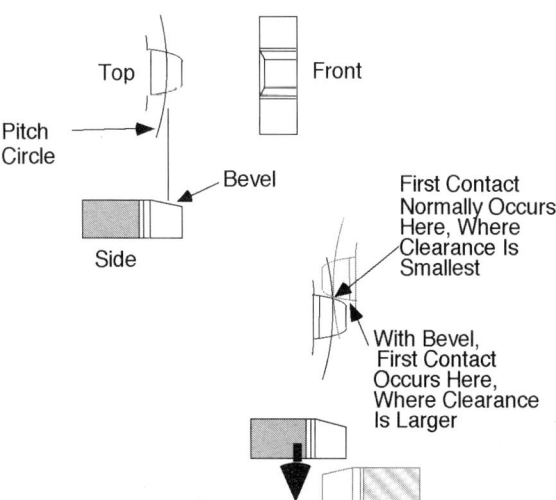

FIGURE 10-41. Bevelling Gear Teeth to Aid Mating.

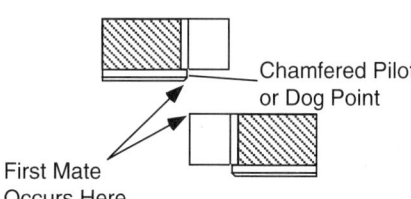

FIGURE 10-42. The "Dog-Point" Gear.

designed so that when they are mated, with the pitch circles tangent, there is little or no clearance between adjacent teeth. When gears are inserted along the spin axis, the pitch circles are typically already approximately tangent. This method therefore depends on the teeth mating under conditions in which there is little or no clearance between them. The arriving gear may simply come to rest on top of its mate and spin without mating, especially if the pitch circles overlap slightly.

People typically make such mates by either (a) waiting until a random chance mates the gears or (b) rocking the arriving gear, tilting its spin axis away from parallel to its final orientation, in order eventually to tilt the tip of a tooth into the space between two teeth on the other gear. These random and unpredictable methods cannot be used by automatic machinery without their being equipped with extra degrees of freedom and sensors. The method also fails to have a predictable completion time, making it an awkward one to include in an otherwise well planned and rhythmic production line. In short, the method lacks structure and should be replaced with a better one.

Two solutions are possible. The first is shown in Figure 10-41. Here, a bevel has been cut on one side of the teeth so that when they meet, the touching places will not be on the pitch circle but instead somewhere else; anywhere else will have larger clearance between mating teeth, so the chance of mating will be much larger.

The second solution is shown in Figure 10-42. This idea is similar in spirit to the dog point screw. To make it work well, the chamfered pilot on the gear must be well made so that it fits snugly within the teeth of the mating gear. This fit places the pitch circles close to each other. Spinning the arriving gear usually causes the teeth to mate easily.

This idea is embodied in U.S. Patent 4,727,770, which is illustrated in Figure 10-43.

Both of these solutions to gear mating have the same disadvantage as dog point screws: They add length to the gears. Since the length of a gear tooth's face is carefully calculated to give the gear adequate load capacity and life, one does not shorten the face in order to accommodate either the bevel or the pilot. Instead, one lengthens the gear to provide space for the bevel or pilot while keeping the tooth face the same size. An entire product can become longer if length is added to some of its parts, and the added length can be a problem for other reasons.

The mating of splines is physically similar to mating of gears. Splines are essentially internal gear mates in which all the teeth mate at once since the pitch circles are concentric.

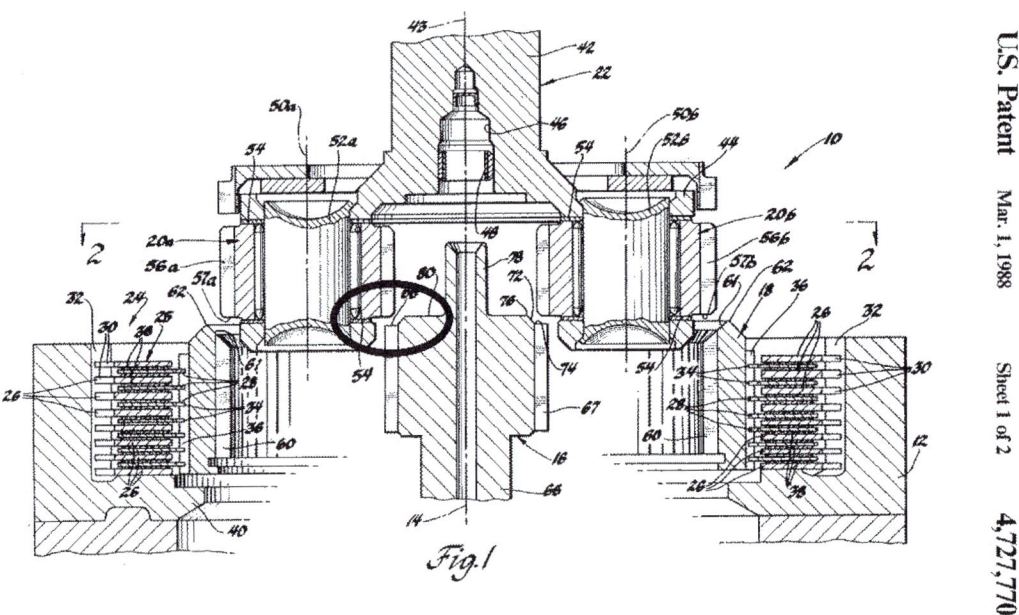

FIGURE 10-43. Richard Ordo's Patent on Multiple Gear Mating. This patent uses the dog-point gear approach. The patented feature is inside the ellipse. The mating situation shown here is similar to that shown in Figure 10-40.

10.G. CHAPTER SUMMARY

This chapter has outlined the behavior of compliantly supported rigid parts during the fine motion phase of assembly. The success of mating was shown to depend heavily on the shapes of the parts, the initial errors between them, the friction coefficient, and the compliance of the supporting tools and grippers. Success for chamfered and chamferless peg–hole mates depends on avoiding wedging and jamming. The mathematical conditions for this are shown and derived in Section 10.J. All of the relevant analyses assume that the parts are moving slowly.

Conditions given for successful mating of screws and gears are geometrical since the theory is not well enough developed to provide anything else. However, the conditions for successful assembly of simple peg–hole mates that take account of friction are more restrictive than the purely geometrical conditions. That is, the allowed errors are much smaller. So it is likely that the geometric conditions given for gears and screws are also merely necessary ones and are not sufficient, implying that the true conditions are more restrictive.

10.H. PROBLEMS AND THOUGHT QUESTIONS

1. Take apart a mechanical item (the stapler, a pump, toaster, light fixture, etc.) and classify the part mates as follows:

- Type of mate—peg/hole, press, tab/slot, screw, solder or glue, thermal shrink, bayonet, compliant snap or wedge, chamferless, and so on.
- Direction of approach of mating parts with respect to each other, based on a common coordinate frame attached to any main part of your choice.
- Accumulate the results for several products and create statistics showing such things as percent occurrence of each mate type and mate direction.

2. Using a micrometer or other appropriate measuring instruments, measure the peg and hole diameters of several part mates in different products and accumulate statistics on percent of mates with clearance ratios in the ranges 0.0001 to 0.001, 0.001 to 0.01, and so on. Compare your results to those shown in Figure 10-16 and Figure 10-17.

3. Obtain some close-fitting parts, such as a ball bearing and its housing, which have a clearance ratio in the range 0.001 to 0.003, approximately. Verify first that the fit has clearance and that it is possible to mate the parts without using force. Then clean the parts thoroughly with soap and water. Next, attempt to wedge the parts by pressing on the bearing on one side while it is part way into the hole. Finally, lubricate the parts and try again to wedge them. Record your observations and explain them in terms of wedging theory.

4. Explain in your own words the difference between wedging and jamming.

5. Derive equations Equation (10-5), Equation (10-6), and Equation (10-7). Show all the necessary steps.

6. Derive Equation (10-11). Show all the necessary steps.

7. Explain carefully all the mating conditions all the way around the periphery of the parallelogram in Figure 10-21. Include the dotted vertical lines as well as the solid sloping lines and the four heavy dots at the corners.

8. Prove the claim made in Section 10.C.4.b that the slope of the sloping sides of the parallelogram in Figure 10-21 is approximately μ when λ is small.

9. Derive the coordinates of each of the four heavy dots in Figure 10-21—for example, the dot at $(1/\mu, -1)$. Similarly, derive the four intersections between the sloping lines and the graph axes—for example, the intersection at $(0, \lambda)$.

10. Draw a picture to show why the shaped ends of the pegs in Figure 10-27 will not wedge.

11. Derive Equation (10-21) and Equation (10-22). Show all the necessary steps.

12. Derive Equation (10-24) and Equation (10-25). Show all the necessary steps.

13. Derive Equation (10-45). Show all the necessary steps.

14. Note that the part mating equations in Section 10.J have been derived for the case where the initial errors θ_0 and ε_0 are both positive. (In fact, the computer program listing in Table 10-2 in Section 10.J is valid only for this case and may give meaningless results or error bombs if other cases are tried.) This represents one of four possible cases, the others given by both errors being negative or one of each being positive while the other is negative. Rederive equations Equation (10-21), Equation (10-22), Equation (10-24), Equation (10-25), and Equation (10-45) for each of the other three cases.

15. Note that the part mating equations in Section 10.J have been derived for the case where the peg approaches the hole along the hole axis. Rederive the equations for the case where the peg approaches along its own axis. You will have to take care when defining the initial errors, since the definitions used in the chapter may not be appropriate.

16. In Figure 10-44 is a sketch of a window sash. The frame squeezes the sash with equal friction force on both sides. There is a little side-to-side clearance between the sash and the frame. To open this window most easily, should you push at A, B, C, D, E, or F? Explain with words or equations as you prefer. Ignore the mass of the sash.

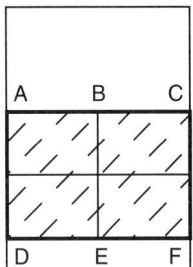

FIGURE 10-44. A Window Sash. The sash moves up and down in the frame.

17. In Figure 10-44, assume the friction force is bigger on the left side than on the right. To open the window most easily, where should you push? Explain with words or equations as you prefer.

18. In Figure 10-45 is a rod supported by a linear spring with stiffness K_x and an angular spring with stiffness K_θ. Write an expression for the total lateral displacement x_2 that relates F, L_g, and the two stiffnesses. [The answer is provided here, but don't peek—use it only to check that you understand the problem.] Also, write an expression for the angle θ.

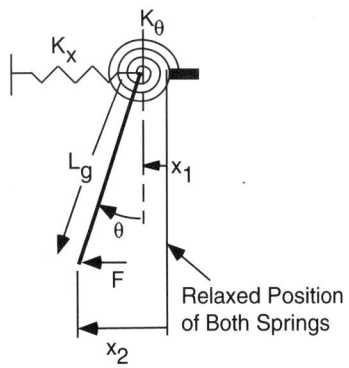

FIGURE 10-45. A Rod Supported by a Linear Spring and an Angular Spring.

Answer to Problem 18:

$$x_2 = F\left[\frac{L_g^2}{K_\theta} + \frac{1}{K_x}\right]$$

19. Continue with Problem 18 as follows:

a. Show on Figure 10-45 where the compliance center is.

b. Explain intuitively and with the aid of the equations how x_2 and θ will behave if K_x is zero or infinity and a force F is applied as shown in Figure 10-45.

c. Similarly, explain how x_2 and θ will behave if K_θ is zero or infinity.

d. Finally, explain how x_2 and θ will behave if L_g is zero.

20. Dan is frugal and brings home from business trips some partially used little bottles of shampoo from hotel rooms. He salvages the shampoo by turning the little bottle up side down and carefully placing its neck in the neck of a large bottle. One time he arranged them as shown in Figure 10-46.

He came back a while later to find the bottles as shown in Figure 10-47. No jostling or vibration occurred to cause this. Use Figure 10-21 and Figure 10-22 as guides to explain what probably happened.

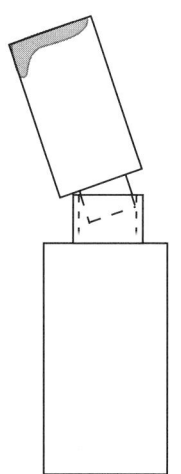

FIGURE 10-46. A Little Shampoo Bottle Balanced on a Large One.

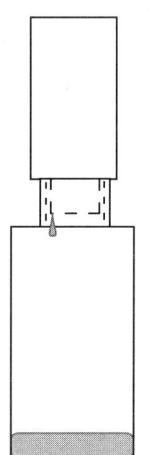

FIGURE 10-47. Configuration of the Two Shampoo Bottles Later.

21. Dan is an observant person who is interested in how people use their hands and how they sometimes get in trouble doing so. He was on an airplane recently and saw that passengers could not open the overhead bins. The harder they pulled on the handle, the more the doors resisted, until they were afraid they would break the handles. See Figure 10-48 and Figure 10-49.[4]

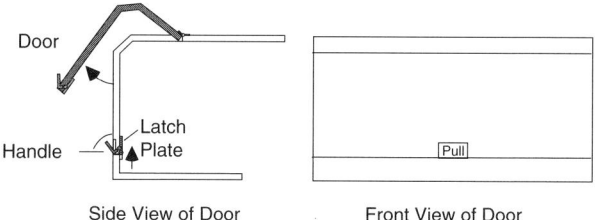

FIGURE 10-48. Side and Front Views of a Luggage Bin Door on an Airplane.

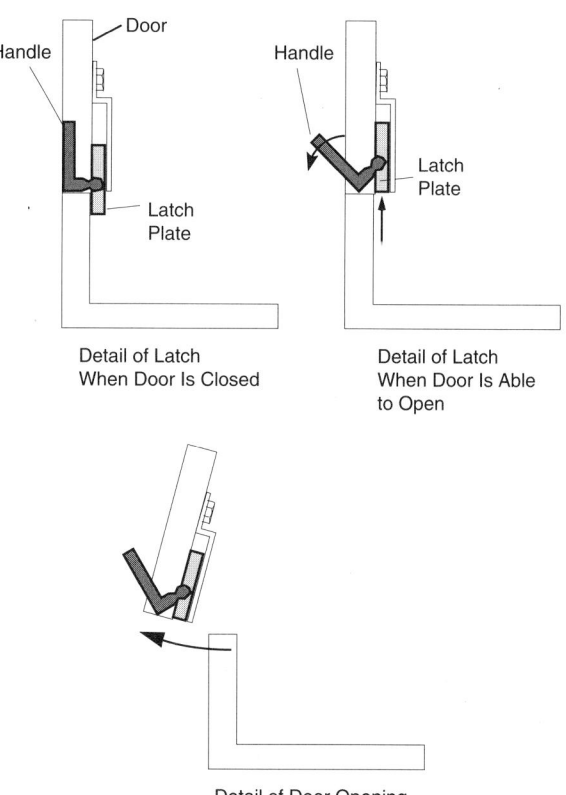

FIGURE 10-49. Detail of Latch When Door Is Closed. When the handle is pulled to the left, the latch plate is supposed to move up.

---

[4]While Dan was studying how the latch worked, another passenger leaned over and said, "Once an engineer, always an engineer!"

Dan got up and did a simple thing that permitted him to open the door effortlessly. Explain

- why the door is hard to open
- what Dan did and why it worked.

To help you answer this question, look at Figure 10-48 and Figure 10-49, which diagram the door, how it is hinged, and how the latch works.

10.I. FURTHER READING

[Arai and Kinoshita] Arai, T., and Kinoshita, N., "The Part Mating Forces That Arise When Using a Worktable with Compliance," *Assembly Automation,* August, pp. 204–210, 1981.

[Baumeister and Marks] Baumeister, T., and Marks, L. S., *Standard Handbook for Mechanical Engineers,* 7th ed., New York: McGraw-Hill, 1967.

[Dunne] Dunne, B. J., "Precision Torque Control for Threaded Part Assembly," M.S. thesis, MIT Mechanical Engineering Department, 1986.

[Gustavson, Selvage, and Whitney] Gustavson, R. E., Selvage, C. C., and Whitney, D. E., "Operator Member Erection System and Method," U.S. Patent 4,324,032, 1982.

[Goto et al.] Goto T., Takeyasu, K., and Inoyama, T., "Control Algorithm for Precision Insert Operation Robot," *IEEE Transactions on Systems, Man, and Cybernetics,* vol. SMC-10, no. 1, pp. 19–25, 1980.

[Nevins and Whitney] Nevins, J. L., and Whitney, D. E., editors, *Concurrent Design of Products and Processes,* New York: McGraw-Hill, 1989.

[Romanov] Romanov, G. I., "Preventing Thread Shear in Automatic Assembly," *Russian Engineering Journal,* vol. 44, no. 9, pp. 50–52, 1964.

[Simunovic] Simunovic, S., "Force Information in Assembly Processes," presented at the 5th International Symposium on Industrial Robots, Chicago, 1975.

[Sturges] Sturges, R. H., Jr., "A Three-Dimensional Assembly Task Quantification with Application to Machine Dexterity," *International Journal of Robotics Research,* vol. 7, no. 4, pp. 34–78, 1988.

[Whitney] Whitney, D. E., "Quasi-Static Assembly of Compliantly Supported Rigid Parts," *Transactions of the ASME, Journal of Dynamic Systems, Measurement, and Control,* vol. 104, pp. 65–77, 1982. This reference contains many other references to part mating theory.

10.J. APPENDIX: Derivation of Part Mating Equations

This appendix sketches the derivations of the basic equations for rigid part mating when the parts are supported by a support with a compliance center. More detail may be found in [Whitney]. The derivations presume that the compliance center is located on the peg's axis an arbitrary distance L_g from the tip of the peg. Chamfer crossing, one-point contact, and two-point contact will be described. The derived equations and computer program treat the case where lateral error and angular error are both positive as shown in Figure 10-9.

10.J.1. Chamfer Crossing

Refer to Figure 10-50, which shows a peg during chamfer crossing and the forces on it.

The compliant support contributes the applied forces, expressed as F_x, F_z, and M at the tip of the peg. The contact between peg and chamfer provides the reaction forces.

The support forces are found by determining how far the compliances described by K_x and K_θ have been deflected. The initial lateral displacement of the support point with respect to the hole's axis is given by U_0:

$$U_0 = \varepsilon_0 + L_g\theta_0 \qquad (10\text{-}15)$$

When $U = U_0$ and $\theta = \theta_0$, both compliances are relaxed. As chamfer crossing proceeds, U and θ are related by

$$U = L_g\theta - z/\tan\alpha + \varepsilon_0 \qquad (10\text{-}16)$$

where α is defined in Figure 10-9. To find U and θ separately, we have to solve for the forces and moments. Writing equilibrium equations between the applied forces and contact forces yields

$$\begin{aligned} f_1 &= f_N\mathbf{B} \\ f_2 &= f_N\mathbf{A} \end{aligned} \quad \text{(contact forces)} \qquad (10\text{-}17)$$

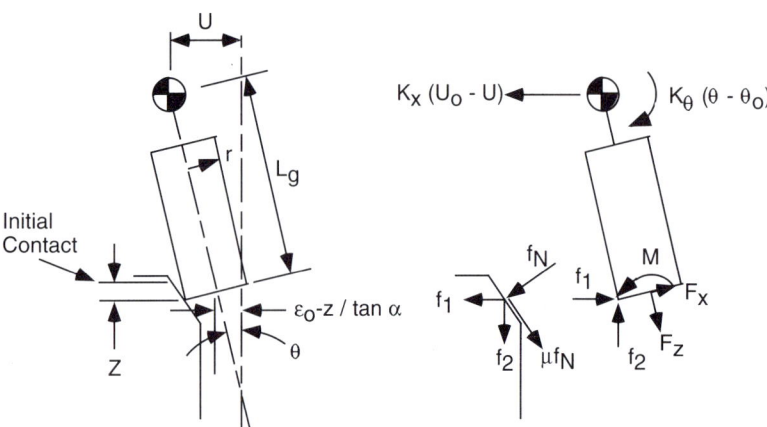

FIGURE 10-50. *Left:* Geometry of Chamfer Crossing. *Right:* Forces During Chamfer Crossing.

where

$$A = \cos\alpha + \mu\sin\alpha$$
$$B = \sin\alpha - \mu\cos\alpha \qquad (10\text{-}18)$$

and

$$F_x = -K_x(U_0 - U) \qquad \text{(compliant support}$$
$$M = K_x L_g(U_0 - U) - K_\theta(\theta - \theta_0) \quad \text{forces)} \qquad (10\text{-}19)$$

$$\left.\begin{array}{l} F_x = -f_1 \\ F_z = f_2 \\ M = rf_2 \end{array}\right\} \begin{array}{l}\text{(equilibrium of contact} \\ \text{forces and compliant support} \\ \text{forces assuming } \theta \text{ is} \\ \text{negligibly small)} \end{array} \quad (10\text{-}20)$$

Combining the above equations yields expressions for U and θ during chamfer crossing:

$$\theta = \theta_0 + \frac{K_x(z/\tan\alpha)(L_g B - rA)}{(K_x L_g^2 + K_\theta)B - K_x L_g rA} \qquad (10\text{-}21)$$

and

$$U = U_0 - \frac{K_\theta(z/\tan\alpha)B}{(K_x L_g^2 + K_\theta)B - K_x L_g rA} \qquad (10\text{-}22)$$

10.J.2. One-Point Contact

The forces acting during one-point contact are shown in Figure 10-51. A derivation analogous to that for chamfer crossing begins with the geometric constraint

$$U = cR + L_g\theta - \ell\theta \qquad (10\text{-}23)$$

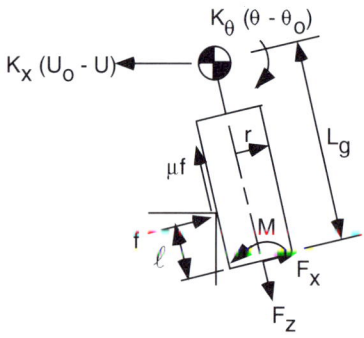

FIGURE 10-51. Forces Acting During One-Point Contact.

and yields

$$\theta = \frac{C(\varepsilon_0' + L_g\theta_0) + K_\theta\theta_0}{C(L_g - \ell) + K_\theta} \qquad (10\text{-}24)$$

and

$$U = U_0 - \frac{K_\theta(\varepsilon_0' + \ell\theta_0)}{C(L_g - \ell) + K_\theta} \qquad (10\text{-}25)$$

where

$$C = K_x(L_g - \ell - \mu r) \qquad (10\text{-}26)$$

and

$$\varepsilon_0' = \varepsilon_0 - cR \qquad (10\text{-}27)$$

10.J.3. Two-Point Contact

Whereas during chamfer crossing and one-point contact we needed to find the forces before we could find U and θ, the reverse is true during two-point contact. We find U and

θ via geometric compatibility

$$R = \frac{\ell}{2}\tan\theta + r\cos\theta \qquad (10\text{-}28)$$

which reduces to Equation (10-2) when θ is small. The relation between U, U_0, θ, θ_0, and ε_0 during one-point contact is obtained by combining Equation (10-15) and Equation (10-23):

$$U_0 - U = \varepsilon'_0 + L_g(\theta_0 - \theta) + \ell\theta \qquad (10\text{-}29)$$

If Equation (10-2) is substituted for θ in Equation (10-29), we obtain the corresponding relation for two-point contact:

$$U_0 - U = \varepsilon''_0 + L_g(\theta_0 - \theta) \qquad (10\text{-}30)$$

where

$$\varepsilon''_0 = \varepsilon_0 + cR \qquad (10\text{-}31)$$

A force analysis based on the right side of Figure 10-18 may be used to determine when two-point contact begins and ends. The result, simplified for the case where $K_\theta \gg K_x L_g^2$ and $K_\theta\theta_0 \gg \mu K_x \varepsilon''_0 r$ is

$$\ell_2 = \frac{cD}{\theta_0} \qquad (10\text{-}32)$$

for the onset of two-point contact, and

$$\ell'_2 \cong \frac{K_\theta\theta_0}{K_x\varepsilon''_0} - \ell_2 \qquad (10\text{-}33)$$

for the termination of two-point contact and the start of line contact. The values of θ at which these events occur may be obtained by substituting and Equation (10-32) and Equation (10-33) into Equation (10-2).

10.J.4. Insertion Forces

Insertion force during chamfer crossing is obtained by substituting Equation (10-20) and Equation (10-21) and into Equation (10-19) and Equation (10-20) to yield

$$F_z = \frac{K_x K_\theta \mathbf{A}(z/\tan\alpha)}{\mathbf{BD} - \mathbf{E}} \qquad (10\text{-}34)$$

where

$$\mathbf{D} = K_x L_g^2 + K_\theta \qquad (10\text{-}35)$$

and

$$\mathbf{E} = K_x L_g r \mathbf{A} \qquad (10\text{-}36)$$

Equations for lateral force and moment are derived similarly.

Insertion force during one-point contact is obtained analogously by substituting Equation (10-23) and Equation (10-24) into Equation (10-19) and Equation (10-20) to yield

$$F_z = \frac{\mu K_x K_\theta(\varepsilon'_0 + \ell\theta_0)}{\mathbf{C}(L_g - \ell) + K_\theta} \qquad (10\text{-}37)$$

Again, lateral force and moment may be obtained analogously.

To derive the forces and moments during two-point contact, we begin by writing the force and moment equilibrium equations between the reaction forces and the support forces expressed in peg-tip coordinates:

$$F_x = f_2 - f_1 \qquad (10\text{-}38)$$

$$F_z = (f_1 + f_2) \qquad (10\text{-}39)$$

$$M = f_1\ell - \mu r(f_2 - f_1) \qquad (10\text{-}40)$$

These may be combined to yield

$$\frac{M}{rF_z} = \pm\lambda - \frac{F_x}{F_z}\mu(1 + \lambda) \qquad (10\text{-}41)$$

where

$$\lambda = \frac{\ell}{2r\mu} \qquad (10\text{-}42)$$

The two equations in Equation (10-41)—one each for the plus sign and the minus sign—form the diagonal lines of the parallelogram in Figure 10-21. Substituting Equation (10-29) and Equation (10-2) into Equation (10-19) and Equation (10-20) yields

$$F_x = -K_x L_g(\theta_0 - cD/\ell) - K_x\varepsilon''_0 \qquad (10\text{-}43)$$

and

$$M = \mathbf{D}(\theta_0 - cD/\ell) + K_x L_g\varepsilon''_0 \qquad (10\text{-}44)$$

Putting these into Equation (10-40) yields

$$F_z = \frac{2\mu}{\ell}[\mathbf{D}(\theta_0 - cD/\ell) + \mathbf{F}]$$
$$+ \mu\left(1 + \frac{\mu d}{\ell}\right)[\mathbf{G}(\theta_0 - cD/\ell) - \mathbf{F}/L_g]$$
$$(10\text{-}45)$$

where

$$\mathbf{F} = K_x L_g\varepsilon''_0 \qquad (10\text{-}46)$$

and

$$\mathbf{G} = -K_x L_g \qquad (10\text{-}47)$$

10.J.5. Computer Program

Table 10-2 contains the listing of a TRUE BASIC computer program that calculates and plots all of the variables discussed in this appendix. This program provided the "theory" lines in Figure 10-24 and Figure 10-25. The following is a brief discussion of how this program works and how the variable names in it correspond to names used in this chapter. The code for this program is on the CD-ROM that is packaged with this book as an "exe" that will run on most PCs.

The first few lines express input data, which may be stored in the program or typed in by the user. Such data include stiffnesses of the supports, clearance ratio between peg and hole, coefficient of friction, location of the support compliance center, and the initial lateral and angular errors. Note that $L_g = 0$ should not be used. To simulate small values for L_g, one may use $L_g = \mu r$. The program

TABLE 10-2. Listing of BASIC Program for Insertion Force

```
1000 REM PROGRAM FOR INSERTION FORCE
1010 REM BASED ON EQUATIONS IN THIS CHAPTER.
1020 REM THIS PROGRAM IS IN TRUE BASIC FOR THE MACINTOSH.
1030 REM VALUES OF COEFFICIENTS AND CONSTANTS ARE METRIC AND
1040 REM CORRESPOND TO EXPERIMENTAL DATA IN TABLE 10-1.
1050 REM
1060 REM PRELIMINARY CALCULATIONS
1065 LET SCALF = 2
1067 LET SCALD = 20
1070 LET SP$ = " "
1080 LET KX = 65
1090 LET KT = 2100
1100 LET D = .46
1110 LET C = .0013
1120 LET MU = .3
1130 LET E0 = .05
1140 INPUT PROMPT "TYPE SP FOR SCREEN PRINT, SG FOR SCREEN GRAPH":AN$
1150 IF (AN$ <> "SP") AND (AN$ <> "SG") THEN GOTO 1140
1160 INPUT PROMPT "INITIAL THETA ":T0
1170 IF AN$="SP" THEN PRINT "L FX FZ M1 F1"
1180 LET LG = 2
1190 LET KL = KX * LG
1200 LET A = KL * LG + KT
1210 LET U0 = E0 + C * D / 2
1220 LET B = KL * U0
1230 LET C1 = - KL
1240 LET AL = KX * (U0 + LG * T0)
1250 LET BE = AL * LG + KX * LG * C * D - AL * MU * D / 2 + KT * T0
1260 LET GA = (A - KX * LG * MU * D / 2) * C * D
1270 LET L2 = (BE - SQR (BE ^ 2 - 4 * AL * GA)) / (2 * AL)
1280 LET L4 = (BE + SQR (BE ^ 2 - 4 * AL * GA)) / (2 * AL)
1290 LET LT = (4 * A + 2 * C1 * MU * D) * C * D
1300 LET LB = 2 * A * T0 + B * (2 - MU * D / LG) + C1 * (T0 * MU * D - C * D)
1310 LET LS = LT / LB
1320 LET L = LS
```

TABLE 10-2. (Continued)

```
1321 let ep = e0-cd/2
1322 LET AA = .707 * (1 + MU)
1324 LET BB = .707 * (1 - MU)
1326 LET FC = KX * KT * EP * AA / ((KX * LG ^ 2 + KT) * BB - KX * LG * D * AA / 2)
1330 GOSUB 2060
1340 LET FM = FZ
1350 IF AN$="SP" THEN GOTO 1500
1360 REM
1370 REM PLOT AXES
1380 SET WINDOW -1,4*L2*SCALD + 2,-4,FM +10
1390 PLOT 0,0; 0,FM+1
1400 PLOT 0,0; 4*L2*SCALD,0
1410 FOR X = 0 TO 4*L2*SCALD STEP 4*L2*SCALD/12
1420   PLOT TEXT, AT X,-1:STR$(INT(10*(X+.5))/10)
1430 NEXT X
1435 LET AX$="INSERTION DEPTH * " & STR$(SCALD)
1440 PLOT TEXT, AT X/2.5, -2: AX$
1450 FOR Y = 0 TO FM + .5 STEP FM/8
1460   PLOT TEXT, AT .2,Y: STR$(INT(100*Y+.5)/100)
1470 NEXT Y
1475 LET AY$ = "FORCE * " & STR$(SCALF)
1480 PLOT TEXT, AT .2,Y+.2: AY$
1490 REM
1500 REM BEGIN MAIN CALCULATION LOOP
1510 FOR L = 0 TO 4*L2 STEP L2/40
1515   if L < 2*L2/40 then plot L*scald,FC*scalf;
1520   IF L >= L2 THEN GOTO 1650
1530   IF L > L4 THEN
1540     PRINT "TWO POINT CONTACT LOST"
1550     GOTO 1800
1560   END IF
1570   LET A1 = KX * (LG - L - MU * D / 2)
1580   LET B1 = A1 * LG + KT
1590   LET EP = E0 - C * D / 2
1600   LET FZ = MU * KX * KT * (EP + L * T0) / (B1 - A1 * L)
1610   LET FX = - FZ / MU
1620   LET M1 = - FX * (L + MU * D / 2)
1630   LET F1 = - FX
1631   LET FZ=FZ*SCALF
1632   LET FX=FX*SCALF
1633   LET M1=M1*SCALF
1634   LET F1=F1*SCALF
1635   LET FZ1=FZ
1640   GOTO 1660
1650   GOSUB 2060
```

(continued)

TABLE 10-2. (Continued)

```
1655    IF FZ < FZ1 AND L < L2 + L2/40 THEN LET FZ = FZ1
1660    IF AN$ = "SP" THEN GOTO 1710
1670    PLOT L*SCALD,FZ;
1680    GOTO 1700
1690    PLOT L,FX;  !CHOOSE THIS ONE TO PLOT FX INSTEAD OF FZ
1700    IF AN$ = "SG" THEN GOTO 1770
1710    LET L = INT (10000 * L) / 10000
1720    LET FX = INT (1000000 * FX) / 1000000
1730    LET FZ = INT (1000000 * FZ) / 1000000
1740    LET M1 = INT (100 * M1) / 100
1750    LET F1 = INT (100 * F1) / 100
1760     PRINT L; SP$;FX; SP$;FZ; SP$;M1; SP$;F1
1770 NEXT L
1780 IF AN$ = "SG" THEN GOTO 2040
1790 REM
1800 REM SUMMARY PRINTOUT OF PARAMETERS
1810 LET LL = LG - C*D/2
1820 LET TC = T0 + KX * LL * EP / (KX * LG * LL + KT)
1830 PRINT "TC= ";TC
1840 LET EPP = E0 + C*D/2
1850 LET LLL = LG - L2 - MU * D / 2
1860 LET T2 = KX * EPP * LLL / (KX * LG * LLL + KT) + T0
1870 PRINT "T2= ";T2
1880 LET AA = .707 * (1 + MU)
1890 LET BB = .707 * (1 - MU)
1900 LET FC = KX * KT * EP * AA / ((KX * LG ^ 2 + KT) * BB - KX * LG * D * AA / 2)
1910 PRINT "FC= ";FC
1920 REM FC IS PEAK CHAMFER FORCE
1930 PRINT "L2= ";L2
1940 PRINT "LS= ";LS
1950 PRINT "FM= ";FM
1960 PRINT "KX= ";KX
1970 PRINT "KT= ";KT
1980 PRINT "LG= ";LG
1990 PRINT "MU= ";MU
2000 PRINT "C= ";C
2010 PRINT "E0= ";E0
2020 PRINT "T0= ";T0
2030 PRINT "D= ";D
2032 PRINT "SCALF= ";SCALF
2034 PRINT "SCALD= ";SCALD
2040 STOP
2050 REM
2060 REM SUBROUTINE TO CALCULATE FORCE DURING TWO POINT CONTACT
2070 LET LD = L / D
```

TABLE 10-2. (Continued)

```
2080 LET TT = LD - SQR (LD ^ 2 - 2 * C)
2090 LET M1 = A * (T0 - TT) + B
2100 LET FX = C1 * (T0 - TT) - B / LG
2110 LET FZ = 2 * MU * M1 / L + MU * FX * (1 + MU * D / L)
2120 LET F2 = M1 / L + FX * (1 + .5 * MU * D / L)
2121 LET M1=M1*SCALF
2122 LET FX=FX*SCALF
2123 LET FZ=FZ*SCALF
2124 LET F2=F2*SCALF
2130 IF (L > L2 + 2) AND (F2 <= 0) THEN
2140   PRINT "TWO POINT CONTACT LOST"
2150   GOTO 1800
2160 END IF
2170 LET F1 = (M1 + .5 * MU * D * FX) / L
2171 LET F1=F1*SCALF
2180 RETURN
2190 END
```

TABLE 10-3

Program Names	Equation Names
KX	K_x
KT	K_θ
MU	μ
EO	ε_0
TO	θ_0
LG	L_g
A	D
UO	ε_0''
B	F
C1	G
L2	ℓ_2
L4	ℓ_2'
LS	ℓ^*
FM	F_z max during two-point contact
A1	C
EP	ε_0'
F1	f_1 (contact force)
F2	f_2 (contact force)
FX, FZ, M1	F_x, F_z, M
TC	θ at end of chamfer crossing
T2	θ when two point contact begins
AA, BB	A, B for $\alpha = 45°$
FC	max insertion force at end of chamfer crossing

asks the user to choose between text and graphical output and to choose an initial angular error in radians.

Next is a short routine that plots axes on the screen.

The next few lines compute ℓ^* and F_m, the depth at which maximum insertion force occur and that force. The values of ℓ_2 and ℓ_2' where two-point contact begins and possibly ends are also computed here.

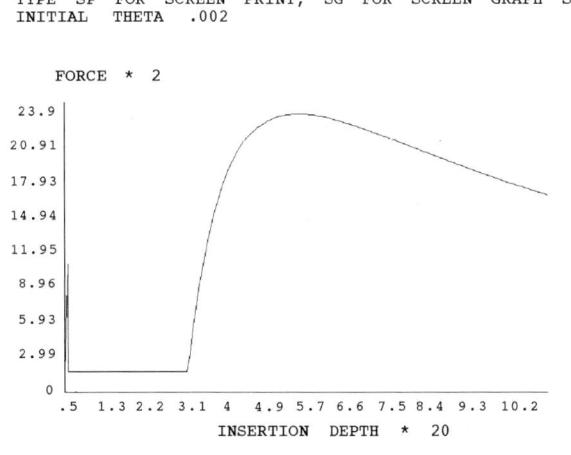

FIGURE 10-52. Sample Output from BASIC Program for Insertion Force.

The main program loop is next, stepping through values of insertion depth ℓ from zero to 4 times the predicted ℓ_2. The first part of this loop calculates insertion forces during one-point contact. When insertion depth exceeds ℓ_2, the corresponding values for two-point contact are calculated. At the end of each pass, values are printed and plotted.

The last part of the program calculates two values associated with chamfer crossing: the value of θ just at the end, where one-point contact begins, and the insertion force at that point. One may assume that chamfer crossing force and angle θ each increase linearly during chamfer crossing, with force starting at zero and θ starting at θ_0.

Finally, there is a summary printout that repeats input data.

Correspondence of variable names is given in Table 10-3.

Figure 10-52 is a sample of graphic output from this program.

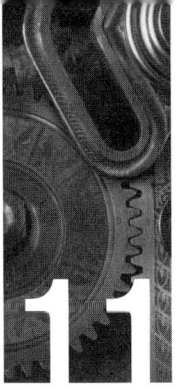

11 ASSEMBLY OF COMPLIANT PARTS

"Mating one pin and socket isn't so hard. Mating 100 at once can require a hydraulic jack."

11.A. INTRODUCTION

Chapter 10 dealt with compliantly supported rigid parts entering rigid holes. It was shown that insertion force and two different failure modes depended on three basic factors: geometry, compliance, and friction. The main design parameter at our disposal was shown to be the location of the compliance center of the rigid part's support.

Different problems and design opportunities arise when at least one mating part is compliant. In particular, we shall see in this chapter that the shape of at least one mating surface can be varied so as to greatly affect the mating force.[1] The first part of the chapter presents analytical models for the main physical phenomena. The second part develops those models for several general cases. The third part focuses specifically on opportunities for designing mating surfaces, while the last part presents experimental verifications of the theory.

Figure 11-1 exhibits numerous applications of compliant parts, including electric connectors, door latches, snap fits, and light bulb sockets. Figure 11-2 shows two simplified geometries that contain the elements considered in this chapter. Compliant sheet metal parts are not treated in this chapter.

11.A.1. Motivation

Compliant part mating is interesting both theoretically and practically. The theoretical issues are similar to those of rigid parts in the sense that the same factors dominate the mating behavior: geometry, compliance, and friction.

However, because the parts are compliant in some places, it is difficult to generate high enough forces to cause wedging to occur. At the same time, it is still possible to observe phenomena similar to jamming. Such events arise during chamfer crossing, when an entering part can become stuck against the chamfer. If the parts are delicate, as is the case with electrical connectors, the insertion force can build up to the point where the parts are damaged or destroyed. Theoretical models of compliant part mating can be used as design guides to achieve desirable assembly features and avoid part damage and excessive mating force.

From a practical point of view, compliant part mating typically involves substantial insertion force. This force can act for good or ill, depending on the situation. Since many compliant part mates are accomplished with bare hands, the amount of force needed cannot be so high that assembly becomes impossible. This can happen with electrical connectors, especially if, as is common, twenty-five, fifty, or more pins must be mated to sockets simultaneously. In electrical connectors, this mating force arises from the need to spread apart portions of the socket elastically because a compressed socket is necessary in order to attain high enough contact force between pin and socket to reduce the contact resistance and allow the connector to function electrically.

Situations like this arise in many compliant part mating situations: Too large mating force will prevent assembly or cause damage, while too small mating force will prevent the item from functioning properly when assembled. Theory can come to the rescue here, permitting the engineer to design the parts so that both needs can be met.

---

[1]This chapter is based on Chapter 6 of [Nevins and Whitney].

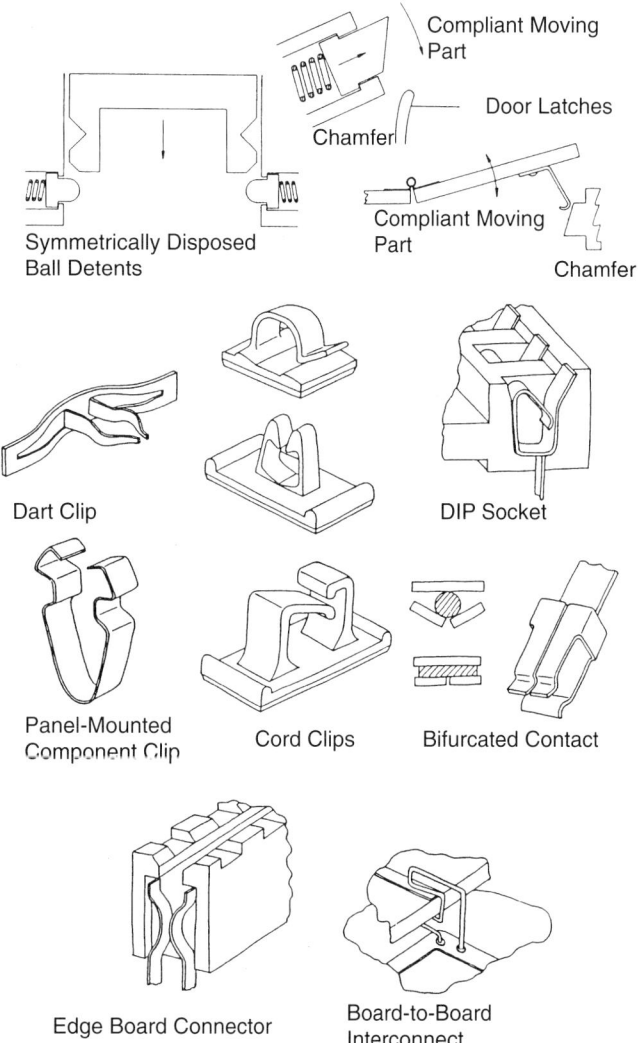

Compliant Moving Part

Door Latches

Chamfer

Compliant Moving Part

Chamfer

Symmetrically Disposed Ball Detents

Dart Clip

DIP Socket

Panel-Mounted Component Clip

Cord Clips

Bifurcated Contact

Edge Board Connector

Board-to-Board Interconnect

FIGURE 11-1. Examples of Compliant Parts. Shown here are door latches, clamps, and electrical connectors. The geometries look superficially different but all can be modeled mathematically, and the equations are similar in all the cases.

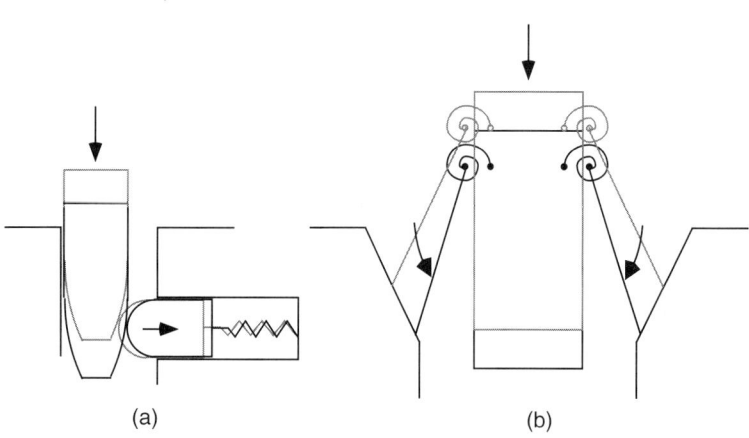

(a)

(b)

FIGURE 11-2. Models of Compliant Part Mating. (a) Rigid peg and compliant hole. A single compliant wall is shown, but both walls may be modeled as compliant if desired. Both peg and hole mating surfaces are shown with shapes that may be represented mathematically. Different shapes give different insertion force behavior. (b) Rigid wall and compliant peg. The peg is modeled as having two compliant sides, but one side may be modeled as rigid if desired. The hole has a straight chamfer shape while the peg's compliant elements are shown as lines that make a point contact with the hole. The chamfer may be given a shape as shown in (a) if desired.

This chapter will explore two theoretical conditions:

a. Rigid peg, compliant hole (Figure 11-2a)

b. Compliant or compliantly held peg, rigid chamfered hole (Figure 11-2b)

Before considering the theory, we will take a look at a practical example.

11.A.2. Example: Electrical Connectors

Figure 11-3 shows some typical shapes of electrical connectors. The pins may have a variety of nose shapes, with spherical being the most common and easy to fabricate. However, as we shall see later in this chapter, tapered noses can greatly reduce the insertion force. Several socket designs are also shown. The compliant element is the electrical contact spring. The socket is manufactured in such a way that the contact spring interferes with the pin as the pin is inserted and deflects the contact spring elastically. This guarantees that there is a residual contact force when assembly is finished. This contact force ensures a good electrical contact. Connectors that carry large amounts of current have thick material sections to reduce their electrical resistance. The contacts typically behave as cantilever beams, and their stiffness is governed by their thickness and length. Thus the contact springs in high-current connectors tend to be quite stiff, and the insertion forces can

be large. A critical design challenge is to achieve high contact force while avoiding high insertion force. Since the same element, the contact spring, is responsible for both phenomena, this goal would appear to be out of reach. In fact, however, it is largely achievable, and we will show in this chapter how to address it.

Figure 11-4 illustrates compliant part mating events for some of the connector pins and sockets shown in Figure 11-3. The insertion force comes from the axial components of contact and friction forces. The friction force is proportional to the contact force, which in turn is proportional to the lateral stiffness and deflection of the contact spring. The amount of insertion force generated also depends on the coefficient of friction and the angle of the surfaces at the pin–spring contact point. Figure 11-4 also illustrates a design in which the insertion process does not go as desired but instead buckles the contact spring.

The remainder of this chapter is devoted to modeling the geometry and force characteristics of parts like these so that their shapes can be designed to achieve desired insertion force and contact force behavior.

Significantly more complex conditions have been solved: compliant peg/compliant hole and minimum energy chamfers in [Hennessey] and a three-dimensional part mating theory in [Gustavson]. The interested reader should consult these references.

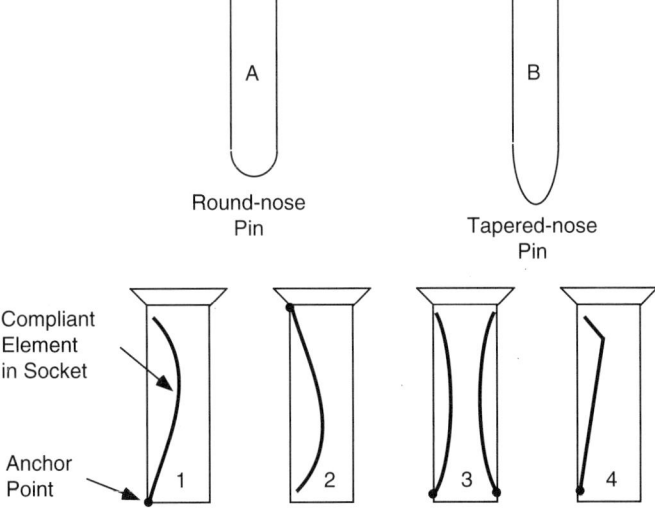

FIGURE 11-3. Examples of Real Compliant Parts. These are schematic drawings of electrical pins and sockets. Each will display very different insertion force versus insertion distance behavior. Some versions of shapes 1, 3, and 4 run the risk of the pin jamming on the chamfer of the compliant element in the socket. Real connector pins have diameters ranging from 1 to 4 mm and lengths ranging from 4 to 10 mm.

A

Round-nose Pin

B

Tapered-nose Pin

Compliant Element in Socket

Anchor Point

1   2   3   4

Sockets with Differently Shaped and Anchored Compliant Elements

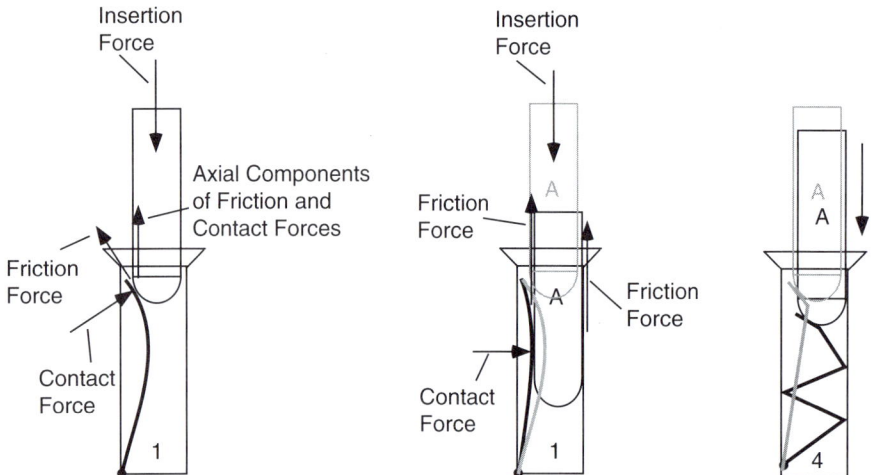

FIGURE 11-4. Schematic Diagram of Insertion Motions and Forces in Electrical Connectors. *Left:* When the pin touches the flexible spring inside the socket, a contact force and a friction force arise. The axial components of these forces are felt as insertion force. *Center:* In a normal successful compliant pin–socket assembly, the pin deflects the spring to one side and enters the socket. The spring is compressed in the final assembled state, giving rise to a contact force that provides firm electrical contact. *Right:* The pin has jammed against the contact spring because the contact force was inside the friction cone. The spring buckled under this axial load and the socket has been destroyed.

11.B. DESIGN CRITERIA AND CONSIDERATIONS

11.B.1. Design Considerations

Compliant parts are designed to perform various functions in various environments. The parts may be mated by hand or machine. They may be delicate or rugged. The designer may, for example, wish easy insertion and difficult withdrawal, or may wish to signal incomplete mating by having the parts pop apart. There are so many criteria that we list only a few, involving the insertion force (force in the direction of insertion) or withdrawal force.

1. Avoid sharp discontinuities in force versus insertion depth.
2. Minimize mechanical work during insertion.
3. Minimize the peak value attained by the insertion force during insertion.
4. Achieve a specific pattern of force versus depth.
5. Achieve a specific ratio of insertion force to withdrawal force.

A number of design features influence insertion force:

1. Peg nose shape
2. Number of springs (compliant members) making up the compliant hole

3. Entry shape of the spring
4. Speed of entry (quasi-static or dynamic)
5. Type of spring deflection (linear, nonlinear)
6. Spring preload
7. Rigid, compliant, or compliantly held rigid pegs
8. Straight or tilted initial entry of the peg into the hole

The most influential feature is the shape of the contacting surfaces. These surfaces are typically the tips of pegs and the mouths of holes.

Four basic types of insertion force behavior have been identified. Each corresponds to a particular type of mating surface shape. The shapes could be on either the peg or the hole.

a. Linear shape (Figure 11-5a)—This is the most common and provides linear force versus depth behavior. The maximum force occurs at the end of insertion and could be very high.

b. Convex shape (Figure 11-5b)—Making the surface convex allows shaping of the force versus depth curve. Various geometrical forms have been tried including circular arcs, parabolas and logarithmic curves.

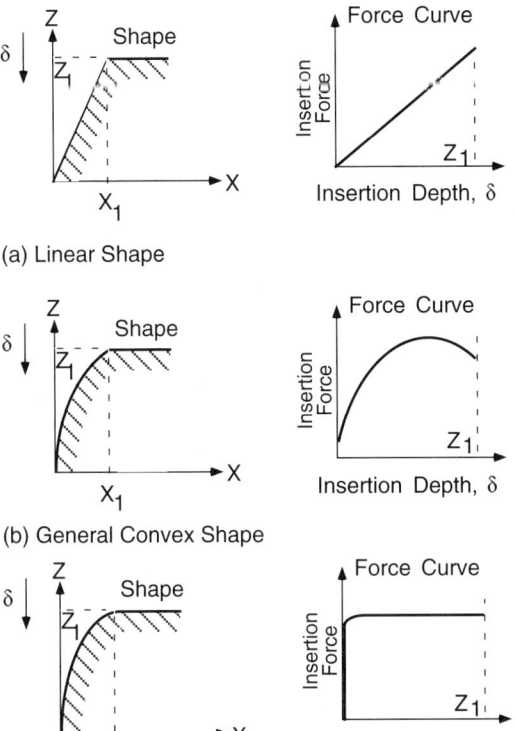

(a) Linear Shape

(b) General Convex Shape

(c) Minimum Peak Force (Constant Force) Shape

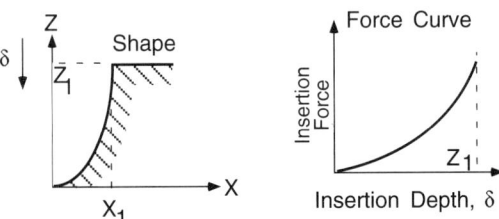

(d) General Concave Shape

FIGURE 11-5. Insertion Force Versus Insertion Depth for Four Generic Chamfer Shapes.

c. Constant force shape (Figure 11-5c)—A particular convex shape with a complex descriptive equation can produce constant insertion force throughout the insertion. This behavior results in minimum peak force.

d. Concave shape (Figure 11-5d)—Reversing the arc provides less force for beginning depths but very large forces near the end.

11.B.2. Assumptions

Table 11-1 lists the assumptions used in this chapter. The comments indicate extensions that might be added.

11.B.3. General Force Considerations

While a significant variety of mathematical models for compliantly held peg into hole or peg into compliant hole can be created, the complexity lies in describing the orientation geometry and the elastic behavior of the compliant elements. If we focus on the peg/compliant member interface at any instant during part mating (see Figure 11-6), we may write the following basic equations for the insertion and lateral forces F_z and F_x (subscript I = insertion, W = withdrawal) acting on the peg during insertion and withdrawal in terms of the normal contact force F_N and interface angle ϕ.

$$
\begin{aligned}
F_{x_I} &= F_N(\sin\phi - \mu\cos\phi) \\
F_{z_I} &= F_N(\cos\phi + \mu\sin\phi) \\
F_{x_W} &= F_N(\sin\phi + \mu\cos\phi) \\
F_{z_W} &= F_N(\cos\phi - \mu\sin\phi)
\end{aligned}
\tag{11-1}
$$

Ratios of F_z to F_x versus ϕ are plotted in Figure 11-7 for the case of insertion and in Figure 11-8 for the case of withdrawal. These figures show that the ratio of insertion force to lateral force during insertion is larger for larger coefficient of friction and smaller for larger interface angle. During withdrawal, the ratio is again smaller for larger angle but smaller for larger friction coefficient. Note that for straight entry shapes, the angle is constant, whereas for curved shapes the angle changes during insertion.

These figures show that three factors control the mating forces of compliant peg–hole combinations:

1. The normal (or contact) force

2. The slope at the interface point where $\phi = \tan^{-1}(slope)$

3. The friction coefficient μ

Establishing and controlling these three factors is fundamental to compliant peg–hole design. They are defined as follows:

11.B.3.a. The Normal Force

The normal force, which produces the insertion force, as given by Equation (11-1), is created by deflection of the compliant member(s). Certain peg–hole combinations

TABLE 11-1. Assumptions for Analysis of Compliant Part Mating

Assumptions	Comment
1. Two-dimensional cases only.	1. Third dimension could be added.
2. Peg travels along the centerline of hole.	2. Lateral and angular misalignments can be added as can inclined approach paths.

3. Peg has prescribed lateral position.
  a. For one compliance—one rigid wall case: peg rides along wall opposite compliance (see Figure 11-9a and Figure 11-12a).
  b. For two equal symmetrical compliance case: peg and "hole" centerlines are coincident (see Figure 11-9b and Figure 11-12b).
	3. Alternately:
	  a. Peg–wall contact may not occur at all.
	
	  b. Compliance may not be shape-symmetric or have equal stiffnesses. Only one compliance may be contacted by peg.

4. Compliant elements are an integral part of the mounting; friction coefficient is uniform on all contacting surfaces.	4. Compliant elements could be separate parts made of different materials requiring specification of two or more friction coefficients.
5. Deflection of the compliance(s) is rigid body motion with respect to a single point, with compliance concentrated at that point.	5. Small deflection beam theory or large deflection theory can be added.
6. Spring has no preload at initial contact with peg.	6. Spring preload can be added.
7. Conditions are quasi-static; motion does not create need for dynamic considerations.	7. Dynamics may play a role in compliant part mating.

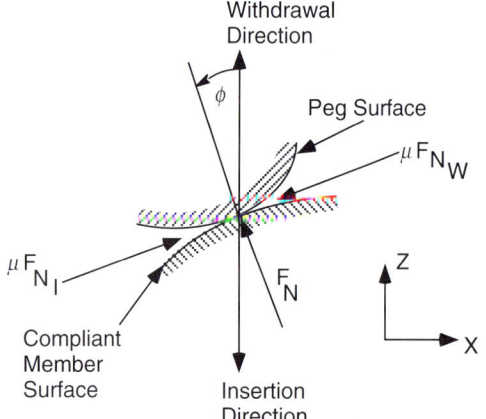

FIGURE 11-6. Definition of Forces and Directions During Insertion and Withdrawal in Compliant Part Mating. During insertion, the contact force generates the friction force μF_{N_I} on the compliant member, whereas during withdrawal it generates the friction force μF_{N_W}.

contain a spring whose action can be analyzed only by large-deflection (nonlinear) theory; they are not included here. For "small" deflections, two types of behavior are possible:

1. Elastic deflection

2. Rigid body motion with respect to a single point where all the compliance is concentrated

This chapter analyzes only the second type; the mathematics is considerably less complex while agreement with experimental results is good.

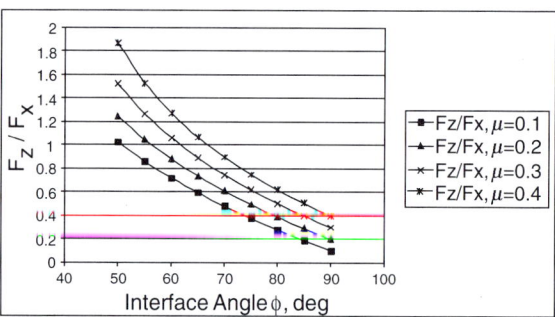

FIGURE 11-7. Ratio of Insertion Force to Lateral Force During Insertion Based on Equation (11-1). As the coefficient of friction μ increases, the ratio of insertion force to lateral force rises. The ratio falls as the interface angle ϕ rises.

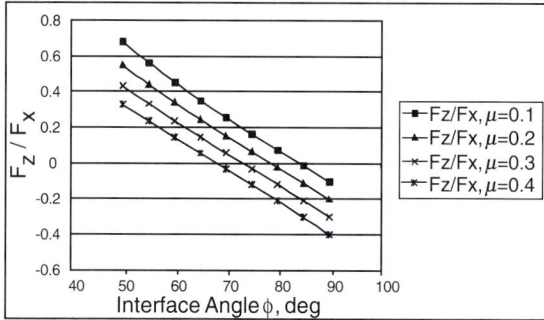

FIGURE 11-8. Ratio of Insertion Force to Lateral Force During Withdrawal Based on Equation (11-1).

11.B.3.b. Slope at the Peg/Spring Interface

The normal force at the point of contact is oriented along the line joining the center of curvature of the peg's surface and the center of curvature of the compliant member's surface; the slope angle at the contact point is ϕ. Depending upon what insertion force versus depth characteristics are required, the designer may alter the interface slope by increasing or decreasing the angle ϕ, or making it variable by using a curved interface surface. Numerous easily definable shapes may be investigated: straight lines, circles, conic sections, parabolic curves, logarithmic curves, and so on. In general, the slope exerts a highly nonlinear effect on insertion force, as may be seen from Figure 11-7. The

reason for this will become apparent when we derive the equations in the next section.

11.B.3.c. Friction

Friction interacts with slope to produce nonlinear effects, as may be seen in Figure 11-7. It can be controlled by lubrication or surface plating. It is modeled very simply for our purposes as a constant throughout a mating event; this has been found experimentally to be quite accurate.

11.C. RIGID PEG/COMPLIANT HOLE CASE

11.C.1. General Force Analysis[2]

The mating of a rigid peg and a compliant hole is analyzed first. The analysis is based on the assumptions in Table 11-1. Two fundamental compliant wall conditions (see Figure 11-9) are considered. Each models the compliant wall as a spring-loaded member with a circular nose, which is constrained to move only in a direction normal to the insertion direction of the peg (assumed to be along the centerline). The nose radius is r and the spring stiffness is K_x. Numerous other models of mating shape and spring behavior could be used by extrapolating from these models.

The end of the peg is modeled as a circular arc with an arbitrary radius (see Figure 11-10). The arc is tangent to the side of the peg. This arc is the primary contact region with the compliant wall(s) during insertion and withdrawal.

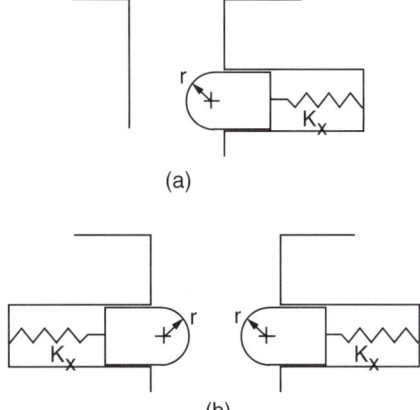

(a)

(b)

FIGURE 11-9. Models of Compliant Walls. *Top:* One rigid wall, one compliant wall. *Bottom:* Two symmetric compliant walls.

11.C.1.a. Forces on Compliant Member

We begin the force analysis by showing the forces on the compliant member. The peg contact point (see Figure 11-11) occurs at angle ϕ which requires the normal force to pass through the center of the radius at the end of the compliant member. The friction forces will be in the direction shown for insertion or in the opposite direction for withdrawal. Thus, as long as friction is present, the insertion normal force is different from the withdrawal normal force. The normal force arises due to the deflection $(L_0 - L)$ of the spring. For insertion we find that the normal force is given by Equation (11-2):

$$F_{N_I} = \frac{K_x(L_0 - L)}{(1 - \mu^2)\sin\phi - 2\mu\cos\phi} \qquad (11\text{-}2)$$

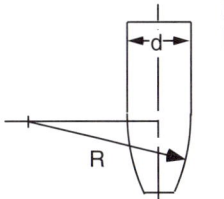

FIGURE 11-10. Definition Shape Parameters of Peg.

while for withdrawal, normal force is given by Equation (11-3):

$$F_{N_W} = \frac{K_x(L_0 - L)}{(1 - \mu^2)\sin\phi + 2\mu\cos\phi} \qquad (11\text{-}3)$$

These normal forces are fundamental to the determination of the insertion and withdrawal forces for compliant part mating. Each equation can be obtained from the other by reversing the sign of μ.

---

[2]Section 11.I contains derivations of the equations in this section.

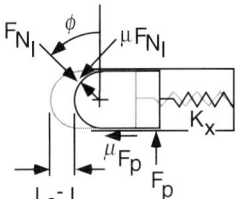

FIGURE 11-11. Forces on Compliant Member During Insertion.

These depend on the geometry of the parts and vary in general during mating. This topic is discussed next.

11.C.1.c. Geometry of Mating Parts

To determine $(L_0 - L)$ and ϕ, we must deal with the one-compliant-member case and the two-compliant-member case separately.

From either Figure 11-13 or Figure 11-14, we find

$$\cos \phi = \frac{Z}{R + r} \qquad (11\text{-}4)$$

From Figure 11-13, we find that the deflection of the compliant member is

$$L_0 - L = L_0 - (D - d) - R + (R + r) \sin \phi \quad (11\text{-}5)$$

for the first case, and from Figure 11-14 it is

$$L_0 - L = L_0 - \frac{D - d}{2} - R + (R + r) \sin \phi \quad (11\text{-}6)$$

for the second case. In both cases, L_0 is an independent feature of the geometry and must be given in advance.

Two other conditions of particular interest occur—initial contact condition and steady-state condition:

11.C.1.b. Forces on Rigid Peg

For present purposes, the peg is assumed to be rigid and constrained to move along a straight path which is coincident with the centerline of the "hole." Since either one or two compliant members could exert force on the peg, there are two separate cases to be defined (see Figure 11-12). The insertion and withdrawal forces are given in Table 11-2.

Although the terms in the table could be divided by $\cos \phi$ making the expressions easier to read, problems would arise in computation when $\phi \to \pi/2$ whereas the equations shown above are always usable.

In order to evaluate the insertion and withdrawal forces, separate calculations for $(L_0 - L)$ and ϕ are needed.

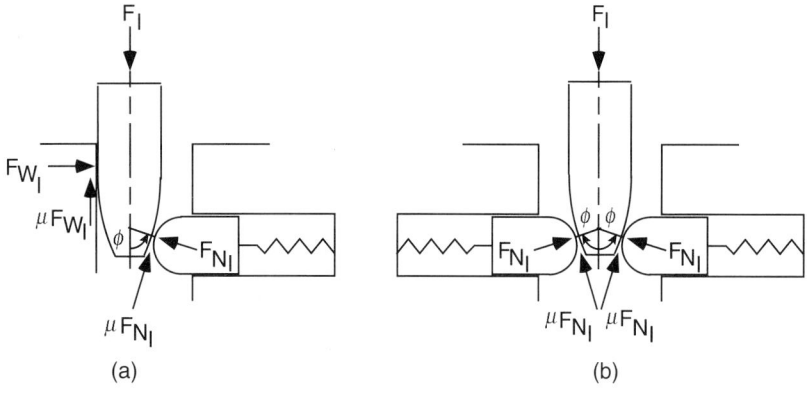

FIGURE 11-12. Forces on the Peg During Insertion. (a) One rigid wall, one compliant wall. (b) Two symmetric compliant walls.

(a)　　　(b)

TABLE 11-2. Formulas for Insertion Force and Withdrawal Force for the Single-Compliant-Member and Two-Compliant-Member Cases

	Insertion Force	Withdrawal Force
One rigid wall, one compliant member	$\dfrac{K_x(L_0 - L)[(1 - \mu^2)\cos \phi + 2\mu \sin \phi]}{(1 - \mu^2)\sin \phi - 2\mu \cos \phi}$	$\dfrac{K_x(L_0 - L)[(1 - \mu^2)\cos \phi - 2\mu \sin \phi]}{(1 - \mu^2)\sin \phi + 2\mu \cos \phi}$
Two symmetric compliant walls	$\dfrac{2K_x(L_0 - L)(\cos \phi + \mu \sin \phi)}{(1 - \mu^2)\sin \phi - 2\mu \cos \phi}$	$\dfrac{2K_x(L_0 - L)(\cos \phi - \mu \sin \phi)}{(1 - \mu^2)\sin \phi + 2\mu \cos \phi}$

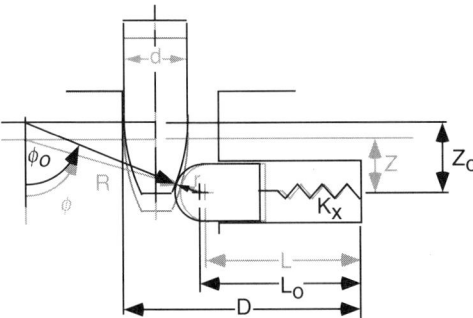

FIGURE 11-13. Geometry of Mating Parts for One Rigid Wall, One Compliant Member Case. The configuration shown in black is the case where the peg first touches the compliant member. The peg's insertion depth at this point is Z_0, and the deflection of the compliant member is L_0. The configuration shown in gray is the case where the peg is part way into the hole. The insertion depth is Z, and the deflection of the compliant member is L.

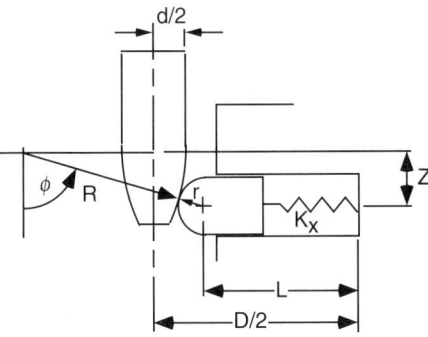

FIGURE 11-14. Geometry of Mating Parts for the Two Symmetric Compliant Member Case. Only one compliant member is shown for clarity. Here is shown only the case where the peg is part way into the hole. For this case, Z_0 and L_0 are defined analogously to the case shown in Figure 11-13.

1. Initial contact occurs when the peg, traveling along its path, first encounters the compliant member's surface, which has a prescribed initial location due to specification of L_0.

2. Steady-state condition occurs when the straight portion of the peg far from the tip contacts the compliant member (i.e., movement of the peg up or down does not cause changes in the magnitude of the forces.) This special case begins when $\phi = \pi/2$. In this state, the insertion force is constant.

Table 11-3 summarizes these two conditions.

11.C.1.d. Critical Geometry

The critical moment in mating in the geometry of Figure 11-12 is the first contact when the normal force has the largest component opposite to the insertion direction. This is the moment when $\phi = \phi_0$. The friction cone surrounding the contact force is most likely to contain the normal force at this instant. This event corresponds roughly to the event of jamming in rigid part mating. If this happens, the denominator of Equation (11-2) goes to zero. To avoid this, we need to satisfy

$$\phi > \phi_{min} = \tan^{-1}\left(\frac{2\mu}{1-\mu^2}\right) \qquad (11\text{-}7a)$$

or

$$\mu < \mu_{max} = \frac{-1 + \sqrt{1 + \tan^2\phi_0}}{\tan\phi_0} \qquad (11\text{-}7b)$$

If $\phi_0 = 30°$, for example, then Equation (11-7) yields $\mu_{max} = 0.268$. It is best to give this limit a wide margin, because the insertion force can become very large if the friction limit is approached.

TABLE 11-3. Equations for Initial Contact and Constant Force for Two Types of Compliant Part Mating

Geometry	Figure	Initial Contact Condition	Steady-State Force Condition
One rigid wall, one compliant member	Figure 11-13	$\phi_0 = \sin^{-1}\left[\dfrac{D - L_0 - d + R}{R + r}\right]$ $Z_0 = (R + r)\cos\phi_0$	$L_F = D - d - r$
Two symmetric compliant members	Figure 11-14	$\phi_0 = \sin^{-1}\left[\dfrac{\left(\frac{D-d}{2}\right) - L_0 + R}{R + r}\right]$ $Z_0 = (R + r)\cos\phi_0$	$L_F = \left(\dfrac{D-d}{2}\right) - r$

Note: Subscript 0 indicates initial contact. Subscript F indicates final constant force. Z_0 and ϕ_0 are the initial insertion depth and contact angle, respectively. L_F is the final deflection of the spring.

11.C.1.e. Numerical Example

A spreadsheet program can be used to represent Equation (11-2) through Equation (11-6) and the equations in Table 11-2 and Table 11-3. Example geometry is shown

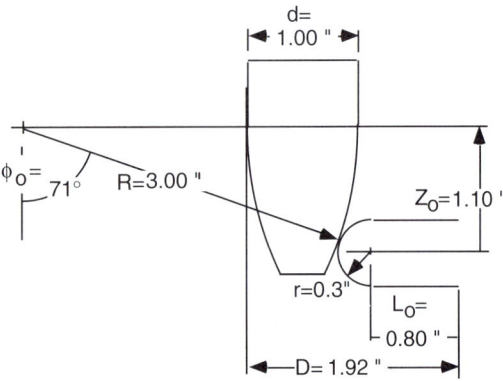

FIGURE 11-15. Example Peg–Hole Geometry.

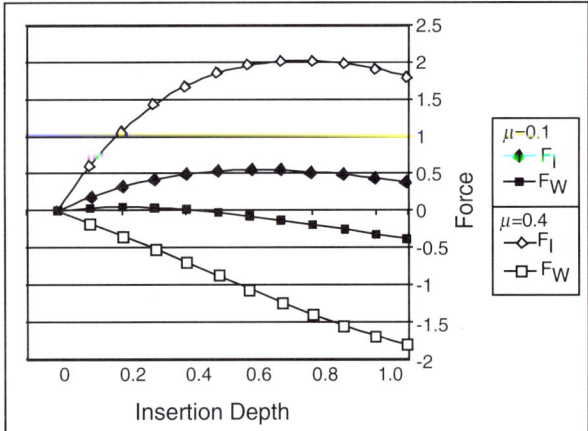

FIGURE 11-16. Insertion and Withdrawal Forces for the Geometry in Figure 11-15 for $\mu = 0.1$ and $\mu = 0.4$. For this geometry, the maximum value of μ is 0.706. Note that a positive value for withdrawal force means that the hole is effectively expelling the peg. This occurs for small μ when the peg is slightly inserted.

in Figure 11-15. The results are shown in Figure 11-16. Here we can see the effect of increasing the coefficient of friction. If R is increased, the peak insertion force is reduced. However, there is often limited space in the axial direction, and increasing R increases the distance in the axial direction between first encounter and steady-state conditions. Thus the design of each compliant mating situation involves a tradeoff among the variables describing geometry, compliance, and friction. The next section of this chapter investigates the problem of designing the entry shape (either on the peg or on the compliant member) so that low insertion force is obtained without increasing insertion depth or reducing steady-state contact force.

11.C.1.f. Electrical Connector

As a real example, let us study a small electrical connector. Drawings and a photo appear in Figure 11-17 through Figure 11-19. It is made by stamping intricate shapes into thin sheet metal and forming the shape into a three-dimensional object. Figure 11-20 shows approximately what the flattened shape of the socket is. These pins and sockets are so tiny that typical charms on a charm bracelet are huge

FIGURE 11-17. Photo of One Pin–Socket Pair of Example Electrical Connector. (Photo by the author.)

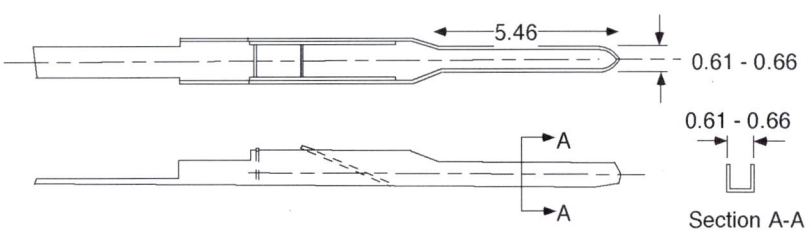

FIGURE 11-18. Top and Side View of Connector Pin. The pin is U-shaped and the tolerance range of dimensions shown applies across the pin in both directions. Dimensions are in millimeters. The material is 0.1 mm thick.

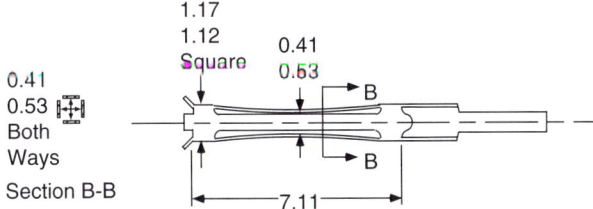

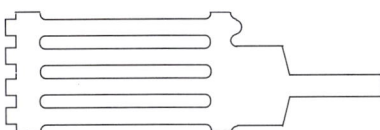

FIGURE 11-19. Drawing of Socket. The socket is square and the tolerance range of dimensions shown applies in both directions. Dimensions are in millimeters. The material is 0.1 mm thick.

FIGURE 11-20. Shape of Stamped Metal of Socket Before Being Formed into Three-Dimensional Shape.

by comparison. Amazingly, they are made by progressive die stamping machines at rates up to 1,000 per minute.

The pins and sockets are typically mounted in injection-molded plastic plugs. Wires are soldered or wire-wrapped onto the contacts. A typical plug has as many as 70 pins or sockets. The manufacturer of these plugs wants to know what the insertion force is, including how it is affected by fabrication variations. Typical tolerances are shown in the figures, but actual measurements on real pins and sockets reveal that some have dimensions that exceed the tolerances by 10–20%. The coefficient of friction is also not known with certainty, but several tests yield a value near 0.14.

Measurements of insertion force were made on plugs having seventy pins and sockets. A typical insertion force history is shown in Figure 11-21. In addition, the equations in Table 11-2 were used to predict the insertion force behavior of a single pin–socket pair, modeling them as a rigid pin entering a compliant hole with two walls. This result was doubled to represent the actual situation. Figure 11-22 is the result. Multiplying this by seventy yields the plot in Figure 11-23. Clearly the actual data do not agree in terms of the shape of the curve, although the total force at the end is in good agreement. The reason for this discrepancy can be traced to the variations in individual pin and socket dimensions. A thought question at the end of the chapter asks the reader to investigate this issue.

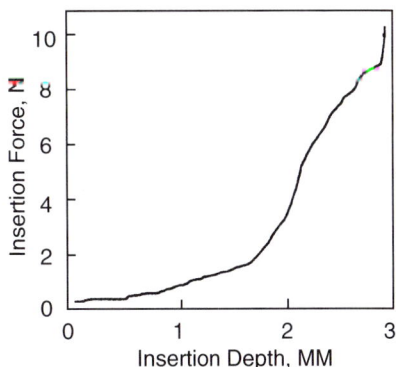

FIGURE 11-21. Actual Insertion Force Data from a Connector Consisting of Seventy Pin–Socket Pairs Like Those in Figure 11-18 and Figure 11-19.

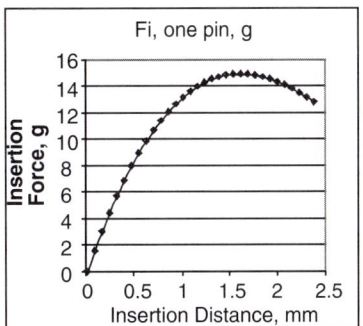

FIGURE 11-22. Insertion Force from One Pin, According to Lower Left Entry in Table 11-2. While no sensor existed to permit verification of the shape of this curve, tests with a gram balance verified the peak value of approximately 15 g.

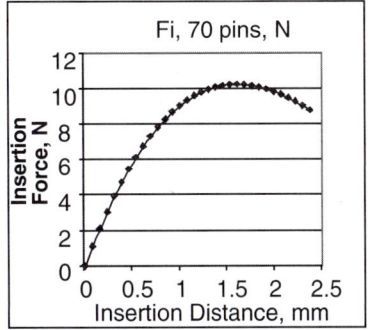

FIGURE 11-23. Insertion Force from Seventy Pins, According to Lower Left Entry in Table 11-2. The peak force value is very similar to that shown in the actual data (Figure 11-21), but the shape of the curve is quite different.

11.C.1.g. Remarks

Figure 11-12 may be used to understand a fundamental point about part mating. The geometry illustrated in the figure shows that the friction force μF_{N_I} points substantially parallel to the insertion direction and thus is a strong contributor to insertion force. By contrast, the normal force F_{N_I} points substantially normal to the insertion direction and thus is a weak contributor to insertion force. If angle ϕ were smaller, then the situation would be reversed, and F_{N_I} would be the strong contributor. If $\mu < 1$, as is common, then $\mu F_{N_I} < F_{N_I}$. Thus a design that exhibits small insertion force is one that has a large ϕ, ensuring that the smaller of these two forces is the one that resists insertion. Since ϕ arises from the shape of the pin and socket at the points where they touch, the design of this shape becomes the focus of attention in the quest to reduce insertion force. This is the topic of the next section of this chapter.

11.D. DESIGN OF CHAMFERS

11.D.1. Introduction

The chamfer is the hero of mechanical parts assembly. It guides parts together when they are laterally or angularly misaligned. Since misalignment is almost inevitable, chamfers are called into play all the time. Yet they are often carelessly designed, routinely chosen to be 45° if straight, given a pleasing shape if curved. Furthermore, many surfaces act as chamfers unbeknownst to the designer. The result is that parts are often much more difficult to assemble than need be.

This section presents analyses and design guidelines for the best shape for chamfers used when the part engaging the chamfer, or its supports, deforms during insertion. The models highlight the relationships between the force-deformation characteristics of the compliant piece, the shape of the chamfer, and friction. These analyses may be combined with one- and two-point contact analyses in Chapter 10 to obtain complete part mating histories of compliantly supported rigid parts.

11.D.2. Basic Model for Insertion Force

The problem of inserting a compliant or compliantly held part across a rigid chamfer can be modeled as shown in Figure 11-24, which corresponds to one symmetric half of Figure 11-2b. Variations on the model can represent other common geometries. The model assumes that the chamfer is rigid and lies between ($X = 0$, $Z = 0$) (the "root"), and ($X = X_1$, $Z = Z_1$) (the "top"). The compliant part is modeled as a rigid, straight thin piece with length L. Its supported end is held by a frictionless bearing that travels along the Z axis, the insertion direction. The compliance is represented by a torsional spring of stiffness K_α at the

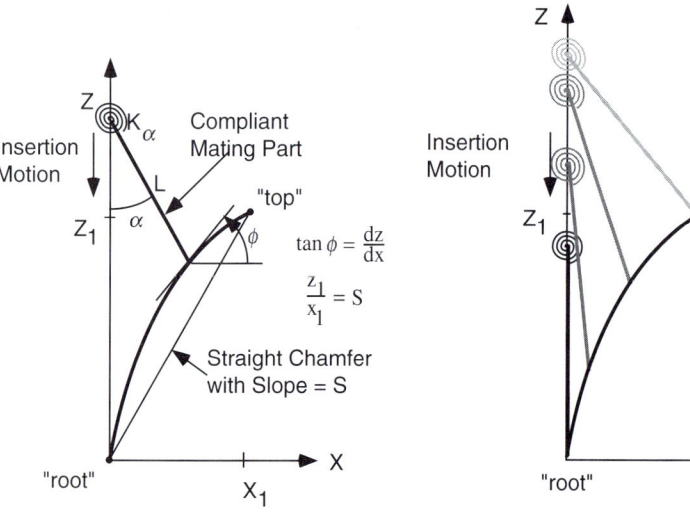

FIGURE 11-24. Simple Model of Compliant Part Mating with Rigid Hole and Compliant Mating Part.

bearing. The contacts of dual in-line packages (DIPs) or the contact springs of electrical connectors may be modeled this way. The other end (the tip) of the compliant part contacts the chamfer first at the point (X_1, Z_1) and is the only portion of the compliant part to touch the chamfer. The tip travels from the top of the chamfer to the root during insertion. Friction coefficient μ acts between the tip and the chamfer surface and is assumed constant. The insertion force F_I is assumed to be positive along the insertion direction (negative Z).

The insertion force is derived using the methods illustrated in Section 11.I and is given by

$$F_I = \frac{K_\alpha(\alpha_0 - \alpha)(\cos\phi + \mu\sin\phi)}{L(\sin(\phi - \alpha) - \mu\cos(\phi - \alpha))} \qquad (11\text{-}8a)$$

where

$$\alpha_0 = \sin^{-1}(x_1/L) \qquad (11\text{-}8b)$$

Figure 11-25 provides notation for Equation (11-8) and sketches the derivation. The force exerted by the compliant part on the chamfer arises from the moment $M = K_\alpha(\alpha_0 - \alpha)$. All other parameters relate that moment to the insertion force. Since the slope of the chamfer at any point equals $\tan\phi$, Equation (11-8) can be revised to show that it is the slope and not the shape that determines insertion force.

In the next section, we will apply this formulation to the problem of designing chamfer shapes. The analysis is aided by simplifying Equation (11-8) in two steps. If α is small and L is large compared to X_1, then

$$F_I \cong \frac{(K/L^2)(X_1 - X)(1 + \mu Z')}{Z' - \mu - X(1 + \mu Z')/L} \qquad (11\text{-}9)$$

where $Z' = \tan\phi$ and $X \cong \alpha L$ and the subscript α has been omitted from K_α.

If, in addition, L is very large, the last term in the denominator of Equation (11-9) can be ignored, yielding

$$F_I \cong \frac{(K/L^2)(X_1 - X)(1 + \mu Z')}{Z' - \mu} \qquad (11\text{-}10)$$

We can observe several things about the insertion force models in Equation (11-9) and Equation (11-10). Both equations can be separated into two factors. The factor $K(X_1 - X)/L^2$ represents the force exerted by the compliant member on the chamfer in the direction perpendicular to the insertion direction. The remaining factor relates this force to the insertion force. Examination of this second factor shows that

- Insertion force depends on chamfer slope Z' but not on chamfer shape.

- There exist combinations of Z', μ, and α that make the denominator small, which in turn makes the insertion force large; the design should be formulated so that such situations do not occur.

- If the chamfer is a straight line (Z' is constant), then insertion force versus insertion depth will also be a straight line.

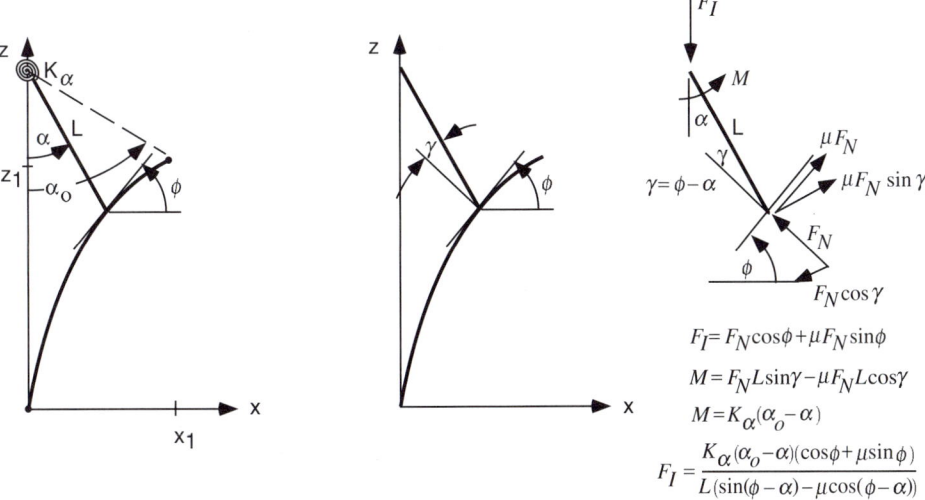

FIGURE 11-25. Notation for Derivation of Equation (11-8).

- If the chamfer is convex, the denominator will be smallest when the contact force is smallest (when the parts first touch each other), and both will increase as insertion proceeds.

- If the chamfer is concave, the numerator will increase and the denominator will decrease as insertion proceeds.

11.D.3. Solutions to Chamfer Design Problems[3]

Three types of solutions will be given. Each has potential uses, advantages, and drawbacks. The three types involve integral criteria, point criteria, and synthesized shapes. The first utilizes calculus of variations to encompass the entire insertion-force history in the criterion and to recommend a chamfer shape. The second deals with one point on the force history, while the third is a family of shapes parameterized by a single variable.

11.D.3.a. Minimum Insertion-Work Chamfers

Calculus of variations has been used to determine the shapes of chamfers that minimize the insertion work:

$$\min \int_{Z=0}^{Z=Z_1} F_I(X, Z') \, dZ \qquad (11\text{-}11a)$$

or, equivalently,

$$\min \int_{X=0}^{X=X_1} F_I(X, Z') Z' \, dX \qquad (11\text{-}11b)$$

Section 11.J contains the derivation of the optimum shape $Z(X)$:

$$Z = \mu X + \sqrt{1+\mu^2} \left[\sqrt{X_1(X_1 - C')} \right.$$
$$- \sqrt{(X_1 - X)(X_1 + C' - X)}$$
$$\left. + C' \log_e \left| \frac{\sqrt{X_1 + C'} - \sqrt{X_1}}{\sqrt{X_1 + C' - X} - \sqrt{X_1 - X}} \right| \right] \qquad (11\text{-}12)$$

where $\log_e$ is the natural logarithm and C' must be determined to match the boundary condition at (X_1, Z_1). The

slopes of this shape at the top and root, respectively, are (see Section 11.J)

$$Z'(X_1) = \mu \qquad (11\text{-}13a)$$

and

$$Z'(0) = \mu + \sqrt{\frac{1+\mu^2}{1 + C'/X_1}} \qquad (11\text{-}13b)$$

To find C', we must employ numerical techniques because Equation (11-12) is transcendental. We will express the shape $Z(X)$ in terms of $Z'(0)$ and the baseline slope, S, defined as

$$S = Z_1/X_1 \qquad (11\text{-}14)$$

We also define an auxiliary variable R, the shape factor, as

$$R = Z'(0)/S \qquad (11\text{-}15)$$

where R represents the ratio of the root slope to the baseline slope. Thus, for straight chamfers $R = 1$, for convex chamfers $R > 1$, and for concave chamfers $R < 1$. Substituting (X_1, Z_1) into Equation (11-12) and using

$$d = C'/X_1, \qquad (11\text{-}16)$$

we obtain

$$S = \mu + \sqrt{1+\mu^2} \left[\sqrt{1+d} + d \log_e \left| \frac{\sqrt{1+d} - 1}{\sqrt{d}} \right| \right] \qquad (11\text{-}17)$$

Equation (11-15), Equation (11-16), and Equation (11-17) can be solved simultaneously to produce a table of R, given μ and S. Figure 11-26 shows sample results and demonstrates that

1. For given μ, S is allowed to have a limited range of values.

2. At each end of the range, $R = 1$.

3. Maximum values of R vary from about 1.1 for $\mu \approx 3$ to 1.25 for $\mu \approx 0.25$, approaching $R = 1.5$ as μ approaches zero, and $R - 1$ as μ approaches infinity.

These observations will be discussed in turn.

First, the lower limit on the value of S is $S = \mu$. If $S < \mu$, then somewhere along the chamfer we would have $Z' < \mu$, which would cause the denominators of Equation (11-9) and Equation (11-10) to pass through zero, causing infinite insertion force.

---

[3]This section of the chapter is based on [Whitney, Gustavson, and Hennessey].

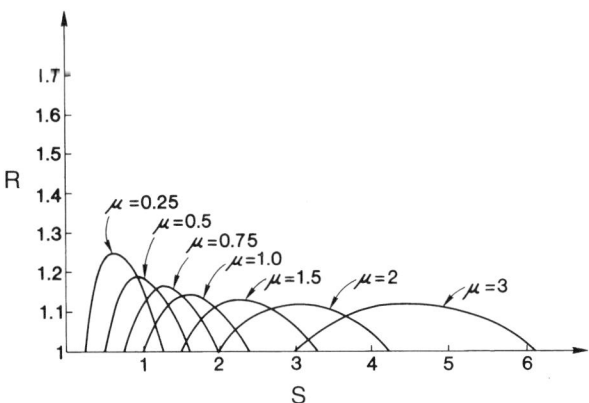

FIGURE 11-26. Curves of R Versus μ and S for Minimum Insertion Work Chamfer.

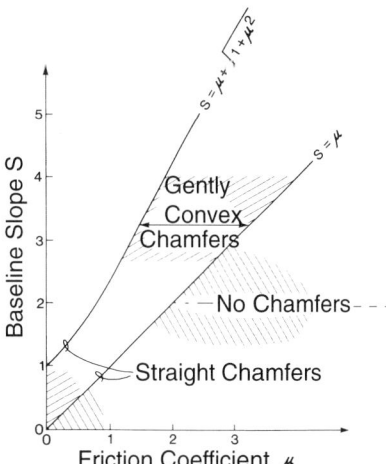

FIGURE 11-27. Parameters of Allowed Minimum Insertion Work Chamfers Expressed as Allowed Values of S and μ.

The upper limit on the value of S is

$$S \leq \mu + \sqrt{1 + \mu^2} \qquad (11\text{-}18)$$

The insertion work is in fact smallest when S has this value. The work is given by Equation (11-19):

$$W^* = 0.5K \left(\frac{X_1 S}{L} \right)^2 \qquad (11\text{-}19a)$$

where the optimum value of S is

$$S = S^* = \mu + \sqrt{1 + \mu^2} \qquad (11\text{-}19b)$$

This does not mean that steeper chamfers cannot be made or used, but only that for each μ, chamfers with $S > S^*$ will not be optimal with respect to insertion work.

Second, at each end of the allowed range of S, the value of R is unity. That is, the slope of the chamfer at $Z = 0$ is the same as the baseline slope. When $S = \mu$, the slope at the root is μ. Since the slope at the top is also μ, the chamfer is straight with slope equal to μ. We can show that when $S = S^*$ the chamfer is also straight.

Third, the fact that R ranges from about 1 to about 1.5 as μ ranges from very large to very small values shows that for S between the allowed limits the chamfers are gently convex. Criteria discussed below yield generally larger values of R, ranging from 3/2 to 10 as μ ranges from 0 to 1. Such chamfers are more sharply arched, yield higher insertion work, and require lower peak insertion force.

Taken together, these results mean that the minimum-insertion-work "curved" chamfer with given μ and optimally chosen S is in fact straight, having W given by Equation (11-19). If S is fixed rather than being chosen optimally, and $\mu < S < S^*$, then a gently convex chamfer

with higher insertion work is obtained. If a continuous-slope chamfer with $S > S^*$ is desired, one must design one's own; it will yield larger work. The calculus of variations will also return a chamfer with $S > S^*$, but its slope will be discontinuous. The chamfer will consist of a straight line of slope S^* between $(0, 0)$ and (X_1, SX_1) followed by a straight vertical line from (X_1, S^*X_1) to (X_1, SX_1). Since there is no insertion force associated with the vertical segment, the work is again given by Equation (11-19) and in effect we have a straight chamfer of slope $S = S^*$. These results and relationships are summarized in Figure 11-27.

In addition to chamfers that obey Equation (11-11), chamfers that obey

$$\min \int_{X=0}^{X=X_1} F_I \, dX \qquad (11\text{-}20)$$

are derived in [Whitney, Gustavson, and Hennessey]. This criterion has no particular physical basis. It is simply a convenient vehicle for generating convex chamfers that are slightly more arched than those derived from Equation (11-11). They also yield lower peak insertion force. Using Equation (11-10), we obtain the following 3/2 power law for the shape:

$$Z(X) = \left(\frac{X_1 - X}{X_1} \right)^{3/2} (\mu X_1 - Z_1) - \mu(X_1 - X) + Z_1$$

$$(11\text{-}21)$$

The top and root slopes are, respectively,

$$Z'(X_1) = \mu \qquad (11\text{-}22a)$$

and

$$Z'(0) = 3S/2 - \mu/2 \qquad (11\text{-}22b)$$

Since $S > \mu$ is required of all chamfers in order to prevent jamming, Equations (11-22a) and (11-22b) show that chamfers described by Equation (11-21) are not concave. If we use Equation (11-9), we obtain a similar but more complex shape formula. The top slope in this case is given by

$$Z'(X_1) = \frac{\mu + X_1/L}{1 - \mu X_1/L} \qquad (11\text{-}23)$$

If we recall that $\tan \alpha_1 \cong X_1/L$ when X_1/L is small, then we can use the familiar formula for $\tan(a + b)$ to show that the top slope is approximately the sum of μ and α_1. Using slopes larger than this will avoid jamming the compliant member against the chamfer.

11.D.3.b. Minimum Peak Force Chamfers

The force models in Equation (11-9) and Equation (11-10) were attacked numerically to determine chamfers whose maximum peak value of insertion force anywhere during insertion was minimized. The result was constant force substantially below the peak-force values of straight chamfers or those obeying the above integral criteria.

Turning the problem around, we can solve for constant-force chamfers analytically. Two cases are considered separately, $\mu = 0$ and $\mu > 0$. Using Equation (11-10) with $\mu = 0$, we easily obtain

$$Z = (2X - X^2/X_1)S \qquad (11\text{-}24)$$

which is a parabola. The value of S is

$$S = \frac{KX_1}{2CL^2} \qquad (11\text{-}25)$$

where C is the constant force value.

The top and root slopes are, respectively,

$$Z'(X_1) = 0 \quad \text{and} \quad Z'(0) = 2S \qquad (11\text{-}26)$$

meaning that $R = 2$.

Using Equation (11-9), we obtain another parabola

$$Z = \left(\frac{X}{X_1}\right)[(2X_1 - X)S + (X - X_1)(X_1/L)] \qquad (11\text{-}27)$$

whose top and root slopes are, respectively,

$$Z'(X_1) = X_1/L \quad \text{and} \quad Z'(0) = 2S - X_1/L \qquad (11\text{-}28)$$

To obtain solutions for $\mu > 0$ requires a technique similar to that used for minimum-insertion-work chamfers. The resulting shape equation $Z(X)$ for the force model in Equation (11-10) is

$$Z = -X/\mu + \frac{L^2 C(1 + \mu^2)}{K\mu^2} \log_e \left[\left| \frac{L^2 C/K\mu - X_1 + X}{L^2 C/K\mu - X_1} \right| \right] \qquad (11\text{-}29)$$

where C, still to be determined, is the value of the constant insertion force. The top and root slopes are, respectively,

$$Z'(X_1) = \mu \qquad (11\text{-}30a)$$

and

$$Z'(0) = \frac{X_1 + \mu L^2 C/K}{L^2 C/K - \mu X_1} \qquad (11\text{-}30b)$$

To determine the constant insertion force C, we put Equation (11-15) into Equation (11-30b) to obtain

$$C = \left(\frac{\mu RS + 1}{RS - \mu}\right)\left(\frac{KX_1}{L^2}\right) \qquad (11\text{-}31)$$

Then, put the boundary condition (X_1, Z_1) in Equation (11-29) to obtain

$$\frac{\mu(\mu S + 1)(RS - \mu)}{(1 + \mu^2)(\mu RS + 1)} = \log_c \left| \frac{\mu RS + 1}{1 + \mu^2} \right| \qquad (11\text{-}32)$$

This can be solved numerically to yield tables of R versus μ and S (see Figure 11-28).

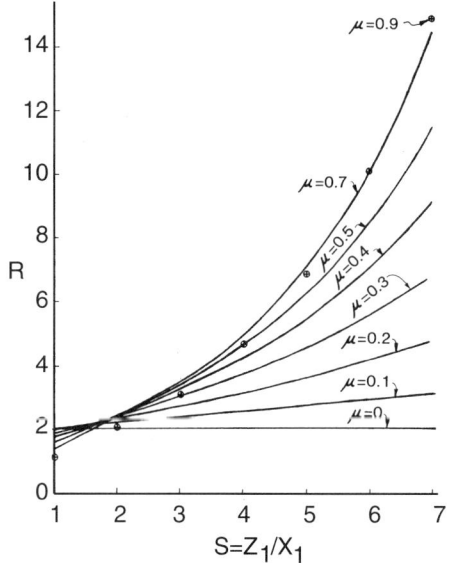

FIGURE 11-28. Shape Factor *R* Versus Baseline Slope *S* and Friction Coefficient μ for Constant (Minimum) Force Chamfers.

TABLE 11-4. Values of Shape Factor R for Different Types of Optimum Chamfers Compared to R for a Straight Chamfer

Straight Chamfer	min $\int F_I\, dZ$	min $\int F_I\, dX$	Constant (Minimum) Force Chamfer
$R = 1$	$1 \le R \le\ \approx 1.5$	$R \approx 1.5$	$R = 2$ if $\mu = 0$ $R > 2$ if $\mu > 0$ (value depends on S)

If we use Equation (11-9), we get a chamfer shape formula which is similar to, but more complex than, Equation (11-29). The top and root slopes are simpler than Equations (11-30a) and (11-30b)

$$Z'(X_1) = \frac{\mu + X_1/L}{1 - \mu X_1/L} \cong \tan(\mu + \alpha_1) \qquad (11\text{-}33a)$$

and

$$Z'(0) = RS \qquad (11\text{-}33b)$$

The convex chamfers may be summarized conveniently by their top slopes and their root slopes. If we require that the denominator of Equation (11-8a) must always be positive, then we can show that the top slopes of all the chamfers discussed here are given by Equation (11-33a). This criterion is more conservative than the others and avoids jamming the compliant member against the chamfer. The root slopes are given by Equation (11-33b), where R is different for each type of chamfer, as shown in Table 11-4. Reading from left to right in the table, we can see that the chamfers have larger values of R, meaning that they become more arched or convex.

It is interesting to compare the various solutions. Figure 11-29a shows three chamfers having $S = 3$ and $\mu = 0.3$, while Figure 11-29b shows their force versus insertion-depth behavior and the value of their normalized insertion work. The minimum-work chamfer has a discontinuity because $S = 3$ is larger than the optimum $S^* = 1.22$ corresponding to $\mu = 0.3$.

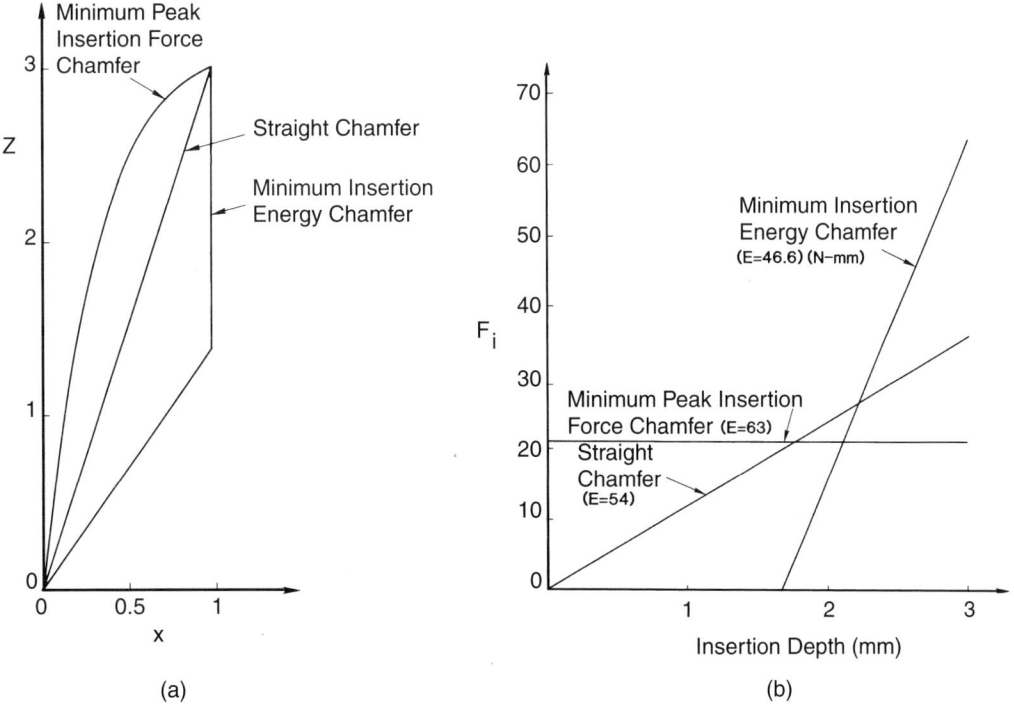

(a)   (b)

FIGURE 11-29. (a) Comparison of Shapes of Straight Chamfer, Constant Force Chamfer, and Minimum Insertion Work Chamfer for $S = 3$ and $\mu = 0.3$. (b) Comparison of Insertion Force Versus Insertion Depth for the Same Three Chamfers.

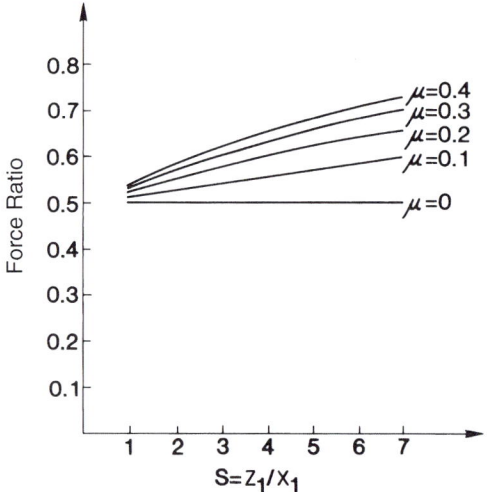

FIGURE 11-30. Ratio of Insertion Force of Constant (Minimum) Force Chamfer to the Peak Insertion Force of a Straight Chamfer Having the Same S and μ.

Figure 11-30 shows the ratio of the minimum constant force to the peak force of a straight chamfer having the same S and μ. The ratio plotted is

$$r = \left(\frac{\mu RS + 1}{RS - \mu} \right) \bigg/ \left(\frac{\mu S + 1}{S - \mu} \right) \qquad (11\text{-}34)$$

which is the ratio of the constant (minimum) force C in Equation (11-31) to the same for a straight chamfer where $R = 1$.

From Figure 11-30 we see that we cannot reduce the insertion force below half that of a straight chamfer (for the same S and μ) no matter what chamfer shape we use. The advantage of the minimum-force chamfer is reduced as S and μ increase.

A heuristic way to design near-constant force chamfers is as follows:

- Determine the value of μ that applies.
- Establish the top and root locations, thereby choosing a value for S.
- Draw a tangent line for the root having slope RS, where R is chosen with the aid of Figure 11-28.
- Draw a tangent line for the top having slope given by Equation (11-33a).
- Draw a smooth curve between the root and the top that matches the slopes at those end points.

11.D.3.c. Synthesizing Chamfer Shapes

A simple formula with a single parameter will yield a family of chamfers containing members similar to each of the above analytically determined shapes. Similar force characteristics are, of course, also obtained. This gives an easy though approximate design technique. We will use it here to design an approximately constant force chamfer.

The model is based on exponential curves of the form

$$Z = a(1 - e^{-bX}) \qquad (11\text{-}35)$$

The root and top slopes are, respectively,

$$Z'(X_1) = abe^{-bX_1} \qquad (11\text{-}36a)$$

and

$$Z'(0) = ab \qquad (11\text{-}36b)$$

For convenience, let $X(1) = 1$. Then we have

$$Z'(1) = abe^{-b} = \frac{\mu + 1/L}{1 - \mu/L} \qquad (11\text{-}37a)$$

and

$$Z'(0) = ab = RS \qquad (11\text{-}37b)$$

where we have used Equation (11-33), which is valid for all the chamfer shapes considered in this section. Also

$$S = Z(1) = a(1 - e^{-b}) \qquad (11\text{-}38)$$

Combining Equation (11-38) with Equation (11-37b), we can write

$$\frac{b}{R} = 1 - e^{-b} \qquad (11\text{-}39)$$

This can be solved for b for any value of R. The result is shown in Table 11-5.

The design procedure for an approximately constant force chamfer is then as follows:

- Pick S and μ ($S = 3$, $\mu = 0.3$)
- Find R from Figure 11-28 ($R \cong 3$)
- Find b from Table 11-5 ($b = 2.824$)
- Find a from Equation (11-38) ($a = 3.18933$)
- Find L from Equation (11-37a) ($L = 4.944$)

TABLE 11-5. Values of b Given R

R	b
1	0.029
1.25	0.465
1.5	0.872
2	1.595
2.5	2.234
3	2.824
4	3.921
5	4.965

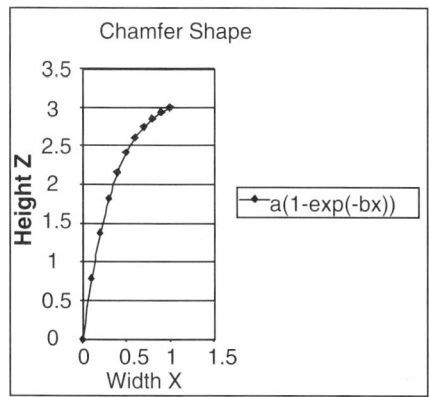

FIGURE 11-31. Approximate Constant (Minimum) Force Chamfer Based on Exponential Shape.

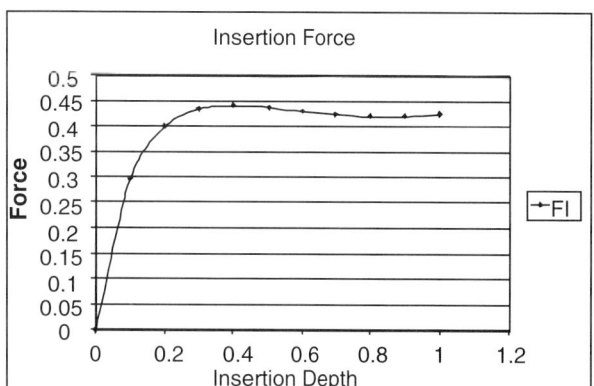

FIGURE 11-32. Insertion Force Versus Insertion Depth Behavior for Approximate Constant (Minimum) Force Chamfer in Figure 11-31. The value of the insertion force is based on scaling K to be numerically equal to L^2. Thus the force value is less important than the shape of the curve. A different value of force can be obtained by choosing a different value for K.

Inserting these values into Equation (11-35) yields the shape shown in Figure 11-31. The root slope of this shape is $RS = 9$, as it should be. Its top slope is $abe^{-b} = 0.5347$.

Using Equation (11-10), the insertion force versus insertion depth behavior for this chamfer is as shown in Figure 11-32. For the purposes of this graph, K is

numerically equal to L^2. The point of the figure is to show how close to constant force has been obtained.

11.E. CORRELATION OF EXPERIMENTAL AND THEORETICAL RESULTS

Three chamfer shapes shown in Figure 11-33 were NC milled to obey the equations for constant insertion force

behavior. Shape PAR07 [constant-force parabola obeying Equation (11-24)] was intended to produce constant

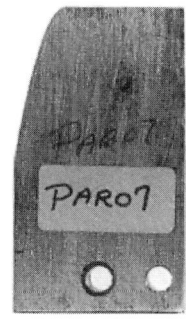

FIGURE 11-33. *Left:* Apparatus Used for Experiments. *Right:* Experimental Chamfers: All three have $S = 3$. PAR07 is a parabolic constant insertion force chamfer based on assuming $\mu = 0$. LOG07 is a constant force logarithmic chamfer with $R = 2.4$ based on $\mu = 0.1$. EXP is an exponential chamfer with $bX_1 = 1$. (Photos courtesy of C. S. Draper Laboratory.)

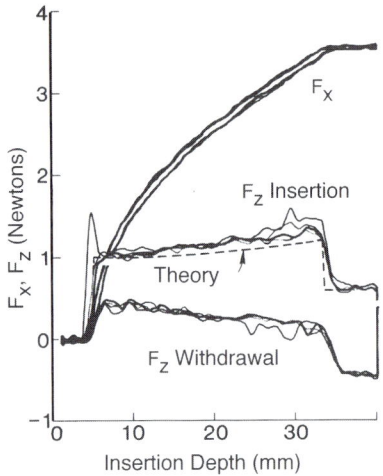

FIGURE 11-34. Comparison of Theory and Experimental Results for the PAR07 Shape (Constant Force if $\mu = 0$). The theory line is based on $\mu = 0.14$, which is the actual value.

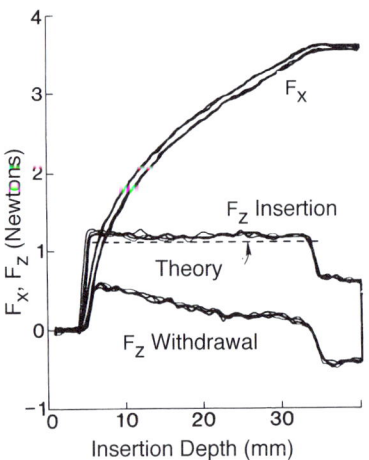

FIGURE 11-35. Comparison of Theory and Experimental Results for the LOG07 Shape (Constant Force if $\mu = 0.1$). The theory line is based on $\mu = 0.15$, which is the actual value.

insertion force if $\mu = 0$. Shape LOG07 [constant-force logarithm obeying Equation (11-29)] was intended to produce constant insertion force if $\mu = 0.1$. Since $\mu \cong 0.14$ to 0.15, neither experimental result yields constant force but is close to what theory would predict for those shapes at that value of μ. Shape EXP N = 1 [exponential shape obeying Equation (11-35) with $bX_1 = 1$] was also used.

Experimental results for these chamfers are shown in Figure 11-34, Figure 11-35, and Figure 11-36. They are in good agreement with the theory, considering that the coefficient of friction assumed when making them was not quite what was obtained in the experiments.

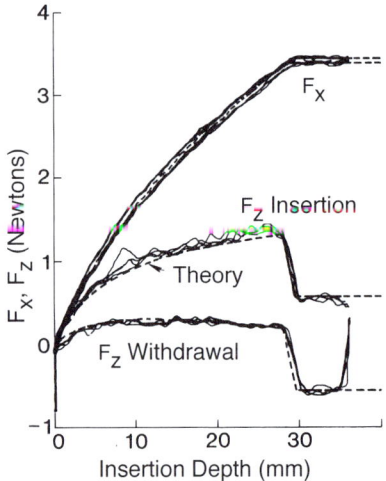

FIGURE 11-36. Comparison of Theory and Experimental Results for the EXP $bX_1 = 1$ Shape. The theory line is based on $\mu = 0.15$, which is the actual value.

11.F. CHAPTER SUMMARY

In this chapter, compliant part mating was studied theoretically and experimentally. Compliant part mating is very common, and there is an opportunity to influence the mating behavior by altering the shapes of the mating surfaces, even if one has little control over other factors such as coefficient of friction. In fact, small changes in shape can make large changes in insertion-force behavior, a statement that is not true of stiffness or friction, except in some limiting cases.

While most chamfers are straight, it appears that, if fabrication costs permit, slightly convex chamfers are preferable. If the chamfer must be straight and if space permits, long chamfers are preferable to short ones.

As is the case for rigid part mating, agreement between theory and experiment is quite good. This means that one can use the equations here with high confidence. However, since small changes in shape can make large changes in insertion force, fabrication variations can cause serious problems. This can have severe consequences in the case of tiny connector pins and sockets like those shown in Figure 11-17.

11.G. PROBLEMS AND THOUGHT QUESTIONS

1. Equation (11-1) describes the ratio of insertion force to lateral force for different coefficients of friction μ and different interface angles ϕ. Use these equations to determine the values of μ and ϕ during insertion that will cause the pin to jam against the compliant element. Your answer should be in the form of an equation relating these two variables. Is a jamming condition possible during withdrawal? Prove your answer using the equations.

2. Take apart a mechanical item containing compliant part mates (a toaster, light fixture, breadbox, electric socket, etc.) and classify the part mates as follows:
- Type of mate: sliding compliance with separate spring, integral part and spring combination, single compliant element working against a rigid one, two compliant elements working against each other, compliance used for electrical contact, compliance used for mechanical contact, compliance used to create a certain "feel" during mating with actual mating or latching accomplished by another feature of the parts, and so on.
- Direction of approach of mating parts with respect to each other, based on a common coordinate frame attached to any main part of your choice.
- Accumulate the results for several products and create statistics showing such things as percent occurrence of each mate type and mate direction.

3. At least one compliant part mate occurs during the operation of the stapler. What is it and how does it operate? Write the equations that describe it and explain what conditions are necessary in this mate in order for proper operation of the stapler to occur.

4. Prove the statement in Section 11.D.3.b that constant force chamfers are parabolas if the coefficient of friction is zero.

5. Figure 11-5 shows that insertion force versus insertion depth curves look much like the chamfers that produce them. Explain why this is so in your own words.

6. It has been pointed out in this chapter that most "preferred" chamfer shapes are convex, some are straight, and none are concave. Explain in your own words why convex chamfers are preferred and none are concave.

7. Attempt a more rigorous answer to Problem 6 by sketching the behavior of the forces generated at the contact between a compliant part and a convex curved chamfer, such as that shown in

Figure 11-25. Draw vectors that show contact force normal to the chamfer surface, friction force tangent to the chamfer surface, and insertion force opposite to the insertion direction. Draw these vectors for three cases: when the part first touches the chamfer, when the part is fully mated, and at some point halfway between the other two. Note how contact force is initially small with a large projection along the insertion direction. Observe, too, that when the mate is complete the contact force is large but has a small projection along the insertion direction, while the much smaller friction force has a large projection along the insertion direction. Explain verbally how this "exchange" of forces projecting along the insertion direction could result in a constant insertion force.

8. Using the same approach as in Problem 7, explain what would happen if the chamfer were concave. Could a concave chamfer possibly yield a constant or approximately constant insertion force characteristic?

9. The text following Equation (11-8) and Equation (11-9) contains several bullet points concerning the properties of those equations. Use the equations to prove each of the bullet points.

10. Figure 11-5a shows a straight chamfer shape and indicates that the insertion force versus insertion distance will also be straight. Assume that an electrical connector is constructed of twenty-five identical mating parts having straight chamfers. Presumably the insertion force will be twenty-five times the force generated by one mating part, as shown in Figure 11-37. However,

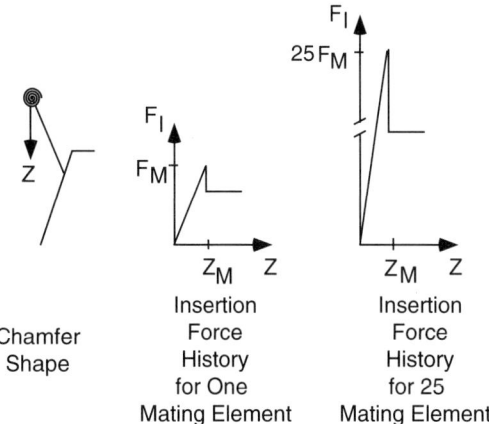

FIGURE 11-37. Theoretical Insertion Force History for One and for Twenty-Five Identical Mating Elements.

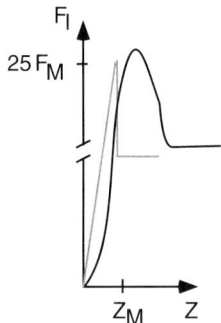

FIGURE 11-38. Observed Insertion Force History for Twenty-Five Nominally Identical Mating Elements. The theoretically expected insertion force history from Figure 11-37 is shown in gray.

when a real connector made of such mating elements is plugged in, the actual insertion force history is an S-shaped curve as shown in Figure 11-38. Explain why this happens. [*Hint:* The mating elements are not identical. Some are a bit longer than nominal and will engage the socket earlier than expected, while others are shorter and will engage later. Some are a bit fatter and will deflect the contact spring a little more, resulting in higher insertion force, while others are thinner and will generate less. Some contact springs are a little stiffer, and so on. Show how the individual insertion force

histories could differ on the basis of these differences. Then sum up these different histories to get the total insertion force history for all of them.]

11. The coefficient of friction is usually measured by measuring normal force and insertion force and calculating the ratio. In a closed connector such as that shown in Figure 11-17, the normal force cannot be measured because it is counterbalanced by an equal and opposite normal force inside the socket. However, the coefficient of friction can be estimated using only measurements of the insertion and withdrawal force if the shapes of the pin and socket can be modeled reasonably accurately. Show how this can be done using the insertion and withdrawal force equations in Table 11-2. [*Hint:* If we know the shape, we can predict what the force would be at any point during insertion or withdrawal. The equations then can be solved for the remaining unknown μ. Force measurements should be taken at several identical values of insertion distance during insertion and withdrawal and the results averaged.]

12. Derive Equation (11-24) and Equation (11-25).

11.H. FURTHER READING

[Dwight] Dwight, H. B., *Tables of Integrals and Other Mathematical Data,* 4th ed., New York: Macmillan, 1961.

[Gustavson] Gustavson, R. E., "A Theory for the Three-Dimensional Mating of Chamfered Cylindrical Parts," *ASME Journal of Mechanisms, Transmissions, and Automation in Design,* June 1985.

[Hennessey] Hennessey, M. P., "Compliant Part Mating and Minimum Energy Chamfer Design," S. M. thesis, MIT Mechanical Engineering Department, September 1982.

[Hildebrand] Hildebrand, F. B., *Advanced Calculus for Applications,* Englewood Cliffs, NJ: Prentice-Hall, 1962.

[Nevins and Whitney] Nevins, J. L., and Whitney, D. E., editors, *Concurrent Design of Products and Processes,* New York: McGraw-Hill, 1989.

[Whitney] Whitney, D. E., "Part Mating Theory," in *International Encyclopedia of Robotics: Applications and Automation,* Dorf, R., editor, New York: Wiley, 1988.

[Whitney, Gustavson, and Hennessey] Whitney, D. E., Gustavson, R. E., and Hennessey, M. P., "Designing Chamfers," *International Journal of Robotics Research,* vol. 2, no. 4, pp. 3–18, 1983.

11.I. APPENDIX: Derivation of Some Insertion Force Patterns

11.I.1. Radius Nose Rigid Peg, Radius Nose Compliant Wall

For this case, the situation is illustrated in Figure 11-39.
  Forces on the peg:

$$F_I = F_{N_I}(\cos \phi + \mu \sin \phi) + \mu F_{W_I}$$
$$F_{W_I} = F_{N_I}(\sin \phi - \mu \cos \phi) \tag{11-40}$$

Combining these equations yields

$$F_I = F_{N_I}[(1 - \mu^2)\cos\phi + 2\mu\sin\phi] \tag{11-41}$$

Forces on compliant member:

Horizontally: $\quad F_{N_I}\sin\phi - \mu F_{N_I}\cos\phi - \mu F_p$
$$- K_x(L_0 - L) = 0 \tag{11-42}$$
Vertically: $\quad \mu F_{N_I}\sin\phi + F_{N_I}\cos\phi - F_p = 0$

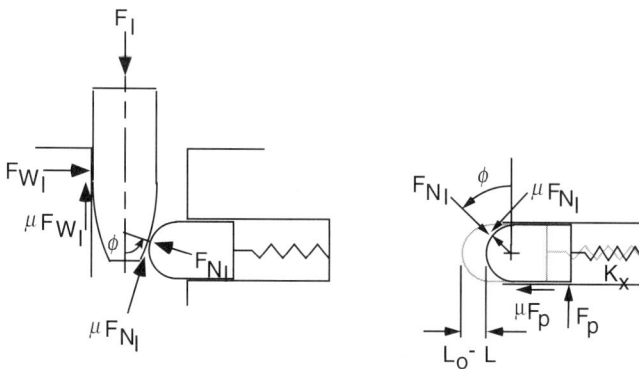

FIGURE 11-39. *Left:* Forces on Peg. *Right:* Forces on Compliant Wall Member.

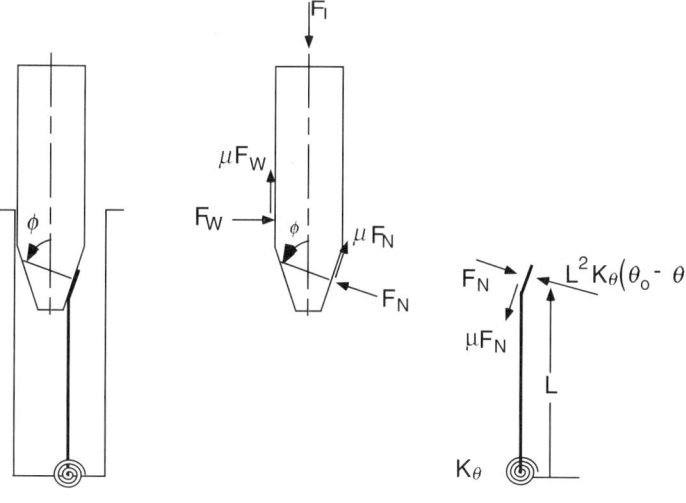

FIGURE 11-40. *Left:* Geometry of Straight Side Peg Entering Hole with Cantilever Beam Compliant Member. *Center:* Forces on Peg. *Right:* Forces on Compliant Member.

Combining these equations yields

$$F_{N_I} = \frac{K_x(L_0 - L)}{(1 - \mu^2)\sin\phi - 2\mu\cos\phi} \qquad (11\text{-}43)$$

Combining Equation (11-41) and Equation (11-43) yields

$$F_I = \frac{K_x(L_0 - L)[(1 - \mu^2)\cos\phi + 2\mu\sin\phi]}{(1 - \mu^2)\sin\phi - 2\mu\cos\phi} \qquad (11\text{-}44)$$

which is the upper left entry in Table 11-2.

Note that when the peg is inserted far into the hole so that its straight side is in contact with the compliant member, the insertion force is constant. This constant value is obtained from the derivation of Equation (11-44) by noting in Equation (11-42) that $F_p = 0$. In that case, we

obtain

$$F_I = \frac{K_x(L_0 - L)[(1 - \mu^2)\cos\phi + 2\mu\sin\phi]}{\sin\phi - \mu\cos\phi} \qquad (11\text{-}45)$$

But $\phi = 90°$ so that Equation (11-45) reduces to

$$F_I = 2\mu K_x(L_0 - L) \qquad (11\text{-}46)$$

11.I.2. Straight Taper Rigid Peg, Cantilever Spring Hole

For this case, the situation is illustrated in Figure 11-40.

The equation for the forces on the peg is the same as in the case above, namely Equation (11-41). The forces on the compliant member are obtained by assuming that the tapered end of the compliant member is parallel to the tapered end of the peg and ignoring the angle at which the compliant member may be deflected:

$$F_{N_I} = \frac{L^2 K_\theta (\theta_0 - \theta)}{\sin \phi - \mu \cos \phi} \qquad (11\text{-}47)$$

Combining Equation (11-41) and Equation (11-47) yields

$$F_I = \frac{L^2 K_\theta (\theta_0 - \theta)[(1 - \mu^2) \cos \phi + 2\mu \sin \phi]}{\sin \phi - \mu \cos \phi} \qquad (11\text{-}48)$$

Because the tapered side of the peg is straight, the deflection $(\theta_0 - \theta)$ is a linear function of insertion depth z. Therefore, the insertion force rises linearly with insertion depth. When the peg is inserted far enough that its vertical side is in contact with the cantilever spring, the steady-state force is obtained by setting $\phi = 90°$ to yield

$$F_I = 2\mu L^2 K_\theta (\theta_0 - \theta) \qquad (11\text{-}49)$$

In typical cases, the insertion force in the steady state will be less than the peak insertion force just before steady

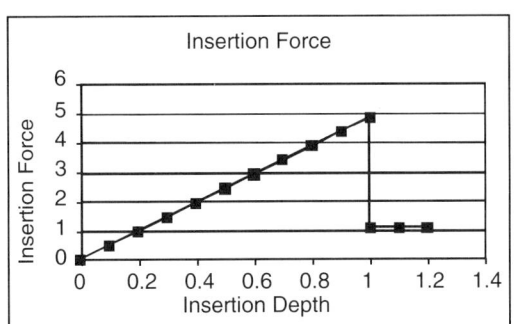

FIGURE 11-41. Insertion Force Versus Insertion Depth for a Straight Side Peg Like That Shown in Figure 11-40. Equation (11-48) and Equation (11-49) are plotted for the case where $L = 1$, $K_\theta = 10$, $\mu = 0.1$, $\phi = 1.0$, and z ranges from 0 to 1 while the compliant element is in contact with the sloping nose of the peg. For insertion depth > 1, the compliant element is in contact with the vertical side of the peg.

state begins. The insertion force behavior will then resemble that shown in Figure 11-41.

11.J. APPENDIX: Derivation of Minimum Insertion Work Chamfer Shape

We wish to find the chamfer shape $Z(X)$ that satisfies

$$\min \int_Z F_I(X, Z') \, dZ \qquad (11\text{-}50)$$

This is a standard problem in calculus of variations ([Hildebrand]). The Euler equation that satisfies Equation (11-50) is

$$\frac{\partial}{\partial Z'}(Z' F_I) = C \qquad (11\text{-}51)$$

where we will use Equation (11-10) to model F_I. Substituting Equation (11-10) into Equation (11-51) yields a quadratic equation in Z' that can be solved to give

$$Z' = \mu + \sqrt{\frac{A\mu(1 + \mu^2)}{A\mu - C}} \qquad (11\text{-}52)$$

where

$$A = (K/L^2)(X_1 - X) \qquad (11\text{-}53)$$

Equation (11-52) can be written as

$$Z' = \mu + \sqrt{1 + \mu^2} \frac{\sqrt{X_1 - X}}{\sqrt{X_1 - X + C'}} \qquad (11\text{-}54)$$

where

$$C' = CL^2/\mu K \qquad (11\text{-}55)$$

Equation (11-54) is a simple integral and is solved using entries 195.04 and 195.01 in ([Dwight]) to yield Equation (11-12). Note that $Z'(X_1) = \mu$ when $C \neq 0$ and $Z' = \mu + \sqrt{1 + \mu^2}$ when $C = 0$. Also, μ is the minimum value that Z' can have anywhere along the chamfer, in view of the denominator of Equation (11-10). Second, we can conclude that the baseline slope $S \geq \mu$. If $S < \mu$, then, by the mean value theorem, Z' would have to be less than μ somewhere along the chamfer. These facts are true of all chamfers obeying Equation (11-10).

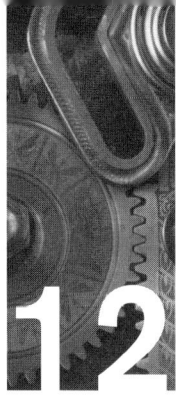

12 ASSEMBLY IN THE LARGE: THE IMPACT OF ASSEMBLY ON PRODUCT DEVELOPMENT

"Just because you can make something doesn't mean you can manufacture it."

12.A. INTRODUCTION

This chapter begins the third major part of the book, dealing with assembly in the large. By this we mean all the issues surrounding designing and manufacturing a mechanically assembled product. The theme of this part of the book is to show how assembly influences, and is influenced by, product development and overall manufacturing strategy. As we saw in Chapter 1, assembly is more than putting parts together. It is a process of deciding how to deliver quality and functionality at the assembly level by designing and producing parts and subassemblies in a top-down fashion.

The topics we will deal with in this and subsequent chapters include

- Analyzing a product or product design from the point of view of assembly in the large

- Understanding how an existing product works in detail

- Product architecture and its influence on product design and manufacturing strategy

- Design for assembly in practice and as an aspect of product architecture

- Design of assembly systems and workstations

- Economic analysis of assembly systems

In addition, Chapter 19 (on the CD-ROM packaged with this book) is a completely worked example based on analysis of an aircraft wing.

Our goal in this chapter is to lay out the issues broadly. We will consider the importance of concurrent engineering (coordinating product and assembly process design), different technological ways of implementing product functions and the assembly implications of the choices, and a step-by-step process for looking at a product from the point of view of assembly in the large. Subsequent chapters will take up these topics in more detail.

12.B. CONCURRENT ENGINEERING

Assembly in the large is a set of design activities that takes place within the larger context of product design and development. There are many books on this subject, which is too broad of us to cover in detail here. One aspect, however, is important enough to deserve our attention. This is the integration of all the different demands on any product that arise from marketing, financial, engineering, manufacturing, assembly, after-market service, upgrading, and recycling. Manufacturers have become increasingly aware of the need to bring these constituencies together during product design and to balance their often conflicting needs.

No sure-fire method of accomplishing this has emerged, and companies constantly revisit the problem and revise their product development methodologies in order to do better ([Smith]).

In 1989 the author and his colleagues wrote a book ([Nevins and Whitney]) on the subject of concurrent engineering. The premise of this book was that the design steps associated with assembly could be used as an impetus toward integrating the many aspects of product development. The following few paragraphs quote directly from pages 198–202 in that book on the subject of combining different constituencies interested in the design of a product. As of this writing, no major changes seem necessary.

"Consider, for example, the role of the Purchasing Department. The heart of a high tech product we recently studied is an infrared detector. Purchasing switched to the lower-cost of two detector vendors, with disastrous consequences for production. Subtle differences between detectors cannot be found until the product is partially assembled with optics, power supplies, and so on. For ruggedness and cost reasons, the unit is glued together, making disassembly to replace detectors very expensive. Naturally, the product could be redesigned to make detector replacement easier. But the product is a single-use weapon; its shelf life is several years, during which it is ignored because it is too complex for field repair; it must work the first time; its useful life is ten seconds. Repair is simply not 'in character' for this product.

"The point is that a seemingly minor decision, made to optimize a corner of a company's operations, can have a pervasive effect on how a product is made or used, with severe consequences for operating costs or the customer's perception of the company. These decisions can completely defeat the designer's intentions. Top Management, Engineering, Purchasing, Personnel and Manufacturing can each contribute to the success or failure of a product.

"Converting a concept into a product is an involved procedure consisting of many steps of refinement. The design requires a great deal of analysis, investigation of basic physical processes, experimental verification, complex tradeoffs and difficult decisions. The initial idea never quite works as intended, or does not perform as well as desired. The designers must therefore make many modifications to the original concept. Along the way, they make increasingly subtle choices of materials, fasteners, coatings, adhesives, and electronic adjustments. Expensive analyses and experiments may be carried out to verify portions of the design. In many cases, the choices become more and more difficult as the design gradually works its way toward acceptability. Furthermore, the choices become more interdependent, and take on the character of an interwoven historical chain in which later choices are conditioned on or forced by ones made previously. The earlier decisions have the most influence on the later course of the design.

"Imagine that a manufacturing engineer comes into this increasingly detailed debate late in the process and begins asking for changes. It is likely that, if the product designers accede to his requests, a large portion of the design will simply unravel, and many difficult choices will have to be made all over again. Where some close calls went one way, they now may go another, in view of the new criteria which the manufacturing engineer brings to the table. New analyses and experiments may be needed.

"As an example, a research scientist at a large chemical company spent a year perfecting a process at laboratory scale. His process operated at atmospheric pressure. When a production engineer was called in to scale it up, he immediately asked for higher pressures because atmospheric pressure would require huge pipes, pumps, and tanks. Unfortunately, the researcher's process failed at elevated pressures and he had to start over.

"In other cases, the manufacturing engineer's requests might not be possible to grant, resulting in an awkward or non-robust process. . . .

"How can problems like this be avoided? If manufacturing and assembly engineers are participants in the design debate from the start, their criteria can be given weight as the difficult choices are being made, and the design process could turn out differently. If repair engineers, purchasing agents, and other knowledgeable people are represented, a better, more integrated design would result, on the basis of a similar debate. Again, the design would represent an interconnected web of decisions, but more parties would make the web better balanced.

"There is at present no perfected method for designing products so that all of the constituencies can have their say, much less get everything they want. It is unlikely that any such method will ever exist. There will always be tradeoffs and compromises between the designer who asks 'What good is it if it doesn't work?,' the production engineer who asks 'What good is it if I can't make it?,' and the marketer who asks 'How can I sell it if it costs too much?'

". . . At the present, the state of the art in concurrent engineering is the team approach. . . . No single designer can have all the knowledge needed to carry out such a comprehensive activity alone. Neither do we have super intelligent computer programs that can design products and manufacturing processes. For the time being, at least, we must rely on teams of specialists to pool their knowledge to create superior products and manufacturing systems.

"Engineers are taught early on that design is an iterative process, but rarely are they taught about the iterations between design and production, or between production and marketing. Perhaps this is the cause of the traditional time separation between product design and manufacturing system design. The real concurrency of concurrency team work cannot be overemphasized. It is not too early to begin the process before there are engi-

neering prototypes, because the essence of a sophisticated design can depend on careful choice of tolerances, materials, or novel fabrication methods that cannot be separated from the design of the manufacturing process."

12.C. PRODUCT DESIGN AND DEVELOPMENT DECISIONS RELATED TO ASSEMBLY

Product design and development involves determining customer needs and deciding how to meet them. This includes the technical aspects of designing the product as well as business issues such as determining how many varieties will be offered, who will manufacture the product, where it will be made, how many units will be made per year, and whether the product is a member (even the founding member) of a family that shares parts, processes, suppliers, and business practices ([Ulrich and Eppinger]).

A sketch of the early phases in a typical product development process is given in Figure 12-1. This figure illustrates a market-driven process, in which means are sought to meet a set of existing or emerging customer

needs. This is commonly called *demand pull*. Not shown is the complementary process of *technology push*, in which a visionary product is developed with the intention of generating a need for it. Demand pull is usually associated with known products such as cars or furniture. Technology push is usually associated with previously unknown products, such as personal computers. However, visionary cars and furniture can or do get developed. In either case, the needs must be converted into engineering specifications, following which (or together with) concepts are generated.

In the next few sections we introduce the most important of these matters, leaving the details for later chapters.

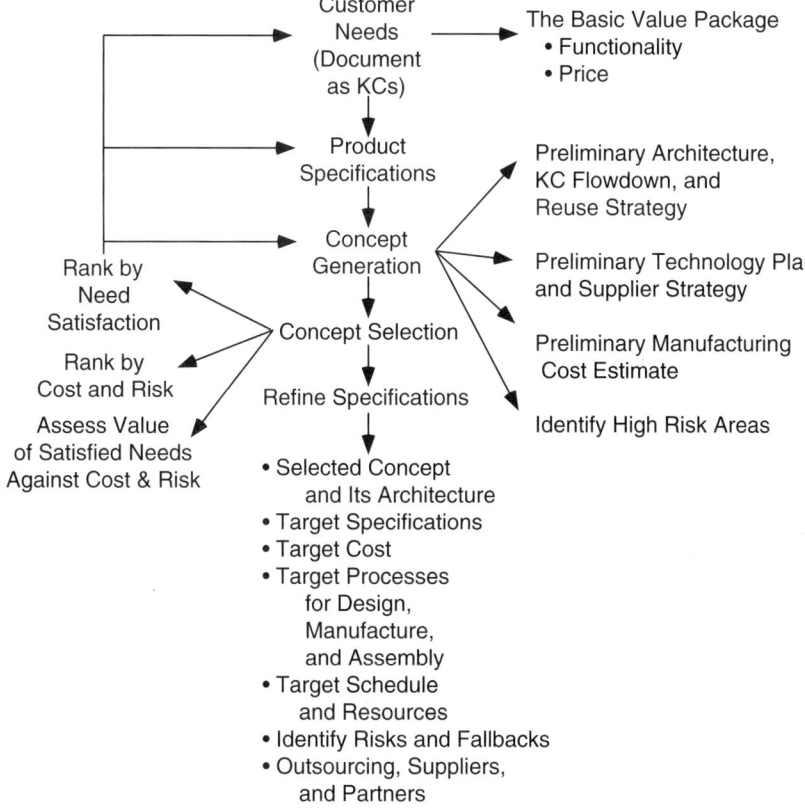

FIGURE 12-1. Important Decisions in Early Product Design. These decisions define the product's basic functions, operating concept, architecture, and plan for manufacture and assembly. Technical and business issues are involved. Subsequent steps of detailed design, realization of the manufacturing plan, including design for manufacturing and assembly, plus integration of all these plans, are not shown.

12.C.1. Concept Generation

Concept generation comprises many creative and exploratory steps that are intended to produce candidate implementations, flesh out the required technologies, demonstrate that the requirements can be delivered, and identify sources of necessary knowledge and capabilities. Costs and risks also need to be identified. Each concept is judged on its ability to deliver the requirements as well as its costs and risks. Concepts may be eliminated, added, or combined as this process moves ahead.

Figure 12-2 shows four ways to print in conceptual form, as well as in historical order. Each concept implements in strikingly different ways the transfer of pigment from a reservoir to a receptor medium in a certain pattern. Different sources of power, number, choice and arrangement of degrees of freedom, and gradual substitution of information technologies for mechanical ones characterize the evolution from design a to design d. Designs a and b contain predefined forms for the images that will be put on the medium, while designs c and d can create any pattern. Design d can put an image on nonpaper media, which the other designs cannot do.

12.C.2. Architecture and KC Flowdown

Architecture involves deciding how physically to implement the products' functions and how to distribute the

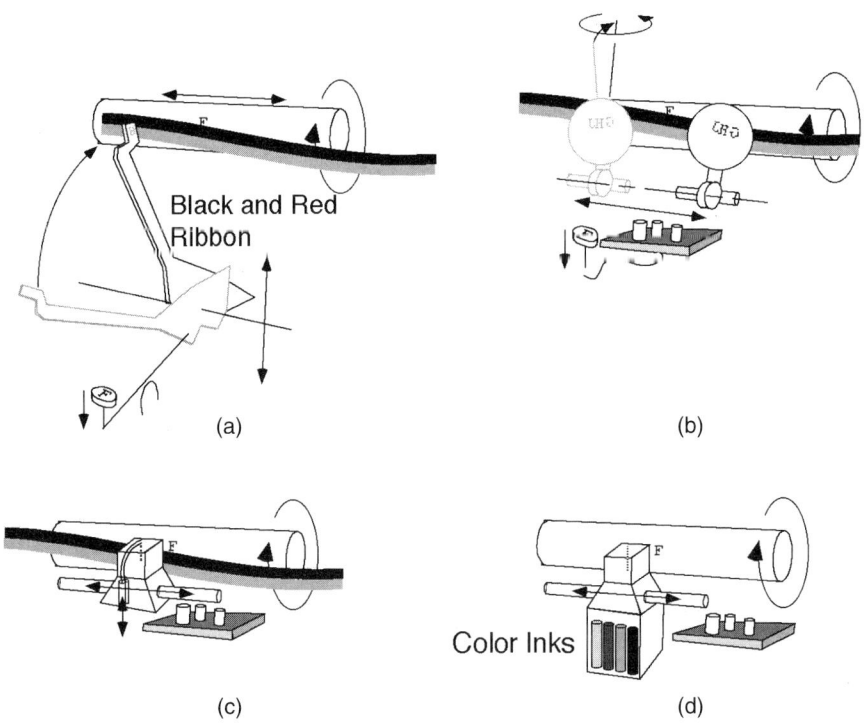

(a)   (b)

(c)   (d)

FIGURE 12-2. Four Ways to Print. (a) Original typewriter design. Only one key is shown. It has one degree of freedom (dof). Additional keys add another dof each. The block carrying the keys has one dof (to allow shifting between lower and uppercase letters). The carriage carrying the paper also has two dof. The total number of dof is in the hundreds. (b) Ball-head design. The schematic circuit board represents electronics. The ball head has three dof while the carriage (now called the platen) has one, for a total of four plus one for each key. In this and the following drawings, there is an actuator for each of the first four dof that is not shown. (c) Dot matrix design. The head contains six to twelve fine wires, each driven by its own magnetic actuator. The head has one dof, as does the platen, for a total of eight to fourteen. There are no keys because the print data come from a computer that has the keys. (d) Ink jet design. The head has one dof, as does the platen, for a total of two. In spite of having the fewest dof, it can print much faster with more colors at finer resolution. The tolerances on these few dof are much tighter than those on the much larger number of dof in designs a and b.

physical elements in space. Architecture also specifies the functional and physical relationships between elements in the product. This means that every concept has an architecture. That architecture can be the result of deliberate choices or it can emerge in the creative process. Either way, architectural decisions have great impact on how the product is designed, built, and used. Product architecture is the subject of Chapter 14.

Architectural decisions are clearly related closely to the definition and flowdown of key characteristics. Architecture provides the basic technologies and subassemblies, and these in turn provide the routes along which KC delivery paths and DFCs will flow. The choice between implementing a function mechanically or electronically, for example, has obvious implications for KC delivery and assembly.

These points are illustrated nicely in Figure 12-2. The four designs represent a transition from mechanical to electronic implementations as well as a great reduction and redistribution of degrees of freedom. The only element that survives throughout the evolution from design a to design d is the curved surface on which the receptor medium rides. The internal forces in designs a and b require that they be made almost entirely of strong metal. Design c contains some important plastic parts while design d is almost completely made of precision injection molded plastic parts.

Without much change in the basic functional allocation or physical layout of design d, it can be inserted into larger system concepts that include more or fewer colors, finer or coarser resolution, optical scanning or computer/electronic input of the required image, much larger size of output medium, remote operation, and so on. A thought question at the end of the chapter asks for definition of some important KCs and DFCs for each of these concepts.

12.C.3. Platform Strategy, Technology Plan, Supplier Strategy, and Reuse

As different technologies and suppliers are considered, large and long range strategy decisions are in the balance. If the product is to be the founder of a family, then these decisions will have to endure as markets and technologies evolve. Subsequent products will presumably be built to fit the family, which usually means reusing some percentage of the parts, along with the necessary design and manufacturing knowledge or suppliers of them. If suppliers are to be relied on for important technologies in the product or its manufacture, careful consideration must be given to the strategic implications of becoming dependent on these suppliers for crucial knowledge. Such topics are beyond the scope of this book. Sources for more information include [Prahalad and Hamel], [Fine and Whitney], and [Fine].

At the assembly design level, a family strategy means reusing KC delivery strategies and DFCs, as well as maintaining compatibility between new and carryover parts at the places where they mate. If different family members are to be made simultaneously on the same production lines, then compatibility must be maintained between the parts and their manufacturing and assembly equipment.

12.D. STEPS IN ASSEMBLY IN THE LARGE

Architectural and technology choices like those discussed in the previous few sections have large implications for how a product will be assembled. Any analysis of the product from a broad assembly-driven point of view will have to take these choices into consideration. Implicit in the discussion about concurrent engineering, however, is the idea that assembly analysis can reveal problems or opportunities that need to be considered when architecture and technology choices are being made. It is a two-way street. Product analysis that takes assembly in the large into account therefore needs to move back and forth between a wide range of issues. The discussion needs to take place early in product development, when basic business, architectural, and technological decisions are being made. This section discusses these issues and presents a step-by-step process plan.

The steps are business context, manufacturing context, assembly process requirements, assembly system design, and product design improvements. These are taken up in turn.

12.D.1. Business Context

The business context for a product includes the expectations of the market for the product as well as the expectations of the company producing it. Among the issues to consider are the following:

- "Character" of the product: Is this a single use item or one that will be used, refilled, repaired, or updated? Different assembly methods, materials, and fasteners will be relevant. Hand grenades and fire extinguishers are different from sewing machines and personal computers.

- What sales volume is anticipated? Small items made at high production rates can be made economically by machines or robots, while large items at almost any production rate are made only by people with the possible aid of equipment of various kinds.

- How many versions of the product are anticipated, and will more be added in the future? First of all, this affects the production volume (units made per year) of any one version. Second, it determines the need to anticipate different versions of the same part or subassembly, necessitating long-range planning of parts interfaces. It also requires defining manufacturing and assembly methods that can be adapted or switched over from one version to another without a lot of downtime between versions.

- Who will make the parts and subassemblies, and how far away (in terms of delivery delays) will they be? If the product has only one version and demand is expected to be steady, then long lead times may not matter. But if several versions are made available at once and there is some uncertainty as to absolute sales demand and the relative popularity of the different versions, then lead times must be kept short or else expensive and risky investments in inventory will be needed. If a typical supply chain evolves, in which suppliers have suppliers, these chains can become quite long without anyone realizing it.[1]

Management of a supply chain also involves responsibility for defining specifications and flowing them down the chain. The DFC is intended to provide an avenue for doing this with respect to the geometry of an assembly, but many other issues are involved, such as colors, material compatibility, surface finishes, electrical and chemical behavior, and so on.

- Will all of the assembly be done at the manufacturer's facility or at upstream suppliers, or will some be done by others farther downstream in the distribution channel or even by the user? The latter strategy is appropriate when demand for the base product is predictable but demand for options is not.

- What are the relative costs of doing business in various regions of the world and implementing production processes that involve people or equipment? Currencies fluctuate, political situations can be volatile, and local laws can affect the available choices. Products sold in different countries may be subject to local content laws, meaning that some percentage of the parts or subassemblies must be made locally. In some countries, one cannot operate a third shift without government permission. In other countries, people cannot work next to a machine whose operating rate controls the person's operating rate. Labor costs in some foreign countries are attractively low, but distance and skill differences usually require expensive supervision and frequent travel, raising overhead costs.

- What are the revenue targets and cost limits imposed by management? Automatic assembly may look attractive, but the money to buy the equipment may not be available. If the market demands a product that sells for ten cents, then the total cost of labor, materials, and overhead (energy, inspectors and supervisors, insurance, shipping, etc.) will likely have to be about three cents. Can it be done?

An underlying theme in many of these considerations is flexibility. This word is used in many contexts to mean many things. In practice, it means being able to change a previously established situation in response to changing conditions. No one can predict the future with regard to demand for a product, costs of materials, or actions of governments. A commitment to one design, one supplier, one market, or one way of manufacturing is inherently risky, although it is less complex and may be more efficient. Companies constantly seek flexibility as a kind of

[1]Currently, the pace-setter for managing such a supply chain is Dell Computer. Not only does it keep its inventories down to the next few hours of orders, but Dell makes it profitable for its suppliers to do the same. Naturally, the buck has to stop somewhere, and that is at the first point where basic process times take more than a few hours. This is usually where physical components are manufactured. Thus the component suppliers are the constraint on the supply side of the equation. In order to be able to sell whatever the customer orders in spite of not having everything in stock or in the supply chain, Dell has become very skilled at managing demand. Customers are strongly encouraged to buy what is available, and sales staff are kept informed almost hourly of what is in stock or can be made in a few hours.

insurance against being wrong. It does not come without cost.

12.D.2. Manufacturing Context

The manufacturing context includes the legal factors mentioned above but focuses more specifically on the factory or factories where the product will be made. Issues to consider include the following:

- What is the work force like? Employees can be highly skilled and motivated, or they may not be. They may belong to a union that has negotiated a restrictive contract with many work rules, or they may be flexible and willing to do many different jobs, regardless of whether they are unionized. If turnover is high, jobs must be simple enough that they can be learned quickly.

- What products has the factory made in the past? The product under consideration may be similar to the factory's experience or it may be quite different. Many details must be mastered in order for a product to be manufactured and assembled reliably at high volume. These take months or years to learn. Seemingly small changes such as shifting from metal to plastic, or from large fasteners to small ones, can cause serious problems.

- Are there facility constraints? It used to be common for automobile assemblers to work in pits under the line when adding underbody parts. This is now often forbidden for work quality and safety reasons. Some companies lift the cars on rails or tilt them sideways while such operations are done. This requires high ceilings or special equipment. Assembly sequences may have to accommodate such restrictions. Similarly, paint shops have to be near ventilation systems. It is unlikely that the paint operations can be relocated to satisfy a new assembly sequence.

Underlying these factors is the need to consider each manufacturing option and location as an individual with different characteristics. This can be especially disconcerting when the same product will be made at several different facilities or when there are multiple suppliers for some of the parts or subassemblies. There is much to be gained in terms of knowledge and economies of scale by standardizing processes. At the extreme, this is called "copy exactly" by Intel as it attempts to control the complex and tricky

process of semiconductor manufacturing. There is also much to be gained by concentrating supply in one source and concentrating production in one place. As discussed above, this is the riskiest and least flexible policy.

12.D.3. Assembly Process Requirements

Design of assembly processes requires understanding the KCs as well as understanding alternate assembly methods. It also involves taking into consideration the issues mentioned above, such as model mix, volume fluctuation, facilities, and people. Often a similar product is already in production, or pilot production will be done. Either case allows for some rehearsal of assembly methods, although such rehearsals can be very misleading.

Here are some factors to consider:

- If the product, or one like it, is currently being assembled manually, then study the manual process. This will reveal difficult spots that may return in the next product, or it may suggest design or process improvements. But care should be taken because the manual process may be distorted for any of several reasons. First, it could be tailored to the idiosyncrasies of the particular factory or design decisions of past engineers. If the product is being assembled by a test crew or skilled technicians, their methods will not be representative of regular assembly operators, who probably have less skill as well as less time to do the work.[2] Also, the processes in use may have evolved and been elaborated beyond anything documented and thus may not represent something that can be easily reproduced on a different product or by a different set of people. The story about ladies "inspecting" fiber in Chapter 1 is an example of this.

- Once any existing process is thoroughly understood, it should be ignored and discarded as a basis for a new process. Instead, the engineers should begin with the product concept, its performance requirements, and its KCs and should design a process that meets those requirements. This may involve requesting some design changes to the product. Such changes should not be requested unless their effect on the product's

---

[2]The quote at the beginning of the chapter refers to this issue. Technicians can always make something work, but that is not the same as manufacturing it. Manufacturing means grinding them out day after day.

TABLE 12-1. Clustering of the Issues in Assembly in the Large into Local and Global Categories Related to Product and Process Decisions

	Global	Local
Product Considerations	Economic and market targets	Assembly sequences
	Volume growth	Types of operations
	Model varieties	Geometric constraints
	Design volatility	Characteristics of parts
	Quality, reliability, safety	Tolerances and clearances
	Make or buy decisions	Tests and inspections
Assembly Process and System Considerations	Cost and productivity goals	System layout
	Interface to the rest of the factory	Equipment choice
	Labor policies	Task assignment
	Failure modes and repair policies	Part logistics and feeding
	Space constraints	Buffers
		Inventory control

performance is thoroughly understood. If assembly engineers do not understand the product and make suggestions that compromise its performance, their suggestions will be rejected then and in the future.

- The product's architecture and outsourcing strategy will dictate, among other things, some decompositions of the product into subassemblies. The boundaries between subassemblies provide clues to subdivisions in the assembly process. Such subdivisions can be opportunities to partition the assembly plant into manageable units dedicated to doing a set of coherent tasks, such as building a portion of the product that does a testable function. The last process step is then likely to be a test of that function. Such boundaries are also rational points at which to outsource.

- The principles of design for assembly (DFA) should be applied throughout the process of generating product concepts and assembly methods. DFA involves simplification of parts handling and assembly processes. It also involves reducing the number of parts, different fasteners, and wasted motions. Reducing part count is an architectural decision that involves changing processes and materials and increasing the number of functions per part. Such decisions cannot be made in isolation of the other factors that influence architecture. DFA is the subject of Chapter 15.

It has been said that the above constitutes "design of assembly."[3] In this sense it transcends DFA.

[3]Steven LeClair, then of the U.S. Air Force Materials Laboratory, Wright Patterson Air Force Base, OH, personal communication.

12.D.4. Product Design Improvements

An analysis as thorough as the one presented above will inevitably generate additional suggestions for design improvements to make assembly easier, more robust, safer, more flexible, more economical, and so on. Some improvements can be made by changing the assembly sequence, but a small design change may be necessary to permit the more desirable sequence. In some cases it may be advisable to add a part to stabilize a subassembly, while in others it may help to consolidate some parts to improve performance, handling, tolerances, or to meet some other goal.

Naturally, few changes can be made to a product that has gone through most of the development process, as discussed above. Thus the process and assembly analyses discussed above must be done quickly and as close to concurrently with product concept development as is practical.

Outsourcing part manufacturing, assembly, or design of the equipment and lines for performing these processes obviously must be done with great care. Close and early involvement of suppliers is essential. Otherwise the processes may be done in possibly uneconomical, inflexible, or clumsy ways that could have been avoided.

12.D.5. Summary

The factors discussed above can be systematized by categorizing them according to whether they apply to the product or to the assembly system and according to whether they can be considered local or global (Table 12-1). Local generally refers to decisions that primarily impact the item at hand, or which can be made without bringing in a wide range of constituents from other organizations. Global generally refers to decisions that have implications beyond the item at hand or which have to be coordinated with similar decisions being made about other items, or which must conform to larger goals or policies set elsewhere.

12.E. CHAPTER SUMMARY

Assembly in the large differs from assembly in the small in that economic, business, and institutional issues share an equal place with technical issues. Many aspects of product design and development are strongly related to assembly or make themselves felt when assembly-related issues are brought into the product design process. The most important of these is product architecture, which defines the physical relationships between elements of the product and relates them to the product's functions. A suitable architecture is an enabler of many important processes from product development to management of variety.

The reader should be careful not to give the name "design for assembly" to the issues discussed in this chapter. This chapter's topics might better be named "design *of* assembly," as noted above, but even that characterization is too narrow. Instead, the overriding issue is concurrent design of the product, its architecture and method of delivering its functions and KCs, and the method of assembling it to meet a variety of needs in the market.

The next chapter shows how to look at a product in detail, part by part, following which we will take up product architecture, DFA, and design of assembly systems.

12.F. PROBLEMS AND THOUGHT QUESTIONS

1. Consider the four concepts for printing shown in Figure 12-2. List the functions that each one performs. Identify major parts and subassemblies for each. Make a table that lists the functions across the top and in each row below indicate the parts or subassemblies that perform or share in performing each function in each concept.

2. For each concept in Figure 12-2, identify the important KCs and draw the DFCs for them. Discuss the accuracy requirements and indicate what the manufacturer's options are, utilizing the table of methods for control of variation that appears at the end of Chapter 6.

3. Continuing with the concepts in Figure 12-2, consider the following rationales for defining subassemblies: each subassembly performs a defined testable function; each is fully constrained; each comprises a single technology that can be purchased conveniently from a single supplier; each has simple connections to other parts and subassemblies; each represents a point at which the customer could exercise choice, requiring different functionality. Use these different rationales to define subassemblies for these concepts. Are the resulting subassemblies similar across the rationales for any of the concepts or different? Is it wise to generalize from these results? If one set of subassemblies is recommended for easy functional testing while another is recommended for simple assembly to other subassemblies, which set of subassemblies should be employed in the design?

4. Two-color typewriter ribbons (black and red, typically) almost always run out of black before they run out of red. Many people are annoyed to throw away what appears to be a half-used ribbon. Designers of color-capable ink jet printers have to decide how to package the ink and face similar potential customer dissatisfaction. One option is to provide three color cartridges, generating black as a combination of colors. Another is to provide one three-color cartridge. A third is to provide two cartridges, one with three colors and one with black. And so on. What kinds of customers or markets would be best matched by each of these methods? (Consider grandparents, real estate agents, doctors, and lawyers as possible customers.) How much ink should there be of the different colors, relatively speaking, including black? Do not forget that ink and cartridges move as the printer operates, and print speed and accuracy could be affected by the forces needed to accelerate and decelerate the mass of these items.

5. High-quality ink jet printers operate at small dot pitch. If, as usual, there are different cartridges for different colors, these must align with each other. If customers replace empty cartridges, all at once or singly, then the customer becomes a product assembler and completes some important KC chains in the process. What are the design alternatives to ensure that the printer prints blacks and colors in proper alignment after the customer intervenes in the KC delivery process?

6. Explain as many assembly differences as you can that differentiate a single-use product like a hand grenade or a fire extinguisher from a multi-use, longlife product like a sewing machine or bicycle.

12.G. FURTHER READING

[Fine] Fine, C. H., *Clockspeed,* Reading, MA: Perseus Books, 1998.

[Fine and Whitney] Fine, C. H., and Whitney, D. E., "Is the Make-Buy Decision a Core Competence?" in *Logistics in the Information Age,* Moreno, M., and Pawar, K., editors, Padova: Servizi Grafici Editoriali, pp. 31–63, 1999. Also available at http://web.mit.edu/ctpid/www/Whitney/papers.html.

[Nevins and Whitney] Nevins, J. L., and Whitney, D. E., *Concurrent Design of Products and Processes,* New York: McGraw-Hill, 1989.

[Prahalad and Hamel] Prahalad, C. V., and Hamel, G., "The Core Competence of the Corporation," *Harvard Business Review,* May–June, pp. 79–91, 1990.

[Smith] Smith, R. P., "The Historical Roots of Concurrent Engineering Fundamentals," *IEEE Transactions on Engineering Management,* vol. 44, no. 1, pp. 67–78, 1997.

[Ulrich and Eppinger] Ulrich, K., and Eppinger, S. D., *Product Design and Development,* 2nd ed., New York: McGraw-Hill/Irwin, 2002.

13 HOW TO ANALYZE EXISTING PRODUCTS IN DETAIL

> "Dr. Whitney, the product we're studying has this part that doesn't seem to do anything."

This chapter lists the steps for analyzing a product, gives advice on how to take a product apart and figure out how it works, and presents some examples. The topics in this chapter are of interest any time one is contemplating improving an existing product or changing the way it is assembled. But similar considerations arise during the design of a new product. The information and methods in this chapter and Chapter 12 are important preparation for understanding the larger issues of product architecture, design for assembly, and design of assembly systems and workstations.

13.A. HOW TO TAKE A PRODUCT APART AND FIGURE OUT HOW IT WORKS

Anyone who has assembled a toy the night before Christmas or had to fix a broken one while a distraught child looks on knows much of what is said here, and that is how the author learned most of it. [Ulrich and Pearson] discusses a number of things that can be learned about a product by taking apart several models or generations of a product and comparing them. [Otto and Wood] presents a method for analyzing products that is used in part in this chapter. The literature on design for disassembly devotes attention to this topic as well.

Taking a product apart should be a systematic process whose goal is to understand its functions and assembly. Each of the steps below should be followed for each major subassembly, starting with the main product and ending when only single parts are left:

- Identify the main functions of the assembly, subassembly, or part. Identify all its degrees of freedom (translations, rotations).

- Document the product, its subassemblies, and its parts. As you take it apart down to the next subassembly level, make an exploded view drawing indicating how the parts join. Take pictures of subassemblies before and after disassembly. Some part-to-part relationships may be disturbed by disassembly and will not be obvious when reassembly is attempted. You only get one chance to observe the correct relationships during disassembly, so document them while you can. Make a liaison diagram[1] of the assembly consisting of nodes representing parts and lines representing the fact that two parts join.

- Make a parts list as you go, including each subassembly. Under each subassembly, list its parts or subassemblies in indented outline form. For each item, note how many there are, what material(s) it is made of, its classification (see below), its function(s), its manufacturing method, the name and location of the manufacturer, and whether it is a standard part like a screw or small motor, or is designed to suit this product. This information will give you a picture of how the designers approached the problem of designing this product. Each part or subassembly represents a solution to some problem and gives a clue as to how the original designers' minds work. Their solutions may be elegant or crude, and one learns a little something each time one notes them.

---

[1]The liaison diagram is defined in Chapter 2.

- Classify the items as follows:
    i. Main function carriers (carriers of important forces, motions, material flows, energy, or information[2]; conveyors or blockers of fields like electricity or heat; locators of main geometric relationships)
    ii. Functional supports (user adjustments, user access, seals, lubricants, vents)
    iii. Geometric supports (brackets, barriers, shields)
    iv. Ergonomic supports (handles, labels, safety items, indicators, warnings, finger guards)
    v. Production supports (test points, adjustment points, measurement points, fixturing or gripping surfaces)
    vi. Fasteners (reversible, irreversible)
- Keep track of dependencies between things, such as alignments, subassembly boundaries, or places where several things must line up for proper function.
- Note any cases where the product has multiple states such as on/off, locked/unlocked, forward/reverse, low-speed/high-speed, and so on. These may be associated with parts that have different positions or mating configurations in the different states.
- Keep track of all the tools needed, all the difficult steps, and any special care or consideration needed.

Take the product apart in stages and ensure at each stage that it can be reassembled from that stage.[3] This is especially important any time the disassembler suspects that energy may be stored in the product. Hidden springs are a typical hazard; they can go flying away unexpectedly and may never be found again. It is a good idea to separate items partially, peek inside if the items are covers, and try to see if any surprises are in store.

Look for clues as to how it comes apart. These include parting lines and the direction from which fasteners appear to insert. This will give an indication of the product's architecture and overall design. Some products are obviously contained within an outer housing which must be separated before internal parts can be seen and further disassembled. A typical example is an electric screwdriver. Other products do not have this kind of architecture. An example is typical clock or watch works, in which the top and bottom plates together provide location and alignment for many other parts. As soon as one plate is removed, the other parts can spontaneously separate from each other. A third architecture is represented by a car engine block. Typically over two hundred parts are fastened to its outside by screws. Inside the block and head are an additional hundred or so parts. But there is no outer cover which, when removed, reveals the remaining parts.

You may encounter parts or features whose purpose cannot be explained. We call these "mystery features." Features cost money and are rarely without purpose. Figuring them out can be educational. Possibly they are of use on a different model of the product and are put there via a parallel production process[4] like molding. It may be cheaper to make all the parts the same than to make a separate mold for each version. On the other hand, the mystery feature may perform an important function, in which case the analyst must determine what it is. Examples are in Section 13.C.4.

It is always useful to have a magnifying glass handy so that small details on parts can be observed. These include surface finish quality, molding methods such as location of risers, dates or location of manufacture, and so on. In a product made in China for export, we found assembly instructions in Chinese molded into the insides of several parts. One can also assess fabrication quality, such as the quality of solder joints.

13.B. HOW TO IDENTIFY THE ASSEMBLY ISSUES IN A PRODUCT

Analysis of a product from the viewpoint of assembly requires addressing many levels of detail. Here we emphasize the lower levels, but it is important to remember that as a whole, we recommend a top-down approach, beginning with functional, physical, and economic requirements, and then proceeding to deal with the supporting details, as outlined in Chapter 12. Top-down is an admirable goal, but

---

[2]These functional categories were developed in [Pahl and Beitz].

[3]This is analogous to "woodsmanship" advice to look over one's shoulder periodically while hiking so that the way back will look familiar.

[4]A parallel process creates all the part's features at once. A serial process, such as machining, creates the features one or a few at a time.

it is not always possible or even feasible. In many cases, one is confronted with an existing design which is being modestly modified. In fact, "reuse" of previous parts or subassemblies is becoming mandated at many companies in the interest of saving development and verification time and cost. Therefore, we begin by listing the steps for analyzing a product in detail:

- Understand each part, its material, shape, surface finish, and so on.
- Understand each assembly step in detail, including all necessary motions, intermediate states, in-process and final checks for completeness.
- Identify high-risk areas.
- Identify necessary experiments to reduce uncertainty about any step.
- Recommend local design improvements.

It is important that these analyses be performed by a group of people working together who collectively have the skills and background to consider a wide range of technical and nontechnical issues. This will ensure that the parts are subject to a broadly based set of eyes and criteria and that interactions between parts and among opportunities for improvement are recognized. This may well be the only time when all the parts are considered at the same time for the same reason. This important opportunity for integration should not be missed.

Analyzing an existing product requires taking it apart. Pointers for doing this and for looking carefully are given in Section 13.A. We now take up each of these steps.

13.B.1. Understand Each Part

Assembly analysts have the responsibility for understanding not only what each part is but also what it does. If its function is not understood, then redesign recommendations may make the part incapable of performing its function. On the other hand, some recommendations listed below seek to combine parts. Again, the required function must never be compromised.

This analysis must include understanding how each part is made, why its material was chosen, what surface finish and tolerances it has, and how these might influence how it will be assembled. As discussed in Chapters 10 and 11, size, shape, surface finish (as it influences friction) and clearance to a mating part heavily influence success or failure during part mating. To help in this process, one

may make drawings of the parts either on paper or in a computer. These drawings are useful in step 2 where each assembly action is studied.

This is the time to recognize and understand mystery features.

13.B.2. Understand Each Assembly Step

In order to begin this step, it is necessary to have either the parts or the drawings made in step 1. Each part mate should be studied in detail. Each surface on a part that will or could contact a surface on a mating part should be identified. Possible mismated states should be noted, along with possible ways that the parts could become mismated. Two such states, called *wedging* and *jamming* respectively, are analyzed in detail in Chapter 10. Find all the places on each part where it might be gripped or fixtured. Keep in mind that only one or a few of these feasible places will actually be possible to use, for a variety of reasons.

First, depending on the assembly sequence, a candidate grip or fixture location could be obscured or in use already as a mating feature to another part. Second, and much harder to see just by looking at the parts, the relationship between the gripped point and the mating feature on the part may not be adequately toleranced. The result of this is that if machine or robot assembly is being used, the mating point may not be in the correct location in space at the moment of assembly even if the gripped point is. The influence of tolerances and the relationships between features within and between parts are discussed in Chapters 2 through 6.

Rehearse or imagine each assembly step occurring before your eyes. "Watch" the parts move through space and meet each other. Try to anticipate how things could go wrong, including collisions with neighboring parts or between parts and tools, grippers, or fixtures. One may be able to use simulation software to aid this part of the analysis. This analysis may turn up many situations where parts could damage each other. For example, soft items like seals could be cut by sharp metal edges. All such edges should be found and targeted for softening or chamfering. Another example is a situation where a part could be assembled the wrong way.

It is often surprising how much one can learn doing one of these analyses, and how often an outsider can learn things that the product's designers or current assemblers do not know. As noted in the Preface, the author spent many years with colleagues analyzing commercial products for assembly. We learned repeatedly that people do not understand their own processes. Once we hired a new employee who accompanied us on his first visit to a client whose product we were assessing for possible robot assembly. We scheduled a one-hour meeting with the line supervisor to learn in detail about the existing manual assembly processes. The meeting quickly extended into three hours and was not completed before we had to depart for the airport. We found that in many cases a step in the "official computer printout" of the process proved impossible. For example, one part could not be assembled in the official sequence because it would obscure an adjusting screw on a previously assembled part. As we identified each such disconnect in the process, the line supervisor became more concerned and perplexed, being reduced finally to making a long list of action items to check the next time he visited the line. As we were approaching the car in the parking lot, well out of earshot of our host, our new colleague asked, "Is it always like this?" We answered in unison: "Yes, it's always like this!"

13.B.3. Identify High-Risk Areas

High-risk areas are those parts of the process that could go wrong, cost a lot, damage parts, injure employees, or cause an assembly station, whether manual or mechanized, to fail too often.

First priority goes to identifying "showstoppers," those events that stop a machine from working, or which violate regulatory or safety standards. Such events get their name from the high likelihood that there is no solution. One example involved the need to apply a small amount of a low-viscosity adhesive to parts that would eventually spin at a high rate. The slightest excess of this material would be instantly sprayed all over the inside of the assembly, ruining it. A redesign was proposed that provided a well in which any excess would be trapped.

Another tipoff that a step has high risk is that only one person on the line can perform it. Once we observed a line that had *two* such steps, each done by a different person. "Don't let those two carpool!" one of us said. This kind of situation leads naturally to the conclusion discussed at length in Chapter 1, namely that if we can't explain a task to another person, we won't be able to explain it to a machine.

Any step where part damage is likely is automatically high risk. In one product we studied, the parts were extremely fragile ceramic insulators, shipped to the line immersed in sawdust. Clearly the objective of the assemblers was to keep from breaking them, well above any requirement to assemble them, since they were very expensive. Similarly, for some parts, even miniscule surface contamination by particles or chemicals will ruin them. Semiconductor wafers are a familiar example. An 8-inch-diameter wafer with 100+ Pentium chips on it represents $30,000 or more value at retail, and particles even smaller than 1 μm will ruin a chip.

A less obvious risk area is one where no available assembly sequence is suitable, although an attractive one is just out of reach for some reason. Perhaps a small redesign will make that attractive sequence feasible, but unless that redesign is accepted, the process contains risk. In one case, we recommended *adding* a part to a subassembly so that it became stable and could be inserted as a unit without complex tooling. Note that this violates the desire expressed above and in Chapter 15 to reduce part count.

Still less obvious but very important for eventual mechanization of an assembly process is risk caused by variable process time. An example is calibration, which can take more or less time depending on how far off the desired setting the assembly is when it arrives at the calibration station. In one case, Denso eliminated most of the task time uncertainty by correlating the final calibrated setting of thirty or so previous assemblies with the initial error observed prior to starting calibration. The first step in the calibration was then selected from the correlation table, and nearly every calibration was finished in two steps, a predictable time.

13.B.4. Identify Necessary Experiments

Experiments are costly and time-consuming and thus should be performed only when really necessary. Simulations are becoming increasingly realistic and should be tried first. Nevertheless, no simulation can anticipate every problem, and some problems are notorious for arising as a result of something that is on the parts but not in the design. Examples include small burrs, sharp edges, springy parts with minor residual shape distortion, or surface contamination from cleaning processes.

Experiments can be directed at confirming either technical or economic feasibility. While the former is the most obvious application, the latter can be tested by finding out how long it really takes to do a task without making a lot of errors, or how much things really cost to make or buy. Sometimes, as indicated in Chapter 18, it is only the product of time and cost that matters, and a slower but cheaper process may be the economic equivalent of a faster but more expensive one. Sometimes the slower alternative is less complex and more reliable, tipping the balance in its favor.

In case of technical feasibility evaluation, it is essential to identify at the outset what are the criteria for successful assembly in terms of time, error rate, tolerable forces exerted on the parts, and so on. Any successful process will contain designed-in poka-yoke that prevents the standard errors and, if possible, signals if any of them occurs.

Finally, a real physical experiment reveals potential undocumented sources of trouble. These can arise from undocumented features on parts or unexpected behaviors of people or equipment. Only by trying them out can such problems be revealed. An example of this was cited in Chapter 1, namely that of the ladies who were "cleaning" fiber.

13.B.5. Recommend Local Design Improvements

All the above analyses and studies will generate suggestions for improvements. These can range from adding or removing a detail from a part to adding or removing parts. The highest priority items address the high risk areas, especially the showstoppers. Others improve technical or economic feasibility. Improvements of this kind address distinctly local issues and are unlikely to affect strategic matters such as how many different product styles can be accommodated or what the platform strategy for a product family will be. These strategic issues are the province of assembly in the large.

The next section gives several examples of product analysis: an electric drill, a toy (surprisingly complex), a camera, and some mystery features.

13.C. EXAMPLES

13.C.1. Electric Drill[5]

An MIT student group took apart and carefully analyzed an electric drill. They listed every part, noted its material, measured key dimensions at places where they joined each other, and enumerated the motions needed to put them together. Figure 13-1 is a photo of the drill with the top cover off. Figure 13-2 is an exploded view. Table 13-1 is the parts list. Table 13-2 lists several part mate dimensions.

The next few paragraphs detail the assembly steps, noting the gross motions of part movement and fine motions of part mating.

13.C.1.a. Transmission Subassembly
13.C.1.a.1. Step 1. This step inserts a small shaft (14) and a pinion gear (13) into the middle mount (12) containing several bearings. See Figure 13-3. Features on parts where assemblers can grip are cylindrical surfaces and gear teeth. The orientation of the assembly is from up to downward against gravity. Jamming can occur in the peg–hole assembly. This process needs two hands, because the assembler should hold the gear to fit the shaft to the hole. If we use

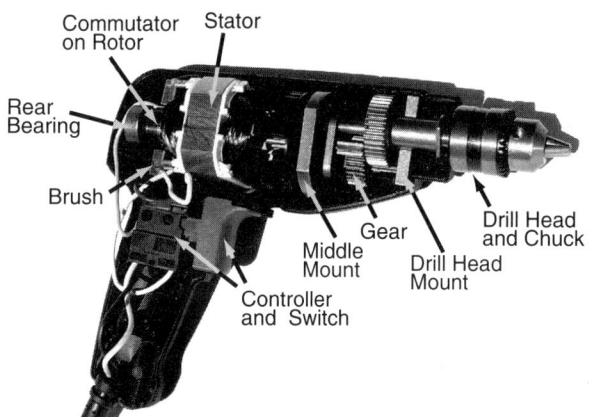

FIGURE 13-1. Electric Drill.

a fixture to fix the mounting plate, it will mate the plate's cylindrical surface.

13.C.1.a.2. Step 2. This step adds the drill head subassembly (15) to the subassembly built in step 1. The drill head's shaft mates to plate (12) and its gear mates to the pinion (13). See Figure 13-4. Features on parts where the assembler can grip are cylindrical surfaces. The subassembly made in step 1 is very loose, because no fasteners are used. So, it can fall apart if we are not careful about holding it with the gear facing upright. If we think about automatic

---

[5]This material was prepared by MIT students Young J. Jang, Jin-Pyong Chung, and Nader Sabbaghian. The drill is also discussed in Chapter 14.

TABLE 13-1. Parts List for Electric Drill in Figure 13-2

Part Number	Part Name	Part Description
1a	Top plastic casing	Plastic casing placed on top of the bottom casing after the insertion of drill subassemblies.
1b	Bottom plastic casing	Plastic casing used to house the drill subassemblies.
2	Stator	Houses the rotor and connected to electromechanical controller and switch.
3	Controller/switch	Variable-speed plastic switch with electrical connectors to power cord and stator.
4	Power cord	Connected to switch, provides connection to 120-V, 60-Hz AC power.
5a	Left brush housing	Brass component connected to wiring from switch, used to hold a brush and spring.
5b	Right brush housing	Same as left brush housing (5a).
6a	Left spring	Spring mechanism used for the placement of the brush in the casing.
6b	Right spring	Same as left spring (6a).
7a	Left brush	Rectangular block of carbon interfacing with the motor and switch.
7b	Right brush	Same as left brush (7a).
8	Thin washer	Plastic washer placed at the back end of the rotor. It is used to prevent lateral movement of the rotor.
9	Thick washer	Same as 8. Possibly selected from several available thicknesses.
10	Rotor	Rotor component equipped with radial fan blades and front gear.
11	Spring washer	Metallic washer used to facilitate the insertion of the subassembly into the plastic casing and keep the rotor from rattling laterally.
12	Middle mount	Used as an interface between the back part of the assembly (rotor) and the front part (drill head).
13	Pinion Gear	Used for the transfer of motion from the rotor to the drill head via the middle mount.
14	Gear shaft	Used to connect the pinion gear to the middle mount.
15	Drill head and chuck	Equipped with gear which interfaces with part 13. Its back shaft is housed in the middle mount and is equipped with a small thrust bearing.
16	Drill head mount	Semicircular structure supporting the drill head, placed inside the bottom casing; supports gear shaft.
17	Rear bearing	Made of powder metal bronze impregnated with lubricant. A locking mechanism prevents it from rotating once placed in the plastic casing.
18	Screws (8)	Fasten top and bottom casings together.

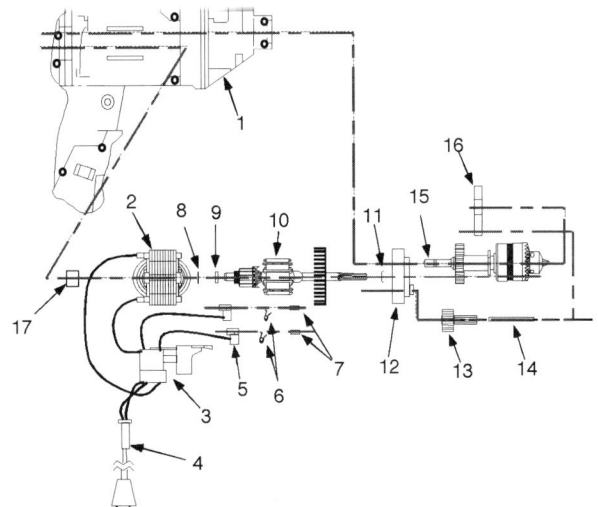

FIGURE 13-2. Exploded View of Sears Craftsman Drill.

TABLE 13-2. Part Dimensions Related to Joints Between Parts

Mating Parts	Clearance (inches)	Clearance Ratio
8 to 10	0.013	0.040
9 to 10	0.013	0.040
17 to 10	0.008	0.025
11 to 10	0.033	0.096
12 to 10	0.008	0.025
12 to 15	0.001	0.003
12 to 14	0.005	0.040
13 to 14	0.005	0.040
16 to 15	0.01	0.016

Note: The clearance ratio is defined as the clearance between two parts at a feature where they join, divided by the size of the feature. For example, in a pin–hole joint, the clearance ratio is the diametral clearance divided by the diameter. This concept is discussed in Chapter 10, where its influence on ease of assembly is quantified.

assembly, the gear teeth between the two gears can collide if not properly positioned during assembly.

13.C.1.a.3. Step 3. This step joins the rotor (10) and drill head mount (16) to the subassembly made in step 2. To make this happen most easily, the subassembly from step 2 should be reoriented in the horizontal direction (see Figure 13-5). This is due to the fact that it is not easy to assemble the rotor shaft vertically into the mounting plate while holding the washers (8 and 9) and journal bearing (17) at the other end. Even when it is reoriented, it is difficult to hold everything without any gripper or fixture. So, the assembler must use his or her whole palm and fingers to assemble these parts. This could present a challenge for the assembler and potentially increase the assembly time. If we use a gripper, it will be easier to perform this step. However, this means introducing an additional step in the process, that of attaching the gripper to the gear-train subassembly.

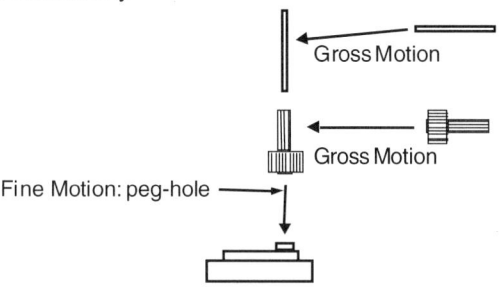

FIGURE 13-3. First Step in Assembling the Transmission Subassembly of the Drill.

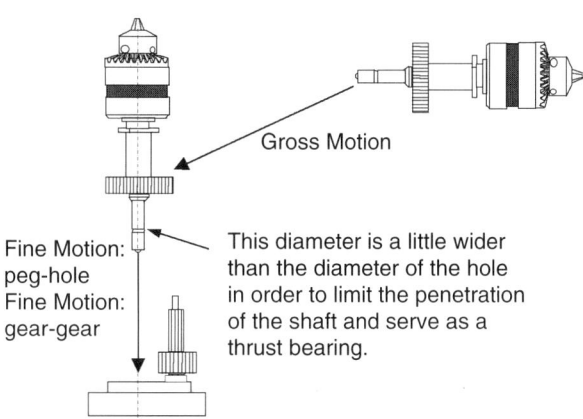

FIGURE 13-4. Second Step in Assembling the Transmission Subassembly of the Drill.

A little grease might be used to hold the bearing onto the end of the shaft temporarily, but this will clog the bearing and keep the impregnated oil from emerging later. Another possible solution is to put the bearing in the bottom casing instead of onto the shaft. But once this is done, it is impossible to mate the shaft with it. In any case, this does not solve the problem of keeping the washers on the shaft.

13.C.1.b. Power Generation Subassembly

The power subassembly (parts 2–7) consists of the motor, switch, and wires, plus brushes and their springs (see Figure 13-6). Except for the brushes, all joints in this unit are pre-assembled and fastened. So, it is easy to handle. But the lengths of the wires are not optimized and are unnecessarily long. It is also very hard to insert the springs that hold the brushes in the rectangular holes. This consists of a spring-locking mechanism that keeps the brushes tightly inserted in the brush holders, yet allows them to be released once assembled to the armature and pressed against

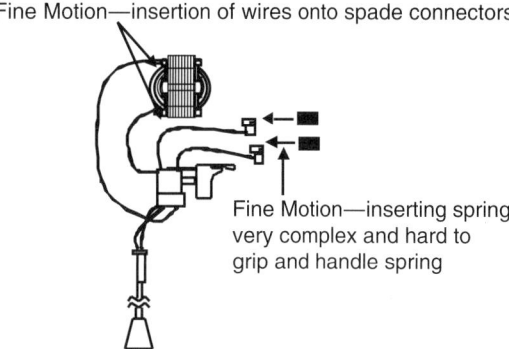

FIGURE 13-6. Assembly of the Power Generation Subassembly.

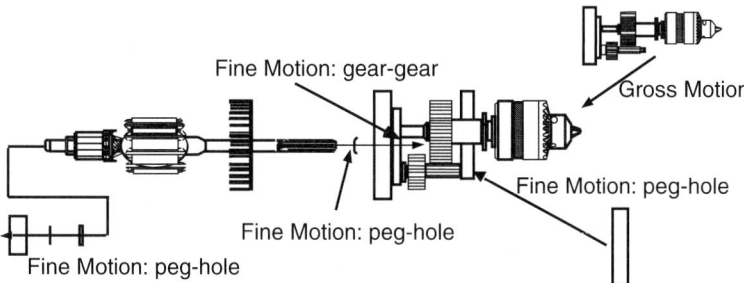

FIGURE 13-5. Third Step in Assembling the Transmission Subassembly of the Drill.

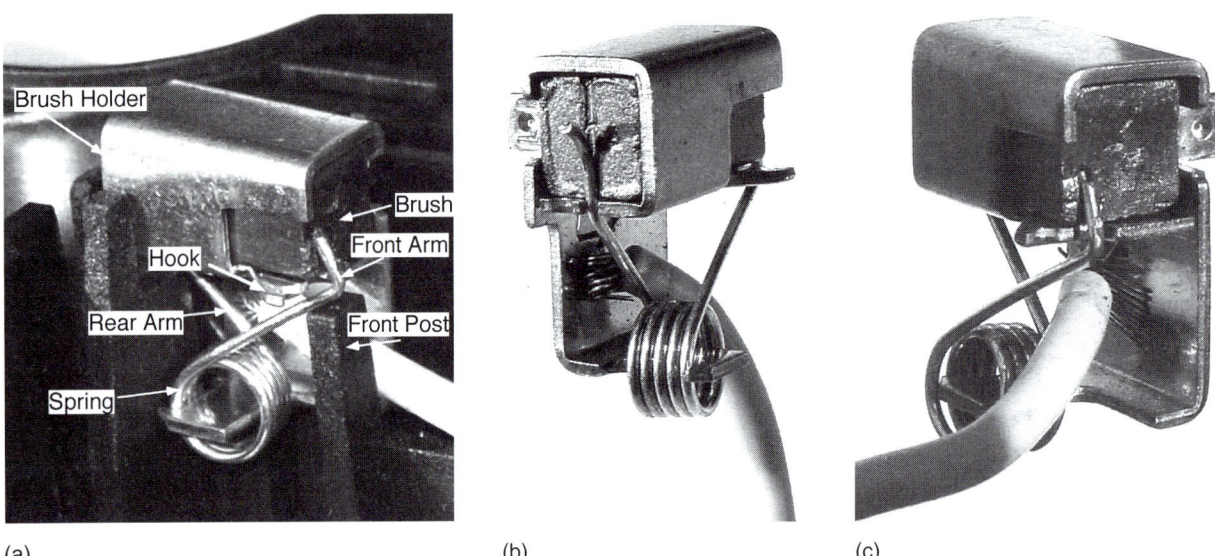

(a)   (b)   (c)

FIGURE 13-7. Photos of Brush Holder, Spring, and Brush Subassembly. (a) Brush and holder partially inserted into the casing. (b,c) Detailed views of brush and holder. This clever subassembly has two states. Before being inserted into the casing, it is cocked: The coil portion of the spring is placed on a pin on the holder with its rear arm inside and its front arm outside. The brush is placed in the holder, and the front arm is carefully stretched and placed on the face of the brush as shown in the detail photos. This pushes the brush back inside the holder. The photo above shows the cocked subassembly after it has been inserted part way into its final position in the bottom case. (Normally, the rotor would be installed before this step, but it has been removed to permit the photo to show the situation.) When the subassembly is inserted all the way, the front post dislodges the front arm of the spring from the face of the brush. The front arm snaps back until it rests on the hook. The rear arm of the spring then can push the brush forward into contact with the rotor. When the drill was first disassembled, the hook was a mystery feature. (Photos by Karl Whitney.)

it.[6] These parts are shown in Figure 13-7 and Figure 13-8. They can be assembled at this stage, or this step can be delayed until after the power subassembly and transmission subassembly have been mated to the bottom casing during final assembly.

13.C.1.c. Final Assembly

To assemble the entire unit, the armature of the transmission sub-assembly should be inside the stator of the power generation subassembly (see Figure 13-9). The joints between the casings and the parts of this subassembly are very tight fitting in order to prevent rattling and wear while transmitting high torque. It is very difficult to hold these two subassemblies together and perform the

gross motion to the plastic casing. In the difficult fine motion between the plastic casing and two subassemblies, many parts must assemble simultaneously into tight clearances. The parts can be tilted relative to each other during the assembly process, because of the clearances between shafts and holes. This can keep the middle mount, drill head mount, and drill head from assembling to the bottom casing.

During the assembly process, manual feedback control in fine motion is needed to adjust the angles of shafts and the middle mount horizontally and vertically. The transmission and power generation subassemblies are only loosely joined, and it is therefore necessary for the assembler to grip the entire subassembly in two locations (one on the transmission and one on the power generation part) to ensure that the overall subassembly maintains its proper alignment for insertion into the plastic casing. The alignment and free motion of the gears and the clearance between the armature and the stator should be checked before the closing of the top plastic casing. The joint between

---

[6]Getting spring-loaded brushes into operating position in contact with commutators is a generic problem in motor assembly. There are many clever solutions, most of which require that the rotor be in place first and the springs activated later.

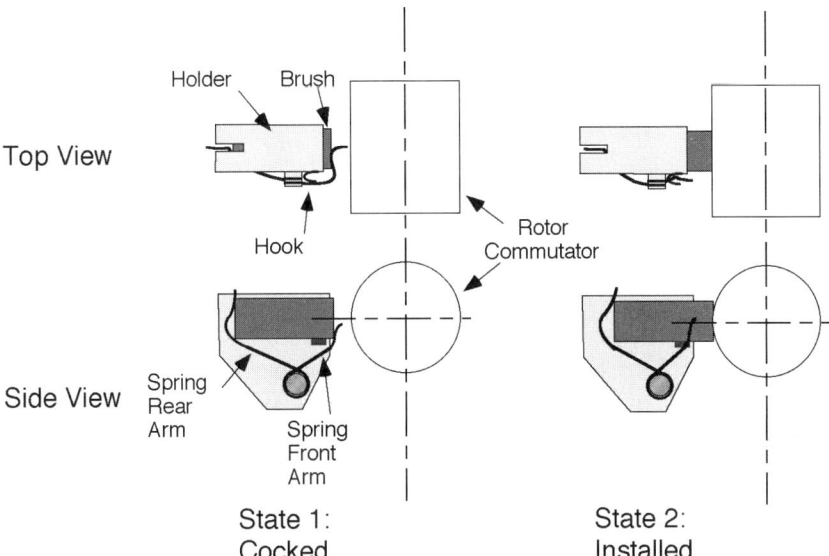

FIGURE 13-8. Illustrating the Two States of the Brush-Holder Subassembly.

Top View

Holder   Brush

Hook

Rotor
Commutator

Side View

Spring
Rear
Arm

Spring
Front
Arm

State 1:
Cocked

State 2:
Installed

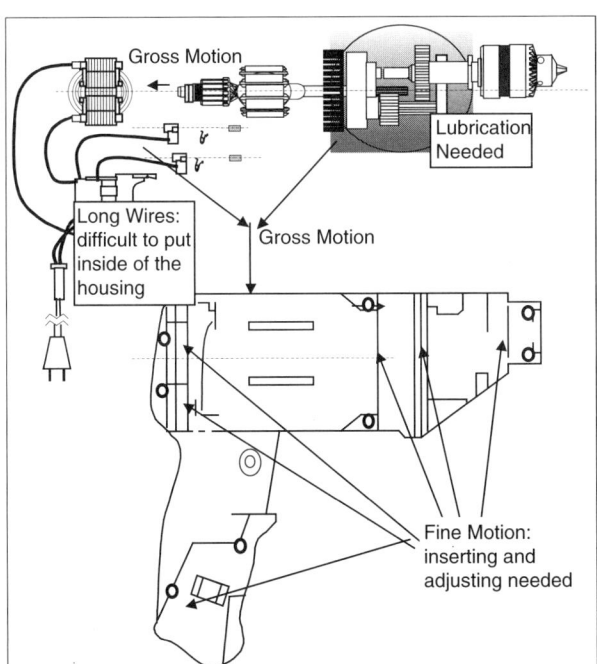

Gross Motion

Lubrication
Needed

Long Wires:
difficult to put
inside of the
housing

Gross Motion

Fine Motion:
inserting and
adjusting needed

FIGURE 13-9. Final Assembly of the Drill.

middle mount and the drill-head's shaft is the one most likely to jam during this final step.

After these parts are installed, the brushes are installed into their housings and the springs cocked, if this was not done before. Then each brush holder is pressed into its pocket in the bottom casing, releasing the brush. This is an awkward motion. If it is done incorrectly, the brush could fly out under spring action.

The wires must be routed carefully and tucked away from the joint between the top and bottom casings. This, too, is an awkward step.[7]

Eight screws are used as fasteners to assemble the two housings.

13.C.2. Child's Toy

Let us examine another example, a low-cost toy. The electric "robot dog," illustrated in Figure 13-10, is operated by a small control box containing two batteries and two buttons. Pushing one button causes the dog to walk, while pushing the other causes the head to bob and the dog to emit a squeak. The dog's tail wags, its ears swing, and lights in its head and tail blink while it is walking. It costs $5.99 retail and is made in China. It is one of a family of four similar toys with similar functionality and the same price and target market.

---

[7]The author had an older drill whose casings were metal. One day he felt a tingling in his hands while using this tool. Upon opening it, he found one of the wires crushed between the casing halves and the conductor exposed, creating an electrical path to his hands. Newer tools must obey double insulation regulations, so this hazard will not occur.

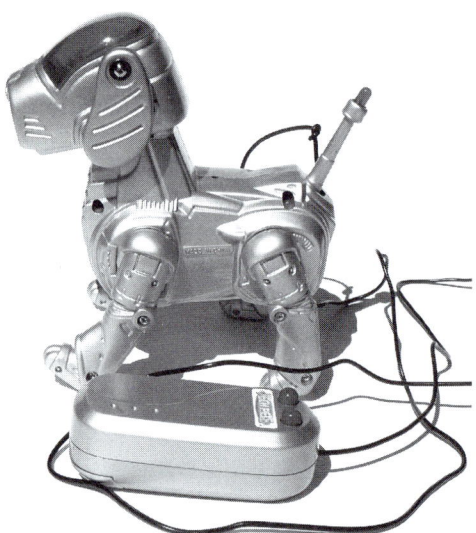

FIGURE 13-10. "Robot Dog" Toy with Control Box. (Photo by the author.)

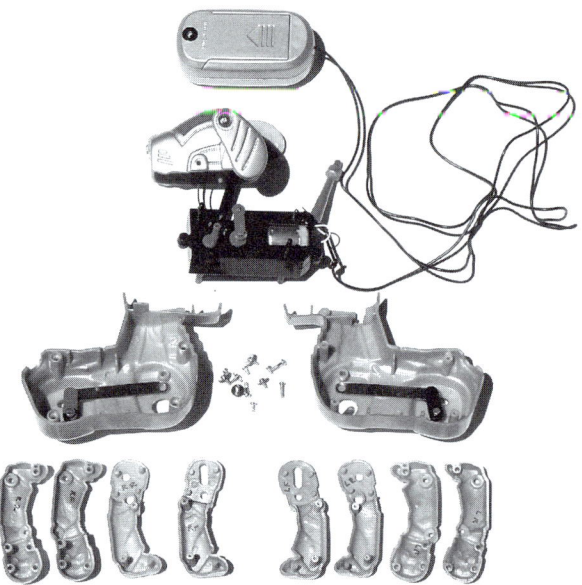

FIGURE 13-11. "Robot Dog" Disassembled Down to the Gearbox Subassembly. (Photo by the author.)

The toy is made almost completely from fair quality plastic injection molded parts. Partially disassembled, it appears in Figure 13-11. The main parts are the head with two ears and a diaphragm that emits a squeaking sound, a two-part body held together with four screws, four two-

part legs each held together with two screws, and a central gearbox and motor subassembly.

The gearbox, shown in Figure 13-12, contains a motor, a right angle power takeoff gear, five other reduction and drive gears, and four levers for driving the left and right leg pairs, the head, and the tail respectively. Table 13-3 lists the parts, their quantities, and materials.

One interesting feature of this toy is the gearbox. It is a separate subassembly. The motor is very small and delivers its power at high speed. Speed reduction and torque enhancement is attained through a right angle drive gear that engages the pinion on the motor shaft. Several reduction stages reduce the speed further. The lowest speed drives the legs while intermediate speeds drive the head and tail. Power is delivered directly to the front legs while individual levers transfer power from them to the rear legs on each side.

The gearbox is completely assembled before the power wires are soldered to the motor. This can be seen by close inspection of the plastic gearbox material near the motor terminals, where it is easy to see melted areas caused by the soldering iron. In turn, this means that the gearbox assembly cannot be tested until it is assembled and the wires attached, and it cannot be disassembled without either unsoldering or cutting the wires. Wires linking the tail and head lights to the power source are soldered to the motor terminals as well, meaning that the entire assembly is tied together permanently inside by wiring. This is typical of small low cost toys.

Another interesting feature of this product is the fact that it is assembled completely with small Philips head screws. It is obvious from the awkwardness of many of the assembly steps that all these screws are installed manually, probably with hand-held power screwdrivers. In fact, it is clear that the whole product is assembled manually because the parts are too awkward for automatic part feeding or assembly. A few of the screws could have been replaced by snap fits, especially where the outer leg parts join the inner leg parts. But such replacement would have required higher-quality molds and plastic material than might have been justified in such a product. In other locations, screws are probably unavoidable and better than most alternatives.

Even though this is a simple toy, it has a remarkable number of parts and functions. It shares many design elements with much more sophisticated products such as cameras and tools: lots of injection molded parts, screws, motors, and wires. It demonstrates that such simple

FIGURE 13-12. Gearbox, Tail, and Head. The gearbox has been opened and some of the gears have been removed. The leg drive gear and shaft is a two-part assembly that passes completely through the gearbox. One half of the shaft must be assembled to the other half after the gearbox is assembled. Head and tail are linked to the gearbox by wires and drive levers that have not been separated from the gearbox in this photo. (Photo by the author.)

TABLE 13-3. Part Statistics for "Robot Dog"

Part Name	Material	Quantity
Body, left and right	Plastic	One each
Leg, outer half	Plastic	Four each
Leg, inner half	Plastic	Four each
Head	Plastic	One
Face in head	Plastic	One
Ears	Plastic	Two
Leg drive arm	Sheet metal	Two
Tail	Plastic	One
Small lights or LEDs	Multiple materials	Three
Tail drive arm	Metal rod	One
Head drive arm	Metal rod	One
Leg drive lever	Plastic	Two
Gearbox body	Plastic	Two halves
Motor	Multiple materials	One
Gears and drive shafts	Plastic, or plastic with metal shafts molded in	Seven
Spring	Steel	One
Remote control body	Plastic	Two halves
Control buttons	Plastic	Two
Electric contacts	Metal	Two
Screws, Philips head	Metal	Four for leg assembly, seven to attach legs to drive linkages, three for gearbox assembly, two for ears, two to attach head to body, four for body assembly, two for remote control assembly; total: 24
Wires	Metal and plastic	Six
Batteries	Multiple materials	Two
Total: 48 plus screws		

TABLE 13-4. Part and Fastener Statistics of a $100 Canon Camera

Fastener Type and Count	Part Type and Count
6 metal rivets	20 springs
2 glue joints	30 plastic gears
2 press fit studs	8 magnets
A few snap fits	40 metal stampings
A few retaining rings	10 lens optical elements
60 screws	10 major plastic molded parts
	1 light pipe
	1 motor
	1 flash unit (bought as a subassembly)
	3 printed circuit boards, both rigid and flexible
	2 relays
	6 switches
	50 electrical components
	20 wire crossovers on circuit boards
	100 other parts not easily classified

Note: This camera has over 350 parts.

products can be interesting and instructive from a design and assembly point of view.

13.C.3. Statistics Gathered from a Canon Camera

Greg Blonder, formerly of AT&T Bell Laboratories (now Lucent Technologies), took apart a Canon camera as part of a study of the design of Japanese consumer electronic products.[8] He carefully took note of the number of parts, type of parts the materials they were made of, the joining methods, and the quality of parts and joints. These are summarized in Table 13-4.

Blonder made several astute comments about this camera and other similar products. First, such products have a remarkable number of complex parts and perform many sophisticated functions, yet they are very modestly priced. (The camera cost $100 in 1990.) Second, a large number of the parts are complex plastic injection moldings. This represents a growing trend in which polymers are

becoming more and more like metals in their ability to support a large number of intricate features and relatively fine tolerances. Third, the molded parts do not have any flash—that is, wisps of material left over from the molding process. Flash often is caused by molten material leaking into gaps between separable parts of the mold. Absence of flash indicates that great care is taken in maintaining the molds. (The plastic parts in the "robot dog" are not high quality by comparison and have considerable flash and poor feature definition.) Fourth, screws are the predominant fastening method, as they are with the "robot dog." They are strong and can be installed with great reliability. Adhesives are rarely used except to hold parts of similar materials where strength and close alignment are not needed.

The point here is not necessarily that these are good product design practices, although some of them may be. The point is that one can learn a great deal by looking very closely at a product or family of products.

13.C.4. Example Mystery Features

A challenging example of mystery features arises in cordless appliances whose rechargeable batteries are soldered to the drive motor. Such batteries typically are uncharged at the time of assembly and remain that way (to extend their shelf life) until purchased. Inside one such product, a small vacuum cleaner, we found a wire with a small metal tab soldered to it, apparently leading nowhere (see Figure 13-13). The analysts (the author and a group of students) noticed that the tab was assembled to a place where it was accessible from outside the product through a small hole. It then became clear that this hole, together with a contact at the battery charger receptacle, permitted the product to be tested after assembly through an electric circuit that bypassed the uncharged batteries.

On a second such product, a cordless screwdriver, a mystery hole was observed in the on–off switch. Close observation revealed that if the switch was pushed to the on position, a small probe could be inserted through the hole and made to contact one side of the motor circuit. Since the other side of the motor circuit could be accessed through the charger receptacle, a test path was again made available. On a third such product, a different brand of cordless screwdriver whose batteries were in a removable pack, no such mystery feature was found since direct access to the motor circuit was available through the contacts used by the battery pack.

---

[8] *Design for Assembly,* video of a presentation by Greg Blonder at Lucent Technologies, January 16, 1990. Given to the author by Greg Blonder.

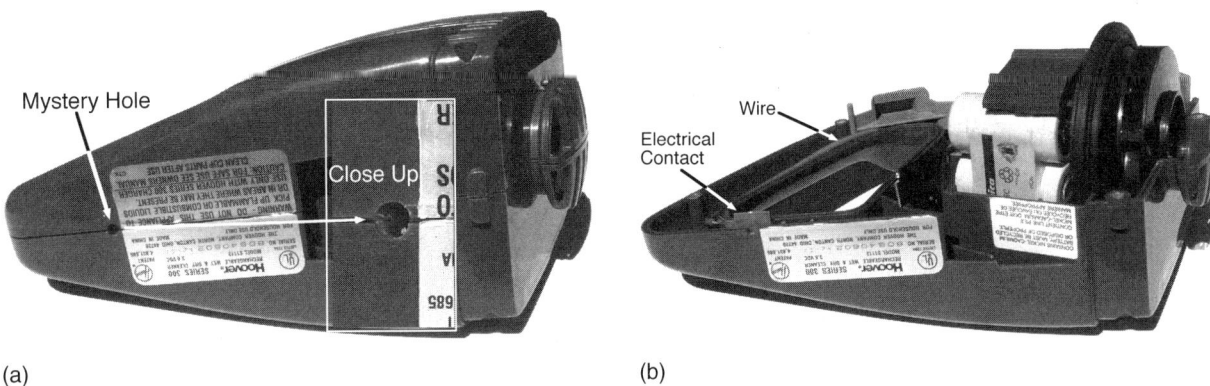

FIGURE 13-13. A Product with a Mystery Part. This product is a small vacuum cleaner. Only the motor end is shown. In part (a) can be seen a small hole whose purpose was initially unknown. When the unit was opened (see part (b)) an electrical contact was found behind the hole, from which a wire led back to the motor.

This example shows several things. First, it is not easy to test cordless products whose batteries are permanently wired in because test current could be diverted into the uncharged batteries instead of into the motor. Thus some kind of workaround is needed. More generally, testing may be difficult for a variety of reasons, and products may contain special nonfunctional features that support testing and only testing. Third, to repeat a point made earlier, there is much to be learned by looking carefully at all details of a product.

13.D. CHAPTER SUMMARY

In this chapter, we discussed how to look at a product in detail, how to take it apart and understand how it works, and how to look for potential assembly problems. Along the way we identified a number of concepts such as part mating failure, design for assembly tradeoffs, product architecture, and economic analysis. These topics are treated elsewhere in this book in detail.

13.E. PROBLEMS AND THOUGHT QUESTIONS

1. Suppose you take apart a product and find that holding the case together are six screws, of which four are long and two are short. Does this represent good or bad design? How could you tell which? What information would you need?

2. On a cordless screwdriver, the handle end is held together by snaps while the screw-driving end is held together by four screws. Why? Perhaps the designer could not make up his mind whether to obey DFA recommendations to eliminate screws or not. Perhaps there is a better reason.

3. The example products discussed in this chapter are of the type where internal parts are packaged by a pair of outer casing parts. This is commonly called a "clamshell architecture." Look around at other products and identify those that have clamshell architectures and those that do not. Try to understand why the designers of these products chose their architectures.

4. Simple consumer products increasingly are being made from injection molded plastic. This applies especially to the outer casings of drills, can openers, food mixers, coffee makers, and so on. The materials are stiff and can be molded with surprising accuracy and high complexity. Discuss how the availability of such processing methods affects assembly.

5. Following on Question 4, it has been noted that simple consumer products of the type mentioned are increasingly being made in low-wage countries and exported to the industrialized countries. Yet the availability of complex molding methods clearly permits a great deal of part consolidation, sharply reducing one of the main requirements for assembly labor. Why isn't the manufacture of such products repatriated to the United States if assembly labor, admittedly more costly here, is almost unneeded, while shipping costs are clearly larger for imported products?

6. See if you can identify mystery features in a product that can only be explained by product variety (that is, the features are used in some other version of the product but not the one you have just taken apart). See if you can figure out what the other version

would use that feature for, or, failing that, obtain another version and see if the mystery feature is used. Discuss the possibility that the feature is not used at all by *any* version of the product, and provide some reasons why it is there anyway.

7. Note any difficult assembly steps in a product you are analyzing and ask yourself if simple tools, holders, clamps, or presses would make the assembly easier. If not, what portions of which parts should be redesigned?

13.F. FURTHER READING

[Boothroyd, Dewhurst, and Knight] Boothroyd, G., Dewhurst, P., and Knight, W., *Product Design for Manufacture and Assembly,* New York: Marcel Dekker, 1994.

[Otto and Wood] Otto, K., and Wood, K., *Product Design: Techniques in Reverse Engineering and New Product Development,* Upper Saddle River, NJ: Prentice-Hall, 2001.

[Pahl and Beitz] Pahl, G., and Beitz, W., *Engineering Design,* 2nd ed., New York: Springer, 1996.

[Ulrich and Pearson] Ulrich, K. T., and Pearson, S., "Assessing the Importance of Design Through Product Archaeology," *Management Science,* vol. 44, no. 3, pp. 352–369, 1998.

14. PRODUCT ARCHITECTURE

"We took apart our car and their car and found that our parts were as good as their parts, or better. But they have a better car and we don't understand how it happened."

14.A. INTRODUCTION

Product architecture is about the relationships between the whole product, its parts and subassemblies, how those items are arranged in space, and how they work together to provide the product's functions. Product architecture is widely discussed and studied because it has such a strong influence on how the product is designed, manufactured, sold, used, upgraded, repaired, and recycled. It is therefore not surprising that it is also widely debated, and no single acceptable definition has emerged that captures all of its influences and nuances.

In this chapter, we will discuss product architecture in general, to show how it influences the product and to show how architecture issues interact with assembly. We will find that, while architecture affects different phases of the product's life, the decisions, once made, are implemented during assembly, affect assembly, or provide or limit the degree to which users and other downstream players assemble or disassemble the product. Product architecture is therefore a major force in assembly in the large.

Product architecture links many technical and nontechnical issues in product design and production, so much so that different constituencies in the product development process may want the product to have radically different architectures. Sorting out the implications for different architectural choices before they are made is extremely important. Among the issues we will take up in this chapter are:

- Integral or modular architecture
- Product families, platforms, and variants
- Commonality, carryover, and reuse
- Management of variety
- Production flexibility and responsiveness to changes in customer demand

These will be illustrated by a variety of examples: consumer products, cars and aircraft, medical devices, power tools, office copiers, and tape players.

14.B. DEFINITION AND ROLE OF ARCHITECTURE IN PRODUCT DEVELOPMENT

We will begin the chapter by defining architecture and discussing its influence on product development. Then we will look at the associated issues listed above. Finally, we will show the many ways that architecture and architectural decisions affect product development and assembly design.

14.B.1. Definition of Product Architecture

A useful definition of product architecture is adapted from [Ulrich and Eppinger]:

Product architecture is the scheme by which the functional elements of the product are arranged into physical chunks and the scheme by which the chunks interact.

When a product architecture is decided, several crucial questions are addressed:

- What subfunctions are needed to carry out each function?
- What technology will be used to implement each function or subfunction?
- How should each physical embodiment be divided into chunks (also called modules) within the constraints imposed by choice of technology?
- How should the chunks be arranged with respect to each other in space?
- How will they need to interact?
- How should the interfaces that provide these interactions be defined and implemented?

While each of these questions appears to be technical, we will see very quickly that the forces that drive the answers are equally technical and nontechnical, involving a variety of business strategy and operational issues.

In terms of assembly, the functional definition appears in the form of KCs which have to be delivered. The chunks are sets of parts assembled together and possibly acting together. The interfaces are obviously assembly features which carry segments of the DFC from one part to another.

Figure 14-1 illustrates some of these points with two different architectures for car power trains, namely, the rear wheel drive and the front wheel drive. What we see

here is a number of physical elements that each carry out a distinct function: engine, transmission, universal joints, drive shafts, differential, and wheels. However, each architecture arranges those elements differently. The rear wheel drive spreads them out, while the front wheel drive packs them all together under the hood, where there is precious little space. The weight of the car is distributed differently, creating different handling and braking characteristics. The components of the front wheel drive are often smaller, so such cars generally have lower power. The management of the product development process is definitely more difficult in the front wheel drive situation due to the need to allocate space much more carefully and to mediate many arguments over how much space is allocated to each function and chunk. [Walton] provides a vivid look at such issues. Finally, assembly is completely different, with the front wheel drive car often built via a subassembly that includes *everything* shown below at the front end except the wheels.

14.B.2. Where Do Architectures Come From?

Several forces drive the creation and form of product architectures, as illustrated on the left in Figure 14-2:

- Technical—architectures emerge from opportunities afforded by new technologies and the engineering design process that implements concepts using particular technologies. Compare, for example, the different layouts and degree of freedom allocations in the four ways of printing discussed in Chapter 12.

- Nontechnical—architectures emerge in response to the need to address a product to particular markets or market segments (by making it in different variants), to design it efficiently (via outsourcing or parallel development of different subassemblies), to manufacture it economically (again via outsourcing or subdivision into subassemblies), to make it easy to recycle (via choice of materials and fastening methods), to respond to various risks and uncertainties related to technological change or customer preferences (via part or module substitution), and so on. (The remarks in parentheses are examples of many possible techniques.)

A company can respond to these forces in many ways. Some of these ways are shown at the right in Figure 14-2. From top to bottom, these responses commit the company

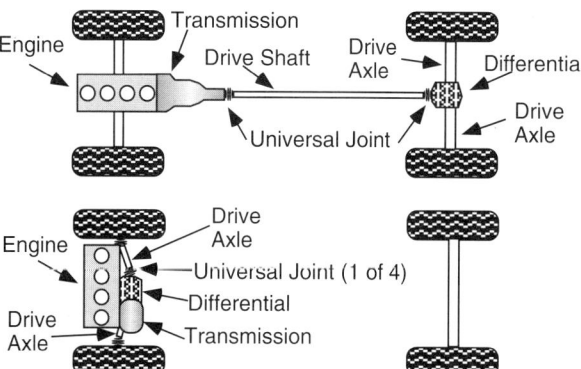

FIGURE 14-1. Two Architectures for Car Power Trains.
Front and rear wheel drive cars have the same items in their power trains, but they occupy different places and are connected to each other differently.

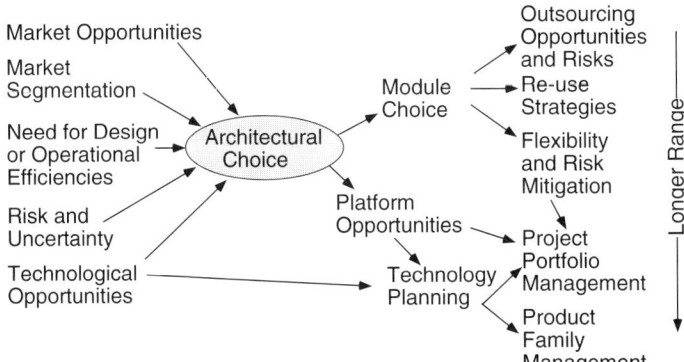

FIGURE 14-2. The Role of Architecture in Product Development.

farther and farther into the future. In the short term, the company can redefine modules within an existing product architecture and thereby change how it makes or outsources different items to suppliers. Different module choices permit different parts and subassemblies to be reused in a series of versions of the product.

In a larger sense, architectural choices affect the company's ability to defend itself against various risks by providing flexibility to rapidly upgrade or redesign the product, or to generate new versions for new markets. This becomes inefficient unless there is some general plan. A common kind of plan is a platform strategy, which combines a basic product design and manufacturing methods with an architecture that permits new versions to be created more easily by building on the platform rather than totally redesigning the product each time. Such a strategy commits the company to a number of product and process technologies, requiring a long view of how these are likely to evolve.

Architecture is also a way to deal with many kinds of complexity and uncertainty. If a product can be divided into segments and each segment can be dealt with separately and recombined later, a reduction in complexity can be achieved. Among the ways of subdividing the product are the following:

- Separate the product into a relatively stable portion and a relatively variable portion; in the variable portion might be items that customers can choose or for which demand may be hard to predict, or items whose technology is changing; in the stable portion may be items that involve costly tooling, long lead times, processes with long learning curves or long setup times, less variable customer demand, more stable technology, and so on.

- Separate the product into base sets of technologies, materials, design and manufacturing methods, and implementation techniques for basic product functions, and then use these bases to generate specific products quickly in response to changing market conditions or new market segments.

- Separate the product into portions whose functions are relatively independent; assign different suppliers or internal engineering groups to design or even build each portion, and retain in the originating company only final assembly and distribution.

- Separate the product into portions that must be designed specifically to meet the requirements and other portions that can be bought as more or less standard items; utilize the standard interfaces on the standard items when interfacing them to the items in the other portion.

It is important to take account of the degree of inherent stability in the industry or the underlying technologies when making these choices. In the technical domain, architectures can remain stable as long as technology remains stable. But technology always changes, so architectures have to change or else products become technologically obsolete. In the nontechnical domain, new market segments emerge or can be created by novel products, new suppliers arrive with novel production techniques or subassemblies, and economic conditions can change, causing costs or prices to change, again causing changes to the architecture.

Researchers such as Abernathy, Clark, and Utterback have documented patterns of evolution of industries and types of products. They point out that novel products are subject to a great deal of exploration as many companies enter the industry and customers experiment with their very diverse offerings. Gradually a consensus emerges around what is called a "dominant design," following which most of these companies fail while a few survive into a mature phase of the industry. As the dominant design takes hold, product innovation tends to slow down and is replaced by process innovation as the survivors compete on price and quality. Customers know what they want and companies know what they have to do. This reduces much of the technical uncertainty and makes it much easier to evolve a relatively stable architecture. Within that architecture, individual modules often undergo considerable innovation ([Erens]). Table 14-1 gives several examples.

In Table 14-1, it is interesting to note two different patterns. One is evolution from decentralized or separate things (airplane wings made of cloth, wire and struts) into a single thing (metal wing). The other is evolution from a centralized thing (central film processing or mainframe computer) into physically or geographically separate things (instant film, drugstore film processing labs, or personal computers). While no trend can be expected one way or the other, it is true that it is easier to make changes when things are separate. Thus in the exploratory phase of an industry or technology, things may be separate, but as the industry matures, some of these things may merge. Examples include the airplane wing and the automobile body. Better materials, improved processes, and

TABLE 14-1. Architectural Evolution of Several Products

Product	Exploratory Phase	First Dominant Design	Subsequent Developments, Some of Which Are Available at the Same Time While Others Drive Out Previous Forms
Airplane (1900–2000)	Two cloth skin wings; struts and wires between wings for stiffness; wings separate from fuselage	One stressed metal skin wing separate from fuselage; separate stiffeners inside skin	• Blended wing and fuselage or flying wing with no separate fuselage; separate skin and stiffeners • Composite graphite and epoxy structures that combine skin and stiffeners • Delta wing for supersonic flight; hybrid wing-fuselage for near sonic flight
Automobile (1880–2000)	Wood body mounted on a separate frame; electric, steam, and gas engines; left, center, right, front or rear steering wheel or tiller	Wood body on frame; gas engine; steering wheel; wheel in front on right or left; rear wheel drive	• Metal unibody mounted on separate frame • Metal unibody integrated with frame • Front wheel drive for small cars • Electric front wheel drive; electric drive with a motor on each wheel (?)
Computer hardware (1940–2000)	Multiple central processors or one processor; separate memories for program and data or same memory	One central processor; same memory for program and data	• Integrated circuit processor with separate memory • Integrated circuit processor with cache memory on processor chip • Multiple PCs networked together for solving large problems • Multiple hand-held devices with docks to computer network, or wireless
Computing service (1950–2000)	One mainframe computer operated by specialists; one user at a time	Time-shared mainframe operated by specialists; user has a terminal; many users at a time	• Sets of minicomputers requiring no specialists; timeshared by many users or one user at a time • Microcomputer; each user has one; specialists on help desk • Client-server; each user has a computer that is connected to a server for networking or storage • Thin client; user has terminal; server does processing, storage, and networking (?)
Camera (1840–2000)	Dark box, lens, one rigid glass or metal plate for each picture	Picture on flexible material that can be rolled up; many pictures on one roll; roll built into camera; user sends camera to central film processing plant (Kodak)	• Separate cassette holds film; customer sends cassette to central processing plant • Film and processing chemicals integrated (Polaroid) • Small decentralized processing machines permit one hour processing • Digital cameras eliminate film and processing; users e-mail photos or print them using PCs

Note: Each of the rows represents approximately 50 to 150 years of development. The "?" indicates a proposed architecture that has not so far been economically significant but may be in the future.

more time to think all contribute to gradual integration of a product. But the opposite trend can also be observed: As industries mature, markets and market segments become better understood, different kinds of customer needs are discerned, and there is a need to keep things separate, variable, adjustable, or substitutable in order to cater to these different sets of needs.

Design and production processes also have to evolve: When a dominant design emerges, one product can be designed and made in huge quantities to suit all customers. An example is the DC-3 airplane or the Ford Model T car. As the industry matures and customer needs begin to fragment, it becomes necessary to design variants faster and to produce them economically in smaller quantities. Globalization connects companies to more distant and varied customers, requiring dispersed design, supply chain, manufacturing, and distribution systems.

Thus there is a constant tension between technically based pressures to integrate and business-based pressures to keep things separate.

14.B.3. Architecture's Interaction with Development Processes and Organizational Structures

Architectures evolve slowly, but when they mature they represent a complex set of relationships that extends well beyond the product itself. As modules are related to each other, so are the design groups or companies that make them. Thus product architectures and company organizations become correlated. For example, current car architectures are divided into bodies, interiors, chassies, and power trains. So most car companies have body, interior, chassis, and power train departments. But if future cars have one electric motor at each wheel that provides motive power and braking, then there will be no exhaust system and no brakes, and thus no departments for them. Power train might even become part of chassis while a new computer algorithm department might develop integrated motor drive and braking controls.

The companies involved in maturing industries develop a set of routines that can harden into habits along with a set of costly investments in methods, equipment, materials, and knowledge. If a new technology or market emerges that demands a new architecture, some companies may be unable to respond because they do not recognize that the architecture is changing. In addition, even if they recognize the change, they can be reluctant to acknowledge

and adopt it for fear of losing existing customers and methods.

When a major change in architecture occurs, the new one is often initially modular to facilitate the necessary experimentation. However, it is difficult at first for companies to write clear specifications for the modules or even to decide the correct modularization, so they tend to do all the design and manufacturing themselves. As the dominant architecture is clarified and new technologies are better understood, outsourcing becomes easier, and the modules can be provided or even designed by specialist suppliers.

These issues are the subject of research in the management sciences ([Henderson and Clark], [Christensen], [Fine], [Fine and Whitney]).

14.B.4. Attributes of Architectures[1]

One reason why architecture is difficult to define is that it displays many different attributes. These interact with each other strongly and have a huge influence on design and operational choices, including assembly. This section discusses a number of these attributes: integrality and modularity; the relation between modules and systems; physical constraints on module choice; families, platforms, and variants; commonality, carryover, and reuse; and intended and unintended consequences.

14.B.4.a. Integrality and Modularity

An important aspect of architecture decisions involves the degree to which functional elements are intended to be independent of each other, and similarly the degree to which physical chunks are designed to be independent of each other as they carry out their assigned functions. One kind of distinction is as follows: Some architectures in the limit are called *modular* while others in the limit are called *integral*. A purely modular architecture, if such a thing existed, would be one in which each function and subfunction were assigned to its own individual physical element. At the limit, each element could be designed and manufactured independently of all the others, and the product could be produced simply by plugging these elements together at their predefined interfaces. By contrast, a purely integral architecture would have a single part that performs all the functions. Most real products are somewhere in between these extremes.

---

[1]Portions of this section are based on [Ulrich].

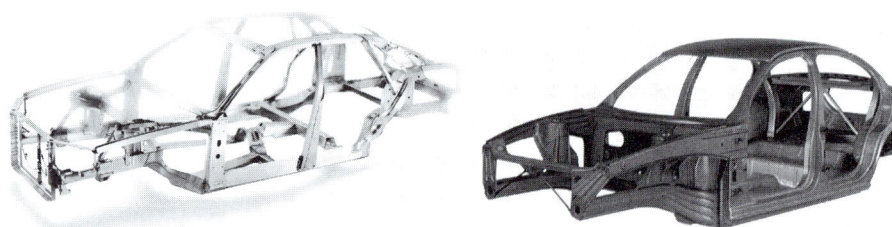

FIGURE 14-3. Two Architectures for Car Bodies. *Left:* A primarily modular aluminum design, where the parts shown function exclusively to provide structural shape and rigidity. The exterior panels provide no rigidity and are added later. (Courtesy of Audi. Used by permission.) *Right:* A mixed modular-integral steel design in which some panels contain both interior structural and exterior appearance portions which share in providing structural rigidity. (Courtesy of the American Iron and Steel Institute. Used by permission.)

An example modular architecture is a printed circuit board together with the components attached to it. The interconnections are provided by the board while the individual circuit functions are provided by separate elements that are made elsewhere and assembled to the board via standard interfaces. A microprocessor is an example integral architecture. It is the integral counterpart to a printed circuit board in which all the individual items and their interconnections are made essentially at the same time in their final assembled locations in one physical entity. This entity has interfaces to other entities in the computer.

Another example is illustrated in Figure 14-3, which shows two architectures for automobile bodies. On the left is an aluminum design that employs a space-frame comprising ribs joined at their intersections. The ribs are extrusions and the joints are castings into which the ribs are plugged and then arc welded or glued. This portion of the car delivers only the interior structure and strength. No large exterior styling surfaces are part of this structure. Instead, these are separate non-load-bearing pieces, often aluminum but sometimes polymers with final color molded in. Separation of structure and appearance marks this design as primarily modular. A major goal of this design is lower weight, which is purchased at the cost of more expensive materials. The tinker-toy structure is used because no good way of welding aluminum exists that does not reduce strength in the region around the weld.[2]

By contrast, on the right in Figure 14-3 is a steel design. Here the panels are spot welded together and some

of them, especially the panel that extends from the rear door area back over the rear fender, comprise a mix of interior ribs and exterior finish surfaces all within a single part. In the sense that structure and appearance are normally separate, their inclusion in a single part marks this design as being somewhat integral. In addition, the exterior portions of some of these panels provide some structural rigidity as well, a function that is provided in the aluminum body exclusively by the frame. The functions that are shared within some of the steel parts thus include appearance, exterior surface, rib-type stiffening, and shell-type stiffening. Some of the weight advantage of aluminum is offset in this design because appearance parts provide some of the stiffness along with the fact that a high strength steel is used, permitting thinner sheet. Rigidity is also provided by box-beam construction of each rib, which requires stamping and welding together a number of pieces that appear in the aluminum design as single extrusions.

As of this writing, it is not clear if the aluminum modular design will replace the integral steel design. In airplane wing design, the old modular design using cloth aerodynamic surfaces with ribs and struts for stiffness has been totally replaced by load-bearing skins contributing shell-type stiffness to an interior rib and spar stiffener system. Cells in this system double as fuel tanks. Most parts and subassemblies thus have three major functions, and their design and construction take these into account.

A deeper understanding of the differences between integral and modular is provided by Table 14-2.

When we compare the implications listed in Table 14-2, we see that integral designs are favored when performance is the highest priority. Such designs are

---

[2]Friction stir welding is a promising process for aluminum, but at present it is too slow for high-volume products like cars.

TABLE 14-2. Comparison of Some Implications for Integral and Modular Designs

Modular	Integral
Generally there are more chunks.	Generally there are fewer chunks.
Chunks may be integral inside but are independent from each other functionally and physically.	Chunks may be integral inside and interdependent among each other.
Standard, predesigned interfaces can be used that can remain the same even if internal characteristics of a chunk change.	Interfaces are tailored to the chunks and are dependent on the functional behavior of the chunk and its surroundings.
Modules can be designed independently to provide their individual contributions to overall function, and sometimes they can be used interchangeably.	Chunks are tailored to their application and surroundings and cannot be interchanged without requiring changes to other chunks.
Unpredictability of module choice requires overdesign to accommodate possible mismatches.	Chunk design can be optimized for a predictable set of functions and implementations.
Standard interfaces are physically separate from the module and thus waste other design resources such as space or weight.	Interfaces can be integral to the chunk, saving space or weight.
Interface management, if planned properly, can provide flexibility during production, use, or recycling.	Interface management occurs entirely during design and is frozen; it is not aimed at flexibility after design.
Business performance may be favored.	Technical performance may be favored.

Source: Adapted from [MacDuffie] with additions.

likely to be more efficient in their use of space, weight, and energy because they can be optimized to a known combination of chunks and can contain their own interfaces. Many costs are increasing functions of the number of parts, regardless of part complexity, so an integral design might cost less per unit to design and manufacture.[3] Modular designs are more difficult to optimize in these ways because allowances have to be made for the size and weight of separate interfaces such as plugs or mounting flanges. In addition, modules are often intended to be substituted for each other in order to create product variety. Since we do not know which modules might find themselves in the same product unit or what future modules might be designed and added to the ensemble, some modules may have to be overdesigned to accommodate these uncertainties. Unexpected failure modes might also arise. However, many business goals are served by modularity, such as outsourcing, independent design, customization, multiple suppliers, and so on. The degree of modularity of each actual product is the result of considerable debate among different constituencies in a company representing performance or business goals, respectively.

It should be noted that integral designs buy their efficiency at the possible cost of flexibility. The stamping dies that make the integral sheet metal parts in Figure 14-3 take a long time to design, and the presses that use them are long-life investments. In a quite symmetric way, modular designs provide flexibility of many kinds but at the cost of efficiency in such domains as space, weight, or the logistics of handling many parts during design and manufacture. Flexibility and efficiency are often at odds, and this is a good example. We shall see later in the examples, particularly in Section 14.C.2.b, that this is not always the case.

By contrast, modular designs often buy their flexibility at the cost of reliability. Such designs have more interfaces, and interfaces are notorious sources of failure. An important example is solder joints in printed circuit boards. Imagine building a computer processor with 10 million transistors, each requiring three solder joints. It is highly unlikely that millions of such processors could be made economically, each having 30 million perfect solder joints. Microprocessors are made in such a way that all 30 million of those joints are made at once by a more reliable process. The chip itself requires a few hundred solder joints to connect it to the rest of the system.

Even simple products must deliver many customer requirements. It was noted in Chapter 8 that many parts in an assembly cooperate to deliver each requirement. It is not surprising, then, that there may be as many requirements as there are parts, perhaps more, and this trend increases if the product is more integral. It is therefore inevitable in typical products that some parts will be involved in delivering more than one KC. Four possible situations are enumerated and named in Table 14-3. The most complex situation listed in Table 14-3 is clearly the chain-integral architecture. It is likely that not all KCs in a chain-integral assembly can be achieved independently.

---

[3] A detailed discussion of this important point is in Chapter 15.

TABLE 14-3. Possible Relationships Between Parts and the Number of KCs to Which They Deliver or Contribute

	One Part Delivers or Helps Deliver …	Many Parts Deliver …
One Function or KC	Modular architecture	Chain architecture
Many Functions or KCs	Integral architecture or function sharing	Chain-integral architecture

Note: The table is read vertically down a column and then across to the left. For example, one part delivering many KCs is said to be involved in function sharing and an integral architecture.

Source: [Ulrich], [Ulrich and Ellison], [Cunningham and Whitney].

Table 14-3 enriches the concepts of integral and modular and shows that assemblies occupy the most difficult cell in this table.

14.B.4.b. Systems and Modules

Modules are identifiable portions of a product or system that do some valuable function but do not do everything that the product or system does. Modules can be considered separately for the purpose of design, manufacture, assembly, and use, but they are not independent in these domains except at the ideal extreme of complete modularity. The items that perform a function need not be contiguous and self-contained but could conceivably be distributed physically in the product. It may seem inappropriate to call such items modules. In general there is no requirement that systems be contiguous and self-contained. Distributed systems are common.

The concept of "module" occurs not only in the context of integral and modular designs but also in the context of systems and system engineering. The basic idea of a system is that it is an organized collection and connection of things that together exhibit some behavior that no subset of these things can perform by itself. Systems can be quite complex and exhibit complex behaviors even when the modules are relatively few and simple. The complexity can appear as unpredictable behavior, behavior that varies over time, or behavior that is so different from that of any single module that it is surprising.

Assemblies are systems whose modules are subassemblies or parts. Among their surprising behaviors are the complex ways that variation at the part level propagates to the KCs. We have a chance to master such complexity if we are careful when the DFC is designed, and especially if we make the final assembly and all its subassemblies properly constrained. Overconstraint creates interdependencies between parts that are in many cases unintended

and have surprising consequences. Even if the assembly is properly constrained, it can be quite difficult to understand assembly behavior because the variations can combine in so many ways, given their statistical nature.

From a practical point of view, the problem in designing a system is to decide how to divide it into modules. This is the process of creating an architecture. The possibilities are illustrated by the car bodies in Figure 14-3, where the same functions are clustered differently in the two designs. Here the decisions are driven in part by the materials and the forming and joining methods that can be used on them. In other instances, the decisions can be driven by, or take advantage of, other considerations. The examples later in the chapter make this clear.

The two car power trains compared in Figure 14-1 are rather different but not because the functions have been assigned differently to the modules. In fact, the modules do the same things in each design. The differences between these systems are expressed in terms of different connections between the modules or in different relative physical locations.

Modules can be quite complex internally. One could even say that a module is a system at some level, and the items below it in the system are modules.

Thus we can say that modules, like systems, are clearly defined by the functions they perform, even if they do not perform the whole function of the product. This helps us distinguish modules from subassemblies, which can be defined in a more restricted way as a collection of parts that is regarded all at once and preferably is stable and properly constrained. If it has a function, then it can be tested to see that it performs that function before it is installed in the product. This is desirable but not necessary.

On this basis, modules are potentially of more interest to the designer or user of the product, while subassemblies are of more interest to the manufacturer, supplier, and manufacturing engineer.

14.B.4.c. Power-Handling Products, Information-Handling Products, and Interface Standardization

Over the last forty years, nearly every mechanical device whose real function was to process information at low power, such as calculators, clocks, and multi-dial numerical displays, has been replaced by much faster, cheaper, and more accurate electronic versions. The new versions are highly integral internally but are easy to use as modules in highly interchangeable ways. As a result, a whole technology has arisen around the plug and play principle. It is exploited in electronic components, stereo systems, computer systems and peripherals, and many other applications. Interface standards have been defined to assist this exploitation, including designs of electrical plugs, voltage levels, assignment of certain pins on the plug to certain functions, and so on. In many ways, one can say that the existence of standard interfaces is the main enabler of modularity in many industries. Why is it that this trend has not been extended to mechanical items that carry or operate at high power? Why are typical high-power or high-stress things like airplane wings integral?

In [Whitney], the author argues that the amount of power or the local power density (power concentrated in a given volume) involved in delivering the product's functions severely limits a designer's choices regarding its modularity. High-power items like automobile engines and aircraft wings need to economize on space, weight, and energy consumption while at the same time delivering multiple functions. Modular designs would not do. They would have too many parts, be too big, or weigh too much. Their interfaces are subjected to considerable physical or thermal stress as part of the item's main function. If the interfaces were independent spatially from the item and designed independently, they would be too big or weigh too much.

Information handling products operate at vanishingly small power levels. An important reason why they are easier to modularize than power-handling products is that their interfaces can be standardized. Products like microprocessors exchange and process information, which is expressed as low-power electrical signals. Only the logical level of these signals is important for the product's function. The interfaces are much bigger than they need to be to carry such small amounts of power. For example, the conducting pins on electrical connectors that link disk drives to motherboards are subjected to more loads during plugging and unplugging than during normal operation. Their size, shape, and strength are much larger than needed to carry out their main function of transferring information. This excess shape can be standardized for interchangeability without compromising the main function. This is why different kinds of disk drives can be used by one computer manufacturer in many models of computer. The information itself can also be standardized, with the result that different disk drives (to continue the example) can be substituted functionally as well as physically with few incompatibilities.

Power-handling items cannot easily be functionally substituted because power exchanges between them will not be efficient unless their power delivery and consumption characteristics are coordinated. This is called impedance matching. Information-handling items exchange so little power that impedance matching is unnecessary. The interfaces of power-handling items carry such large loads that there is little design slack left over to divert to interface standardization.

It is debatable whether microprocessors carry out a single function, and the large power densities in microprocessors cause their internal elements to interact strongly, making their design difficult to modularize. Nevertheless, the majority of information-handling items do one or a very few functions that can be clearly separated from each other internally and externally. Designers of these items have considerable freedom to add or subtract functions. This freedom is not often available in power-handling products because the higher power levels bring with them side effects like vibration, crack growth, and heat radiation that cannot be avoided. More design effort typically goes into predicting and mitigating these side effects than goes into determining how to deliver the main functions. Obviously, side effects cannot be standardized, and this is another reason why power-handling items cannot easily be substituted functionally.

In summary, modularity in many applications is enabled by standardization of interfaces, which in turn is enabled when

- The interfaces carry low power or stress.
- They do not deliver a main function or affect performance.
- They do not consume major design resources like space.
- Economy of scale exists for their manufacture.
- They can be defined and designed independently of the items they join.

14.B.4.d. Families, Platforms, and Variants[4]

Along with the terms integral, modular, module, and system, we have the terms family, platform, and variants. Product families are sets of products that share some major characteristics and typically consist of a platform and variants. Platform is another term with many definitions

---

[4]Portions of this section are based on [Erens].

and uses. Establishing the structure of a platform is an architectural decision: One has to decide which parts or functions are part of the platform. In addition, one also has to consider whether implementation of a function would differ depending on whether it is in the platform or not.

[Lehnerd and Meyer] define a product platform as "a set of subsystems and interfaces that form a common structure from which a stream of derivative products can be efficiently developed and produced." This definition emphasizes the aim of allowing development of related products while requiring less effort in design and less duplication of production facilities. Such a family would have similarities that derive from the platform, but different versions of the product could be quite different without requiring expensive redesign of the whole thing.

The platform definition is coordinated with a set of distinct markets as well as a set of matched product and process technologies. This is illustrated schematically in Figure 14-4. Market segments could be geographic or could differentiate types of users. Market tiers could

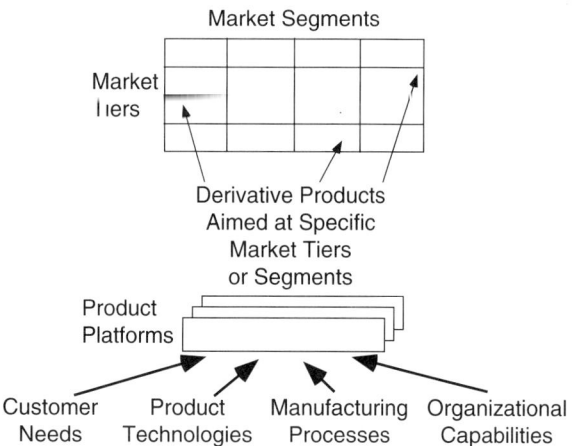

FIGURE 14-4. Lehnerd and Meyer's Concept of Product Platforms. In this concept, product platforms arise from a common set of building blocks comprising capabilities and a recognized set of customer needs. Target markets are identified and divided into segments and tiers. The platform has to be planned in advance with the capabilities, needs, segments, and tiers in mind, so that it will be efficient to develop individual products targeted at each of the segment/tier combinations that are deemed attractive. (Printed with the permission of The Free Press, a Division of Simon & Schuster, Inc., from [Lehnerd and Meyer]. Copyright © 1997 by The Free Press.)

represent sizes, quality levels, or different amounts of features or options. A segment for portable tape recorders might be Japan or the United States. Different tiers might contain mono, stereo, sporty look, and so on. For office copiers, segments might be home office, small company, large corporation, or graphics service industry. Tiers could be divided by range of copy speed, black–white versus color, combination of copying with fax or digital networking, and so on. Each variant product built on the platform is coordinated so that it efficiently reuses the techniques, common parts or modules, equipment, and knowledge while addressing the markets and tiers distinctly and without giving rise to confusing and inefficient overlap and internal competition.

The essence of platforms is reuse. That is, some portions of the product or its design/production infrastructure are reused in multiple products or product versions. Among the classes of things that can be reused are parts and subassemblies, enabling technologies, manufacturing methods or equipment, standard items, and knowledge of design methods or other skills.[5]

A more general definition of a platform is as follows: "a portion of a product (or set of products, or products and their design and manufacturing systems) that is totally divided from the rest of the product by a set of interfaces such that portions of the product on either side of the dividing line can be altered with minimal effects on the other side."[6] An example is a computer operating system. It provides a platform for developers of application software and supports a consistent user interface for all the applications that use that operating system. In addition, the operating system performs some generic functions for all applications like opening and saving documents, printing, and driving the display.[7]

Platforms are of interest when flexibility and economy are sought across a set of products even if they are not related in any functional ways. One often sees products that are divided into a portion that is expected to stay the same (the platform) plus other portions that could be

---

[5]The importance of reuse in understanding platforms was pointed out to the author by Christopher Magee.

[6]This definition is adapted from one created by a committee of the MIT Engineering Systems Division in May 2001.

[7]In the DOS operating system, each application did its own printing and contained its own printer drivers. Installing the application involved setting up its connection to the printer. This is no longer necessary in Windows and was never necessary on the Macintosh.

changed for a variety of reasons. Those portions that remain the same should be isolated from the product's main functions so that the functions can be modified across the family without disrupting the platform. Alternately, whatever functions are delivered by the platform portion should be the same for all family members.

Family members may differ by scale in some way, such as motors of different power level or electrical controllers of different wattage. These may be scaled versions of each other, with the internal parts simply getting bigger as the main scale is increased. For several reasons, such simple scaling is not always possible, and one sees different implementations of the same function in entire sub-ranges of the scale. An example is plastic gears for low-torque applications and metal ones for higher torques. Another is coil springs for low stiffness and Bellville washers for high stiffness.

Platforms are also of interest when they can be the basis of an industry standard. In the software, information, and communication industries, standardization of operating systems (Windows by Microsoft), programming languages (JAVA by Sun), encoding methods (Stuffit by Aladdin), and bandwidth compression techniques (CDMA by Qualcomm) has been used to convey market power to the company that owns the standard. These standardized items perform, or are vital to, the product's main functions. This is far different from standardization of interfaces discussed in Section 14.B.4.c, which do not play a large role in delivering the main functions of the products they are in.

Table 14-4 gives examples of several product families. It states or estimates the family's purpose and distinguishes what stays the same and what varies. Several purposes may be achieved. Some platforms may be intended to be utilized repeatedly over time, such as successive generations of Sony Walkmen. It can be a great competitive advantage to be able to generate new models quickly, especially if sales depend on styling and fickle customer preferences. Other purposes may be utilized unpredictably, such as being able to bring a second car line into an existing body shop if demand for that car grows beyond the capacity of its original factory. Platform design may also permit an existing car factory to be used with minimal capital investment to make the next generation car. The money saved can be hundreds of millions of dollars. The design standardization needed is so trivial that it hardly interferes with the car's main functions at all.

For example, Figure 14-5 shows a simplified view of the power tool product platform and family structure developed by Black and Decker in the 1970s. The platform comprises product design commonality such as the same motor design and manufacturing methods, a single motor diameter, and a stack architecture for all the products. Details about this platform are in Section 14.D.7.

TABLE 14-4. Example Product Families with Definition of Platform Portion and Variant Portion

Product Family	Purpose of Family	What Stays the Same	What Varies
Ford cars; Toyota cars	Reuse body shop equipment for the next car model; permit different cars to be made in the same factory at the same time	Underbody main locators; body shop fixtures; body assembly sequence	The rest of the car
Volkswagen cars; Chrysler cars	Reuse chassis; bring new cars to market faster for less money	Chassis and portions of drive train	Upper portions of car, interior and exterior
Xerox digital copiers	Sell to several different kinds of customers	The idea that it is a digital copier, along with all the supporting technologies	Black–white versus color; slow copy rate versus fast; operating software
Black and Decker small power tools	Present a coordinated product line; enjoy economies of scale especially in small motors	Motor diameter, motor housing	Business end, handle end; length of motor, hence motor power; details of housing where it mates to handle or business end
Sony Walkmen	Present a coordinated product line; bring new styles to market quickly and see if they catch on	Hard-to-design tape handling mechanisms	Exterior parts, styling, and user interface that can be changed quickly
Boeing aircraft	Bring new passenger capacity models to market less expensively	Fuselage diameter, major assembly fixtures, engines, main controls and cockpit	Fuselage length, wing length, fuel capacity, number of seats, range

Source: Based on information from Christopher Magee, Ford, Maurice Holmes, Xerox, [Lehnerd and Meyer], [Sanderson and Uzumeri], and the author's experience.

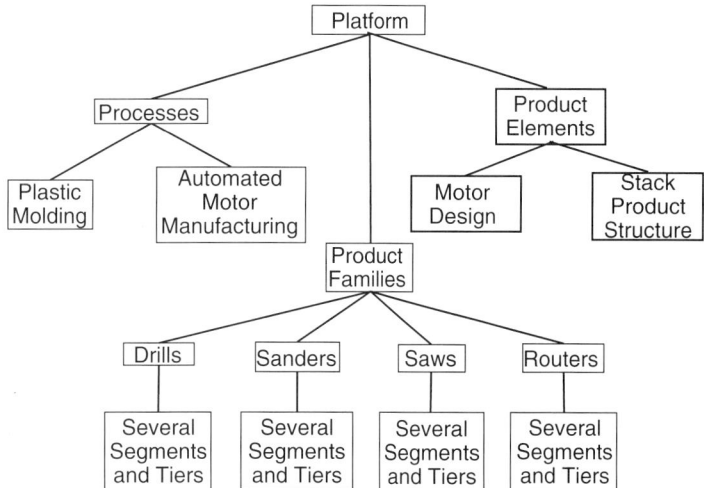

FIGURE 14-5. Simplified Structure of Black and Decker Power Tools. The platform is made of several product and process elements. These are common to several product families. Each family contains several products that differ according to the market segment or quality and performance range to which they are targeted. ([Lehnerd and Meyer])

Figure 14-15 (in Section 14.D.1) shows the tape recorder mechanism for the Sony Walkman product series. This mechanism plays the role of a platform for many models of the Walkman. It is inside many products whose exteriors look completely different. Some look businesslike, others look like toys, still others are waterproof. These exteriors are injection-molded plastic. This permits them to be tough as well as colorful. Even for the same mold design, different colors may be had by changing the plastic. Other molds can be designed relatively quickly. On the other hand, the tape mechanism represents several years of design as well as design of the assembly system to put it together.

Figure 14-6 illustrates an automobile body platform concept aimed at reusing body shop fixtures for the next generation car as well as for reusing body and body shop design principles and best practices. The platform consists of the constraint and locator scheme for delivering body assembly and welding accuracy, plus consistency of arc welding lines in the underbody. Standardizing these items hardly affects the car's main functions at all. At Ford, cars are given size designations like A, B, and so on, with each one in alphabetical sequence being longer, wider, and taller than the previous one. Within a size group, cars can differ somewhat in length by having more overhang in the front and rear structures plus longer floor pan in the middle (front floor) structure. Small changes in width can be

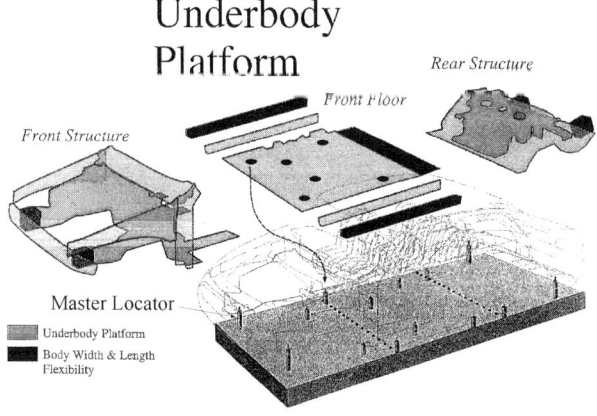

FIGURE 14-6. Car Body Platform. The platform consists of the car underbody locator system and weld line location plus the pallet for carrying the body through the body shop. The underbody parts themselves can differ within prescribed ranges as long as the main locators stay in the same places relative to each other. In the Ford scheme, bodies in a family can vary in length but not much in width. In the Honda system, they can vary substantially in both length and width. (Courtesy of Ford Motor Company. Used by permission.)

obtained by using different rocker panels (stiffeners along the sides of the floor pan).

Main assembly of the car body is accomplished by building the separate underbody subassemblies shown in Figure 14-6, joining them using a fixture similar to the pallet shown in the figure, then using this pallet to carry the body through the rest of the process as side frames and roof are put on and welded in place. Then doors, hood, and trunk lid are added. Because each pallet is slightly different due to how it was built or how it wears, each car could inherit variation that is unique to the pallet it was built on. For this reason, some car companies prefer to use locator pins attached to each workstation instead of pallets. Simple conveyor belts just carry the car body to the next station and place it on the pins.[8]

14.B.4.e. Commonality, Carryover, and Reuse

Commonality, carryover, and reuse are important aspects of platform design. Generally they mean sharing of parts, equipment, or knowledge across products or in subsequent similar products. This is done usually to save money and time. It does not necessarily involve deliberate declaration of a family or definition of a distinct platform, but carrying it out involves many of the same kinds of decisions and methods of implementation. Although the idea has recently been rediscovered, it is at least as old as the 1920s, when it was implemented at General Motors ([Sloan], pp. 156–159). When GM decided to design a new Pontiac, the decision was made to invest only in a new engine but to reuse as much of the previous *Chevrolet's* chassis as possible. Sloan says, "Physical coordination in one form or another is, of course, the first principle of mass production, but at the time it was widely supposed, from the example of the Ford Model T, that mass production on a grand scale required a uniform product. The Pontiac, co-ordinated in part with a car in another price class, was to demonstrate that mass production of automobiles could be reconciled with variety in product. . . . If cars in the [lower volume] higher price class could benefit from the volume economies of the lower-price classes, the advantages of mass production could be extended to the whole car line."

In order for parts or subassemblies to be reused in other current or later products, it is of course necessary to standardize the interfaces as well as the tolerances on those interfaces. In this way, as in many other aspects of architecture, assembly is the point in the process where the strategy is implemented and either succeeds or fails.

Decades after Sloan, it was discovered that Toyota could design and build cars with half the number of people needed by U.S. firms ([Cusumano], p. 199). The reason for the difference in the 1970s and early 1980s was that Toyota outsourced much of its design and manufacturing, especially of commodity items like small parts, lights, door handles, and so on. In the late 1980s and early 1990s, Toyota extended its advantages by using as much as 60% of a car's "invisible" parts in subsequent models. Later in the 1990s, entire car design projects were overlapped so that both engineers and their parts were applied to follow-on programs while the previous ones were still being designed. Naturally, reuse must be done with care because many "invisible" parts are members of systems. Each system has its own requirements and each part in a system is designed to play its role in that system. Mixing parts from different systems just to accomplish reuse, on the assumption that it doesn't matter because the customer cannot see them, ignores the possibility that the customer will feel or hear the difference anyway.[9]

14.B.4.f. Intended and Unintended Interactions

[Ulrich and Eppinger] point out that when an architecture is defined, the engineer not only assigns functions to technologies and geometric space, but he or she also defines relationships between the physical entities. These are called intended interactions; they serve to carry out or aid in those product functions that require more than one entity. It is inevitable, however, that other interactions will arise. These are called unintended interactions or side effects.

In an electrical system operating at low power but with high frequencies, electromagnetic interference can occur. This is possible in cellular telephones where miniaturization places the radio-frequency components very close to the digital logic components, making the latter difficult to design and debug.

In a mechanical system operating at high power, vibrations can occur and be transmitted as motions or noise to other parts of the system. An example is a car engine, whose vibration is transmitted to the driver through the steering column. The car's body engineers and vibration specialists try to design the steering column and the surrounding body so that they do not resonate at frequencies generated by the engine, especially when it is idling.

---

[8]Individual pallets provide some flexibility to route the work to different stations, or to increase or lower production rate by adding or removing pallets.

[9]Another quite unexpected risk is that older cars which contain parts used in newer cars will be subjects of theft. This apparently has happened to Toyota Camrys with model years 1988–1992. "Stop Thief! That's My Camry." *Business Week*, April 23, 2001, p. 14.

However, most cars are offered with a choice of engines, each of which idles at a different frequency. In all, there are so many vibration frequencies that it is almost impossible to defend against them all. Here we see clearly the difficulties that can arise due to product variety and an architecture in which the engine is a customer choice module but the body is part of a common platform.

Architecture and company organization can interact in unintended ways as well. Once the author was told by a manager at an auto company, "We use the same engine on thirty-eight different car models." "Good," said the author. "Bad," said the manager. "Every time we want to change a screw, we have to get permission from thirty-eight different program managers."

14.C. INTERACTION OF ARCHITECTURE DECISIONS AND ASSEMBLY IN THE LARGE

Architectural decisions are made at every stage in product development, but, except for highly integral products, these decisions have their impact during assembly. Every physical decomposition generates an assembly interface; every interchangeable module has to have the same assembly interface as every other module it could be interchanged with; every option that the customer could order later and self-install must be easy to assemble properly and quickly; the specifications for each outsourced item must include strict requirements on interfaces to items not made by that source; any function or item in the product that could be upgraded later needs to be identified early in the design process so that it can be provided with an interface even if it is not used right away. This part of the chapter deals with these issues in a general way while the next section provides several examples.

14.C.1. Management of Variety and Change

The main nontechnical impact on product architecture is the need to accommodate variety and change.[10] Variety involves changes over short time spans that apply to a single design. Change involves longer-term evolutions of a design. Both are related to architecture. The main goal in constructing a product architecture for the purpose of accommodating variety and change is to provide as much variety and change as the market can absorb with as little effort and investment as possible.

It is possible to deal with variety and change by adopting certain operational methods such as careful management of inventories, logistics, scheduling, and data processing. Such methods, while necessary, will be greatly enhanced if the product or product family are designed specifically to enable flexible operations. The main way this is done is by careful choice of architecture and platform, leading to flexible assembly operations.

14.C.1.a. Benefits and Costs[11]

The main benefits of being able to offer variety in a product is that more customers can be attracted even though their wants are not exactly alike. If the process is managed correctly, the customer will get the product quickly in spite of being able to choose from many varieties. The manufacturer's goal is to do this by making minimal changes during design or production, so that the cost of providing the variety will be low and the manufacturer will be able to get almost as much economy of scale as if only one product were being made. Another main benefit of having variety or the capacity to generate variety is to be able to follow unpredictable shifts or swings in demand. It may be that aggregate demand will be roughly predictable but options chosen by customers will not be. Switching between these options should therefore be as easy and fast as possible. Alternately, styles or preferences may change, and one variety will stop selling forever while others will see rising demand.

The main costs of accommodating variety and change are that extra resources are required, in addition to which the right resources may not be available when or where they are needed. The product will have more parts and more internal interfaces. Extra design effort is required, and extra tooling or other facilities must be acquired and kept ready. Items made but not sold may have to be

---

[10] A manager at a manufacturer of large home appliances once said, "Marketing wants them in seven colors and manufacturing wants them all to be white."

[11] "We spend most of our time building cars for which there are no buyers, making customers wait a long time to get the car they want, and then losing money on incentives to get rid of dealers' unsold inventories."

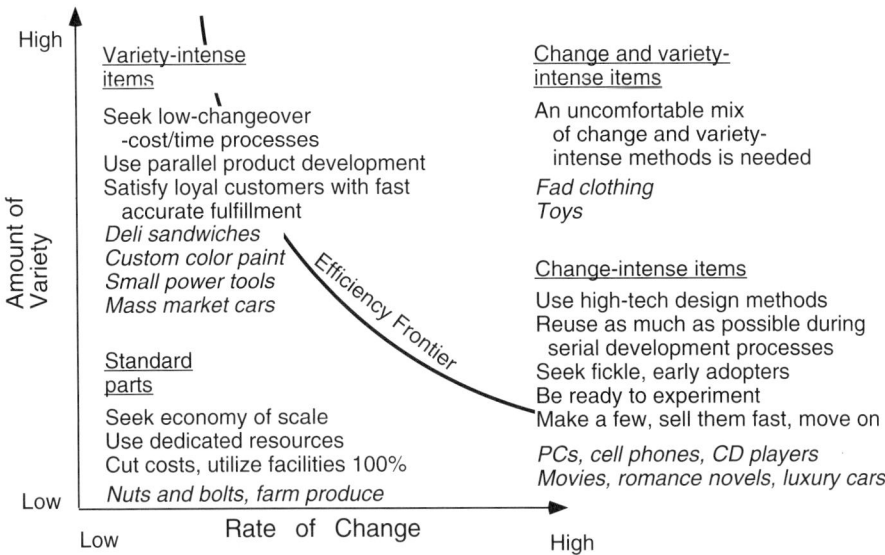

FIGURE 14-7. The Variety–Change Tradeoff Space. Strategies exist for operating at the extremes on each axis but not in the middle of the plane where change and variety are both intense. [Sanderson and Uzumeri] postulates the existence of an efficiency frontier, shown as a curve in the figure. The management styles, kinds of design and production methods, and company organizations needed to be successful at each extreme are so different that no one company can operate a product line in both domains at once. (Adapted from [Sanderson and Uzumeri]. Copyright © 1997 McGraw-Hill. Used by permission.)

scrapped, while customers who order beyond inventory or production capacity will be kept waiting or will get impatient and buy something or somewhere else. If some facilities are dedicated to one version and that version goes out of production, those facilities will be worthless. As discussed elsewhere in this book, dedicated facilities can be very economical, whereas flexible facilities cost more but may survive a change in demand. Other costs include the sheer effort of keeping track of all the varieties, scheduling production activities, managing orders to suppliers, making sure that different options are compatible with each other, avoiding mistakes while configuring the product to the customer's specifications, and so on.

14.C.1.b. Variety–Change Tradeoffs[12]
Variety in a current product line is different from change impacting that product. Some products never change but must be provided in enormous variety, such as nuts and bolts at a hardware store. Other items change rapidly even if variety is low. High-tech products or those in the immature stage of industrial development undergo rapid change.

Examples include laptop computers, personal digital assistants, and cellular telephones.

The approach of [Sanderson and Uzumeri] to managing product families is structured around the variety–change tradeoff, illustrated in Figure 14-7. Variety increases on the vertical axis while changes occur more often to the right on the horizontal axis. A few example products of each type are shown, along with a few basic strategies. It should be noted that in the high-variety low-rate-of-change region, modular innovation is possible but architectures will probably stay the same. At the high-rate-of-change extreme, there may be no dominant design and neither architectures nor modules will be stable. Architectures may be able to stabilize if the rate of change in product design or technology is not too high. [Sanderson and Uzumeri] hypothesizes that one company can be active at either extreme while making the same product, but not at the same time. This gives rise to the efficiency frontier shown in the figure. In addition, some companies will address a product by operating at one extreme while other companies will address the same product by operating at the other extreme. In particular, every time there is a major change in technologies in an industry, most of the companies retreat to a conservative position far from either extreme and then

---

[12]Portions of this section are based on [Sanderson and Uzumeri].

venture forth toward one or the other. Each cycle of this kind has the potential to push the frontier farther out into the plane.

If a company chooses to operate a product line at one extreme, then its entire design and production process for that product line must be constructed consistently to do so. This includes the suppliers and the distribution chain.

It appears that it is easiest to operate near the origin in Figure 14-7, next more difficult to operate in the variety-intense region, next more difficult in the change-intense region, and most difficult in the combined change-variety intense region. In both the cases of Toshiba in laptop computers and Sony in Walkmen, [Sanderson and Uzumeri] shows that both companies dominated their respective industries in the 1980s and early 1990s by establishing an architecture and then moving quickly to create huge variety. No even moderately successful company in either industry tried being change-intensive or could afford to operate there for long, except for IBM.

14.C.1.c. Manufacturing Strategies and Decoupling Points[13]

It should be clear by now that architectural decisions cannot be made just based on how the functions of a product will be implemented. In addition, the strategy for how the product will be sold and distributed must also be taken into account. Two extreme strategies can be distinguished:

- Build to stock and wait for customers to order; ship immediately from stock; make more of what is bought and try to keep the unbought items from spoiling (examples include tooth paste, lamb chops, fast food, airline seats, common hardware items, and low-cost houses built on speculation).

- Design and build to order; nothing is wasted but the customer has to wait while the order is made (examples include highways and bridges, some office buildings, power plants, and custom-built expensive homes).

Between these extremes are several intermediate strategies including

- Build to order from stock designs (mid-range houses, restaurant meals, high-end automobiles, custom man's suit).

- Build variations onto a standard design (commercial aircraft with different seating arrangements, mass production automobiles in different colors and options, men's suits off the rack).

- Assemble a custom version from available standard subassemblies (deli sandwich, Denso panel meters, custom color paint).

- Program standard physical items to order electronically (EPROM, home alarm or climate control system, user-configurable software).

- Design the product so that the customer makes his own from standard parts (salad bar, Lego toy, component stereo).

- Engage in risk-sharing partnerships with suppliers or retailers who hold inventory at various stages of assembly.

- Manage demand so that customers order what is in stock or what can be built quickly from stock items in a platform product (Dell Computer, dealer incentives and discounts).

Common to all of these strategies, in addition to obvious architecture and interface issues, is the concept of the decoupling point. (See [Erens], which cites a number of sources for this idea. Also see [Ulrich et al.]). Two kinds of decoupling point have been identified: the design decoupling point and the production decoupling point.

The design decoupling point is the point in the architectural decomposition below which existing technologies, platforms, or subassemblies are carried over from previous designs, and above which something new will be designed. The deeper in the decomposition this point is, the more thorough the redesign is, or the more profound the innovation is. The vast majority of product development is redesign at a relatively shallow level in the decomposition, thus preserving the main product, process, and business architectures. For example, a new car design may be created every ten years while a refresh consisting of revised sheet metal and interior styling may occur as often as every two years. A new commercial aircraft design may occur every twenty years while variants within the family may occur every three to five years. Suppliers of major subassemblies may change along with the new design. A new prime mover technology for cars or airplanes may be attempted every fifty years. For most high-power or high-stress items like buildings, bridges, cars, and aircraft, major changes in primary structural materials occur

---

[13]Portions of this section are based on [Erens].

TABLE 14-5. Design and Production Decoupling Points for Automobile Components and Customer Choices

Item	What Is Chosen	Last Opportunity to Choose
Car body	Styles to be made	When stamping dies are designed (~2 years before car production starts)
Gasoline engine	Main operating characteristics, to suit the car it will be used in	When the engine is designed (~5 years before engine production starts)
Diesel engine	Main operating characteristics, to suit the car it will be used in	When the engine is designed, but important changes can be made by reprogramming the engine control module (~1 year before car production starts)
Passive suspension system	Main ride characteristics	After linkages have been designed or even during production: by changing elasticity of ball end inserts and shock absorber mounts
Many customer options	Color, body style, seats, roof rack, etc.	Volvo 21-day car: For customer, 21 days before delivery. For suppliers, 12 hours before car is finished. For Volvo, 8 hours before car is finished.
Active suspension system	Main ride characteristics	Customer can choose any time during use

Note: Reading down the table, the decoupling point is closer to the customer. The last opportunity to choose may in practice be exceeded but only at a very high cost. The Volvo 21-day car is described in Chapter 16.

TABLE 14-6. Various Design and Production Decoupling Points in Automobile Components Depend on How Integral the Item Is

Item	How Integral?	Last Opportunity to Choose Major Characteristics
Transmission	Coupled to engine and drive train characteristics	When main driveability characteristics of the car are chosen (usually early in design)
Wire harness in instrument panel[14]	Must connect to every electrical item in the panel; these items comprise a set of options from which customers can choose	During design: when instrument panel option set is determined During production: when customer order is received; once it is installed in the panel, that set of choices must be installed
Instrument cluster type (analog or LCD, for example)	Can be plugged into opening in instrument panel assembly	A few hours before final assembly of instrument panel; the right wire harness must have been installed first
CD player in the car or not	Can be plugged into a slot in the instrument panel	During assembly of instrument panel or after buying the car
A CD in the CD player	Not at all	Any time the customer wants to play a CD

Note: Reading down the table, the item in the left column is less integral and uses less power.

extremely rarely. Whole industries arise or disappear in response.

The production decoupling point is the point in the product delivery process where the item being made is designated as being for a particular customer who provided a set of requirements. This point can be anywhere along the delivery chain, at the very beginning for design and build to order strategies or at the very end if the customer makes his own. The decoupling point is the stage after which customer preferences and choices have to be recorded and responded to. In [Simchi-Levi et al.] this is called the push–pull boundary because production upstream can proceed by the push method whereas downstream it

proceeds by the pull method. Thus design, interfaces, architecture, manufacturing, information systems, and assembly all have to be capable of responding to the customer's choice downstream of the decoupling point but not upstream of it. The decision of where to put the decoupling point is thus both a technical and a business decision, which heavily affects the architecture of the product.

These two kinds of decoupling points are not independent of each other, and within the same product there can be different decoupling points for different modules or subassemblies. This is illustrated for automobiles in Table 14-5.

It is interesting to see how the design decoupling point varies according to some of the characteristics of integral and modular items listed in Table 14-2 when applied to some of the items in Table 14-5. Table 14-6 shows that

---

[14] According to Porsche, reported to the author by Henning Rudolph, a student at Technical University of Berlin.

design and production decoupling points can blur into each other: The same item, such as the optional CD player, can be the subject of a design or a production decoupling point, due to its modularity and the fact that it is a low-power item with standard interfaces.

A study of the mountain bike industry by [Ulrich et al.] shows that different manufacturers of the same product can have different decoupling points depending on what kind of variety they want to offer to customers. Of the four manufacturers studied, one offers frame geometry choices, another frame material, another add-on components, and the fourth color. Some of these choices require different distribution, sales, or customer service strategies while others require different manufacturing strategies. The manufacturer that offers different frame geometries had to design its frame manufacturing process to be very responsive to customer orders, while the others can make a few standard frames and provide variety in other ways. These latter manufacturers make their frames on rigid fixtures by welding together standard tubes whose ends have been prepared for welding by die cutting using standard dies that require long changeover times. This process has lower unit costs as more identical frames are made on one setup. The manufacturer that customizes frame geometry cuts its tubes to custom lengths using numerically controlled flame cutters, and provides a tab in slot joint geometry. A simple clamp holds these parts together while they are welded. This process can make any size frame in any quantity quickly for about the same cost. In terms of our vocabulary, the tab–slot joint enables the frame to be a Type 1 assembly while the fixtured designs are obviously Type 2 assemblies. Generally speaking, Type 1 designs enable more flexible assembly operations. A thought question at the end of the chapter asks the reader to consider when and why this might not be true.

14.C.1.d. Delayed Commitment and Other Variety-Management Strategies Based on Assembly Sequence

If the assembly sequence for a product can be chosen appropriately, the decoupling point can be placed advantageously during the assembly process. As a general rule, it is better to place the decoupling point as close to the customer as possible. The reason is that most of the assembly process can be accomplished in the same way regardless of what the customer orders. Items can be made at a rate similar to expected orders without much fear that a great deal of customizing will be done for items that have no customers.

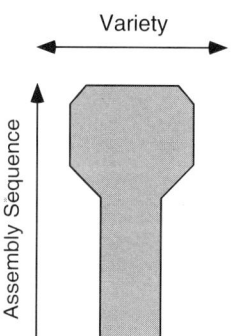

FIGURE 14-8. Mushroom Product. The vertical direction indicates progress in assembling the product. The horizontal direction indicates the amount of variety possible at a given stage of assembly, with wider meaning more variety. In the mushroom product, variety is concentrated near the end of the process. ([Mather])

Furthermore, most of the assembly process is the same, permitting assemblers to become proficient and avoiding confusing changeovers. Since special treatment is needed for an order downstream of the decoupling point, more adaptable workers (more scarce and costly) are needed for those tasks. Finally, less work is needed to complete the order, so it can be shipped sooner, shortening the time the customer has to wait.

One of the most common assembly sequence strategies for dealing with variety is called *delayed commitment* ([Lee]), also known as the *mushroom product* ([Mather]). A similar but slightly more complex strategy is called *plain vanilla box* ([Swaminathan and Tayur]). "Commitment" means adding the parts that differentiate the product for a particular customer. "Delayed" means placing the decoupling point close to the customer.

The mushroom product idea is illustrated in Figure 14-8. The motivation in this case is to provide for long lead time items by adding them at the end of the process. This permits the process to begin without them if necessary and still finish on or nearly on time.

Martin, Hausman, and Ishii give more quantitative ways of evaluating different assembly sequences in order to achieve a mushroom-like product structure ([Martin, Hausman, and Ishii]). They studied a car instrument cluster with 10 parts, plus a final test point, and 18 possible final versions. If the decoupling point were at the first assembly station, there would be 198 different valid subassembly types (18 at each of 10 assembly stations plus 18 at final test). If decoupling could be delayed until the final test station, there would be only 28 different subassemblies (one in each of the first 10 assembly stations plus 18 varieties at the last station). The original assembly sequence gave rise to 113 different subassemblies. Due to assembly precedence relations, it is impossible to

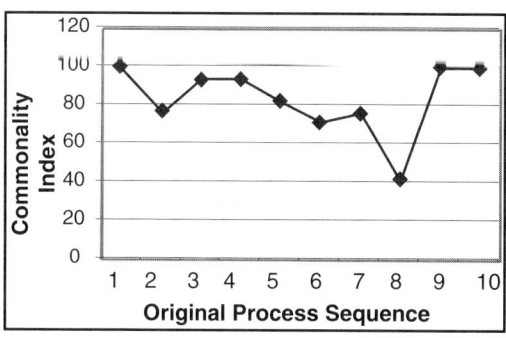 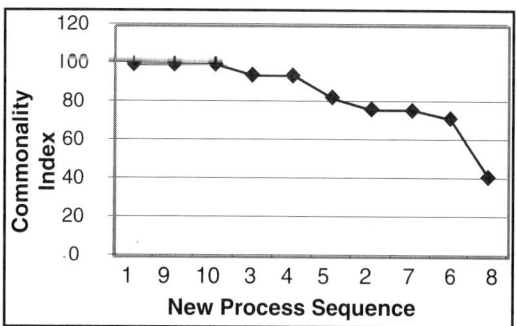

FIGURE 14-9. Commonality Index Versus Assembly Sequence for Two Sequences. The sequence on the left gives rise to 113 possible subassemblies while the one on the right gives rise to 72. ([Martin, Hausman, and Ishii])

delay differentiation to the last station, but the best feasible sequence results in 72 different types of subassemblies.

Martin, Hausman, and Ishii calculate a commonality index C_j:

$$C_j = 1 - \frac{u_j - 1}{v - 1}, \qquad 0 \le C_j \le 1 \qquad (14\text{-}1)$$

where u_j is the number of unique part numbers at step j and v is the final number of product varieties.

If C_j is small, then many different versions of a part are used at step j, whereas if C_j is large, then only a few are used. In a mushroom or delayed commitment product, C_j falls as the assembly process advances. For the instrument cluster, the original assembly sequence shows C_j falling and rising erratically, whereas for the improved sequence, C_j falls almost monotonically as assembly proceeds. This is illustrated in Figure 14-9.

Kota, Sethuraman, and Miller compared part commonality across entire product lines of Walkman-like products ([Kota, Sethuraman, and Miller]). They compared the actual use of the same part for the same function to the potential reuse, where "same" included occupying the same amount of space, employing the same manufacturing processes, and using the same assembly and fastening methods. After normalizing for the number of parts, the Sony family had a reuse score of 80.3% while two competitors' families had scores of 44.1% and 24.4% respectively. The lowest scoring family also had twice as many parts in each product unit as the others.

A more sophisticated analysis is provided by [Lcc], who incorporates the value of inventory. Manufacturers hold safety stock in order to avoid running out when orders arrive. The more variation there is in orders, the larger the safety stock must be.[15] Any method that reduces the variation or reduces the amount of inventory that is subject to uncertainty reduces the safety stock that has to be held, saving money. The strategy of delayed commitment is aimed at this kind of savings.

In the typical delayed commitment strategy, a product is built in generic form that would satisfy any customer except for certain features or parts. Those parts are added only when a customer's order arrives. Since demand for the generic items is relatively predictable, safety stock of them can be small. The more generic parts there are (that is, the closer the decoupling point is to the customer), the more parts there are that are subject to low variety and small safety stock. Also, if fewer parts are subject to customization, or if those parts are cheaper, the value of the safety stock that is at risk due to high variety and uncertainty is smaller. However, the more parts are added to a generic product, the more value it has, raising the investment in inventory. The optimum place to put the decoupling point to minimize generic inventory holding cost has to be evaluated individually for each case.

An example given by Lee concerns a manufacturer of computer printers. Customers in different countries use electricity of different voltages, requiring different power supplies. Making printers with power supplies in advance of actual orders is risky because orders from different

---

[15] Safety stock is an amount of inventory that is held for the purpose of avoiding running out. It is like a design margin or design safety factor in the sense that it represents an investment that will almost never be used but has to be made to avoid something worse. If we had perfect knowledge of all things, then safety stocks and design margins would not be needed.

countries can vary a lot. Building printers to order makes customers in distant countries wait while their printer is shipped. So the company made printers without power supplies and shipped them to local distributors near each customer country. Along with these incomplete printers were shipped large numbers of different and relatively low cost power supplies, along with instructions on how to install them and test the printer. Note that this method, like the mushroom product, requires that the assembly sequence permit the customizing steps to be last or nearly last. It may be necessary to modify the product's design to permit this. In addition, the steps carried out by the distributors have to be foolproof because the printer manufacturer has no control over the distributors' hiring and training. In the case of the printer, the power supply was a separate module with a wire that simply plugged into the back of the printer.

The plain vanilla box strategy ([Swaminathan and Tayur]) is even more sophisticated than delayed commitment. In this case, the manufacturer may construct several generic versions containing different subsets of the final product's parts. These different subsets each can be built into one or several versions of the product, as illustrated later in Figure 14-10. In this case the issue is to balance the production and inventory holding cost of building vanilla boxes that might not be ordered for a while, the cost of not being able to fill an order in time, and the cost of assembling different kinds of components into vanilla boxes or end products. In the case shown in Figure 14-10, it may not be economical to build all three kinds of vanilla boxes.

As in the other strategies, each vanilla box must have a feasible assembly sequence permitting each intended end product to be built.

14.C.1.e. Combinatorial Implementations

In some cases, wide variety in a product line can be achieved by substituting modules, each of which does a particular function that the customer recognizes. It may be that the combinations are subject to few constraints on how they are chosen, combined, and assembled. An example is sectional furniture or office partitions. On the other hand, it may be that while many versions of each module are available, only certain combinations are possible due to interference or other kinds of incompatibility. It may be that in a car one cannot choose the highest-capacity air conditioner and the biggest radio/CD player because there is not enough space for them both. In these cases, the manufacturer or distributor must weed out orders that contain incompatible combinations, a task that often requires computer support.

When the modules can only be assembled in specific ways, the product is called a *configuration product* ([Mittal and Frayman], quoted by [Erens]). In a configuration product, there is a limited number of choices for each module, they must be assembled to each other in specific ways, and the design process itself involves sophisticated classification of functions, subfunctions, and alternative ways of carrying them out. In the extreme, a new class of products is defined this way. Examples include the Denso Panel Meter and various kinds of industrial and medical equipment such as X-ray machines. Both of these are discussed below.

14.C.2. The DFC as an Architecture for Function Delivery in Assemblies

14.C.2.a. Product Family DFCs

We saw in previous chapters that the DFC expresses design intent for an assembly by identifying the KCs as well as chains of mates between parts that place the parts properly in space for the purpose of delivering the KCs. If a particular kind of product is to be designed over and over in different styles, shapes, or even materials, but with the same KCs, it may make sense to standardize the DFC across the entire family. This permits an important aspect of the design to be reused, including validation of KC delivery, tolerance analysis, assembly sequences, and so on. This idea is consistent with the philosophy of assembly

End Products

Plain Vanilla Boxes

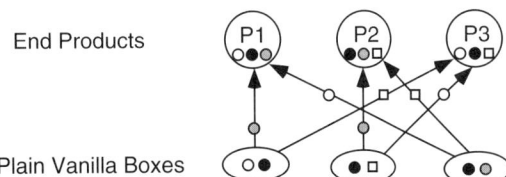

FIGURE 14-10. Plain Vanilla Box Strategy. In this schematic, there are three end products, each with three components. Also shown are three possible vanilla boxes together with the components built into them and the component that needs to be added to create one of the end products. The cost of each strategy depends on demand variability for each combination of parts in a plain vanilla box, the amount of safety stock needed for each kind, and the cost of finishing assembly from the vanilla box stage. In general, this is a difficult problem to solve. ([Swaminathan and Tayur])

design in this book, namely that connectivity is the root from which the assembly's design springs, and details of part shape are determined later to fit the connectivity design. Product family DFCs capture common aspects of design knowledge about the family, and they permit later entries in the family to be consistent in important ways with the rest of the family.

14.C.2.b. Denso Panel Meter

The Denso Panel Meter illustrates the idea of the Product Family DFC very well. Figure 14-11 shows the meter in detail and schematically illustrates different family

members. Figure 14-12 contains the same information using our symbols for KCs and DFCs. Note that detailed part shapes are not needed to convey the information. The architecture that governs the configurations is based on the two parallel centerlines on which the different parts stack. One of these centerlines is the main datum of the assembly. Any new member of the family must be designed so that it conforms to this structure, so that assembly and fastening methods and equipment can be used unchanged. Similarly, a new version of any one of the parts must also conform by adhering to standardized interfaces between parts.

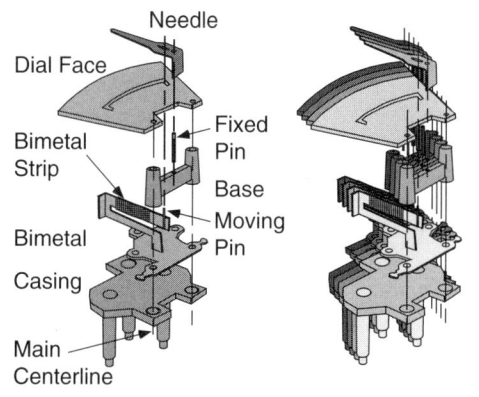

One Instance of the Panel Meter

Several Instances of the Panel Meter

FIGURE 14-11. Denso Panel Meter Family. *Left:* One member of the panel meter family. The part called "bimetal" contains a bimetal strip with a small pin at its free end. This pin connects to the slot in the needle. A fixed pin mounted to the part called "base" interfaces to the hole in the needle. Different signals change the temperature of the strip, causing its pin to move left or right. This causes the needle to rotate on the axis of the fixed pin. *Right:* Different family members are represented by parts of different shades of gray but the same shape. In general, different versions of the same part will not have exactly the same shape but will be the same at key points such as mates to other parts and places where assembly fixtures and grippers interface to them.

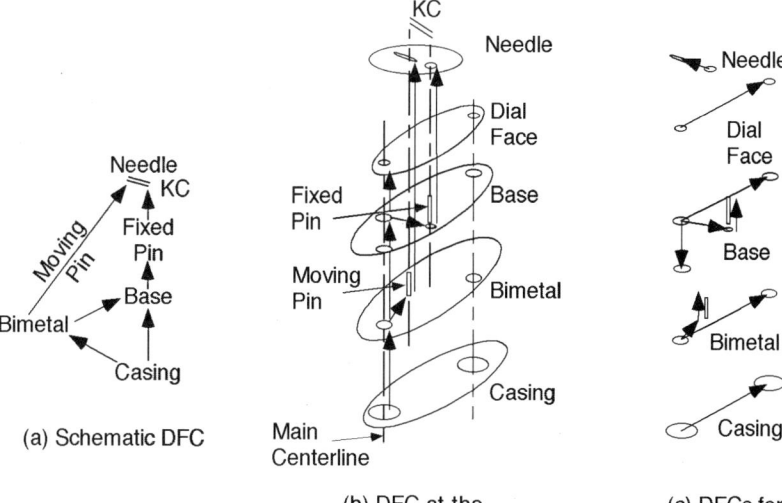

(a) Schematic DFC

(b) DFC at the Assembly Level

(c) DFCs for Each Part

FIGURE 14-12. KC and DFC of Denso Panel Meter Family. (a) A sketch of the DFC, showing the main parts by name and the KC. The KC is proper spacing and alignment of the moving pin and the fixed pin. (b) The DFC follows the main centerline up from the casing to the dial face and needle. It branches to the moving pin. (c) DFCs inside each part follow the flow set by the assembly level DFC. That is, the dimensional datum inside each part is the mate feature that aligns with the main centerline. Any new part added to the family must conform to this dimensioning scheme or else it risks destroying the ability of the family to function properly.

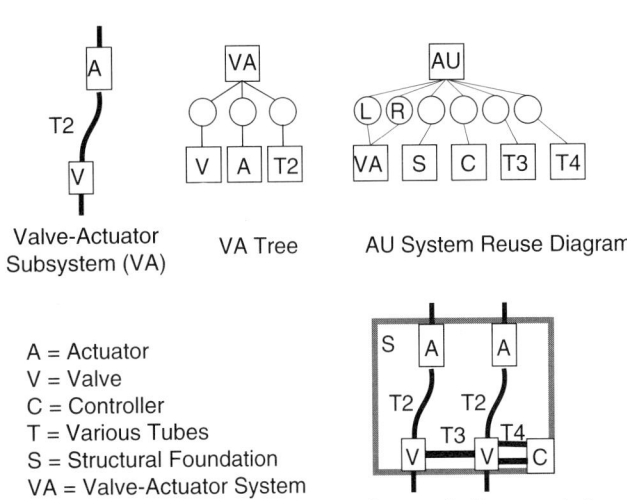

Valve-Actuator
Subsystem (VA)   VA Tree   AU System Reuse Diagram

A = Actuator
V = Valve
C = Controller
T = Various Tubes
S = Structural Foundation
VA = Valve-Actuator System
AU = Actuator Unit

Geometric Representation

FIGURE 14-13. A Nonduplicative Model of Assemblies That Encompasses Reuse. Valve–actuator systems comprise certain stock parts such as valves, actuators, controllers, tubes, and a structural foundation. Once geometric and connective models of these primitive elements have been defined, systems made of different numbers of them can be diagrammed efficiently using the reuse diagram. The system shown geometrically at the lower right is diagrammed in the reuse diagram at the upper right which in turn uses the tree in the upper middle which in turn uses the geometric model at the upper left, plus others not shown. ([Callahan]. Copyright ©1997 by Association for Computing Machinery. Used by permission.)

14.C.3. Data Management

In Chapter 3 we saw that we could model an assembly by noting names and coordinate frames of parts and their assembly features. This assembly model is sufficient if there is only one version of the product. If there are multiple versions, or if the product comprises a platform and variants, then a more sophisticated data model is needed. This model has to be able to represent many versions at the same time so that it will not simply be a bulky agglomeration of every possible individual version. In addition, it is desirable to represent the options to the customer in the form of functions rather than part names, because dozens or hundreds of unfamiliar part names might be needed, and they might be distributed over several subassemblies.[16] Finally, it has to represent forbidden combinations.

[Callahan] suggests a modeling scheme comprising two linked diagrams. One diagram is a physical decomposition tree made without regard to duplicate usage. The other is called a reuse graph. The decomposition can be used to store generic relationships between parts and the subassemblies they belong to. The reuse graph contains physical models of parts and subassemblies including their assembly features and can compactly represent multiple occurrences of subassemblies simply by referring to their tree representations.

Suppose, for example, that in a hydraulic actuator system every actuator has a valve but that different subsystems can have one, two, or three valve/actuator assemblies and these can occur in left-hand or right-hand configurations (see Figure 14-13). The figure shows that it is more compact to diagram each specific system with a shorthand that calls out specific instances of the generic valve/actuator subassembly with a short parameter list to indicate how many and whether they are lefties or righties.

[Erens] proposes an extension of the familiar bill of materials to handle some kinds of product variety. It is built on the idea of a choice tree, an example of which, applied to X-ray systems, appears in Figure 14-14. Each X-ray system has certain required elements, but several of these come in different sizes or functional capabilities. First of all, the system can be for heart or blood vessel examination (cardio or vascular). Depending on the customer's use requirements, some X-ray tubes are valid choices while others are not. Other combinations are similarly allowed or forbidden. Backing up these functional choices are many physical subassemblies that need not concern the customer.

A good way to set up a choice tree like that in Figure 14-14 is to imagine the customer's view of the product and the functional aspects or operational steps the customer would use while trying to decide which elements

---

[16]Some industries are much better at doing this than others. One need only compare choosing options in a car versus choosing PCI cards for a computer.

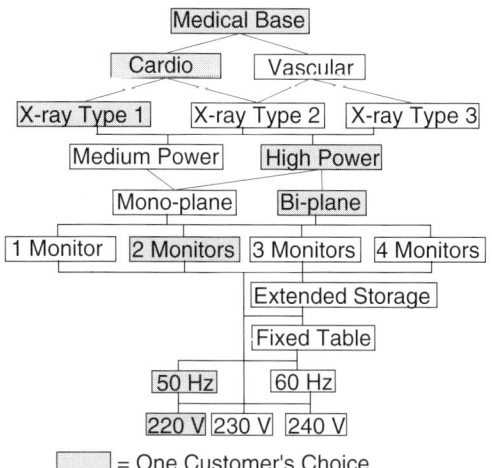

FIGURE 14-14. A Choice Tree for X-Ray Equipment. ([Erens])

to buy. The electrical controller company Telemechanique set up entire product catalogs this way, providing different chapters to indicate the functional choices and leading the customer through them using color coding. Essentially, Telemechanique "designed the buying experience" first, identified the modules that meant the most to the customer, and then subdivided the physical elements of the product family into subassemblies that lie entirely within the respective modules so that the customer would always choose something that was geometrically compatible ([Morelli]).

Telemechanique then built an assembly system to make these products to order using combinations of standard modules. Manual assemblers were helped to choose the right components by video monitors that showed them what to do. Barcodes on the product signaled the video system for which display to show.

14.D. EXAMPLES

In this section are many specific examples where product architecture has been chosen for technical, nontechnical, or a mix of these reasons.

14.D.1. Sony Walkman

The Sony Walkman is one of the most successful product families ever produced. Introduced in 1979, versions of it are still in production. It emerged from the need of Japanese commuters to have a pocket-sized tape player to use while riding the train to and from work for hours at a time. It depended on three Sony component innovations: very small and light earphones that produced excellent audio quality (1979), a thin "chewing gum" battery (1982), and a steady RPM pancake motor (1984). These elements created a small high-quality product. Following the Lehnerd–Meyer scheme, important market segments were the United States (large units with good quality and high price) and Japan (small units with excellent quality and small units at low price). The basic product relied on the earphones and formed the basis for a low cost but long-lived family. Battery and motor innovations were used to launch second and third main families that were smaller and/or had higher audio quality. Once these three families were established, literally hundreds of versions were produced, some involving only cosmetic changes. A common internal module shown in Figure 14-15

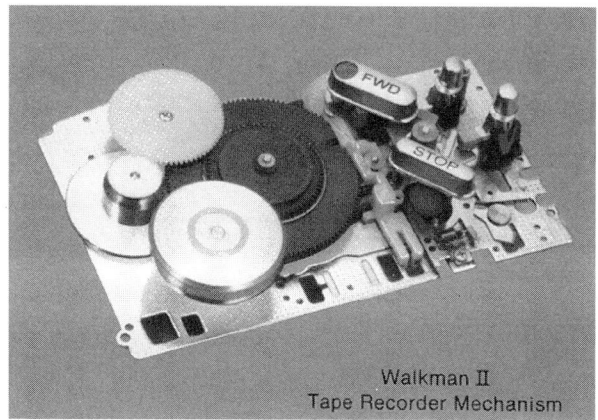

FIGURE 14-15. The Walkman II Recorder Chassis. In this version of the product, the circuit board is a separate unit. In later versions, after surface mount electronics were introduced, the mechanical chassis and the circuit board became the same unit. This process improvement permitted much thinner units to be made. (Photo courtesy of Sony FA.)

underlay the different versions. Supporting all this variety was Sony's strong capability in assembly robot design, production, tooling, and programming. The robots and their parts feeders (see Chapter 17) were easy to reprogram and retool to make each new version. Even though each component technology could be or was copied, no other manufacturer produced so many models or became so identified in the public's mind with dedication to this kind of product ([Sanderson and Uzumeri]).

Even though most companies rely on suppliers for many key components and production technologies, Sony developed and exploited this product family using its own component and production know-how and products. The most interesting production know-how was embodied in a family of assembly robots with easily reprogrammable part feeders. These are described in Chapter 17. This technology permitted Sony to switch rapidly as different versions of the Walkman were introduced. For this reason, Sony product designers and marketers did not have to worry about their products' entry into the market being delayed by assembly constraints as long as they adhered to some basic rules about design for assembly, the topic of the next chapter.

14.D.2. Fabrication- and Assembly-Driven Manufacturing at Denso—How Product and Assembly Process Design Influence How a Company Serves Its Customers

Denso's panel meter has been used in different ways in this book to illustrate different things. Here we use it to show how Denso designed it to meet some important global requirements. We will also see that Denso implemented a rather different strategy for manufacture than is customary.

The meter comes in many varieties, and Denso can predict only roughly what the demand will be for each one. The challenge is that Toyota will make a fairly predictable number of cars each year and each one will need two or three meters. Thus Denso could potentially face demand for hundreds of thousands per year of model 22 and 100 for model 33, or it could face demand for hundreds of thousands of model 33 and 100 for model 22. There is no economical way to prepare capacity to meet demand for hundreds or for hundreds of thousands for each of 40 varieties. The only way is to make one plant with total capacity equal to Toyota's anticipated needs and switch the factory from variety to variety as demand shifts.

Furthermore, demand can shift rapidly, perhaps several times during a single day. When the author visited the plant where the meters were made (they are no longer in production), he saw that shipments were packaged and sent out every hour, and kanbans (order tickets in the Just-in-Time

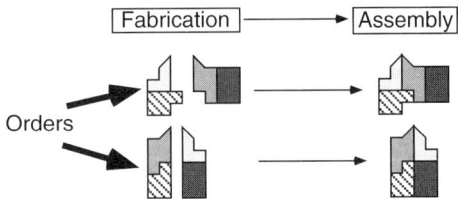

Relies on Fabrication to Express Model Mix and Achieve Flexibility:

In Response to Orders, Complex Parts Are Made and Then Assembled into Final Items.

This Is a Low Bandwidth Method Because Fabrication Takes So Long.

FIGURE 14-16. Fabrication-Driven Manufacturing. The factory makes several different product versions, of which two are shown. Orders arrive at the fabrication end. To fulfill each order, different parts are fabricated. Once made, they are assembled. Such a process can respond to changes in demand as fast as the fabrication processes can be changed over from making one set of complex parts to another. It usually takes a long time to do this.

system) arrived by return truck. This is faster than fabrication equipment or supplier orders can typically be changed over and in some cases is faster than parts can be made.

The typical manufacturing process operates by taking orders in at the fabrication end and sending them out from the assembly end. This can be called fabrication-driven manufacturing and is illustrated in Figure 14-16. It is difficult to switch such a process rapidly from one product version to another if changes come in faster than the fabrication shop can change over and make new varieties of parts. If demand shifts rapidly and significantly, the company will run out of some items, delaying order fulfillment and annoying or losing customers. A costly way out is to build up and hold inventory of many kinds of complex parts in the (possibly vain) hope that someone will eventually order them.

An alternative to fabrication-driven manufacturing is assembly-driven manufacturing, shown in Figure 14-17. Orders arrive at the assembly factory, where they are put together by means of different combinations of ready-made relatively simple parts. The speed of response of the fabrication shop is not a factor in the ability to fill an order quickly. Instead it is the speed and flexibility of the

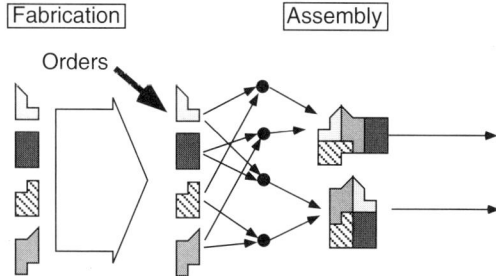

Relies on Assembly to Express Model Mix and Achieve Flexibility:

Simple Parts Are Made to Statistical Trends.
In Response to Orders, Items Are Assembled.

This Is a High Bandwidth Method Because Assembly Happens So Quickly.

FIGURE 14-17. Assembly-Driven Manufacturing. In this process, parts are made without specific orders for them being in hand. The risk in this is low because most parts are used in a high percentage of the versions that are likely to be ordered. Such a process can respond to changes in demand as fast as the assembly personnel or equipment can arrange to substitute different parts. This can usually be done in a few seconds or minutes.

assembly process that matters. It should be easy to see that the Denso panel meter fits this description. The flexibility of the process is determined during design when the assembly interfaces of the product were standardized across all versions of each part. This strategy depends on holding inventory in the assembly area, but, as discussed in Chapter 1, the parts are simple and low cost, and they are each used in so many versions of the product that there is little risk that any of them will be unused for long. Another reason why inventory is held at assembly is that Denso cannot manage demand the way Dell can.

In Chapter 15 we learn that a fundamental principle of DFA is part count reduction, which could make parts more complex. The strategy of assembly-driven manufacturing depends on keeping the parts simple, at least at the places where they join. It also requires keeping them separate in the sense that aspects subject to choice by customers need to be on separate parts so that satisfying the choices can be accomplished with simple substitutions. Alternately, parts can be made programmable. Electronics are the most obvious example.

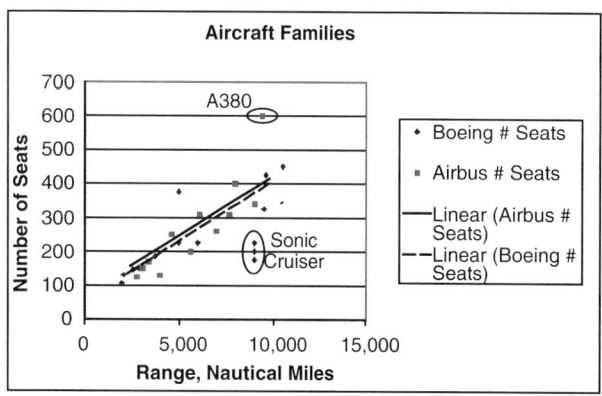

FIGURE 14-18. Range-Seats Chart of Boeing and Airbus Aircraft, Circa 1997. A range-seat chart is the typical way passenger aircraft manufacturers represent their product lines. Prior to 2000, Airbus did not have a long-range high-capacity aircraft to compete with the Boeing 747. It is currently building the A380 to fill this gap. Meanwhile, Boeing announced the Sonic Cruiser. Each of these planes occupies a totally different region of range-seat space.

14.D.3. Airbus A380 and Boeing Sonic Cruiser

Boeing and the Airbus consortium are the only surviving manufacturers of large commercial passenger and freight aircraft. Each firm makes a variety of aircraft and typically differentiates them according to combinations of seating capacity and range. Figure 14-18 shows this combination for Boeing and Airbus as of the late 1990s. The figure shows that Airbus lacked an aircraft in the region occupied by the Boeing 747. Accordingly, Airbus decided in late 2000 to launch the A380.

Boeing said all through the late 1990s that there was not a big enough market for a plane as big as the A380. Instead, it promoted and sold many of its 777s, which carry fewer passengers but have long range. More recent members of the 777 family have extended range rather than more seats. Airbus marketed the A380 as a way for long-haul airlines to serve major hubs. Boeing claimed that hub and spokes arrangements were inefficient and predicted that airlines would prefer to fly smaller planes to smaller airports using point-to-point routing.[17] The latter is more

---

[17]Improved air traffic control based on global positioning systems could also play a role in enabling efficient point to point flight plans.

convenient for passengers and opens up many more air-ports to international travelers while relieving congestion at major hubs. Airbus predicts that the A380 will be so at-tractive to long-haul airlines that they will encourage any needed upgrades of airports. Additionally, Airbus intends to make many structural parts from composites rather than metal to reduce weight.

Thus Boeing and Airbus each have a different view of the architecture of future air travel although their views of how to structure a family of aircraft are basically the same.

In the late 1990s, Boeing queried customers about a redesigned 747 but did not get a strong response. In April 2001, it announced a new concept aircraft called the Sonic Cruiser.[18] This plane was intended to have a range of 9,000 miles and seating capacity of 175 to 250. This put it be-low the 777 in seat capacity but at a similar range. The novelty of the Sonic Cruiser was that its cruising speed would have been 0.95M or 95% the speed of sound, mak-ing it 10% faster than current airliners that fly 0.85M. This would cut one to three hours off transcontinental flights ([Aviation Week]). Higher speeds and direct flights to many airports could be attractive to both passengers and airlines.

An aircraft that can fly so close to the speed of sound (called transonic) cannot look like conventional aircraft with a simple fuselage and wing architecture. Various al-ternate designs are possible, including a delta wing. Tran-sonic and supersonic fighter aircraft do not have constant fuselage diameter. High drag is experienced as the air en-counters the increased cross-sectional area of wing and fuselage. In the early 1950s, Richard Whitcomb devel-oped the so-called "coke-bottle" shape for the fuselage, effectively reducing the cross-sectional area in the region around the wings ([Carlson]).

A rough sketch of the concept appears in Fig-ure 14-19. The sketch shows a blended wing, similar to the delta wing, plus two small forward wings called canards. It has a constant diameter fuselage. "The coke-bottle fuse-lage is really hard," said a Boeing executive ([Aviation Week]). "The body has no constant section, it's struc-turally inefficient, hard to build, the number of passen-ger aisles varies, and you can't stretch it." The last point refers to Boeing's traditional way of generating variants on a product family, namely to lengthen the fuselage.

---

[18]This aircraft did not survive beyond the concept stage.

FIGURE 14-19. Artist's Rendition of Boeing Sonic Crui-ser Concept. (Courtesy of Boeing. Used by permission.)

The design challenge to Boeing was to get sufficient lift to drag ratio for this aircraft. Alternate designs have much larger wings in proportion to their fuselages, with the extreme being a flying wing. That design is more inte-gral than the conventional aircraft in the sense that every square inch of exterior surface contributes lift. In con-ventional aircraft there is a separate fuselage that carries passengers but contributes no lift. Passengers in a fly-ing wing would probably go faster but would have no windows. Thus the Sonic Cruiser was a compromise ar-chitecture that tried to combine speed, range, passenger comfort, manufacturability, family extensibility, and con-formance to a projected future architecture of airline routes and operating methods that is quite different from that of Airbus.

14.D.4. Airbus A380 Wing

Until 2000, Airbus aircraft were made by the Airbus Consortium, a partnership of companies in four Euro-pean countries: England, France, Germany, and Spain. The work of making current Airbus aircraft was care-fully divided to satisfy political and technical criteria, and each member of the consortium sold its subassem-blies to the consortium. Aircraft subassemblies are made in three countries: wings in England, fuselage sections in Germany, and tail sections in Spain. These are flown to Toulouse, France in huge transport aircraft, where they are fitted out with engines and interiors, assembled, and

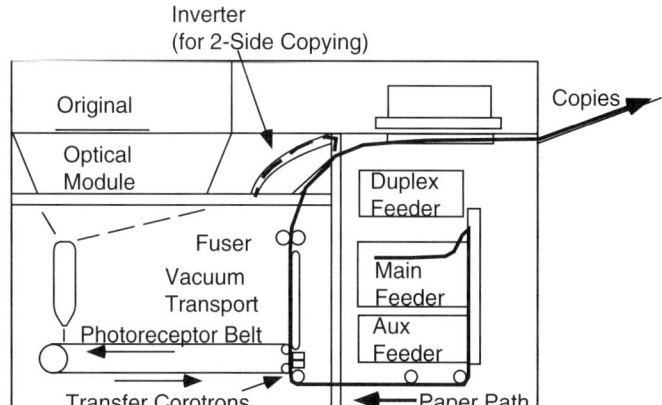

Inverter
(for 2-Side Copying)

Original

Optical
Module

Copies

Fuser

Vacuum
Transport

Photoreceptor Belt

Duplex
Feeder

Main
Feeder

Aux
Feeder

Transfer Corotrons

Paper Path

FIGURE 14-20. Optical Copier Architecture. Light gathered from the original is passed through an optical path to the photoreceptor belt, which carries it to charged wires called corotrons. These charge the paper and make the black toner powder stick. The fuser fixes the image onto the paper. If two-sided copies are needed, the paper is diverted through the inverter and sent to the duplex feeder. From there, it is sent through the image transfer process again. Much of the interior of the machine is taken up by the optical path and the photoreceptor belt.

delivered to customers. Each time a new aircraft was added to the Airbus family, the work content of each partner was painstakingly negotiated to accord with its percentage share of the consortium. Profits from aircraft sales were distributed to the partners in proportion to their share. Thus the architecture of European politics, the aircraft, and the consortium aligned.

A new company, EADS, was formed to carry on this activity and to build the A380. The A380 will have wings that are too large to be transported in final form by air from England to where-ever the plane is finally assembled from the other subassemblies. As the plane was being designed and the consortium was negotiating its way to becoming EADS, there was considerable speculation as to where final assembly would take place, given the difficulty of transporting the wings and the difficult political issues surrounding allocation of jobs. As of this writing, the assembly process is: wings made in England, fuselage sections made in Germany, tail sections made in Spain, preliminary assembly in Toulouse, aircraft flown to Germany for final fitting out of interiors. EADS plans to transport wings and fuselage sections by large barges and trucks.

Boeing avoids such problems by conducting final aircraft assembly in the same building where wings are built. An overhead crane carries wings and fuselage sections to the final assembly area. Boeing and Airbus assembly processes are compared in Chapter 16.

This example and the previous one show how the architectures of technologically and commercially important products can be influenced by matters of national security and pride to the extent that production operations can be deeply affected. Architectural choices that literally bet the

company involve predicting the existence of future market segments, the operating choices of other companies, and the purchase choices of future consumers.

14.D.5. Office Copiers

Office copiers have been revolutionized in the last five years by the conversion from optical to digital image acquisition. Copiers originally had high-power lights and lenses that obtained an image and transferred it via a complex optical path to a photosensitive belt, from which it was transferred to the paper. An example of such an architecture is shown in Figure 14-20.

By contrast, a digital copier obtains its image via a scanner or network connection to a computer. This image, possibly processed by a computer inside the printer, is transferred electronically or by fiber optics to a laser diode which exposes the photosensitive drum. An example of this architecture is shown in Figure 14-21. Note that the image acquisition portion of the machine, similar to a scanner, does not have to be physically attached to or near the copying portion of the machine. Also, it can act as a printer on a computer network or as a fax machine.

Copier design involves a number of interlocking architectural decisions.[19] In an optical copier, the paper must not interrupt the optical path, so the paper path must run on the opposite side of the photoreceptor drum or belt from

---

[19]Maurice Holmes, formerly of Xerox Corp. provided these examples and called the interlocking nature of their design process "architectural flow."

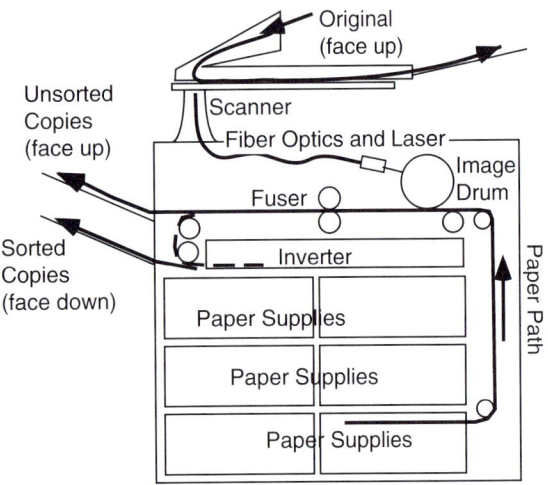

FIGURE 14-21. Architecture of a Digital Copier. The optical portion is very small, allowing for a smaller machine that has a larger paper supply and possibly a simpler paper path.

the optical path. Since the optical path must run from the top of the copier, where originals are inserted, to the drum or belt, the paper must be low in the machine when it receives the image. This usually results in a long and circuitous paper path, which in turn can lead to more frequent paper jams than a short straight path.

If entire documents are to be copied, a major design choice is whether they are to be inserted face up or face down. If face up and fed from the top of the stack, then they will be copied in page order and thus the copies must be delivered face down in order to be delivered in page order. If the blank paper passes over the transfer drum, the image will be face down and it will be easy to deliver the pages in page order. If the paper passes under the transfer drum, as it must in an optical machine, then it will be face up as it receives the image and thus must be inverted before moving to the exit tray in order to be in page order.

If documents are to be placed face down, then either they must be fed off the bottom (not easy) or else they will be copied in reverse page order. It will be difficult to deliver the originals in page order unless they are inverted before being passed to the user. Copies can then be delivered face up and will be in page order. They can thus easily be passed under the transfer drum. Flexibility to run the paper above, under, around, or anywhere else inside the machine is enhanced in the digital design compared to the optical design.

Note that the machine in Figure 14-21 treats individual unsorted copies differently from stacks of sorted copies. The former have a simple paper path while the latter must pass through the inverter. A thought question at the end of the chapter asks the reader to critique this choice.

Office copiers are extremely complex, often containing software with millions of lines of code. Their technology and architecture are currently in a tumultuous transition which is accompanied by various kinds of technological and market convergence. Newer architectures that exploit new technology have very different internal space constraints, higher reliability, and enhanced functions. They also enable different work processes in offices.

14.D.6. Unibody, Body-on-Frame, and Motor-on-Wheel Cars

At least three major architectures exist for wheeled land vehicles: body on frame (the original architecture of wagons and buggies for centuries: all suspension and drive train parts attach to a frame, while the body is a separate unit bolted to the frame), unibody (suspension parts and drive train are attached directly to a body shell), and motor on wheel (usually a version of body on frame). Body on frame is suitable for large vehicles with large engines and rear wheel drive, and it was the most common architecture for American cars until the 1970s. It is modular in the sense that the body can be redesigned without requiring much redesign of the frame, and vice versa. In a frontal collision, the frame will absorb the collision energy, so the body's design is not deeply affected by this issue. Body on frame sometimes includes a front axle that extends the full width of the car, requiring the engine to be high off the ground. This naturally raises the entire vehicle. A high body also facilitates passage of the drive shaft from the engine in front to the rear wheels.

In the 1970s, smaller and more fuel-efficient vehicles were needed, and front wheel drive, originating in the 1920s, was revived. Since it does not have a full-width front axle, the engine can be placed low, between the front wheels. This lowers the entire vehicle, a move that is not blocked by a drive shaft extending to the rear. Since a frame would introduce extra material between the body and the ground, a unibody makes sense in this case. However, the body now must be designed to absorb all the frontal collision energy. These factors naturally tend to

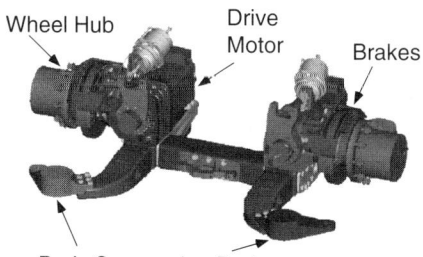

FIGURE 14-22. Drive Train Module for Airport Bus. Each wheel has its own electric motor. Between the motors is a space for the bus' floor. The resulting vehicle has very low ground clearance, making it easy for passengers to get on and off. (Courtesy of ArvinMeritor, Inc. Used by permission.)

integrate body, suspension, and power train design, making changes in body design difficult without redesigning portions of the other two.[20]

Figure 14-22 shows the drive train of a motor on wheel, body on frame architecture used for airport buses. Electric motors drive individual wheels through a right angle drive. The bus body rests on the suspension pads. As a consequence of the low-slung middle region of the suspension module, the floor of the bus can be quite low. This is important for allowing passengers easy access on the tarmac where there are no stairs or curbs to help them up.

This example shows how different the architectures for apparently similar products can be, depending on differences in scale or use that appear small at first glance.

14.D.7. Black and Decker Power Tools

In the 1970s, Alvin Lehnerd oversaw the redesign of Black and Decker's entire line of hand power tools—saws, drills, routers, sanders, and jigsaws ([Lehnerd and Meyer]). It was the first time he had the opportunity to create a product platform. The impetus was a new Federal

---

[20]Company X makes shock absorbers for unibody cars made by company Y. Company Y complained to company X about suspension noise and vibration. Company X asked for design data and computer models of the front end sheet metal of the body where the shock absorbers attach, but Company Y refused, saying, "The problem is with your shocks. Fix them." After a year, Company Y agreed to provide the information, following which Company X did some computer simulations that assessed the interactions of flexible suspension and flexible sheet metal, and designed a simple fix. Company X understood that the body/shock/suspension system was integral, whereas Company Y thought it was modular, probably because it was made of separate parts that were screwed together.

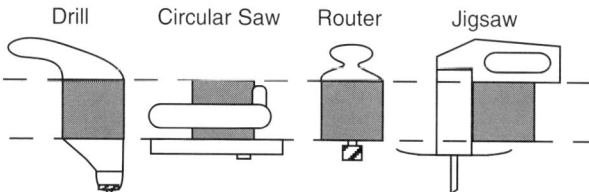

FIGURE 14-23. Power Tool Platform with Common Motor Module. Each product comprises three main modules: handle, motor, and "business end." Within each product type, many varieties are offered, each having different motors with different power.

regulation regarding double insulation, which required redesigning the motors. Lehnerd decided to define a family of products based on the redesigned motors. The essence of the physical architecture of this family is shown in Figure 14-23.

The structure of this platform should be clear from Figure 14-23: All the products have a handle end, a business end, and a motor in the middle. All the motor housings are the same across the entire product family, which requires all motors to be the same diameter regardless of their output power but saves the cost of designing different injection molds for different plastic motor housings and the parts that mate to them. Tools with more power therefore require longer motors. However, the length of the motors will not change drastically as their power rises, as shown in Figure 14-24. This requires some clever design, plus strong understanding of motor fundamentals. The motors

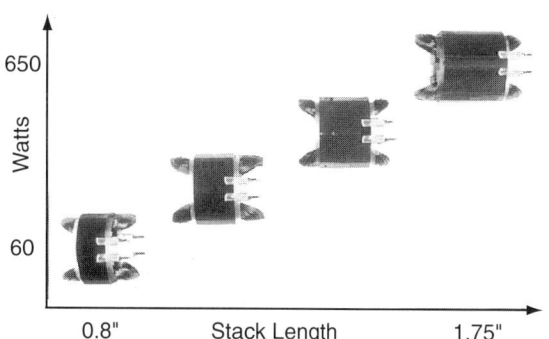

FIGURE 14-24. Motors for Power Tool Platform. These motors differ by nearly a factor of 10 in power but only a factor of two in length. They are the same diameter. (Printed with the permission of The Free Press, a Division of Simon & Schuster, Inc., from [Lehnerd and Meyer]. Copyright © 1997 by The Free Press.)

were simplified in other ways to enhance automation of their manufacture. The first products put on the market under this approach had longer motor housings than they needed so that the same housing could be used on later versions that had more power. Lehnerd and Meyer call this "building degrees of freedom into the platform."

Once all members of the platform used the same motor design and mostly the same motor parts, huge economies of scale in motor production became possible, dropping the price of motors. This, along with other design and production modifications, permitted Black and Decker to underprice its competitors, generate a series of new products quickly, and take the market for these products within a few years.

The structure of the products in Figure 14-23 is a stack arrangement. Assembly can be done by placing the business end down in a fixture, dropping shafts and gears in, then adding the motor module, then the handle end. Automatic assembly, though not necessarily economical, is at least technically feasible. An alternate structure for small power tools is shown in Figure 14-25. It is called the clamshell architecture. A drill designed this way is discussed at length in Chapter 13. In this arrangement, assembly is accomplished by assembling the power train, comprising motor, brushes, wires, shafts, and gears, into a semi-stable subassembly, and then laying this subassembly down into one half of the shell. The last step is to place the other half of the shell on top and insert the fasteners. This is a dexterous operation that is suitable for manual assembly but not for automatic assembly.

This example shows how pervasive product architecture can be even for apparently simple products, touching family design, technical essentials of a central module, and choice of manufacturing and assembly method.

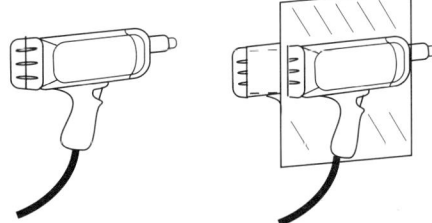

FIGURE 14-25. "Clamshell" Architecture for Hand Tools. *Left:* The assembled tool. *Right:* An imaginary vertical plane separates the left half and the right half of the tool. Each case half resembles half a clam shell, giving rise to the name for this architecture.

14.D.8. Car Air–Fuel Intake Systems

Car air–fuel intake systems have evolved from an assembled set of cast iron and aluminum parts to an integrated injection molded polymer part, as a result of improvements in materials and molding methods. The functions of this system have grown apace and now include air filtering and throttling, fuel injection (the injectors and their parent fuel rail are molded or assembled into the polymer part), air and temperature sensing, and distribution of the air to the cylinders. Thus the architecture has evolved from modular to integral. Correspondingly, looking at this system from the point of view of the engine, the modern part qualifies for the name "module" because it is designed and manufactured by a supplier and simply bolts onto the engine in a few seconds. A typical system is shown in Figure 14-26.

14.D.9. Internal Combustion Engines

Automobile engines used to be made of cast iron, but fuel economy standards call for weight reduction. In response, manufacturers have turned to aluminum, a lighter but less stiff material. The architecture of engines has changed accordingly. Figure 14-27 shows a typical cast

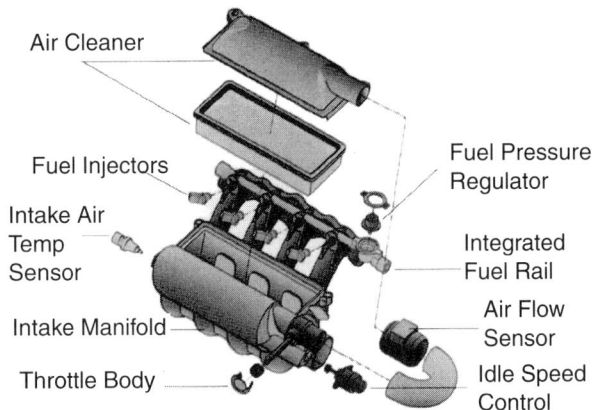

FIGURE 14-26. Integrated Air Intake System. In this design, a number of previously separate parts, such as the manifold, the filter pack, the fuel rail, and the throttle body, are integrated into one unit. The main enablers for this are development of heat-resistant polymers and a molding process that uses low-temperature metal forms to create the internal passageways. These forms are melted out later with boiling water, and the metal is reused. (Courtesy of Siemens VDO Automotive. Used by permission.)

iron block engine. The block is so stiff that it can act as the foundation for the rest of the engine. Many other parts are simply bolted onto it. Note that the oil pan is a simple sheet metal piece, which is adequate for its main function of retaining the oil. Compare this with the all-aluminum engine in Figure 14-28. The block requires additional stiffness contributions from the other parts as well as from the very long tie bolts. The sheet metal oil pan has been replaced by a substantial cast aluminum lower crank case

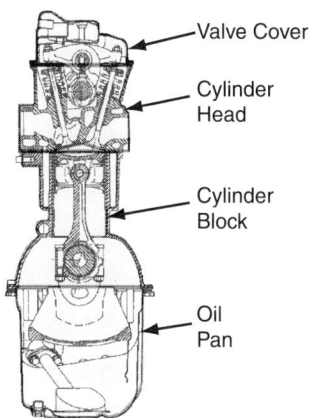

FIGURE 14-27. Cast Iron Block Engine. The block is massive and stiff, providing all the structural rigidity that the engine needs. Other parts are just screwed on. Note the sheet metal oil pan, just stiff enough to retain the oil and resist puncture by objects thrown up from the road. ([Taylor]. Copyright © 1985 by MIT Press. Used by permission.)

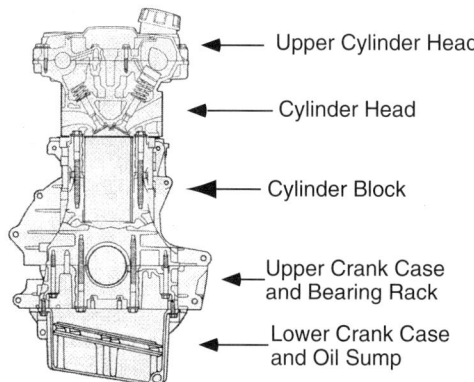

FIGURE 14-28. Aluminum Engine. All the main cast parts shown are aluminum. Each of them is needed to contribute to the structural rigidity of the assembly. Note how long the tie bolts are, how many there are, and in how many directions they work, compared to those in Figure 14-27. ([Larsson et al.])

and oil sump. The construction of this engine is reminiscent of prestressed concrete, with the long tie bolts playing the role of the reinforcing bars. All cast parts other than the block have dual "engine" and structural functions that the corresponding parts in the cast iron block engine do not have. Thus engines have evolved from integral to modular in the sense that the newer ones have more parts, but several parts have evolved from performing one function to performing two or more.

14.D.10. Car Cockpit Module

Car cockpits (see Figure 14-29) consist of some foundation pieces, a main exterior dashboard, and many attached submodules. The [Sako] study of modularization trends in the world automobile industry showed that different companies buy cockpit assemblies with widely different complements of these submodules attached, and even the same cockpit used on the same model car built in a different factory will have different submodules attached to it when it arrives at the plant. Comparing seven car makers, Sako found that the only parts that were consistently part of the incoming assembly were the cross-car beam assembly, the dashboard, and the glovebox. All the other submodules were on half or fewer of the cockpits studied. This is especially surprising for the case of the wire harness, given how difficult it is for a person to lie on his back and install it after the cockpit is in the car.

Figure 14-30 shows the cockpit's liaison diagram. Two kinds of relationships are shown in this figure. One is a traditional liaison diagram, which indicates that there are three "hubs" from which spokes radiate: the cross-car beam, the dashboard, and the wire harness. Each of these hubs functions as a sort of integrator for the items attached to it. The other kind of relationship, shown by rectangular regions, is functional, consisting of structure, steering region, user interface for electronics, and dashboard. Note how the functional clusters overlap or underlap the liaison hubs. The fact that the physical and functional links and clusters do not correspond is typical and indicates that it is difficult to achieve pure modularity in both functional and physical domains.

14.D.11. Power Line Splice

High-voltage power lines are thick cables. When they break, they must be spliced back together. Breaks can

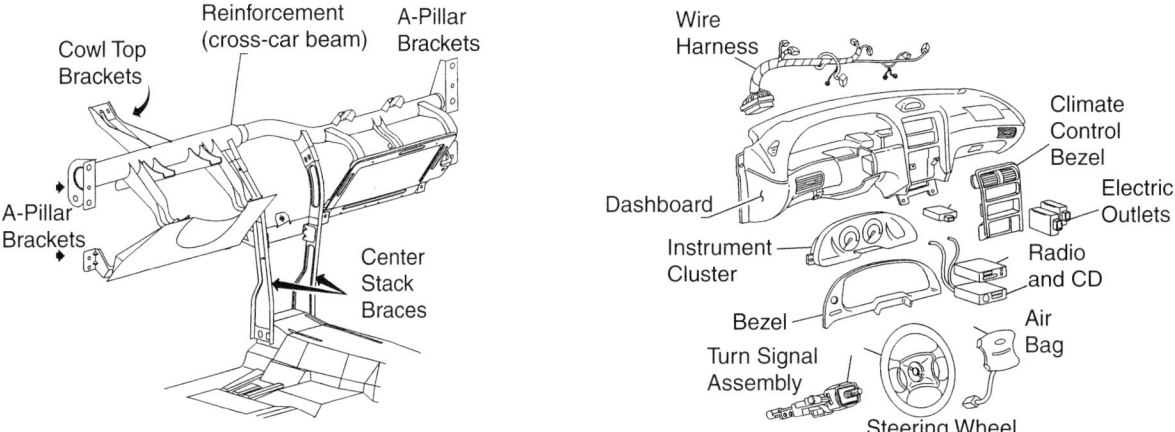

FIGURE 14-29. Car Cockpit Module. *Left:* The underlying structure of the cross-car beam: center stack braces, reinforcement structure, cowl top brackets. *Right:* The main submodules that are assembled to the cross-car beam: wire harness, dashboard, instrument cluster, instrument cluster bezel, radio, CD player, electric outlet, climate controls, fuse box, steering wheel, turn signal assembly, climate control bezel, and airbag. Depending on the manufacturer and customer options, this module can contain brake and clutch pedals, brake booster, and navigation system. The glovebox and air conditioning (HVAC) module are not shown, but these are always (for the glovebox) and sometimes (for the HVAC) part of the purchased cockpit. (Courtesy of Ford Motor Company. Used by permission.)

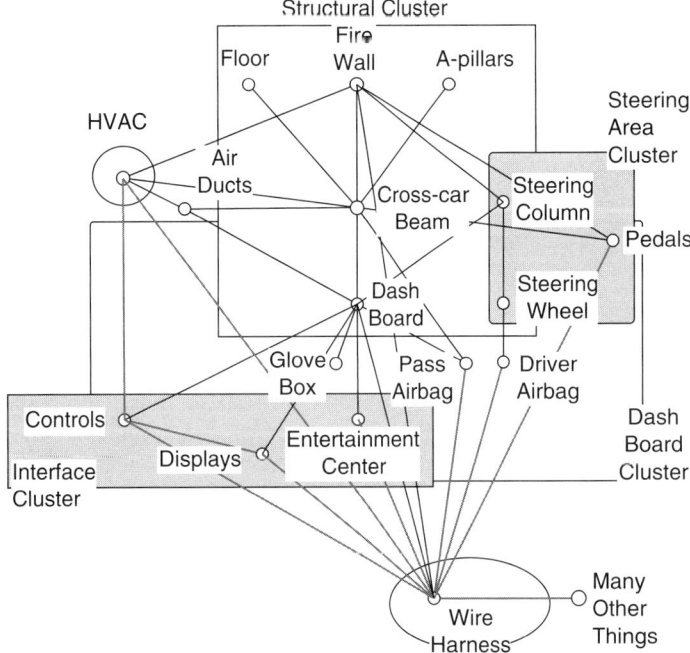

FIGURE 14-30. Cockpit Liaison Diagram. This simplified liaison diagram contains at least three hub–spokes arrangements. In addition, several functional clusters are indicated by rectangular regions. Items such as the dashboard that lie on overlapped regions should be interpreted as belonging to all such regions. The ratio of arcs to nodes (see Chapter 7) is higher than is typical for assemblies. Black arcs represent mechanical joints while gray arcs represent wiring connections.

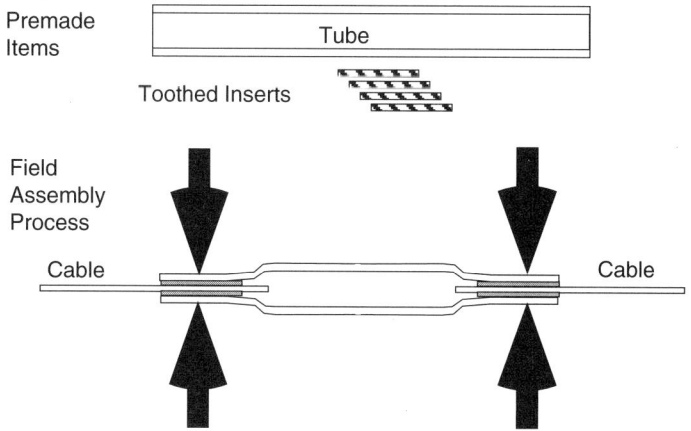

FIGURE 14-31. Typical Power Line Splice. The pre-made items consist of an aluminum tube and a set of toothed inserts. Tube and insert sizes must be chosen to suit the cable that needs splicing. The inserts are placed in the tube surrounding the cable, and the tube is compressed around the cable by a hydraulic press. The press and its power generator must be transported to the site where the splice is made.

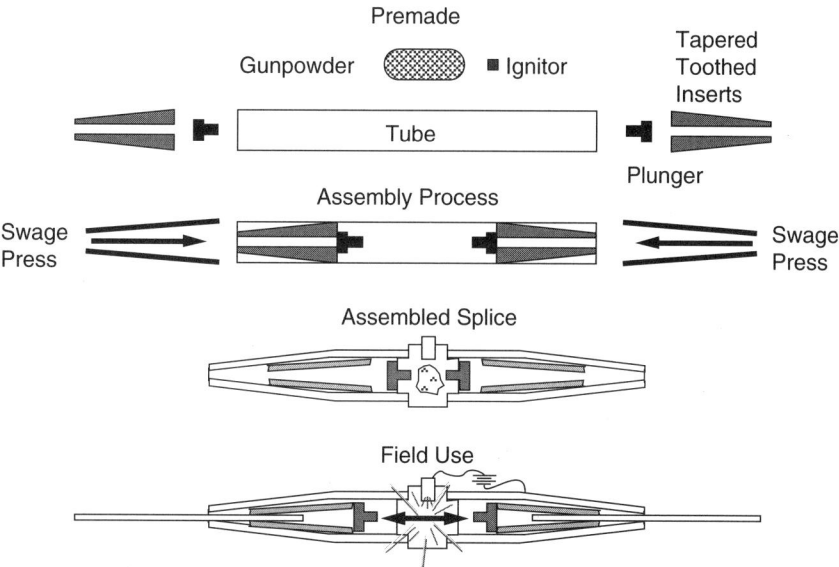

FIGURE 14-32. New Concept for Power Line Splice. In this concept, the pre-made items consist of a tube, as before, but the toothed inserts are tapered and there are additional parts comprising plungers, gunpowder, and an ignitor. The ignitor contains a small gunpowder charge and a fuse wire. Assembly involves selecting and assembling the correct tube, plungers, and inserts, then tapering the ends of the tube in a swaging press, then inserting the gunpowder and ignitor. In the field, the cables are inserted into the ends of the assembly. A lantern battery is applied to the ignitor, melting the fuse wire, setting off the ignitor's powder, which sets off the main powder charge, firing the plungers into the inserts and trapping the cable firmly.

occur in inaccessible places like jungles or mountains but nevertheless they must be fixed. Typical splices are aluminum tubes plus separate inserts that grip the cable with teeth on their contact surfaces. See Figure 14-31. These pieces must be chosen to be the correct size for the particular cable that needs splicing. A power company will typically buy a variety of sizes and have them on hand when a cable breaks. The required pieces are then taken, sometimes laboriously, to the site of the break, along with a portable power generator and a hydraulic press.

Total cost for such a splice is about $5.00, plus installation costs.[21]

The splice manufacturer came up with a new design, shown in Figure 14-32. This design accomplishes splicing by using gunpowder to fire toothed inserts into a tapered tube, trapping the cable in a way similar to the Chinese

---

[21]This story dates from 1980, and the prices quoted are as of that time period.

finger trap. As with the original design, tube and inserts must be chosen to suit the cable that needs splicing. However, no heavy and costly power generator and hydraulic press are needed. A simple lantern battery sets off an ignitor, which blows the gunpowder charge and drives the inserts into the cable trap. The manufacturer needed a hydraulic swaging press to put the taper on the tube after the correct inserts were placed inside. The price of this product was $100. The manufacturer thought it would appear to be a bargain to power companies in view of its very simple and low cost method of use, compared to hiring pack animals or a heavy lift helicopter to ferry a generator and hydraulic press into the jungle.

Figure 14-33 compares the architectures of these two designs. The figure shows the functions on the left and the physical elements that accomplish them on the right. It also shows which actions occur in the factory and which occur in the field, as well as where major sources of energy are.

Both of these products are designed to implement a delayed commitment strategy. Because the size of the broken cable cannot be known in advance, and because there are many sizes of cables, only part of the manufacturing and assembly process can be done in advance, unless one wants to accumulate a large inventory of splices in the hope that

they will eventually be used. Note that when the swaging press tapers the tube in the new design, the product really is committed in the sense that it cannot be disassembled and resized for another size cable. The new design places the decoupling point closer to the customer.

Unfortunately, the new design was not a success. Not only did power companies resist paying $100 for something that they thought should cost only $5, but the manufacturer was unable to operate the swaging press consistently successfully. Instead of forming a nice taper, it often crumpled the ends of the tube. The reason is that each size of tube requires a different swaging die, different amounts of dry lubricant, and different pressing speed. These items were not known with sufficient accuracy, or the process was not sufficiently repeatable, or both. Yet the delayed commitment strategy depended on being able to make the press operation work correctly the first time. There was no way to make money if five or ten tubes had to be crumpled while the press operator experimented to find the correct settings.

This problem might have been possible to overcome if the product could have been designed to accept the tapered toothed inserts after the tube had been swaged. Then the manufacturer could have made a supply of swaged tubes

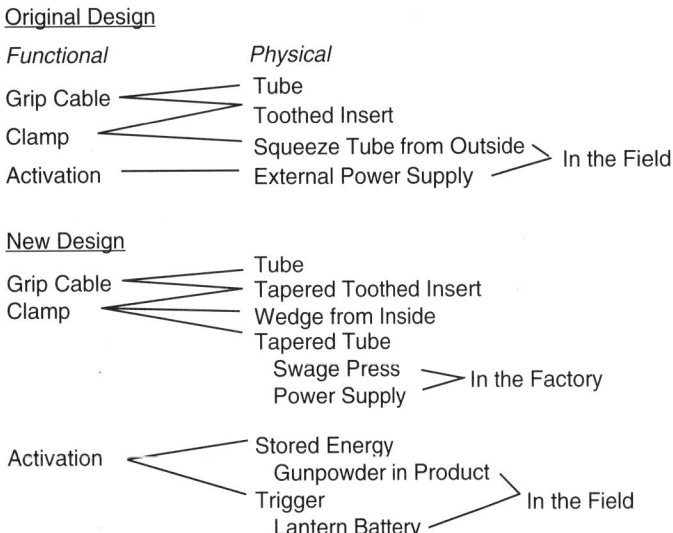

FIGURE 14-33. Functional and Physical Architectures of Two Power Line Splices. In the original design, there are a few simple parts. Important "manufacturing" steps occur in the field, requiring large energy provided by power supplies and presses. In the new design, there are more parts and some are more complex. A large energy supply is required in the factory, where all actual manufacturing occurs. Because of the high energy stored in the gunpowder, only a lantern battery is needed in the field.

of various sizes, perhaps crumpling a few at the beginning. This obviously would place the decoupling point even closer to the customer and avoid having a low repeatability process downstream of the decoupling point. The author attempted to create such a design but the manufacturer feared that if the inserts could be put in after tapering the tube, then accidental discharge of the gunpowder would shoot the inserts out of the tube like bullets. This could not have happened in the design shown in Figure 14-32.

The uncertainty of the swaging process and the high selling price caused the manufacturer to take the product off the market.

14.E. CHAPTER SUMMARY

This chapter deals with issues that transcend assembly. In some sense, assembly is a subset of architecture while in another sense assembly enables architecture by creating parts, subassemblies, interfaces, assembly sequences, decoupling points, and many other process features. These in turn are the tools of the many strategies enabled by architectural choice, such as management of variety and change, mitigation of risk, design of platforms and families, standardization of interfaces or whole functionalities, and so on. Several of the interacting factors are listed in Table 14-7, which focuses on module definition but repeats nearly all the items discussed in this chapter.

We have seen examples of architecture driven by such diverse forces as market segments, continental and regional politics, evolution of materials, attempts to dominate an industry or technology, and product selling schemes. Such architectures can be primarily modular or integral and can evolve over time from modular to integral or in the reverse direction.

TABLE 14-7. Considerations that Bear on Choice of Module Scope

Technical Considerations	Nontechnical Considerations
Enable an attractive assembly sequence.	Enable a substitution strategy for managing variety.
Create a testable unit comprising a functional cluster.	It is readily available from a number of suppliers.
Easy to add to the final product, with simple fasteners and few interface points.	Define module so that production work is balanced between different internal shops or factories.
Easy to fit into shipping containers, not awkward, fragile, or hard to nest.	Distribute risks of various kinds over different items.
The last step in assembly seals the item (success requires that the item be tested before sealing).	Module boundaries are optimized to balance labor and material costs versus shipping costs.
The item moves relative to things to which it attaches.	A supplier has a patent, trade secret, or other intellectual property, so the supplier defines the scope of the module.
The item must be physically, electrically, or thermally isolated from its neighbors.	Local content laws encourage contracts to local companies to make the item.
Its operating characteristics and boundary conditions can be specified precisely.	The buyer hopes that the supplier will do all the worrying and bear all the cost and risk.

Note: The factors are divided into technical and nontechnical domains. Most of them have been discussed in this chapter. It should be obvious that module choice is complex and subject to conflict. (The notion of functional cluster is discussed in [Pahl and Beitz] and [Otto and Wood].)

14.F. PROBLEMS AND THOUGHT QUESTIONS

1. Figure 14-4 shows a scheme for developing product platforms that depends on dividing the market for a product family into segments and tiers. Consider the following products: automobiles, airplanes, credit cards, dogs, copying machines, and digital cameras. For each one, list the following: a set of market segments, a set of market tiers, a set of functions or modules that would be in a product platform, and a corresponding set that would be in a product platform, and a corresponding set that would be in a nonplatform portion suitable for being designed to suit a market segment and tier or suitable for being customized on an individual basis.

2. Identify a product that appears to be built by a platform or family strategy and provide information for it in the format of Table 14-4.

3. Figure 14-6 shows a car body manufacturing strategy based on building the car on a pallet. An alternate strategy provides locating pins at each workstation and a conveyor that simply carries bodies from station. The text provides some arguments in favor of the workstation approach. What are some arguments in favor of the pallet approach?

4. The caption of Figure 14-12 states that failure to design a new part in conformance with the DFC plan shown would risk "destroying the ability of the family to function." Why is the family at risk and what function or functions are involved?

5. The digital copier shown in Figure 14-21 flows single unsorted copies through in a simple straight path but requires sorted copies to traverse a more convoluted path through the inverter. Is this the best choice? Consider what to do if single copies are more likely than copies of stacks, and vice versa. Consider, too, the cases when one-sided or two-sided copies, respectively, are more likely. What market segments and tiers are more likely to want single pages copied, stacks, one side, and so on?

6. Figure 14-14 shows the choice tree for an X-ray system. Draw the choice tree for the Denso panel meter assuming that there are three kinds of casings, four terminals, four bimetals, three voltage regulators, one base, and two fixed pins.

7. It was stated at the end of Section 14.C.1.c that Type 1 assemblies enable a more flexible assembly process. Discuss the conditions under which this would be true, and discuss counterexamples or mitigators that could make Type 2 assemblies flexible as well.

8. Consider the four ways of printing described in Chapter 12. Make a table listing each method and fill in the following information about each: the technical need that it fulfills; the technology used; the business case for the manufacturer; the product architecture; the main technology used to fulfill the need; the structure of product families, if any.

9. In addition to the four types of printing implementation described in Chapter 12, there exist others, such as the laser printer, the daisy wheel printer, and the chain printer. See Figure 14-34 for sketches of the latter two. A daisy wheel has a character near the tip of each thin petal of a plastic wheel that looks just like a daisy flower. The wheel spins rapidly between an electromagnetic hammer and the ink ribbon. The spin axis is perpendicular to the paper. The wheel rotates to the correct position and the hammer strikes the petal just behind the character, printing it. Then the wheel–hammer assembly translates to the next character position and the action repeats. A chain printer has a continuous chain that carries all the characters, much like a chain saw chain with characters instead of teeth. The chain runs parallel to a line being printed on the paper and carries the letters along the line extremely rapidly. Along the full length of the chain, each character appears three times. Behind each column position on the paper is a fixed electromagnetic hammer, 120 hammers in all for a 120-column printer. Hammer action is timed to strike the chain exactly when the desired letter passes by the column position where that letter needs to be printed. Prepare and fill in a row for an extension of the table built in Problem 8 for each of the laser printer, the daisy wheel printer, and the chain printer.

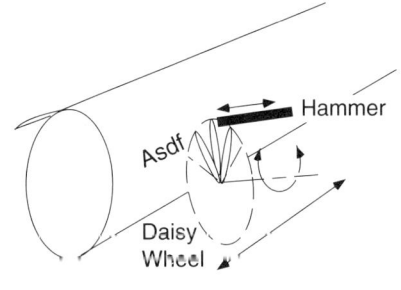

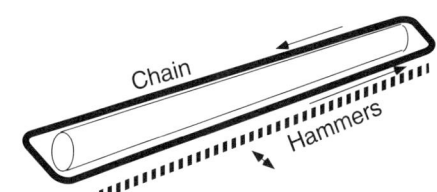

FIGURE 14-34. Daisy Wheel Printer and Chain Printer.

10. Consider the electric drill discussed in Chapter 13 (or any similar product). Disassemble it, draw its liaison diagram, and cluster the parts into functional groups along the lines of Figure 14-30. Comment on the alignment of physical clusters and functional clusters.

14.G. FURTHER READING

[Abernathy and Clark] Abernathy, W., and Clark, K. B., "Innovation: Mapping the Winds of Creative Destruction," *Research Policy,* vol. 14, no. 1, pp. 3–22, 1985.

[Abernathy and Utterback], Abernathy, W., and Utterback, J. M., "Patterns of Industrial Innovation," *Technology Review,* vol. 80, no. 7, pp. 40–47, 1978.

[Aviation Week] "It's Boeing's Time for Something New," April 2, 2001, pp. 32–33, and "Boeing Banking on Speed, Not

Size," April 2, 2001, pp. 34–35; "Business Case Will Decide Future of Boeing's New Faster Aircraft," April 23, 2001, pp. 60 61, and "Mach 0.95 Design Origins," April 23, 2001, pp. 60–61; correspondence by Carl Ehrlich, April 23, 2001, p. 9.

[Callahan] Callahan, S., "Relating Functional Schematics to Hierarchical Mechanical Assemblies," in *4th Symposium on Solid Modeling and Applications,* Hoffmann, C., and Bronsvort, W., editors, New York: ACM Press, 1997.

[Carlson] Carlson, W. D., "Development of the Transonic Area-Rule Methodology," NASA Langley Aero Research Center webpage: http://larcpubs.larc.nasa.gov/randt/1993/RandT/SectionB/B17.html.

[Christensen] Christensen, C. M., *The Innovator's Dilemma: When New Technologies Cause Great Firms to Fail,* Boston: Harvard Business School Press, 1997.

[Cunningham and Whitney] Cunningham, T. W., and Whitney, D. E., "The Chain Metrics Method for Identifying Integration Risk During Concept Design," DETC paper DETC98/DTM-5662, ASME DETC, Atlanta, September 1998.

[Cusumano] Cusumano, M. A., *The Japanese Automobile Industry,* Cambridge, MA: Harvard University Press, 1985.

[Erens] Erens, F. J., *The Synthesis of Variety: Developing Product Families,* Ph.D. thesis, University of Eindhoven, published by the Eindhoven University Press, 1996.

[Fine] Fine, C. H., *Clockspeed,* Reading, MA: Perseus Books, 1998.

[Fine and Whitney] Fine, C. H., and Whitney, D. E., "Is the Make-Buy Decision a Core Competence?" in *Logistics in the information age,* Moreno, M., and Pawar, K., editors, Padova: Servizi Grafici Editoriali, pp. 31–63, 1999. Also available at http://web.mit.edu/ctpid/www/Whitney/papers.html

[Henderson and Clark] Henderson, R. M., and Clark, K. B., "Architectural Innovation: The Reconfiguration of Existing Product Technologies and the Failure of Established Firms," *Administration Science Quarterly,* vol. 35, pp. 9–30, 1990.

[Kota, Sethuraman, and Miller] Kota, S., Sethuraman, K., and Miller, R., "A Metric for Evaluating Design Commonality in Product Families," *ASME Journal of Mechanical Design,* vol. 122, pp. 403–410, 2000.

[Larsson, et al.] Larsson, T., Pettersson, J., Rydquist, J. E., and Tunander, H., "The New Volvo 5-Cylinder Modular Engine," *Technology Report: A Joint Technical Magazine for Renault and Volvo,* vol. 1, no. 1, pp. 34–63, 1992.

[Lee] Lee, H., "Effective Inventory and Service Management Through Product and Process Redesign," *Operations Research,* vol. 44, no. 1, pp. 151–159, 1996.

[Lehnerd and Meyer] Lehnerd, A. P., and Meyer, M. H., *The Power of Product Platforms,* New York: The Free Press, 1997.

[MacDuffie] MacDuffie, J.-P. "Automotive 'Build to Order': The Modularity—E-Business Link," presentation to IMVP Sponsors' Meeting, Cambridge, MA, September 27, 2000.

[Martin, Hausman, and Ishii] Martin, M., Hausman, W., and Ishii, K., "Design for Variety," in *Product Variety Management: Research Advances,* Ho, T.-K., and Tang, C. S., editors, Boston: Kluwer, pp. 103–122, 1998.

[Mather] Mather, H., "Logistics in Manufacturing: A Way to Beat the Competition," *Assembly Automation,* vol. 7, no. 4, pp. 175–178, 1987.

[Mittal and Frayman] Mittal, S., and Frayman, F., "Towards a Generic Model of Configuration Tasks," 11th International Conference on Artificial Intelligence, 1989.

[Morelli] Morelli, A. (Director of Automation and Productivity, Telemechanique R & D Laboratory, Paris), personal communication, 1992. The complete report is available at http://web.mit.edu/ctpid/www/Whitney/Europe/telemechanique.html.

[Otto and Wood] Otto, K., and Wood, K., *Product Design: Techniques in Reverse Engineering and New Product Development,* Upper Saddle River, NJ: Prentice-Hall, 2001.

[Pahl and Beitz] Pahl, G., and Beitz, W., *Engineering Design,* London: Springer, 1991.

[Redford and Chal] Redford, A., and Chal, J., *Design for Assembly,* London: McGraw-Hill Europe, 1994.

[Sako] Sako, M., "Modularity and Outsourcing: Main Issues and Trends," presentation to IMVP Sponsors' Meeting, Cambridge, MA, September 2000.

[Sanderson and Uzumeri] Sanderson, S. W., and Uzumeri, M. *The Innovation Imperative: Strategies for Managing Product Models and Families,* New York: McGraw-Hill/Irwin, 1997.

[Simchi-Levi et al.] Simchi-Levi, Kaminsky, D. P., and Simchi-Levi, E., *Designing and Managing the Supply Chain: Concepts, Strategies, and Case Studies,* 2nd ed., New York: McGraw-Hill, 2002.

[Sloan] Sloan, A. P., Jr., *My Years with General Motors,* New York: Doubleday, 1963 (1990 paperback edition).

[Swaminathan and Tayur] Swaminathan, J. M., and Tayur, S. R., "Managing Broader Product Lines Through Delayed Differentiation Using Vanilla Boxes," *Management Science,* vol. 44, part 2 of 2, pp. S161–S172, 1998.

[Taylor] Taylor, C. F., *The Internal-Combustion Engine in Theory and Practice,* Cambridge, MA: MIT Press, 1985.

[Ulrich 1993] Ulrich, K., "The Role of Product Architecture in the Manufacturing Firm," *Research Policy,* vol. 24, pp. 419–440, 1993.

[Ulrich and Ellison] Ulrich, K. T., and Ellison, D. J., "Customer Requirements and the Design-Select Decision," University of Pennsylvania Wharton School Working Paper 97-07-03,

Automotive Engineers Aerospace Technology Conference, 1998.

[Ulrich and Eppinger] Ulrich, K. T., and Eppinger, S. D., *Product Design and Development,* New York: Irwin, 2000.

[Ulrich et al.] Ulrich, K. T., Randall, T., Fisher, M., and Reibstein, D., "Managing Product Variety," Chapter 9 in *Product Variety Management: Research Advances,* Ho, T.-H., and Tang, C. S., editors, Boston: Kluwer, 1998.

[Utterback] Utterback, J. M., *Mastering the Dynamics of Innovation,* Boston: Harvard Business School Press, 1994.

[Walton] Walton, M., *Car: A Drama of the American Workplace,* New York: Norton, 1997.

[Whitney] Whitney, D. E., "Why Mechanical Design Cannot Be Like VLSI Design," *Research in Engineering Design,* vol. 8, pp. 125–138, 1996.

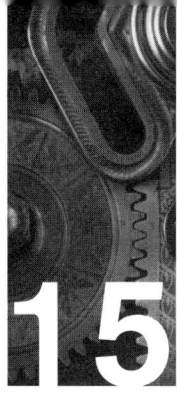

15 DESIGN FOR ASSEMBLY AND OTHER "ILITIES"

"There were no screws. It all went together with snap fits. We used high-speed movies to watch it come apart when we dropped it and kept refining the design until it just bounced."

–Chief Engineer for Polaroid SX-70 Camera, 1972.

"Word came down that we couldn't use screws. So we used snap fits. Then word came down that it had to pass a drop test. So we dropped it and it fell apart."

–Desktop copier design engineer, 1991.

15.A. INTRODUCTION

Design for assembly (DFA) is one of several DFx's, where each "x" is a characteristic of the product, its production, or its life cycle that is important to someone or in some context. In addition to DFA, there is design for manufacturing (DFM) as well as design for disassembly, repair, recycling, upgrade, and so on. Each DFx represents a body of knowledge, procedures, analyses, metrics, and design recommendations intended to improve the product in the domain "x." These "x's" are sometimes called "ilities." In this chapter we will look at DFA and some of the other DFx's, learn the basic principles behind them, and place them in the context of both assembly in the small and assembly in the large.

We will approach this topic by dividing the methods in use into two categories:

- Methods or process steps that can be applied to one part at a time by an engineer working alone (which we will call DFx in the small)

- Methods or process steps that involve consideration of all the parts in an assembly at once and that may need many people to interact (which we will call DFx in the large)

What we will discover is that this topic does not reward us with nice, clean answers. Instead we will find, for example, that some DFA recommendations conflict with some DF-recycling recommendations. More generally, recommendations arising from DFx in the small are less likely to encounter conflict with each other while those arising from DFx in the large, especially when they affect product architecture, are more likely to encounter conflict. An important goal of this chapter is therefore to help understand when DFx recommendations can be applied by an engineer working alone and when the interests of others, both technical and nontechnical, must be considered.

DFx methods provide the most benefit when they are applied early in the design process when changes are relatively easy to make. If DFx is delayed until detailed designs are well under way or finished, there will be too little money or time to make more than cosmetic changes. The irony is that, as we shall see, much of DFx deals with details that are unclear early in design. Most DFx methods attempt to deal with this paradox in one way or another, usually by devising scoring systems that alert designers to DFx problems without requiring too much detailed information.

While advantageous in many ways, these scoring systems can cause problems. Too strong reliance on the score can lead to incorrect design decisions. A score-based system can also lead people to think that experience is not needed to get good results.

15.B. HISTORY

15.B.1. DFM/DFA as Local Engineering Methods

As noted in Chapter 1, assembly has been studied seriously for only the past 30 years or so, in contrast to basic fabrication processes such as metal cutting, casting, grinding, and forging. The latter processes involve large amounts of power and high investments and thus attracted the attention of factory managers and researchers almost as soon as they came into use. Assembly has traditionally been performed by people, and their ingenuity makes up for many shortcomings in product and assembly process design. Only when automatic machinery and especially robots began to be considered for assembly did it become obvious that more attention needed to be paid to assembly itself.

The basic goals of DFM and DFA are to make fabrication and assembly easier, less costly, simpler, more reliable, and on. To achieve these goals, engineers must often modify their designs and expand their focus when designing so that factors other than product performance are seriously taken into account. While this sounds admirable, it faces two formidable barriers: Many engineers think that considering manufacturing or assembly will compromise product performance, and often they do not know enough about manufacturing and assembly to make the appropriate changes in their designs in order to achieve DFx goals. If engineers can carry out their own DFx analyses, they can protect product function and will probably learn that there is little chance that function will be seriously impaired. Systematic DFx methods seek to enable engineers to do their own analyses, but often the issues are too complex and other knowledgeable people must be involved.

One of the first manufacturers to deliberately focus design attention on the assembly process was Henry Ford, whose early cars had simpler designs and fewer parts than many of his competitors' ([Hounshell]). Design simplification and standardization were extremely important to Ford, whose methods were widely adopted in the United States, though less so in Europe. During World War II, U.S., Russian, and British defense contractors adopted simple standard designs for aircraft, jeeps, tanks, trucks, and rifles, and ground them out in huge quantities. This contrasted sharply with the methods of German industry, which kept "improving" the designs of many of these items under pressure from the military, with the result that logistics, training, and field repair became difficult ([Overy]).

DFA was first systematized in the 1960s by Geoffrey Boothroyd and his colleagues Alan Redford and Ken Swift at the University of Salford, England. Other contributions to DFA research and applications include [Andreasen et al.], [Miyakawa and Ohashi], [Redford and Chal], and [Sturges]. Research first focused on methods of feeding parts by means of vibratory and other mechanical techniques (results summarized in [Boothroyd, Poli, and Murch]). Attention turned in the 1970s to classifying parts and assembly tasks in an effort to provide a simple way for engineers to judge the assembleability of their designs. Hitachi developed a set of assembleability evaluation methods at this time as well. It is these part-focused assembleability methods that we call DFx in the small.

The basis for the approach is "classification and coding," a technique common in Europe in the domain of group technology. The aim of group technology is to cluster things into groups so that common process techniques, fixtures, or methods can be applied to them even though they are not identical. This permits the same machine, process, or setup to handle more parts, increasing economies of scale, reducing time spent switching setups, and increasing the utilization of machines. For example, all shafts whose ratio of length to diameter (L/D) is more than three might be put in one group for the purpose of rough machining, while those with L/D less than three would be in a different group. Other details about the shafts are presumably not important for rough machining. The differences in these details are thus safely ignored or might even be changed later without causing an item to be moved to a different group.

For each group, a process plan or cost and time prediction is developed, usually by conducting experiments or analyses of typical parts that fall into that group. These plans or predictions will not be exactly right for every part but will be good enough for predictions. The benefits are that (1) predictions can be made ignoring many details that may not be known early in the design process and (2) time and money can be saved by using an off-the-shelf process and tooling.

To apply this method, one needs to know what differences are important and which are not for the purpose at hand (rough machining, assembly, etc.). Then a way of dividing the items according to the important differences is needed (must shafts be split into five different L/D ranges, or will three suffice?). Finally, a simple coding scheme is needed to label the items so that each class can be treated appropriately.

Typically, several characteristics have to be isolated before a complete classification is done (L/D, number of different diameters on the same shaft, and material, for example). Each characteristic gets its own code number or letter; "321" might be a shaft with $L/D > 3$, two different diameters, made of steel, while "234" might be a shaft with $L/D < 3$, three different diameters, made of brass.

The result of this process is that an item (a shaft to be machined or a part to be assembled) can be coded and assigned to a class, and process rules or predictions for time and cost applicable to the entire class can just be read from a table and applied with reasonable accuracy to that item.

For example, all shafts with code 321 might be sent to shop 1 while those with code 234 might go to shop 2. Within each shop would be machines of the appropriate size to accommodate the shafts it received, plus people who were used to dealing with parts of this type.

Finding out what differences matter for classification purposes may be simple or it may require considerable research and experimentation. In the case of assembly, Boothroyd identified feeding, orienting, handling, and inserting as focus areas for DFA. His research showed that part shape, size, weight, L/D, and symmetry were things that mattered, and classifications were developed to differentiate parts according to those features. Hitachi identified nonvertical direction of insertion, number of extra operations in addition to insertion, and others. Denso identified ease of switching from one version of a part to another, as well as the cost of presenting parts to robot assembly stations.

Once a part is classified according to these metrics, it receives a score or a time estimate, with smaller time or higher score being better. A lower score implies more time needed to assemble it, or more difficulty feeding it in an automatic feeder, or more chance of assembly error. Low-scoring parts are therefore targets for redesign.

In the 1980s, Boothroyd encapsulated his classifications and scoring techniques into software for the PC and now sells it through a company called Boothroyd Dewhurst, Inc. This software asks the engineer a series of questions about each part and then returns scores and makes other recommendations that are discussed later in this chapter. Hitachi also sells its software. A number of companies use internally developed DFA or DFx software

but do not sell it. These include Sony, GEC (UK), Lucas, Fujitsu, Denso, and Toyota. Some of these companies' methods will be described briefly in this chapter. A comprehensive comparison of DFA techniques, including descriptions of the Lucas, Boothroyd Dewhurst, and Hitachi methods, may be found in [Redford and Chal].

It is important to understand that this method works by taking the parts out of context and ignoring many details about them. The reward is that important facts can be learned without considering the details, and the analysis can be done by someone who does not know much about assembly. This simplifies the analysis and permits it to be done early in the design process when many details remain unclear. It also permits an engineer to judge the parts in isolation, that is, just by looking at them one at a time. Since many details are missing and the results are frankly estimates, the process is best used to compare design alternatives. The downside, as we shall see later (and reflected in the second quote at the beginning of the chapter), is that the context can be important, even overwhelming, in which case it cannot be ignored. In such cases, a lookup table corresponding to simple codes will not do the job.

15.B.2. DFM/DFA as Product Development Integrators

Experience has shown that the context is important in a large fraction of cases. In Chapters 12 and 14 we learned how varied and interlocked these contexts can be. As a result, many companies have decided to forgo the advantages of allowing engineers to judge their own parts according to a classification and coding procedure. Instead they acknowledge that no single person can know enough about all the relevant processes or have command of all the contexts in which the part or assembly may be involved.

Researchers have also studied alternate ways of enriching design decisions to include manufacturing and assembly. Favored approaches include use of expert system technology and other similar computer-aided methods ([Hayes-Roth, Waterman, and Lenat]). In both the research and industrial communities, attempts to build truly general computerized analyses have not so far borne fruit. Instead one sees specific implementations that focus on particular kinds of parts used in particular product applications, or a focus on specific process technologies. For a review of the theory and some applications of this approach, see [Tong and Sriram]. Examples include predicting flow of molten plastic in injection molds, bending of sheet metal in stamping dies, assembling a particular automotive component on an existing assembly line with particular dimensional limits, and design of molds for plastic car tail-light bezels to meet regulations. The amount of knowledge that has to be accumulated and coded into these systems in order to make them competent for real products and processes is very large.

A typical noncomputerized approach in companies is to hold meetings of people likely to have an interest in the design, such as representatives from the engineering, manufacturing, assembly, and field service departments.[1] This is commonly called *concurrent engineering*. It has the salutary effect of acquainting people with each other and generally improving communication. Topics other than those on the original meeting agenda inevitably come up, permitting other problem areas to be discovered and addressed. Thus DFM and DFA become vehicles for improving integration of the product design overall. Books that focus on the use of assembly to drive integration of the product development process include [Andreasen et al.] and [Nevins and Whitney].

15.B.3. DFA as a Driver of Product Architecture

When Boothroyd first addressed DFA in the late 1960s and early 1970s, he and many others believed that assembly accounted for 30–50% of manufacturing cost ([Boothroyd], [Boothroyd and Redford], [Boothroyd, Poli, and Murch]). This conclusion was based on the observation (still valid) that most of the people in a factory are doing assembly work or helping them assemble by bringing them parts, taking away empty containers, and so on. Relatively few people are involved in fabrication due to the mechanized nature of those processes. Thus it seemed logical to try to reduce the number of parts in an assembly as a way to reduce this cost. In order to systematize the search for parts that could be eliminated from an assembly, Boothroyd developed criteria for flagging *theoretically* unnecessary parts along with a metric called *assembly efficiency,* which measures the degree to which theoretically unnecessary parts have been eliminated. Details of this technique are described later in this chapter.

When Boothroyd Dewhurst, Inc. accumulated industry experience with this technique in the 1980s, it was realized that, while costs were indeed reduced, the reason was not because of reduction in assembly cost. Indeed, [Boothroyd, Dewhurst, and Knight] does not state that assembly cost is 30–50% of manufacturing cost. Instead, it cites data showing that assembly is a very small fraction of cost. The parts themselves are the most costly items, and the major savings accrue from eliminating *them* rather than eliminating their assembly.[2] DFA, as presented in both [Boothroyd, Dewhurst, and Knight] and [Andreasen et al.], is thus associated with the process of simplifying product structure, what we call product architecture.

It is because part count reduction is really aimed at product architecture that we call methods of this type DFx in the large. It should be clear that efforts of this type cannot be conducted out of context by an engineer studying the parts one at a time. Instead, all the issues associated with product architecture must be considered.

In summary, we can observe that DFA has matured from an initial focus on single parts, along with the belief that assembly was a major cost center, to a richer view with two phases. The first phase considers all the parts at once and adds assembly process criteria to the search for a good product architecture, while the second phase looks carefully at the surviving parts to see how their fabrication (if we include DFM) and assembly can be improved. Many "ilities" must be considered beyond manufacturing and assembly, so we will refer to the entire set of concerns as DFx.

Sections 15.C, 15.D, and 15.E consider these two phases in more detail and present some examples.

15.B.4. The Effect on DFM/DFA Strategies of Time and Cost Distributions in Manufacturing

We learn in Chapter 18 that cost analysis is a complex topic that involves many tradeoffs. Among them, the most basic are those between labor and capital and between labor and materials. Labor costs dominate the total rolled-up costs of a product over the whole supply chain, but materials costs dominate within any one company in the chain. We will learn later in this chapter that when products

---

[1] Along the way, some of these companies decide not to use DFA software any more ([Chung]).

[2] This conclusion is discussed and supported by data in Chapter 18.

TABLE 15-1. Recommended DFA/DFM Strategy When Time Is a Consideration

Across: Production Scenarios Down: Development Time Scenarios	Low Lifetime Production Volume	High Lifetime Production Volume
Development Time Is Critical	*Example Products:* • High-performance Computers • Telecommunications equipment *DFM/DFA Strategy:* • Avoid long lead time tooling • Use standard components • Minimize production risk	*Example Products:* • Notebook computers, toys *DFM/DFA Strategy:* • Minimize complexity of most complex part • For complex parts, use processes with fast tool fabrication methods **Apply traditional DFM/DFA to less time-critical parts**
Development Time Is Not Critical	*Example Products:* • Machine tools • Electrical distribution equipment *DFM/DFA Strategy:* • Avoid expensive tooling • Use standard components • Other issues are likely to dominate	*Example Products:* • Blank videocassettes • Circuit breakers *DFM Strategy:* • Combine and integrate parts • Consider automatic assembly **Use traditional DFM/DFA**

Note: In this table, "traditional DFM/DFA" refers to the methods of Boothroyd and his colleagues, which emphasize part count reduction and assembly simplification. The research behind this table included a study of plastic injection molded parts that contained as many as 1100 dimensioned and toleranced features and whose molds took as long as 4 months to design and debug. This time was the longest single time in the product development schedule for the product using these parts. The conclusion is that DFM and DFA should be used when the time they take to apply and exploit is short compared to the time available to design the product, and the product will be manufactured for a long time, increasing the accumulated amount of time and money saved.

Source: [Ulrich, Sartorius, Pearson, and Jakiela].

are redesigned to save labor, the redesign often includes changing materials and part manufacturing processes. For this reason, most mature DFA methodologies include a strong element of design for manufacture and supply chain implications, so that decisions can be evaluated across as many cost impacts as possible.

In addition, different DFA choices are affected by time in several ways. Table 15-1 compares two different manufacturing scenarios with two different design scenarios to develop appropriate DFM strategies. An interesting conclusion from this analysis is that DFA and DFM as described in the literature are most relevant to only one of the four situations shown, and only partially relevant to one other situation. For the remaining two, totally different approaches are recommended ([Ulrich, Sartorius, Pearson, and Jakiela]).

15.C. GENERAL APPROACH TO DFM/DFA[3]

[Redford and Chal] places DFA within the larger scope of product design as follows:

a. Design the product to achieve the functions and "ilities."

b. Pay attention to cost.

c. Decide the best fabrication and assembly process and method for each part.

d. Design the part to suit that process and method.

Processes and methods can be classified as shown in Table 15-2. More detail on the different kinds of assembly methods and processes and their relative economic attractiveness is in Chapter 16 and 17.

Only a few products are technically or economically feasible to assemble by means other than people. So, it is reasonable to conduct DFA on the assumption that the product will be assembled manually. In addition, a product that is easy to assemble manually will usually be easy to assemble by machines, although this is not always true for every design decision or assembly action.

[3]This section is based in part on [Redford and Chal], pp. 23–60.

TABLE 15-2. Appropriate Assembly Methods and Processes

Across: Processes Down: Methods	People	Dedicated Automatic Machines	Programmable Machines (Robots)
One or a Few Workstations, Lots of Work at Each Station	Will be successful if there is not too much work for each person because people cannot learn too many tasks; appropriate for smaller production rates	Cannot use this process for this method because each dedicated equipment station does only one task	Will not be economical at very low or very high production rates; larger number of tasks per station does not impose a learning burden but does impose tool change time losses
A Line of Stations with Very Little Work at Each Station	Efficient; boring for the people Required if parts are large	The most economical by far for high production rate assembly of small parts, but perhaps only 10% of all products fit this class	Probably will not be economical; small parts are assembled more economically by dedicated machines, while large parts are too heavy for robots; car body welding is the big success for this process

Note: This classification is sufficient for a rough determination of appropriate processes and methods. Detailed analysis of technical and economic feasibility is required in each case.

Source: [Redford and Chal].

Regardless of how the product will be assembled, assembly cost can be saved if the number of assembly operations is reduced. This will happen, of course, if the number of parts is reduced, but some parts will require more ancillary operations than others, or could be designed to require fewer. In general, assembly of a part requires getting the part, orienting it if necessary, confirming that it is the correct part and that it is of good enough quality to be used, carrying it to the insertion point, inserting it (possibly using a tool that must be acquired separately and/or using another hand), fastening or tightening it, or applying something like adhesive or solder to hold it in place, and finally testing that it has been properly inserted and (possibly) that it functions correctly. Each new tool requires time to change tools. Each part that is of insufficient quality takes time to assess and repair or discard. Each extra action like gluing, lubricating, or riveting, takes time and extra apparatus.

Engineering and process planning time or cost can be saved if standard parts are used rather than designing parts to suit the product. If all the fasteners put in at a given workstation are the same, then there can be no confusion about which fastener to use where, or time lost switching from one screwdriver to another to suit each fastener. As noted below, however, standardization can affect function adversely by limiting the engineer's choices and the product's function.

Part feeding and presentation are not too critical if people are doing the assembly. However, they are critical for machines. For example, it is relatively easy, though somewhat time-consuming, for a person to pick a part from a jumbled heap and reorient it dexterously in one hand while moving it to the insertion point, whereas these actions are essentially impossible for any machine in existence today. If parts are to be assembled by machines, they must be presented automatically to the insertion apparatus in the correct orientation. Design for easy feeding and orientation therefore become proportionately more important. Most of the mechanized ways of separating, orienting, and presenting parts work only for small parts, say less than 10 cm long in the longest direction. One reason is that the feeding equipment is usually a factor of ten or more larger than the part being fed. The other is that the rate at which parts can be fed is usually an inverse function of the length of the part, and at some point the feed rate becomes too small to support the production rate. Formulas for understanding these factors are derived in [Boothroyd and Redford].

Parts should be presented so that they are easy to grasp firmly so as to facilitate insertion. This is not as critical for people as it is for machines. We note in Chapters 10 and 17 that the gripper of an assembly machine or workhead is an essential element in the DFC for the process of insertion. The point at which the part is gripped (an assembly feature) is the point at which the coordinate frame of the gripper is transferred to that of the part.

Elsewhere on the part is the feature that will be mated to another part during insertion. Thus a part that will be assembled by a machine must be designed so that the assembly DFC is designed properly. People do not assemble parts by navigating a DFC but by using their sense of touch, to a lesser degree their vision and, occasionally, hearing. Thus it is more important for people to have space for their fingers or assembly tools and to be able to see the assembly action than it is for the grip point of the part to be properly designed for assembly accuracy.

People learn assembly tasks more quickly if the work is simpler and there is less of it. Henry Ford divided the work on his assembly line into the smallest possible elements. These could be learned quickly, compensating for the high turnover among his employees caused by the pace and boredom of short assembly cycles. Nissan manufacturing engineers estimate that it takes a manual assembler about 3,000 cycles or about a week to become really proficient at an assembly task. Kilbridge, quoted in [Jürgens], says that it takes about 1,000 trials if the work takes one minute, and proportionately longer if the work takes longer.

Ever since Ford designed the first assembly line, there has been debate about how much work each person should do. Ford used team assembly of cars before turning to the assembly line. Team assembly requires workers to get the parts, while the line brings the parts to them. Volvo experimented for many years with team assembly of cars, as discussed in Chapter 16. Toyota adopted the Ford method and elaborated on it by figuring out how to use every second of the assembly cycle. Workers on a Toyota assembly line look like they are in a ballet, with every second occupied and every motion applied to getting parts, repositioning one's body, arranging tools, or performing the assembly itself.

The academic and popular literature abound with rules of thumb for making things easy to assemble. A sample of

TABLE 15-3. Rules of Thumb for Easy Assembly

Some DFA Rules of Thumb[a]	Attributes of a Product that Is Easy to Assemble[b]
Minimize part count	Can be assembled one-handed by a blind person wearing a boxing glove
Provide ample space for insertion tools or fingers	Is stable and self-aligning
Design parts so that they do not tangle with each other	Tolerances are loose and forgiving
Provide chamfers and other alignment aids	Few fasteners
Insert parts from above	Few tools and fixtures
Make it obvious how to assemble	Parts easy to grasp and insert

[a] Adapted from Table 14.1 in [Otto and Wood].
[b] Supplied by Peter Will, ISI, partly tongue in cheek.

these rules appears in Table 15-3. Boothroyd and others point out that a systematic method is needed in order to find out which, if any, of these rules should be applied and which should be ignored in any given situation.

Within this general framework, different researchers and practitioners of DFA have developed their own often proprietary and well-guarded methods for simplifying part feeding, orienting, and inserting. Recent textbooks give general descriptions of DFA and DFM and provide examples of how to estimate manufacturing costs ([Otto and Wood], [Ulrich and Eppinger]). Detailed cost quotes usually require special knowledge of an individual process as well as the capabilities of the supplier. General macroeconomic conditions and negotiating skill also play a role in determining the price paid for a part or assembly.

15.D. TRADITIONAL DFM/DFA (DFx IN THE SMALL)

DFx in the small focuses on simplifying the feeding, orienting, and inserting of individual parts. It does this by various means that involve classifying the parts or the assembly actions required, and then scoring or timing them approximately according to the classification. The primary method in use today is that of Boothroyd. This method and others addressing individual parts or operations are described below.

15.D.1. The Boothroyd Method

Boothroyd and his colleagues Swift and Redford first analyzed automatic parts feeders such as vibratory bowls.

Design of these items is more an art than a science, and Boothroyd realized that some parts are harder to feed automatically than others for reasons that could be avoided if part designers had more information. He then turned to manual assembly and identified two main phases of single part assembly, namely handling (which includes grasping and orienting) and insertion. Each of these is also affected by part design.

The part features that affect manual handling are listed in Table 15-4. These features are used to assign the part to a handling difficulty classification and give it a two-digit code ranging in value from 00 to 42, for a total of 27 classifications. A portion of the manual handling time

(a) For parts that can be grasped and manipulated with one bare hand

		No handling difficulties			Part nests or tangles		
		Thickness > 2 mm		< 2 mm	Thickness > 2 mm		< 2 mm
		Size > 15 mm	6 mm < size < 15 mm	Size > 6 mm	Size > 15 mm	6 mm < size < 15 mm	Size > 6 mm
Symmetry $(S = \alpha + \beta)$	Code	*0*	*1*	*2*	*3*	*4*	*5*
$S < 360°$	*0*	1.13	1.43	1.69	1.84	2.17	2.45
$360° < S < 540°$	*1*	1.5	1.8	2.06	2.25	2.57	3
$540° < S < 720°$	*2*	1.8	2.1	2.36	2.57	2.9	3.18
$S = 720°$	*3*	1.95	2.25	2.51	2.73	3.06	3.34

(b) For parts that can be lifted with one hand but require two hands to manage

	$\alpha \leq 180°$		$\alpha = 360°$
	Size > 15 mm	6 < Size < 15 mm	Size > 6 mm
Code	*0*	*1*	*2*
4	4.1	4.5	5.6

FIGURE 15-1. Selected Manual Handling Times in Seconds. Parts (a) and (b) are mutually exclusive. Both apply to small parts within easy reach, that are no smaller than 6 mm, do not stick together, and are not fragile or sharp. Symmetry is measured by summing angles α and β; α is the number of degrees required to rotate the part about an axis normal to the insertion axis in order to return it to an identical configuration, and β is the same with respect to an axis about the insertion axis. The code to be assigned is the combination of the row and column headings in *italics*. For example, a part coded "12" has handling time 2.06 sec. (Courtesy of Boothroyd Dewhurst, Inc. Copyright © 1999.)

table appears in Figure 15-1. Each code is accompanied by an estimated handling time in seconds, ranging from 1.13 seconds to 5.6 seconds. These times were developed over a period of years by means of experiments and are applicable to small parts.[4] Individual companies have also developed their own time estimates. Boothroyd also provides guidelines for scaling the times for larger parts.

The assembly conditions that affect assembly time are listed in Table 15-5. A portion of the manual insertion time table appears in Figure 15-2. There are 24 code numbers with insertion times that range from 1.5 seconds to 10.7 seconds. As with the numbers in Figure 15-1, these

[4]MIT students who have used these times for handling and assembly report that they are accurate within about 10%. However, it is important to recall the information cited above that it takes 1,000 to 3,000 trials to become really proficient at an assembly task, whereas the MIT student data are based on ten or twenty practice runs at most.

TABLE 15-4. Part Features that Affect Manual Handling

- Nesting, tangling, fragility
- Need to use two hands or more than one person
- Need to use tools
- Size, thickness, and weight
- Flexibility, slipperiness, stickiness
- Need for mechanical or optical magnification assistance
- Degree of symmetry of the part

Source: [Boothroyd, Dewhurst, and Knight].

times apply to small parts and must be scaled up for larger ones. For example, a person assembling cell phones might install several complex-shaped metal shields over a circuit board to block radio-frequency interference during a cycle time of 15 seconds or less. By contrast, on an automobile final assembly line, station times are typically 45 to 60 seconds, during which one large item like a seat, roof, hood, or battery might be obtained and installed. Sometimes, two people work together to handle the larger items. Often there is no time to install and tighten fasteners, so another person does this at the next station.

In support of the time estimates in these tables, [Boothroyd, Dewhurst, and Knight] presents several detailed explanations for the sources of the estimates, including empirical formulas and graphs. These include:

- The influence of symmetry or asymmetry on the time a person needs to orient something correctly starting

TABLE 15-5. Conditions that Affect Manual Insertion Time

- Whether the part is secured immediately or after other operations
- Accessibility of the insertion region
- Ability to see the insertion region
- Ease of aligning and positioning the part
- A tool is needed
- Whether the part stays put after being placed or whether the assembler must hold it until other parts or fasteners are installed
- Simplicity of the insertion operation

Source: [Boothroyd, Dewhurst, and Knight].

(a) Part inserted but not secured immediately, or secured by snap fit

		Secured by separate operation or part				Secured right away by snap fit	
		No holding down required		Holding down required			
		Easy to align	Not easy to align	Easy to align	Not easy to align	Easy to align	Not easy to align
	Code	0	1	2	3	4	5
No access or vision difficulties	0	1.5	3.0	2.6	5.2	1.8	3.3
Obstructed access or restricted vision	1	3.7	5.2	4.8	7.4	4.0	5.5
Obstructed access and restricted vision	2	5.9	7.4	7.0	9.6	7.7	7.7

(b) Part inserted and secured immediately by power screwdriver. Note: add 2.9 seconds to get power tool.

		Easy to align	Not easy to align
	Code	0	1
No access or vision difficulties	3	3.6	5.3
Restricted vision only	4	6.3	8.0
Obstructed access only	5	9.0	10.7

(c) Separate operation times for solid parts already in place

	Screw tighten with power tool	Manipulation, reorientation, or adjustment	Addition of nonsolid materials
Code	0	1	2
6	5.2	4.5	7.0

FIGURE 15-2. Selected Manual Insertion Times (Courtesy of Boothroyd Dewhurst, Inc. Copyright © 1999.) Parts (a) and (b) are mutually exclusive, while Part (c) contains times that are added to times in the other two tables when required. Times in Part (a) apply to small parts where there is no resistance to insertion.

from a random orientation (time rises approximately linearly regardless of detailed part cross-sectional shape from a base of 1.5 seconds to a peak of 2.7 seconds as the required number of degrees of rotation rises)

- The influence of part size and thickness (size greater than about 15 mm does not impose any handling time penalty, while thickness greater than 2 mm does not cause any handling time penalty; these conclusions obviously do not apply to parts the size of car seats)

- The influence of part weight (for small parts, the time rises linearly with weight, and a part weighing 20 pounds imposes a penalty of 0.5 seconds plus any additional time associated with walking)

- The influence of clearance ratio (see Chapter 10) on insertion time (time penalty is inversely proportional to the log of the clearance ratio and ranges from 0.2 to 0.5 seconds depending on whether there is a chamfer or not)

In addition to the time estimates provided in [Boothroyd, Dewhurst, and Knight], one can use standard time handbooks such as [Zandin]. These handbooks use standard work actions like "reach," "grasp," and so on, without taking the design of the part or the assembly operation into account. However, they contain data that applies to larger parts, walking time, and time to position equipment to aid assembly.

These time estimates do not take account of variations due to fatigue or time of day. In many factories, assembly line workers can adjust the speed of the line during the day as long as they make the total number of assemblies required by the end of the day. This approach is satisfactory for a line that feeds a warehouse but not for one that feeds another line unless additional measures are taken to ensure that the downstream processes receive assemblies when they need them.

Several general guidelines are also offered:

- Avoid connections, or make them short and direct. Items like pipes that join different parts or assemblies could be made shorter, straighter, or even eliminated if the parts were closer to each other or otherwise better arranged. A guideline like this can run into conflicts if the parts in question must be replaced for maintenance or are subject to design revision or customer options. Conflict can also arise if the parts must be kept separate in order to allow cooling air to

pass between them or to reduce the effect of radio-frequency interference, for example.

- Provide plenty of space to get at the parts and their fasteners during assembly. This guideline often conflicts with the need to make products small even as they become more complex. Car engine compartments, cell phones, and cameras are typical examples. In such cases, assemblers need tools, magnifiers, dexterity, and extra time.

- Avoid adjustments. Adjustments take time, hence the guideline. Sometimes, as discussed in Chapter 6, it is not economical to make parts of sufficient accuracy to avoid adjustments. In other cases, the customer makes the adjustments in the normal course of using the product. The user of a sewing machine adjusts thread tension to accommodate different thread materials with different coefficients of friction.

- Use kinematic design principles. As noted in Chapter 4, overconstraint makes the assembly strategy operator-dependent and thus makes both time and quality operator-dependent.

[Redford and Chal] notes that the classification method, while not explaining in detail what to do if a part or operation takes longer than desired, nevertheless places it in the table next to other classification possibilities that are better or worse. Thus the engineer can see what kinds of improvements might be made in a given case: Would the part be better if it was thicker, had a chamfer, didn't tangle, was a little more symmetric, and so on? How much time will that save? And so on.

[Boothroyd, Dewhurst, and Knight] notes that design changes for ease of assembly, like those that reduce part count (discussed below) cannot be made without knowing their impact on the cost of making the part. Thus [Boothroyd, Dewhurst, and Knight] also contains chapters on design for sheet metal, injection molding, machining, and other manufacturing processes, as well as robot assembly.

The information in the tables for handling and insertion times is encapsulated in software available from Boothroyd Dewhurst, Inc., Kingston, Rhode Island.

15.D.2. The Hitachi Assembleability Evaluation Method

The Hitachi Assembleability Evaluation Method (AEM) belongs to a class of "points off" methods ([Miyakawa,

Ohashi, and Iwata]). In these methods, the "perfect" part or assembly operation gets the maximum score, usually one hundred, and each element of difficulty is assigned a penalty. There are twenty different operational circumstances, each with its own penalty. Each circumstance is accompanied by a simple icon for identification, permitting the method to be applied easily with little training. The AEM is part of a larger suite of tools including the Producibility Evaluation Method (PEM, [Miyakawa, Ohashi, Inoshita, and Shigemura]), the Assembly Reliability Evaluation Method (AREM, described below), and the Recyclability Evaluation Method (REM).

The method is applied manually or with the aid of commercially available software. When a part or operation is fully evaluated, all the penalties are added up and subtracted from one hundred. If the score is less than some cutoff value, say eighty, the operation or part is to be subjected to analysis to improve its score. The penalties and time estimates have been refined based on the experience of the entire Hitachi corporation, which makes a wide range of consumer and industrial goods such as camcorders, television sets, microwave ovens, automobile components, and nuclear power stations. All the evaluations are based on comparing the current design to a base design that is either "ideal" or represents the previous design of the same or a similar product. Because of the depth of the underlying dataset and the ratio technique of evaluation, the method is especially useful for designing the next in a series of similar products over a period of years. Repeated use of the method on the same product line relentlessly drives out low scoring operations.

The evaluation takes place in two stages. First, each operation is evaluated, yielding an evaluation score E_i for each operation. If several operations are required on one part, an average score E is calculated. The score for the entire product is either the sum of all the individual part

scores or the average of the part scores. In either case, it is possible that an assembly with fewer parts will have a higher score simply because fewer penalties are available to reduce it. In this case, the method clearly states, "reduction in part count is preferable to better score." However, the method does not include a systematic way of identifying which parts might be eliminated.

Examples of the penalties and use of the method appear in Figure 15-3 and Figure 15-4.

15.D.3. The Hitachi Assembly Reliability Method (AREM)

The Hitachi Assembly Reliability Evaluation Method ([Suzuki, Ohashi, Asano, and Miyakawa]) extends the AEM beyond cost and time into the domain of assembly success and product reliability. The impetus for this method arises from several trends: the rise in product liability suits, the introduction of new product and process technologies resulting in production uncertainties and long ramp-up times, shorter product development time resulting in design mistakes, and the degree to which outsourcing makes a manufacturer dependent for quality on the work of other companies. The method has proven useful for products that must achieve very high reliability, products that change drastically from one model or version to the next, complex products, ones that are assembled at multiple sites around the world, and products containing many parts and subassemblies from suppliers. The basic logic of the method is shown in Figure 15-5.

The method is similar in style to the AEM in the sense that each operation is evaluated and compared to a standard, resulting in a penalty. In addition, the method contains a scale factor called the basic shop fault rate, based on data from a given factory, that permits the failure rate at that factory to be estimated based on the product's design.

Category	Basic Element Example		AEM Symbol	Coefficient
Movement	Downward Movement	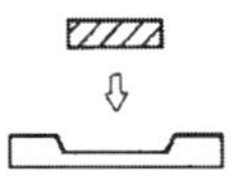	↓	1.0
Joining	Soldering	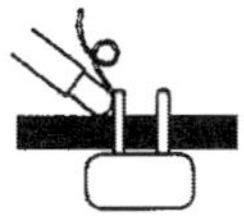	S	2.2

FIGURE 15-3. Examples of AEM Symbols and Penalty Scores. ([Miyakawa, Ohashi, and Iwata]. Hitachi, Ltd. Used by permission.)

Examples	Product Structure and Assembly Operations		Part AEM Score $_aE_i$	Product AEM Score $_aE$	AEM Cost Ratio $_aK$	Part to Be Improved
Structure 1 [before improve-ment]	C(↓ ↻) B(↓…) A(↓-)	1. Set base A.	100	76	100	C
		2. Bring down block B and hold it to maintain its orientation.	79			
		3. Fasten bolt C.	50			
Structure 2	B(↓…) A(↓-)	1. Set base A.	100	88	Approx. 0.8	B
		2. Bring down and press-fit block B.	75			
Structure 3	A(↓-)	No assembly.	100	100	0	—

FIGURE 15-4. Assembleability Evaluation and Improvement Examples. ([Miyakawa, Ohashi, and Iwata]. Hitachi, Ltd. Used by permission.)

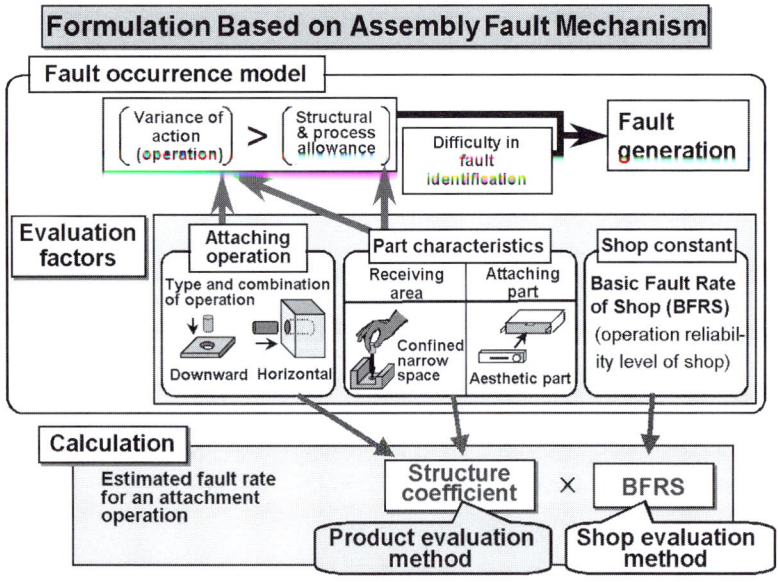

FIGURE 15-5. Hitachi Assembly Reliability Evaluation Method. (Hitachi, Ltd. Used by permission.)

On this basis, one can decide either to improve the product or to improve the factory in order to increase the score.

The basic assumption behind the method is that if the assembly reliability is low, either the product is at fault (resulting in a product structure penalty) or there is some variation in the assembly process (resulting in an operational variance penalty). Product structure factors that influence assembly faults include dimensional variation, flexibility or fragility of parts, lack of sufficient access to the assembly point, too much force needed to ensure complete insertion, and so on. Operational variance factors include not positioning a part accurately enough, applying too much force or not enough force, not driving screws all the way in, cutting a wire, and so on.

These factors are to some extent represented in the Boothroyd handling time and insertion time tables but are associated with time rather than failure to perform the assembly correctly. In addition, other kinds of mistakes are

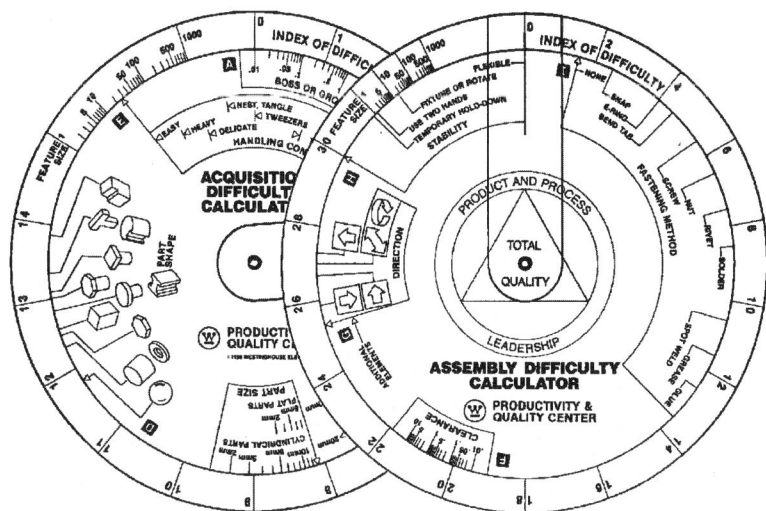

FIGURE 15-6. The Westinghouse DFA Calculator. The calculator is a rotary slide rule. It consists of a large disk with a slightly smaller disk and a transparent cursor on each side. The smaller disks can be rotated independently of the large disk and the cursors. Difficulty starts at zero and accumulates as the topics marked A, B, C, and so on, are addressed in turn. (Reprinted from [Sturges] with permission from Elsevier Science.)

possible, as discussed in Chapter 16. The most frequent of these are using the wrong part and using a damaged part. No DFA method deals directly with these issues, although general guidelines include warnings about helping the operator to distinguish between similar parts.

15.D.4. The Westinghouse DFA Calculator

Sturges developed a rotary calculator at Westinghouse for estimating handling and insertion difficulty (Figure 15-6). On one side the user calculates a handling difficulty index that is interpreted as seconds required. On the other side the same kind of calculation is done to estimate assembly time. Factors such as part shape, symmetry, size of features to be grasped or mated with, direction of insertion, clearance, and fastening method are assessed and added up by repositioning the disks and the cursor.

15.D.5. The Toyota Ergonomic Evaluation Method

Most DFA methods are designed to evaluate assembly of small parts. In the auto industry, final assembly of the product involves relatively large and heavy parts. Here, ergonomics, the science of large-scale human work and motion, is applicable. Toyota has determined that the product of the weight of a part and the time it must be supported by a worker is a good indicator of physical stress ([Niimi and Matsudaira]). In addition, the worker's posture is important: standing, slightly bending, or bending

deeply are each more stressful than the one before for the same weight and duration. Thus Toyota has developed a stress evaluator called TVAL (Toyota Verification of Assembly Line) to prioritize assembly operations for improvement to reduce physical stress. The form of TVAL is

$$\text{TVAL} = d_1 \log(t) + d_2 \log(W) + d_3 \quad (15\text{-}1)$$

where d_1, d_2, and d_3 are constants and t and W are the time and part weight, respectively. For example, installing a lightweight grommet onto a car door requires standing for 30 seconds and has a TVAL of about 25. By contrast, installing a rear combination lamp involves bending forward deeply for over 60 seconds and has a TVAL of 42. Before TVAL was applied to a section of assembly line, TVALs ranged from 30 to 48. After redesigning the worst stations, TVALs range from 22 to 35.

15.D.6. Sony DFA Methods

Sony has a unique way of involving its engineers in the DFA process. The engineers must prepare exploded view drawings of all concepts. This forces consideration of assembly even before detailed design begins. This is illustrated in Figure 15-7. A DFA analysis is done on the concept, based on the exploded view, using Sony's own DFA software. The DFA score is included with other criteria in judging the merit of each concept.[5]

---

[5]This process was explained to the author during two visits to Sony in 1991.

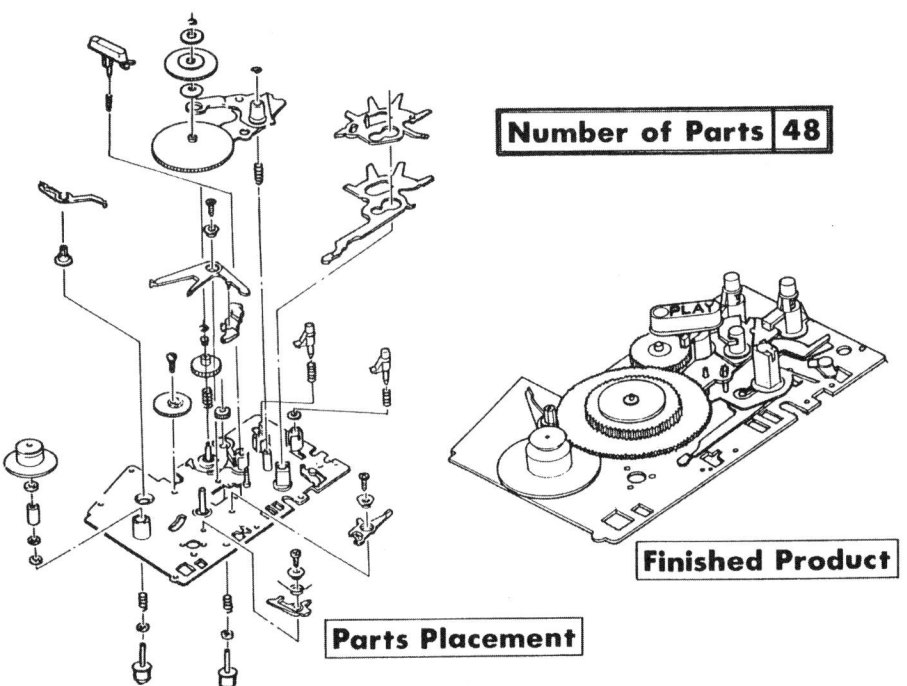

Number of Parts 48

Finished Product

Parts Placement

FIGURE 15-7. Exploded View Drawing of Sony Walkman Chassis. Drawings like this are made by design engineers for every design concept. (Used by permission of Sony FA.)

15.E. DFx IN THE LARGE

DFx in the large deals with issues that require consideration of the product as a whole, rather than individual parts in isolation, and likely will require consideration of the context in the factory, supply chain, distribution chain, and the rest of the product's life cycle. We take up such issues here. Our focus will be on (a) product structure and its relation to product simplification and (b) design for disassembly, repair, and recycling.

15.E.1. Product Structure

Product structure involves many of the issues normally associated with product architecture, but the focus is on the structure more than on its influence on architecture issues. That is, one reads about products that are built in stacks or in arrays, or about consolidating parts, in the context of simplifying assembly rather than about "integrality" or "modularity." Nevertheless, one of the first books to deal with design for assembly, [Andreasen et al.], clearly recognizes the close connection, not only between DFA

and product structure, but between these two topics and the larger issue of product development processes themselves. Early consideration of assembleability inevitably turns to opportunities for restructuring the product, and this cannot be done except early in the design process. A design process that does not permit early consideration of assembly issues will therefore be a very different process from one that does, and the resulting product will be different as well. Furthermore, the differences will extend beyond the local issue of assembleability.

15.E.1.a. Styles of Product Structure and their Influence on Ease of Assembly
Several architectural styles have been identified in assemblies. These are the stack and the array. Examples of these are shown in Figure 15-8.

In general, arrays present the fewest constraints on the assembly process. Printed circuit boards are the most obvious example. These are usually made by high speed machines that select parts from feeders each of which

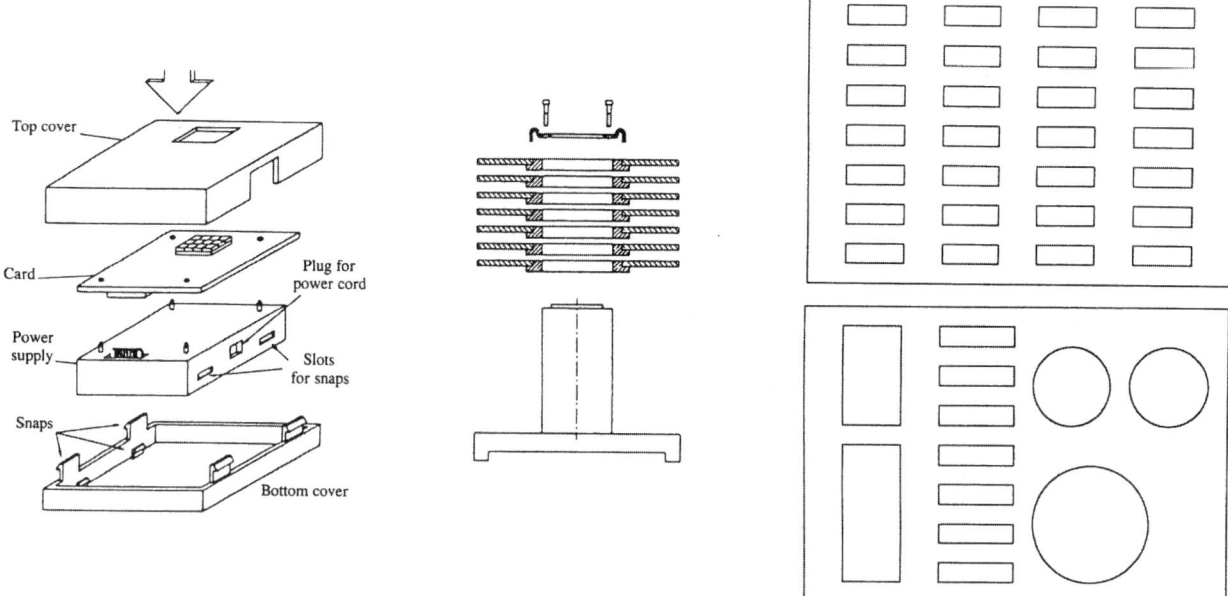

FIGURE 15-8. Examples of Stack and Array Product Structures. Both stacks and arrays come in two generic varieties: the parts are mostly the same or mostly different. ([Redford and Chal]. Copyright © Alan Redford. Used by permission.)

presents one part (100K resistor, a particular integrated circuit, etc.) Because this product structure is so simple, the assembly sequence can be optimized to suit selection and insertion of the different kinds of parts. The factors involved include how far the insertion head has to travel to get each kind of part, how many of each kind are needed, how close together on the board they are, and so on. Optimization algorithms have been developed to find the best insertion sequence.

The main justification for a stack architecture is that gravity aids the insertion process. If locating features are provided, a part will stay put once it is placed. In Figure 15-8, two types of stacks are shown, namely, those with identical parts and those with different parts. In the former case, there are ample opportunities for alternate assembly sequences, such as preparing a separate subassembly comprising the stack of the identical disks. When the parts are quite different, as suggested by the illustration, their individual properties and mating features may create assembly sequence constraints.

Most products are combinations of the generic structures illustrated above. [Kondoleon] conducted a survey of a dozen varied products, including consumer and industrial items, noting which assembly operations were needed

and the directions along which they occurred. The results appear in Figure 15-9 and Figure 15-10. They show that there are two dominant insertion operations and two dominant directions. The implication is that these products appear to have a major axis of insertion and perhaps of operation as well. Perpendicular to this axis is the direction in which fasteners are installed. These observations probably reflect the Cartesian nature of the architectures of the machine tools used to make the parts.

15.E.1.b. Simplification Methods

As noted earlier in this chapter, a major effort of DFA is product simplification. Simpler products have fewer parts, which means fewer assembly operations, workstations, factory space, and workers. In addition, each part represents design effort and overhead. Whether simpler/fewer always means less expensive is a separate issue discussed below.

While most researchers and practitioners of DFA understand the desirability of reducing the number of parts, only the Boothroyd method presents a systematic approach to doing this. The idea is to subject each part to three criteria that might justify its inclusion in the product, and eliminate any part that fails the criteria.

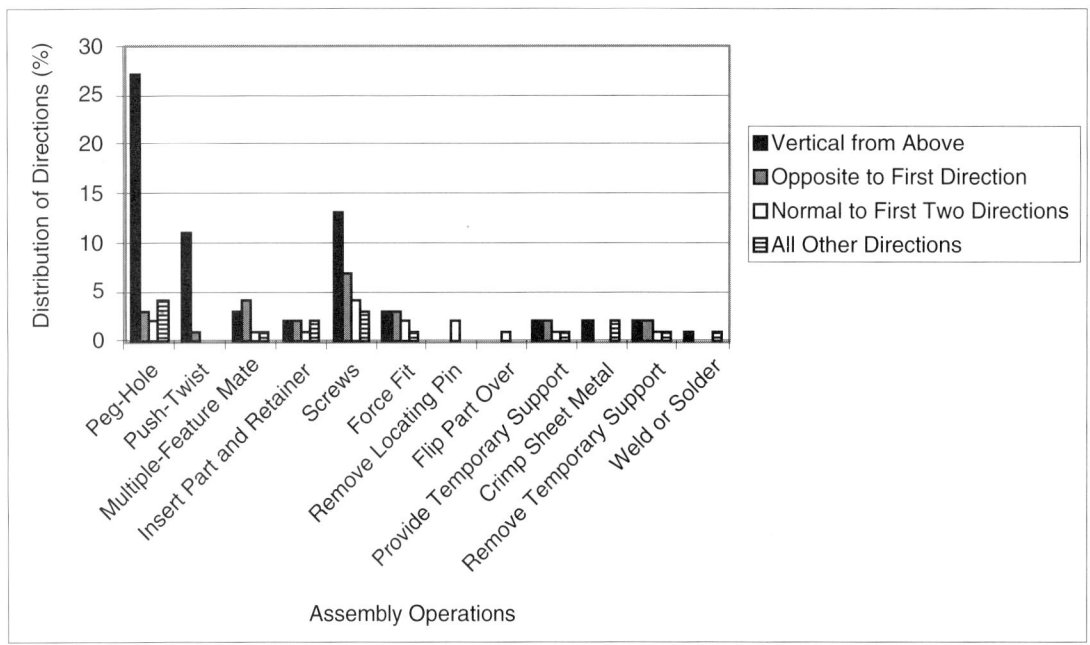

FIGURE 15-9. Census of Assembly Operations and Their Directions. The conclusions to be drawn from these data must be tempered by the fact that they were gathered in the middle 1970s. Product design methods and product materials have changed greatly since that time but no study comparable to this has been repeated since. ([Kondoleon])

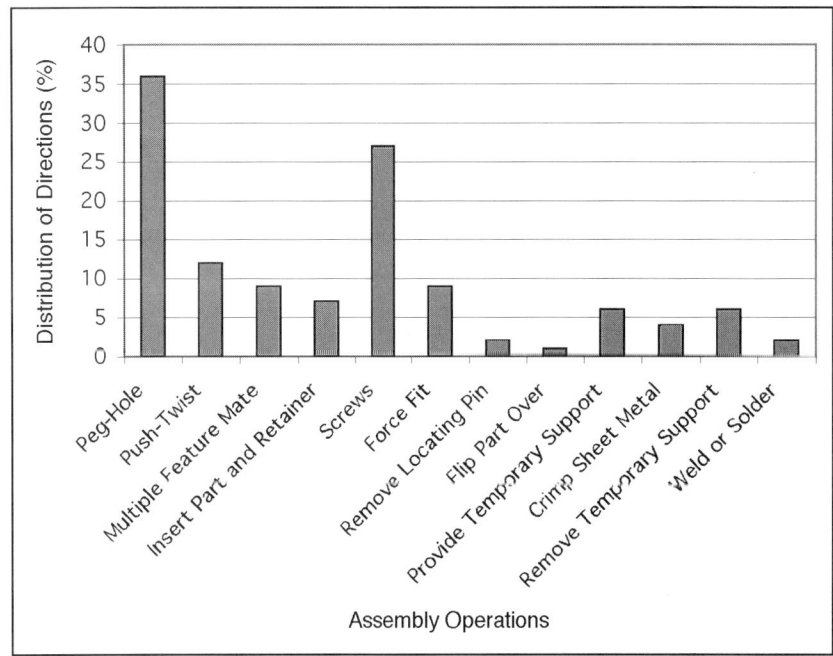

FIGURE 15-10. Summary Census of Assembly Operations. ([Kondoleon])

The three criteria are as follows ([Boothroyd, Dewhurst, and Knight]):

1. During operation of the product, does the part move relative to all other parts already assembled? Small motions that could be accommodated by flex hinges integral to the parts are not counted.

2. Must the part be of a different material or be isolated from all other parts already assembled?

3. Must the part be separate from all other parts already assembled because otherwise the assembly or disassembly of other separate parts would be impossible?

Unless at least one of these questions can be answered "yes" for a part, that part *theoretically* can be combined with another part or eliminated entirely. This criterion is applied ruthlessly using main product functions as the focus. Thus, for example, all separate fasteners are automatically flagged as theoretically unnecessary. The effect of part consolidation on part cost is evaluated separately using DFx in the small. It is not expected that the theoretically unnecessary parts will really be eliminated because other criteria for performance or manufacturability might be affected. The purpose of the exercise is to focus attention on necessity.

The *assembly efficiency* metric is calculated as follows:

$$\text{Assembly efficiency} = \frac{\left(\begin{array}{c}\text{theoretical minimum} \\ \text{number of parts}\end{array}\right) * 3 \text{ sec/part}}{\begin{array}{c}\text{estimated assembly time} \\ \text{including all parts}\end{array}} \quad (15\text{-}2)$$

In this metric, an assembly time of 3 seconds per part is assumed, based on an ideal assembly time for a small part that presents no difficulties in handling, orienting, and inserting. Thus the numerator represents an ideal minimum assembly time for a relatively simple manually assembled product that contains only those parts that survive the three questions listed above. The denominator represents the actual assembly time of the current or modified design. Typical products that are ripe for part count reduction often have assembly efficiencies on the order of 5% to 10% while efficiencies after reduction analysis or redesign are typically on the order of 25%. An assembly efficiency of or near 100% is unlikely to be achieved in practice. This finding implies that other valid reasons beyond those listed in the three questions above intervene to prevent parts from being eliminated. Considering the issues raised in Chapters 12 and 14, this should be no surprise.

FIGURE 15-11. Plastic Injection Molded Part. This part goes into a domestic hot water heating system and has dozens of features on it. It is about 1.5″ high. Its mold clearly took a long time to develop. It utilizes "hollow core" molding, which involves folding and moving mold core parts. Such a part will not be economical unless it is made in very large quantities. (Poschmann Industrie-Plastic GmbH & Co KG. Photo by the author.)

Some of the products used as part-count-reduction examples in the DFA literature may appear ridiculous at first sight. These typically are rich in threaded fasteners, including washers and nuts. Each screw/washer/nut set counts as three parts that are automatically eliminated, driving down the assembly efficiency. As Boothroyd points out, some of these products look like model shop prototypes that were put directly into production without any attempt to design them for production efficiency. Anecdotally, fasteners seem to account for low assembly efficiency in many cases.[6] Eliminating them is thus an easy way to boost the score. The pros and cons of modifying or eliminating fasteners are discussed later in this chapter.

Today, many products exhibit evidence of careful attention to structure and part consolidation. As reflected in the examples in Section 15.F, even quite modest consumer products contain injection molded or stamped parts of high quality, exquisite tolerances, and complex features. See Figure 15-11 for a picture of one such part. This is the result of recent progress in development of stamping methods as well as of new polymer materials having high strength, low shrinkage, and high-dimensional stability over time. Examples include the casings of electric drills

---

[6]Ken Swift, University of Hull, personal communication.

and screwdrivers, covers of cell phones and computers, and interior components of automobiles. Design of these items and their molds is supported by three-dimensional CAD models and simulation of the flow of molten plastic into the molds.

Nevertheless, there are many case studies across a range of industries that show an average of about 45% part count reduction ([Boothroyd, Dewhurst, and Knight], [Swift and Brown], [Swanstrom and Hawke]).

15.E.1.c. Tradeoffs and Caveats

Application of DFx in the small subjects each part to scrutiny separately while DFx in the large summarizes the appropriateness of the product as a whole through metrics like that in Equation (15-2). Blind adherence to the "rules" and "metrics" of DFA, however, is not recommended. Instead, these methods should be used in combination with other criteria. In a number of situations, the right thing to do is not what the DFA analysis recommends. This section contains some comments and examples.

15.E.1.c.1. General Considerations. We noted at the beginning of the chapter that parts costs greatly exceed assembly costs. For this reason, DFA must be accompanied by DFM. Naturally, any choice of manufacturing method and material must deliver the required functionality, reliability, durability, and appearance. DFM is a huge topic with a rich literature that we cannot address in this book. [Boothroyd, Dewhurst, and Knight] presents methods for estimating the cost of making cast, molded, stamped, and powder metal parts. [Ostwald and McLaren] gives methods for estimating process costs for a variety of processes based on given hourly operating costs for machines. [Hu and Poli] describes a method of comparing the cost of stamped, molded, and assembled parts based on guaranteeing functional equivalence feature-by-feature. [Esawi and Ashby] describes the Cambridge Process Selector, which searches for good candidate processes based on preliminary part information early in design. Extensive analysis and testing are required to compare different materials and processes for making "the same" part.

The Boothroyd method as presented above applies to manual assembly. If automatic assembly is contemplated, then a different set of criteria, codes, and operation times must be used. These are available in [Boothroyd, Dewhurst, and Knight]. Unfortunately, it is often difficult to know which method of assembly will be used. A new product might begin life assembled manually and could be switched to automatic assembly if it becomes a market success. Rarely is there an opportunity to redesign it at this stage because the effort is directed at getting as many units of the original design out the door as possible.

Second, any DFA method requires that a nominal assembly sequence be chosen, because assembly difficulty often depends on which other parts are present when a given part is to be installed. In many cases, little guidance is provided regarding how to select an assembly sequence. Often one is advised to "pick the base part." It may not be obvious which part this is, although [Redford and Chal] recommends that special design effort be devoted to being sure that every product has one. Properties of a good base part include being wide enough to provide stable support and being well enough toleranced to function as an assembly fixture. The casing of the Denso panel meter meets these criteria. On the other hand, as we saw in Chapter 7, many quite attractive assembly sequences begin with parts that would not be chosen as the base, such as the rotor nut on the automobile alternator. This part was chosen in order to permit vertical assembly with no reorientation of the product during assembly.

Third, assembly difficulty is not easy to predict, and many ways exist to reduce it. As noted above, people get better as they practice, and a difficult task can often be made easy with the provision of a simple tool. Until one actually has the parts in hand and is able to try assembling them, it is difficult to know for sure what will be easy and what will be difficult. Furthermore, many operations that are easy for people (turning the assembly over or quickly determining if a part is suitable for use) are difficult, expensive, or impossible for machines. Similarly, many operations that are difficult for people are easy for machines, such as picking up little parts with tweezer-like end effectors, placing integrated circuits to within 0.01" tolerance at the rate of 6 per second, and tightening fasteners to an exact torque every time. Thus, DFA analyses are predictions at best. Recent research, such as [Gupta et al.], applies virtual environments to help predict assembly problems.

In addition, we saw in Chapter 8 that eliminating and consolidating parts can deprive the assembly process of needed adjustment opportunities. Depending on the industry, its cost structures, the skill of its assemblers, the variation in the parts, and the time available for each assembly operation, it may be of advantage to permit adjustments or it may not be. Each case needs to be evaluated carefully.

Existing Arm Bracket Assembly

Side Arm Assembly

Top

Side Arm

Rivets

Clinch Nut

Pivot

Munro
&
Associates, Inc.

New Arm Bracket Assembly

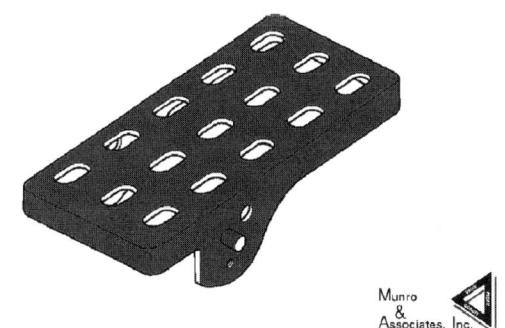

Munro
&
Associates, Inc.

Lean Design Idea Implemented

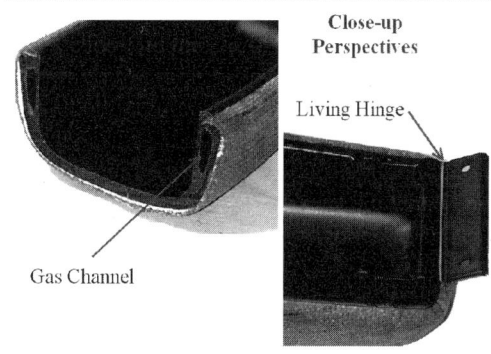

Close-up
Perspectives

Living Hinge

Gas Channel

FIGURE 15-12. Redesign of Automobile Interior Arm Bracket. *Top left:* Existing design, requiring several parts, fixtures, and assembly operations. *Top right:* New bracket. *Bottom:* New armrest with bracket molded in. (Courtesy of Munro and Associates. Used by permission.)

15.E.1.c.2. Does Consolidating Parts Really Save Money? More deeply, it is clear that consolidating parts makes them more complex. Boothroyd and other practitioners of DFA, including Sandy Munro of Munro and Associates, are firmly convinced that fewer but more complex parts add up to a less expensive product due to lower parts costs and lower assembly costs. The true conditions must be evaluated individually for each part and product.

Sometimes the consolidated design is totally different from the original. Developing it requires intimate knowledge of materials and process technologies. An example appears in Figure 15-12. Not only is the metal bracket transformed to a single stamping, but the armrest itself is molded integrally with the bracket. Its shape is created in part by injecting gas into the sides during molding.

[Boothroyd, Dewhurst, and Knight] presents equations permitting one to estimate the number of hours needed to fabricate an injection mold. Factors that influence the time include the area and volume of the part, the number of features such as surface patches, holes and depressions, and tolerances and surface finish. Most of these factors affect mold development time linearly, but complexity in terms of features is estimated to increase mold development time by the power of 1.27. Figure 15-13 shows the results of calculating mold development time for the following problem: Given some number of separate parts of given complexity, is it better in terms of mold development time to make a separate mold for each part or to combine the parts and make them with one mold? Naturally, the combined part is more complex. If the individual parts are sufficiently complex themselves, the nonlinear factor

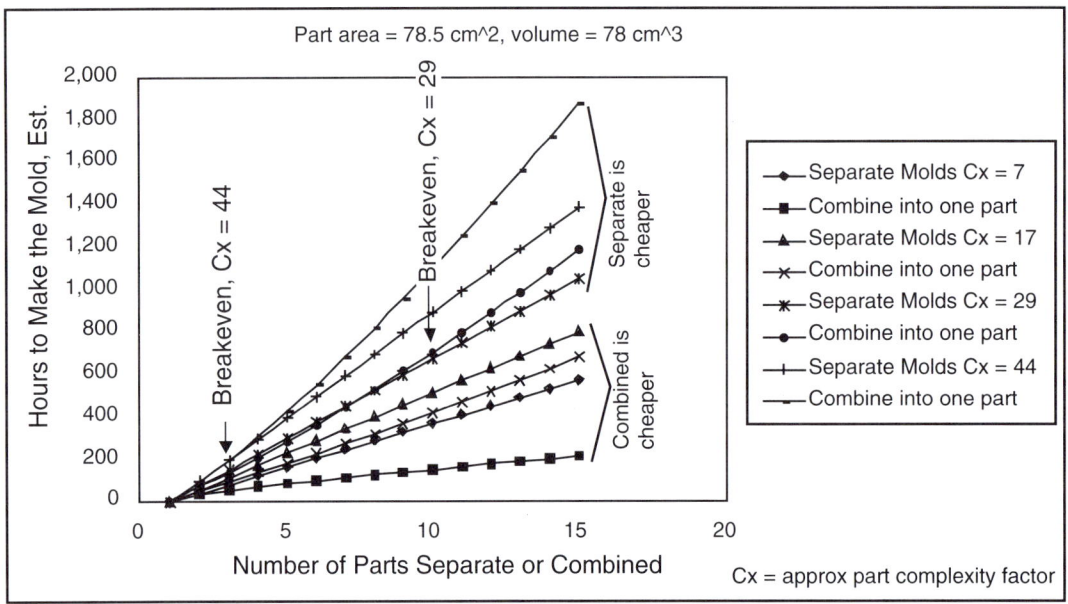

FIGURE 15-13. Cost Versus Complexity of an Injection Mold. Hypothetical parts with different degrees of complexity are considered as candidates for consolidation, and the number of hours to develop the mold is calculated using equations in [Boothroyd, Dewhurst, and Knight]. These equations include factors for estimating the complexity of a part. Each pair of lines in the chart compares time to make separate molds versus time to make one mold that makes a combined part. If the parts are not very complex, then it is always better (in terms of mold development time) to consolidate them. If they are complex, then there is a maximum number that should be consolidated, above which it is better to create separate molds for each one. This number is lower when the parts are individually more complex. The chart is illustrative only, and a similar analysis would have to be made in any real situation.

by which complexity influences development time will sooner or later make the combined-part mold take longer than separate molds. The study in Figure 15-13 does not include alternate strategies like combining only some of the parts, but rather only compares all versus none. It is illustrative only, and each real case must be evaluated on its own merits.

[Hu and Poli] presents a more refined cost model that includes material and assembly costs for the parts as well as tool development cost. This part fabrication model estimates total cost by summing the cost of creating each feature on the part. The model is linear and does not contain an explicit measure of complexity. It concludes, contrary to Figure 15-13, that there is always a number of parts to be combined above which it is cheaper to combine them than to make them separately and assemble them.

[Fagade and Kazmer] expands the scope of analysis to include time to market and long term profit. This model is based on statistical analyses of price quotes and delivery times from mold vendors on a variety of parts. While each proposed consolidation must be evaluated on its merits, the research concludes that the three criteria for part consolidation given by Boothroyd must be augmented. Plastic injection-molded parts may be consolidated unless

- The consolidation does not reduce the number of tools,
- The parts have vastly different quality requirements,
- The design process is not certain of delivering the product and there is significant sales cost sensitivity, and
- The manufacturing processes are not capable of delivering high yields of complex products

These conclusions are consistent with those of [Ulrich, Sartorius, Pearson, and Jakiela].

If the product must meet criteria for repair, recycling or reuse, then other factors must be considered. For example,

FIGURE 15-15. **Low Melting Point Bismuth Alloy Lost Core.** (Courtesy of Poschmann Industrie-Plastic GmbH & Co KG.)

15.E.1.c.3. Is It DFA or Product Redesign? More deeply still, it may not be obvious where modifying product structure stops being a DFA activity and begins to resemble redesign. As an example, consider the two pump designs shown in Figure 15-16. This figure illustrates use of the Lucas/University of Hull DFA method. It is similar in many ways to the Boothroyd method in that it calculates a number of metrics based on deciding which parts are really functionally necessary and which are not. The metrics compare time or effort devoted to "unnecessary" parts relative to that devoted to the necessary ones.

But this figure shows something else, namely that the two pump designs do not operate the same way. The path taken by the pumped fluid is different, the style of valve is different, and external piping and packaging arrangements are different. In one case the volume above the piston fills with fluid while in the other case it does not. This means that the seals around the piston rod are crucial in one design and negligibly important in the other design. Each design is likely to exhibit different failure modes. This is not to say that the new design is not a good one but rather to point out that much more differentiates the two designs than mere application of DFA rules and metrics. In general, DFA must take its place among all the other pressures exerted on product design, and DFA recommendations must be weighed against other factors.

15.E.1.c.4. The Role of "Product Character". Finally, it is likely that consumer and industrial products will provide different opportunities for DFx. Consumer products like food mixers and can openers are subject to much less stringent performance and durability requirements than are industrial components like automobile transmissions

FIGURE 15-14. **Glass-Filled Nylon (PA-66) Injection-Molded Parts for a Home Hot Water System.** These parts are members of a product family that allows a heating contractor to customize a home hot water system to the customer's needs. The parts share common exterior and interior diameters as well as axes where fasteners are inserted. (Courtesy of Poschmann Industrie-Plastic GmbH & Co KG. Photo by the author.)

parts that are subject to wear should be separate, low cost, and easy to remove and replace.

An example of real parts with real mold time and cost data, consider the parts in Figure 15-14. These parts go into home hot water systems and are designed so that they can be combined in many ways to configure custom systems. The diameter across the fastener diagonal is 10 cm. These parts sell at wholesale for about $2.00 to $6.00 each. They are very complex, including curved internal passages that are created using a low temperature melting point bismuth alloy mold insert (see Figure 15-15) that is later washed out of the finished part using hot oil. The molds take 6 to 8 weeks to design and 4 to 12 months to bring to their final state, able to deliver parts with tolerances around ±0.2 mm. One of these molds can cost from $100,000 to as much as $500,000.[7]

---

[7]Information provided in 2000 by Andreas Meyer of Poschmann Industrie-Plastic GmbH & Co KG of Germany, the company that makes the molds and the parts. Meyer estimates that doubling the number of features on a part can triple the design and tryout time for a mold.

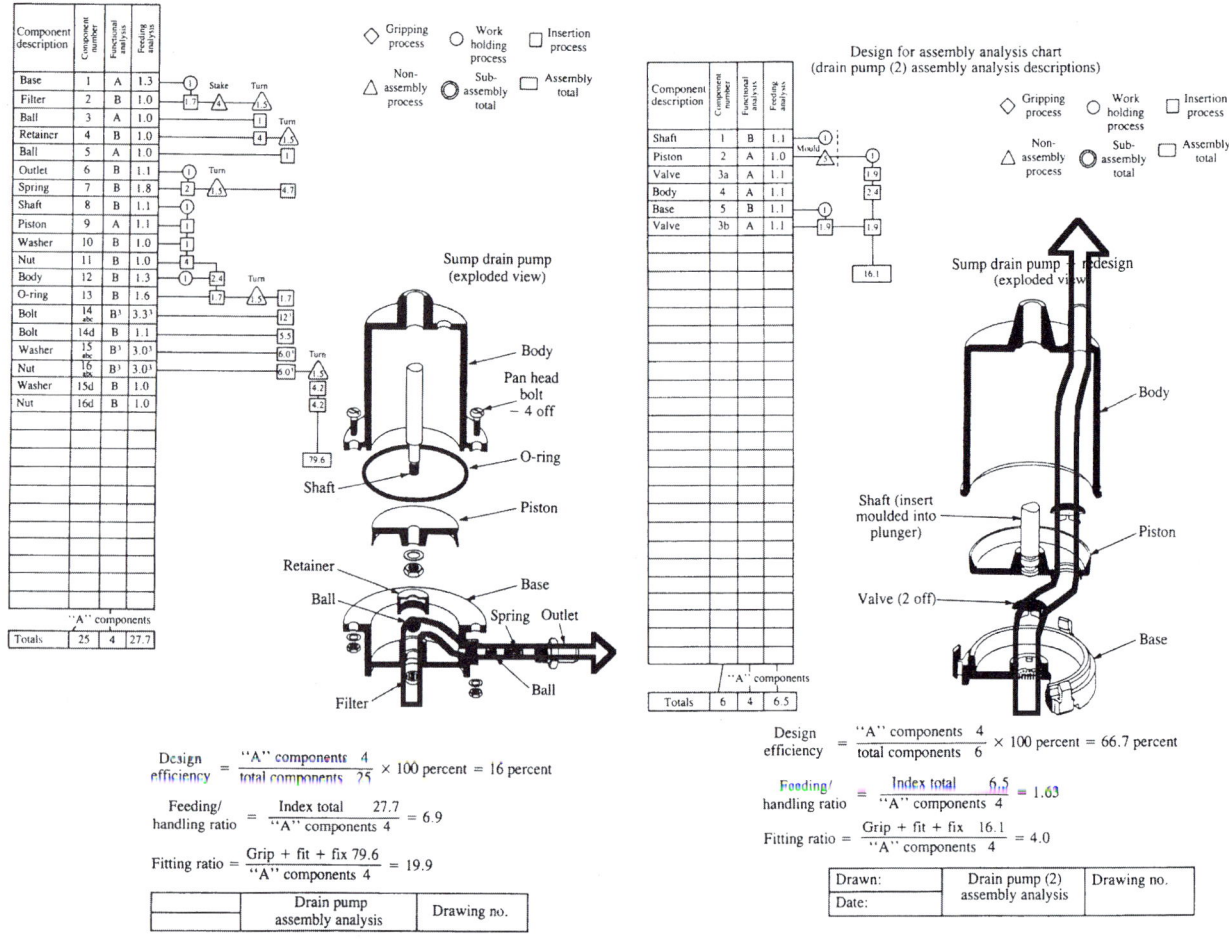

Before   After

FIGURE 15-16. Pump Redesign Example. Close inspection of the before and after pump designs shows several functional and application differences. For example, the fluid paths, shown by heavy hollow arrows, are different in the two designs. This example illustrates use of the Lucas/University of Hull DFA methodology. This methodology judges the value of keeping a part in the product based on three different metrics: design efficiency (similar to the Boothroyd assembly efficiency, the ratio of the total number of parts to the number of "A" parts, the latter being the functional minimum), feeding and handling ratio (ratio of total feeding effort to that needed to feed only the A parts, and the fitting ratio (ratio of time needed for all assembly operations to that needed for the A parts). ([Redford and Chal]. Copyright © Alan Redford. Used by permission.)

and aircraft engines. A home handyman's electric drill will get as much use in a year as a professional carpenter's drill will get in a single day. For such reasons, designers will choose materials, part boundaries, and fasteners much more carefully for an industrial product. The result is that opportunities for part consolidation and elimination of fasteners will be fewer.

Table 15-6 summarizes several factors to consider when deciding whether or not to consolidate parts.

15.E.1.d. Fastening Choices

As noted above, fasteners are the chief targets of part count reduction. One of the motivations for this was the belief in the 1970s that threaded fasteners took a long time to

TABLE 15-6. Factors to Consider Regarding Parts Consolidation

Consolidation	Differentiation
Supports functional drivers requiring integrity, absence of interfaces, absence of fasteners	Supports business drivers requiring substitution, differentiation, and modularity
Complex design process: KCs must be achieved by means of fabrication process design and execution	Supports adjustment to achieve KCs
Material selection is crucial	Permits multiple materials and other opportunities for design refinement
Design must accommodate the most demanding requirement	Each part can meet its own requirements, including need for periodic replacement or support for low cost reuse
Larger, heavier parts	More parts, longer assembly line
Fewer assembly steps, more reliance on fabrication processes to create quality	More assembly steps, more reliance on assembly processes to create quality
Fabrication tooling is more expensive and takes longer to develop	Lower-cost fabrication tooling may be attractive for low volume production
Complex fabrication processes	Fabrication and assembly steps can be interspersed on the assembly line to achieve differentiation, adjustment, better tolerances
Many features are created at once	
Requires care in defining location and orientation of split planes	Each feature can have its own material, fabrication process, surface finish, tolerance, etc.
Reduces "fixed" design and management costs like parts management, logistics, contracts, etc., that grow with the number of parts regardless of their complexity	Saves on costs associated with part complexity such as time to design and prove out complex production tooling
Process yield is crucial for large or complex parts; failure creates expensive scrap (microprocessors, thermoset aircraft assemblies) or inconvenient meltdown (metals, thermoplastics)	Process yield risk is distributed over many parts; high risk can be concentrated on one or a few parts, and assembly can use tested good parts

install and that installation was error-prone. Whether that was true or not at the time, it is less true today. Automatic screw insertion machines and powered hand tools are available that are very fast and reliable, and usually include automatic feeding of each screw and sensing of the correct torque. Even though it is still a good idea to examine a design to see if parts can be eliminated or consolidated, fasteners may be a less tempting target than they were in the past.

Many fastening alternatives exist. These include screws, screws with washers already attached, self-tapping screws (especially useful for fastening plastic and soft metal parts), rivets, adhesives, welding, crimping, heat or ultrasonic staking, and snap fits. Each of these has its advantages and disadvantages, some of which are listed in Table 15-7.

Generally, welding, screws, and rivets are preferred for joints subjected to large loads in products such as machinery, autos, aircraft, bridges, and buildings. Attempts to reduce the number of fasteners in major machinery joints in the name of DFA have been known to cause catastrophic failure.[8] A typical screw joint in a simple consumer

product might have four screws at 90° intervals while one in a machine tool, aircraft engine, or construction crane will have fasteners densely spaced no more than two or three fastener diameters apart around a bolt circle. The purpose of this tight spacing is to avoid large differences in contact stress between material under the bolt heads and material between them.

Properly spaced screws and rivets, or adhesives, may be superior to other joining techniques if there is a need to keep two surfaces flat and tight against each other. This can be necessary to seal against leaks or to prevent buzzing noises caused by vibration inputs.

In summary, choice of fastening method is influenced by many factors, only one of which is assembly time or cost. It is not always possible to consolidate parts, so fasteners will be with us for the foreseeable future.

15.E.2. Use of Assembly Efficiency to Predict Assembly Reliability

The Hitachi AREM predicts assembly reliability by examining and evaluating individual assembly operations. [Boothroyd, Dewhurst, and Knight] reports data from Motorola showing that products with higher assembly efficiency have fewer defects per million parts.

---

[8] Ken Swift, University of Hull, personal communication.

TABLE 15-7. Advantages and Disadvantages of Different Kinds of Fastening Techniques

Fastening Method	Enablers or Advantages	Cautions or Disadvantages
Threaded Fasteners	Strong	Can take 3 sec each
	Strength or tightness can be tested on each one as it is installed	Installing several at once requires a special tool
	Reversible (good for recycling and repair)	Can vibrate loose
Rivets	Strong	Must be drilled out or otherwise destroyed to enable disassembly
	Will not vibrate loose	
Welding	Same as rivets	Requires same material on both parts
		Does not work well on some materials (e.g., aluminum)
Heat staking, ultrasonic welding	Strong, good for thermoplastics	Must be destroyed to enable disassembly
Adhesives	Strong	Must be destroyed to enable disassembly
	Works on dissimilar materials	Could be deteriorated by chemical attack or time
Snap fits	Fast assembly	Suitable for small products not subject to large loads
	Requires parts made of flexible material	May take up more space than screws for the same required strength
	User cannot disassemble easily but repair or recycling is easy with the right tools	

[Beitler, Cheldelin, and Ishii] expands on a method from [Hinckley] that predicts the fraction of defective products using the same data as used to calculate assembly efficiency. While Hinckley used the Westinghouse DFA calculator ([Sturges]) to obtain assembly operation time estimates, the basic idea is the same.

Hinckley discovered that a factory's defect fraction could be predicted by calculating the *difference* between the actual assembly time and the theoretical assembly time based on the *actual* number of operations. [Recall that assembly efficiency as defined by Equation (15-2) is the *ratio* of theoretical assembly time using only *theoretically necessary* parts to the actual assembly time.] Hinckley called this the complexity factor and calculated it as

$$CF = TAT - TOP * t \qquad (15\text{-}3)$$

where *CF* is the complexity factor, *TAT* is the actual assembly time, *TOP* is the total number of assembly operations, and *t* is some nominal ideal operation time.

Hinckley took data at several factories on a number of different assemblies and their defect rates. When each factory's data were plotted on a log–log scale, he found a good straight line with a slope of about 1.3. An illustrative chart of such data appears in Figure 15-17.

Hinckley used a baseline value of *t* = 1.4 sec for each individual operation, but any convenient value will do.

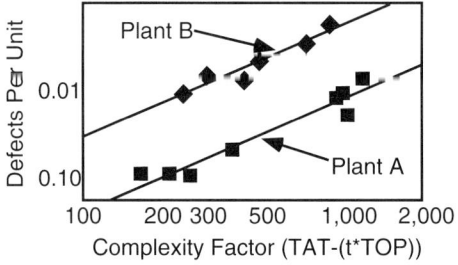

FIGURE 15-17. Notional Data Illustrating Equation (15-3). Each square or diamond represents one factory's defect rate on an assembly with the complexity factor shown. ([Beitler, Cheldelin, and Ishii])

The idea is that if operations take longer than the baseline time, there is an increased chance that a mistake will be made. While this is not in itself surprising, the consistency of the data is surprising and potentially useful.

If one wants to predict the defect rate of a new assembly in a given factory whose defect rate on other similar assemblies is known, then this method allows such a prediction. If one wants a better defect rate than that factory can deliver, one can use a different factory with a better defect rate for such assemblies, or one can attempt to redesign the product to reduce the lengthy operations.

15.E.3. Design for Disassembly Including Repair and Recycling (DfDRR)

Disassembly for repair or recycling was always an important part of DFA but has risen in importance in recent years. In some European countries, laws require manufacturers to take back their products at the end of their useful life and recycle them. Regardless of legislation, it is economical to recycle many products, and industries exist to do this. By weight, 85% of automobiles are recycled.

However, the practicality, scope, and economics of recycling are heavily affected by choice of materials and fastening techniques. Thermoset plastics cannot be melted down for reuse, and thermoplastics with imbedded fibers, such as used in the parts in Figure 15-14, cannot be recycled many times because recycling involves chopping the parts into little pieces. Each cycle cuts the fibers, and at some point they are too short to function properly. Liquid aluminum easily dissolves a number of impurities that spoil its ability to function. Therefore, aluminum cannot be recycled unless all other materials are separated out first. Snap fits are convenient for assembly but can cause problems for disassembly. Rivets and adhesives are even more problematic. Not only are they difficult to reverse, but they are often used to join dissimilar materials that cannot be recycled unless they are separated. If parts joined this way are ripped apart, the parts themselves may rip, and portions of one will remain attached to the other. For these reasons, design for disassembly and recycling is subject to many more conflicting forces than is conventional DFA.

One approach to DfDRR is classification and coding in the spirit of DFA. [Kroll and Hanft] is typical of this approach. It identifies four cost drivers in determining if a product is easy to disassemble:

- Accessibility—is there clearance to insert the necessary tool or hand?
- Positioning—how accurately must the tool or hand be positioned in order to remove the part (grabbing and yanking is easier than positioning a tool)?
- Force—how much force is needed (less is better and some fastening methods require more force to reverse them than others)?
- Basic disassembly time—each operation has its own estimated time.

The basic operations are unscrew, remove, hold/grip, peel, turn, flip, saw, clean, wedge/pry, deform, drill, grind, cut, push/pull, hammer, and inspect.

The difficulty of each task is estimated for each part in the product, based on the difficulty presented by each of the cost drivers. Difficulty scores range from 1 to 10, and the time to do a disassembly operation in seconds is estimated to be $1.04 * (difficulty - 1) + 0.9 * (number of hand and tool operations)$. A reference score can be calculated by estimating the time to disassemble the product if Boothroyd's rules for consolidating parts are applied. Thus both the time to disassemble and the sources of "unnecessary" disassembly time can be estimated.

Another approach to DfDRR is process-based. [Kanai, Sasaki, and Kishinami] describes a method for graphically representing the process alternatives for disassembling, shredding, and recycling a given design. It is based on an extended assembly model of the type described in Chapter 3. A comprehensive model of this type permits computer algorithms to make evaluations of the kind described in [Kroll and Hanft] as well as to search for the lowest cost process combinations.

The model represents the following issues:

- What the parts are made of
- What fastening methods are used
- Whether any part or subassembly can be reused

This information is used to decide if further disassembly of any item is needed or whether the resulting item can be reused whole or shredded whole. This question is asked recursively, starting from the whole product and working down. Feasible disassembly sequence and method choices are evaluated based on whether the disassembled items would be rendered unusable or unrecyclable by a chosen process. The logic of the search is diagrammed in Figure 15-18.

A search routine is used to look for a sequence of disassembly steps that maximizes the "weight fit ratio" defined as

$$Weight\ fit\ ratio = \%\ of\ parts\ by\ weight\ that\ can\ be\ treated\ properly$$

using the minimum total number of operations, where "treat" means reuse, recycle, or dispose, and "properly" means, for example, that a reusable part can really be reused and is not recycled or dumped instead. "Operations" include disassembling, sorting, and shredding. A cross-plot of weight fit ratio versus total number of operations reveals the desirability of a process plan, as indicated in Figure 15-19.

Figure 15-20, Figure 15-21, and Figure 15-22 illustrate the use of this method on a simple Sanyo electric shaver. If all parts must be reused, then the best process that can be found has a weight fit ratio of about 50% after about 150 operations. Further operations cannot improve the ratio. This means that the product is not well suited for reusing every part, even though that is a desirable goal. By contrast, a less ambitious goal is simply to recycle every part. In this case, the weight fit ratio rises to 80% within about 150 operations. Finally, a more nuanced goal

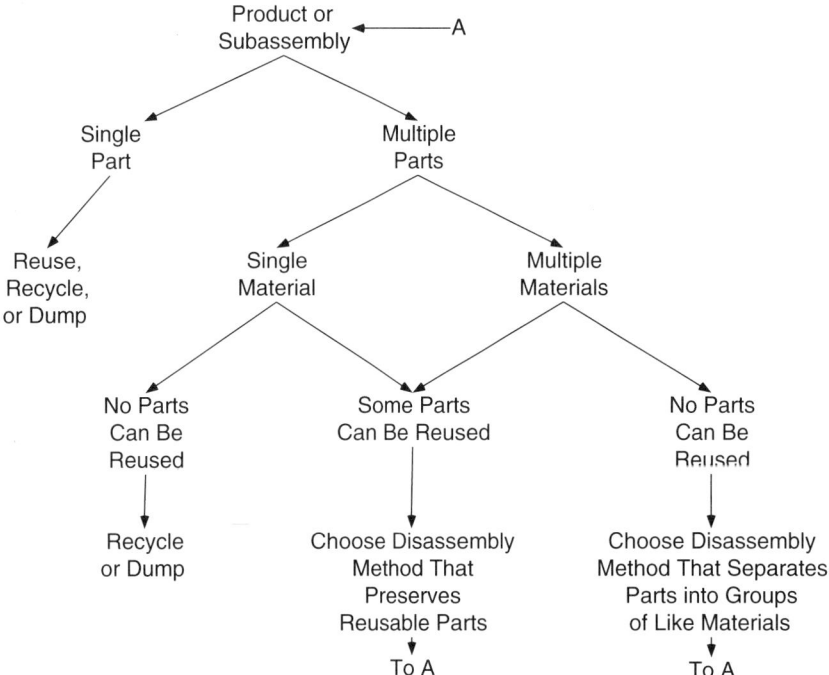

FIGURE 15-18. Options for Planning the Reuse, Recycling, and Disposal Process.

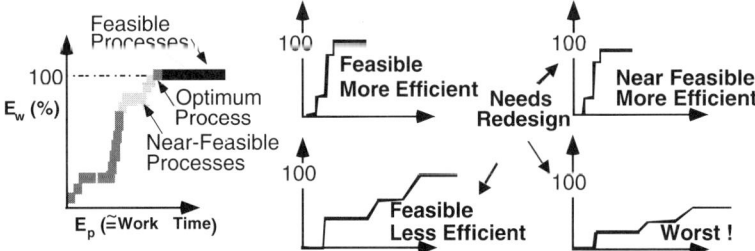

FIGURE 15-19. Cross-Plot of Weight Fit Ratio E_w and Number of Operations E_p. Better process plans have a steeply rising cross-plot that approaches 100% while worse ones have a slowly rising plot that falls short of 100%. The former succeeds in deploying each part to its desired final state in a small number of operations while the latter spends time but sends many parts to the wrong destination (recycled or dumped instead of reused, for example). ([Kanai, Sasaki, and Kishinami]. Courtesy of S. Kanai. Used by permission.)

that calls for reusing the motor and battery, recycling all polymer parts and all metal parts over 10 g in weight, and disposing of the rest scores 82% within about 140 operations.

This method can be used to compare process goals or product designs and can indicate, based on the cross-plot, which parts or operations are responsible for keeping the process from efficiently meeting the goals.

FIGURE 15-20. Shaver Used for DfDRR Example. ([Kanai, Sasaki, and Kishinami]. Courtesy of S. Kanai. Used by permission.)

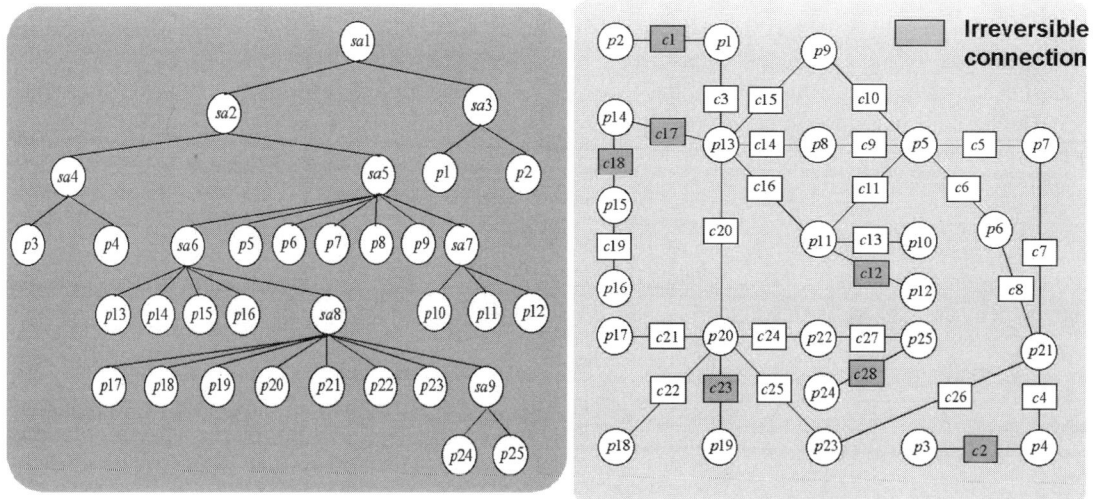

FIGURE 15-21. *Left:* Product Structure of Shaver. *Right:* Liaison Diagram of Shaver. "sa" denotes a subassembly, "p" denotes a part, while "c" denotes a liaison. ([Kanai, Sasaki, and Kishinami]. Courtesy of S. Kanai. Used by permission.)

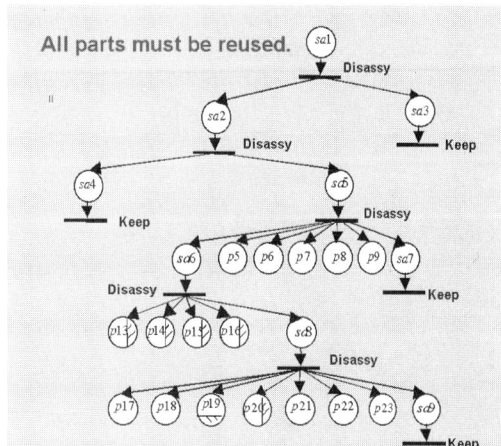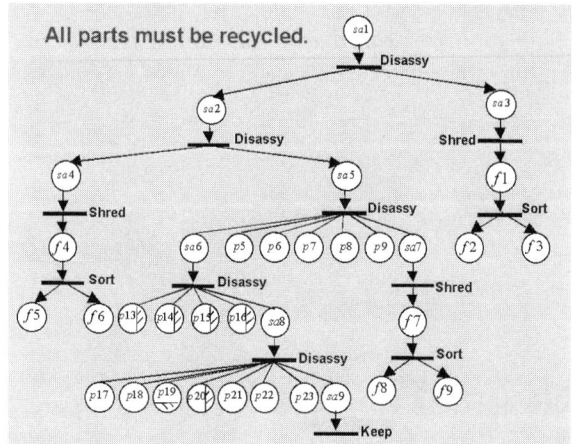

FIGURE 15-22. Two Process Plans for Disassembling the Shaver. *Left:* All parts must be reused. In this case, the weight fit ratio is about 50%. *Right:* All parts must be recycled. In this case, the weight fit ratio is about 80%. "sa" denotes subassembly, "p" denotes part, and "f" denotes fragment of a part. Fragments result from shredding or arise when parts are ripped apart, leaving a fragment of one attached to the other. ([Kanai, Sasaki, and Kishinami]. Courtesy of S. Kanai. Used by permission.)

[Harper and Rosen] provides metrics for assessing a design in terms of refurbishing and remanufacturing. In remanufacturing, a product is disassembled, some of its parts are replaced, and others are repaired, still others are simply cleaned and refinished. Among the factors that they identify for evaluating the recyclability of a product design are those given in Table 15-8.

15.E.4. Other Global Issues

Some companies develop their own DFA methodologies and in so doing are able to emphasize factors that particularly affect their operations. A good example is Denso, a company that must deal with high volatility in its production schedules and high variety in its products ([Whitney]). As we have seen, Denso deals with these challenges during the design process and executes its solutions during assembly.

Figure 15-23 shows how Denso approaches one aspect of DFA. Like other rational approaches, Denso's begins with a cost analysis. The cost shown in this figure is that

TABLE 15-8. Factors Entering Refurbishing Criteria

• Number of theoretically necessary parts	• Number of tests
• Disassembly time	• Testing time
• Assembly time	• Cleaning score
• Number of parts	• Number of refurbished parts
• Number of replaced parts	• Number of key parts

Source: [Harper and Rosen].

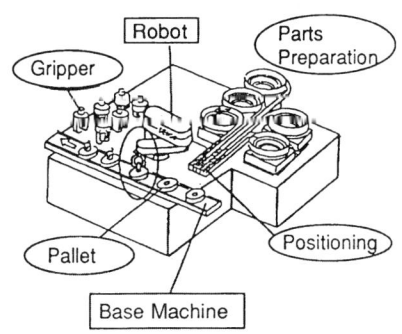

Flexible Assembly System

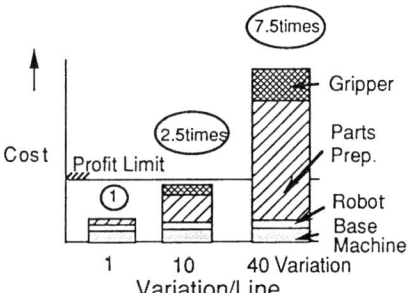

FIGURE 15-23. Cost Analysis of Assembly of High Variety Products. This figure shows that parts preparation (feeding and orienting) costs rise faster than other costs as the number of variants of a product rises. Denso's production is particularly highly influenced by the need to handle many variants. Thus its DFA methodology scores parts according to the ease with which feeders can switch from one version to another. (Courtesy of Denso Co., Ltd. Used by permission.)

(Ex.) High speed Coiling of Stator Core Adaptable to Three types

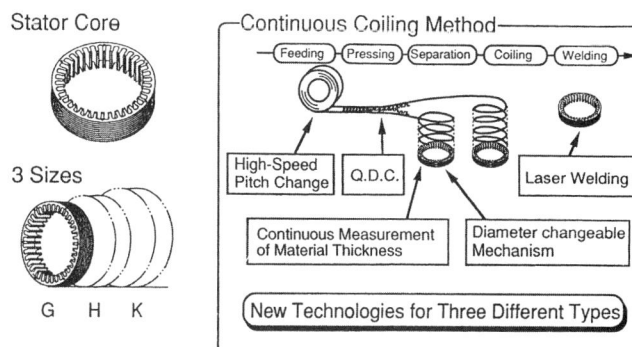

FIGURE 15-24. Denso Method of Making Alternator Stators in a Variety of Diameters and Lengths. Denso makes alternators in three different sizes. Both diameter and stack height can be varied easily. Instead of stamping flat stator core plies and stacking them up, Denso winds them helically from straight stock. Fewer fixtures are needed, changeover is fast, and much less scrap is generated. This is one of several process innovations used by Denso to permit flexible manufacture and assembly of many varieties of alternator on one set of machines in response to orders from Toyota. (QDC means quick die change.) Compare this method of dealing with different power levels with that of Black & Decker discussed in Chapter 14. Here, Denso varies both diameter and stack length whereas Black & Decker varies only stack length. (Diagram courtesy of Denso. Used by permission.)

of a robot assembly station, presented as a function of how many variants of the product this station will have to accommodate. The figure shows that the cost of parts presentation grows faster than any other cost component, and in this example it is unprofitable to deal with more than

10 variants. Denso therefore includes in its DFA evaluation the cost of switching from one version of a part to another. This cost might be reduced by redesigning the part or by redesigning the product as a whole. An example of the latter is shown in Figure 15-24.

15.F. EXAMPLE DFA ANALYSIS[9]

This product is a staple gun made by the Powershot Tool Company, Florham Park, NJ, and sold under the Sears and Powershot names. See Figure 15-25. This is a heavy-duty product with rugged and finely made stamped metal operating parts and well finished plastic exterior parts. It retails for $29.95 and is made in the United States at an estimated annual production volume of 500,000. It is assembled manually. Figure 15-26 shows an exploded view of the staple gun while Table 15-9 is a parts list.

The Boothroyd design for assembly (DFA) method was selected for the analysis of the staple gun.

Given its design and several difficult assembly operations, the appropriate assembly method for the current

staple gun design is manual assembly. The analysis below assumes a series of assembly stations, simple transfer lines and simple assembly fixtures. Manual assembly includes the gross motions of part selection and the fine motions of part insertion or positioning. Parts are classified using the terms alpha and beta to establish the end-to-end and rotational symmetry. Parts are evaluated for the ease of handling relative to jamming, tangling, size, flexibility, and slipperiness/sharpness. These parameters are used to classify the part handing and fastening type. Each classification has an associated time with penalties added for difficulty. The assembly labor costs can then be determined by using the standard assembly hourly rate. Each of these analyses is described below.

15.F.1. Part Symmetry Classification

Parts are classified by alpha and beta symmetry. Alpha is the angle through which a part must be rotated about

[9]The material in this section, except the analysis of the low-cost staple gun, was prepared by MIT students Ben Arellano, Dawn Robison, Kris Seluga, Tom Speller, and Hai Truong, and Technical University of Munich student Stefan von Praun. They used a previous version of Boothroyd Dewhurst software and code numbers that do not align completely with those in Figure 15-1 and Figure 15-2.

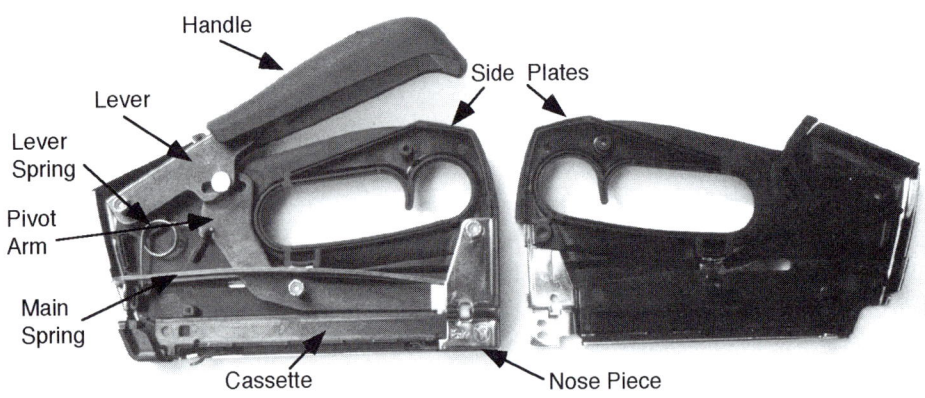

FIGURE 15-25. Photo of Powershot Staple Gun. (Photo by the author.)

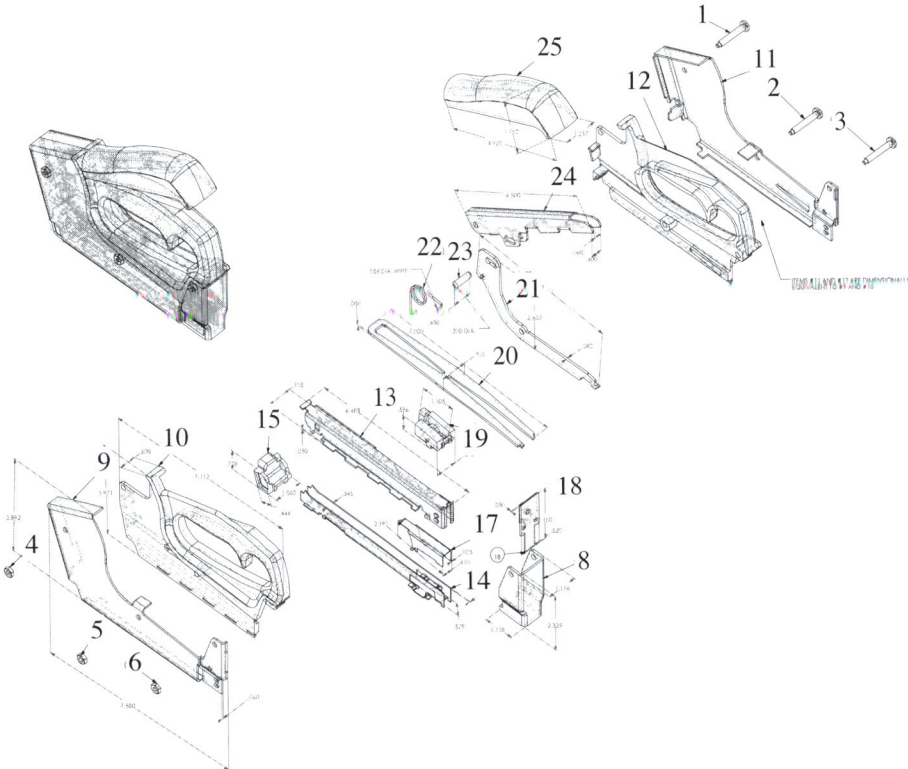

FIGURE 15-26. Staple Gun Exploded View.

an axis perpendicular to the insertion axis to orient it correctly for insertion. Beta is on the angle through which a part must be rotated about the axis of insertion to orient it correctly for insertion. Table 15-10 shows the symmetry categorization of each part of the staple gun. The sum of alpha and beta determines the effect of symmetry on orientation time.

15.F.2. Gross Motions

Gross motions can be defined as the selection and handling of the part to the assembly fixture. They can be performed quickly and do not require accuracy.

Table 15-11 shows the type of gross motions associated with each part assembly step in the staple gun.

TABLE 15-9. Parts List for the Staple Gun

Description	Part Number
Shoulder bolt-rear	1
Shoulder bolt-center	2
Shoulder bolt-front	3
Nylon lock nut-rear	4
Nylon lock nut-center	5
Nylon locknut-front	6
Self-tapping screw	7
Nose piece	8
Right side plate (metal)	9
Right handle body (plastic)	10
Left side plate (metal)	11
Left handle body (plastic)	12
Cassette	13
Staple guide	14
Staple guide handle	15
Staple advance spring	16
Staple advance bracket	17
Anvil	18
Anvil guide (plastic)	19
Main spring	20
Pivot arm	21
Lever spring	22
Dowel pin	23
Lever (metal)	24
Lever handle (plastic)	25
Self-tapping screws (2)	26
Staples	27

TABLE 15-10. Alpha and Beta for Staple Gun Parts

Description	Part	Alpha	Beta
Shoulder bolt-rear	1	360	0
Shoulder bolt-center	2	360	0
Shoulder bolt-front	3	360	0
Nylon lock nut-rear	4	180	0
Nylon lock nut-center	5	180	0
Nylon locknut-front	6	180	0
Self-tapping screw	7	360	360
Nose piece	8	360	360
Right side plate (metal)	9	360	360
Right handle body (plastic)	10	360	360
Left side plate (metal)	11	360	360
Left handle body (plastic)	12	360	360
Cassette	13	360	360
Staple guide	14	360	360
Staple guide handle	15	360	360
Staple advance spring	16	360	360
Staple advance bracket	17	360	360
Anvil	18	360	360
Anvil guide (plastic)	19	360	360
Main spring	20	360	360
Pivot arm	21	360	360
Lever spring	22	360	360
Dowel pin	23	180	0
Lever (metal)	24	360	360
Lever handle (plastic)	25	360	360
Self-tapping screws (2)	26	360	0
Staples	27	90	180

Note: Alpha and Beta are summed to determine the total reorientation restrictions on each part for the purpose of coding handling difficulty.

15.F.3. Fine Motions

Fine motions can be defined as the final orientation and placement of the parts. They need a high level of accuracy and are likely to be much slower than gross motions. Fine motions are small compared to the part and are typically a series of controlled contacts with closed feedback loops for reorientation.

Table 15-12 shows the type of fine assembly motions associated with each part assembly step in the staple gun.

15.F.4. Gripping Features

In the staple gun, all gripping of parts is done by hand (manually) and in a location that is perpendicular to the axis of insertion.

15.F.5. Classification of Fasteners

There are two types of fasteners: (1) self-tapping screw and (2) shoulder bolt with nylon lock nuts. There are three self-tapping screws. In addition, there are two shoulder bolts of the same length and one that is slightly longer.

15.F.6. Chamfers and Lead-ins

The only chamfers are located on the dowel pin. The cassette has lead-ins on the keyway tabs to help manual positioning. The right- and left-hand plates have a type of chamfer that helps to guide the plastic guide handle locator. It is surprising that the shoulder bolts do not have chamfers to help locate them during the assembly process. This should be considered as an assembleability design improvement.

15.F.7. Fixture and Mating Features to Fixture

The staple gun can be categorized as a Type 1 assembly. However, fixturing is recommended to aid in preloading of the main spring and to fix firmly the unsecured parts during the assembly process. Otherwise the

TABLE 15-11. Codes for Manual Handling Gross Motions for the Staple Gun

Code	First Digit: Symmetry	Second Digit: Difficulty
00	Each part beginning with 0 is of nominal size and weight, can be grasped without tools, and can be maneuvered with one hand. Part symmetry is < 360°	Easy to grasp with thickness > 2 mm and size > 15 mm
04		Easy to grasp with thickness < 2 mm and size < 6 mm
08		Difficult to grasp with thickness < 2 mm and size > 6 mm
09		Difficult to grasp with thickness < 2 mm and size < 6 mm
30	Each part beginning with 3 is of nominal size and weight, can be grasped without tools, and can be maneuvered with one hand. Part symmetry = 720°	Easy to grasp with thickness > 2 mm and size > 15 mm
36		Difficult to grasp with thickness > 2 mm and 2 mm < size < 15 mm
38		Difficult to grasp with thickness < 2 mm and size > 6 mm
80	Parts severely nest or tangle but can be grasped with one hand	No additional grasping difficulties: $\alpha < 180$, size > 15 mm
88		Additional handling difficulties: $\alpha = 360°$, size > 6 mm

Note: The codes in this table correspond to an earlier version of the Boothroyd method and thus do not match the codes in Figure 15-1. Nevertheless, the handling times are similar.

TABLE 15-12. Codes for Manual Assembly Fine Motions for the Staple Gun

Code	First Digit	Second Digit
00	Part is added but not secured immediately and is easily maneuvered into position	No holding required, easy to align, no resistance to insertion
02		No holding required, easy to align, no resistance to insertion
08		Holding required, not easy to align, no resistance to insertion
12	Part is added but part can't easily reach desired location	No holding required, not easy to align, no resistance to insertion
30	Part is secured immediately, can reach desired location easily and tool can be operated easily	No screw operation, easy to align, no resistance
38		Screw tightening, easy to align, no torsional resistance
39		Screw tightening, not easy to align, resistance
40	Part secured immediately, location cannot be reached easily due to obstruction or blocked view	No screw operation, easy to align, no resistance
44		Plastic deformation, not easy to align, resistance
98	Separate operation after parts are in place	Nonfastening process (manipulation of parts, grease)

Note: The codes in this table correspond to an earlier version of the Boothroyd method and thus do not match the codes in Figure 15-2. Nevertheless, the assembly times are similar.

assembly tends to spring apart before it can be completed. Figure 15-27 is a CAD drawing of the staple gun in the proposed assembly fixture, highlighting the locating pin configuration.

In Figure 15-27, part number 11, the left side plate, locates the assembly to the plane of the fixture. The pins are designed to prevent the parts from moving and provide alignment during assembly. The locator pins on the bottom and right hand side have clearance so as not to over constrain the assembly in the fixture. The top four pins provide a resisting force against the force required to preload the main spring, while the left-hand side pins resist the force required to attach the nose piece. The fixture also contains fixed Philips head screwdriver tips to hold shoulder bolts 1 and 2 while the nuts 4 and 5 are tightened.

A clamp is required to hold the subassembly down during the assembly of the subsequent parts and the mainspring loading process. The clamp is shown in Figure 15-28. A test was conducted to confirm that this clamp will secure subassembly 1 during the loading of the spring and the attachment of the nose piece.

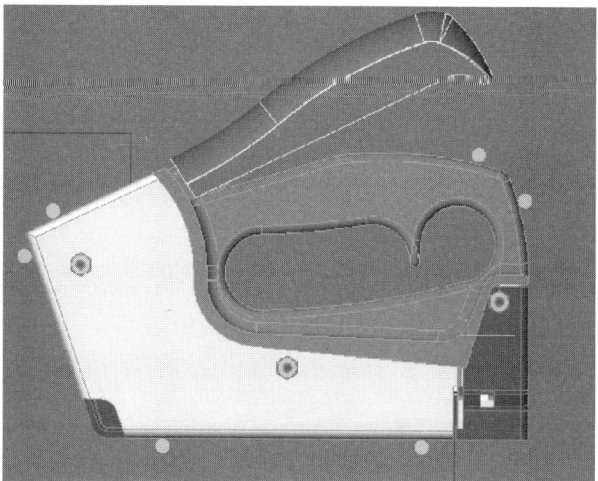

FIGURE 15-27. Method of Fixturing the Assembly.

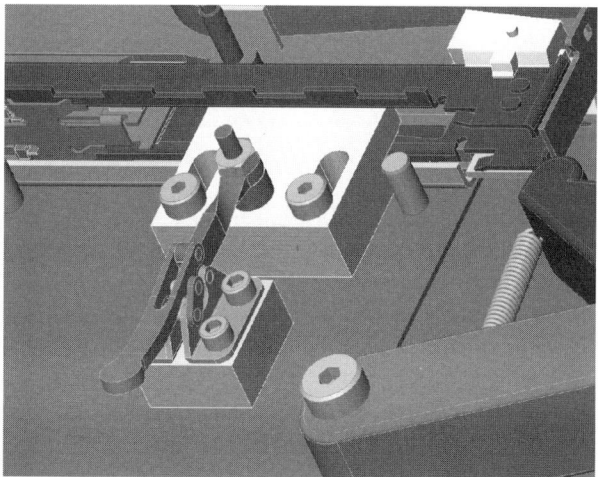

FIGURE 15-28. Clamp for Holding the Assembly in the Fixture in Figure 15-27.

15.F.8. Assembly Aids in Fixture

Figure 15-29 shows two assembly aids. The left lever arm closes a clamp that aids in the preloading of the main spring. The right lever locates the nose piece to the assembly.

15.F.9. Auxiliary Operations

Two auxiliary operations are required. They are greasing and the quality control check.

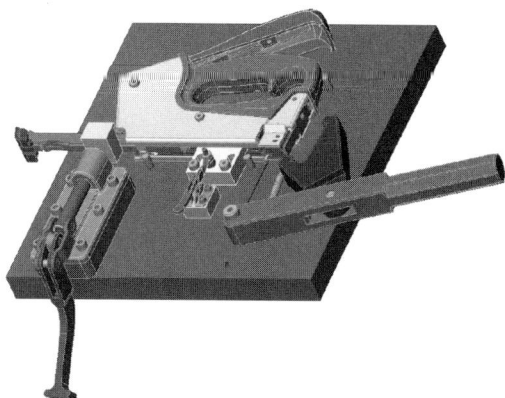

FIGURE 15-29. Assembly Aids Attached to the Fixture in Figure 15-27.

15.F.10. Assembly Choreography

The assembly choreography is:

- Create subassembly 1. Parts 11 and 12 are joined together to form subassembly 1.
- Shoulder bolts 1 and 2 are located in the holes of subassembly 1.
- These four parts are located in fixture 1. The shoulder bolts must be rotated until they seat in the fixed Philips head locators in the fixture.
- Grease is added to the area where the anvil slides along subassembly 1.
- Subassembly 4, the staple gun delivery subassembly, is created using Parts 13–19 and joining them using the following sequence: Part 14 to 16, Part 17, Part 15, Part 13, Part 18, Part 19.
- Install subassembly 4 and leave it in the open position.
- The clamp in Figure 15-28 is closed to secure subassembly 4 to the base parts and fixture.
- The pivot arm 21 and the mainspring 20 are brought together simultaneously. Each part is gripped in one hand and the pivot arm is inserted in the slot contained in the mainspring. The pivot arm is then placed on shoulder bolt 2 while the main spring is mated to the rectangular holes in the anvil 18.
- Another clamp aids the assembly worker in preloading the mainspring and putting it in its final position.

- The lever spring 22 is then preloaded and mated to the handle body through the main spring and attaches to the slot located in the lever arm. This operation is difficult as it requires two hands and is obstructed.
- Create subassembly 2—the staple gun handle assembly. Parts 9 and 10 are joined together to form this subassembly. This subassembly is then located on shoulder bolt 1.
- The dowel pin 23 is inserted through the handle subassembly and the pivot arm into the handle body.
- Create subassembly 3 by joining parts 9 and 10.

- Grease is applied to subassembly 3 and then it is added to the other parts.
- The rear and center nylon nuts, parts 4 and 5 are installed and tightened to secure subassemblies 1 and 3.
- Part 7, the self tapping screw, is threaded into the assembly.
- The nose piece 8 is attached to another assembly aid, which is then used to push the nose piece against the assembly.
- All clamps are opened and the staple gun is removed.

TABLE 15-13. Time and Cost Analysis of Staple Gun Assembly

Task	Number of Items	Manual Handling Code	Handling Time per Item	Manual Insertion Code	Insertion Time per Item	Total Operation Time	Total Operating Cost
Create subassembly 1—Parts 11 and 12	2	30	1.95	30	2.00	7.90	$0.032
Create subassembly 2—Parts 24, 26, and 25	3	04	2.18	38	6.50	26.04	$0.105
Create subassembly 3—Parts 9 and 10	2	30	1.95	30	2.00	7.90	$0.032
Create subassembly 4—Parts 13–19							
Part 14 to 16	2	36	3.06	40	4.50	15.12	$0.061
Part 14 and 16 to 17	1	30	1.95	30	2.00	3.95	$0.016
Parts 14, 16, and 17 to 15	1	30	1.95	30	2.00	3.95	$0.016
Parts 14, 15, 16, and 17 to 13	1	30	1.95	30	2.00	3.95	$0.016
Parts 13, 14, 15, 16, and 17 to 18	1	33	2.51	00	1.50	4.01	$0.016
Parts 13, 14, 15, 16, 17, and 18 to 19	1	30	1.95	30	2.00	3.95	$0.016
Final assembly							
Subassembly 1 to part 1	2	00	1.13	00	1.50	5.26	$0.021
Subassembly 1 to part 2	1	00	1.13	00	1.50	2.63	$0.011
Put in fixture 1	1	00	0.00	98	9.00	9.00	$0.036
Grease	1	00	0.00	98	9.00	9.00	$0.036
Add subassembly 4	1	80	4.10	08	6.50	10.60	$0.043
Put in fixture 2	1	00	0.00	98	9.00	9.00	$0.036
Add parts 20 and 21	2	88	6.35	12	5.00	22.70	$0.092
Put in fixture 3	1	00	0.00	98	9.00	9.00	$0.036
Add part 22	1	38	3.34	44	8.50	11.84	$0.048
Add subassembly 2	1	30	1.95	12	5.00	6.95	$0.028
Add part 23	1	08	2.45	12	5.00	7.45	$0.030
Grease	1	00	0.00	98	9.00	9.00	$0.036
Add subassembly 3	1	30	1.95	02	2.50	4.45	$0.018
Add part 4	1	09	2.98	39	8.00	10.98	$0.044
Add part 5	1	09	2.98	39	8.00	10.98	$0.044
Add part 7	1	09	2.98	39	8.00	10.98	$0.044
Add part 8	1	30	1.95	00	1.50	3.45	$0.014
Put in fixture 4	1	00	0.00	98	9.00	9.00	$0.036
Add part 3	1	80	4.10	00	1.50	5.60	$0.023
Add part 6	1	09	2.98	39	8.00	10.98	$0.044
Add part 27	1	00	1.13	00	1.50	2.63	$0.011
Quality check	1	00	0.00	98	9.00	9.00	$0.036
Total time in seconds						267.25	

		Min	Cost	Cost/hr
Total time in minutes and cost in $		4.4542	$1.083	$14.58

Note: The codes in this table correspond to an earlier version of the Boothroyd method and thus do not match the codes in Figure 15-1 and Figure 15-2. Nevertheless, the handling and assembly times are similar.

- The last (longer) shoulder bolt 3 and nylon nut 6 are attached to the gun.
- Staples 27 are added to the gun and the final quality check is performed.

15.F.11. Assembly Time Estimation

Table 15-13 shows the evaluation of the staple gun using the Boothroyd DFA method. It includes all the handling and insertion tasks, their codes, estimated assembly times, and estimated assembly costs.

15.F.12. Assembly Time Comparison

The team performed the assembly sequence 10 times with the resulting average assembly time of 4.012 minutes, as shown in Table 15-14. This appears to be reasonable and consistent with the times that were calculated using the Boothroyd method.

15.F.13. Assembly Efficiency Analysis

Table 15-15 shows the assembly efficiency analysis for the staple gun as is. This analysis assumes that parts have been eliminated rigorously according to the Boothroyd method, with the understanding that all such removals will be reviewed later for engineering acceptability. On this basis, the staple gun has an efficiency of nearly 17%. This is well above typical values cited in [Boothroyd, Dewhurst, and Knight], indicating that some DFA has probably already been done on this product.

Table 15-16 continues the analysis by calculating the impact of actually eliminating all or just some of the parts identified as candidates for removal. Two concepts are presented, one that risks performance and the other that does not. The first achieves efficiency of over 30% while the other achieves over 25%. The latter figure is typical of what is to be expected following DFA analyses, according to [Boothroyd, Dewhurst, and Knight].

15.F.14. Design Improvements for the Staple Gun Design for Assembly

The ideal assembly plan brings all parts down vertically, but the insertion of the shoulder bolts requires an operation on one side of the gun and then threaded nuts must be torqued on from the other side. Assembly would be easier if at least one of these bolts were replaced by a feature on

TABLE 15-14. Data on Manual Assembly Times

Actual Assembly Time (minutes)	Average Assembly Time	DFA Calculated Operation Time
4.21	4.012	4.45
4.43		
4.27		
3.83		
3.93		
4.11		
3.8		
4.1		
3.62		
3.82		

TABLE 15-15. Assembly Efficiency Analysis for Staple Gun

Number of liaisons	57
Number of parts	27
Ratio	2.11
Number of fasteners	8
Number of other joinable parts	8 (11, 12, 18, 20, 21, 23, 24, 25)
Total "unnecessary" parts	12 (all fasteners plus half the others; note functional risk in eliminating some of them)
Total assembly time	267.25
Minimum number of parts	15
Theoretical assembly time	45
Assembly efficiency	16.84%

Note: This analysis rigorously applies the Boothroyd analysis and eliminates all fasteners plus a few other parts with the understanding that this could create some risk to proper function.

TABLE 15-16. Assembly Efficiency Analysis Based on Two Part Consolidation Concepts

Total necessary parts for reliable function	19
Theoretical assembly time	57
New total time	225.41
New assembly efficiency	25.29%
Total necessary parts with functional risk	15
New total assembly time	148
Theoretical assembly time	45
New assembly efficiency	30.41%

Note: If we eliminate all the parts that we conceivably could, taking some risk with function, we obtain a part count of fifteen and an efficiency of over 30%. If we are more conservative and only eliminate some of the candidates, we end up with nineteen parts and an efficiency of over 25%.

part 12 that performed the function of holding the handle and/or the main spring. Separate fasteners could be used to hold the clamshell style sides together. This would add parts but would make assembly go much faster.

The current design does not permit, at least not easily, automatic locating of the pivot arm in the main spring,

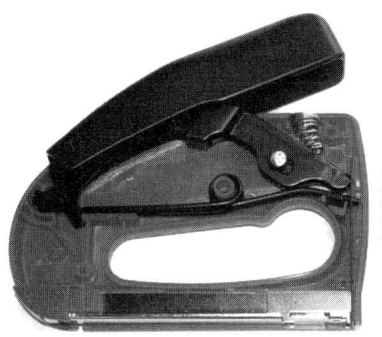
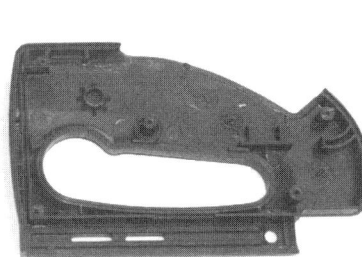

FIGURE 15-30. Photo of a Lower-Cost Staple Gun Made by the Same Manufacturer as the One Shown in Figure 15-25. This staple gun exhibits several of the DFA suggestions made above. It has 18 parts compared with 27 for the rugged staple gun, including staples. (Photo by the author.)

loading of the main spring, or locating of the coil spring in the two small pin holes in a preloaded condition. Because of the necessity and difficulty of preloading the main spring and to a lesser extent the coiled handle spring, the assembly becomes unstable. To stabilize and constrain the assembly during the spring preloading operation, a fixture clamp must be used resist the rotational and translational movements in the nose of the gun until the final side plate and nosepiece are attached, fully constraining the final assembly.

Using the three shoulder bolts, the current design must be made by hand with the assistance of a fixture base to stabilize the assembly during the initial locating operations and then another fixture to resist the twist in the unconstrained subassembly prior to preloading the main spring. Furthermore, once the nose clamp is assembled, the remaining assembly work must continue at a single station, creating a bottleneck.

Shoulder bolts 1, 2, 3 should be the same length for DFA but they are not—one is longer. The three shoulder bolts can be commonized to minimize the complexity and cost of storage.

15.F.15. Lower-Cost Staple Gun

The Powershot's manufacturer also makes a lower-cost version that sells for $14.99. A photo of it appears in Figure 15-30. Interestingly, it exhibits several of the DFA suggestions listed in the previous sections. The suggestions were made without knowledge of this lower-cost version.

Comparing Figure 15-30 and Figure 15-25, we can see that the latter is much more rugged. It is bigger and drives bigger staples. It has 50% more parts, as indicated in Table 15-17. These parts are bigger and thicker or are made of stronger materials. Whenever one considers part count reduction or material substitution, one has to be careful

TABLE 15-17. Comparison of Parts in Low-Cost Staple Gun and Rugged Staple Gun

Parts in Low-Cost Staple Gun	Corresponding Parts in Rugged Staple Gun[a]
Five screws	1 through 3
Left side	11 and 12
Right side	9 and 10
Staple carrier	14
Staple carrier handle	15
Staple pusher	17
Pusher spring	16
Handle and lever	24 and 25
Anvil	18
Main spring	20
Pivot arm	21
Lever spring	22
Dowel pin	23
Staples	27

[a]Rugged staple gun parts eliminated completely: 4–7, 8, 13, 19, 26. Rugged staple gun parts combined with others: 9, 11.

not to compromise performance. The low-cost staple gun is obviously less able to withstand the shock of driving staples and is likely to wear out faster or break after driving fewer staples than the rugged version.

A DFA analysis was conducted on the low-cost staple gun.[10] The results are in Table 15-18 and Table 15-19. The low-cost item takes less than half the time to assemble as the rugged version, and its assembly efficiency is a respectable 31%. This result confirms the design change suggestions made above for the rugged stapler and indicates that little further improvement in efficiency can be achieved. However, there are several difficult assembly steps that could be made to go faster with some design changes. These include finding a way to retain the cock

---

[10]This analysis was performed by the author. He is responsible for any discrepancies relative to the analysis of the rugged staple gun.

TABLE 15-18. DFA Analysis of Low-Cost Staple Gun

Task	$\alpha + \beta$	Number of Items	Manual Handling Code	Handling Time per Item	Manual Insertion Code	Insertion Time per Item	Total Operation Time	Cost	Part Can Be Eliminated?
Create subassembly 1— parts 3, 4, 5, 6	720	1	30	1.95	00	1.5	3.45	$0.014	
Part 3 to part 4	720	1	30	1.95	00	1.5	3.45	$0.014	
Add part 6	180	1	33	2.51	21	6.5	9.01	$0.036	4
Add part 5	720	1	30	1.95	00	1.5	3.45	$0.014	
Put left side 1 into fixture	720	1	30	1.95	00	1.5	3.45	$0.014	
Insert anvil 8	540	1	24	2.85	00	1.5	4.35	$0.018	
Insert main spring 9	720	1	30	1.95	34	6	7.95	$0.032	
Insert pivot 12	180	1	03	1.88	00	1.5	3.38	$0.014	
Insert cock lever 10	720	1	30	1.95	06	5.5	7.45	$0.030	
Insert handle 7	720	1	30	1.95	03	3.5	5.45	$0.022	
Insert cock level spring 11	360	1	10	1.50	19	10	11.50	$0.047	11(?)
Insert subassembly 1	720	1	30	1.95	06	5.5	7.45	$0.030	
Install right side 2	720	1	30	1.95	19	10	11.95	$0.048	
Insert screw 13	360	5	00	1.13	92	5	30.65	$0.124	13
Insert staples	360	1	00	1.13	00	1.5	2.63	$0.011	
		Parts					Time	Cost	Parts eliminated
Totals		19					115.57	$0.468	7

lever spring, which easily flies out, plus guide chamfers that permit the right side to mate with the anvil and the pivot more easily. A snap fit assembly might eliminate the screws, but this would have to be tested carefully in view of the possibility that the shock exerted by the stapling action might dislodge the snaps.

TABLE 15-19. Assembly Efficiency Analysis of Low-Cost Staple Gun

Theoretically necessary parts	12
Theoretical assembly time	36
Actual assembly time	115.57
Efficiency	31%

15.G. DFx'S PLACE IN PRODUCT DESIGN

DFx in the small is reasonably easy to separate from other design processes, but DFx in the large is hard to separate from product architecture and product design overall. Figure 15-31 attempts to compare these different topics and to lay them out in approximate temporal order (with the understanding that there is usually a lot of iteration among them as a product is being designed). This figure indicates that there is a phase when it is necessary to add parts in order to achieve technical and business goals, followed by various attempts to reduce the number of parts while balancing several competing goals.

Ironically, as the product enters production, there is a tendency for the number of parts, or the number of varieties of some of the parts, to increase again. Several forces are at work. These include evolution of markets and

customer preferences, the tendency of individual product development teams to make their product unique, and unawareness by top management of the loss of production efficiency caused by increasing the number of options and varieties of a product. These forces often act very slowly and their effects accumulate before anyone notices.[11]

---

[11]This phenomenon is especially evident in automobile engine plants. Engine designs and plants are typically used for a decade or more, but different engine displacements and varieties are added every year. This increases the frequency with which tools and fixtures must be changed, reducing the amount of time available for production. One engine plant manager told the author, "We cook slowly like the frog."

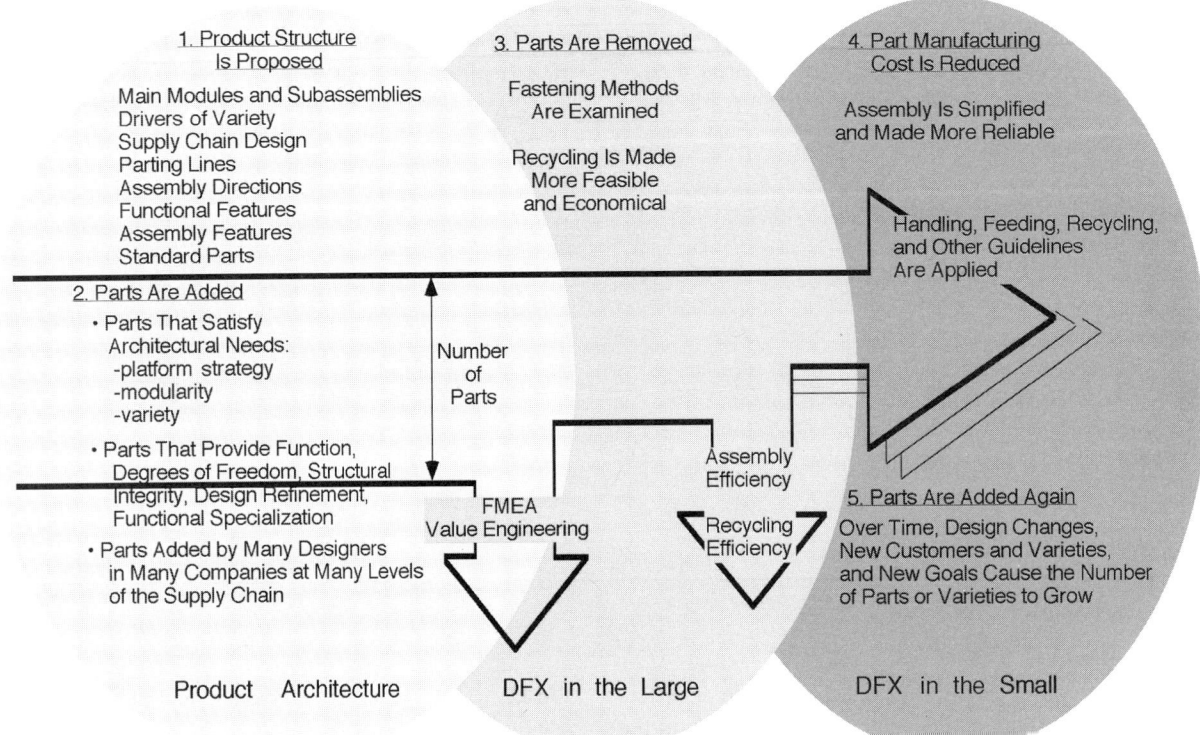

FIGURE 15-31. Relationship Between Product Architecture, DFx in the Large, and DFx in the Small. When product architecture is defined, a structure for the product is proposed and parts are added through a variety of mechanisms and for a variety of reasons. Value engineering and DFx in the large tend to reduce the number of parts, while DFx in the small seeks to lower their cost and make their assembly and eventual disassembly more economical. However, as time goes on during the product's life, various forces tend to increase the number of parts or the number of varieties of some of them.

15.H. CHAPTER SUMMARY

Over the past thirty years, DFx has become a more important and sophisticated element of product development, beginning with DFA in the 1970s and extending into DFM and DfRRD in the 1980s and 1990s. When products were simpler, when designers had more manufacturing knowledge, and when a larger fraction of the product was made by the company that designed it, it was probably easier to include DFx directly in the design process. Today, products are very complex and comprise many different technologies, materials, and processes. Many specialist companies cooperate to design and make them. Under these conditions, it is very difficult to know when a product's design is effective from the point of view of all the x's. Many of these constituencies want opposing things, and everyone wants low cost, high performance, and high quality.

When DFA was first proposed, most products were made of machined parts fastened together with screws. They were ripe for part count reduction and increased assembly efficiency. Today, designers can exploit advanced materials and forming processes to create increasingly intricate, complex, and beautiful parts comprising hundreds or even thousands of features. Even very unassuming consumer products like food mixers and staplers exhibit sophisticated design, careful choice of materials, and relatively high assembly efficiencies. However, industrial products subject to exacting performance requirements and large loads need design refinement more than do consumer products. This obstructs efforts to consolidate their parts and reduces DFA opportunities.

Assembly was the process that got people thinking about all these issues, but assembly cannot be the only or even the dominant criterion. Assembly is too small an element of product cost and production time. Architecture trumps all the x's, as indicated in Figure 15-31, since it seeks the broadest view of the contending issues. But only assembly has the integrative power to force these issues into the open where they can be debated and resolved.

15.I. PROBLEMS AND THOUGHT QUESTIONS

1. Take apart a product like the desktop stapler, a small food mixer, a can opener, a shaver, or a drill. Classify all of the parts into one of the following categories: main function carriers, parts that support the main function carriers, fasteners, and others.

2. Return to the product you considered in Problem 1. Take one of the main function parts and determine what it does and why it is in the product. Then ask yourself what other materials it might be made out of, what alternate processes might be used, and how these choices would affect the ability of the part to function as needed, to be fastened to its neighbors, and to be separated from them when the product is recycled. Comment on how well it would function if consolidated with some of its neighbors or if it was separated into two or more different parts of possibly different materials.

3. Do a complete DFA analysis on the product in Problem 1. You may calculate the handling and assembly times based on a classification and coding process or you may simply assemble the product several times and average the durations. Using the three criteria in Section 15.E.1.b, determine which parts are theoretically necessary. Calculate the minimum assembly time using 3 seconds per theoretically necessary part and determine the assembly efficiency. Do you think substantial improvements in the assembly efficiency are possible? If so, how?

4. A low-cost electric drill has a plastic injection-molded case and a clamshell architecture. The bearings are oil-impregnated brass. The reduction gears run in bearing blocks that are set directly in the plastic case. A higher-cost drill has a stack architecture with a metal nosepiece that contains all the bearing blocks for the gears. Long tie bolts parallel to the drill bit axis hold the assembly together. Ball bearings are used everywhere and the motor brushes are replaceable. Explain these differences in terms of function, durability, material cost, and assembly efficiency.

5. Researchers interested in DfRRD and integrality/modularity developed an algorithm that will recommend material changes in an assembly. The algorithm balances disassembly cost and sorting cost. The researchers applied the algorithm to the console of an automobile. This assembly, illustrated in Figure 15-32, sits on the floor of the car between the driver and the passenger, and contains the transmission shift lever, handbrake lever, ashtray, cup holder, electric outlet, and other convenience items.

The liaison diagram of this assembly has a hub and spokes structure. The hub piece embodies the large hump shape typical of these assemblies and is made of a plastic called ABS. Speculate on why the algorithm recommends that almost all the other parts be made of ABS as well ([Coulter, McIntosh, Bras, and Rosen]).

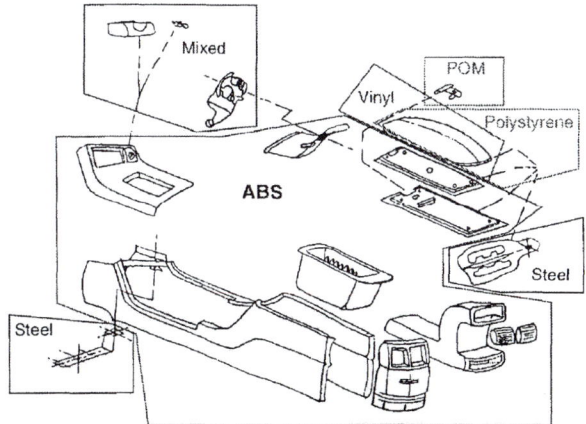

FIGURE 15-32. Figure for Problem 5. (Copyright © 1998 ASME. Used by permission.)

15.J. FURTHER READING

[Andreasen et al.] Andreasen, M. M., Kähler, S., and Lund, T., *Design for Assembly,* London: IFS Publications, 1983.

[Beitler, Cheldelin, and Ishii] Beitler, K. A., Cheldelin, B., and Ishii, K., "Assembly Quality Method: A Tool In Aid of Product Strategy, Design, and Process Improvements," Proceedings of the 2000 ASME Design for Manufacturing Conference, Baltimore, MD, paper no. DETC2000/DFM-14020, September 2000.

[Boothroyd] Boothroyd, G., *Assembly Automation and Product Design,* New York: Marcel Dekker, 1992.

[Boothroyd, Dewhurst, and Knight] Boothroyd, G., Dewhurst, P., and Knight, W., *Product Design and Manufacture for Assembly,* New York: Marcel Dekker, 1994.

[Boothroyd, Poli, and Murch] Boothroyd, G., Poli, C., and Murch, L. E., *Automatic Assembly,* New York: Marcel Dekker, 1982. Portions of this book repeat contents of *Mechanized Assembly,* by G. Boothroyd and A. H. Redford, which is out of print.

[Boothroyd and Redford] Boothroyd, G., and Redford, A. H., *Mechanized Assembly,* London: McGraw-Hill, 1968.

[Chung] Chung, E., "History of Design for Assembly," B. S. thesis, MIT Mechanical Engineering Department, May 2001.

[Coulter, McIntosh, Bras, and Rosen] Coulter, S. L., McIntosh, M. W., Bras, B., and Rosen, D. W., "Identification of Limiting Factors for Improving Design Modularity," Proceedings of the 1998 ASME Design Engineering Technical Conferences, Atlanta, GA, paper no. DETC98/DTM-5659, September 1998.

[Esawi and Ashby] Esawi, A. M. K., and Ashby, M. F., "Cost Estimation for Process Selection," Proceedings of the 1999 ASME Design for Manufacture Conference, Las Vegas, NV, paper no. DETC99/DFM-8919, September 1999.

[Fagade and Kazmer] Fagade, A. A., and Kazmer, D. "Optimal Component Consolidation in Molded Product Design," Proceedings of the 1999 ASME Design Engineering Technical Conferences, Las Vegas, NV, paper no. DETC99/DFM-8921, September 1999.

[Gupta et al.] Gupta, R., Sheridan, T., and Whitney, D., "Experiments on Using Multimodal Virtual Environments for Design for Assembly Analysis," *PRESENCE: Teleoperators and Virtual Environments,* vol. 6, no. 3, pp. 318–338, 1997.

[Harper and Rosen] Harper, B., and Rosen, D. W., "Computer-Aided Design for Product De- & Remanufacture," Proceedings of the 1998 ASME Design Engineering Technical Conferences, Atlanta, GA, paper no. DETC98/CIE-5695, September 1998.

[Hayes-Roth, Waterman, and Lenat] Hayes-Roth, F., Waterman, D. A., and Lenat, D. B., editors, *Building Expert Systems,* London: Addison-Wesley, 1983.

[Hinckley] Hinckley, M., "A Global Conformance Quality Model—A New Strategic Tool for Minimizing Defects Caused by Variation, Error, and Complexity," Ph.D. dissertation, Stanford University, 1993.

[Hounshell] Hounshell, D., *From the American System to Mass Production, 1800–1932,* Baltimore: Johns Hopkins University Press, 1984.

[Hu and Poli] Hu, W., and Poli, C., "To Injection Mold, to Stamp, or to Aassemble? A DFM Cost Perspective," *ASME Journal of Mechanical Design,* vol. 121, pp. 461–471, 1999.

[Jürgens] Jürgens, U., "Rolling Back Cycle Times: The Renaissance of the Classic Assembly Line," in *Transforming Auto-* *mobile Assembly,* Shimokawa, K., Jürgens, U., and Fujimoto, T., editors, Berlin: Springer, pp. 258–259, 1997.

[Kanai, Sasaki, and Kishinami] Kanai, S., Sasaki, R., and Kishinami, T., "Representation of Product and Processes for Disassembly, Shredding, and Material Sorting Based on Graphs," Proceedings of the IEEE International Symposium on Assembly and Task Planning, Porto, 1999.

[Kondoleon] Kondoleon, A. S., "Application of Technology-Economic Model of Assembly Techniques to Programmable Assembly Machine Configuration," MIT Mechanical Engineering thesis, May 1976.

[Kroll and Hanft] Kroll, E., and Hanft, T. A., "Quantitative Evaluation of Product Disassembly for Recycling," *Research in Engineering Design,* vol. 10, pp. 1–14, 1998.

[Miyakawa and Ohashi] Miyakawa, S., and Ohashi, T., "The Hitachi Assemblability Evaluation Method (AEM)," Proceedings of the 1st International Conference on Product Design for Assembly, Newport, RI, April 1986.

[Miyakawa, Ohashi, Inoshita, and Shigemura] Miyakawa, T., Ohashi, T., Inoshita, S., and Shigemura, T., "The Hitachi Producibility Evaluation Method (PEM)" (in Japanese), *Kikai-sekkei,* vol. 33, no. 7, pp. 39–47, 1989.

[Miyakawa, Ohashi, and Iwata] Miyakawa, S., Ohashi, T., and Iwata, M. "The New Hitachi Assemblability Evaluation Method (AEM)," Proceedings of NAMRI, published by SME, 1990.

[Nevins and Whitney] Nevins, J. L., and Whitney, D. E., editors, *Concurrent Design of Products and Processes,* New York: McGraw-Hill, 1989.

[Niimi and Matsudaira] Niimi, A., and Matsudaira, Y., "Development of a New Vehicle Assembly Line at Toyota: Worker-Oriented, Autonomous, New Assembly System," in *Transforming Automobile Assembly,* Shimokawa, K., Jürgens, U., and Fujimoto, T., editors, Berlin: Springer, pp. 82–93, 1997.

[Ostwald and McLaren] Ostwald, P. F., and McLaren, T. S., *Cost Analysis and Estimating for Engineering and Management,* Upper Saddle River, NJ: Prentice-Hall, 2003.

[Otto and Wood] Otto, K., and Wood, K., *Product Design: Techniques in Reverse Engineering and New Product Development,* Upper Saddle River, NJ: Prentice-Hall, 2001.

[Overy] Overy, R., *Why the Allies Won,* Chapter 6, New York: W. W. Norton, 1995.

[Redford and Chal] Redford, A. H., and Chal, J., *Design for Assembly: Principles and Practice,* Maidenhead, Berks, UK: McGraw-Hill Europe, 1994.

[Sturges] Sturges, R. H., "A Quantification of Manual Dexterity: The Design for Assembly Calculator," *Robotics and Computer-Integrated Manufacturing,* vol. 6, no. 3, pp. 237–252, 1989.

[Suzuki, Ohashi, Asano, and Miyakawa] Suzuki, T., Ohashi, T., Asano, M., and Miyakawa, S., "Assembly Reliability Evaluation Method (AREM)" 4th IEEE International Symposium on Assembly and Task Planning, Fukuo ka, Japan, May 2001.

[Swanstrom and Hawke] Swanstrom, F. M., and Hawke, T., "Design for Manufacturing and Assembly: A Case Study in Cost Reduction for Composite Wing Tip Structures," 31st International SAMPE Technical Conference, Chicago, October 1999.

[Swift and Brown] Swift, K. G., and Brown, N. J., "DFM Experiences and Strategies for Implementation," International Forum on DFMA, Newport, RI, June 2001.

[Tong and Sriram] Tong, C., and Sriram, D., *Artificial Intelligence in Engineering Design,* three volumes, New York: Academic Press, 1992.

[Ulrich and Eppinger] Ulrich, K., and Eppinger, S., *Product Design and Development,* 2nd ed., New York: McGraw-Hill, 2000.

[Ulrich, Sartorius, Pearson, and Jakiela] Ulrich, K., Sartorius, D., Pearson, S., and Jakiela, M., "The Influence of Time in DFM Decision-Making," *Management Science,* vol. 39, no. 4, pp. 429–447, 1993.

[Whitney] Whitney. D. E., "Nippondenso Co. Ltd.: A Case Study in Strategic Product Design," *Research in Engineering Design,* vol. 5, pp. 1–20, 1993.

[Zandin] Zandin, K. B., *MOST Work Measurement Systems,* New York: Marcel Dekker, 1990.

16 ASSEMBLY SYSTEM DESIGN

"Stoppage? What stoppage?"

16.A. INTRODUCTION

This chapter deals with assembly system design.[1] It lays out the basic issues, the choices that system designers must make, and some approaches to making these choices systematically. The topics in this chapter are generic to most manufacturing systems, although the emphasis here is on assembly. Here we will discuss the basic decisions that need to be made, take a look at different kinds of assembly equipment and methods, describe two system design techniques, and study some examples from industry that cover a range of products and assembly methods. The next two chapters deal with workstation design and economic analysis.

Manufacturing system design is not a science, even though several aspects of it are supported by well-developed computer aids. There is still a great deal that is subject to expert judgment, arbitrary decisions, and lack of information about future conditions that the system may face. No design can cope with all future events and still retain adequate efficiency. No single technology can do all jobs, much less all jobs well.

For these reasons, our approach to system design emphasizes careful specification of the information needed for good design decisions. It also encourages the development of hybrid systems made up of suitable mixes of specialized or fixed automation, flexible automation, and people. Even though we present the topics in a particular sequence, it should be kept in mind that the actual process is highly iterative.

16.B. BASIC FACTORS IN SYSTEM DESIGN

Manufacturing system design can begin when a candidate product design is available along with the requirements for each process step and a candidate assembly sequence. It is vital not to wait until the product design is "done" because product and process comprise a coherent system that must be designed as a unit. All decisions are subject to change as the design effort proceeds. The process is illustrated schematically in Figure 16-1. It comprises these steps:

1. Analyze the product and the necessary fabrication and assembly operations. Determine alternate fabrication methods, fabrication and assembly sequences, and candidate subassemblies. Determine fabrication and assembly process requirements. Assess the maturity of these processes and estimate process yield. Identify flexibility requirements such as batch sizes and model mix. Identify problematic assembly steps and suggest product modifications.

2. Select an assembly sequence for use in assembly system design.

3. Determine the production capacity required of the system, taking into account factors like downtime, time to switch models, employee breaks, process yield and other factors that effectively reduce capacity.

4. Tabulate feasible fabrication and assembly techniques (equipment or people) for each operation and estimate the cost and time for each.

---

[1] This chapter is based in part on Chapter 10 of [Nevins and Whitney].

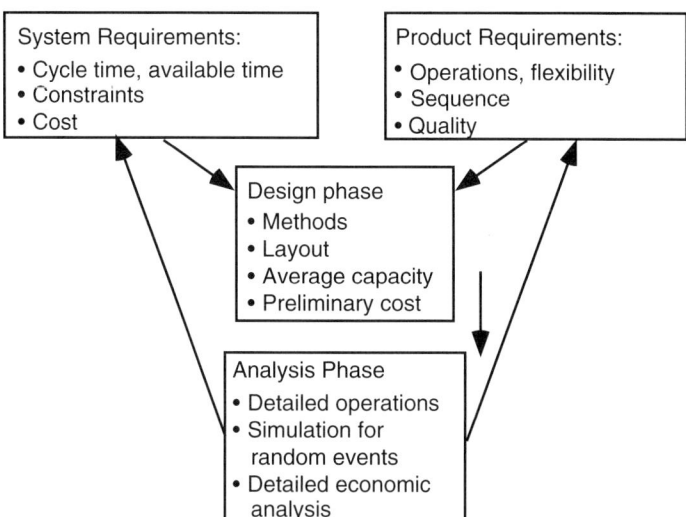

FIGURE 16-1. Basic Steps in Designing an Assembly System. The system and the product both provide constraints to the design process. Some modifications to the product's design may be desirable or necessary.

5. Using either intuitive techniques or the computerized methods described later in this chapter, select a set of equipment or people that can make the product at the required rate for a reasonable cost.

6. Either make preliminary economic analyses or proceed to detailed workstation designs and then perform economic analyses.

At any stage in this process, economic or technical evidence may appear that forces a reconsideration of product design, selection of subassemblies or assembly sequence, timing requirements, and so on. If all of the required information is not available, or if system design reveals knowledge gaps, then additional product or process design effort, engineering, or experiments may be necessary. The alternative is a system design with less robustness and predictability than desirable. Product quality, delivery, or cost may suffer, or the time to reach full production may be prolonged, as a result.

Even if the analysis is incomplete, performing it has great benefits. Visibility into the lack of robustness of processes or product design gives management the evidence to decide whether a product is ready to be manufactured.

Assuming that at least preliminary information is available, system design deals with the following topics, though not necessarily in this order:

16.B.1. Capacity Planning—Available Time and Required Number of Units/Year

Capacity planning requires providing the system with the ability to deliver the required number of correct parts or assemblies per hour, day, and year. If there are several models, then the batch size and changeover frequency are also part of capacity planning information. The number of units needed per year is an estimate based on marketing surveys, sales of similar products, and economic conditions. Required capacity establishes a minimum operating speed for each workstation expressed in time per part or assembly. Figure 16-2 shows that much less than 24 hours/day is available to deliver good assemblies ([Chow]). The words "utilization," "uptime," "availability," and "capacity" are often misused and confused. A chart like Figure 16-2 is probably the only way to make sure everyone is using and understanding the terms in the same way. Note that some kinds of scheduled downtime can be considered good in the sense that they either support higher level strategies (variety change supports customization, while tool change supports good surface finishes in machining) or should improve uptime overall (preventive maintenance). Section 16.I discusses how the Toyota Production System accommodates high variety assembly. Plant personnel often overlook short downtime

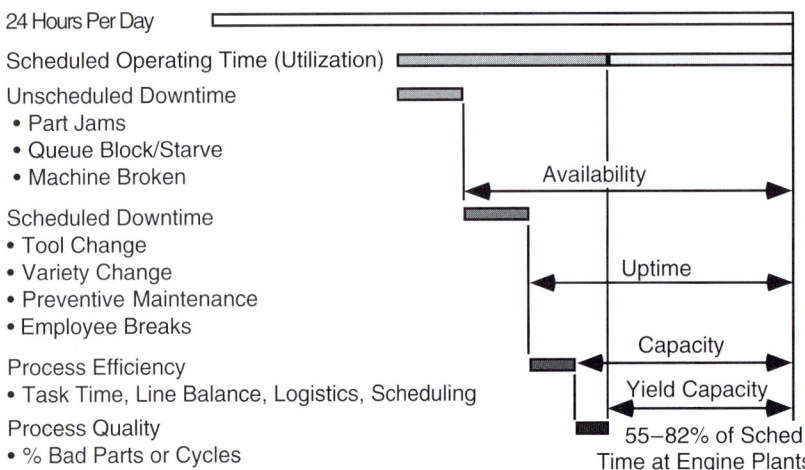

24 Hours Per Day

Scheduled Operating Time (Utilization)

Unscheduled Downtime
- Part Jams
- Queue Block/Starve
- Machine Broken

Scheduled Downtime
- Tool Change
- Variety Change
- Preventive Maintenance
- Employee Breaks

Process Efficiency
- Task Time, Line Balance, Logistics, Scheduling

Process Quality
- % Bad Parts or Cycles

Availability

Uptime

Capacity

Yield Capacity

55–82% of Sched.
Time at Engine Plants

FIGURE 16-2. Where Does All the Time Go? A plant may schedule one-, two-, or three-shift operation. Many factors reduce the actual time available. These factors are grouped into unscheduled downtime, scheduled downtime, and process losses. According to [Peschard and Whitney], automobile engine plant machining shops have yield capacity ranging from 55% to 82%. (Terminology adapted from [Chow] and [Peschard and Whitney].)

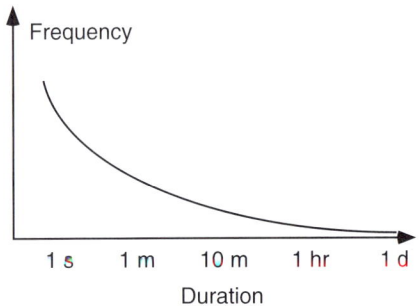

Frequency

1 s   1 m   10 m   1 hr   1 d

Duration

FIGURE 16-3. Illustrative Histogram of Downtime Durations in Assembly. Most downtime events are quite short, so short that people may not notice them. The total time lost is proportional to the area under this curve, and it can be quite large because there are so many short stoppages. (Not to scale.)

periods,[2] but they are the majority, as indicated schematically in Figure 16-3. Section 16.G.3 shows how to estimate how much capacity is lost when assemblies have to be reworked. Section 16.H discusses how to design buffers into assembly lines to reduce the effect of machine stoppages.

16.B.2. Assembly Resource Choice

A great deal of judgment is involved in choosing a suitable way to do each assembly operation. "Resources" could include people, and the design methods in Sections 16.K and 16.L allow this choice wherever the designers say that it is feasible. Mechanical equipment may have to be designed specifically for some steps, making it unlikely that there will ever be a universal equipment data base and totally automated assembly system design. Often, a company outsources the design of its assembly lines and is at the mercy of the vendor regarding types of equipment.

The choice of technology is governed by the technical and economic factors discussed in Section 16.E. Production volume, size of the product, and difficulty of the assembly tasks are the main determining factors. For example, high volume, a small number of small parts per product unit, and simple unidirectional assembly tasks make fixed automation a possible candidate. The Denso panel meter is an example. On the other hand, large parts dictate manual assembly often aided by lifters or other equipment. Automobile final assembly is an example.

It may be discovered that some operations must be done manually due to inadequate design of the product. If so, and if machine assembly is desired, the relevant parts must be redesigned. Even if manual assembly is desired, it may be discovered that some operations are too difficult or prone to mistakes[3] or human injury. It would be better if consideration of alternate assembly methods and sequences were made part of concept design of the product, as advocated by followers of DFA and concurrent

---

[2]A manager went to a line to find out for himself the sources of downtime. After a short stoppage was repaired, he asked the foreman the cause of the stoppage and got the answer quoted at the beginning of the chapter.

[3]The word "mistake" is used in this book to mean a procedural error on the part of a person or machine. The word "error," while commonly used to mean the same thing, is reserved in this book to mean lateral and angular misalignments between parts. This is done to avoid confusion.

engineering, but this does not always happen, especially if an outside vendor is assumed to be involved in providing assembly methods and equipment.

16.B.3. Assignment of Operations to Resources

Operation assignment involves deciding which tasks in the system should be done by which resources. Several alternatives for each operation might be feasible, and some kinds of resources—people and flexible automation, for example—may be able to do several operations in a row on each product unit. As long as the available equipment options are equally feasible and capable, the matter of resource choice and task assignment becomes dominated by cost and time. A systematic design procedure must therefore integrate these first three steps, choosing equipment and assigning tasks to take advantage of equipment versatility, so that the operations are done fast enough to meet the capacity specification and the cost meets suitable economic criteria.

It should be clear that these first three choices are interrelated, and making them is difficult. The size of the product and number of units needed per year (called the annual production volume) have a decisive effect on the choices that are technically and economically feasible. People are flexible, dexterous, and adaptable, but cannot work at extreme speeds or on very large objects without help from machinery. Small items (cigarettes, ball point pens, razor blades) have few parts and are made in high volumes, while large items (aircraft, ships) are made in low volumes. The layout of the facilities and the freedom of the operators to vary their methods may be restricted by equipment needed to do specialized tasks or manipulate large parts. The decision path often follows the pattern shown in Figure 16-4.

Figure 16-5 shows example unit assembly cost curves plotted against annual production volume. The general shapes of these curves are discussed in Section 16.E, while the assumptions and details are explained in Chapter 18. For now, it is sufficient to note that the three different methods shown (manual assembly, dedicated or fixed automation, and flexible or robotic automation) exhibit very different cost versus production volume behavior. For the labor and equipment costs shown in the figure, manual assembly is by far the lowest cost unless production volumes are quite large.

It should also be clear that product design and assembly sequence heavily affect system design. A different sequence presents different task assignment opportunities, for example, or may require extra steps like turning a subassembly over, requiring different or extra equipment, while at the same time making it easier to discover or diagnose quality problems.

16.B.4. Floor Layout

Floor layout requires arranging the fabrication, assembly, part feeding, and material handling equipment into a compact, efficient, and effective layout on the factory floor. Layout and operation sequence can sometimes affect each other if certain operations must be located at designated places on the floor (painting near ventilation, for example).

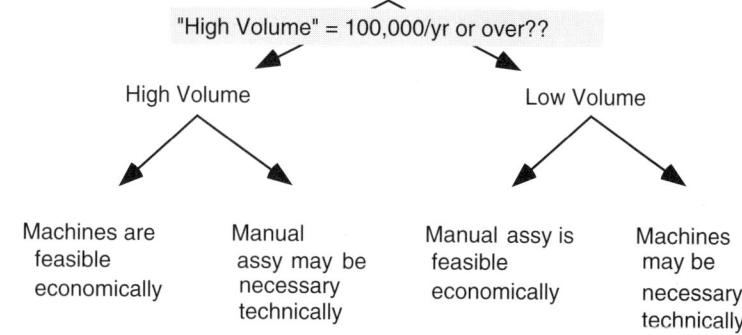

FIGURE 16-4. The Basic Decision Tradeoffs Between Manual and Mechanized Assembly. As a rule of thumb, usually confirmed by detailed economic analysis, assembly operations divide into high and low volume at around 100,000 units per year. The technical capabilities of people and machines and their relative costs determine what is feasible and what is necessary.

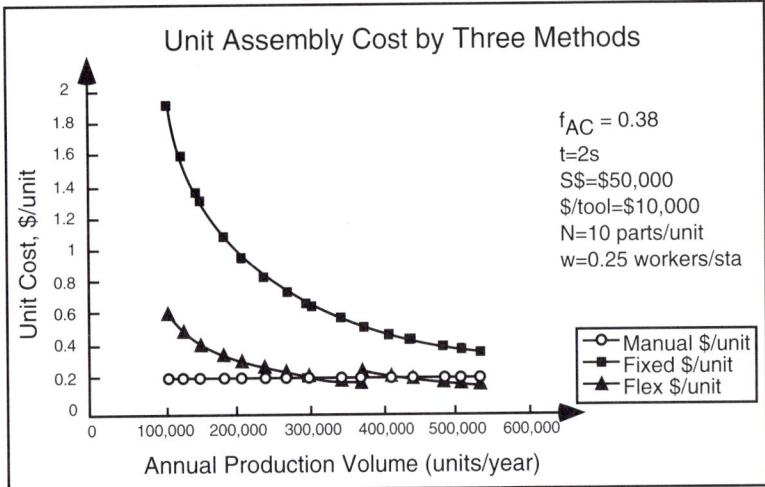

FIGURE 16-5. Example Unit Assembly Cost by Three Generic Assembly Methods Versus Annual Production Volume. Legend: f_{AC} is the fraction of equipment cost charged each year. t is the time to do one operation, L_H is the hourly labor cost, $S\$$ is the cost of one fixed automation or flexible workstation.

16.B.5. Workstation Design

Each station must be designed to do the tasks it must perform. The designer must insure that parts can be reached, tools can move rapidly and accurately enough, and assembly can be done reliably. This is the subject of Chapter 17.

16.B.6. Material Handling and Work Transport

The designer must decide how to convey assemblies and parts within the system. This and floor layout are often determined together, with emphasis on meeting the capacity requirements. Assembly happens very rapidly, so transport is needed frequently. A factory making 300,000 units per year with 10 parts each must convey 3 million parts per year. Transport speed, number of transporters and parts fixtures or pallets,[4] and distances traveled are the important variables. In addition, capacity and efficiency are affected by the number of assemblies, or the number and kind of parts, handled by one transporter or pallet, as well as by the space available for queues of pallets in the system. Flexibility of the process, including the ability to change the priority of individual assemblies, is affected by the method chosen to transport assemblies. A fixed conveyor offers little opportunity for such adjustments, while

free-ranging automated guided vehicles offer a lot more but may be much more expensive.

16.B.7. Part Feeding and Presentation

How will parts be fed to equipment or people? Options include bulk vibratory feeders, cassettes, or feeder strips for small parts, and pallets or kits for large parts. Sometimes these options interact heavily with other aspects of system design; other times part provisioning decisions may be made independently. Different methods differ greatly in speed, reliability, and capability to handle odd-shaped parts. This topic is discussed as part of workstation design in Chapter 17.

16.B.8. Quality: Assurance, Mistake Prevention, and Detection

Assembly happens very rapidly, and it is easy to overlook a problem or mistake. Parts need to be designed for easy assembly and to make mistakes hard to commit.[5] Different assembly sequences may make mistake detection or repair easier or more difficult.

16.B.9. Economic Analysis

Economic analysis involves deciding if the proposed system design meets economic criteria. Since any system will require an initial investment, typical economic criteria seek to measure whether this is a good investment or not.

---

[4]A pallet is a work-holder that carries partially completed assemblies between stations and is used to position them accurately in the station. Pallets are not always used, especially for manual assembly, where accurate location is not needed and the product may be able to support itself. In the Denso Panel Meter assembly machine, the part called casing acts as a pallet.

[5]These issues are discussed in Chapter 17.

The basis for judgment is usually a comparison of the required investment and the amount and timing of either (a) the savings generated by a replacement system compared to the one being replaced or (b) the net revenues generated by sale of a new product made on a new system. This is the topic of Chapter 18.

16.B.10. Documentation and Information Flow

Every assembly operation requires instructions, especially if different versions of the product are assembled on the same line. Many operations require data to be recorded, such as tightness of fasteners, results of tests, and so on. Reasons for machine downtime and time to repair are sometimes recorded, but not in enough factories. Production schedules are distributed or posted so that everyone knows what to make when, or information is passed between stations for the same reason. The pace of the line, the number made so far that shift, and the number remaining are also often posted prominently. All this adds up to an enormous flow of information that must be organized, managed, kept up to date, and displayed.

16.B.11. Personnel Training and Participation

Assembly is very repetitive work, and enormous concentration is required to do each step exactly the same correct way every time. As discussed below, machines are better at this than people, but people have their advantages, too, and by far the vast majority of assembly is done by people. It is essential that they understand what they are doing, how they contribute to the success of the product, and what they can do to improve the operation.[6] Some of the issues are discussed in Chapter 12.

16.B.12. Intangibles

In addition to the usual system performance specifications, there are others which are desirable and which may be costly to provide, but whose contribution may be difficult to measure. The latter fact does not diminish their importance but may cloud decisions if typical economic criteria are applied. Among the intangibles are flexibility, responsiveness, and quality.

Flexibility and responsiveness are discussed from the point of view of product architecture in Chapter 14. Here we have space only to say that the required flexibility must be made part of the design specifications for the assembly system and its resources. Workstations, transport, and part feeding are especially impacted by these issues.

Quality may be defined as the degree to which the product's original specifications are met by each unit. The cost and quality of a product are determined mainly during design and are rarely or slightly improved during production. Workpieces and the processes for making and assembling them are ideally designed to be robust, well understood, and reproducible. Taguchi calls this off-line quality control ([Taguchi]). Quality control in the factory can only ensure that the design intentions are still being achieved but cannot economically correct problems caused by poorly designed products and poorly operated systems. This is called on-line quality control.

The above attitude toward quality involves extra expense during design and, possibly, more expensive parts and processing equipment. Justifying such extra expenses has often been difficult in the past because the benefits of higher quality are hard to measure quantitatively. As companies and their customers become more conscious of quality, the justifications are becoming easier. One approach is for the company to make quality one of the specifications for the product. Units that fail are reworked or discarded, an action whose cost can be measured. The consequences of different designs for product and system can then be measured, including the quality costs. The company can also view quality as one of its marketing strengths. The cost of lost sales can sometimes be measured and applied as above.

16.C. AVAILABLE SYSTEM DESIGN METHODS

In spite of the availability of good texts on assembly line design ([Chow], [Nof et al.], [Scholl]), many assembly

systems are designed without detailed analysis. Also, in all too many cases, the products are designed without a lot of attention to assembly, and the task of designing assembly processes and equipment, and operating them, falls to factory personnel. Past methods are often repeated

---

[6]A body inspector at a vehicle plant said, "I wouldn't want that mark on my truck."

because their performance is known, and often products are designed the same way again and again, preserving continuity and knowledge but also repeating the same mistakes.

Furthermore, too many companies outsource the design of assembly lines to vendors of such equipment. Many of these companies are small and have good knowledge of only one or two kinds of assembly methods. This means that the assembler does not get a wide range of process choices. In some cases, each subassembly in a complex product will be built on equipment from a different vendor, which is the result of devolving vendor choices onto the equipment purchasing department. Only after these different systems are installed and operators try to connect them do a variety of incompatibilities appear. These may include different kinds of motors and controllers, causing the assembler to stock many kinds of spare parts and to have to train its maintenance personnel on many kinds of equipment. Operating incompatibilities include different philosophies concerning surge capacity, batch

sizes, and rework patterns. Companies that take charge of this process can save a lot of money and time, even if they outsource detailed design and construction of the equipment.

The most common design method is a kind of trial and error in which a design is proposed and then tested using discrete event computer simulation. (See Section 16.J for a short discussion of this important topic.) This is "design by analysis," and often the scope of different system designs that can be studied is limited because different simulations usually only adjust parameters of the original design without greatly changing its architecture.

Sections 16.K and 16.L discuss systematic heuristic and analytical techniques, respectively, for designing assembly systems. These techniques provide considerable scope for exploration while at the same time applying some mathematical rigor. These methods must be augmented by simulation to cover stochastic properties of the system that no design method can currently take into account.

16.D. AVERAGE CAPACITY EQUATIONS

The first responsibility of the assembly system designer is to provide a system that can make the required number of assemblies per year. As discussed in Chapter 14, great uncertainties surround this information. The product may be a market success or a failure. The number of versions made may have to change, and the rate at which different versions are needed may change as well. Over the life of the product, demand will rise, plateau, and fall. In any case, a number must be chosen for the required capacity, and then a system must be designed. Systems concepts capable of changing capacity are discussed in Sections 16.N.3 and 16.N.4.

The required effective rate at which the system must operate is based on the work content of the assembly. This is expressed as the total number of operations required to complete one unit. The total number of operations required per year is then:

$$\text{\# operations/year} = n * Q \qquad (16\text{-}1)$$

where n is the number of operations per unit and Q is the number of units per year.

This is converted into operations per second by calculating how many seconds are in an operating year. The

calculation below is based on 280 operating days per year (but this varies by industry and country) and that there is a choice of one, two, or three 8-hour shifts per day:

$$\text{\# operations/sec} = \text{\# operations/year} * (1/Y) \qquad (16\text{-}2)$$

where Y is the number of seconds per year (28,800 sec/shift $* n$ shifts/day $* 280$ days/year, $n = 1, 2,$ or 3) and it is assumed that each shift is 8 hours long.

If we allow for scheduled downtime, then shifts are effectively shorter, perhaps only 7 hours. This will increase the number of operations that must be done per second. The same effect occurs if we work fewer than three shifts per day or fewer than 280 days per year. If required production volume is uncertain, it makes sense to plan for less than three-shift operation, leaving the option to grow capacity by adding shifts.

The number of operations per second can be easily converted into the required operation time or number of seconds available to the system to perform each operation:

$$\text{available operation time} = \text{sec/operation}$$
$$= \frac{1}{\text{\# operations/sec}} \qquad (16\text{-}3)$$

It is important to remember that the number of operations can exceed the number of parts by a factor of two or more. Additional operations may be needed to apply lubricant, check that the part was actually inserted or fastened properly, and so on.

Against this must be put the operating capability of each assembly resource assigned to do an operation. This is expressed as needed operation time t in Equation (16-4). Equation (16-4) includes factors that increase the time a resource will need, while Equation (16-2) and Equation (16-3) include factors that decrease the time available. Also included in Equation (16-4) are any net savings in tool change time and station in–out move time that result from loading several product units onto one pallet and having the resource repeat the operation on all these units. Therefore, these equations permit the designer to realistically estimate which resources are technically feasible in the sense that they have time to do the work assigned to them.

Operation time needed $= t$

$$= \frac{1}{\varepsilon} \left[\text{pure assembly time} + \frac{\text{in–out time}}{\text{units/pallet}} \right.$$
$$\left. + \frac{\text{tool change time} * \text{\# changes/unit}}{\text{\# units/tool change}} \right] \quad (16\text{-}4)$$

where operation time needed is the average time this resource needs to do one assembly operation; pure assembly time is the average time to do an operation based on fetching and inserting the parts; in–out time is the time for moving finished work out and moving new work in (also known as station-station move time or just station move time); units/pallet is 1 or more, depending on the design of the pallet; # changes/unit is the number of tool changes needed to do the assigned work on one unit; # units/tool change is the number of units worked on before a tool change occurs (cannot be larger than the number of units/pallet); and ε is the fractional uptime of the resource (a number between 0 and 1).

Some methods of estimating operation time capability of resources are discussed in Chapter 17 and the references cited there. Typical operation times are shown in Table 16-1 for different kinds of resources.

Once we know the average operation time capability of a resource, we can calculate the makespan S, the total time that resources of this type will need on average to make one complete product unit:

$$S = nt \quad (16\text{-}5)$$

TABLE 16-1. Typical Operation Time Capability for Different Assembly Resources

Resource	Application Example	Typical Operation Time
Person	Small parts assembly	3–5 seconds
Robot	Small parts assembly	2–7 seconds
Fixed automation	Small parts assembly	1–5 seconds
Person plus lifting aids and tools	Automobile final assembly	One minute
Person plus lifting aids and tools	Aircraft final assembly	Several minutes to an hour or more

Then we can calculate N, the total number of such average resources that will be needed to make the required quantity per year Q during the available time per year Y:

$$N = \left[\frac{nt\,Q}{Y} \right],$$

where $[\cdots]$ means largest integer in $\cdots$ (16-6)

Equivalently, we can think of N as the number of equal length time chunks into which creation of one product unit must be divided. Corresponding to N is the amount of working time τ in each time chunk.

$$\tau = Y/Q \quad (16\text{-}7)$$

Time τ is called the cycle time or the takt time—that is, the time interval at which product units must be finished in order to meet demand.

These idealized equations assume that a resource will take the same time to do all operations. In reality, some operations may take less time while others take more. If $t = \tau$ for an operation–resource combination, then one such resource will just be able to do the operation. If $t < \tau$ for several operations in a row, then the resource may be able to do more than one operation during the time chunk, assuming it is technically capable and enough time remains in the cycle, including any required tool changes and in–out move times.[7] If $t > \tau$ for an operation, then multiple resources will be needed in order to finish it in the required

---

[7]Remember that tool change times and in–out station move times are often similar to the assembly operation times themselves and therefore can consume a significant fraction of the cycle time. In machining systems, operation times can be several minutes, and in–out and tool change times are often negligible.

time. Alternatively, the required τ (and available Y) can be effectively increased by scheduling more shifts per day, adding overtime, and so on. Or, a faster (often costlier) resource may be selected. A decision process like this underlies the manual and algorithmic system design techniques discussed in Sections 16.K and 16.L, except that actual operation times for each operation–resource combination are used instead of averages.

It is important to understand that these equations give average operation time capability. Several stochastic effects usually reduce the capability further. These are usually investigated using simulations and may include unusual bunching up of equipment downtimes or the effect of having a buffer between stations fill up or empty completely, causing one or more stations to stop while others continue.

16.E. THREE GENERIC RESOURCE ALTERNATIVES

Three basic types of assembly resources can be distinguished: people, fixed automation, and flexible automation. These are discussed in terms of assembly but similar considerations apply to fabrication. The discussion is divided into technical and economic characteristics for each type of resource.

Factory operating costs are generally classified as fixed or variable. Fixed costs are the same regardless of how many units are made. The factory building itself is an example, as is any machinery whose capability is defined by fixed factors such as the number of parts in the assembly. Variable costs depend on how many units are made. Examples include raw material, energy, labor, lubricants and tool bits, and so on. These concepts will be used in the following discussion and are treated in detail in Chapter 18.

16.E.1. Characteristics of Manual Assembly

16.E.1.a. Technical
Manual assemblers are the most flexible, adaptable, innovative, dexterous, and responsive to challenging tasks. They are also the least efficient and most variable and sometimes cannot resist "improving" an operation and thereby introducing variability, contamination, or damage that they are unaware of. These attributes contrast sharply with those of machines, which are relatively more efficient and less variable, adaptable, dexterous, and so on. The above characterizations must be tempered by the fact that the task to be done sets the conditions that must be met. The "same" task may require more or less dexterity, for example, depending on its design, physical layout, variability of incoming parts, or a host of other reasons.

For example, a circuit board "needs" to be assembled manually if some of its components are of odd shape so that no mechanical means of presenting them to an assem-

bly machine is available. A differently shaped component, with the same electrical characteristics, or one made by a vendor who will package it in a cassette for easy feeding, could be selected by the designer, permitting machine assembly. Thus design choice strongly affects the dexterity needed and the feasibility of an assembly method.

People can also do several operations at once, something that machines usually cannot. For example, a person can move a part while simultaneously reorienting and inspecting it. This saves time and makes the required cycle time shorter.

People are especially necessary for managing poorly understood processes and for dealing with assembly steps that require tests, adjustments, and decisions about many complex measurements. While it is sometimes possible to redesign the product to tighten tolerances and thus eliminate such adjustments, it is often more economical to make low tolerance, low cost parts and adjust the assembly into operation. The problem with adjustments is that they take a different amount of time on each assembly. This makes total assembly time a variable, which makes capacity planning difficult.

Unlike machines, people need to rest and eat. Typically, there is a rest break of 10 to 15 minutes in the morning and afternoon, along with a lunch break of 20 to 30 minutes in the middle of the day. Second and third shifts have breaks of similar duration. If a shift is nominally 8 hours, then only 7 to 7.25 hours are scheduled for work. Correspondingly, machines need regular maintenance and occasional repairs.

16.E.1.b. Economic
To estimate the cost structure of manual assembly, we note that a person operates at a given speed, and if more production is needed, more people may be employed. If less production is needed, they may given more operations to do or be employed elsewhere, due to their flexibility. Thus

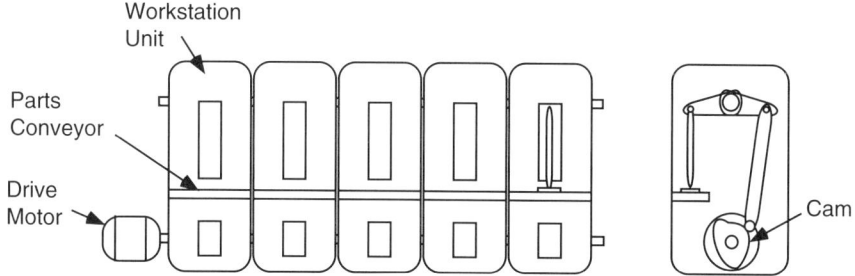

FIGURE 16-6. Typical Fixed Automation Assembly Machine. *Left:* Front view. *Right:* End view. The machine is made of a series of identical base workstation units connected by a synchronous transfer belt or conveyor. Assemblies move from one station to another synchronously and stop while work is performed. The motor at the left drives a common shaft with a cam specially designed for each workstation. The cam operates one or more levers or other mechanisms that actually perform the assembly operations. A similar cam operates the work transfer mechanism. Each assembly station has a part feeder, which is not shown.

labor cost is primarily a variable cost.[8] Often the tools or fixtures that cannot be redeployed are a small cost component. They represent a fixed cost. Since the main cost component, people, is directly proportional to the annual production quantity, the assembly cost per unit is approximately independent of production volume. For this reason, the line representing manual unit assembly cost in Figure 16-5 is constant regardless of production volume.

16.E.2. Characteristics of Fixed Automation

16.E.2.a. Technical

"Fixed automation" typically describes machines built to do one task without frequent changes, or for which changes require a significant amount of reconstruction. As indicated in Chapter 1, fixed automation has been under continuous development for nearly two hundred years, with increasing speed, accuracy, and range of task capability being the result. A fixed automation assembly machine reminds one of a diesel engine, with arms, wheels, and levers moving in perfect synchrony, often remarkably slowly. Slow speed is feasible because suitable products usually are small, and good design of product and machine keep motion distances small as well, perhaps only a few inches. Each station of such a machine installs or tests just one part. All stations operate simultaneously, driven by a common shaft and cams to run each workhead, so the

FIGURE 16-7. A Small Parts Fixed Automation Assembly Machine. In the foreground are parts feeders that operate by vibration. The parts crawl up the helical track and are picked up at the top. Also visible are several work heads that can make simple up–down or in–out motions. They are operated by air pressure controlled by valves operated by a central controller. (Photo courtesy of *Assemblagio*.)

speed of a station is the speed of the machine. Figure 16-6 illustrates such a machine schematically. Its architecture is linear. Figure 16-7 is a photo of a similar machine that is operated by air pressure.

For smaller parts, or for situations requiring shorter cycle times, rotary dial machines are often used. A drawing of one appears in Figure 16-8.

---

[8]Some observers note that union contracts and other obligations often make it impossible to eliminate people, making them effectively a fixed cost to the company, if not to a particular assembly line at a given time.

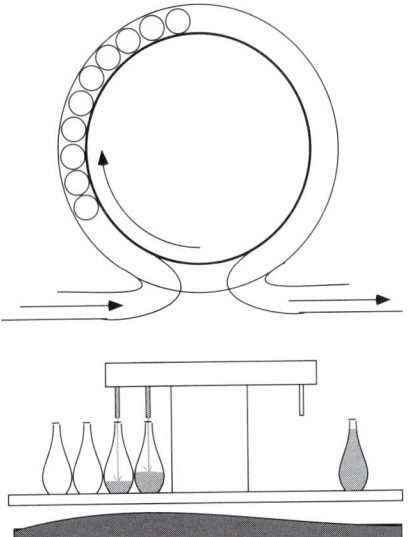

FIGURE 16-8. Typical Rotary Dial Machine. *Above:* Top view. *Below:* Side view. This machine accepts incoming units from any kind of conveyor and places them in pockets on a rotating table. There is one work head per pocket dedicated to each unit. The work heads rotate in synchrony with the rotating table, and are operated by a stationary cam (indicated schematically by the gray contour in the side view). This kind of machine can do more operations per unit time than an in-line machine like the one in Figure 16-6 because many units are being worked on at once and no time is lost while units stop and start. The production rate is the RPM of the table times the number of pockets, even though the workheads can work at a relatively leisurely pace. The application shown, filling bottles, is a common use of this kind of architecture but it has also been used for assembly of small items like razor blade cartridges.

Machines of the type shown in Figure 16-6 and Figure 16-8 are most suitable when simple operations requiring few degrees of freedom are sufficient to do the necessary work. Quite clever engineering is often involved in making the product capable of being assembled on this kind of machine, and quite clever engineering is usually required of the machine's designers.

Most machines require people to tend them, refill parts feeders, and clear jams. When these people go on rest or meal breaks, the machines can often continue running but some companies turn them off for safety reasons.

16.E.2.b. Economic

A dedicated or fixed automation machine, almost by definition, must have individual workstations for each operation needed. The number of workstations is thus approximately proportional to the number of parts in the assembly. This is a fixed amount and does not vary with production volume. Thus fixed automation represents a fixed cost. Labor associated with the machine, such as operators who clear part jams or do repairs, are a variable cost because they are not needed unless the machine is running. This cost is usually small.

Consider a machine for assembling automobile cigarette lighters. If the machine's cycle time (time for each station to act, and thus the time between completed assemblies) is 3 seconds, then if it operates 8 hours a day, 5 days a week for a typical working year of 280 days, it will produce 2,688,000 lighters in a year. If we allow for 20% downtime for maintenance and part feeding jams, then at most five such machines working one shift or three working two shifts could make over 9,000,000 lighters per year, enough to satisfy the needs of all the domestic car manufacturers and still have some capacity left over for the replacement market.

The conclusions to draw from this and other examples are that such machines are often one or few of a kind, can easily produce large quantities of small items, require engineering design to focus their general capability onto a specific product, have one station per part or operation required by the product, and will stand idle if demand for the product falls. Consequently, they are most economical for high-volume products that will be made for several years. When demand falls, the machine's fixed cost is spread over fewer units, and the cost per unit rises proportionately. The overall result is that the cost to assemble one unit is approximately proportional to the number of parts in one unit and inversely proportional to annual production volume. This fact accounts for the hyperbolic shape of the fixed automation unit assembly cost curve in Figure 16-5.

A key to the economical use of such machines is to keep them running,[9] because when one station stops, they all stop. Some companies go to extraordinary lengths to accomplish this while others seem not to understand the economic rewards. One need only look at the wide range of uptime statistics mentioned in the caption of Figure 16-2 to realize this.

Such machines are made economically by companies that specialize in a particular machine architecture and

---

[9] Assuming, of course, that all their output is needed.

limited size range of products, and which have created innovative basic machine concepts. These concepts minimize the amount of engineering required to adapt the machine's standard foundation and actuators to a specific product. Modern "fixed" automation assembly machines are also often capable of preplanned production flexibility, based on the ability to switch feeders or skip operations at given stations depending on model mix requirements.[10] However, since these machines usually operate synchronously, it is difficult to include a station such as an adjustment whose cycle time may be variable.

16.E.3. Characteristics of Flexible Automation

16.E.3.a. Technical

The third type of assembly equipment we take up is flexible automation, commonly thought of as robots. In general, however, flexible automation may be defined as any automation which is reconfigurable to do a range of tasks. Such machines are characterized by several moveable or controllable degrees of freedom so that different (often arbitrary) shapes, paths, angles, and different (within limits) sizes, forces, directions, and so on can be encompassed. They can change tools. Much of their flexibility is based on computer control. There is strong potential for adaptation of behavior (different path or force, etc.) based on sensor inputs and computed alterations in behavior, although this is only beginning to be applied. There may be some setup time penalties, and some people may be needed in attendance. In addition to robots, other kinds of flexible automation include knitting, weaving, welding, and tube bending machines, as well as entire arrays of actuators and welders or riveters that assemble and fasten ribbed structures for ships or aircraft.

16.E.3.b. Economic

The main economic feature of flexible equipment is its ability to do more than one task. One may interpret this as the ability to be turned to a different application after a period of years, but more frequent and more important is the ability to turn to a different task after a few seconds or minutes. A typical assembly robot can assemble two different parts in a row, whereas a fixed automation assembly machine requires two workstations to do the same thing. The cost difference can be large: the cost of a second station compared to the cost of another gripper. (Sometimes even the same gripper can be used.) Like fixed automation and unlike a person, a robot can also work 16 or 24 hours per day. The economic consequences are that the cost of a robot assembly system does not have to grow strictly in proportion either to the required production volume, as manual assembly cost does, or in proportion to the number of parts in the product, as fixed automation does. Instead, one buys as many robots as their cycle time permits and the production rate requires, and at most as many tools as there are assembly operations, and runs the system as many hours as needed. For this reason, flexible systems' costs are a combination of variable (the number of robots needed) and fixed (the number of tools and part feeders needed).

Polaroid Corporation used Sony assembly robots for many years to assemble its cameras. These were very intricate mechanisms whose design changed frequently. During the 1980s, the same robots were reused to assemble four different designs of shutter mechanism during one five year period. Each time, different programming and different tools had to be designed, but the pace of changeover to a new shutter was limited by the availability of new parts, not redesign or reprogramming of the line, which typically took 8 to 10 weeks. Among the uses for flexibility that Polaroid exploited included experimenting with different assembly sequences to get the best balance of assembly times on each station, and insertion of extra tests and measurements while the system was being brought up to speed, operations that were omitted when everything checked out.[11]

16.F. ASSEMBLY SYSTEM ARCHITECTURES

Architecture of assembly systems refers to the spatial arrangement of the workstations. Many varieties exist.

Several are discussed below together with some advantages and disadvantages.

---

[10]The Denso panel meters were assembled on such a machine. See Section 16.N.1.

[11]Norman Ward, former Director of Automation, personal communication, April 22, 1993.

16.F.1. Single Serial Line (Car or Airplane Final Assembly)

The serial assembly line, in which a conveyor brings the work to the operators, was invented by Ford in 1913. It is by far the most efficient for high volume assembly but is used for products of all sizes and production rates. Dell uses it for assembling computers, GM and other car makers use it for final assembly of cars as well as assembly of smaller components, and it is used as well in other industries.[12] When the production rate is high, the amount of work done by one person is very limited and the work can be boring.

Figure 16-9 shows two ways of setting up a line to do five operations. In the series arrangement, each station does one operation while in the parallel setup each station does all five. While nominally one would think that each arrangement has the same capacity and cost structure, this is not true. A thought question at the end of the chapter asks the reader to think about this.

Figure 16-10 shows design variations on the simple assembly line. One variant shows what to do if one operation takes much longer than the others: Add more resources in parallel. The other variant shows what to do if several operations in a row take a short time and a flexible resource is available that can do them all. Clearly, this option depends on a favorable assembly sequence and design of the product.

Finally, Figure 16-11 shows a flexible architecture in which parts or assemblies have access to multiple resources. This can be an advantage when resources have high downtimes or when task durations are variable: the work can go to the next available resource in the sequence. However, a disadvantage is that if mistakes are found at the end of the line, it can be difficult to diagnose the problem unless a record is kept about the actual route that each assembly took.

16.F.2. Team Assembly

In a team assembly architecture, a group of operators works together to do a large number of operations. Volvo uses this technique for final assembly of engines in high variety and relatively low volume. The operators can choose a variety of ways to divide the work, including

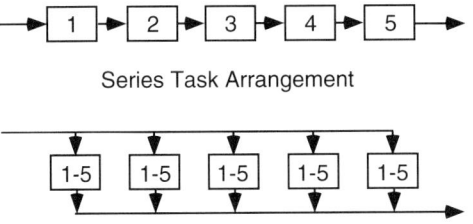

Series Task Arrangement

Parallel Task Arrangement

FIGURE 16-9. Two Generic Ways of Arranging an Assembly Line.

having several engines in process at once worked on by smaller subteams. Volvo also used team assembly for final assembly of cars, a technique Ford abandoned as too inefficient. [Engström, Jonsson, and Medbo] presents data that show that team assembly at Volvo's Uddevalla plant was some 20–30% more efficient in terms of person-hours per car than a comparable conventional assembly line at Torslanda. However, quality suffered because operators occasionally forgot one of the many operations during the 90 minutes they spent assembling a car.[13] Opinion about the advantages of team assembly was divided inside Volvo, and ultimately the team assembly plants were closed.

16.F.3. Fishbone Serial Line with Subassembly Feeder Lines

A fishbone line structure consists of a main backbone final assembly line with several side lines delivering finished and tested subassemblies at the points where they are needed on the final assembly line. The side lines are directly connected to the main line by conveyors. This architecture is not practical for large products like automobiles, in which case the subassemblies are delivered by independent transport means such as fork lifts. The fishbone line works well for smaller items like automobile transmissions and desktop computers. Fishbone lines offer many chances to test the product systematically at the subassembly level and are preferred for complex products. A modular design whose subassemblies perform well-defined functions is a prerequisite to reaping this advantage.

Fishbone lines offer an opportunity to apply the method of delayed commitment at the subassembly level. Each

---

[12]Ford got the idea from meat packing plants.

[13]"How on earth could we forget this?" ([Engström, Jonsson, and Medbo], p. 214)

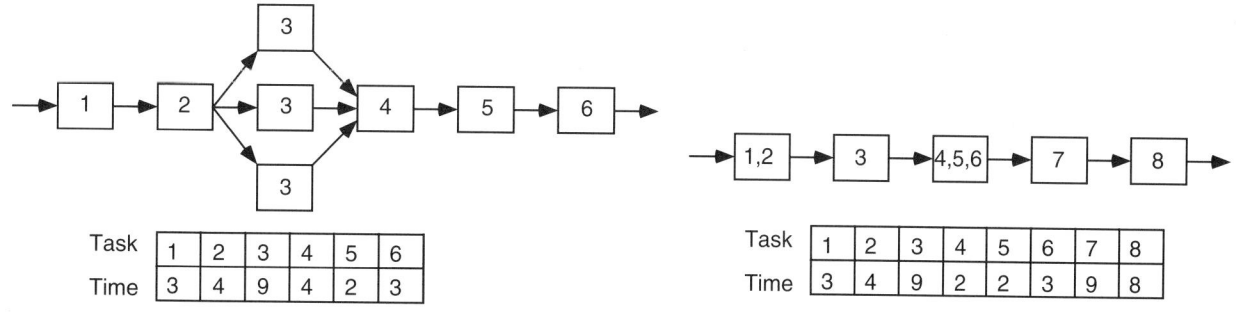

Task	1	2	3	4	5	6
Time	3	4	9	4	2	3

(a) Three Copies of Station 3 Are Needed Because Its Task Takes So Long

Task	1	2	3	4	5	6	7	8
Time	3	4	9	2	2	3	9	8

(b) Grouping Work at Stations Improves Balance of Station Times

FIGURE 16-10. Two Variations on the Simple Assembly Line. (a) One operation takes much longer than the others, so three copies of the station are employed. (b) A series of operations together takes such a short time that one (manual or flexible) station has time to do them all. In each case, the lines are reasonably well balanced in terms of work time.

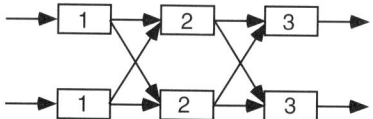

There Are Six Possible Paths

FIGURE 16-11. A System with Multiple Paths.

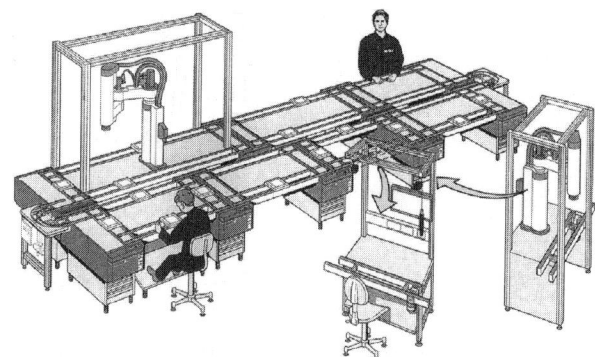

FIGURE 16-13. A Modular Assembly System Based on a Loop Architecture. This system consists of a standard conveyor loop plus options consisting of manual or robotic workstations. (Photo courtesy of *Assemblaggio*.)

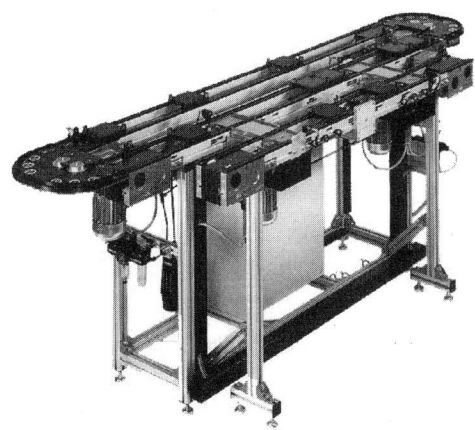

FIGURE 16-12. Typical Loop Architecture for Small Parts Assembly. (Photo courtesy of *Assemblaggio*.)

subassembly can be customized for its eventual customer near the end of the line that assembles it. A hierarchical delayed commitment strategy can then be implemented.

16.F.4. Loop Architecture

A common architecture for assembly systems is the closed loop. This is convenient when the line uses pallets

because the empty pallets return easily to the front of the line to begin the process again. Figure 16-12 is a photo of a loop machine foundation with its conveyor. Assembly workheads will be attached to it. An example system is shown in Figure 16-13.

In some cases, the loop encloses an open area. This area can be used for places where the operators sit or stand. This arrangement simplifies replenishment of the parts, which can arrive at the outside of the line and roll down chutes to the operators. Unfortunately, this arrangement tends to trap the operators inside the loop and require them to mount stairs or duck in order to get out. If the operators are on the outside, then part replenishment becomes awkward. Forklifts run by at close proximity and must approach the operators to one side or the other instead of opposite.

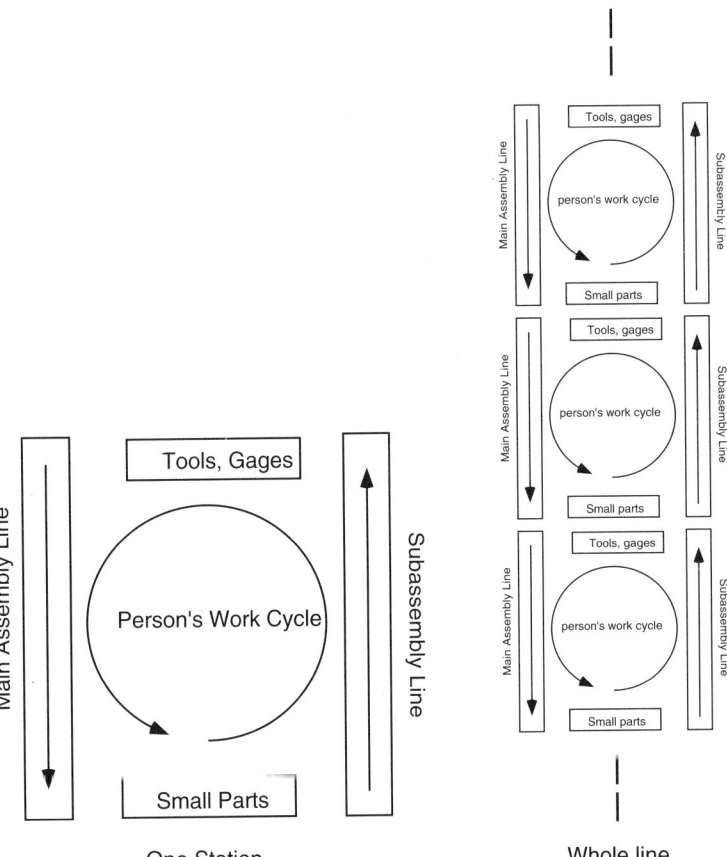

One Station

Whole line

FIGURE 16-14. A Cellular Assembly System. This architecture, shown on the right, consists of many cells, one of which is shown at the left. The final assembly line is continuous and runs down the left side of the string of cells. A person is inside each cell. He walks from point to point, picking up small parts from trays in the cell and larger subassemblies from the conveyor at the right. He works as he walks, and he places his finished unit on the main line or onto a partially finished assembly on that line. Production rate can be varied by changing the number of people in the cell.

Extra space along the line is needed to accommodate both operators, fork lifts, and boxes of parts. Chutes are next to the operators instead of opposite.

16.F.5. U-Shaped Cell (Often Used with People)

A U-shaped architecture is a variation on a line and can be used if the line is short. All the operators are near each other and can communicate. Especially important is the fact that the end and beginning are near each other so that problems discovered at the end but originating earlier can be discussed and fixed quickly.[14]

---

[14] A famous cartoon shows two assembly line workers. One says, "Tomorrow I'm retiring after thirty years at this workstation. I'm going to visit the end of the line and find out what we make here."

16.F.6. Cellular Assembly Line

Assembly or fabrication cells employ one operator to do several tasks. These tasks are often associated with a coherent set of operations, such as all the machining steps on a part or all the assembly steps in a subassembly. An advantage of this arrangement is that its production rate can be varied by adding or removing people. Another advantage is that one person has cognizance over the coherent set and can better understand and diagnose problems. However, people in such cells are constantly on the move. As with team assembly, there is more walking time per unit of assembly time than on a line. Like a line, people can be paced by the equipment and often have little or no time to rest. Figure 16-14 is a drawing of such a line, used for assembling automatic transmissions at a plant in Japan.

16.G. QUALITY ASSURANCE AND QUALITY CONTROL

Many authors have studied the general problem of the cost of lost assembly cycles due to failed assembly operations. For example, see [Boothroyd]. Boothroyd includes the tradeoff between more costly parts that cause fewer failures and the cost of lost assembly time and equipment utilization. Another way an assembly system can lose capacity is through the need to rework some assemblies. The lost time is costly over and above the possible cost of scrapped parts because the resources, which must be paid for continuously, do not produce good assemblies on each cycle. Here we discuss two aspects of this problem: (1) calculating the cost of rework and (2) determining a strategy based on assembly sequence choice for minimizing the amount of disassembly needed once a fault is found.

16.G.1. Approaches to Quality

The "quality" of a product is an important determinant of the product's potential for market success. "Quality" has many components, some quantifiable, some inherently not; some subjective, others quite objective.

The subjective quality components include styling, appearance in a showroom, presentation, certain simple comparisons with competing products, some aspects of fit and finish, and aspects of the buyers' and users' expectations. These matters are vitally important to a product's success, but cannot concern us here.

One of the major achievements of lean manufacturing was getting it right the first time ([Womack, Jones, and Roos]). A metric for judging this is to see how big is the rework area at the end of the line, where smaller is better.[15] Many steps are involved in attaining good first-time quality. These include simplifying the tasks, carefully planning and simulating each assembly task, providing proper tools and training, and allowing enough time to complete the assembly operations.

The quantitative and objective components of quality may show up in whole batches or populations of a product, or in occasional individual items. The quality of whole batches may be determined by the adequacy of the original design, the materials, or the production processes, whereas a single item's quality is usually due only to production adequacy. Quantitative components include other aspects of fit and finish; whether a product works when it is delivered; how many of a population do not work at various stages of production and what is the distribution of problems; comparisons of performance indicators of finished products with performance standards; how often products must be serviced for other than planned maintenance; and the like. Many of these matters are of concern to us here.

Several ways of obtaining high quality are known. Broadly speaking, these are as follows:

1. "Designing quality in," that is, choosing materials, dimensions, tolerances, and procedures so that the likelihood of the desired outcome is very high. Often this approach leads to expensive products, but Taguchi has shown that systematic statistical techniques can be used to choose among items or options of similar cost so as to obtain higher quality.

2. Monitoring production processes to ensure that they continue to perform as expected. A statistical approach may be taken here as well, by improving the process until no systematic or repeating error remains. The residual nonsystematic error is acknowledged to be purely random and uncontrollable; if it is still too large, a different process must be tried. If it is small enough, occasional sampling is sufficient to detect deviations that require correction. This approach is broadly referred to as statistical process control and is discussed in Chapter 5.

3. Testing each product unit at one or more points during its production. The possibility of testing raises many questions, and the set of answers to these many questions constitutes a "test strategy." Matters to be addressed include what faults could occur and what could cause them; what tests are possible, and how many of them to do; where in the assembly sequence various faults become testable, and which of several testable states for each fault to choose; whether to modify an assembly sequence to accommodate or enhance testing; and so on.

This section is an introduction to several aspects of inspection and testing of assemblies and briefly examines the tradeoff between more costly but higher quality parts and extra testing and repair of lower quality ones.

---

[15] Size of the rework area has to be normalized to the production rate, so a reasonable metric is the fraction of a shift's production that can fit in this area. One or two percent is considered good.

16.G.2. Elements of a Testing Strategy

Testing serves as one means toward the delivery of a fault-free product. A proper test program screens for common product faults and directs faulty units or subassemblies to rework. Testing also documents the statistics of faults, providing basic data for establishment or revision of the strategy.

Manufacturing and assembly process improvement, on the other hand, directly reduces the frequencies of fault occurrences. It is a more fundamental means toward quality control and improvement. Taken to the limit, it can pre-empt various, or even all, parts of a testing strategy. We may distinguish two opportunities for process improvement, namely part fabrication and part assembly.

We may divide assembly considerations into two classes, those involving single part mates and those covering the entire assembly sequence.

Single part mates are subject to six mistakes: wrong part, damaged part used anyway, missing part, incorrect assembly action (not tight enough, not fully mated, etc.), dirt or other contamination, or damage caused by the mating action itself. Both manual and automatic assembly are subject to each of these. People tend to make more random mistakes and rarely make the same mistake again and again once their training is complete. Automatic assembly tends to behave in the opposite way: A jammed feeder may result in twenty assemblies in a row with the same missing part. The potential for extremely high quality seems better for machines because they "tolerate" the extravagant vigilance that is necessary better than people do. Automation technology can also improve assembly processes by mechanizing assembly steps in which the details of motion, part trajectory, or technique are critical to the success of the step or avoidance of damage. Attention to many details of design for assembly also improve the chances that the correct part will be used and will be inserted properly. All these single mate considerations affect the probabilities of failures and thus affect the scope and shape of the testing strategy.

Beyond single part mates lies the domain of the assembly sequence as a whole. Assembly sequence is a major basis for a choice of test strategy. A test strategy is the list of tests that will be performed, chosen from the generally larger list of all tests which can be performed. A potential fault cannot be tested for until the requisite parts have been mated. Design often constrains the sequence and we have no options. One can choose from many examples: A

case must be assembled and closed before it can be tested for leaks; soldering of components must be done before a printed circuit board can be checked for solder bridges and performance of major functions; valve train clearances of a pushrod overhead valve engine cannot be checked until the head is bolted to the specified tightness.

In other situations, there are choices between assembly sequences that offer different test opportunities. One sequence may create a subassembly that can be tested while another will leave out a crucial part until the subassembly has already been mated to the final product; one sequence will permit a likely fault to be detected early in the assembly while another will delay detection.

Summarizing, assembly sequence determines fault viability, viability determines whether a test can exist, and the strategy choice is made from the set of possible tests.

In addition, there are often better and worse assembly sequences in terms of the potential for damage during assembly, directly affecting fault occurrence. One sequence will permit parts to be securely jigged or gripped on well-toleranced surfaces, reducing the likelihood of wedging and jamming; another will not. One sequence will permit each mate to involve securely gripped single parts while another will require bulky groups of loosely stacked items to be mated simultaneously.

16.G.3. Effect of Assembly Faults on Assembly Cost and Assembly System Capacity

When an assembly fault is detected, the cost of assembly goes up in several ways. Time is spent repairing the assembly, parts may be scrapped, or the entire assembly may be scrapped. In addition, the assembly system, its equipment, and its operators will have spent time on the original assembly without delivering a good one. This reduces the effective capacity of the system.

As a simple example, suppose the assembly system consists of one workstation with final test, plus one rework station. Reworked assemblies are sent back to the workstation where they must be completely reassembled and tested again. This extreme assumption will be relaxed shortly. If q units are processed in this fashion and m of them fail and must be reworked, then the yield fraction y of the system is

$$y = \frac{1+q}{1+q+m} \qquad (16\text{-}8)$$

If the cost of the parts is M, the cost of assembly and test is P, and the cost of rework is $R + \alpha M$ (α is the fraction of parts that must bc rcplaccd during rcwork), then if one repair cycle is sufficient, the cost including rework divided by the cost if no rework were needed is called C_r:

$$C_r = 1 + \frac{(1 + (R + \alpha M)/P)m}{(1 + M/P)q} \qquad (16\text{-}9)$$

m can be eliminated from Equation (16-9) using Equation (16-8) to yield

$$C_r = 1 + \frac{(1 + (R + \alpha M)/P)}{(1 + M/P)} \left(\frac{1}{y} - 1\right)\left(\frac{1}{q} + 1\right) \qquad (16\text{-}10)$$

A more general analysis, described next, allows assemblies to be reworked more than once if necessary, and it permits more complex networks of assembly architectures to be analyzed. Consider the assembly network shown in Figure 16-15. It makes two subassemblies and joins them. Each subassembly is tested and possibly reworked. The final assembly is also tested with three possible outcomes as shown. Subassemblies and final assemblies can circulate inside this system as long as necessary. The question is, What is the average cost of assembly and the effective capacity of the system compared to one where every assembly passes the tests the first time?

A network equivalent to Figure 16-15 is shown in Figure 16-16.

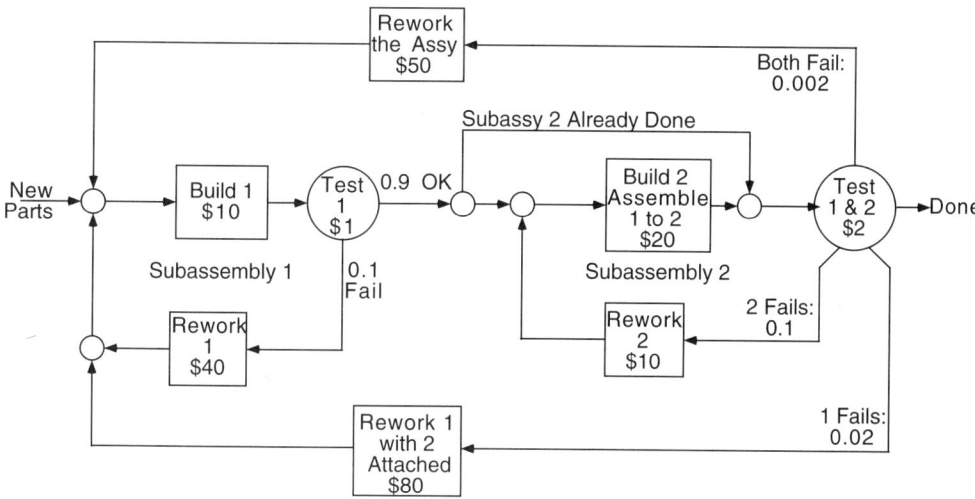

FIGURE 16-15. An Assembly Network with Several Rework Loops. The cost of each step is shown, along with the probability of rework at several test points. The cost of assembly and test is $33 if nothing fails. For the probabilities shown, the actual cost averages $44.76, or 36% more. The branch labeled "subassy 2 already done" covers the situation where the whole assembly fails the final test but only subassembly 1 is found to need repair.

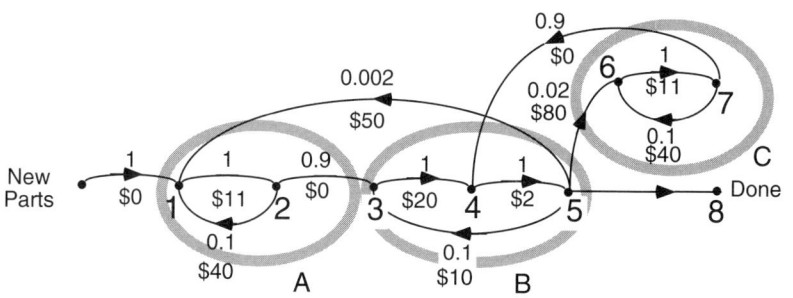

A Build/Repair Subassembly 1 and Test It
B Build/Repair Subassembly 2 and Test Both
C Repair/Rebuild 1 While Attached to 2

FIGURE 16-16. Network Equivalent to Figure 16-15. The arc in Figure 16-15 marked "subassy 2 already done" is replaced by segment C in the network to permit separate calculation of the flows of assemblies that follow that path.

To find the effective yield of this system, we need to find the fraction of assemblies that pass through the rework loops. Equivalently, we want to know the average flow rate of products on each branch of the network given that the flow in is unity. The average flow out of node 8 divided by the average flow into node 1 in Figure 16-16 is the yield fraction of the system.

The problem of finding average or steady-state flows in a network is quite easy to solve. One simply writes a conservation of flow equation at each node where branches merge or separate. The sum of all flows in must equal the flows out. The flow out of node i is called y_i. If the probability of going from node i to node j is called p_{ij} (known quantities), then the flow from node i to node j is f_{ij} given by

$$f_{ij} = y_i p_{ij} \qquad (16\text{-}11)$$

where we must have $\sum_j p_{ij} = 1$ for each i.

To conserve flows at any node j, we write

$$y_j = y_j p_{jj} + \sum_{k \neq j} y_k p_{kj} + x_j \qquad (16\text{-}12)$$

where x_j represents any new assemblies entering the network and p_{jj} is the probability that an assembly will immediately reenter the node it just left. We rule this out by setting $p_{jj} = 0$ for all j. Arranging all the p_{ij} into a matrix called P, all the y_j into a vector called Y, and all the x_j into a vector called X, we may rewrite Equation (16-12) as

$$Y = P^T Y + X \qquad (16\text{-}13)$$

whose solution is

$$Y = [I - P^T]^{-1} X \qquad (16\text{-}14)$$

The equilibrium flows f_{ij}, grouped into a matrix called F, may be found from Equation (16-11) as

$$F = box(YY, P) \qquad (16\text{-}15)$$

where YY is a square matrix whose columns are each vector Y and $box(A, B)$ is a matrix whose i, j entry is $A_{ij} * B_{ij}$.

If c_{ij} is the cost of moving the assembly from node i to node j, then the cost of processing an assembly through the system is

$$cost = \sum_i \sum_j f_{ij} c_{ij} \qquad (16\text{-}16)$$

MATLAB code for these calculations for the example network in Figure 16-16 appears in Table 16-2.

Suppose we wanted to improve this network and explored the result of spending an extra \$2.00 to cut the

TABLE 16-2. MATLAB Code for Cost of Rework for the Network in Figure 16-16

```
»P=zeros(8)

»C=zeros(8)

»P(1,2)=1
»P(2,1)=.1;
»p(2,3)=.9;
»P(2,3)=.9;
»P(3,4)=1;
»P(4,5)=1;
»P(5,3)=.1;
»P(5,1)=.002;
»P(5,6)=.02;
»P(5,8)=1-P(5,6)-P(5,3)-P(5,1);
»P(6,7)=1;
»P(7,4)=.9;
»P(7,6)=.1;
»X=[1 0 0 0 0 0 0 0];
»X=X'

»Y=inv(eye(8)-P')*X
```

TABLE 16-2. (Continued)

```
Y =

   1.1136
   1.1136
   1.1162
   1.1390
   1.1390
   0.0253
   0.0253
   1.0000

≫YY=[Y Y Y Y Y Y Y Y]

≫F=box(YY,P)

F =
0.0000 1.1136 0.0000 0.0000 0.0000 0.0000 0.0000 0.0000
0.1114 0.0000 1.0023 0.0000 0.0000 0.0000 0.0000 0.0000
0.0000 0.0000 0.0000 1.1162 0.0000 0.0000 0.0000 0.0000
0.0000 0.0000 0.0000 0.0000 1.1390 0.0000 0.0000 0.0000
0.0023 0.0000 0.1139 0.0000 0.0000 0.0228 0.0000 1.0000
0.0000 0.0000 0.0000 0.0000 0.0000 0.0000 0.0253 0.0000
0.0000 0.0000 0.0000 0.0228 0.0000 0.0025 0.0000 0.0000
0.0000 0.0000 0.0000 0.0000 0.0000 0.0000 0.0000 0.0000

≫C(1,2)=11;
≫C(2,1)=40;
≫C(3,4)=20;
≫C(4,5)=2;
≫C(5,1)=50;
≫C(5,3)=10;
≫C(5,3)=10;
≫C(5,6)=80;
≫C(6,7)=11;
≫C(7,6)=40;

≫cost=sum(sum(box(C,F)))

cost =

  44.7608

≫FF=sum(sum(F))

FF =
  5.6720

≫EX=FF/5

EX =

  1.1344
```

Note: The total cost of assembly and rework is $44.76. If there were no rework, the cost would be $33.00. FF is the sum of all elements in matrix F—that is, the total flow of product in the system. If there were no rework, we would have $FF = 5$. The excess of FF over 5 is the amount of excess product flow in the system due to rework. This amounts to 13.44% excess load on the system, which can be interpreted as a loss of capacity, diverted capacity, or extra capacity that the system must be built to provide that would not be needed if there were no rework. The CD-ROM that is packaged with this book contains this code in a file called "reworkloops."

failure rate of subassembly 1 in half. The total cost of assembly, including the extra $2.00, falls to $42.40 from the original $44.76, on which basis we may conclude that the extra investment was worth it. On the other hand, if it costs $5.00 to cut subassembly 1's failure rate in half, we find that the total cost rises to $45.30, indicating that test and rework is less expensive than attempting to prevent rework in the first place. Issues like this come up all the time.

16.H. BUFFERS

16.H.1. Motivation

Most manufacturing and assembly processes require a large number of operations. While we have seen that there are many possible architectures for assembly systems, the great majority involve a series of workstations, with the output of one being the input to the next. In the ideal situation, each station has exactly the same amount of work, in which case we would expect partially finished assemblies to move simultaneously from one station to the next, and no workstation would be overworked or underworked. In this case, the line is called perfectly balanced.

Two issues intervene to make this ideal situation unrealistic. The first is that assembly operations typically do not take the same length of time. If there is so much work that each workstation has time for only one operation, then the longest operation will set the pace for the entire line, and all other workstations will have idle time. The line is then said to be unbalanced. If there is less work required, it may be possible to put several operations at one station. This presents the opportunity to achieve better balance by grouping operations. Team assembly achieves better balance not only by grouping operations but also by permitting cross-trained operators to help each other if one falls behind ([Engström, Jonsson, and Medbo]). Different feasible assembly sequences offer different grouping opportunities, as discussed in Chapter 7.

Assembly operations can take different amounts of time for several reasons. Human operators naturally vary from cycle to cycle as well as during the day as a result of fatigue. Even mechanical operations like calibration can take unpredictable amounts of time. Finally and most commonly, when there is model mix on the line, different versions of the product will require more parts or different parts or different assembly methods, again leading to variations in process time for the "same" task. At Toyota's Georgetown, Kentucky, assembly plant, which assembles several versions and many customer options, work content in 19 final assembly stations varied as much as 30% from the simplest to the most complex car in one sample ([Mishina]). The same study showed that in September 1992 the whole Toyota corporation made 197,000 cars expressing 38,633 of the over one million version combinations possible. 23,010 of these varieties were represented in only one car that month. That is, 60% of the cars made that month were unique, although the differences did not necessarily entail major changes in assembly time.

The second reason why assembly operations can take different amounts of time is the likelihood that not every assembly operation will be successful. Either there was a mistake, or a bad part, or a malfunction in the operation itself. If the station is a machine, the machine may break or simply come to a stop for a simple reason that is easily fixed. Typical stoppages include jammed part feeders or a part that fails to assemble due to variation. Figure 16-3 indicates that most stoppages are very short, but most assembly operations are also very short, so even a few stoppages can cause many assembly cycles to be lost.

If an assembly system is built so that each station feeds the next one directly, then in effect the stations will have to operate in lockstep because they cannot operate unless they have a partially completed assembly available. This in turn means that if one station stops, even for a second, or finishes late for any reason, the whole line will stop and the clock will keep ticking, but no assemblies will be worked on or finished. For this reason, most lines are built with buffers between some or all of the stations. A buffer is an empty space where partially completed assemblies can wait after leaving a station before entering the next one.[16] These buffers decouple the stations from each other and, in theory at least, permit the system to keep operating even if a station has stopped or is late finishing. The stations downstream of the stopped/late one can work on the items in their input buffers until those buffers are exhausted, at which time those stations are said to be starved. The

---

[16]Sometimes, a buffer is called a queue. Sometimes, the set of waiting parts or assemblies is called a queue. The size or capacity of the buffer is the number of items it can hold, while the size of the queue is the number of items actually waiting at any one time. Most mathematical theory about buffers assumes that they have infinite capacity, but this is unrealistic.

stations upstream of the stopped/late one can also continue to work until their output buffers fill up, at which point they are said to be blocked. A thought question at the end of the chapter asks the reader to think about whether the stations that can keep working are really accomplishing anything in terms of the whole system's output or not.

Some kinds of assembly machines are incapable of accepting buffers. The rotary dial machine is one of these. Usually the assemblies are carried directly by the dial from one station to the next as the whole dial rotates. Cam-operated assembly machines also typically do not have buffers because the workstations operate synchronously, making it impossible for one station to stop and others to continue. Buffers can be placed between dial or linear cam machines if several of them are in series.

Other machine architectures permit buffers because they consist of independent stations connected by a conveyor. If the conveyor operates by gravity or friction, the assemblies can ride it until they enter the next station or bump into another assembly waiting to enter.

The main objection to buffers is economic: assemblies in buffers represent work in process inventory, which has value. The company pays for those parts, and the more parts are sitting in buffers, the more money is tied up. The assembly line is bigger and thus takes up more valuable space in the factory. Nevertheless, stoppages are frequent, and if the whole assembly process stops for lack of buffers, all those parts are sitting there doing nothing, and the same economic argument can be applied, although the details are beyond the scope of this book. See [Boothroyd] and [Nof, Wilhelm, and Warnecke] for discussions of this topic.

So we are faced with a tradeoff: If the cycle time is set to the average, longer operations will fall behind. On the other hand, if the cycle time is lengthened until there is always time for the most pessimistic operation time, then equipment or people will be idle everywhere else on the line. This, too, represents a waste of resources. In Section 16.I we discuss Toyota's approach to this tradeoff.

16.H.2. Theory

If it is agreed that some buffers are necessary, then how big should they be? The following is an approximate analysis, based on [Boothroyd]. It addresses time lost due to machine breakage and repair, but other reasons for an unusually long assembly time can be treated the same way.

Suppose we have an assembly machine with s identical stations. That is, each station operates with the same cycle time t and stops, for whatever reason, during or after the same fraction of operations x, and it takes time T to fix the problem and get the station started again. Equivalently, we can imagine that a fraction x of the time, the station will take T longer to finish its work. Suppose that we decide to put identical buffers between the stations, sized to hold up to b partially completed assemblies. What should b be? Boothroyd shows that, under these assumptions, the total downtime d on any station while building N assemblies is

$$\frac{d}{Nx} = T + [2T - bt] + [2T - 2bt] + [2T - 3bt] + \cdots \tag{16-17}$$

where the series terminates at the last positive $[\cdots]$. This may be rewritten as

$$\frac{d}{Nx} = T + \sum_{i=1}^{\frac{s-1}{2}}(2T - ibt) = sT - bt\frac{(s^2 - 1)}{8} \tag{16-18}$$

where s is the number of workstations in the line or the largest integer that satisfies $2T > bt(\frac{s-1}{2})$.

Each additional $[\cdots]$ in Equation (16-17) represents the effects of stations farther from each other, and at the point where the $[\cdots]$ becomes negative, the stations are so far apart that they can be fixed before the effects of their stoppage propagate to distant stations. The inequality following Equation (16-18) expresses the same phenomenon.

The fraction of time that the entire system is down is given by

$$D = \frac{d}{Nt + d} = \frac{x\left[\frac{T}{t} + \sum_{i=1}^{\frac{s-1}{2}}\left(2\frac{T}{t} - ib\right)\right]}{1 + x\left[\frac{T}{t} + \sum_{i=1}^{\frac{s-1}{2}}\left(2\frac{T}{t} - ib\right)\right]} \tag{16-19}$$

Table 16-3 evaluates $uptime = 1 - D$ using Equation 16-19 for $x = 0.01$ and two values of T/t.

Table 16-3 shows that adding buffers of size 2 to a synchronous line ($b = 0$ in the table) greatly improves the uptime. However, stoppages must be fixed within a very few cycles. For a high-speed machine with a cycle time of 5 seconds, this means, for $T/t = 5$, that workers have only 25 seconds to recognize that the line has stopped, walk (not run) to the stopped station, diagnose the problem and fix it, and then push the button to restart the station. This example shows why uptimes rarely reach 85%, which is

TABLE 16-3. Assembly System Uptimes

T/t	Buffer Size	Number of Coupled Stations	Uptime
5	0	They are all coupled; uptime based on 10 stations	0.666666666
	2	10	0.798403194
	4	6	0.888888889
	6	4	0.91954023
	8	3	0.934579439
	10	1	0.952380952
10	0	They are all coupled; uptime based on 20 stations	0.33333333
	2	20	0.49937578
	4	10	0.66445183
	6	7	0.74626866
	8	5	0.79365079
	10	4	0.82474227

Note: If $b = 0$, there are no buffers and the machine operates synchronously. For larger buffers, the number of stations whose stoppages affect other stations drops, and the uptime improves. These results compare reasonably well with the much more sophisticated models in [Enginarlar et al.].

considered good by factory operators. It also shows why it is recommended that synchronous machines be limited to a few stations, say 10. If more stations are needed, separate small machines should be used, linked by buffers.

16.H.3. Heuristic Buffer Design Technique

The analysis above gives us clues as to how big buffers should be. A full analytical treatment is beyond the scope of this book. See [Gershwin]. It would have to deal not only with average station failure rates x but also with the probability distribution of values of x. Usually, that is better done with discrete event simulation, which is discussed briefly in Section 16.J.

A simple heuristic design technique is based on the idea that we want the buffers not to become empty or jammed full during the time a station is down. Since a buffer could just as likely be upstream of the stopped station as downstream, both starving and blocking are equally likely. Thus the ideal steady state for a buffer is to be half full. The number of assemblies (or the number of empty spaces) in the buffer represents the number of assembly cycles available to the repair crew before the buffer empties (or fills completely). If T/t is the number of cycles needed to fix a stopped station, then we need

$$\frac{b}{2} = \frac{T}{t} \quad \text{or} \quad b = \frac{2T}{t} \qquad (16\text{-}20)$$

When b takes this value, the very first $[\cdots]$ in Equation (16-17) is zero and the inequality in Equation (16-18) is satisfied for $s = 1$, indicating that all the stations are

effectively independent, and the line has the same uptime as each individual station. But this is a very conservative design and may take up too much space. Shorter buffers couple the system and may degrade performance. A simulation is usually used to study this issue.

16.H.4. Reality Check

The above analysis does not cover every possibility. First, assembly and fabrication stations rarely have the same operating time, except for machines that are designed to operate synchronously and have no buffers. Usually one station takes the longest and is called the bottleneck station. A cycle lost on the bottleneck station is a cycle lost forever. Faster stations can catch up if they are stopped for a time, but the bottleneck cannot. For this reason, one of the principles of the Theory of Constraints ([Goldratt]) is that the buffer upstream of the bottleneck must never be allowed to become empty and the buffer downstream must never be allowed to become blocked. So our real focus should be on the sizes of those two buffers more than on any others.

Naturally, if there are no buffers anywhere, then if any station stops, the bottleneck will stop. But if we put just one buffer in the line, the probability that a stop on another station will immediately stop the bottleneck is cut from 1.0 to 0.5. If there are buffers between all the stations, then the likelihood that a stop elsewhere will stop the bottleneck is greatly reduced. This implies that buffers at the other stations are not there for the benefit of those stations at all but rather to keep the bottleneck running.

If the buffer upstream of the bottleneck starves (or the one downstream blocks), then the time to fill it back up halfway (empty it half way) is proportional to the ratio of the top speed of the station upstream (downstream) to the speed of the bottleneck. If those speeds are similar, it could take a long time to reestablish ideal buffer contents. Until then, the bottleneck is at risk of being blocked or starved.

Second, downtimes at stations vary in duration according to some probability distribution, and they could in theory be quite long. Buffers have finite size and thus could become starved or blocked even if the average calculations from the previous section tell us that the buffers are large enough to prevent this. In general, extra length must be added to buffers in proportion to the standard deviation of any uncertainty that affects buffer size, such as variation in operator time, variation in time to fix a broken machine, or variation due to different models being assembled.

Finally, if the factory undertakes a program of continuous improvement, then efforts will naturally focus on the bottleneck station. If its speed is improved, then some other station will become the bottleneck and the operators must focus on the buffers adjacent to it. Therefore, any station that has the potential to become the bottleneck needs to be treated as such during the design of the system.

16.I. THE TOYOTA PRODUCTION SYSTEM[17]

16.I.1. From Taylor to Ford to Ohno

As discussed in Chapter 1 and Chapter 5, manufacturing has passed through several stages of development. Three recent stages are associated with individuals: Frederick Taylor, Henry Ford, and Taiichi Ohno. Taylor invented scientific analysis of manual labor. He carefully timed workers and advocated dividing work into the smallest chunks possible for both analysis purposes and workplace design. Ford arranged these small chunks into a serial assembly line and efficiently produced identical or nearly identical cars for many years. Much of the efficiency of his factories stemmed from heavy investment in fixed automation. Later auto manufacturers adopted this technique to a greater or lesser degree. It flourished in an environment of little model mix on the assembly line. Ohno of Toyota faced the twin problems of educated workers who would not tolerate short assembly cycle times and forced paced assembly along with educated consumers who wanted more choice in the products they buy. The Toyota Production System (TPS) is the result of accommodating those problems. The TPS is a complex topic, and space prevents more than a brief discussion. See [Monden] for the authoritative description. The role of the TPS in the development of the Japanese car industry is described by [Cusumano]. The TPS also comprises a complex set of social, managerial, and cultural patterns ([Spear and Bowen]).

16.I.2. Elements of the System[18]

16.I.2.a. Recognition and Elimination of Waste

The basic motivator for the TPS is identifying and removing waste. In the late 1940s, Toyota, like other Japanese companies, was very short of money. Ohno decided that cutting inventory would save money. This meant reducing the size of buffers and thus reducing the variations in process time and model change effects that usually force larger buffers to be used. If process time variations could be reduced, idle time of operators could be reduced as well. This led to a focus on efficient utilization of the parts and the people, even if it meant less efficient utilization of the equipment. In the investment-heavy Ford system, the focus is often on efficient utilization of the equipment. But it represents a small part of the cost. At the same time, a focus on keeping it busy makes it difficult to cut production if demand falls, and often difficult to change it over if demand shifts to different models. The TPS is people-heavy but the ability to shift people around is encouraged by cross-training, and some reduction in capacity in Japanese plants is accomplished by having about 30% temporary workers who bear the brunt of layoffs.

16.I.2.b. Just-in-Time Production

To ensure that there is little waste of parts, the TPS tries to build only what has been ordered. This in turn requires

---

[17]This section is based on [Mishina] and [Monden].

[18]Space limits prevent discussing every element. This section presents a selection.

pulling work from upstream stations based on customer orders rather than pushing work downstream based on a pre-planned schedule. The aim is to produce what is wanted, when it is wanted, and where it is wanted.

To accomplish this, the system is run by passing orders upstream in the form of "kanbans." Kanban is the Japanese word for ticket, but the kanbans act like money in the sense that they are used by downstream stations to buy parts from upstream stations. For this reason, if the orders are entered at the very end of the line, a signal representing what was just made will propagate upstream, causing the same things to be made over and over. In order to guarantee that the actual mix of incoming orders is reflected upstream, and to combat the variations caused by model mix, Toyota employs production smoothing or load leveling, which are discussed next. Furthermore, as discussed in Section 16.I.3, the order stream may be inserted in the middle of the line instead of the very end.

16.I.2.c. Production Smoothing or Load Leveling

Orders from customers do not arrive in the best sequence for production. Suppose the plant makes car A and car B, among others. Assume car A takes much less than the average time to make, while car B takes much longer. If 10 orders each for car A and car B arrive, it may disrupt the line to schedule them each in a solid batch. If the factory operates at a standard pace, operators working on a solid batch of 10 A's will have time left over and nothing to do. On the other hand, operators working on a solid batch of 10 B's will fall behind. It is better to interleave these orders as ABAB . . . so that over these 20 cars the operators will take about the average time.

Another kind of smoothing is also pursued. Suppose the plant receives orders for sedans, hardtops, and wagons in the following proportions: 50% sedans, 25% wagons, and 25% hardtops. If these different cars use some different parts, then demand for the parts will vary. As in other respects, a goal of TPS is to reduce variation and thus reduce the need for buffer stocks that absorb that variation. On this basis, one should not make all the day's sedans first, then all the wagons, and then all the hardtops. Instead, one should interleave them in a pattern like SSWHSSWHSSWH. . . ([Monden], pp. 68–69).

Naturally, these two formulae for sequencing the cars cannot both be obeyed, although one can approach both goals. Toyota actually favors the second kind of smoothing and gives it priority when solving its sequencing problem each day ([Monden], p. 254). If time for W is longer than

for S and H is shorter, one might then make the above cars in the sequence SHSWSHSWSHSW. . . if that smoothed the different station times better.

16.I.2.d. Short Setup Times

Since the TPS involves mixing the different orders rather thoroughly in order to keep variation in demand down, some upstream processes, particularly machining and stamping operations, have to change over frequently. This will never be economical unless changeovers can be done quickly. This is a topic of its own, exemplified by the single minute exchange of dies process (SMED) ([Shingo]).

16.I.2.e. Single Piece Flow

In the TPS, individual orders are treated individually, so that large batches of parts and assemblies are not made. This is sometimes called single piece flow. Among the advantages are short waiting times for parts of a particular type, low work in process inventories, and quick discovery of mistakes. If 5,000 of part A are made before any of part B are made, products that need part B will wait while all 5,000 A's are made, or else a large (wasteful) supply of B's parts must be held in inventory. If a mistake is found in the 500th A, all 5,000 may contain the mistake and have to be reworked or scrapped. Single piece flow supports another element of the TPS called the visible control system, in which it is easy to see what is happening to every part. [Linck] reports that automobile component plants that use single piece flow have lower mistake rates and can make more units with fewer employees in less floorspace than batch process plants making the same components.

Single piece flow is accomplished in machining operations by creating a cell architecture. A few operators walk individual parts from machine to machine. The parts follow their required machining sequence but the operators visit the machines in the sequence in which they finish and need a new part. The operators make the parts called for by the kanbans. If demand falls, fewer operators are assigned to the cell and fewer kanbans arrive.

The alternative to single piece flow is batch processing. Batch size is governed by the economic lot size formula, which balances cost and time for changeovers with cost of holding the batch as work in process inventory. According to this formula, shorter changeovers make smaller batches economical, although this forces transport events to happen more often and may require more resources to carryout these events.

In industries like aircraft, where the products are large, there is no alternative to single piece flow.

In addition to the advantages of single piece flow discussed above, batch processing requires investments in transport equipment that can carry a whole batch or a large fraction of it. This can create problems of its own in the form of a transport department with its own procedures and costly equipment.[19]

16.I.2.f. Quality Control and Troubleshooting

In order for a low work in process inventory system to operate successfully, there must be very few assembly mistakes. The TPS emphasizes mistake reduction by several means, including foolproofing operations and empowering operators to inspect their own work. Reduction in inventories also makes problems appear rapidly because workers are affected quickly when their buffers run out. Ohno called this "lowering the water so you can see the rocks." It is the reverse of the strategy of using buffers as protection against unforeseen events.

16.I.2.g. Extension to the Supply Chain

It took Toyota a number of years to discover that the TPS had to be extended to its suppliers in order to gain full advantage. The basic issue is the need to reduce costs all down the supply chain. The TPS recognizes waste in the form of idle labor and idle parts or assemblies. The cost of production at any stage in the supply chain is mostly the cost of parts and assemblies purchased from the stage below. Labor (and equipment depreciation) is a small proportion of the cost. But, summed over the entire chain, labor is the largest proportion, as discussed in Chapter 18. Thus, if a company looks only at its own operations, it will focus more on the materials and less on the labor. But if it looks at the whole chain, it will focus on labor. Since Toyota knew how to make efficient use of both labor and materials in its own plants, it undertook to teach its suppliers to do the same. It also taught its suppliers how to get along with less fixed equipment and to be able to cut costs when demand fell.

---

[19] A car engine plant visited by the author consisted of separate machining lines linked by transport vehicles that brought several parts at once. When a line lacked parts, its operators blamed the transport department. The transport department blamed the upstream line for not notifying it when parts were ready to ship. The problem was solved by directing the downstream operators to get the parts themselves.

16.I.3. Layout of Toyota Georgetown Plant

Toyota's design for the Georgetown, Kentucky, plant shows a sophisticated mix of pull- and push-type production (Figure 16-17). As described in [Mishina], final orders are smoothed as described above and sent to the beginning (not the end) of the line just after the press shop. The line runs as a conventional push-type conveyor from that point forward. However, the subassembly feeder lines and supplier lines operate on a pull basis and supply parts according to what is consumed by the main line. Since the main line is sequenced to represent the average flow of orders, the supplier and subassembly lines produce versions according to that average or use the concept of delayed commitment to modify their output at or near the end of their sub-lines in order to satisfy each individual order. A small amount of inventory in the form of a "convenience store" is held at the ends of these lines as well.

16.I.4. Volvo's 21-Day Car

Volvo has built a factory in Ghent, Belgium, that delivers a car to a customer twenty-one days after it is ordered. Typical delivery intervals are six to eight weeks in most countries. A variety of techniques, many of them similar to Toyota's, contribute to Volvo's ability to deliver this quickly. Unlike the Denso panel meter, where product design and assembly process design were crucial enablers, Volvo's process uses largely standard part design and fabrication processes and depends instead on carefully managed logistics. Volvo has decided carefully where and when to make each subassembly (make ahead and keep in stock, make only when the customer orders, make at lineside, make at supplier, etc.). The elements of the approach are illustrated in Figure 16-18.

Like Denso, Volvo presents customers with a limited amount of variety from which to choose, although the range is still generous. Three body styles and twenty colors are available. The customer can choose seat coverings, interior colors, and any or none of the following: roof rails, air conditioning, cruise control, electric windows, and electric mirror. Several engine options are also available, as are transmission options.

The strategy includes partitioning these items according to their value and the time it takes to make them. High-value long-lead items like engines, transmissions, seats, and instrument panel assemblies are made at nearby

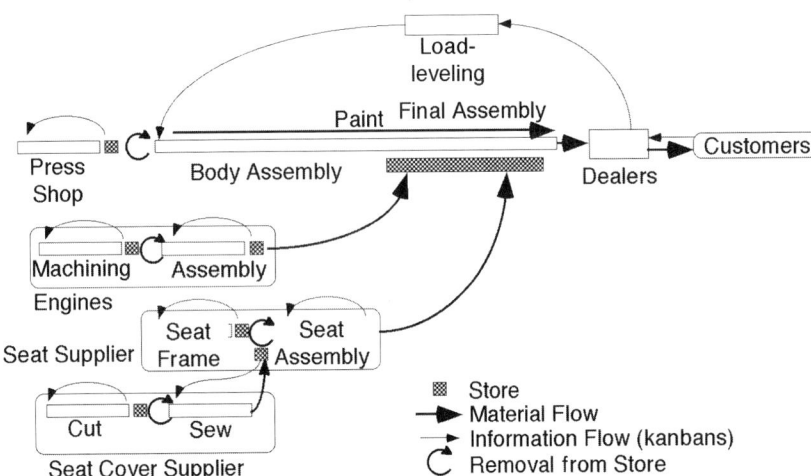

FIGURE 16-17. Layout of Toyota Georgetown Plant as of 1992. This figure shows an in-house supplier for engines, a first-tier supplier of seats, and a second-tier supplier of seat covers. One or two hours of parts from suppliers not shown are arrayed along the assembly line in what Mishina calls "stores." Press shop, engine shop, seat supplier, and seat-cover supplier operate pull systems. Final assembly starting in the body shop is a push system. According to this layout, finished engines are drawn from a store rather than being built to match a particular car. At an auto plant in Germany, the engine assembly line is notified 4.5 hours before an engine is needed by the adjacent assembly plant. Since it takes 3 hours to assemble an engine from finished parts, there is no need for a store at the end of the engine line. However, blocks are machined in large batches, and it takes three weeks to generate all the necessary varieties. (Observed by the author in 1996.) In the Volvo 21-day car system described in the next section, orders enter at the output of the paint shop buffer. This, too, permits engines to be assembled to suit each car. (Adapted from [Mishina]. Copyright © 1999 Ashgate Publishing Ltd. Used by permission.)

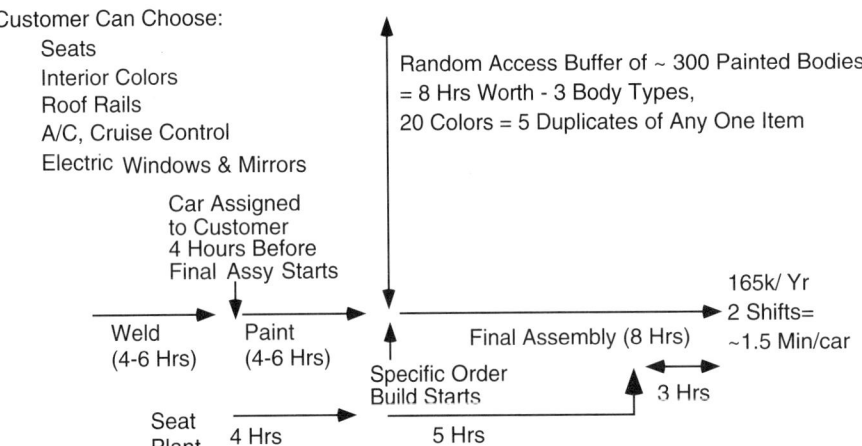

FIGURE 16-18. Volvo's 21-Day Car. The customer orders the car and many parts are marshaled in the time leading up to assembly day. A fixed variety of body styles and colors is made almost regardless of orders. Due to the possible unreliability of paint processes, cars are not painted to order. Instead, painted cars are stored in a buffer and a specific order begins to be built when one of these bodies is assigned to a customer. Many items, such as seats, are built in nearby plants to match the order and are ready at the time they are needed on the final assembly line. (Information provided by M. Etienne DeJaeger of Volvo.)

plants. Basic engines are standard and made in Sweden, but accessories can be added quickly in the final assembly plant to meet a customer's needs. Seats are similar, with power motors and fabric coverings being matters of customer choice. Medium value items with short process times like steering columns are built in the final assembly plant from standard parts that are small and not too valuable. There are big stocks at lineside of low cost small parts.

A big ballet of signals, conveyor lines, and trucks mesh these items together during an eighteen-hour period that begins with welding together stamped body parts and painting them. (Eighteen hours is typical for this overall process at most car plants.) Three body types and twenty colors makes sixty customer choices, and a buffer of three hundred vehicles ahead of final assembly thus contains five of each possible type, ready to pick when a customer's order becomes active. Seat and engine plants are notified after welding but before painting, giving them between four and nine hours notice that a particular item will be needed. A finished car rolls off the line every 1.5 minutes, two shifts a day.

16.J. DISCRETE EVENT SIMULATION[20]

An important step in the design of many manufacturing systems is the simulation of system operation. Simulation may be incorporated in the design process for specifying system characteristics or it may be used to verify the performance of a proposed system after the specification process is complete. Simulation of the type described here, called discrete event simulation, is a very powerful tool in operations research and is widely used for such problems as route and equipment scheduling for transportation systems. Consequently, numerous software tools and languages exist for system simulation. It is beyond the scope of this text to cover any particular simulation software package in depth or even to list all the available packages. Rather, the purpose of this section is to describe, in a general sense, how and when simulation may be effectively applied to the design of manufacturing systems. For a more detailed description of simulation and the available tools, the reader is referred to the references ([Pooch and Wall], [Fishman]).

Simulation is the operation of computer models of systems for the purpose of studying deterministic and stochastic phenomena expected to occur in those systems. Simulation is instrumental in the design process because it allows the engineer or analyst to:

1. Study the performance of systems without building them.

2. Study the impact of different operational strategies without implementing them.

3. Study the impact of major external uncontrollable events such as component failures without requiring them to occur.

4. Expand or compress time to study phenomena otherwise too fast or too slow to observe.

5. Realistically represent random events and nonlinear effects like finite buffer sizes that are difficult to capture mathematically.

The key to any simulation effort is the formulation of a model of the system under study. The results obtained through simulation can be only as accurate as the underlying model. The model is an abstract representation of a system or part of a system. The model describes, in some convenient way, how the system will behave under all conditions that it is likely to experience.

All discrete event simulation tools share a common modeling viewpoint—that of entities, activities, and queues. The model is a network of activities and queues through which the entities flow. The essence of constructing the model is to specify the network and the logic that governs that flow. Entities are objects that flow through the system or resources that reside in the system. Examples of entities are workers, robots, machine tools, and production parts. Activities are the productive elements of system behavior and require the participation of one or more entities in order to occur. Examples of activities are the machining of a part or the replacement of a machine's cutting tool. Finally, queues are places where entities collect when not participating in any activity. Queues may represent real aspects of the system such as inventories of materials, or they may represent fictitious quantities such as raw materials that have not yet entered the system or machines

---

[20]This section is based in part on Chapter 15 of [Nevins and Whitney].

in the idle state ready to be assigned work. In some cases, the behavior of queues may be of specific interest because the size of an inventory queue or time that machines are idle are important aspects of system performance.

Each activity has a duration, which can be a random number. The simulation starts by finding all the activities that can start because they have all the entities they need. The simulator then advances the clock until the next event, which is caused by completion of the ongoing activity that has the shortest time-to-go. The simulator distributes its entities to different queues according to the model and then looks to see if any other activities can start or finish. The simulation continues in this way until a time limit is reached or for some reason no activities can start.

The concepts of entities, activities, and queues are illustrated by a simplified model shown in Figure 16-19. This figure, called an activity cycle diagram, depicts the various activities as rectangles, the queues as circles, and the "flow" of entities as connecting lines. The flow of entities along the connecting lines is instantaneous; at all times, every entity must be either involved in an activity or waiting in a queue. The connecting lines represent the possible state changes for each class of entity. Two classes of entities are included: pallets and a cutting tool. The pallets can move between the activities and queues defined by the network paths shown by solid lines. The cutting tool is constrained to the network paths shown in dashed lines. The process that this model simulates can be described as follows:

- A part is loaded onto an empty pallet.
- The part is machined using the cutting tool.
- The finished part is removed from the pallet, which returns to the beginning of the system.
- Provision has been made for the cutting tool to be replaced when worn or broken. While the tool is being replaced, no machining can occur.
- Similarly, if there are no empty pallets, parts cannot be fed into machining.

Two features illustrated in the figure are especially important to discrete event simulation: cooperation and branching. Machining cannot occur without the cooperation of a pallet and the cutting tool. The cutting tool may branch from queue "sharp tools" to either activity "machining" or activity "replace tool." The model must specify some logic for determining which branch to follow. This model could be used to study how in-process storage requirements change when activity durations and tool replacement strategies are varied.

Commonly, simulation is used to do the following:

1. Determine resource utilizations to identify bottlenecks in system performance and to fine-tune the line balance. In the above example, simulation would have shown that machine utilization was less than expected because of the idle time caused by waiting for a sharp tool.

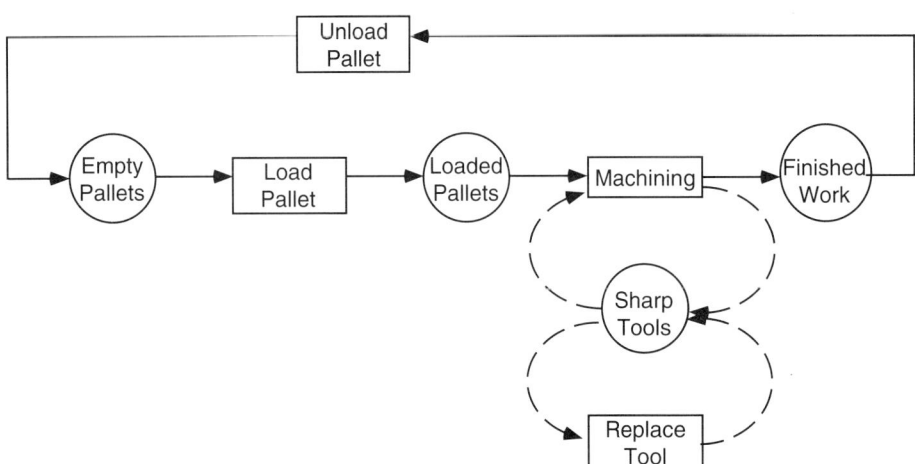

FIGURE 16-19. Example Activity Cycle Diagram.

2. Investigate scheduling strategies. System performance is often affected by changing the scheduling and priority of activities. For example, simulation could show that a system's throughput could be improved by giving highest priority to the repair of the machines with the highest utilization.

3. Determine inventory levels. These may be inventory levels or buffer sizes that result from operation of the system in a prescribed manner, or the inventory or buffer sizes required to achieve system performance unconstrained by the effects of finite buffer size.

4. Investigate the impact of different batching strategies for batch-process systems.

The usefulness of the simulation to the system designer relies on the use of other tools such as economic analysis. Without proper interpretation of its results, simulation would be merely a trial and error process. Simulation will yield the characteristics of a single point in design space: It is the responsibility of the designer, using other methods such as those described elsewhere in this chapter, to optimize the system within the design space.

Discrete event simulation is a valuable tool in the design and specification of manufacturing systems. It is not, however, a substitute for analytical methods. It is useful when a system is complex or subject to random behavior and as a means of verifying results obtained by an analysis based on unproven assumptions. A rough analysis is always a prerequisite for formulating a simulation model.

16.K. HEURISTIC MANUAL DESIGN TECHNIQUE FOR ASSEMBLY SYSTEMS

This section and the next one deal with specific steps in designing an assembly system for the base case where one or a few versions of a product are to be assembled. This section describes a manual design method while the next shows how to use a computer algorithm to help with part of the process. Some of the steps in this process are illustrated with the staple gun[21] whose DFA is considered in Chapter 15. Five hundred thousand of these items are made each year.

16.K.1. Choose Basic Assembly Technology

In this manual method, it will be assumed that one dominant assembly method will be used: manual, fixed automation, or flexible automation. The computer algorithm described in the next section chooses the most appropriate technology for each operation or group of operations and generates mixed-technology designs.

16.K.2. Choose an Assembly Sequence

We learned in Chapter 7 how to generate and select assembly sequences. Different sequences may favor different assembly technologies. For example, if the assembly

sequence requires turning the product over many times, manual assembly (or manual operation of the turnover steps) may be the best choice. A product whose different versions require different part counts or different sequences may be feasible via a fixed automation machine that allows stations to be skipped if their part is not needed by that version. More often, such products are assembled by robots or people.

16.K.3. Make a Process Flowchart

A process flowchart is a diagram that follows the pattern of the assembly sequence, indicating separately each subassembly that is built and introduced to the line. The flowchart also includes all nonassembly operations that require attention, time, or equipment, such as inspections, lubrication, or record-keeping.

Figure 16-20 is the process flowchart for the staple gun.

16.K.4. Make a Process Gantt Chart

Gantt charts are commonly used in scheduling any kind of work sequence. An example appears in Figure 16-21. Time runs along the horizontal axis, while the tasks from the process flowchart are arrayed down the vertical axis in sequence from first to last. Times for tasks that occur in series must be placed end to end in the chart. Operations on subassemblies that can be done in parallel are shown going on at the same time as other tasks. An estimate of the time required for each task should be calculated using

---

[21]The staple gun example is based on work by MIT students Benjamin Arellano, Dawn Robison, Kris Seluga, Thomas Speller, and Hai Truong, and Technical University of Munich student Stefan von Praun.

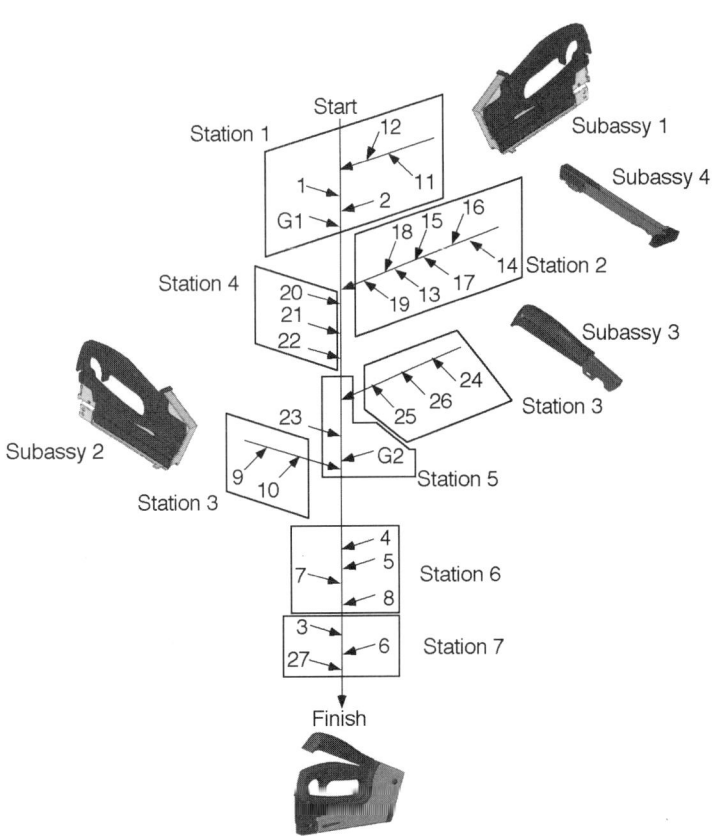

FIGURE 16-20. Process Flowchart for the Staple Gun. G1 and G2 are greasing operations.

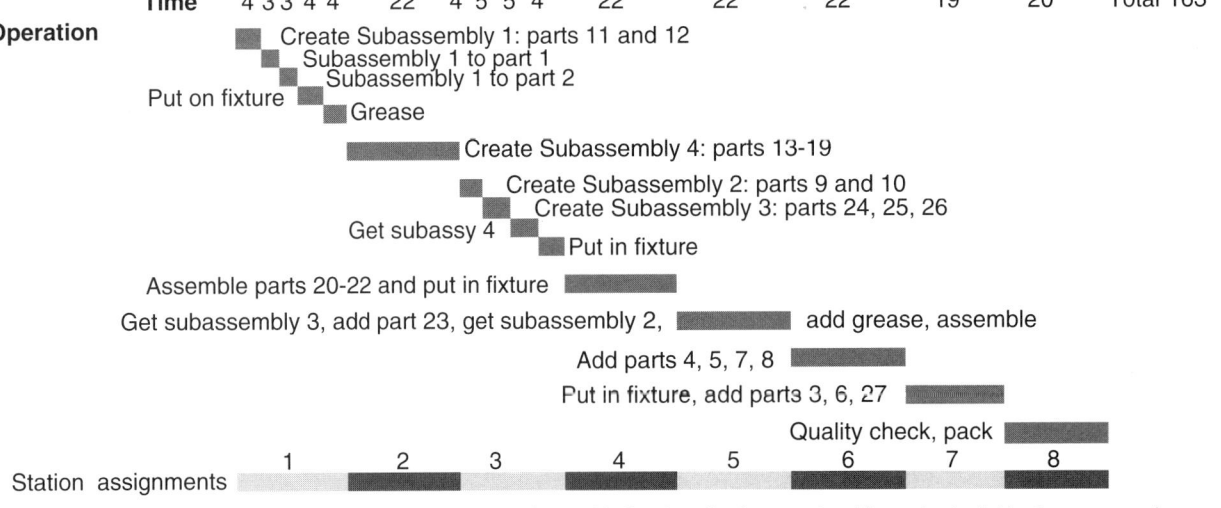

FIGURE 16-21. Assembly Gantt Chart for the Staple Gun with Station Assignments. Times for individual steps are shown for stations 1 and 3, while aggregate times are shown for the others. Two seconds transport time between stations is not shown. Also not represented is any downtime loss, which the designers of this system assumed would be 15%. The makespan without these effects is 163 seconds.

Equation (16-4) or some other suitable method. A time appropriate to the resource being used must be chosen. The total time (makespan) needed to assemble one unit can then be read off the chart.

16.K.5. Determine the Cycle Time

Assuming that the number of assemblies needed per year is known, the required cycle time can be computed using Equation (16-7). This cycle time reflects an assumption about how many shifts will be needed. It is easiest to start by assuming one shift operation.

16.K.6. Assign Chunks of Operations to Resources

Equation (16-6) tells us how many equal-sized time chunks are needed to do all the operations. The longest time chunk (called t in Equation (16-4)) should not be longer than the cycle time (τ in Equation (16-7)). Our goal is to assign chunks of operations to resources so that all the work gets done and each resource has about the same amount of work to do.

In Figure 16-21, the number of chunks is eight. In this case, several time chunks are longer than the operation times in those chunks, so one manual or flexible resource can do several tasks.

In general, the operation time may exceed the cycle time, may be about the same, or may be much less. Each case is handled differently.

First, see if some operations take much longer than others. If so, consider providing additional stations in parallel to do those operations, as shown in Figure 16-10a. Keep doing this until those operations can be done in approximately one cycle.

Next, look for operations that take much less time than the others and see if they can be clustered into one workstation so that their total time is approximately one cycle. An example is shown in Figure 16-10b. This option is feasible only if the resource can do more than one task; this is inapplicable to fixed automation, whose operation times by definition are the same for each step in the assembly and consist of one step only.

At this point, one may have a line of stations which, operating in series, can produce the assemblies at the required rate.

Even after chunking the operations into approximately equal time clusters, there still may not be enough time to make all the needed assemblies unless a very large number of parallel stations is used. This would be unwieldy and take up a lot of space. Instead, consider adding a second or even a third shift of operation. Equivalently, consider simply building more than one identical system. Either approach effectively multiplies the required cycle time by two or three over that calculated at first and may enable the system to finish the needed assemblies in the available time. Naturally, adding shifts will affect the economics (discussed below) in different ways, depending on whether the system is manual or not. The reason is that adding a second shift doubles the labor cost while the same machines can be used on any number of shifts without buying them again. Only the people needed to tend the machines must be paid for a second (or third) time. Duplicating the system means buying additional machines as well as hiring additional people.

The plan for the staple gun shown in Figure 16-21 can deliver the required 500,000 units per year if it is operated for two shifts per day. Its cycle time of 22 seconds plus 2 seconds station move time permits just over 1,000 units to be made per shift at 85% uptime.

16.K.7. Arrange Workstations for Flow and Parts Replenishment

The above steps create a list of stations and identify the time sequence of their operation, or equivalently the sequence in which assemblies must visit the stations. The next step is to arrange these stations into a floor layout, perhaps using one of the layout types discussed in Section 16.F. In doing so, the designer must account for space for people to work and move about, space for the assembly equipment and work tables, and access paths and storage space for incoming parts and finished assemblies. Buffers between stations must also be considered, especially on either side of the slowest station. Areas for rework following test operations must also be provided. The floor area must be arranged so that paths of transport vehicles do not cross each other and present safety problems or traffic jams. If the system contains robots or fixed automation, good practice is to leave plenty of space between stations for people to stand in if a station is broken for an extended period.

Figure 16-22 shows the assembly system for the staple gun. The station times shown here include an extra 2 seconds for passing the work from one station to the next, in addition to the process times shown in Figure 16-21.

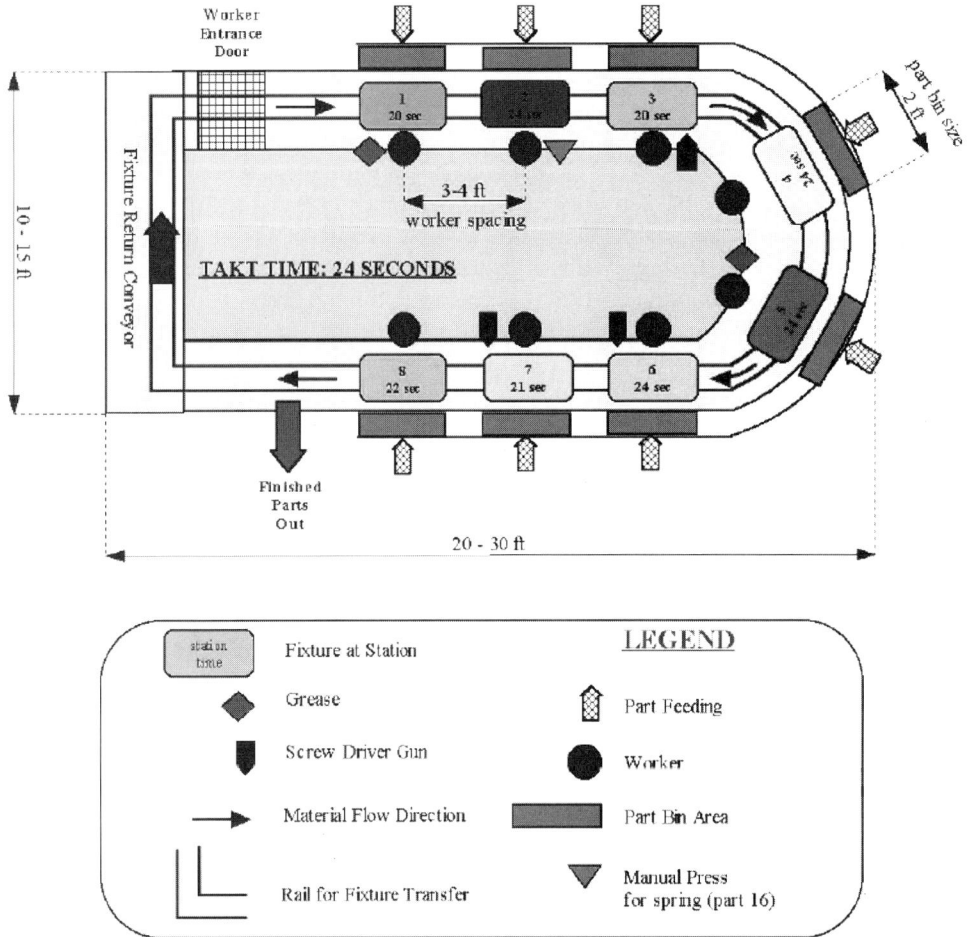

FIGURE 16-22. Assembly System Design for the Staple Gun. This system is estimated to require an investment of $32,000 and yield a unit assembly cost of $0.90 counting only direct labor at $15/hr.

Note that the operators are inside this loop while parts arrive from the outside. A door is provided to permit the operators to enter and leave.

Table 16-4 shows the parts supply strategy for this system. Based on the size of the parts and the rate at which they are consumed, different delivery schedules are appropriate for the parts needed at each station.

16.K.8. Simulate System, Improve Design

The above design process creates a system that is sufficient to meet average demand under average operating conditions. Many sources of variation will affect its operation, usually negatively. For this reason, it is necessary to

make a discrete event simulation of the proposed design to see how it works. As discussed in Section 16.J, the result could be addition of buffers, enlargement of buffer space, improvement in anticipated machine downtimes, hiring of additional repair or part replenishment people, and so on.

16.K.9. Perform Economic Analysis and Compare Alternatives

The above procedure creates an assembly system based on assuming a given assembly technology, together with its costs. These consist of investment in equipment plus the ongoing cost of labor. In some situations, floor space is assigned an overhead cost or even taxed as real estate

TABLE 16-4. Parts Supply Schedule for the Staple Gun

	Different Parts Supplied	Bulk Bins in Rack at Any Time	Size of Bin	Trays in Rack at Any Time	Size of Trays	Maximum Supply Interval	Recommended Supply Interval
Station 1	4	2	$17 \times 7 \times 2.5$	8	$17 \times 7 \times 5$	1 hr	0.5 hr
Station 2	7	5	$17 \times 7 \times 2.5$	5	$17 \times 7 \times 5$	1 hr	0.5 hr
Station 3	5	1	$17 \times 7 \times 2.5$	15	$17 \times 7 \times 5$	1 hr	0.5 hr
Station 4	3	1	$17 \times 7 \times 2.5$	2	$17 \times 7 \times 5$	1.15 hr	0.5 hr
Station 5	1	1	$17 \times 7 \times 2.5$	1	$17 \times 7 \times 5$	4 hr	3 hr
Station 6	4	3	$17 \times 7 \times 2.5$	3	$17 \times 7 \times 5$	1 hr	0.5 hr
Station 7	2	1	$17 \times 7 \times 2.5$	1	$17 \times 7 \times 5$	8 hr	8 hr
Station 8	1	0	$17 \times 7 \times 2.5$	5	$17 \times 7 \times 5$	1 hr	0.5 hr

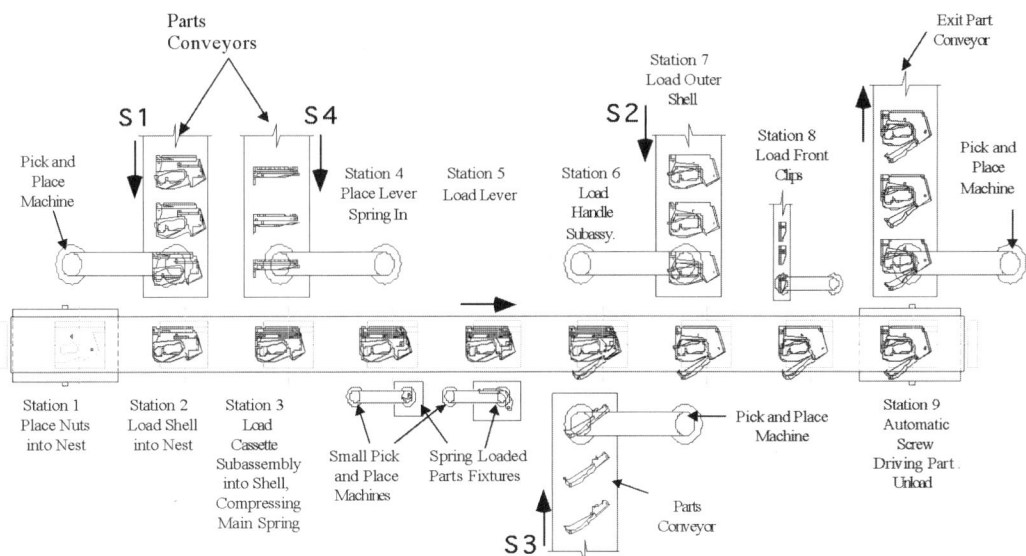

FIGURE 16-23. Robotic Assembly System Proposed for Staple Guns. This system can make 500,000 units per year operating one shift. Each station operates in 10 seconds, and 2 seconds are allowed for station–station move time. It is estimated to require an investment of $1.26 million. There are nine automated stations plus four manual stations (not shown) that prepare subassemblies S1 through S4. Each unit bears about $0.59 to repay this investment at prevailing interest rates.

by the surrounding municipality. To see if the proposed system is the most economical, an economic analysis of it must be made. Following this, a different design must be created and subjected to all of the above steps so that its performance and cost may be compared to the first one. This process is repeated as many times as the designer has imagination or time, until a satisfactory design is obtained. Naturally, if design of the system is outsourced to a vendor, the vendor will do all this tedious work but will most likely choose the assembly methods it is most familiar with and prepared to deliver.

In the case of the staple gun, an alternate design consisting of fixed automation and robots was designed and compared to the manual line described above. It is shown in Figure 16-23. Economic analysis, as explained in more detail in Chapter 18, shows that it would cost slightly more to assemble one staple gun on this system than on the manual system, even though it would make all the needed staple guns in one shift. It also faced considerable technical challenges in accomplishing the more difficult assembly tasks.

16.L. ANALYTICAL DESIGN TECHNIQUE

One of the more difficult steps in the manual design process is to choose among different resources for each task so that the work is done within the cycle time and the whole assembly system has minimum cost. In this section, an algorithm for doing this is briefly described, along with software that carries it out. The algorithm is described in [Graves and Holmes-Redfield] and [Cooprider]. The software, originally written in QBASIC by Curt Cooprider, was corrected and ported to Microsoft Visual Basic by Michael Hoag with help from David Whitney.

16.L.1. Theory and Limitations

The Holmes–Cooprider method assumes that the assembly system will be implemented as a single line with no incoming sub-lines and no recirculation for rework. All station times are assumed to be deterministic. The annual cost of a resource is assumed to consist of a fraction[22] of any long-term investment plus the annual operating cost, primarily direct and indirect labor. Each resource that can do an assembly task is described by the time it takes to do that task, a tool number, and the cost of that tool. If a resource is technically incapable of doing a task, no data are entered. Each resource also has a tool change time that applies to any tool used by that resource. Each resource also has a characteristic uptime fraction and a characteristic number of people needed to keep it running. The assembly system as a whole has a characteristic time to move work from one station to the next.

In addition to the above, input data include the number of shifts to use, the number of operating days in a year, and the number of assembled units required per year. The costs of direct and indirect labor are also provided. Data are prepared on a chart shown in Table 16-5.

The algorithm operates by creating a network of node pairs representing the assembly tasks, along with arcs joining nodes that represent assignment of a resource to a group of tasks. An example network is shown in Figure 16-24. Theoretically, if there are n nodes, there are $n(n - 1)/2$ arcs for each kind of resource allowed, but an explosion in the number of arcs is avoided for several

FIGURE 16-24. Task Node Diagram. There are three tasks in this assembly sequence. The arcs show that there exists at least one resource that can do task 1, at least one that can do task 2, at least one that can do both tasks 1 and 2, and at least one that can do task 3.

reasons. First, if several types of resources can satisfy one arc (i.e., they have time to do all the assigned tasks), only the lowest-cost type is chosen. Second, many arcs are inactive because the designer has deemed the resource technically incapable. Other arcs are eliminated because the designer has set an upper limit on how many duplicate resources of a given type can be assigned to a set of tasks.

The cost and time of an arc are based on the tasks and the resource. If more than one tool is required, tool cost is added to resource cost, and tool change time is added to task time. If the last tool used is different from the first one, then one more tool change time is added unless it is shorter than the station-to-station move time, in which case station-to-station move time is added. All times are inflated to reflect uptime less than 100%, and the result is again compared to the available cycle time. If one such resource cannot do the work in the required time, additional identical resources are added (up to the limit specified by the designer) until they all can do the work on that arc working in parallel.

The resulting network consists of time-feasible arcs with different annual costs. A shortest path algorithm then finds the least cost path. This path is a list of resources together with the tasks assigned to them. Since this path runs from the first node to the last, all the tasks are assigned.

16.L.2. Software

The software is called SelectEquip. It is written in Visual Basic and runs on PCs with Office 2000 or higher. An executable version is on the CD-ROM packaged with this book, along with instructions and the data file for the example in Section 16.L.3. The opening window appears in Figure 16-25. Different sub-windows may be opened to permit information about resources and tasks to be entered.

---

[22]This fraction (called f_{AC} in Figure 16-5) depends on the number of years that the investment is expected to be productive, as well as prevailing interest rates and other factors. It is explained in Chapter 18, along with detailed cost equations for each kind of resource.

TABLE 16-5. Task-Resource Matrix for SelectEquip for IRS Rear Axle

		Station-station move time (s)	5	
		Production Volume	300000	
			400000	
Legend				
	For each resource:			
			When a resource can be used:	
	Short Name			
C	hardware Cost ($)		Operation	Tool
rho	installed cost/hardware cost		time (s)	number
e	% uptime expected			
v	operating/maint rate ($/hr)		Tool cost	
Tc	Tool change time (s)			
Ms	Max # stations/worker			

Resource: ...... Task:	___MAN_ C_2000___ rho__1.5__ e__80___ v__4.00___ Tc__5___ Ms__0.83__	__FXD_____ C__0____ rho_1.5__ e__95___ v__6.00___ Tc__5___ Ms__4___	__RBS_____ C__40000_ rho__2.5_ e__90___ v__6.00___ Tc__5___ Ms__4___	___RBB____ C_80000__ rho__2.5_ e___90___ v__6.00___ Tc__10___ Ms__4___	____TRN___ C_____ rho_____ e_____ v_____ Tc_____ Ms_____
1 Put frame on pallet	15 | 101 15000	10 | 201 75000	■	10 | 401 10000	■
2 Mate body mounts to frame	25 | 102 5000	15 | 202 100000	15 |302 20000	15 |402 20000	
3 Subassy shafts-A arms	60 | 103 30000	15 | 203 300000	25 |303 40000	25 |403 50000	
4 A-arms to frame	30 | 104 2000	15 | 204 300000	■	20 |404 40000	
5 Diff to Frame	15 | 105 15000	10 | 205 150000	■	8 | 405 20000	
6 Mate diff, shafts, & frame	75 | 106 15000	20 | 206 250000	■	35 |406 50000	
7 Arrange brake cables	40 | 107 2000	■	■	■	■
8 Transport Betw Stations	■	■	■	■	5 | 508 258000
9	|	|	|	|	|
10	|	|	|	|	|

Note: Resource data include the purchase cost *C*, the uptime expected, extra labor required for maintenance or operational support, tool change time, and the number of stations that an attending worker can support (charged at the regular labor rate). This figure is less than 1.0 for manual stations to account for scheduled rest and lunch breaks. "rho" is the ratio of engineering cost to resource and tool purchase cost and represents extra cost to design the workstation and install it; rho is larger for more complex resources. Task data include the time the resource needs to do the task, the tool number needed, and the cost of the tool. The cost of fixed automation is all accounted for in the tool cost to reflect the fact that a fixed resource can do only one task.

16.L.3. Example

SelectEquip was applied to an example assembly consisting of an independent rear axle for automobiles. This example was studied in Chapter 16 of [Nevins and Whitney]. The axle and its parts appear here in Figure 16-26.

Table 16-5 lists the data task by task, showing which of four assembly resources can do each task, using what tool, and at what cost. The purchase cost and other data for each resource are listed across the top. Fixed automation resources are listed as individual tools that have no other purchase cost. This forces the algorithm to assign

Basic environmental data

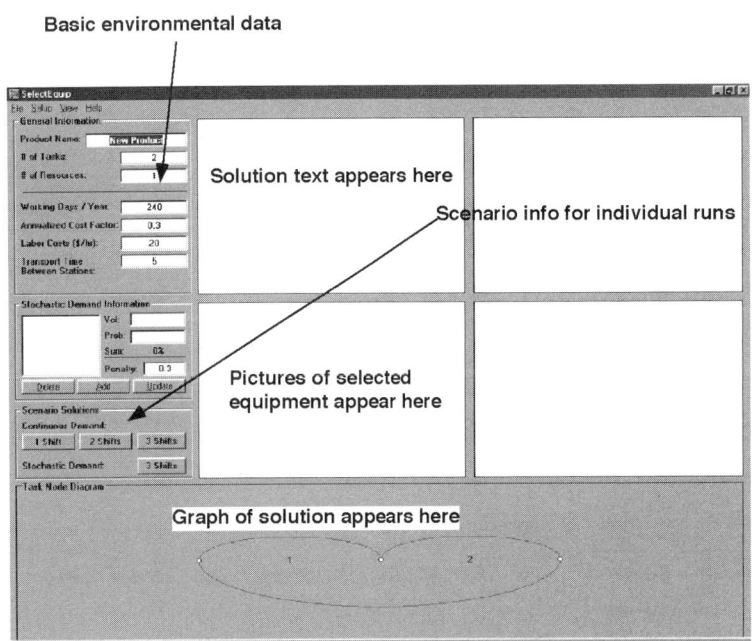

FIGURE 16-25. Opening Window for SelectEquip Software. Different parts of the user interface window are labeled.

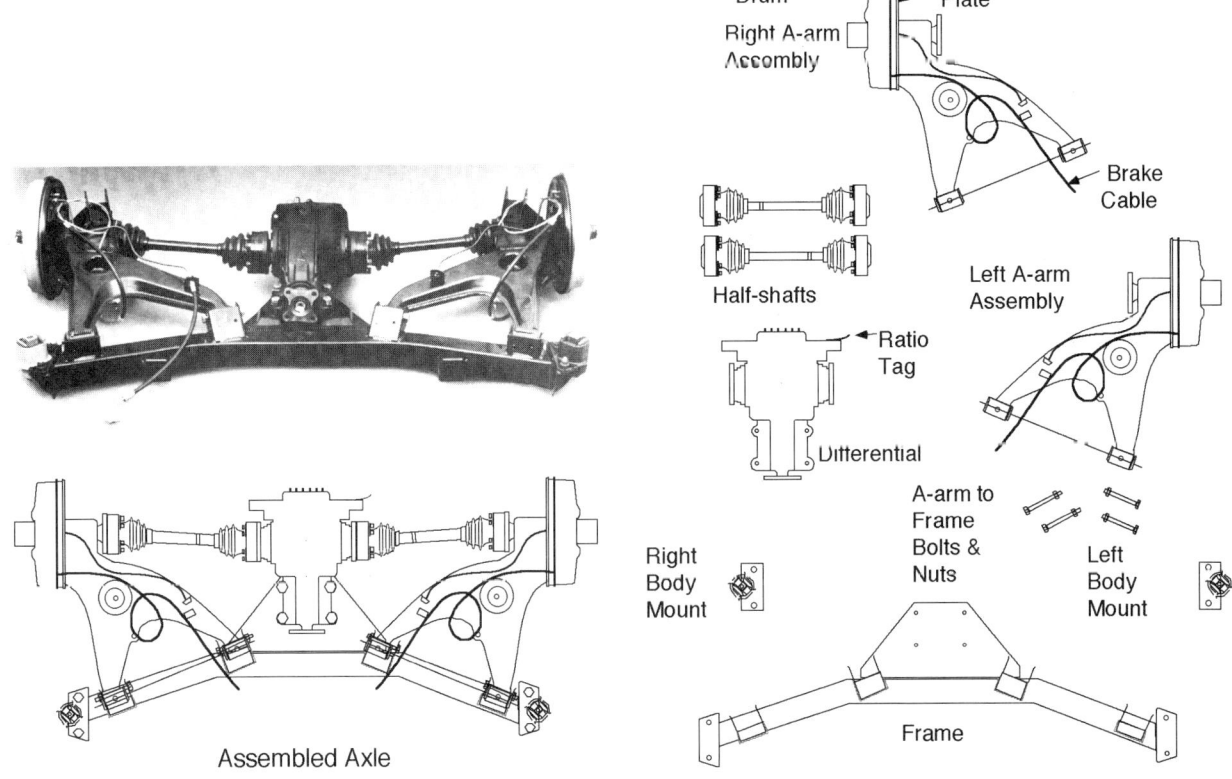

Assembled Axle

FIGURE 16-26. IRS Rear Axle and Its Parts.

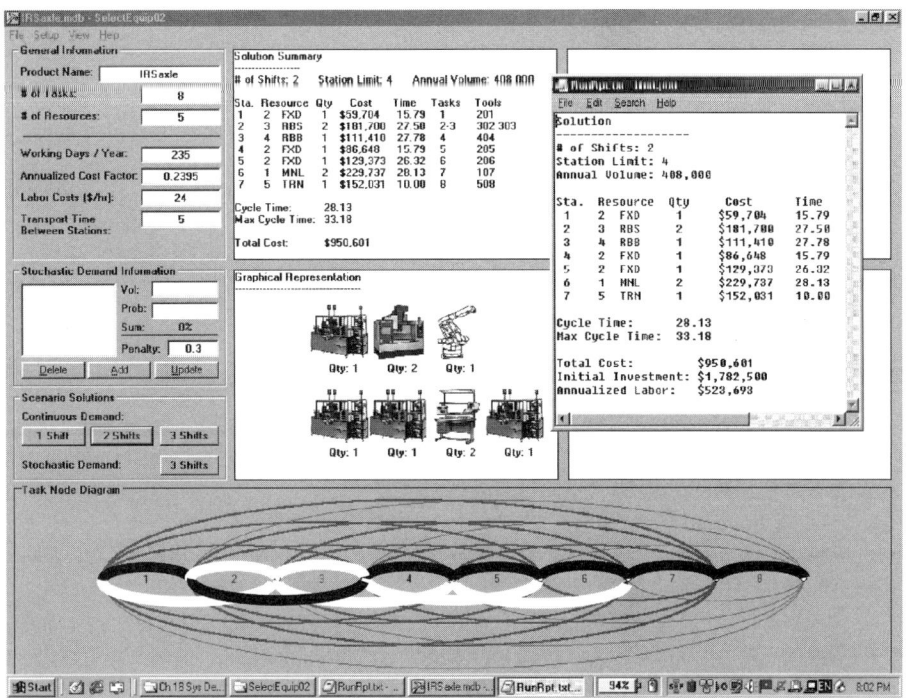

FIGURE 16-27. Example Output from SelectEquip for the Data in Table 16-5. The Notepad run report at the right contains the details of the solution in text form. Pictures and quantities of the required resources appear in the Graphical Representation window. The task node diagram is at the bottom. Each gray arc is mathematically available, but only the white arcs represent resources actually assigned, as noted in Table 16-5. Thin black arcs represent resources that could do the assigned tasks except that more duplicates than the user has allowed would be needed. The thick black arcs represent the optimal solution.

only one task to each fixed resource and to buy it in full for that task only. Blacked-out areas represent tasks that cannot be done by the respective resource. The cost of the entire transport system is lumped into the resource TRN, accompanied by a dummy task called Transport. When the algorithm runs, the 5-second transport time is applied to each inter-station move.

The solution for the IRS Rear Axle, assuming two shift operation and 408,000 units per year, appears in Figure 16-27. It consists of a mix of all available resources.

16.L.4. Extensions

SelectEquip addresses one of many problems in assembly system design. Milner combined a different implementation of the SelectEquip algorithm called ASDP ([Gustavson]) with assembly sequence generation software to find the lowest cost assembly sequence by systematically searching the sequence network diagram ([Milner]). Klein manually generated alternate assembly sequences and used ASDP to find least cost systems ([Klein]). He found unit cost differences of as much as 20% based on saving people, equipment, or tools were found. These savings emerged because the same tool or resource could do several tasks if the assembly sequence permitted them to be done in an unbroken series. These tasks could then be grouped on resources to save buying the same tools or resources multiple times. [Nof et al.] describes a wide range of algorithms for scheduling and balancing assembly lines. [Scholl] relates the problems of sequence design and line balancing and contains an extensive reference list. The interested reader is referred to these sources for more details.

16.M. EXAMPLE LINES FROM INDUSTRY: SONY

Sony designed the FX-1 assembly system in 1981 to accommodate the frequent shifts in design of the Walkman product family. Styling changes occurred as often as every six months. Such a time span is not only too short to recoup the investment in a typical fixed automation assembly machine, but shorter than the time needed to design, build, and debug one. The FX-1 system layout is shown in Figure 16-28. It consists of two separate lines occupied by three programmable assembly stations each. These stations are described in detail in Chapter 17. The assemblies were manually placed in pallets along with the necessary parts, and the pallets were loaded onto the station's worktable. This table was capable of $X-Y$ motions, allowing it to place the assembly under individual tools dedicated to a single operation. The station's architecture permitted new assembly tools to be attached and checked out independently of ongoing assembly operations.

This system was used to assemble the Sony Walkman tape recorder mechanism described in Chapter 14. As originally designed, this chassis had parts on both sides of a central board. Stations A-1 through A-3 took parts from the pallet and put them on one side. Operators then removed the chassis from the system, turned it over, placed it on a new pallet with a new stock of parts, and fed it to stations B-1 through B3. They also installed some parts that were difficult to place robotically.

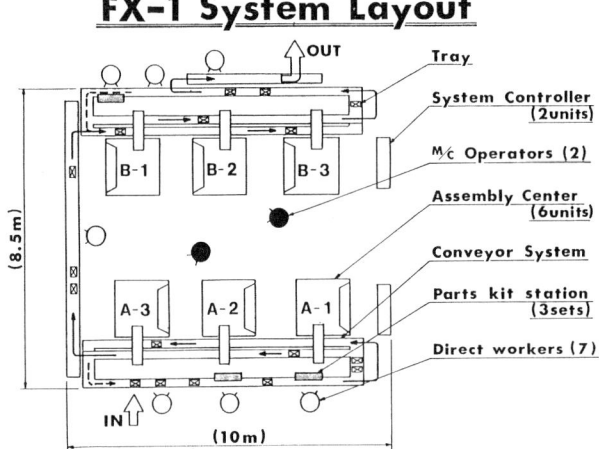

FIGURE 16-28. Sony FX-1 Assembly System for Walkman Products. (Courtesy of Sony FA.)

A few years later, Sony replaced this system and station concept with a straight line of robots. A few FX-1 stations were retained to install press-fit pins in VCR chassis because their rugged construction permitted them to exert the necessary forces. In other ways, these stations proved too expensive and incapable of the reach and speed needed for further applications. Lines of twenty-five or more robots were developed for assembling other Walkman models as well as complex VCR tape changing mechanisms and videocameras.

16.N. EXAMPLE LINES FROM INDUSTRY: DENSO

Denso's main customer for the last fifty years has been Toyota. Denso has learned over this time to accommodate Toyota's high variety and small batches. The three assembly systems described here are sample milestones in Denso's growing capability to conquer production variety ([Whitney]).

16.N.1. Denso Panel Meter Machine (~1975)

The Denso panel meter discussed in Chapter 1 was assembled in arbitrary batch sizes on an essentially ordinary fixed automation assembly machine. This product and its assembly process were one of the first attempts by Denso to merge product design, process design, and company

strategy for dealing with its most important customer. As the following examples show, Denso has evolved a sophisticated technology strategy that has successfully tackled more and more complex problems over the last thirty years. The progression has extended from small products like the panel meter having a few substituteable parts to large products like air conditioning modules whose different versions can have different numbers of parts or can even be of different sizes.

16.N.2. Denso Alternator Line (~1986)

The Denso alternator assembly line comprises twenty robots, designed and built by Denso (Figure 16-29).

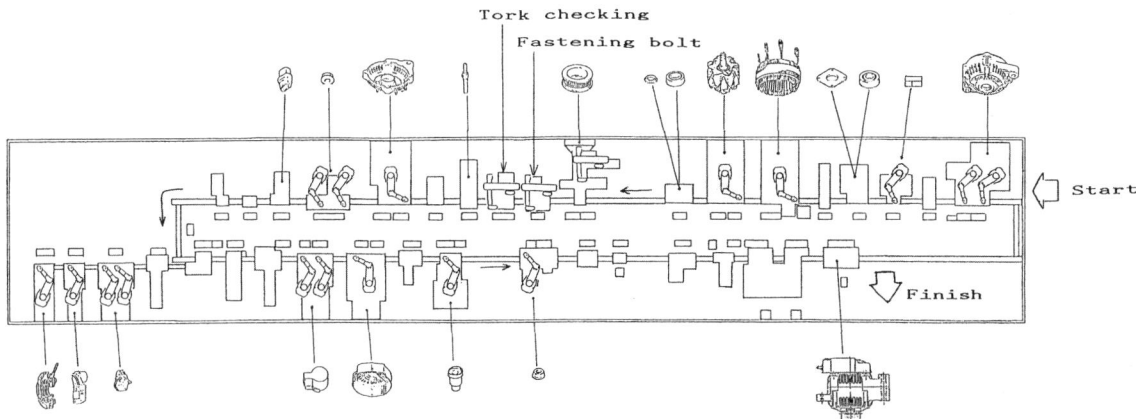

FIGURE 16-29. Denso Robotic Assembly Line for Alternators. This system is arranged in a loop. Assemblies are carried on pallets which return to the start of the line to pick up a new assembly. (Courtesy of Denso Co., Ltd. Used by permission.)

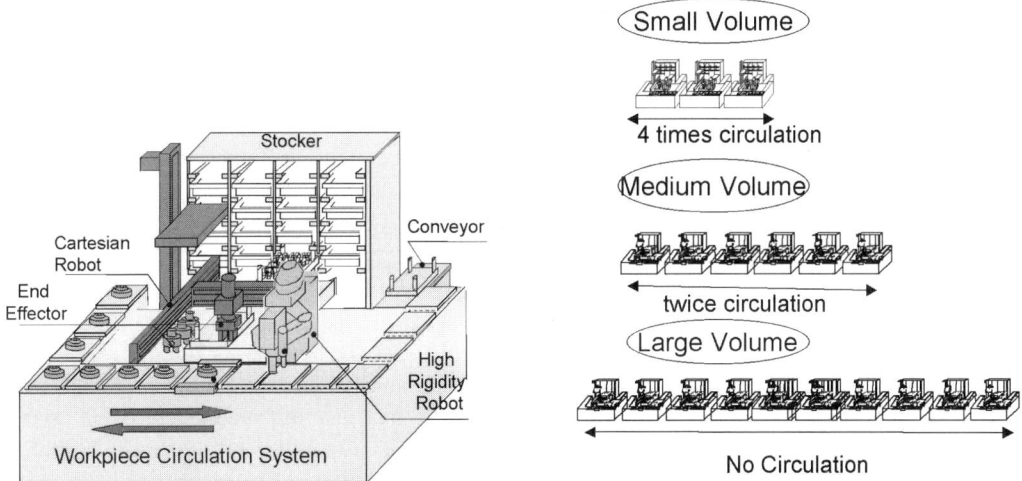

FIGURE 16-30. Denso Variable Capacity Line. The line is made of standard assembly cells consisting of a stock of parts, a robot that retrieves trays of parts from the stocker, a tool-changing Cartesian robot, and a high rigidity SCARA type robot. Different numbers of these cells can be deployed to assemble products at different production rates. Low rates require a few stations, each of which has many tools and assembles many parts onto each assembly. (Courtesy of Denso Co. Ltd. Used by permission.)

Several workstations contain vision systems that permit them to pick up unoriented parts from a tray. An interesting feature of this system is its ability to assemble alternators of different sizes, including both diameter and length variations.

16.N.3. Denso Variable Capacity Line (~1996)

The variable capacity assembly system shown in Figure 16-30 consists of standardized assembly cells that can be placed next to each other in any number. For

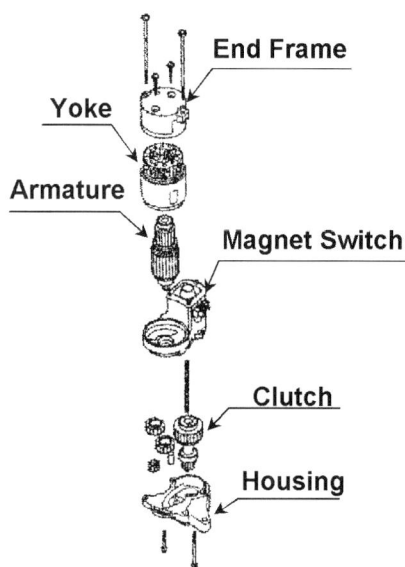

FIGURE 16-31. Denso Roving Robot Line for Starters. In this system, robots are not assigned to a specific assembly station but can cluster under decentralized control at places where excess work has accumulated. The line is similar to cells discussed in Section 16.F.6 in the sense that production rate can be varied by adding or removing robots from the line. The workpieces travel along a conveyor. Parts are fed from the side of the line opposite the robots. The robots carry a suite of tools and pick up the tool needed by the next part at whatever station they are attending. ([Hanai et al.]. Copyright © IEEE 2001. Used by permission.)

low-volume applications, one or a few stations will be used. Assemblies can circulate inside each station, returning to the assembly robots several times as they change tools and add more parts. Also, assemblies can circulate among several stations for the same purpose. Parts are placed in the stocker at the rear of the station ([Hibi]).

16.N.4. Denso Roving Robot Line for Starters (~1998)

The roving robot line shown in Figure 16-31 is capable of adjusting its capacity by addition or removal of robots. These robots can position themselves at any station and can cluster around an overloaded station or a broken robot in order to help each other work off the backlog. This is accomplished by a decentralized control system.[23]

---

[23]The author observed Denso employees helping each other during a visit in 1974. An employee who was ahead ran downstream to help the adjacent employee who had fallen behind, then ran back to work off her own backlog. In 1981, Hitachi described a slightly different roving robot concept in which the robots carried the partially finished assemblies as well as the tools, and they obtained parts from the different stations they visited.

16.N.5. Comment on Denso

Denso designed and built all the foregoing assembly systems in its Production Tooling Department over the past thirty years. Denso makes all its own robots and is fully capable of creating any assembly system it needs. It has also pursued a consistent strategy of advancing its capability in automatic assembly over that period, as described schematically in Figure 16-32. To the author's knowledge, it is the only company that has its own multi-decade manufacturing technology roadmap similar in spirit to the product-process technology roadmap of the semiconductor industry. Each step in the strategy has addressed a new and more difficult problem, such as combinatoric model mix assembly of small parts, model mix assembly of large parts, assembly of products with different size parts in different models, and variable production capacity assembly with low fixed cost. This strategy was described by the author in [Whitney] based on knowledge available in 1992. The company's strategy is still intact as of this writing approximately twelve years later.

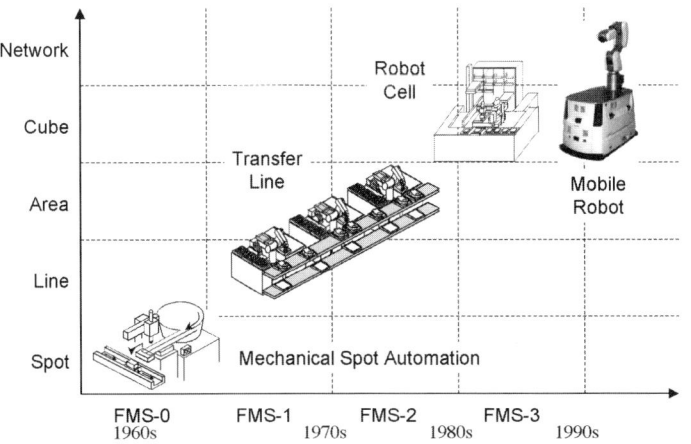

FIGURE 16-32. Denso's Manufacturing Technology Roadmap for Assembly Automation. The panel meter assembly machine belongs to the FMS-1 category, the alternator line belongs to FMS-2, and the cell and mobile robot systems belong to FMS-3. ([Hibi]. Copyright © IEEE 2001. Used by permission.)

16.O. EXAMPLE LINES FROM INDUSTRY: AIRCRAFT

Aircraft are much larger than automobile components, but they are still assembled on a line. This section describes Boeing's method of assembly of the 777. Each station does a particular set of operations over a three-day period. During the third shift every three days, all the assemblies move ahead to the next station. At the beginning of the line, fuselage segments of the type described in Chapter 8 are assembled into complete tubular sections. Wiring and some internals are then installed in each section. On a separate line, wings are built from pieces made by suppliers or in other Boeing plants. Tail sections are similarly assembled nearby. All these parts are brought together at a final body join station. Then landing gear are added and the plane rolls to a final outfitting station. Finished aircraft roll out the door and are flown to their customers. This sequence is shown in Figure 16-33 while the floor layout is shown in Figure 16-34.

For comparison to Figure 16-33, the final assembly process for Airbus aircraft (except for the A380) is shown in Figure 16-35.

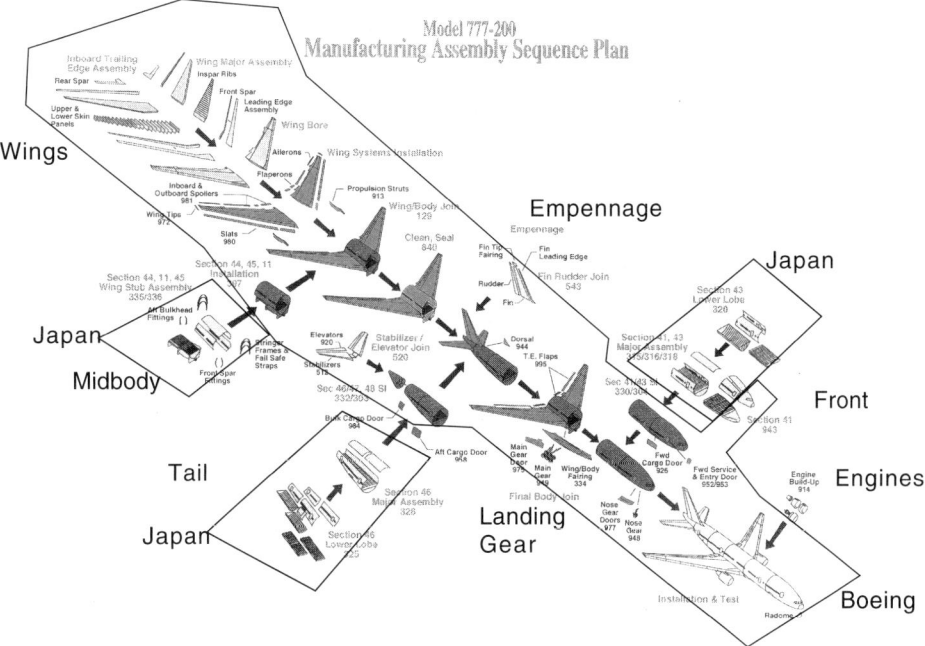

FIGURE 16-33. Assembly Sequence of Boeing 777 Aircraft. Note that main body fuselage section pieces are made in Japan. Wings and empennage are made at Boeing's final assembly plant. Fuselage section pieces are assembled into fuselage sections at Boeing. (Courtesy of Boeing. Used by permission.)

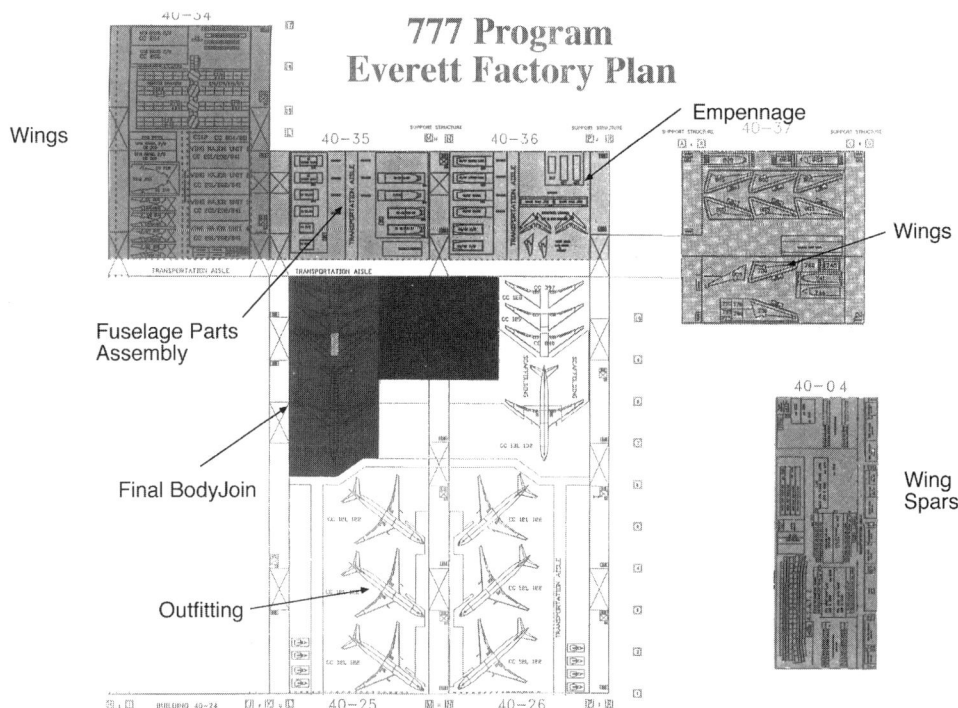

FIGURE 16-34. Boeing 777 Assembly Floor Layout. Wings are made at the left, while fuselage sections are made in the upper middle and tails are made at the upper right. Body join occurs in the middle, while final outfitting is at the bottom. (Courtesy of Boeing. Used by permission.)

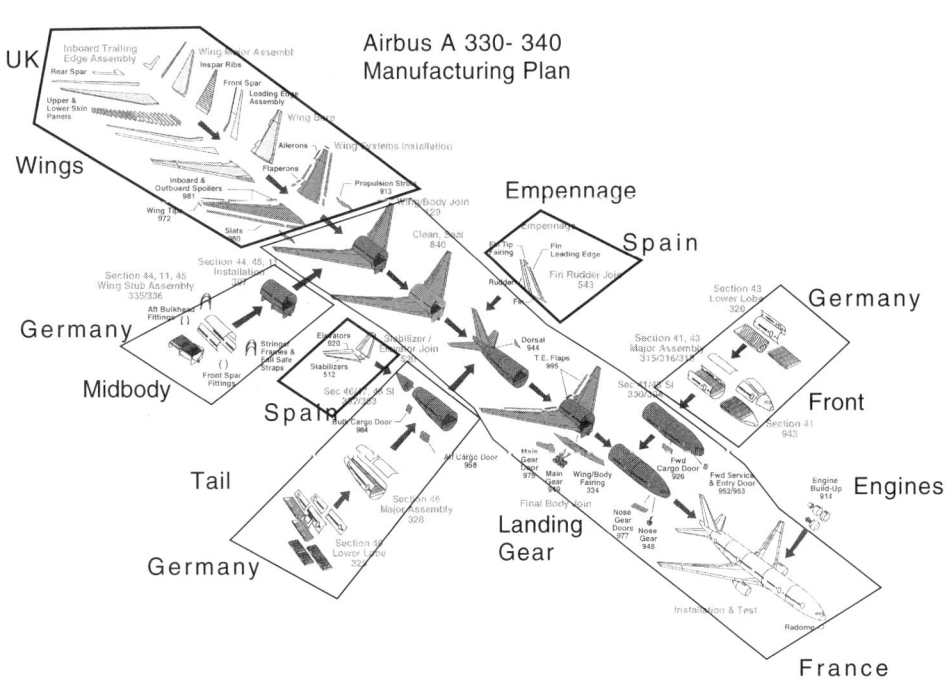

FIGURE 16-35. Final Assembly Process of Airbus Aircraft. Airbus aircraft are assembled in a sequence similar to Boeing's, except that consortium members in other countries do more assembly work before sending pieces to France for final assembly. The A380 will have a somewhat different assembly process. (Courtesy of Boeing. Used by permission.)

16.P. CHAPTER SUMMARY

This chapter deals with design of assembly systems and shows that a system must meet a wide variety of operating conditions and judgment criteria. It must have sufficient capacity, be reliable, produce good products, be a good place for people to work, be responsive to changes in its operating environment, and be capable of improvement over time. Combining this chapter with Chapter 14 and Chapter 15, we can see that product and assembly system design need to be carefully coordinated in order for the maximum benefit to be realized. The leaders in these things appear to be Toyota from the point of view of continuous evolution of operational methods and Denso from the point of view of long term management of technology and product-process coordination.

16.Q. PROBLEMS AND THOUGHT QUESTIONS

1. Figure 16-9 shows two ways to arrange assembly operations. In theory they have identical operating characteristics, but in reality they do not. Identify the differences and comment on which arrangement has the advantage for each.

2. Consider an assembly line with identical workstations and the same size buffers between them. Assume each buffer is half full when the system starts up. If one station stops for a while and the buffer ahead loses pieces while the one behind gains, how long will it take after the station starts working again until those buffers again have the contents they had just before the station stopped?

3. Consider an assembly line with identical stations except for one bottleneck station that runs at 90% of the top speed of the others. Suppose that the stations are separated by buffers with capacity for ten assemblies, and that each buffer has five pieces in it when the bottleneck stops for three cycles. Assume that the other stations can be individually sped up or slowed down by the operators as needed, but not until the bottleneck starts running again. What options do the operators have with their ability to speed up and slow down the other machines? What will happen to overall output of the system if the operators exercise each of these options?

4. Continuing the story from the previous problem, suppose that later the bottleneck stops again for three cycles. What will happen to output from the system, depending on which option the operators chose after fixing the bottleneck the previous time? What options do they have this time?

5. Sketch a simple assembly line with identical stations and identical buffers between them. Assign identical assembly times, probability distributions of breakdowns, and probability distributions of repair time to each station. Perform a discrete event simulation, varying the buffer capacities, and compare the results with the analytical predictions in Section 16.H.

6. Calculate the capacity (product units/unit time) of the Denso panel meter machine if batches of one type contain 1, 2, 4, 8, 16, etc., units. Express your answer as a ratio of the capacity to that of the same machine making exactly one type all the time.

7. Use SelectEquip to design a manual assembly system for the staple gun using task times from the DFA analysis in Chapter 15. Compare it to the one shown in Figure 16-22.

8. Use SelectEquip to design an automatic assembly system for the staple gun for comparison with the one shown in Figure 16-23. Assume that subassemblies S1 through S4 are made manually and that S4 also includes parts 20 and 27. The task assignments in Figure 16-23 are given in Table 16-6.

If you think that some stations in this system are too complex, such as station 3, then break them into distinct tasks, provide lower cost resources, and see what SelectEquip does.

TABLE 16-6. Task Assignments for Automatic Staple Gun Assembly System in Figure 16-23

Station	Parts
1	4, 5, 6
2	S1
3	S4, 20, 27
4	22, 23
5	21
6	S3
7	S2
8	8
9	1–3, 7

9. Should the buffers upstream (downstream) of a bottleneck be half full (empty) or totally full (empty)?

16.R. FURTHER READING

[Boothroyd] Boothroyd, G., *Assembly Automation and Product Design,* New York: Marcel Dekker, 1992.

[Chow] Chow, W. M., *Assembly Line Design,* New York: Marcel Dekker, 1990.

[Cooprider] Cooprider, C., B., "Equipment Selection and Assembly System Design Under Multiple Cost Scenarios," S. M. thesis, MIT Sloan School of Management, June 1989.

[Cusumano] Cusumano, M. A., *The Japanese Automobile Industry,* Cambridge: Harvard University Press, 1985.

[Enginarlar et al.] Enginarlar, E., Li, J., Meerkov, S. M., and Zhang, R. Q., "Buffer Capacity for Accommodating Machine Downtime in Serial Production Lines," *International Journal of Production Research,* vol. 40, no. 3, pp. 601–624, 2002.

[Engström, Jonsson, and Medbo] Engström, T., Jonsson, D., and Medbo, L., "Developments in Assembly System Design: The Volvo Experience," in *Coping with Variety: Flexible Productive Systems for Product Variety in the Auto Industry,* Lung, Y., Chanaron, J.-J., Fujimoto, T., and Raff, D., editors, Aldershot, UK: Ashgate Publishing, Ltd., 1999.

[Fishman] Fishman, G. S., *Discrete Event Simulation,* New York: Springer-Verlag, 2001.

[Gershwin] Gershwin, S. B., *Manufacturing Systems Engineering,* Englewood Cliffs, NJ: Prentice-Hall, 1994.

[Goldratt] Goldratt, E. M., *The Goal,* Great Barrington, MA: North River Press, 1992.

[Graves and Holmes-Redfield] Graves, S. C., and Holmes-Redfield, C., "Equipment Selection and Task Assignment for Multiproduct Assembly System Design," *International Journal of Flexible Manufacturing Sys.,* vol. 1, pp. 31–50, 1988.

[Gustavson] Gustavson, R. E., "Computer-Aided Synthesis of Least-Cost Assembly Systems," Proceedings of the 14th International Symposium on Industrial Robots, Gothenburg, 1984.

[Hanai et al.] Hanai, M., Hibi, H., Nakasai, T., Kawamura, K., and Inoue, Y., "Development of Adaptive Production System to Market Uncertainty—Autonomous Mobile Robot System," Proceedings of the 2001 IEEE International Symposium on Assembly and Task Planning, Fukuoka, Japan, May 2001.

[Hibi] Hibi, H., "Development of Mobile Robot System Adaptive to Sharp Fluctuation in Production Volume," keynote speech and paper in *Proceedings of 2001 IEEE International Symposium on Assembly and Task Planning,* Fukuoka, Japan, May 2001. Additional detail about this system is contained in the companion paper by Hanai et al. cited above.

[Klein] Klein, C. J., "Generation and Evaluation of Assembly Sequence Alternatives," S. M. thesis, MIT Mechanical Engineering Department, February 1987.

[Linck] Linck, J., "A Decomposition-Based Approach for Manufacturing System Design," Ph.D. thesis, MIT Mechanical Engineering Department, June 2001.

[Milner] Milner, J., "The Assembly Sequence Selection Problem: An Application of Simulated Annealing," S. M. thesis, MIT Sloan School of Management, May 1991.

[Mishina] Mishina, K., "Beyond Flexibility: Toyota's Robust Process-Flow Architecture," in *Coping with Variety: Flexible Productive Systems for Product Variety in the Auto Industry,* Lung, Y., Chanaron, J.-J., Fujimoto, T., and Raff, D., editors, Aldershot, UK: Ashgate Publishing, Ltd., 1999.

[Monden] Monden, Y., *Toyota Production System: An Integrated Approach to Just-in-Time,* Norcross, GA: Engineering & Management Press, 1998.

[Nevins and Whitney] Nevins, J. L., and Whitney, D. E., editors, *Concurrent Design of Products and Processes,* New York: McGraw-Hill, 1989.

[Nof et al.] Nof, S. Y., Wilhelm, W. E., and Warnecke, H.-J., *Industrial Assembly,* New York: Chapman and Hall, 1997.

[Peschard and Whitney] Peschard, G., and Whitney, D. E., "Cost and Efficiency Performance of Automobile Engine Plants," available at http://web.mit.edu/ctpid/www/Whitney/papers.html

[Pooch and Wall] Pooch, U., and Wall, J. A. *Discrete Event Simulation: A Practical Approach,* Boca Raton, FL: CRC Press, 1993.

[Scholl] Scholl, A., *Balancing and Sequencing of Assembly Lines,* Heidelberg: Physica Verlag, 1995.

[Shingo] Shingo, S., *A Revolution in Manufacturing: The SMED System,* Stamford, CT: Productivity Press, 1985.

[Spear and Bowen] Spear, S., and Bowen, H. K., "Decoding the DNA of the Toyota Production System," *Harvard Business Review,* September–October, pp. 96–106, 1999.

[Taguchi] Taguchi, G., *Introduction to Quality Engineering: Designing Quality into Products and Processes,* White Plains, NY: Unipub-Kraus International Publications, 1986.

[Whitney] Whitney, D. E., "Nippondenso Co. Ltd: A Case Study of Strategic Product Design," *Research in Engineering Design,* vol. 5, pp. 1–20, 1993.

[Womack, Jones, and Roos] Womack, J. P., Jones, D. T., and Roos, D., *The Machine that Changed the World,* New York: Rawson Associates, 1990.

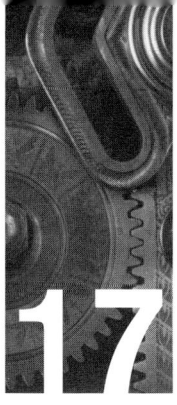

17 ASSEMBLY WORKSTATION DESIGN ISSUES

"If the work must be done in 60 seconds and your robot needs 59 seconds, you get the job. If your robot takes 61 seconds, you don't get the job. It's that simple."

—Joseph P. Engelberger, Unimation, Inc.

17.A. INTRODUCTION

This chapter deals with designing a single assembly workstation.[1] The problem has three major aspects: strategic, technical, and economic. The strategic issues center on choice of method of accomplishing the assembly—manual, robotic, and so on—plus part presentation, flexibility, inspection, and throughput. The technical problems involve detailed technology choice and assurance of proper performance, mainly achieved via an error analysis. Economic analysis is concerned with choosing a good combination of alternative methods of achieving assembly and controlling error.

The information developed during workstation design is used in, and is influenced by, the effort to design an entire assembly system. Choices of assembly sequence or assembly resource will influence what choices are available, economical, or reasonable for the individual stations, and vice versa. The process is typically iterative.

Our objective in designing an assembly workstation is to accomplish one or more assembly operations, in the presence of errors, so as to meet a specification, and to verify the station's performance. The number and identity of the operations to be performed at a station are often tentatively decided during overall system design and may be revised often as station designs are attempted. Typical operations are part mating, application of adhesives, use of tools, application of heat, and measuring. The

errors may arise from parts fabrication, assembly equipment, jigs, fixtures, part feeders, human performance, and so on. Verification must comprise not only the bare minimum—that the parts have been pushed together—but that the work has been accomplished within prescribed tolerances on interpart forces, accelerations, temperature, pressure, cleanliness, or whatever may be of concern.

In creating a workstation design, we have to provide for presenting the parts, providing the tools, transporting assemblies into and out of the station, displaying instructions, recording data, and generally making it possible for the assembly resource to do the job in the available time. The resource must be able to reach everything, do the work efficiently and effectively, and, if it is a person, remain comfortable, confident, and safe.

17.A.1. Assembly Equals Reduction in Location Uncertainty

From a 50,000-ft altitude, we may view assembly as a process by which parts that are far from each other in position and orientation somehow get to the point where they are assembled properly. This is illustrated schematically in Figure 17-1.

This figure illustrates a wide variety of methods. They include

- Having a person do the assembly
- Having a person load a fixture or pallet so that equipment can finish the process

---

[1]This chapter is based in part on Chapters 10 and 11 of [Nevins and Whitney].

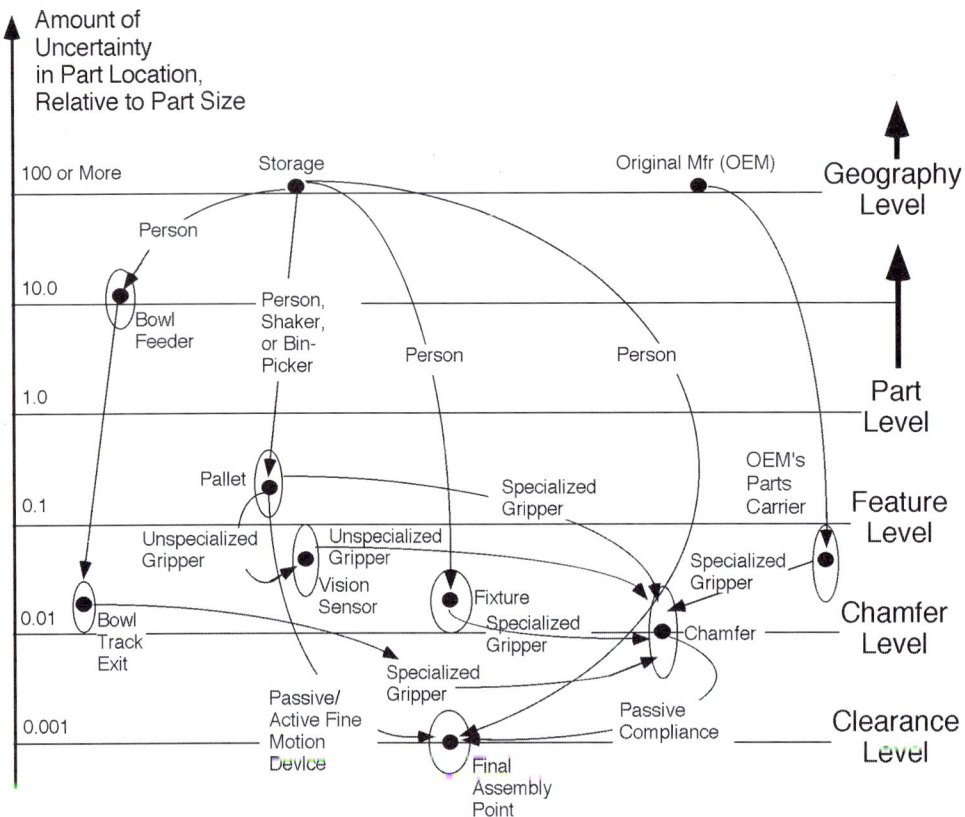

FIGURE 17-1. Different Ways That a Part May Be Brought to Its Final State of Assembly. Removal of uncertainty in relative location and orientation may be done in stages. Different methods are capable of different amounts of relative uncertainty reduction. Each has a different cost, reliability, and speed.

- Having a chain of people or equipment hand off the part

Different approaches demand different amounts of technological sophistication, cost, reliability, and speed.

Some of the steps may occur at the place where the part is made, while the rest occur where assembly occurs. In some cases, the problem of choosing a method may be solved as a shortest path problem using SelectEquip (discussed in Chapter 16).

17.B. WHAT HAPPENS IN AN ASSEMBLY WORKSTATION

Here is a typical assembly cycle. It will be repeated, ideally identically, hundreds of times per shift:

- An incomplete assembly arrives (or several arrive at once).
- Parts to be assembled arrive as single parts or as a subassembly.
- Parts may have to be separated, oriented, given a final check.

- Necessary tools are fetched.
- Parts are joined to the assembly.
- Assembly correctness is checked.
- Tools are set aside.
- Documentation may have to be filled out.
- The assembly is passed on to the next station.

The station designer must accommodate all of these steps. If people are involved, the station's design must be robust against differences in people, such as age, handedness,[2] and, sometimes, gender. Each of these will have an effect on speed, accuracy, mistakes made, and weight that the assembler can be required to lift.[3]

In order to do this properly, the designer must take account of a number of issues: getting the work done in time, adhering to the assembly requirements, and avoiding a variety of mistakes. These issues are discussed next.

17.C. MAJOR ISSUES IN ASSEMBLY WORKSTATION DESIGN

17.C.1. Get Done Within the Allowed Cycle, Which Is Usually Short

In Chapter 16 we learned how to determine the amount of time available in which to perform each assembly operation. We noted that different resources take different amounts of time to do the same task. Thus an important design requirement is to choose a resource that can get the work done in time.

The work steps that occupy the time include:

- Moving work into the workstation. Until the work is settled into position, the resource cannot work on it. (On moving assembly lines in some car companies, workers will walk upstream to meet the oncoming work. Sometimes they will pick up parts or tools on the way and get ready to install the parts while they are still walking.[4])

- Deciding what to do. If different versions are built on the same line, some time is needed to gather information about what the oncoming item is and what parts and operations it needs. There is plenty of opportunity for mistakes at this point.

- Getting ready to work. If the worker is seated, or if the resource is a machine with a fixed location, then the resource must wait for the work but can use this time to fetch a tool and a part. A two-handed person can do each with one hand, but equipment usually fetches the tool first and then moves to the part. In

this and many other ways, people can overlap operations that equipment must do serially.

- Moving to get the part.
- Moving the part to the insertion point.
- Inserting the part. This step, and the two just before it, must be done without doing any damage to the part or the assembly. For large or delicate parts, this can be the most critical phase of the process.

- Checking that assembly was accomplished properly. This usually follows strict instructions. Is the part actually there? Is it secured? Does it operate freely? Did it survive assembly? For a person, this is relatively easy, but for a machine, answering these questions may require special equipment or even a separate workstation.

- Recording information about what was done, how much force was used, and so on. Increasingly, this information is recorded automatically. It is essential for the following: quality; ability to trace problems back to their root causes; training; and improving performance.

- Passing the assembly out of the station.[5]

Methods exist for predicting how long individual assembly operations take. These are discussed in Chapter 15. Here we note that for both people and equipment, every gross motion must follow a pattern of acceleration, steady state speed, deceleration to a creeping state, and finally stopping. In many cases, as illustrated in Figure 17-2, a small percentage, or even zero percent, of the motion will occur at top speed. For this reason, it is unwise to base station operating times on quotes of top speeds. Simulation software, discussed in Section 17.H, usually contains

---

[2] A manufacturing engineer was assigned to find out why exactly half of the assemblies made on a two-shift process had identical assembly mistakes. After eliminating everything else, he determined that the cause was a left-handed assembler on the second shift who could not properly operate the station as originally designed. The assembler was assigned to a different station and the mistakes stopped.

[3] Toyota's method of determining the fatigue impact of an assembly operation, called TVAL, is described in Chapter 15.

[4] At one automobile factory, the author saw workers essentially moving and doing work every second of the assembly cycle, like ballet dancers.

[5] At the end of an assembly line for automobile alternators, the author observed a worker skillfully tossing, underhand, each finished alternator onto the overhead conveyor hook that carried assemblies to the test cell. Only occasionally did he miss. The floor was made of wood blocks, and alternators always passed the test even if they hit the floor on the way.

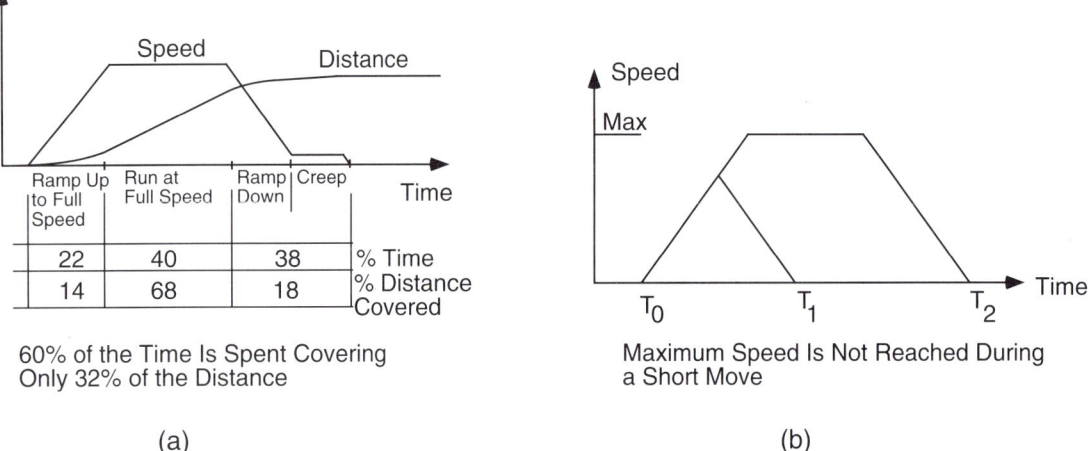

FIGURE 17-2. Patterns of Speed During Gross Motion. (a) Every move comprises ramp-up to full speed, motion at full speed, and ramp-down to a stop. Surprisingly little time, percentage-wise, may be spent going full speed. (b) For short motions, top speed may never be reached.

dynamic models of different assembly resources and can be used to estimate station operating times once a geometric layout of the station is available.

17.C.2. Meet All the Assembly Requirements

To repeat a phrase from Chapter 1, assembly is more than putting parts together. It has to be done correctly or else it is possible that the assembly will not work properly or will not last as long in service as it is supposed to. Typical assembly requirements include the following:

- Using the correct amount of torque on fasteners. Modern fastener installation tools contain torque sensors as well as data recorders. Insufficiently tightened fasteners present severe safety risks in some products like cars and aircraft. Tightening them too tight can be just as bad.

- Applying lubricant. Too little will cause obvious problems. Too much can cause damage or make a mess and make the customer angry.

- Applying adhesive. The same issues arise here as with lubricant.

- Keeping the assembly clean. This is crucial for precision assemblies like optical trains in camera lenses. It is also important in any product that conveys control fluids because orifices or valves can become

clogged and the product will malfunction. Surface contamination can cause adhesives to fail.

- Avoiding scratches, dents, and other cosmetic damage.

It is especially important to be sure the requirements are really required. Some "requirements" are actually evidence that the writer of the specifications is unsure of what is required, so something quite restrictive was written. Such requirements can sometimes make assembly prohibitively expensive. Another problem is requirements that are vague or that assume some common understanding that may not exist. Typical of these are statements like "use a small amount" or "avoid overtightening." These are of no help because there is no certain way to determine if they have been met or not.

17.C.3. Avoid the Six Common Mistakes

Assembly happens very fast, and operators can easily fall into mechanical activities in which they stop paying attention to what they are doing. Six kinds of assembly mistakes are listed in Chapter 16.

In many factories, engineers go to great lengths to prevent mistakes, beyond training and nagging the operators. In Japan this is called poka yoke or mistake-proofing. Examples include designing parts so that there is only one way to insert them, or employing screwdrivers with overload clutches that prevent overtightening. The Hitachi design for reliability method described in Chapter 15 is an example of this approach. The amount of effort needed to achieve part-in-a-million mistake rates can be high indeed. It is common to place lights over bins to tell the operator which part to select. It is less common to place photocells on parts bins to check if the operator selected the correct part. It is even less common to provide a recorded voice to tell the operator which part to select. At one plant, Toyota discovered that only by using all three of these techniques was part-in-a-million accuracy achieved.

On top of all this, a factory is a noisy place, full of moving equipment, tools, and people. Operators need to be shielded from distractions that can cause them to lose concentration and then make a mistake.

It is sometimes said that the justification for assembly automation equipment is that it will operate the same way every time, not get tired, not get distracted. As long as assembly requires rote repetition, this is probably true. But there are many instances where judgment is required, even if it is only to quickly check that a part is suitable for assembly. Such judgments may be relatively easy for a person while being impossible or uneconomical for a machine. Sometimes an appropriate compromise is to assign operators to make kits of known good parts that are then assembled by machines.

17.D. WORKSTATION LAYOUT

The geometric layout of a workstation is dominated by two factors, namely, accessibility and time. Assembly involves the transfer of parts or tools from one location and orientation to another. There may be obstacles, such as other partly done assemblies, tool storage racks, or other parts in the same assembly. Part size and presentation method play a major role. Total task completion time may depend critically on location of elements within the station area.

The major items to be located are the work itself, the parts to be assembled, and the tools or assembly fixtures and aids, if any. Naturally, these should be as close to each other as size and shape permit. Interestingly, however, the relative positions of items visited once per cycle affect only the order in which they are visited but hardly affect total task time.

Whether the workstation is operated by a person or a machine such as a robot, the parts and tools must be laid out within easy reach, preferably in the order in which they are needed. Tools should be stored so that the operator's hand can approach them easily, without awkward twisting that could cause fatigue or injury. Heavy tools are often hung on counterbalances.

Instructions for the operator need to be easily visible. In some factories, each product unit comes with paper instructions, often called a traveler, that tells what the unit is, what version it is, what parts it needs, what tests have already been performed, and so on. The operator will read this information, perform the necessary work, and possibly make an appropriate notation on the traveler. In other factories, all of this will be computerized. The incoming unit will have a bar code or an escort memory in the form of magnetic or other information storage. Items called RF cards attached to products or packages can send a radio signal to the station announcing the unit's arrival and providing a wealth of information about what work is needed, which parts to use, and so on. Work done at the station will generate further information, which can be fed into the card.

Transport and logistics must be arranged at the workstation. The product unit must be transported in. Roller conveyors, belts, carts, automatic guided vehicles, or even smooth surfaces on which the unit slides are all used. Individual parts may arrive in a wide variety of forms. Details about the options are discussed in Section 17.E.2. In general, the options are loose parts in bins or bags, entire kits containing all the parts that are needed at that station, automatic feeders that convert unoriented bulk parts into ones that are oriented or even set up in pallets, and chutes that carry the parts or bins of parts by gravity down to where the operator can reach them.

Figure 17-3 shows a workstation layout for assembly of a sport fishing reel. The designers of the assembly system decided that partially assembled reels were too unstable to be easily transported from one station to the next, so they designed a station where one operator could assemble the entire product. Production requirements were met by providing eight of these stations. Naturally, this complicates the logistics, because parts must be brought to eight different stations, and finished reels must be gathered up from these stations. However, the parts and the product are small and lightweight, so quite a few can be easily transported at once by a single logistics person.

It is important to remember that assembly happens very quickly, so logistic events are needed frequently. In most

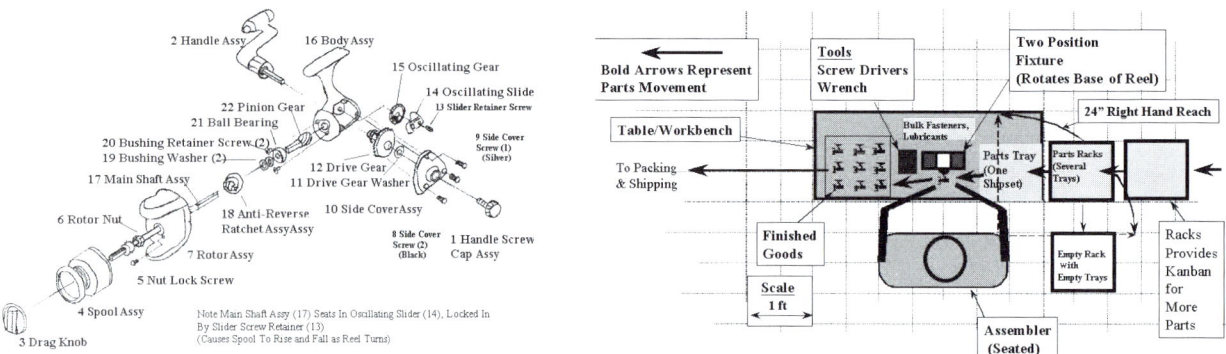

FIGURE 17-3. Assembly Station Layout for Fishing Reel. (Based on work by MIT students Michael Cuppernull, Troy Hamilton, Everardo Ruiz, and Robert Slack.)

assembly systems, parts can be replenished on a schedule that depends on part size, useage rate, and space available near the station. In several of the stations illustrated in this chapter, different parts have different replenishment rates based on such logic. If care is not taken in this part of the design, the operators could run out of parts frequently, or the area around the station could be cluttered with parts and part carriers.

17.E. SOME IMPORTANT DECISIONS

17.E.1. Choice of Assembly "Resource"

Every assembly workstation needs a "resource" that will perform the necessary tasks. Choice of resource is discussed in Chapter 16 because it is a system issue rather than an individual workstation issue. Choices of resources at different stations are interdependent. In addition to the resource, each station needs additional equipment to meet the totality of the requirements. These include tools, part presentation, sensors, transportation for the assemblies, and assembly aids like fluid dispensers, fixtures, and clamps. Most of these are discussed in the various examples in this chapter. The most common, part presentation, is discussed in detail in the next section.

17.E.2. Part Presentation

Part presentation or feeding has one obvious purpose, namely to bring parts to the point where they will be assembled. There is at least one nonobvious purpose, that is, to keep control over the parts so that they stay intact and clean, and so that none get lost or diverted. Choice of presentation method depends on a part's size, shape, and weight. Most of the methods described below do not apply to extremely large parts. However, even manufacturers of cars and tractors have found that nearly all of their parts are smaller than 6" across and weigh less than 5 pounds.

Part feeding methods may be categorized as bulk or individual. Bulk methods take in several or many disoriented parts at once and, by any of several means, transport them a short distance and, more importantly, orient them correctly and present them individually for assembly. Individual feeding methods present prepackaged, preoriented parts individually. Methods include pallets, cassettes, carrier strips, kits, trays, racks, and other arrays or stacks.

17.E.2.a. Bulk Feeding Methods

Examples of bulk feeders include vibratory bowls or hoppers, counterflowing conveyors, and tilting trays. [Boothroyd and Redford] contains detailed information and analyses of these and other kinds of feeders. Vibratory bowls are the most common. They are suitable for small parts whose outer surface can stand some repeated rubbing and impacts with other parts. A typical vibratory bowl feeder for small parts appears in Figure 17-4, while one for larger parts is in Figure 17-5. The feeder works by vibrating rotationally and vertically at the same time, effectively tossing the parts up a helical track that runs up the inside of the bowl. Bowl feeders have the advantages that they work continuously and can be replenished automatically from simple overhead dumpers. On the other hand, they must be designed individually by experts with considerable ingenuity. Their feeding speed is

FIGURE 17-4. Vibratory Bowl Feeder for Small Parts. Above and to the left of the feeder bowl is a hopper that will spill parts into the bowl when a sensor detects that the bowl is nearly empty. A person, or a conveyor or other automatic method, refills the hopper. (Photo courtesy of *Assemblaggio*.)

FIGURE 17-5. Vibratory Bowl Feeding System for Large Parts. This system is similar in many ways to that in Figure 17-4. These parts are about as large as are practical to feed by vibratory methods. (Photo courtesy of *Assemblaggio*.)

inversely proportional to the number of stable states of the part, all but one of which must be eliminated by traps or pockets cut into the track until only parts oriented correctly for assembly arrive at the top.

A disadvantage of vibratory feeders is that their operation often depends on a subtle combination of geometry and friction. Since the bowls wear under continued use, small changes in their shape or friction properties induced by wear may cause them to suddenly and mysteriously stop working properly. A related problem is that it is often impossible to copy a bowl feeder in order to obtain additional feeding capacity. Instead, a new bowl must be made from scratch. Feeder tracks leading from the bowl's exit to the assembly point are also subject to jamming. Finally, parts with very complex shapes may be impossible to feed this way.

An alternative to bowl feeders with mechanical orienting means are bowls combined with vision systems. Parts and bowl tracks are contrasting colors, and the vision system can see if the part is in the correct orientation. If not, it instructs an air jet to blow the part off the track. This approach is less idiosyncratic than mechanical orienting. Due to the cost of vision systems, the method is not widely used but promises to spread as vision systems become less expensive.

Less structured feeding methods are being tried in several companies. Often this consists of manually placing the parts roughly arrayed on a flat surface, not touching, and almost in the correct orientation. A simple vision system can find each part in about one second, permitting a robot to pick it up. This approach is well suited for products that are made in smaller quantities, for which it is not economical to build special pallets or bowl feeders, as well as for larger parts that cannot be fed and oriented using bulk means. This method is used at several workstations in the Denso alternator assembly system discussed in Chapter 16.

The least structured bulk feeding method is a bin or box, with parts lying in it in arbitrary locations and orientations. While there has been research progress on automatic "bin-picking", it is rarely practiced in industry for several reasons. First, it is slower and more costly than the semistructured vision approach. Second, parts grasped from a bin are in an arbitrary location and orientation in the robot (or other) gripper, and must be further analyzed and reoriented. This process takes time and requires additional motion axes, all of which cost money. People are the fastest and cheapest bin pickers and are likely to remain so.

17.E.2.b. Individual Feeding Methods

Individual feeding usually implies pallets or kits, although pallets of small parts may be filled by bulk methods. Pallets usually contain one kind of part or assembly. Kits usually comprise enough of each kind of part to make one unit or subassembly. Kits are used when careful control of parts is necessary; reasons include documentation, cleanliness, prior matching or certification, and so on. The choice between kitting and palletizing often depends on the size of the parts. Most kits are made by people, although kits of small electronic parts can be made by machines.

17.E.2.c. Combined Bulk and Individual Feeding Methods

Bulk methods are often used to load individual feeders with properly oriented parts. Such individual feeders might be pallets or carrier strips. The pallet or strip is then presented to the assembly station. A two-stage feeding system results, with the sometimes troublesome bulk methods accomplished far from the assembly site so that jams do not interfere with assembly.

17.E.2.c.1. Pallets. An example of the pallet method is the Sony APOS (Automatic Positioning and Orienting System) feeding system (see Figure 17-6). Pallets have approximately hundred pockets, each of which will hold one part of a specific kind. Pallets visit one of several pallet loaders, each of which can load several kinds of parts into their respective pallets. Parts are dumped automatically from a hopper onto the pallet, which is vibrated while being held at a slight down slope. Parts that fail to fall into a pocket in the pallet fall instead off the lower end of the pallet and are recirculated. It may take a minute for a pallet to fill up. Vibration speed and tilt angle of the pallet are determined experimentally. This technique is less specialized than vibratory bowl feeders and may take less time to get working; however, the pallets are costly since they must be made by accurately molding or NC machining the individual parts pockets.

A workstation utilizing these pallets is illustrated in Figure 17-7. This workstation contains a robot that assembles parts and exchanges empty pallets for full ones. A control computer sends for a full pallet when one at the station is nearly empty. Clearly there is a limit to the number of such pallets that the station can hold, and some gross motion time is used up swinging the robot over to the more distant pallets. Time is also lost exchanging pallets.

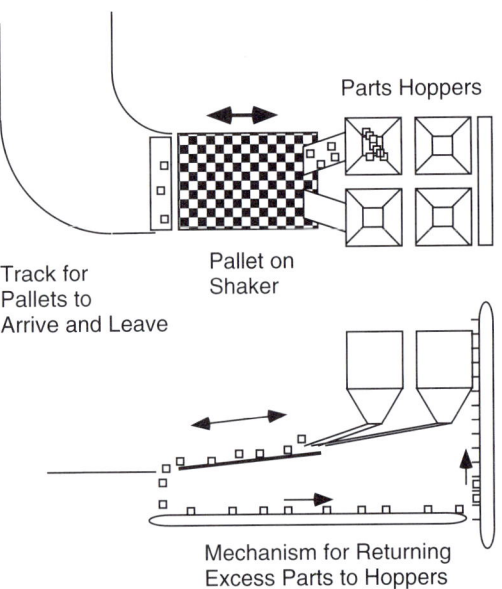

FIGURE 17-6. APOS System for Filling Pallets. This apparatus is part of the Sony robot assembly family of products. Pallets usually fill to about 85%. The robot gripper can detect if a pocket is empty via grip sensors. See [Krishnasamy et al.] for details about the physics of pallet filling.

But this method is quite general and can feed very complex parts. Whatever limitations there may be on feeding speed or feeding reliability are not felt by the assembly station because feeding occurs elsewhere.

17.E.2.c.2. Pallet Arrangement for Large Parts. Figure 17-8 is a sketch of a possible assembly station for a truck automatic transmission. The transmission contains a few quite large parts such as the case and the torque convertor, plus many medium size parts like planetary gear sets and shafts, as well as a large number of small parts. It is awkward to feed many large parts at once, and inefficient to provide only as many small parts as are needed by one transmission. It is probably better to sort the parts by size into two or three classes and devise appropriate feeding and handling methods for each class. This has been done in Figure 17-8. The largest parts are individually presented while the smaller ones are lined up in kits or provided in bulk in sufficient numbers for many transmissions. Large parts must therefore be presented as often as once per takt time while smaller ones may be brought in every 10, 20, or even 50 takt times, depending on their size and rate of consumption.

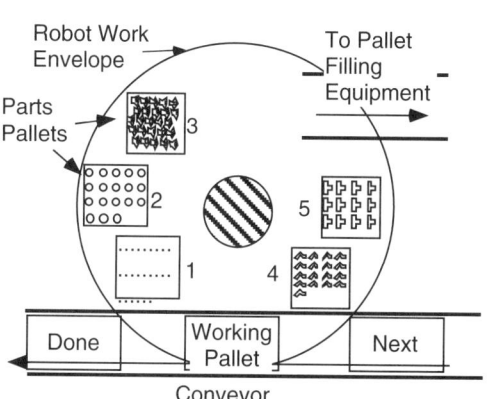

FIGURE 17-7. Workstation with APOS Pallets. In the center of the workstation is a robot. It picks a part from a parts pallet and inserts it in the product mounted on the working pallet. When a parts pallet is empty, the robot places it on the upper conveyor, which takes it to the pallet filling equipment.

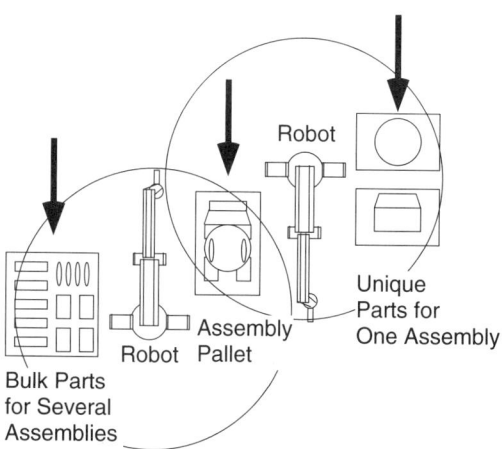

FIGURE 17-8. Conceptual Arrangement of Assembly of a Large Product. The product is an automatic transmission for trucks, which measures approximately 30 × 50 cm. An assembly station might have one or two large robots as shown, or it could have people performing the assembly. Large parts are brought to the station on their own carriers, while smaller ones arrive in kits sufficient for the station's needs for at least one complete transmission, perhaps more.

17.E.2.c.3. Carrier Strips. An example of the carrier strip method is that of electronic connector pins and sockets. They are made from metal strips by a series of stamping processes, and then left in the strip like paper dolls through all subsequent processing (cleaning, plating, and transport) until the instant when they are cut off the strip and pushed into the body of the connector. This process is illustrated in Figure 17-9.

A second carrier strip example ([Kaneko and Saigo]) is quartz watch assembly by Seiko, where special attention was given to feeding speed and reliability. As indicated in Figure 17-10 and Table 17-1, careful study revealed that there is a direct correlation between the failure rate and the amount of structure applied to the part's manufacturing and feeding method. The methods Seiko has used include stamping parts from metal strips as in the previous example, plastically molding them into metal carrier strips, or mechanically inserting them into such strips. In each case, parts are removed from the strip at the moment of insertion into the watch. Figure 17-11 shows examples of these carrier strips while Figure 17-12 shows stacks of them ready to be fed into a robot assembly station.

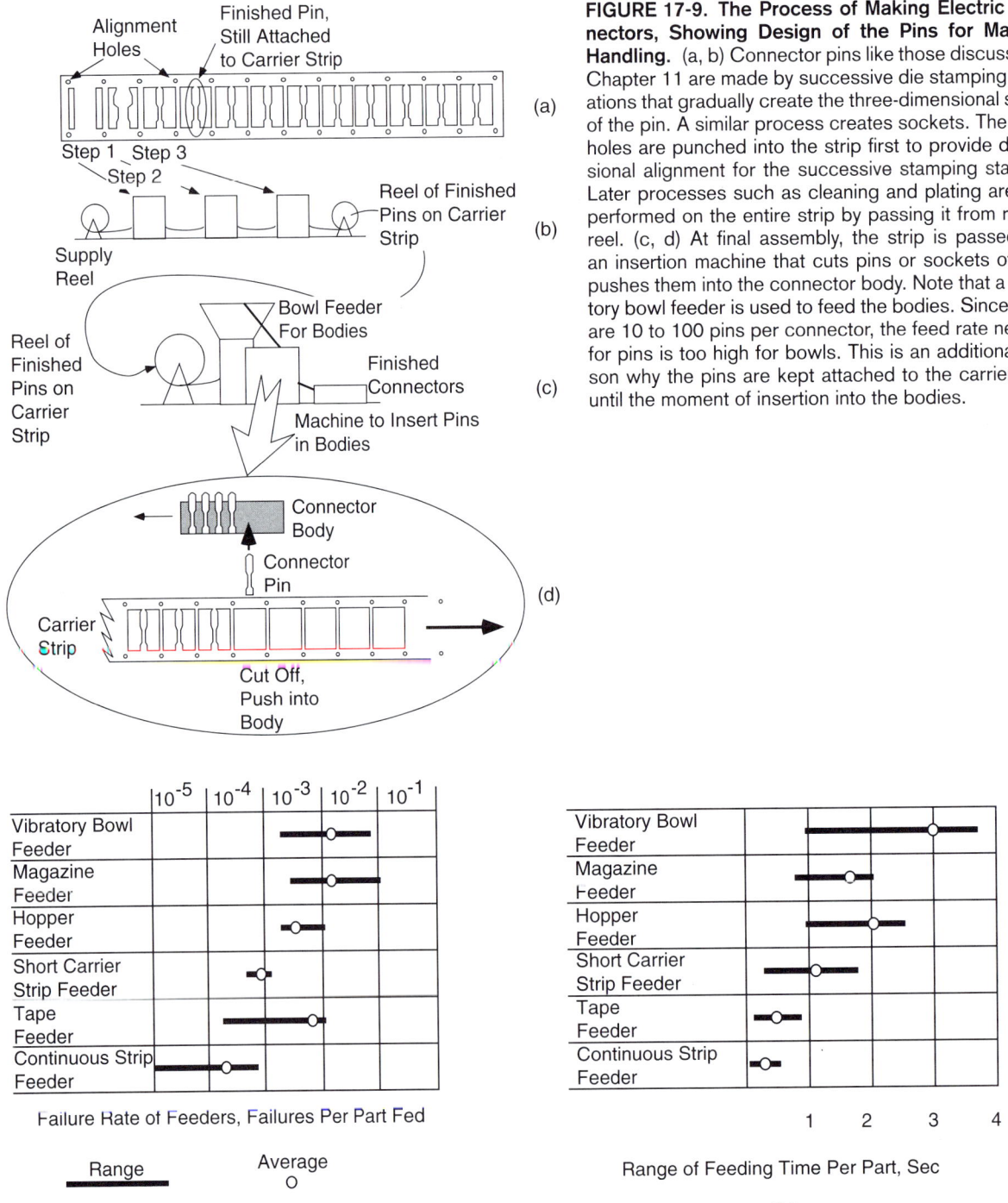

FIGURE 17-9. The Process of Making Electric Connectors, Showing Design of the Pins for Material Handling. (a, b) Connector pins like those discussed in Chapter 11 are made by successive die stamping operations that gradually create the three-dimensional shape of the pin. A similar process creates sockets. The small holes are punched into the strip first to provide dimensional alignment for the successive stamping stations. Later processes such as cleaning and plating are also performed on the entire strip by passing it from reel to reel. (c, d) At final assembly, the strip is passed into an insertion machine that cuts pins or sockets off and pushes them into the connector body. Note that a vibratory bowl feeder is used to feed the bodies. Since there are 10 to 100 pins per connector, the feed rate needed for pins is too high for bowls. This is an additional reason why the pins are kept attached to the carrier strip until the moment of insertion into the bodies.

FIGURE 17-10. Reliability and Feed Rate of Different Small Part Feeding Methods. (a) Experiments on feeding methods for quartz watch parts revealed that feeding parts off continuous strip feeders is at least an order of magnitude more reliable than other methods, but it requires separate equipment to load the strip. An exception is the connector pins in Figure 17-9, which are made directly on the strip. (b) Continuous strip feeders are also faster than other methods. ([Kaneko and Saigo])

17.E.2.d. Other Alternatives

In addition to the above common feeding methods, one may note traveling magazines, traveling pallets, and roving robots. Denso's roving robots are discussed in Chapter 16.

Table 17-2 summarizes several part feeding options according to part size and logistic requirements.

17.E.2.e. Feeding, Tool Changing, and Station Efficiency

Feeding method can have a profound effect on the efficiency of an assembly station. The ideal situation is one in which the parts are ready, in the correct orientation, the moment the resource needs them. No assembly time is wasted. If parts are on a pallet, there may be a loss of time while an empty pallet is moved out of the station and a full one moved in. The actual time lost per assembled unit is less if more parts are on a pallet than the number needed for one product unit, since the in/out time is spread over several units. For large parts, where in/out times could be long, the benefit of even four units per pallet is great. Similarly, the need for tool changes has the same effect on productive cycle time, and a similar approach mitigates the effect: if several product units requiring the same tool are within reach of a resource, it can work on all before changing tools. These tradeoffs are discussed more fully in Chapter 16.

TABLE 17-1. Reliability and Changeover Time of Different Kinds of Part Feeders

	Reliability	Changing Time	Comments
Magazine feeder	99.92%	5 minutes	Preparation takes a lot of time
Short carrier strip feeder	99.93%	7 minutes	It takes relatively more time to change the press die
Tape feeder	99.97%	2 minutes	Very good
Screwdriver	99.96%	4 minutes	Very good

TABLE 17-2. Alternatives for Part Feeding

	Feed Separately to Each Station	Feed to Many Stations
Small parts <1 inch	Bowl, hopper, magazine, carrier strip	Traveling magazine Traveling pallet Kit Roving robot
Medium parts <6 inches	Roller conveyor Stack Manual Ordered layers in bin	Kit of the parts for one unit Pallet of identical parts Roving robot
Large parts >6 inches	Conveyor rack ordered layers	May not be practical

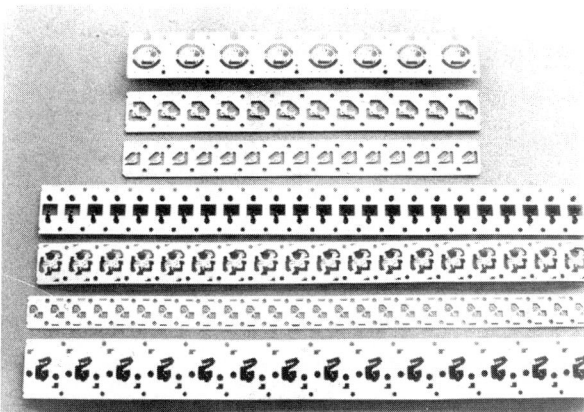

FIGURE 17-11. Quartz Watch Parts on Carrier Strips. Parts may be inserted into these strips by several means: a machine may insert them into pockets in the strip; the parts may be made right on the strip, like the connector pins in Figure 17-9; or the strip may be placed in an injection molding machine and the parts molded into the strip.

FIGURE 17-12. Carrier Strips with Quartz Watch Parts Stacked Up Behind Assembly Robot. A strip is taken off the bottom of the stack at the right and sent to the assembly robots, whose back ends are visible at the left.

17.F. OTHER IMPORTANT DECISIONS

17.F.1. Allocation of Degrees of Freedom

Assembly involves maneuvering two parts relative to each other until they are properly mated. This involves removing incompatibilities in up to six degrees of freedom. Naturally, it is possible to require one part to move while the other stands still, or one part may move in n degrees of freedom while the other moves in $6 - n$ degrees of freedom. People can easily maneuver two small items simultaneously in space and will quickly adopt the easiest way. But the designer of a machine or workstation may not have a wide range of choices, and a person dealing with one large part will need help.

Imagine a large assembly like a car engine, onto which must be put many small external parts like oil lines, spark plug wires, brackets, manifolds, and so on. A likely design for a workstation would provide a support for the engine that permits it to be reoriented about at least two axes. A person or machine that must add a part to the engine will maneuver this part from its orientation in the feeding apparatus to its orientation on the engine. The number of motions needed for the engine obviously depends on how many locations on it will receive parts plus the awkwardness of requiring assembly if no reorientation is provided.

Example workstation designs where allocation of degrees of freedom is important are in Section 17.I.4 and Section 17.I.5.

17.F.2. Combinations of Fabrication and Part Arrangement with Assembly

Sometimes it is convenient to make parts right at the point of assembly rather than make them elsewhere, package and transport them, and feed them to the assembly station. Most fabrication methods position and orient the part precisely, and most packaging methods destroy this

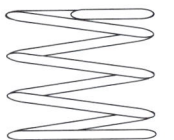 

FIGURE 17-13. Closed (*left*) and Open (*right*) End Springs. The former are unlikely to tangle with each other in a bulk part feeder, while the latter are very likely to do so. Closed end springs use more material and require additional fabrication steps. If they are made at the assembly station one at a time as needed, then it does not matter if they have open or closed ends.

position and orientation. The feeding method must then reestablish it, and many people argue that this is a waste of information, time, and money. Assemblers may require their suppliers to package the parts in a specific orientation that is convenient for assembly, but in some cases this results in poor packing efficiency and increased transport costs.

An alternative is to make the parts at the assembly station. This is commonly done when the parts are made in one or a few simple operations using bulk material. An example is stamping out gaskets from a strip of flexible material. Another is winding springs from a coil of wire. In each of these cases, a difficult feeding problem is avoided: Flexible materials in general are hard to position accurately, while springs tend to tangle unless they are made with closed ends, as shown in Figure 17-13.

In some cases it is convenient to make a subassembly of small parts right at the station where the subassembly will be installed rather than make the subassembly at a distant point. This method may be selected if the subassembly is unstable until it is installed.

Naturally, making parts at the assembly station is not a good idea if the required equipment is large, noisy, or dirty.

17.G. ASSEMBLY STATION ERROR ANALYSIS

Figure 17-14 is a sketch of a robot workstation together with several kinds of errors. Errors arise from parts, grippers, fixtures, interface points where tools attach to the robot, and even in the software that controls the robot and drives it to a commanded point. As pointed out in Chapter 10, relative position and orientation errors must not be so large that the wedging and chamfer crossing criteria are violated.

When a machine performs an assembly operation, we may think of proper relative position and orientation of the parts as a KC and draw a DFC for this KC. Such DFCs appear in Chapter 8. The methods of Chapter 6 may be used to evaluate the probability that these KCs will be delivered. One can go so far as to calculate a C_{pk} for the assembly station.

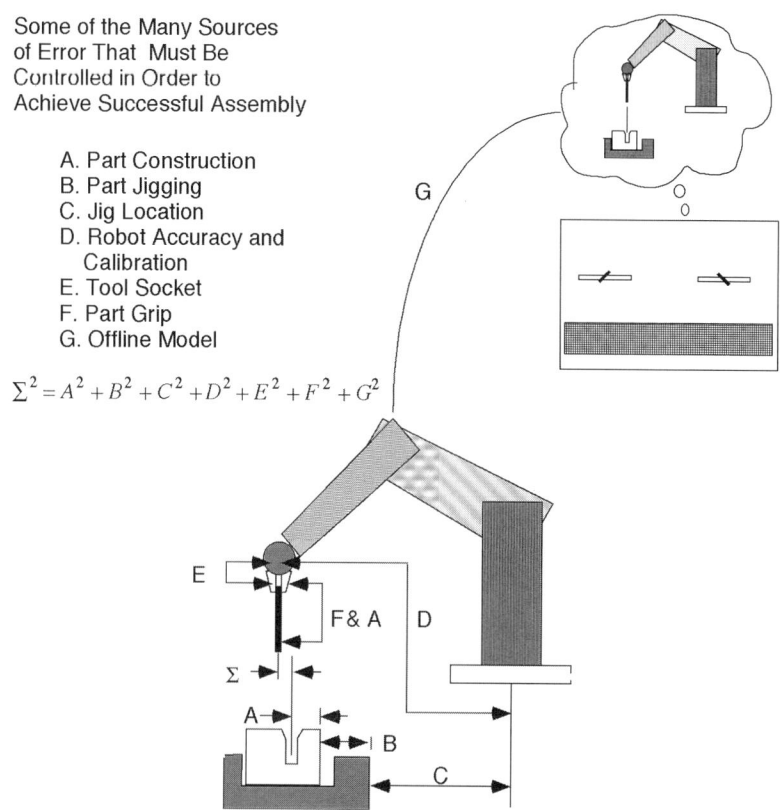

Some of the Many Sources
of Error That Must Be
Controlled in Order to
Achieve Successful Assembly

    A. Part Construction
    B. Part Jigging
    C. Jig Location
    D. Robot Accuracy and
       Calibration
    E. Tool Socket
    F. Part Grip
    G. Offline Model

$$\Sigma^2 = A^2 + B^2 + C^2 + D^2 + E^2 + F^2 + G^2$$

FIGURE 17-14. Sources of Error in a Robot Assembly Station.

17.H. DESIGN METHODS

Generating an assembly workstation, like any other design problem, is a creative act in which time, cost, geometry and a variety of constraints are all managed. There exists no algorithm that will design workstations. Instead there are some aids that will help with certain aspects of the design. These are computer software like simulations and drafting tools, and search algorithms like SelectEquip.

17.H.1. Simulation Software and Other Computer Aids

Simulation software can be a great help in designing a workstation. Examples are shown in Figure 17-15.

    Software of this type usually contains dynamic models of equipment so that cycle time estimates can be made. Cycle time depends on station layout, so the software includes, or links to, CAD software for creating the layout. The simulation can test for accessibility for parts and tools

as well as human hands and arms. The necessary fine motions along the trajectory can also be worked out. Increasingly, these simulated motions can be transferred to the equipment, creating a first pass gross motion program for the robots. Usually, the program must be adjusted on the real equipment because small errors can arise from elastic deflections under static and dynamic loads, backlash in joints, incorrect zero positions on joint angle sensors, and so on. Also, the control program for an assembly station contains many logical tests, signals to actuators like pallet locks, signals from sensors, and other program steps that must be added by skilled programmers.

    Simulations of people are also available. They predict gross motion times as well as possible body stress situations. An example is "Jack" shown in Figure 17-16.

    An example of alternate layouts is shown in Figure 17-17 and Figure 17-18 for the case of robots assembling rear axles of the type discussed in Chapter 7. The assembly step under consideration is installation of the axle shafts into the carrier subassembly. The two drawings are studies of the reach requirements of available assembly robots. Two different pallet designs and robot choices are shown. One design places two axles on one pallet while the other design has four.

FIGURE 17-15. Examples of Computer Simulations of Assembly Workstations. This figure illustrates the capabilities of software from Silma, Inc., a division of Adept Technology, but similar software is available from other companies. At the upper left, the simulation can test to see if the robots will collide. At the lower right, the issue is whether the engine will collide with other parts of the car. The apparatus that carries the engine is not shown, but when it is added, its possible collisions can be tested as well. (Images courtesy of Adept Technology, http://www.adept.com/silma/.)

FIGURE 17-16. Human Simulation Software Jack Performing an Automotive Assembly Task. Jack is sold by EDS, a CAD and factory design company. (Image taken from http://www.plmsolutions-eds.com/products/efactory/jack/. Used by permission.)

17.H.2. Algorithmic Approach

Workstation and system design each are complex problems of choosing among many alternatives. In each case, several steps must be accomplished, and there are several methods available for doing each step. Faced with

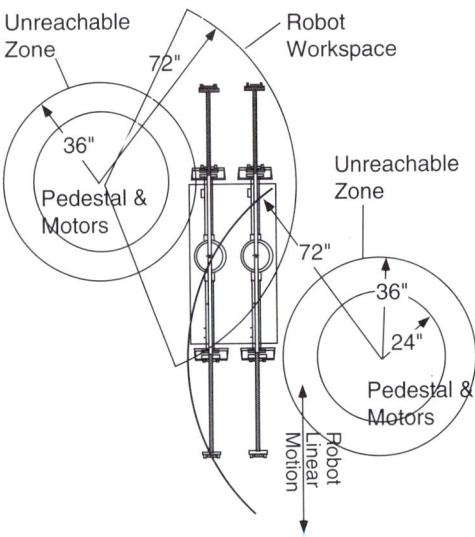

FIGURE 17-17. First Design for Robot Assembly of Axle Shafts into Carrier Subassemblies. This plan view is drawn to scale so that different size robots can be tested against the reach and motion requirements of the task.

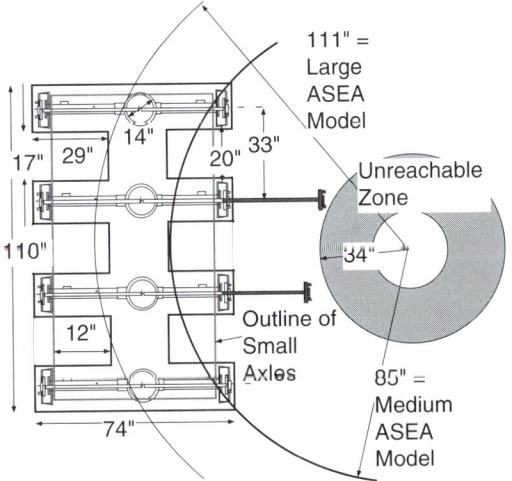

FIGURE 17-18. Second Design for Robot Assembly of Axle Shafts into Carrier Subassemblies. This diagram is also drawn to scale and represents a different approach to the problem illustrated in Figure 17-17.

this complexity, designers often pick what they are familiar with or make arbitrary choices. Here we outline a systematic method that utilizes SelectEquip ([Graves and Holmes-Redfield]).

At present, workstation design and system design are not integrated due to their complexity. Instead, the custom is to design them iteratively. System design as discussed in Chapter 16 is capable of suggesting groupings of tasks at each station. Station design starts with the suggested group of tasks and attempts to create a design that can accomplish them within the required time, error, and cost budgets. If this is not possible, a new system design is created, with different groupings of tasks at stations.

For each operation at a station, we proceed as follows:

1. The operation has several required phases:
   a. Part presentation (feeders, pallets, etc.)
   b. Placement or reorientation of the receiving part or assembly (on a jig, say)
   c. Acquisition of the part (by person, gripper, or tool)
   d. Transportation of the part to the assembly point on the receiving part (by person, robot, gripper, actuator, etc.)
   e. Mating of the transported part to the receiving part

2. For each phase there are different resource choices. Each has an acquisition and operating cost, an operation time, and a contribution to the final error. The effective operating time could be zero if the operation is overlapped with an adjacent operation. The error contribution could be positive, negative, or zero, although final error cannot be negative. Jigs, grippers, pallets, and robots typically contribute positive error, whereas sensors and chamfers typically reduce or absorb error (contribute negative error). Some resources can do more than one phase, although more cost, time, and error might result. Cost and error are related inversely for each resource. That is, given two resources capable of doing the same task, the one that reduces error more or increases it less will cost more. Cost and operating time are also inversely related.

3. To design the workstation to do an operation, we must select resources so that
   a. Each required phase of the operation is done.
   b. Total cost is minimized.
   c. Total time does not exceed a specified maximum.
   d. Total error does not exceed a specified maximum.

SelectEquip will handle (a), (b), and (c) but in its present formulation will not handle (d). Instead, one must keep track of error separately and try different resources or tools if the error is too large.[6]

The design procedure is as follows:

1. The list of M required phases is partitioned into every possible grouping and subgrouping.

2. From the list of N possible resource choices for each phase, each possible assignment of resources to phases or groups of sequential phases is made, subject to the limitation that the total time required to perform each phase or group of phases cannot exceed the cycle time available for performing the whole operation. [In the system design problem, several stations can operate simultaneously, so each must obey the cycle time restraint. In the station design problem, there is only one cycle available and the phases or groups of phases must be executed in succession, so not only must each feasible group take less than one cycle, but the total time taken by all groups must be checked separately to ensure that their total time (and total error) do not exceed the limits imposed.] If more than one candidate resource exists for each phase, the lowest-cost candidate is chosen.

3. The various lowest-cost equipment choices are arranged into a network, where each pair of nodes in the network is a phase and each arc or leg of the network is an equipment choice that does one phase or a group of phases.

4. A shortest path algorithm is used to find a sequence of resource choices of minimum cost that will do all the phases. However, this sequence may not achieve the time and error constraints. If a shortest path algorithm is used that finds not only the shortest, but also the next shortest and next shortest paths, and so on, then one searches through these for the first one that meets both the time and error constraints. Many such algorithms exist ([Dreyfus], [Fox]).

---

[6]Extending SelectEquip to handle error would require treating error the same way that total task time is treated. This is not considered a difficult extension.

The procedure is illustrated by an example in Figure 17-19.

This approach will yield good, but not necessarily optimal, designs. The reason for non-optimality is the arbitrary way we decided between competing candidate equipment choices for each group, in which we took the least cost one. We could have taken the least error or least time one, having no idea what the impact of such a choice would be on the overall outcome. This issue may be moot in any particular example, or the choice between candidates may be obvious to the designer.

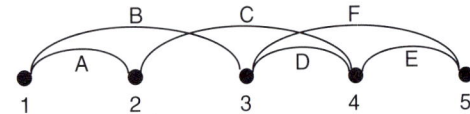

Required Phases

1 - 2: Present parts   3 - 4: Carry part to assembly point

2 - 3: Acquire a part   4 - 5: Mate parts

Feasible Methods

A: Feed part by gravity down
   a track from a bin, using a
   vibratory bowl feeder

B: Pick part directly from bin
   using vision

C: Grip part from end of track
   and carry in gripper to
   chamfer-engagement

D: Use vision to correct
   inaccurate grip while carrying

E: Mate parts from chamfer-
   engagement using compliance

F: Carry parts. Use super-
   accurate parts and vision
   to skip chamfers and mate directly

Method	Cost	Time	Translational Error
A	$10,000	Overlapped	0.02"
B	$35,000	6 sec	0.04"
C	$15,000	4 sec	0.015"
D	$15,000	overlapped with B	—80% of prior error —All prior error if less than 0.04"
E	$9,000	0.2 sec	
F	$15,000	3 sec	—All prior error except 0.005"

Note: Costs, times, and errors are illustrative only.

Comparison:	Cost	Time	Error
Path A-C-E	10+15+9 = $34K	4.2 sec	0.035" (absorbed by chamfer)
Path B-D-E	35+15+9 = $59K	6.2 sec	0.008" (absorbed)
Path B-F	35+15 = $50K	9.0 sec	0.005"

FIGURE 17-19. Example of Algorithmic Workstation Design. *Top:* The network of task phases and the feasible arcs. *Middle:* Time, cost, and error data for each resource choice. *Bottom:* Several possible solutions. Path A–C–E is optimal because it has the lowest cost and error.

17.I. EXAMPLES

17.I.1. Sony Phenix 10 Assembly Station

The Sony Phenix 10 assembly station, shown in Figure 17-20, was used to assemble many versions of the Walkman. As discussed in Chapter 16, six Phenix stations comprised a complete assembly system. Each station con-

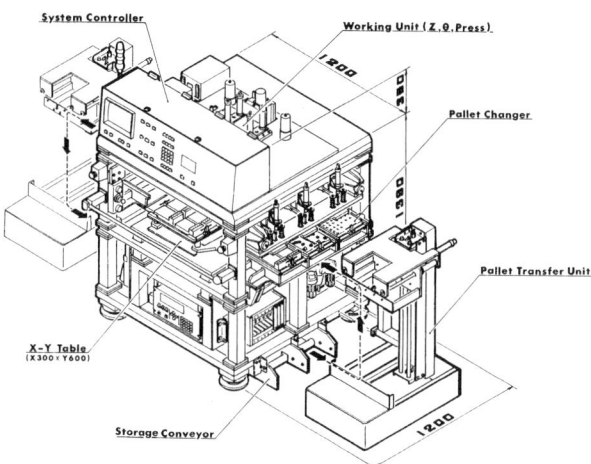

FIGURE 17-20. Sony Phenix 10 Assembly Station. Six of these stations were used to assemble Walkman units. Parts trays illustrated in Figure 17-21 were used. (Drawing courtesy of Sony FA.)

sisted of a computer-controlled $X-Y$ platform that carried a pallet (see Figure 17-21) to tools arrayed around the periphery of the station. These tools moved in the Z direction. The pallet would position a part under an insertion tool, for example, which would take the part from the pallet. Then the pallet would move until the part's insertion point was under the tool, and then the tool would insert the part.

These systems were later replaced by robots equipped with tool turrets. A line of these robots is shown in Figure 17-22, while the turret is shown in Figure 17-23. Part feeding in these robot systems is accomplished using the APOS described in Section 17.E.2.c.

Sony chose to provide multiple tools on a turret rather than have the robot change tools. Tool changing involves swinging the robot to a tool storage rack and executing a change maneuver, which takes time. On the other hand, a turret is heavy because it carries all the tools at once, and this might slow the robot down. In this case, the turret is a better choice because the parts are small and so are the tools, so they are not very heavy.

In other applications where production rate is slow, a workstation might have to assemble ten or fifteen different parts. In this case, Sony employs turret changing, in which the robot exchanges one five-tool turret for another.

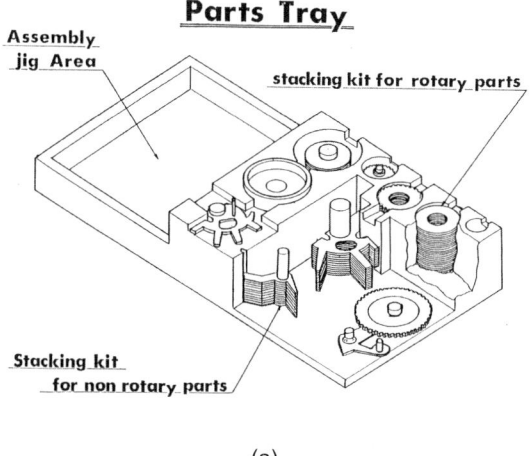

(a)

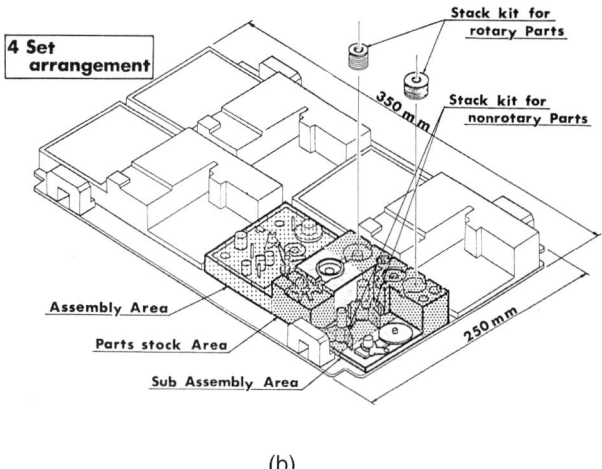

(b)

FIGURE 17-21. Parts Trays Used in Sony Phenix 10 System. (a) A tray carried one assembly and a supply of parts sufficient for many assemblies. (b) Four such trays were placed in each pallet. (Drawing courtesy of Sony FA.)

FIGURE 17-22. A Line of Sony Robots Assembling VCR Recording Heads. Twenty-five of these robots assembled a total of about hundred parts. Each robot was equipped with a turret of up to five tools, shown in Figure 17-23. (Photo courtesy of Sony FA.)

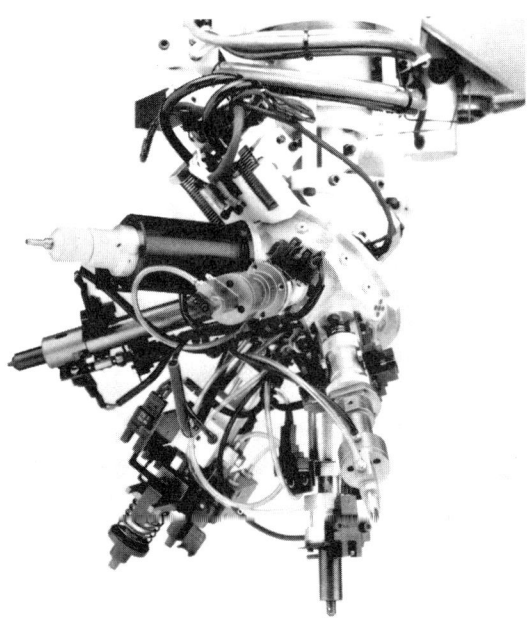

FIGURE 17-23. Turret of Sony Assembly Robot. As the robot swings around to pick up the next part, the turret rotates so that the required tool is in position. The turret is equipped with a gripper that can grasp a parts pallet for the purpose of exchanging an empty one for a full one. (Photo courtesy of Sony FA.)

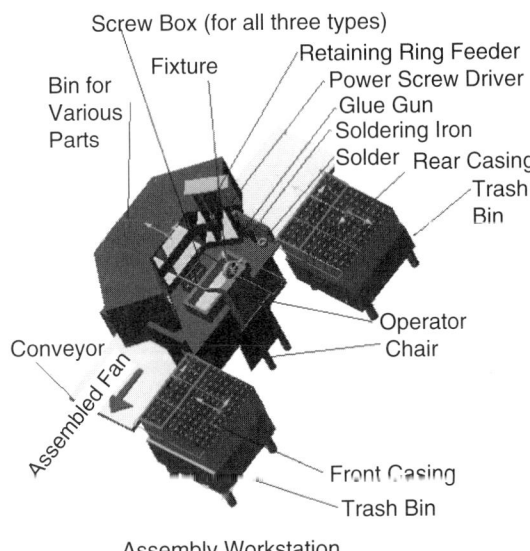

Assembly Workstation

FIGURE 17-24. Assembly Workstation for Twin Window Fan. The entire fan is assembled at this station. Large parts are in the two roller carriers at the operator's right and left. Small parts are in the bins arrayed in front of the operator. Logistics people put bins of parts in from the side opposite the operator and take away finished fans.

17.I.2. Window Fan[7]

The assembly station shown in Figure 17-24 allows a person to assemble a complete twin window fan. This layout solves the problem of providing parts and a work area for a large product that contains a few large parts and many small ones. Different transport, storage, and access methods are required for each size range. The designers chose to provide the large parts in roller carriers, while the small ones were provided in bins. The operator has an unobstructed work area and can easily reach all the parts.

17.I.3. Staple Gun[8]

The assembly system for the staple gun was discussed in Chapter 16. The system comprises a conveyor loop with the operators inside. A typical workstation is shown in Figure 17-25. Parts are fed from the opposite side via bins that glide down on rollers by gravity to the operator. The operator reaches over the conveyor to get the parts. At some stations where heavy springs need to be compressed, the operator has an assembly aid shown in Figure 17-26.

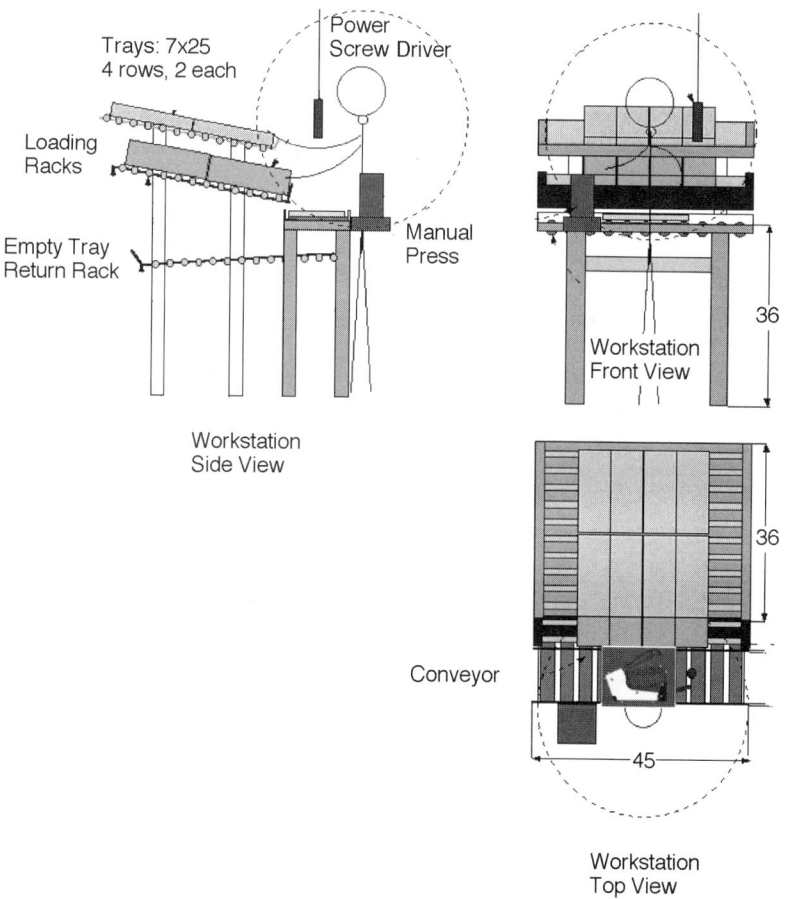

FIGURE 17-25. Assembly Workstation for Staple Gun. This workstation illustrates the method of replenishing parts from the opposite side from where the operator works. Gravity feeds bins of parts down a roller track to where the operator can reach them. Empty bins are placed on the roller track under the workstation, from where they are taken by logistics people to be refilled on a regular schedule.

[7]Based on work by MIT students Aaron Fyke, David Johnson, Bukola Masha, Joshua Pas, Eric Schmidlin, Prabhat Sinha, and Jey Won.

[8]Figure 17-25 and Figure 17-26 are based on work by MIT students Benjamin Arellano, Dawn Robison, Kris Seluga, Thomas Speller, Hai Truong, and Stefan von Praun.

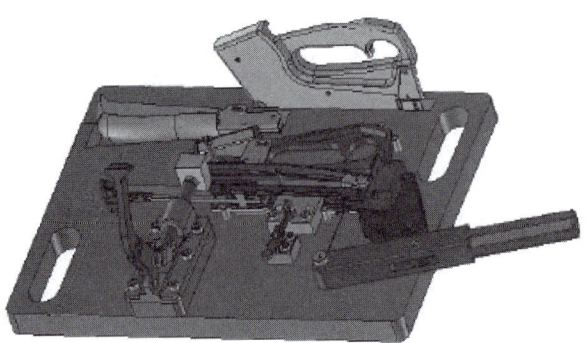

FIGURE 17-26. Aids for Staple Gun Assembly. These fixtures help the operator compress the strong springs and keep the parts from flying apart while fasteners are inserted.

17.I.4. Making Stacks

In the assembly of automatic transmissions, it is necessary to create stacks of clutch plates (illustrated in Chapter 8). Plates of two types, called "a" and "b," arrive at the workstation where they must be merged into ababab... stacks totaling ten plates. A finished stack is needed every 25 seconds. The challenge is to design a robot workstation to do this boring task. Different conveyors bring plates of different diameters, and ideally we would like to be able to make stacks of any diameter with the same equipment or multiples of the same equipment. Here we discuss three different concepts. Each one distributes actions and degrees of freedom differently between part feeders and robot-borne tooling.

The first concept, with two versions, is shown in Figure 17-27. It requires the robot to do everything. In the first version, the robot picks up an "a" plate, puts it on the conveyor, then picks up a "b" plate, puts it on the conveyor, and so on, until the stack is finished. This process takes $15T$ seconds, where T is the time the robot takes to make one gross motion from point to point within the station. To meet the requirement of a stack every 25 seconds, the cycle or takt time must be 1.67 seconds. If $T = 5$ seconds, then we will need four robot stations to provide the stacks at the required rate. The second alternative in Figure 17-27 is a bit faster owing to a more capable gripper that can accumulate the stack in the gripper itself, and it requires only three robot stations to meet the demand.

The second alternative appears in Figure 17-28. This takes advantage of the ring shape of the plates and simply accumulates them onto a spindle. The robot moves the spindle from the "a" conveyor to the "b" conveyor, lifts up

to spear a plate off the end, then repeats this move until a stack of 10 is built. Plates and spindles are deposited on the output conveyor. This is done so quickly that only two stations of this type are needed.

Figure 17-29 shows the third concept. This idea extends the spindle idea with the realization that the robot does not really need to move if the spindle can be made part of the end of the conveyor. Instead of the spindle moving up, the plates fall down, pushed out from under a simple stack that builds up under the end of the conveyor. Furthermore, the symmetry of the situation allows us to build two stacks at once, one on each spindle. This idea is so efficient that only one robot is needed and it can service ten such stack builders.

The main difference between these concepts is what degrees of freedom are put where. In the first concept, the robot has all the degrees of freedom and must make all the moves necessary to build a stack. Even though most of these moves are fine motions, they are given to a large robot whose best capability is for gross motions. Proceeding from concept 1 to concept 2 to concept 3, the fine motion degrees of freedom are successively taken away from the robot and given to more sophisticated tooling placed at the end of the conveyors, until the robot has only infrequent gross motions to make. It thus operates in concept 3 at the center of a rather capable cell that can build all the stacks needed. Since the robot is much more expensive than the other tools, the third concept is much less costly than the first two.

17.I.5. Igniter

The igniter shown in Figure 17-30 activates the power line splice discussed in Chapter 14. When the steel cap is connected electrically to the steel body, the fuse wire heats up, igniting the gunpowder in the igniter, which in turn sets off the main powder charge.

Figure 17-31 shows the original manual assembly process. It involves several subassemblies and intricate maneuvers and requires the parts to be inverted several times. To save money and remove people from possible danger of explosions, an automated process was sought. A completely different assembly sequence and maneuvers were proposed, as shown in Figure 17-32 and Figure 17-33. An important difference between the proposed process and the original is the different distribution of degrees of freedom and the different choices of which parts move and which ones remain stationary during assembly. This

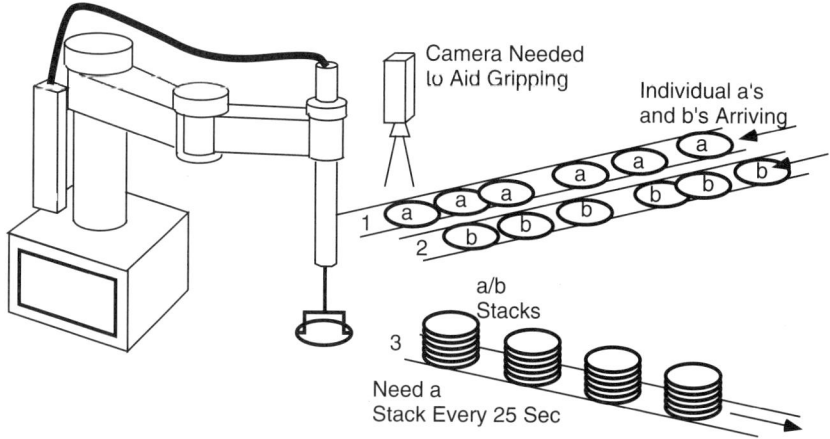

Alternative 1: Robot Grips 1 a and 1 b

    Program Is 1, 2, 3, 1...
    Each Submove Takes T Sec:
    1 to 2, 2 to 3, 3 to 1, 5 Times.
    Time for a Stack of 5 a's, 5 b's = 15*T Sec
    So T Must Be 25/15 = 1.67 Sec.

Alternative 2: Robot Can Grip up to 5 a's and 5 b's

    Program Is (3 to 2), {(2 to 1) 5 Times, (1 to 2) 4 Times Interleaved}
    Plus (1 To 3).
    Time for a Stack of 10 = 11*T Sec.
    So T Must Be 25/11 = 2.27 Sec.

        Since T for a Robot of This Size Is Typically 5 Sec,
        Alternative 1 Would Require 4 Such Stations, While
        Alternative 2 Would Require 3.

FIGURE 17-27. First Concept for Making Stacks. Two alternatives are illustrated. Alternative 1 requires a simple gripper, while alternative 2 requires a more capable gripper but can do the job faster. Plates are ring-shaped, and the gripper grasps them from the inside.

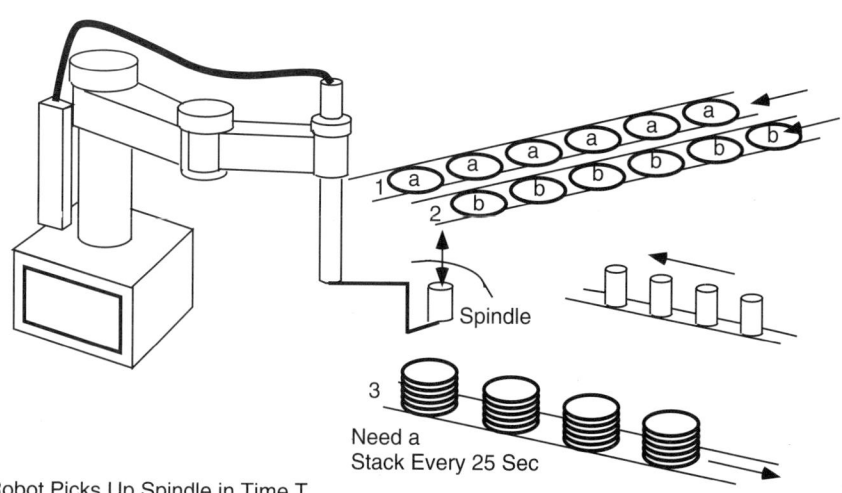

FIGURE 17-28. Second Concept for Making Stacks. This concept dispenses with the gripper and uses a spindle to lift a plate off the end of each conveyor alternately until ten plates are on the stack. Spindles full of plates are then placed on the output conveyor. This concept delivers stacks faster than the first concept.

Robot Picks Up Spindle in Time T.
Robot's Tool Swings Spindle from 1 to 2 in Time 0.7*T.
Robot Deposits Stack with Spindle in Time T.
Total Time = (9*0.7+2)*T = 8.3*T, So T Must Be 3.01 Sec.

On This Basis, Two Such Stations Will Suffice.

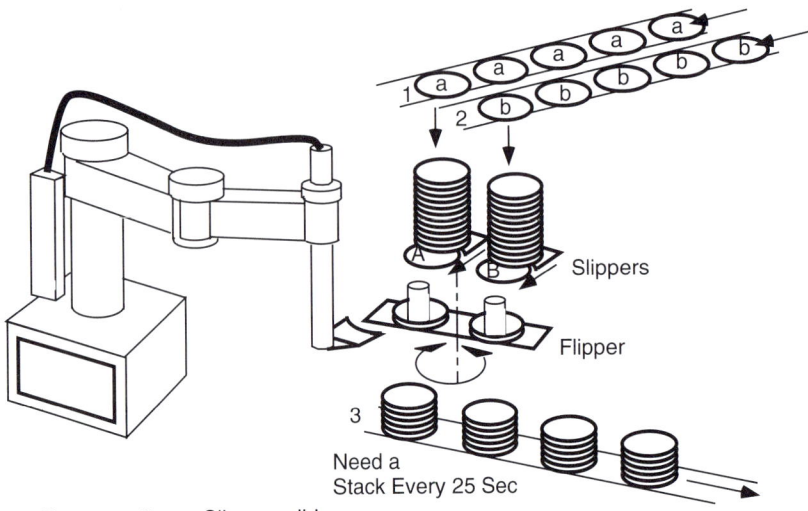

FIGURE 17-29. Third Concept for Making Stacks. This concept dispenses with the spindle carried by the robot and instead comprises a pair of spindles located at the end of the conveyor that effectively builds two stacks at once. The robot's job is simply to pick up finished stacks and place them on the conveyor.

Flipper flips every 2 sec. Slippers slide a disk onto each spindle on the flipper. Two complete stacks are done in 10 sec. Robot transfers them to conveyor in 5 sec. Net time per stack = 7.5 sec

If there are N flipper-slippers, the robot serves each one every 5N sec. So each one produces a stack every 2.5N sec. If each makes a style that is needed once each 25 sec, then a robot can serve 10 flipper-slippers.

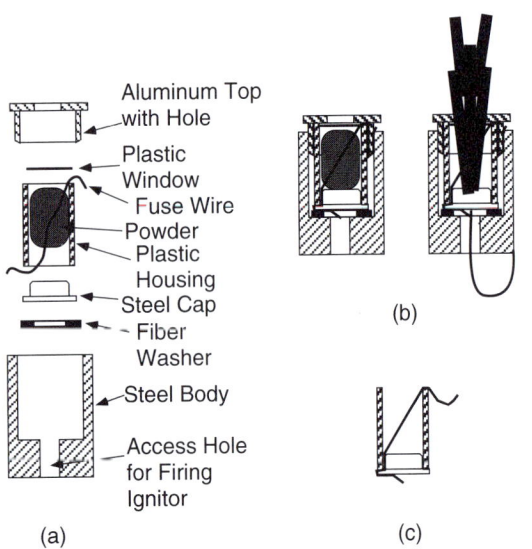

(a)   (b)   (c)

FIGURE 17-30. Exploded View and Method of Operation of the Igniter. (a) The exploded view. (b) Method of operation. The powder is ignited by passing electric current through the fuse wire. (c) Subassembly of plastic housing, fuse wire, and steel cap.

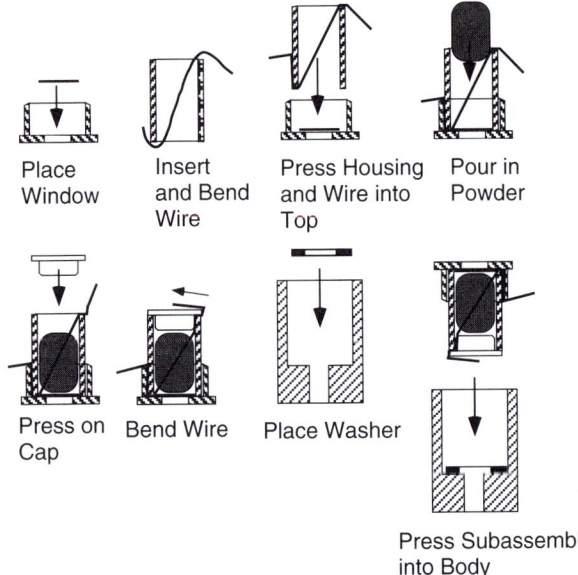

FIGURE 17-31. Manual Assembly Process for the Igniter. The sequence goes from left to right across the top row of figures, then across the bottom row of figures. All steps are done by hand except the three press steps.

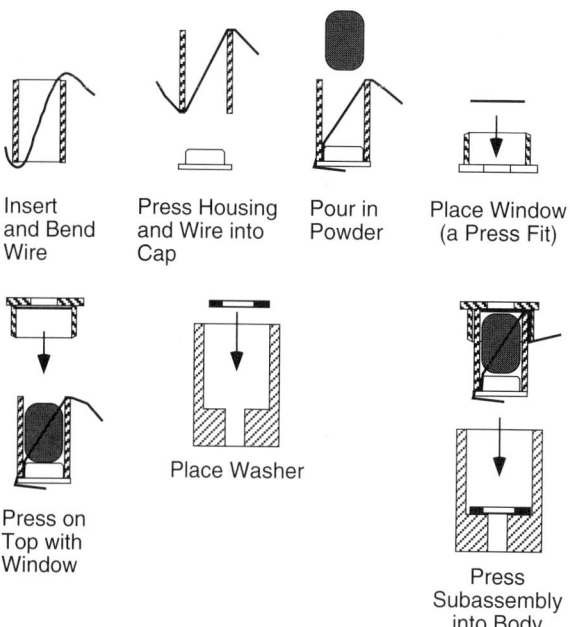

FIGURE 17-32. Proposed Assembly Process for Automatic Implementation. The assembly sequence is quite different from the manual one.

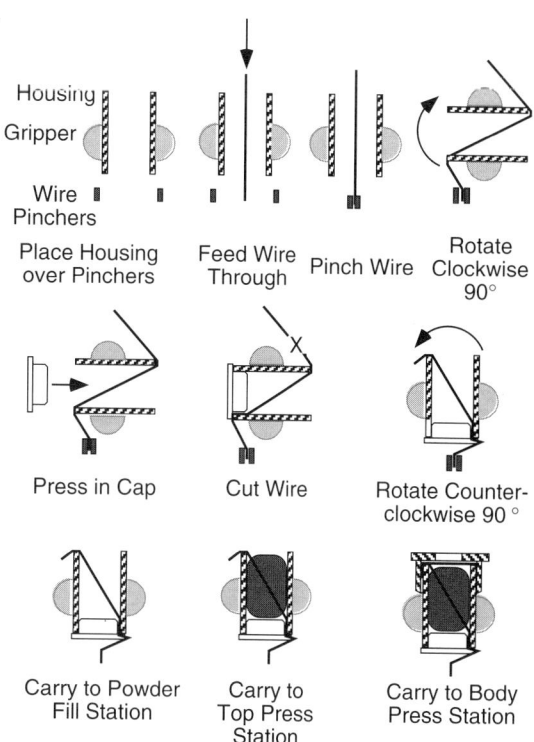

FIGURE 17-33. Modification of Proposed Process to be Accomplished "One-Handed." This process does the difficult wire threading step by inverting the motion relationships between the wire and the housing, compared to the manual method.

difference is required because no reasonable cost method of threading the fuse wire through the plastic housing could be found that duplicated the manual method. Naturally, it is not necessary to duplicate the manual method; what is required is to get the wire through the housing and bend it at each end as shown. The proposed method does this. In principle, it can be done with a robot that can make only a few simple motions.

A workstation concept to do the required operations is shown in Figure 17-34. It consists of a circular table around which are arrayed part feeders and tools like presses. The wire feed and cut station is the most complex. In the center is a simple robot that moves between several fixed points on each of three axes: in–out, rotating about the vertical axis, and rotating about the in–out axis. It also has a gripper that can perform simple open-close operations. Test and rejection of bad units is done by performing the test, then moving to the "not OK" point and opening the gripper if the unit is bad, otherwise moving on to the "OK" point and opening the gripper again.

This system was not built because the parent product was not a success in the marketplace. Assembly systems that accomplish complex operations with simple robots and tooling do exist, however.

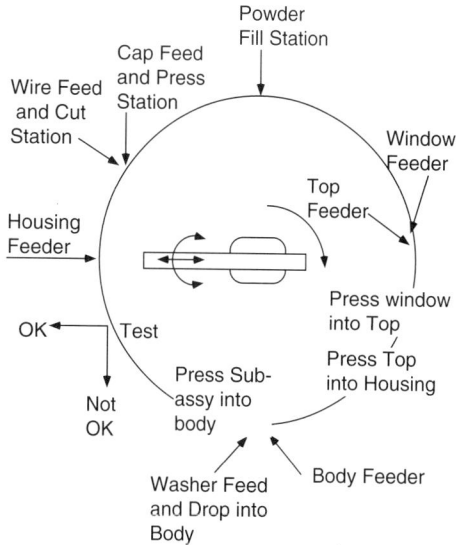

FIGURE 17-34. Top View of Concept Workstation to Build Igniters.

17.J. CHAPTER SUMMARY

Workstation design is a complex problem involving geometric layout, timing, and resource choice. Engineering judgment and knowledge are required, and no algorithmic design method exists except for certain subsets of the total problem. Workstation design interacts with system design, and iteration between these two phases is usually needed. Several examples show that simple equipment can conceivably perform complex operations. In addition, as discussed in Chapter 1, the examples show that machine methods are quite different from manual methods. A satisfactory design cannot be obtained simply by mechanizing the motions and actions of people. Not discussed here but of great importance is reliability and uptime of the station. Great care is required to choose robust methods and to include sensors and checks to see that only good assemblies are passed on to the next station.

17.K. PROBLEMS AND THOUGHT QUESTIONS

1. Use SelectEquip to choose between different part feeding alternatives as illustrated in Figure 17-1. To do this will require estimating costs and times of different methods. A search of the Internet will provide plenty of information.

2. Take apart a small product and decide which part feeding methods might be appropriate for each of the parts.

3. Assume that a person or a robot can accelerate at 0.6 g's maximum and can move top speed 1 m/sec. How long will it take to move the following distances: 0.1 m, 0.2 m, 0.5 m, 1 m, 2 m.

Plot these times versus the corresponding distances. (One g is the acceleration of gravity: $9.8 \ m/s^2$.)

4. If a person is paid $20/hr, is it worth buying a $5000 feeder system to orient small parts if it takes a person 0.5 sec to take the parts from a bin and orient them manually? Assume that the takt time is 10 sec, assembly takes an additional 4 sec, and the calculation applies to one year of operation. What if reorientation takes 5 sec? What if the assembly operation takes 6 sec? (Answer separately for manual reorientation times of 0.5 sec and 5 sec.)

17.L. FURTHER READING

[Boothroyd and Redford] Boothroyd, G., and Redford, A. H., *Mechanized Assembly,* New York: McGraw-Hill, 1968.

[Dreyfus] Dreyfus, S. E., "An Appraisal of Some Shortest Path Algorithms," *Operations Research,* vol. 17, pp. 395–412, 1969.

[Fox] Fox, B. L., "Data Structures and Computer Science Techniques in Operations Research," *Operations Research,* vol. 26, pp. 686–717, 1978.

[Graves and Holmes-Redfield] Graves, S. C., and Holmes-Redfield, C., "Equipment Selection and Task Assignment for Multiproduct Assembly System Design," *International Journal of Flexible Manufacturing Systems,* vol. 1, pp. 31–50, 1988.

[Kaneko and Saigo] Kaneko, K., and Saigo, T., "A Newly Developed Unit Feeder Used in the Analog Quartz Watch Assembly System," Proceedings of CIC, Besançon, pp. 69–72, 1984.

[Krishnasamy et al.] Krishnasamy, J., Jakiela, M., and Whitney, D. E., "Mechanics of Vibration-Assisted Entrapment with Application to Design of Part Feeding Apparatus," IEEE International Conference on Robotics and Automation, Minneapolis, April 1996.

[Nevins and Whitney] Nevins, J. L., and Whitney, D. E., *Concurrent Design of Products and Processes,* New York: McGraw-Hill, 1989.

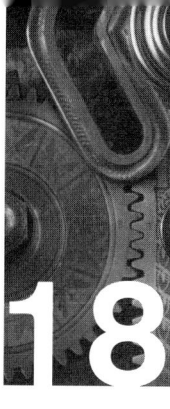

18 ECONOMIC ANALYSIS OF ASSEMBLY SYSTEMS

"Don't ask us how we do investment justification. We just fill out a form and mail it in. After a while, an answer comes back: Yes or No. Usually No."

18.A. INTRODUCTION

Economics in general is about deciding how to allocate scarce resources.[1] If resources were not scarce, there would be no problem. "Resources" in classical economics usually means money, land, and human capital. In our case we deal with time, capital to invest, capable people, and floor space, to name a few. In different situations, we may have more of some and less of others. Frequently, we find that many solutions are possible, but they differ as to which of these resources is favored. One solution may make better use of space but cost more, while another may cost less but take longer to assemble the parts. Thus we find ourselves trading some scarce items for others.

Technology presents us with many choices for performing assembly. Without some means of comparing the alternatives, we have little guidance as to which to choose. Should we invest in designing a simple product so that machines can assemble it in the United States? Should we dispense with simplicity and choose a low-wage country in which to make our product and incur transportation

costs and various communication delays? A strict economic analysis, such as described in this chapter, will not totally solve these problems because not all of the alternatives can be expressed directly in the same metric, such as dollars. As long as we keep this important limitation in mind, the tools discussed here can provide a start on making a rational decision.

It is important to understand that cost is extremely difficult to quantify. One can look at the checkbook and see what the expenditures are, but it is difficult or impossible to find out how the money was actually spent and why. In a manufacturing operation, many costs are distributed or shared because they are associated with support activities. They cannot be directly allocated to a given activity, part, machine, or assembly step. Accounting systems are often designed to help the accounting department balance its books rather than to help management understand cost structures and make improvements. This limitation must also be kept in mind.

18.B. KINDS OF COST

Costs are usually divided up into categories, based on when the money is spent or what it is spent on. The basic categories are fixed, variable, institutional, direct, and indirect. These are not independent categories but in fact overlap.[2]

18.B.1. Fixed Cost

Fixed cost is that group of costs involved in a going activity whose total will remain relatively constant throughout the range of operational activity. That is, fixed costs are more or less the same regardless of production rate, number of shifts, number of people, and so on. Fixed costs usually represent investments that are made in advance of the start of operations. They include buildings, power and waste facilities, machines, conveyors, and so on.

---

[1]This chapter is based on Chapter 12 of [Nevins and Whitney] as well as numerous sections of [Thuesen and Fabrycky].

[2]Some of the definitions in this section are taken from [Thuesen and Fabrycky].

18.B.2. Variable Cost

Variable cost is that group of costs that vary in some relationship to the level of operational activity. Thus variable costs rise as production rate rises because more people are hired, more electricity is used, more tool bits wear out, and so on. Some variable costs may arguably be called fixed, such as buying additional fork lift trucks to support increased production volume. Once bought, they may be hard to sell. This also applies to people, as discussed in Section 18.F.6.

18.B.3. Materials Cost

Every manufacturing operation needs to buy the parts and raw materials that go into the product. In most of the analyses that follow, materials costs will be ignored because they will be the same in spite of different ways of assembling the parts. Materials costs enter when scrap cost is a factor, such as when a different assembly resource will create a different amount of scrap. Other material costs, such as lubricants, are often grouped with variable costs.

18.B.4. Administrative Cost

In order for a company to operate, it must incur a wide variety of other costs, such as a sales force, managers and supervisors, purchasing and human relations departments, design engineers, and so on. In some industries, these costs far outweigh the costs of materials and labor in the factory.[3]

18.B.5. Direct Cost

Direct cost is the cost that is easiest to allocate to the production and assembly process. It includes all the material that goes directly into the product plus all the labor that actually works with the product and the associated immediate supervision. All variable costs are direct.

---

[3]In accounting practice, the cost of materials and labor directly related to manufacturing is usually called the cost of goods sold. Other accounting cost categories include general and administrative costs, sales costs, interest expense, research and development, depreciation, dividends, and taxes.

18.B.6. Indirect Cost

Indirect cost is much harder to quantify and associate to specific products, especially if, as usual, the company makes several things. Indirect costs include general management, engineering, all back office functions like human relations and purchasing, plus facilities like test labs, the lunchroom and company newspaper, and so on. Some of these, as mentioned above, can be quite large.

It is customary to sum up all of the indirect costs and apply them to each item produced by the factory in proportion to the direct labor cost of the item. In this method, the costs are called overhead or burden. The costs of paying for machines bought in the past are also often added to the burden in order to come up with an hourly billing rate for the use of the machines. This method can be misleading for several reasons. For example, one product may be so simple to make that little administrative effort is needed. Yet it will share the overhead costs with other products that make much larger demands on the indirect facilities. The simple product then may look quite expensive to management, who will then investigate outsourcing it when in fact it might be quite profitable on its own.

In order to better understand indirect costs, a method called activity based costing (ABC) has been used ([Cooper and Kaplan]). In order to find out, for example, how much the purchasing department spends to support the purchase of a part for product X, one actually writes down all the activities involved in purchasing, including follow-up phone calls, visits to the supplier's factory, and so on. The cost of performing those activities is then estimated based on time, wage rates of people involved, and so on.

18.B.7. Distribution of Costs in the Supply Chain

Virtually all products are produced in a supply chain. At each stage the basic atoms or molecules are processed or improved in some way by means of the addition of capital, labor, and knowledge. We will see presently that the costs at any one stage are dominated by purchased materials; but if we sum over the whole chain, the dominant component is labor. One result is that economic decisions made at one stage may not be the same as if one could extend the scope of coordinated decision-making over the entire chain. Henry Ford tried to own the chain from rubber plantations and iron mines to final assembly. Toyota

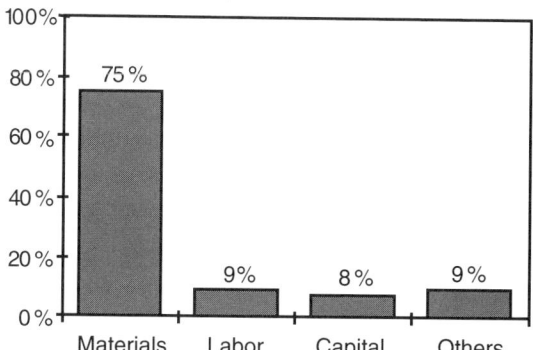

FIGURE 18-1. Cost Distribution in Automobile Engine Plants. This figure shows the components of the cost of a single engine. A typical engine plant can have $500 million in equipment and 500 employees per shift, operating two shifts. "Capital" is all the investment in machines and facilities, allocated to a single engine by sharing the cost among all the engines built over a multi-year period. "Others" includes scrap, rework, tool bits, rags, lubricant, energy, and so on. ([Peschard and Whitney])

Cost Distribution for All Manufacturing (NAICS Sectors 31-33) and for Motor Vehicle Industry (NAICS Sectors 3361, 3362, and 3363) Source: Commerce Department Annual Census of Manufactures

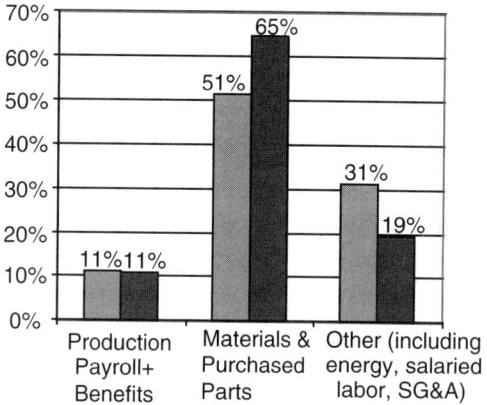

FIGURE 18-2. Cost Distribution for All Manufacturing (NAICS Sectors 31-33) and for Motor Vehicle Industry (NAICS Sectors 3361, 3362, and 3363) for 1999. This chart shows that production labor is a relatively small part of costs within manufacturing enterprises and that materials and purchased parts are the dominant cost. (NAICS: North American Industrial Classification System.)

owns stakes in many of its suppliers and tries to improve their costs and operations. With those and a few other exceptions, the chain is disaggregated, and each level makes its own decisions.

Thus, most companies control only the cost within their four walls. The dominance of materials costs is illustrated in Figure 18-1, which shows cost distributions for a composite of twenty-seven different automobile engine plants. The interesting thing to note about Figure 18-1 is that labor and capital are about equal contributors to total cost. Capital cost is shared by all the engines made over several years according to calculation methods described later in this chapter.

For a view across the entire U.S. economy, consider Figure 18-2, which is derived from the U.S. Department of Commerce Census of Manufactures for 1999. These data are gathered every year and combined with other census data. A similar figure appears in [Nevins and Whitney] with data from 1984. Both charts show the same thing—namely, that labor cost in any one manufacturing firm is about 10% to 15% of sales. This says that for any one firm, the incentive is to first reduce materials and parts costs, and then reduce labor costs.

But increasingly, firms buy from many other firms in a supply chain that stretches back to raw materials extractors and processors. Figure 18-3 shows how these costs

accumulate. If we assume that the same atoms and molecules are simply refined and passed up the chain, then each firm adds labor, logistics, and other costs in the process of converting these atoms and molecules into parts, subassemblies, and products.[4] If we assume that each stage in the chain buys about the same fraction (say 65%) and that there are, say, 5 tiers in the chain, then materials are $0.65^5 = 0.116$ of the cost of the final item. A calculation of this sort is reflected in Figure 18-3.

Once we are sensitive to these issues, we can take a deeper look at costs and place assembly costs in

---

[4]Karl Marx said that all value is attributable to labor. Figure 18-3 shows how much can be attributed to natural resources. Not shown in this figure is the amount attributable to knowledge, which accumulates so slowly compared to labor's contribution that sometimes it goes unnoticed. Also not shown is capital invested, which is usually combined with materials cost.

context. A selection of the issues and options appears in Table 18-1.

All of the above considerations must be kept in mind when economic analysis of manufacturing is undertaken.

18.B.8. Cash Flows

Costs are money that flows out of the company. When products are sold, money flows in. Accountants refer to

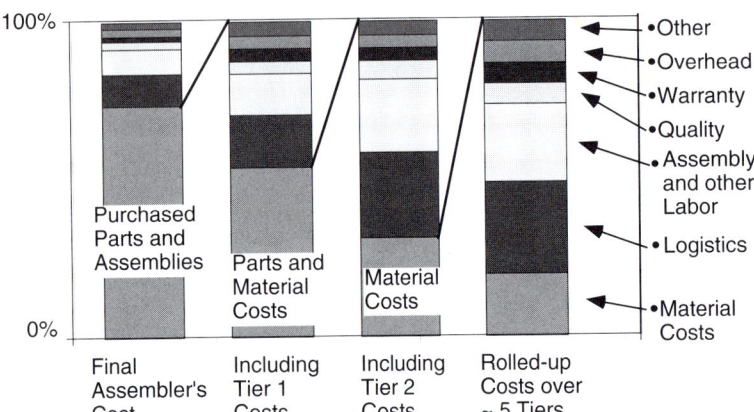

FIGURE 18-3. Estimated Cost Distribution Across a Supply Chain Reflected in the Costs of the Firm at the Top of the Chain. At the left is the cost distribution of the firm at the top of the chain, which is consistent with the data in Figure 18-2. However, this firm buys so much from its first tier suppliers that the costs other than materials costs of those firms really should be included at the top. When they are, the second column from the left is the result. If the chain is roughly five tiers deep, the actual distribution is more like the one at the right. What we see here is the same materials in various states of refinement and shape being worked on by more and more people at more and more companies, each of which applies its knowledge, incurs costs, extracts profits, and pays or charges logistics and shipping costs as items are sent from firm to firm up the chain. Chart provided by Sandy Munro, Munro and Associates, from data accumulated by Daimler-Chrysler. The estimate that logistics can account for as much as 25% of manufacturing cost is confirmed by James Masters, Director of the Master of Science in Logistics program at MIT (personal communication).

TABLE 18-1. Some Cost Drivers in Addition to Labor Plus Some Options in Response

The Cost Driver Is	A Solution May Be	Remarks
Labor	Seek low-labor-cost firms or economic regions	Overhead costs in the form of supervision and communication may rise
	Use automation to reduce labor needs	Requires fabrication methods that are easy to automate and need little skilled labor
	Make high value products with high profit margins	
Transportation	Make the product near customers and/or near suppliers	Local content laws may require this; communication costs may rise
	Encourage suppliers to locate plants near final assembly plant	
	Take care in packaging	Densest packaging randomizes part orientations, requiring costly reestablishment of orientation before assembly can happen
Parts	Reduce the number of parts, choose lower-cost materials and processes	Part complexity may rise, increasing tooling cost and design time
Inventory	Use just-in-time methods to keep inventory low	Works best with predictable order stream and nearby suppliers
Warrantee	Be careful during design and use foolproof assembly methods	

these events as cash flows. An illustrative pattern of cash flows is shown in Figure 18-4. This term will be used repeatedly in this chapter. Presumably the company will make money if income exceeds expenses. But it is not this simple. In the next section we will discuss the fact that cash flows occurring at different times have different values even if the amount is nominally the same. This gives rise to the need for a way to combine them consistently.

18.B.9. Summary

All of the costs discussed above must be taken into account when making an economic analysis and seeking to choose between investment alternatives. It may emerge that the favored alternative wins because of scrap reduction, fewer supervisors, less floor space, better flexibility, or a variety of what might at first appear to be secondary reasons.

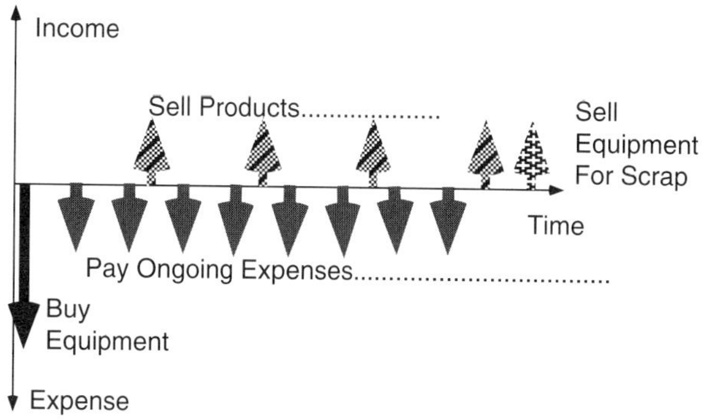

FIGURE 18-4. Illustrative Pattern of Cash Flows. Investment and labor are expenses, while product revenues and sale of used equipment are income. Equipment purchase is a fixed cost that usually occurs before production begins, while wages to people are a variable cost. Sales revenue is a variable income. Variable costs and income usually occur after production begins.

18.C. THE TIME VALUE OF MONEY

We said above that fixed costs such as initial investments and product development are incurred in advance of production startup while variable costs such as labor and materials are incurred if and when operations occur and vary in proportion to the scale of operations. Furthermore, investments are made in the hope of making money, and sales revenues arrive over time but well after initial investments are made. How can the company tell if it is making money? The answer is hard to determine because the cash flows occur at different times, and the value of money is a function of time. The evidence for this is the existence of interest rates.

One can put money in the bank and earn interest. If I promise you $1 a year from now and can earn, say, 3% interest, then I need only 97¢ today. I will put it in the bank and withdraw the dollar in a year. Equivalently, I can just give you 97¢ today. The actual amounts vary according to the interest rate, a factor that will be mentioned many

times in this chapter. Sometimes, the rate in question will be the actual bank or government bond interest rate; at other times, it will be a desired rate of return. The point is that the "same" amount of money can have different values depending on the year when the money is spent or received. Any comparisons between amounts of money, such as subtracting cost from revenue to calculate profit, must be made with comparable quantities. To compare cash flows from different times involves using some kind of interest rate in order to determine what the amounts would have been worth had they come in or gone out at the same time. If they arrived in the past, their value should be increased by the interest they could have earned in the meantime. If they will arrive in the future, then they must be discounted (decreased) by the amount of interest that they could be expected to earn if given the chance to do so. The latter process is called discounting to the present value, and will be described later in this chapter.

18.D. INTEREST RATE, RISK, AND COST OF CAPITAL

In general, people invest in a business in order to make money. The profit they make, in the form of dividends and capital gains, can be thought of as interest paid on their investment. But businesses are not risk-free investments. Risk-averse investors buy risk-free government bonds. Economists often refer to "the social rate of return" to

denote a risk-free interest rate. A value of 3% is often ascribed to this rate, but there is no theoretical reason for any particular value.

When there is risk, as there is in any business, investors always demand a higher rate than the social rate of return, and the higher the risk, the higher return they demand. The excess over the social rate of return is often called the risk premium. How much risk premium to demand is an open question. Certainly, inflation presents a risk of loss of value of future earnings, so the demanded rate of return usually includes an element to account for inflation.[5] Similarly, businesses in new industries face higher risks, as do factory investments in new technologies. How much risk to assume and what rate of return to demand are often matters of personal choice. The methods commonly used to evaluate investments, described here, allow the decision-maker to choose the interest rate or risk premium.

Furthermore, businesses usually must borrow the money they invest in machines and other business activities, by going to banks, selling bonds, or selling stock. In each case, the company can calculate what is called the weighted average cost of capital (WACC). (For a tutorial and representative data, see http://valuation.ibbotson.com.) Certainly, any investment made with that money must return more than the cost of capital or else the company will not make money in the long run.[6] WACC is discussed further in Section 18.G.4.

18.E. COMBINING FIXED AND VARIABLE COSTS

Based on the foregoing, we can state that the total cost of making and assembling something is

$$\text{Total cost} = \text{Fixed cost} + \text{Variable cost} \\ + \text{Material cost} + \text{Other costs} \quad (18\text{-}1)$$

This equation applies over the entire life cycle of the product. As noted above, these costs are incurred at different times, and combining cash flows from different times requires special care. Moreover, fixed manufacturing or assembly cost is usually spent all at once before production starts, while the other costs usually can be identified with individual product units on an ongoing basis. To compare different ways of making something, we need a way to allocate the fixed cost to the individual product units in a way that takes account of time. Several methods exist for doing this. Each one takes the form

$$\text{Cost/Unit} = \text{Variable cost/Unit} \\ + \text{Some } f(\text{Fixed cost and \# units} \\ \text{made in some time period}) \quad (18\text{-}2)$$

This equation converts the initial investment in equipment into an approximately constant expenditure spread out over the same time span as that of the variable costs. This approximation is overcome by a formal cash flow analysis described in Section 18.G.

The variable cost per unit is relatively easy to calculate. For example, if 10 person-minutes are needed to assemble the unit and the wage rate is 15¢/minute, then the unit labor cost is $1.50.

To calculate the contribution of fixed cost, several methods are used.

The payback period method defines a time period during which the fixed cost is said to be paid back by charging each unit an equal share of the fixed cost. If this time period is a number of years P and the number of units made in a year is called Q, then the payback period method determines the unit cost to be

$$\text{Unit cost} = \text{Variable cost} + \frac{\text{Fixed cost}}{PQ} \quad (18\text{-}3)$$

The annual recovery method replaces the $1/P$ term in Equation (18-3) with a fraction that represents the amount that must be paid each year if the fixed cost was like a mortgage with equal principal and interest payments. This method allows us to use a realistic interest rate that includes the social rate of return plus any risk premium we feel is justified.

The fraction paid each year is based on the standard mortgage amortization formula

$$A = I_0 \left[\frac{r(1+r)^H}{(1+r)^H - 1} \right] \left[1 - \frac{v_H}{\rho(1+r)^H} \right] \quad (18\text{-}4)$$

In this formula, A is the annual payment, I_0 is the initial investment or fixed cost, r is the interest rate (expressed as a fraction charged per unit time), and H is the time horizon over which the investment must be paid back. H is similar in spirit to P in the sense that each represents a time period during which the investment is considered to be productive or during which the decision-maker

---

[5]The amount of inflation that people expect over a given future period (say five years) may be estimated by subtracting 3% from the interest paid by good government bonds that mature at the end of that period.

[6]According to http://valuation.ibbotson.com, the weighted average cost of capital for major U.S. manufacturing firms in SIC Class 3XX as of March 15, 2002, is approximately 15.6%.

wants the investment paid back so that the money can be deployed elsewhere. In some cases, the tax law mandates a time period, such as 5 years for computers. If the equipment is taken out of service before period H, it is said to have salvage value v_H, which equals 0 after period H. ρ is the ratio of the cost of the investment to the depreciable hardware cost.

If we define

$$f_{AC} = \frac{A}{I_0} \qquad (18\text{-}5)$$

as the annual capital recovery factor, then the annual recovery method for calculating unit cost is

Unit cost = Variable cost + f_{AC} * Fixed cost/Q (18-6)

The relationship between Equation (18-3) and Equation (18-6) may be seen in Figure 18-5. It shows that f_{AC} declines as H increases and rises as r increases. When

$r = 0$, the annual recovery factor may be calculated using Equation (18-3).

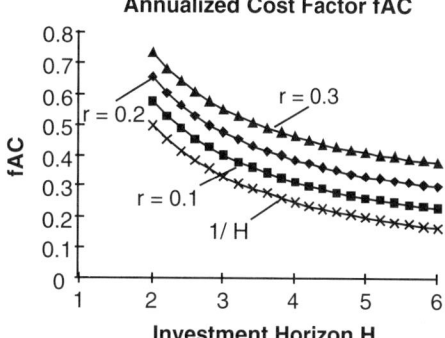

FIGURE 18-5. **Comparison of Annual Recovery Method and Payback Period Method.** The chart is constructed by setting $P = H$. The payback period method is the annual recovery method with $r = 0$.

18.F. COST MODELS OF DIFFERENT ASSEMBLY RESOURCES

In this section we present simple models for the cost of different assembly resources: manual assembly, fixed automation, and flexible automation. We used some of these concepts in Chapter 16 for estimating costs for input to SelectEquip. Here we explain the basis for these estimates in more detail. In particular, we will develop what are called unit costs—that is, the cost of making or assembling one product unit. Often, but not always, the method with the lowest unit cost is preferable. The models are illustrated by charts that use the numerical values in Table 18-2.

18.F.1. Unit Cost Model for Manual Assembly

If we assume that the cost of tools and facilities for manual assembly is negligible, then the cost is entirely accounted

for by labor. The equation for unit assembly cost is then given by Equation (18-7):

$$C_{\text{unit,manual}} = \frac{A\$ * \#\text{People}}{Q} \qquad (18\text{-}7)$$

where

$$\# \text{People} = \left[\frac{TNQ}{2{,}000 * 3{,}600}\right] \text{ and } [x] \text{ denotes rounding } x$$
to the next larger integer

$A\$ $ = the annual cost of a person = $2{,}000 L_H$
$2{,}000$ = number of hours/year
$3{,}600$ = number of seconds/hour

Equation (18-7) is plotted, using the values in Table 18-2, in Figure 18-6.

18.F.2. Unit Cost Model for Fixed Automation

Fixed automation is specialized and inflexible for the sake of efficiency. Thus, a workstation is needed for each part or assembly operation. If we assume that the cost of labor for fixed automation is negligible, and if we use Equation (18-6), then the unit assembly cost is given by Equation (18-8):

$$C_{\text{unit,fixed}} = \frac{f_{AC} * N * S\$ * \#\text{Workstations}}{Q} \qquad (18\text{-}8)$$

TABLE 18-2. **Assumptions Used to Illustrate Simplified Economic Models**

Assembly time per part, T, seconds	5
Labor cost, L_H, $/hr	12
Number of parts/unit, N	10
Annual capital recovery factor, f_{AC}	0.38
Cost of a workstation, fixed or flexible, $S\$ $	50,000
Cost of a tool used by a flexible resource, $T\$ $	10,000
Number of workers/station, w	0.25

where

$$\#\text{Workstations} = \left\lceil \frac{TQ}{3 * 2,000 * 3,600} \right\rceil$$

Equation (18-8) is plotted, using the data in Table 18-2, in Figure 18-7.

18.F.3. Unit Cost Model for Flexible Automation

If we assume that each flexible workstation will need a different tool to handle each part in the assembly, and if we further assume that flexible automation requires some

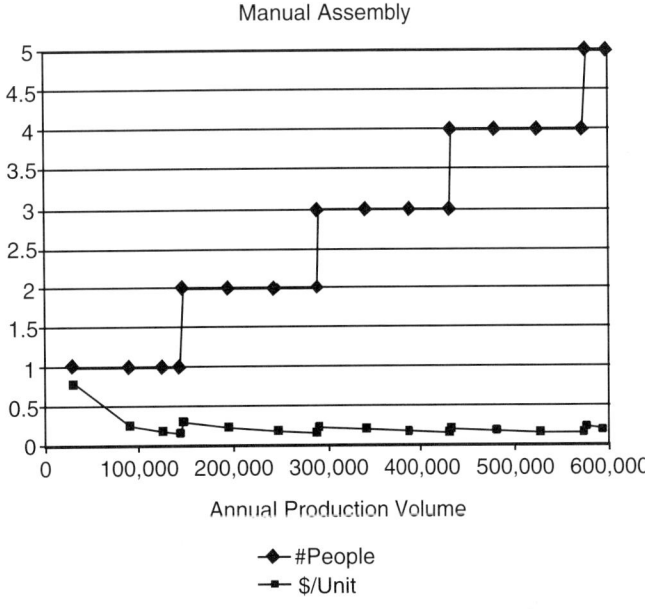

FIGURE 18-6. Simplified Economic Model of Manual Assembly. Each person can do only so many assembly operations in a given amount of time. As the production volume grows, the number of people needed (called #People) grows. Since people are added in proportion to volume, the cost per unit stays almost the same. In this case, assembly cost is all in the category of variable cost.

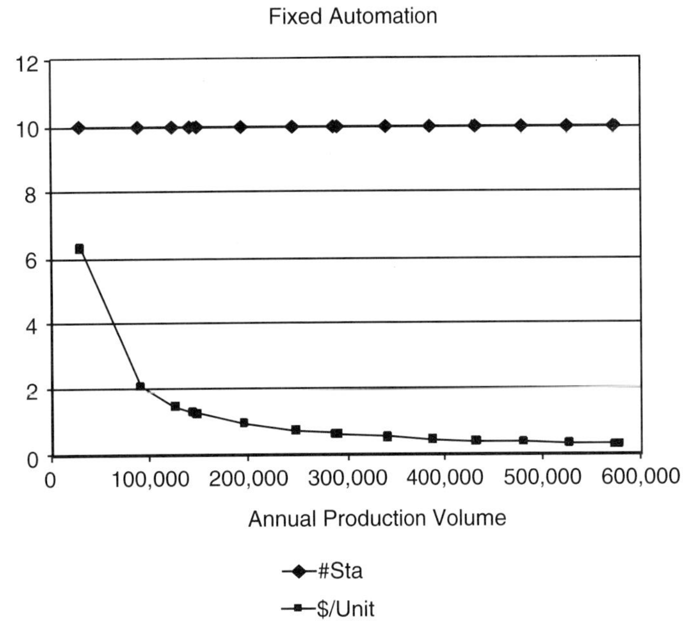

FIGURE 18-7. Simplified Economic Model of Fixed Automation. A fixed number of workstations (#Sta) is needed regardless of production volume since each station can insert only one part. Since we assumed ten parts, the #Sta is always $N = 10$ regardless of production volume. Cost per unit drops steadily as volume increases because there are more units available over which to spread the cost. In this case, the cost of assembly is all in the category of fixed cost.

human attendance, then we may write the unit cost model in Equation (18-9):

$$\text{Cost}_{\text{unit,flexible}} = \frac{f_{AC}I}{Q} + \frac{L\$}{Q} \qquad (18\text{-}9)$$

where

I = total investment in resources and tools
I = #Workstations $* S\$ + $ #Tools $* T\$$

$$\text{#Workstations} = \left\lceil \frac{TNQ}{2,000 * 3,600} \right\rceil$$

$L\$$ = annual cost of labor in the system
$L\$ = w * $#Workstations$ * L_H * 2,000$
#Tools $= N$(one tool for each part)

Equation (18-9) is plotted, using the data in Table 18-2, in Figure 18-8. This equation is approximately given by Equation (18-10):

$$C_{\text{unit,flexible}} \approx \frac{f_{AC} * (S\$ * T) * N}{2,000 * 3,600}$$
$$+ \frac{f_{AC} * T\$ * N}{Q} + \frac{w * T * N * L_H}{3,600} \qquad (18\text{-}10)$$

Equation (18-10) contains a term $S\$ * T$, which is the cost of a flexible resource multiplied by the time it takes to insert one part. This factor is called the price–time product ([Lynch]). Its form indicates that cost and speed can

be traded for each other in the sense that faster resources can be expected to cost more. It can also be used as a metric for such equipment because resources whose price–time product is too high are unlikely to be economically attractive.

18.F.4. Remarks

Some comment on these graphs is in order. For all the resource types, the total cost (though not necessarily the unit cost) is proportional to the number of parts in each product unit. For fixed automation, the number of resources is fixed since it is assumed that a different station is needed for each part. Therefore it is rarely economical at low production volumes. For both manual and flexible automatic assembly, the number of people or flexible resources increases as the production volume grows. At the same time, the number of assembly operations done by each person or flexible resource drops. At the lowest volume, one person or resource has time (but not necessarily the technical capability) to build the entire assembly. To accomplish all these tasks typically requires many tools, tool changing, feeding of many parts, and so on, whose cost and time required are not included in these simple equations. At the highest volume, each station has time for at most one operation. Only one tool and part are used at each station.

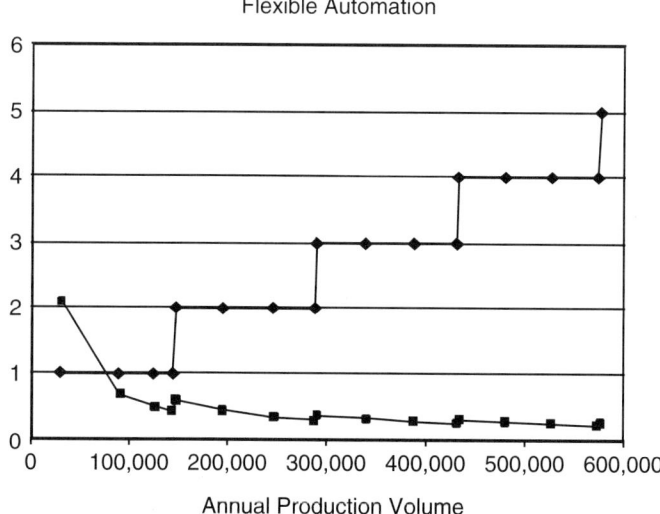

Flexible Automation

Annual Production Volume

→ #Sta
→ $/Unit

FIGURE 18-8. Simplified Economic Model of Flexible Automation. The cost structure here is a mix of fixed and variable costs. The number of workstations (#Sta) needed is proportional to production volume because flexible resources are like people and can do any number of tasks if they have the right tools. But each part needs its own tool regardless of production volume, which contributes a fixed cost. The number of tools (one for each part, or ten) is not shown on the plot.

This represents the extreme in division of labor. If volume exceeds this limit, then the assembly system must run a second or third shift. Alternately, duplicates of the slowest stations can be added. Ultimately, a duplicate assembly system is required.

These equations take account only of the people directly involved in assembling the parts. More comprehensive models account for people who bring new parts and remove finished assemblies. Larger production rates require more such people. Machines need to be tended and repaired. More sophisticated machines need more highly paid people. The number of such people is proportional to the number of machines.

In Figure 18-9 we compare the unit costs for the three kinds of assembly resource shown in the previous three charts. Note the general trend: Manual assembly is the lowest cost of the three methods until production volumes in the range of 750,000 per year are reached, and flexible automation is never the lowest cost. These conclusions are valid only for the numerical values shown in Table 18-2. For other assumptions, such as more expensive labor or faster robots, totally different conclusions may be reached. Flexible assembly may become economical at 200,000 units per year, for example. In order to determine what is best in each case, one must use a more detailed method,

such as SelectEquip, discussed in Chapter 16. In particular, SelectEquip permits design of hybrid systems containing the best mix of manual, fixed, and flexible methods. In addition, it takes into account several important details, such as tool change time, transport time and cost, repair and maintenance staff, and the need for extra shifts or equipment due to saturation of the system or a workstation.

Another comment is in order. Curves like those in Figure 18-9 are similar to other curves whose axes have the same names: unit cost on the vertical axis and production volume on the horizontal axis. These other plots are called learning curves. In spite of the apparent similarity, these two kinds of plots are completely different. A learning curve records the history of improvement in cost as more and more units are produced. The horizontal axis represents cumulative actual production. Cost may fall as more units are produced for many reasons, such as operators getting better, machines breaking down less, suppliers making parts closer to tolerances, and so on. Such improvement is welcome and is sometimes mandated by contracts. The curves in Figure 18-9, by contrast, are predictions about what the unit cost would be if annual production volume were some number. They do not represent history and they assume that all learning improvements have been taken into account.

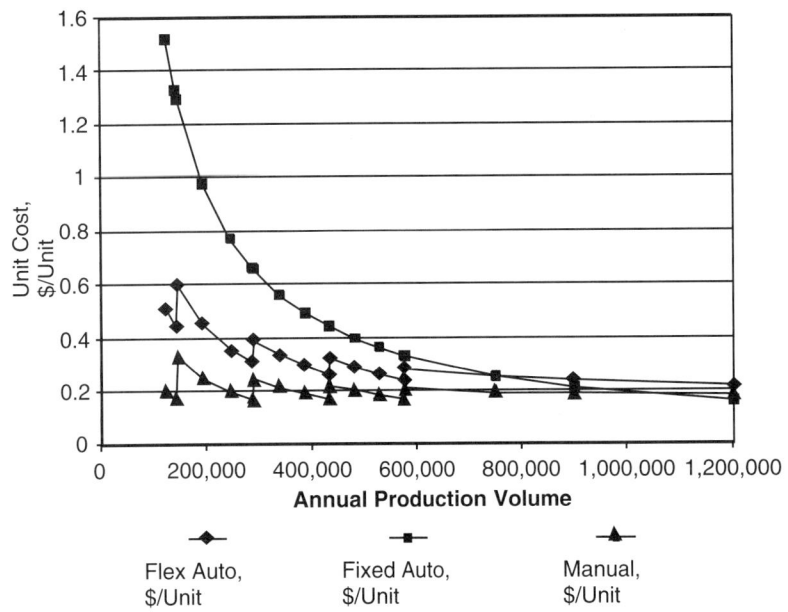

FIGURE 18-9. Comparison of Unit Assembly Cost by Three Kinds of Resource.

18.F.5. How SelectEquip Calculates Assembly Cost

In Chapter 16 we use SelectEquip to design least cost assembly systems. SelectEquip uses unit cost models for each resource based on Equation (18-7), Equation (18-8), and Equation (18-9) with $N = 1$. Tools are assigned based on data supplied by the user of the program, and it may not be necessary to buy one tool for each individual part. The total unit cost of assembling a product is the sum of the individual resource unit costs, and the chosen set of resources minimizes total unit cost. It is important to remember that the number of resources and tools required to assemble a product depends on the assembly sequence and the production volume. Depending on the sequence as well as the design of the parts and tools, tools can be reused at a station or the same tool can be used for two or more operations in a row. The number of tasks clustered at one resource (which creates opportunities to use a tool for different tasks) is inversely proportional to the production volume. Thus tool cost and tool change time depend on assembly sequence and production volume. The cost of assembling each part cannot be calculated in advance without knowledge of these parameters. The net result of this is that total unit assembly cost is a function of the entire sequence.

If the cost of assembling a part could be calculated independently of these factors, then the lowest cost assembly system could be found from the assembly sequence network developed in Chapter 7. One would find the least-cost resource for assembling each part and would put that cost onto the state transition that represents that operation. A simple shortest path algorithm could then find both the best assembly sequence and the best set of resources all at the same time. Some assembly researchers have addressed the problem in this way. But their solutions are unlikely to be correct.

18.F.6. Is Labor Really a Variable Cost?

Most economic analyses treat labor as a variable cost, as we do in this chapter. Strictly speaking, this assumption is valid only if labor can be dialed up or down like water from a faucet. In practice, this is unlikely to be true except under certain circumstances. Many factory employees are protected by union contracts from being laid off arbitrarily. Most companies invest a lot in training their workers and do not like to see that training investment lost. So they are reluctant to lay people off regardless of union contracts. [Mishina] points out that the traditional assembly line creates a workplace for each task, requiring an operator at each workplace regardless of whether the line runs slowly or fast. This, too, suggests that labor is not a variable cost in such a factory.

Several approaches may be taken to address this issue. At a company-wide level, many companies hire fewer permanent employees than they need and fill the gap with temporary workers. These workers bear the brunt of variations in economic activity and the changes in demand that occur as a consequence. At the level of individual products, some companies arrange production into small cells whose production rate can be varied by changing the number of people assigned to the cell. The production rate is then set to the takt time of overall production. People not needed in one cell can be used in another one unless production everywhere has fallen. In that case, the company-wide solution must be considered.

18.G. COMPARING DIFFERENT INVESTMENT ALTERNATIVES

At the beginning of the chapter, we noted that economics deals with allocation of scarce resources. A company always has more ways to spend money than it has money. How should a company decide to invest in one project rather than another? Several methods are used, but each at its core attempts to determine which alternative has the most attractive return. The standard methods include the payback period, internal rate of return, and net present value. Each of these follows the form of Equation (18-1). However, instead of using the equation to compare unit costs as we did in Section 18.E, we will use it to compare rates of return.

Each of the methods must evaluate the attractiveness of a set of cash flows, some negative and some positive, some occurring now, others occurring, or predicted to occur, in the future. Each alternative investment will in general have its own pattern and its own rate of return.

To perform these calculations economically accurately, one must take account of several factors. First, profits are taxed, so only after-tax cash flows should be included.

Second, investments are subject to depreciation. Depreciation is an amount intended to be set aside by the company to replace old equipment some time in the future after it wears out. Total depreciation equals the amount originally paid but is deducted in increments over a period of years rather than being taken as a business expense all at once in the year the investment was made. Each depreciation increment is treated as an expense in the year it is taken and is not subject to taxes. These issues will be illustrated when an actual cash flow example is given in Section 18.G.5.[7]

In the following sections, it will be assumed that there are at least two alternatives called A and B. Each one will have its own pattern of cash flows. For simplicity, we will assume that the time horizon for comparison is the same for both alternatives. If a certain interest rate is demanded, it will be the same for both as well. It can be shown that under these circumstances, the same answer will be obtained if we simply subtract the cash flows of A from those of B and perform the calculation on the difference. If we want to evaluate A as a stand-alone investment, we can say that B represents doing nothing.

When A and B are actual alternatives, then the positive and negative cash flows studied are the differences between the respective items for A and B. For example, the "investment" is actually the difference between the investment required for A and that for B. When B is "do nothing," then all investments and income are those due to A. This situation arises when one is going into business for the first time. The investment is required to set up the factory and equipment, while the savings represent revenue from sales of the goods made by the factory minus the ongoing costs of operation and materials.

18.G.1. Discounting to Present Value

The essential calculation used to compare cash flows occurring at different times is called discounting. This method effectively applies some interest rate to cash flows over the time durations between their arrivals. Although we could pick any time as the reference for calculating the durations, it is common to pick the time when the initial investment is made, and to call that time "now." Then the question becomes, what is the value now (called the present value PV) of all the future cash flows? This is given by Equation (18-11):

$$PV = \sum_{t=0}^{n} F_t(1+r)^{-t} \qquad (18\text{-}11)$$

where

$r =$ the rate of interest expressed as a fraction due each time period
$t =$ an index of time periods
$n =$ the total number of time periods
$F_t =$ the cash flow in period t

Figure 18-10 plots Equation (18-11) for the case where $r = 0.05$ and $F_t = 1$. It should be clear that the discounted PV is substantially less than the undiscounted total. What may be less obvious is that the earlier cash flows contribute much more to the total than the later ones. Equation (18-11) thus multiplies more distant future flows by smaller factors.

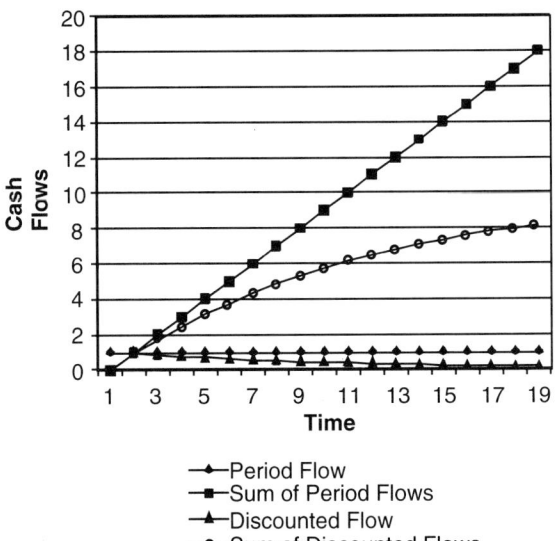

FIGURE 18-10. Illustration of Discounting Future Cash Flows. The early cash flows contribute much more to the discounted total than the later ones.

---

[7]Tutorials on financial calculations are available on many academic web sites. For example, see http://pages.stern.nyu.edu/ ~adamodar/ New_Home_Page/

The different methods described next each use Equation (18-11) in a different way. Some fix the value of n and solve for a value of r. Others fix both n and r and solve for the value of PV, while others set $PV = 0$, fix n, and solve for r. In many cases, the value of n is given by the market duration of the product or the useful life of the investment. However, the value of r must be chosen or solved for. In each case, the items being studied are actual cash coming in or going out during each time period.

18.G.2. Payback Period Method

The payback period method ignores the time value of money and simply seeks the time it takes for the positive cash flows to equal the negative ones. The advantage is that the calculation is simple. One sets $r = 0$ and $PV = 0$ in Equation (18-11) and solves for n. Presumably, one would choose the alternative with the smallest n. But this method will value all investments with the same payback period the same, even if they earn in the aggregate very different amounts of money. It will also value investments whose positive cash flows occur sooner after the initial investment, even if more money could be made in another investment by waiting longer.

18.G.3. Internal Rate of Return Method

The internal rate of return method includes the time value of money in valuing future cash flows. In this method, one chooses a value for n, sets $PV = 0$, and solves for r. This value of r is called the internal rate of return (IRoR). Presumably, one would choose the alternative with the highest IRoR.

18.G.4. Net Present Value Method

In the net present value method, one chooses values for r and n and solves for PV. Presumably, one chooses the alternative with the largest PV. In order to use this method, a value of r must be chosen. This is usually called the minimum attractive rate of return (MARR). Business schools teach that the company should set r equal to its weighted average cost of capital (WACC),[8] although this can be

---

[8]Weighted average cost of capital is calculated by averaging costs from several sources of capital, such as stock or bonds, weighted by the percent of a company's capital derived from each source.

difficult to calculate if most of the capital comes from selling stock.

18.G.5. Example IRoR Calculation

To perform an IRoR calculation, we need to construct what is called a pro-forma cash flow analysis, laying out all the income and expenditures over a period of years. If we are comparing two alternatives, then we use the difference between the two investments as well as the difference between all the subsequent expenses and incomes. If the question is whether to buy equipment in order to save money, then the income comprises the anticipated savings stream resulting from the investment.

Figure 18-11 shows a spreadsheet arranged to find the IRoR for an investment and a savings stream so that the net present value is zero. (This spreadsheet is on the CD-ROM that is packaged with this book.) Several items in this figure deserve explanation. See Table 18-3.

18.G.6. Example Net Present Value Calculation

Figure 18-12 shows a realistic estimate of the cost of developing a new large passenger aircraft, such as the Sonic Cruiser or the A380. In this figure, savings equal profits, which are assumed to range from $40 million to $60 million per plane, for an average of $50 million per plane. It is assumed that production ramps up from 10 planes in year 4 to 60 per year in year 25, for a total of 660 planes. Seventy-five percent of the $10.5 billion development cost is presumed to be undepreciable engineering and testing cost. (This spreadsheet is on the CD-ROM that is packaged with this book.) The figure shows the IRoR for a zero present value. However, the project is unlikely to be approved unless its present value is greater than zero. To find what the PV will be for different interest rates, the spreadsheet is calculated by inserting different interest rates in cell F36 and recording the present value. The result of doing this is shown in Figure 18-13. The IRoR of 11.33% appears in this figure as the point where the curve crosses the $NPV = 0$ axis. If the company's cost of capital is less than 11.33%, the NPV will be positive. Otherwise it will be negative. Capital costs as of this writing are close to this value, so the project is unlikely to be very attractive. It is likely to be approved anyway for strategic reasons, however.

	A	B	C	D	E	F	G	H	I
1			NET PRESENT VALUE CASH FLOW ANALYSIS						
2									
3		7	YEARS ECONOMIC LIFE		0%	SALVAGE VALUE % OF COST AT END OF ECONOMIC LIFE			
4									
5			EXPENSE FORECAST			INCOME FORECAST			
6									
7	YEAR	RATIO	TAX RATE	DEPRECIABLE	SAVINGS	DEPRECIATION	TAX RATE	CREDIT	
8	0	100.00%	34.00%	66.67%					
9	1				$100	14.29%	34.00%		
10	2				$181	24.49%	34.00%		
11	3				$198	17.49%	34.00%		
12	4				$150	12.49%	34.00%		
13	5					8.92%	34.00%		
14	6					8.92%	SUM OF UNUSED YRS		
15	7					8.92%	DEPR=	31.22%	
16	8					4.46%	USED FOR SALVAGE VALUE		
17							OF REMAINING DEPRECIABLE INVESTMENT		
18									
19				TOTAL INVESTMENT		$400		TAX CREDIT IN YR 0 ON	
20				DEPRECIABLE INVESTMENT		$267		UNDEPRECIATED INVESTMENT	
21				INTERNAL RATE OF RETURN		18.41%	RESULT OF		
22						GOAL SEEK			
23						ON CELL G38 = 0			
24				PRO FORMA CASH FLOW					
25									
26	YEAR	INCOME	DEPRECIATION	TAXES	CREDITS	NET	DISC NET		
27	0	($400)		($45)	$0	($355)	($355)		
28	1	$100	$38	$21	$0	$79	$67		
29	2	$181	$65	$39	$0	$142	$101		
30	3	$198	$47	$51	$0	$147	$88		
31	4	$150	$33	$40	$0	$110	$56		
32	4	$83	$0	$0	$0	$83	$42		
33	SALVAGE VALUE								
34	IN YEAR 4								
35									
36	GROSS INCOME	$713	$183	$152	$0	$561	$355		
37									
38	NET INCOME	$313	$183	$106	$0	$206	($0)		

FIGURE 18-11. Spreadsheet for Performing Net Present Value Calculation. This sheet is set up to find the IRoR that yields zero net present value. It does so using the Goal Seek feature, seeking the rate of return in cell F21 that drives the discounted return in cell G38 to zero.

TABLE 18-3. Explanation of Terms in Pro-Forma Cash Flow in Figure 18-11

Term	Meaning
Ratio	How much of the investment occurs in year 0
Depreciable	What fraction of the investment is depreciated over several years; the rest is taken as an expense in year 0. The ratio of total cost to depreciable cost is called ρ. Undepreciable expenses include engineering and installation of the system. They generate a tax credit in year 0.
Savings	The difference between (revenues minus costs) of alternatives A and B during each time period
Depreciation (difference between A and B)	The amount of the depreciable part of the investment that is deducted each year. The pattern is mandated by U.S. tax laws. If the horizon of the investment is less than the eight years shown, the investment is assumed to have a salvage value equal to the sum of the unused depreciation.
Tax rate	This is approximately 34% by U.S. tax law. Taxes are paid on income (savings) less depreciation.
Net income NI (difference between A and B)	$NI_t = (1 - \tau_t)S_t + \tau_t D_t$, where S_t = savings, D_t = depreciation, τ_t = tax rate in period t
Disc net	Net income discounted to year 0 using the IRoR shown in cell F21
Gross income	Sum of rows 28–32
Net income	Sum of rows 27–32

(a)

	Net Present Value Cash Flow Analysis							
	7 Years Economic Life			0% Salvage Value % Of Cost At End Of Economic Life				
	ExpenseForecast			IncomeForecast				
Year	Ratio	Tax Rate	Depreciable	Savings	Depreciation	Tax Rate	Credit	
0	100.00%	34.00%	25.00%					
1				0	14.29%	50.00%		
2				0	24.49%	50.00%		
3				0	17.49%	50.00%		
4				$400	12.49%	50.00%		
5				$600	8.92%	50.00%		
6				$800	8.92%	50.00%		
7				$800	8.92%	50.00%		
8				$800	4.46%	50.00%		
9				$1,000	0	50.00%		
10				$1,000	0	50.00%		
11				$1,000		50.00%		
12				$1,500		50.00%		
13				$1,500		50.00%		
14				$1,500		50.00%		
15				$1,500		50.00%		
16				$1,500		50.00%		
17				$1,500		50.00%		
18				$1,750		50.00%		
19				$2,100		50.00%		
20				$2,100		50.00%		
21				$2,100		50.00%		
22				$2,400		50.00%		
23				$2,400		50.00%		
24				$3,000		50.00%		
25				$3,000		50.00%		
			Total Investment		($10,500)	50.00%	Tax Credit In Yr 0 0n	
			Depreciable Investment		($2,625)		Undepreciated	
			Internal Rate Of Return		11.33%		Investment	
					Goal Seek			
					On Cell G71 = 0			

(b)

			Pro Forma Cash Flow					
Year	Income	Depreciation	Taxes	Credits	Net Income	Disc Net	Sum Of Undisc Net Inc	
0	($3,500)		($3,937.50)	$0	$438	$438	$438	$438
1	-3500	($375)	($1,562.44)	$0	($1,938)	($1,740)	($1,500)	($1,303)
2	-2500	($643)	($928.57)	$0	($1,571)	($1,268)	($3,071)	($2,571)
3	-1000	($459)	($270.44)	$0	($730)	($529)	($3,801)	($3,099)
4	$400	($328)	$363.93	$0	$36	$23	($3,765)	($3,076)
5	$600	($234)	$417.08	$0	$183	$107	($3,582)	($2,969)
6	$800	($234)	$517.08		$283	$283	($3,299)	($2,686)
7	$800	($234)	$517.08		$283	$133	($3,016)	($2,553)
8	$800	($117)	$458.54		$341	$145	($2,675)	($2,408)
9	$1,000	$0	$500.00		$500	$190	($2,175)	($2,218)
10	$1,000	$0	$500.00		$500	$171	($1,675)	($2,047)
11	$1,000	$0	$500.00		$500	$154	($1,175)	($1,893)
12	$1,500	$0	$750.00		$750	$207	($425)	($1,686)
13	$1,500	$0	$750.00		$750	$186	$325	($1,500)
14	$1,500	$0	$750.00		$750	$167	$1,075	($1,333)
15	$1,500	$0	$750.00		$750	$150	$1,825	($1,184)
16	$1,500	$0	$750.00		$750	$135	$2,575	($1,049)
17	$1,500	$0	$750.00		$750	$121	$3,325	($928)
18	$1,750	$0	$875.00		$875	$127	$4,200	($801)
19	$2,100		$1,050.00		$1,050	$137	$5,250	($664)
20	$2,100		$1,050.00		$1,050	$123	$6,300	($542)
21	$2,100		$1,050.00		$1,050	$110	$7,350	($431)
22	$2,400		$1,200.00		$1,200	$113	$8,550	($318)
23	$2,400		$1,200.00		$1,200	$102	$9,750	($217)
24	$3,000		$1,500.00		$1,500	$114	$11,250	($103)
25	$3,000		$1,500.00		$1,500	$103	$12,750	$0
Gross Income	$34,250	($2,624)	$14,937	$0	$12,313	$0		
Net Income	$23,750	($2,624)	$11,000	$0	$12,750	$0		

FIGURE 18-12. Example Net Present Value Calculation for a Large Passenger Aircraft. (a) Net present value cash flow analysis. (b) Pro-forma cash flow.

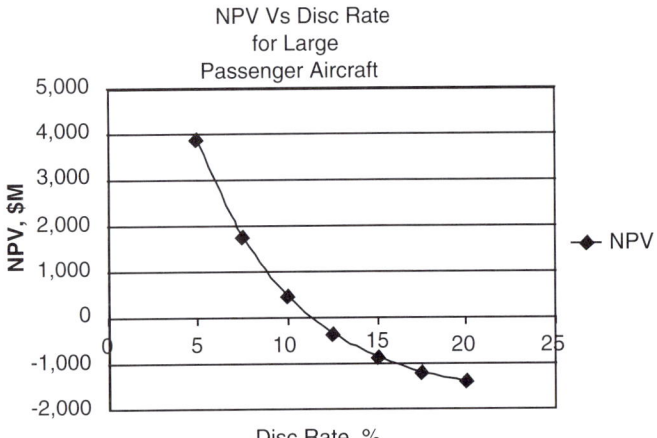

FIGURE 18-13. NPV for Large Passenger Aircraft. NPV is positive for interest rates less than 11.33%, not a very attractive investment on economic grounds alone.

18.G.7. Remarks

The pattern of cash flows shown in Figure 18-4, in which there is one large negative flow at the beginning followed by numerous smaller positive and negative flows thereafter, is typical in the kinds of problems studied here. This kind of cash flow pattern gives rise to the pattern of *PV versus discount rate* behavior shown in Figure 18-13. The methods of comparing investments discussed above are valid when the pattern of *PV versus discount rate* looks like this but may give the wrong answer if it does not.

The NPV method has its critics and there are many ways to interpret the results. Note that the goal of a company is to make money, not to earn a particular rate of interest. Suppose the company has $100 million to invest and has two choices: to invest $90 million for an IRoR of 15% or to invest $15 million for an IRoR of 20%. One investment earns a higher rate of return but the other makes much more money. Thus the results of the calculations must be judged carefully and a decision rule should not be followed blindly.

Another criticism of the NPV method is that it favors short term results and tends not to select projects that will mature over a longer period. While this is true, there are other reasons why a short term view is often taken, even if they are not always good reasons. Capital costs money, and that cost is certain. Profits are in the future and they are uncertain. Discounting is the main way to compensate for the differences in uncertainty.

Another way to take uncertainty into account is to imagine different scenarios for future cash flows. Perhaps one can assign a most likely value, a most optimistic value, and a most pessimistic value. Then it is possible to calculate the mean and standard deviation of the IRoR and *PV*. Investments with a larger mean and smaller standard deviation might be more attractive. In practice, the mean and standard deviation of returns are usually correlated, and one will not find the lowest standard deviation together with the highest mean.

18.H. CHAPTER SUMMARY

This chapter and the two before it comprise a way of looking at assembly (or other manufacturing) systems in a combined economic–technical way. This process begins with the requirement to produce a product or family of products at a certain rate for a certain period of time using some mix of resources. Investments and ongoing costs are involved. A simplified diagram of this process appears in Figure 18-14. It shows that product design (including design simplification), assembly sequence, alternate assembly technologies, and macro- and microeconomic factors all must be considered.

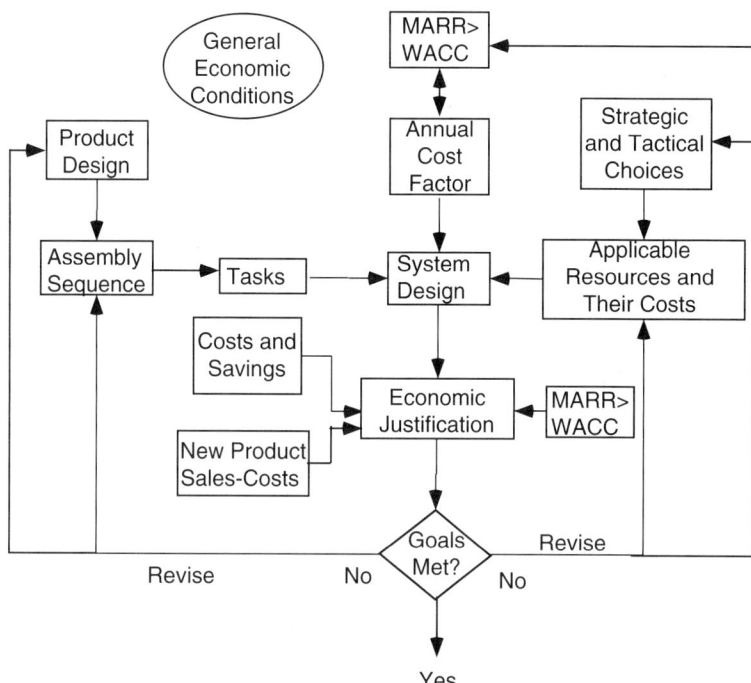

FIGURE 18-14. Logic Diagram for the Creation of Economically-Technically Effective Systems. Most of the factors discussed in earlier chapters are involved in this diagram. "General economic conditions" affect many blocks in the chart, so this block is not linked by arrows to other blocks in the interest of simplicity.

18.I. PROBLEMS AND THOUGHT QUESTIONS

1. Prove that the payback period method of annualizing fixed costs is equivalent to the annual recovery method with $r = 0$. Note that this cannot be proven by substituting $r = 0$ in Equation (18-4). Instead, L'Hôpital's Rule must be used.

2. In Figure 18-6 and Figure 18-8 the unit cost versus production volume plot falls and then rises suddenly, then repeats this pattern several times. However, in Figure 18-7 no such behavior can be seen. Explain why the sudden rises happen in two of the figures but not in the third.

3. Discuss the various terms in Equation (18-10). In particular, discuss possible tradeoffs between robot speed, represented by T, robot cost, represented by $S\$$, and tool cost, represented by $T\$$. For example, a more costly robot could be afforded if some of the cost were devoted to versatility that required fewer tools.

4. The NPV analysis of large passenger aircraft in Figure 18-13 utilizes a tax rate of 50%, appropriate for Europe. If 34% is used, appropriate for the United States, one finds that the NPV is considerably smaller. Explain why this is so.

18.J. FURTHER READING

[Cooper and Kaplan] Cooper, R., and Kaplan, R. S., "Measure Costs Right: Make the Right Decisions," *Harvard Business Review,* September–October, pp. 96–103, 1988.

[Lynch] Lynch, P. M., "Economic-Technological Modeling and Design Criteria for Programmable Assembly Machines," Ph.D. thesis, MIT Mechanical Engineering Department, June 1976.

[Mishina] Mishina, K., "Beyond Flexibility: Toyota's Robust Process-Flow Architecture," in *Coping with Variety: Flexible Productive Systems for Product Variety in the Auto Industry,* Lung, Y., Chanaron, J.-J., Fujimoto, T., and Raff, D., editors, Aldershot, UK: Ashgate Publishing, Ltd., 1999.

[Nevins and Whitney] Nevins, J. L., and Whitney, D. E., *Concurrent Design of Products and Processes,* New York: McGraw-Hill, 1989.

[Peschard and Whitney] Peschard, G., and Whitney, D. E., "Cost and Efficiency Performance of Automobile Engine Plants," available at http://web.mit.edu/ctpid/www/Whitney/papers.html.

[Thuesen and Fabrycky] Thuesen, G. J., and Fabrycky, W. J., *Engineering Economy,* Upper Saddle River, NJ: Prentice-Hall, 2001.

INDEX

A

Accommodation method, 257
Activities in a simulation, 447
Activity-based-costing, 490
Activity cycle diagram, 448
Acyclic graph, 215
Additive processes, 134
Adept Technology, 478
Adhesive bonding, 167
Adhesives, 318, 338, 384, 401
Adjustment, 167, 168, 318, 328, 388, 396, 428
    of an assembly, 23
    of a sewing machine, 9
Airbus, 365, 461
Airbus A380, 365, 461, 501
Airbus A380 wing
    product architecture example, 367
Aircraft assembly, 213
Aircraft engines, 400
Aircraft fuselage
    DFC example, 240
Aircraft product family example, 365
Aircraft structures, 70, 105
Aircraft wing, 344, 346, 349
Aircraft wing subassembly, 251
Aladdin, 351
American National Standards Institute, 114
AND/OR tree, 188, 194
Angular error, 152, 198, 255, 256, 265, 266, 269
    influence on wedging, 271
Angular stiffness, 267
Angular velocity vector, 80
Annual recovery method, 494
ANSI Y 14.5-M, 114
APOS
    vibratory part feeding system, 472, 481
Archimedes system, 194

Architectural flow, 367
Architecture, 348
    attributes, 345
    desktop stapler, 3
    of fixed automation assembly machine, 429
    product, 3, 8
    of product family, 10
    of products and companies, 345
Architectures for automobile bodies, 346
Array product structure, 392
Artificial constraints, 259
ASDP, 457
Assembleability, 380
Assembleability problems, 86
Assemblies
    improperly constrained, 64
    overconstrained, 86
    perform many functions, 2
    properly constrained, 64, 86
    as systems, 16, 348
    types, 34
    underconstrained, 86
Assembly
    definition of, 64
    design process, 1
    feature, 3, 6
    high volume, 11
    history, 12
    importance, 1
    integrative nature, 1, 317, 329
    low volume, 11
    main activities, 11
    nominal, 19
    supports business processes, 2, 6
    time available, 421, 467, 479, 484
    time required, 421, 467, 479
Assembly approach direction, 44
Assembly cost, 190, 396, 479
    increased by rework, 437

Assembly cost analysis, 382, 489
Assembly cycle times
    short, 443
Assembly design intent, 1, 21, 34, 211, 215, 221
Assembly difficulty, 328, 389, 396, 422
Assembly efficiency, 382, 395
    low-cost staple gun example, 414
    related to assembly reliability, 401
    rugged staple gun example, 413
Assembly errors, 254, 263, 479
Assembly feature, 34, 42, 44, 62, 112, 141, 142, 213, 214, 245, 342, 384
    chosen to achieve KCs, 217
    constructed from basic surface contacts, 89
    in a DFC, 216
    Screw Theory model of, 78
    toolkit of, 78, 90, 107
Assembly fitup, 243
Assembly fixtures, 65, 77, 407, 410
Assembly forces, 253
Assembly instructions, 425
Assembly interface, 354
Assembly in the large, 2, 253, 317, 379
    influenced by product architecture, 341
    steps in, 321
Assembly in the small, 2, 253, 379
    steps in, 329
Assembly line, 13, 388, 432, 443, 499
Assembly method, 396, 426, 465
Assembly mistakes, 329, 390, 424, 432, 445, 467, 468
    six kinds, 436
Assembly model, 34, 213
    including product variety, 362
Assembly modeling
    in CAD systems, 213
Assembly motion, 253

Assembly operations, 420
Assembly process capability, 152, 244
Assembly processes, 383
Assembly process requirements, 323
Assembly requirements, 468
Assembly resource, 422, 470
Assembly resource choice, 13, 423, 449,
    465, 470, 479
Assembly robots, 13, 431
Assembly sequence, 2, 180, 221, 246, 330,
    393, 396, 420, 431, 436, 449, 457,
    465, 499
  algorithms, 181
  of automobile alternator, 195
  of automobile transmission, 229
  counting how many, 202
  criteria, 180, 189, 197, 205
  disassembly questions, 184
  feasible, 182, 187, 190
  igniter, 484
  infeasible, 183
  influence on assembly system
    design, 423
  of juicer, 55, 199
  KCs and DFC of, 232
  linear, 183, 190
  of missile seeker head, 53
  of rear axle assembly, 201
  official, 330
  precedence relations, 188
  questions, 183, 184, 187,
    191, 199
  related to decoupling point, 358
  related to delayed commitment, 360
  relation to DFC, 214
  to remove KC conflict, 236, 241
  of sheet metal parts, 167
  state, 189, 202, 203
  transitions, 189, 205
  two-handed, 194
Assembly sequence analysis, 211
Assembly sequence design, 183, 186
Assembly sequence editing, 186, 189
Assembly sequence method
  Bourjault, 184, 190, 207
  cut set, 184, 192
  exploded view, 182
  onion skin, 184
  subset rule, 186
  superset rule, 186
Assembly sequence software
  Archimedes (Sandia), 194
  Draper/MIT, 194
Assembly simulation software, 188
Assembly success, 265, 389
Assembly system, 253
Assembly system capacity, 426

Assembly system design, 317, 327,
    420, 465
  basic factors, 420
Assembly system design methods, 425
  heuristic design method, 449
  systematic, 426, 454
Assembly system performance, 447
Assembly task network, 454
Assembly technology. *See* Assembly
    resource
Assembly time, 497
  actual, 395
  ideal minimum, 395
Assembly time and cost, 420
Assembly time available, 426
Assembly time estimate
  low-cost staple gun example, 414
  rugged staple gun example, 413
Assembly time required, 427, 449,
    451, 454
Assembly tree(s), 205
Assembly workstation, 253
  design methods, 477
Assembly workstation design, 317, 327,
    420, 465
Assembly-driven manufacturing, 364
AT&T Bell Laboratories, 338
Automatic assembly, 55, 198, 322, 332,
    370, 380, 383, 396, 436
  igniter, 484
Automatic assembly machines, 13
Automatic guided vehicles, 469
Automatic screw insertion machines, 401
Automatic transmission, 72
  cellular assembly, 434
  truck, 472
  See also Automobile transmission
Automobile air-fuel intake systems
  product architecture example, 370
Automobile alternators, 396
  robot assembly, 195
Automobile axle and wheel
  DFC example, 217
Automobile body, 344
  product architecture example, 368
Automobile body platform, 352
Automobile cigarette lighters, 430
Automobile cockpits
  product architecture example, 371
Automobile door. *See* Car door
Automobile engine, 76, 342, 349, 353, 476
  KCs, 116
  product architecture example, 370
  selective assembly, 168
  valves inserted by chamferless assembly
    device, 278
  valve train, 117

Automobile engine plant, 422, 491
Automobile transmission, 399, 484
  DFC example, 227
Availability, 421
Average of a sum
  derivation, 140

B

Balance
  of assembly system, 440
Ball-head, 320
Bandwidth
  of fine motions, 260
  of gross motions, 254
Bar code, 469
Base coordinate frame, 36
Base part, 182, 197, 215, 396
Basic dimensions, 121, 122
Batch processing, 444
Batch sizes, 420, 421
Bearing, 333
  self-aligning, 226
Belts
  elastic, 263
Best fit assembly, 166
Bin-picking, 471
Black and Decker, 369, 370
Blame the last part, 16
Blocked workstations, 441
Body on frame
  automobile body architecture, 368
Boeing, 134, 365, 461
Boeing 747, 365
Boeing 747 fuselage, 213
Boeing 777, 115, 134, 461
Boeing Sonic Cruiser, 366, 501
Bolt circle, 120
Boothroyd Dewhurst, Inc., 381
Boothroyd method for DFA, 385, 393,
    396, 399
Bottleneck, 448
Bottleneck station, 442
Bottom-up design, 20
Bourjault, 182
Brushes
  motor, 333
Budd Company, 8
Buffer, 428, 440, 447, 451, 452
Buffer size calculation, 441
  heuristic, 442
Buffer sizes, 449
Build to order, 356
Build to print, 114, 115,
    169, 170
Build to stock, 356

Business context
    for assembly in the large, 321
Business goals
    drivers of modular architecture, 347
Business issues
    influence of assembly, 321
    in product development, 319
Butt joints, 166, 219, 221

C

CAD systems, 20, 34, 45–47, 68, 75,
    213, 215
calibration, 330, 440
Cambridge Process Selector, 396
Canon camera
    example, 338
Capacity, 421, 432, 436
    reduced by rework, 422, 437
Capacity planning, 421
Capital
    cost component, 382
Car body sheet metal, 25, 169
Car door, 27, 116, 235, 353
    KC conflict in, 29, 236
    KC flowdown, 24
    variation, 157, 237
Car door mounting
    Ford method, 31, 237
    GM method, 29, 237
Car floor pan, 221
Car hood, 166
Carrier strip
    for part presentation, 472, 473
Car seat, 69, 76, 104, 113, 245,
    386, 445
Case-hardened armor plate, 71
Cash flows, 493, 500, 504
    comparing, 499
Cellular assembly system, 434, 444, 499
Cellular telephones, 353, 355, 386, 396
Central Limit Theorem, 136
Chain of mates, 217, 221, 239, 245
    including fixture, 221, 222
Chamfer, 152, 263, 272, 304,
    388, 409
    alternate shapes, 276
    influence on assembly, 275
Chamfer crossing, 265, 268, 285
Chamferless assembly, 263, 276
Change, 354, 355
Charles Stark Draper Laboratory,
    Inc., 195
Chevrolet, 353
China, 335
Chinese puzzle, 208

Choice tree
    for configuring a product family
        member, 362
Clamshell architecture, 370
Classification and coding, 380,
    381, 388
    for design for recycling, 403
Clearance, 112
    in form closures, 75
    in hydraulic valve parts, 276
    to relax two-sided constraint, 96
    small, 74
Clearance between feature surfaces, 147
Clearance between parts, 263, 329, 334
Clearance ratio, 154, 270, 271, 288, 332
    effect on assembly time, 388
Closed-loop assembly system
        architecture, 433
Clutches, 227, 484
Coach joint, 221
Coke-bottle aircraft shape, 366
Combinatoric method, 6, 460
Commonality, 359
Complexity, 348
    addressed by architecture, 343
    of injection molded parts, 397
Complex products, 16
Compliance center, 267, 273, 285, 293
    influence of location on assembly, 269
    location when RCC is used, 274
Compliance matrix, 267
Compliance of parts
    influence on assembly mechanics, 265
Compliant parts, 165, 253, 263, 293
    design goals, 296
Compliant support of mating parts, 266
Compliant supports, 263
Composite aircraft parts, 366
Compound assembly feature, 49, 143
    representation of car door hinges, 157
Concept design
    DFA during, 391
Concept generation, 320
Concurrent engineering, 253, 317, 318,
    321, 382
Cone point screws, 279
Connective assembly model, 42, 46, 47,
    62, 112, 141
    including fixtures, 155
    of nominal assembly, 48, 142
    nominal assembly examples, 48
    of varied assemblies, 48, 142
    varied assembly examples, 152
    with variation, for compound
        features, 144
    with variation, for simple
        features, 142

Connector pins
    manufacturing method, 474
Connectors
    electrical, 263
Constant Force, 297
Constraint, 34, 113, 141, 180, 193, 199,
    246, 247
    applied by assembly feature, 46, 62, 76
    and datum surfaces, 121
    delivery by DFC, 221
    DFC role in providing, 214
    one-sided and two-sided, 73
    overconstraint, 62
    proper, 214
    proper, kinematic, exact, 62
    proper, necessary for validity of
        DFC, 245
    provided by fixtures, 155
    single-part, 152
    two-sided, 96
    underconstraint, 62
Constraint analysis, 86
    graphical technique for, 98
    results, 87
Constraint analysis phase, 245
Constraint and mobility
    distinguished, 68
Constraint between parts, 62
Constraint mistakes, 63, 68
Constraint plan, 211, 245, 352
Constraint rule, 244
    algorithm for, 251
Constraint situations
    summarized, 75
Contact force, 265, 295, 297
    during assembly, 268
    effect on part motion, 269
    in electrical connectors, 295
    engineered compliance and, 266, 274
    equations for, 285
    in wedging, 271
Contact rule, 244
    algorithm for, 251
Contacts, 5, 73, 212, 217
    in aircraft assembly, 213
    in automobile transmission, 230
    in car doors, 239
    in Cuisinart, 231
    in sheet metal example, 219
Continuous improvement, 443
Conveyor, 424, 445, 447, 469
Conveyor belts, 353
Cooprider, Curt, 454
Coordinate frame, 36, 62, 79, 142–144,
    147, 152
    of basic surfaces, 87
    for calculating twist, 80

Coordinate frame (*continued*)
  for calculating twist intersection, 87
  for calculating wrench, 82
  in a chain of frames, 41
  on desktop stapler, 36
  element of connective model, 57
  expressed by matrix transformation, 37
  on a feature, 44, 47, 107
  global, 83
  on a hand or gripper, 256
  local, 83
  for motion analysis, 91, 97
  on a part, 44, 45, 47, 91
  world, 45
Coordinate frames
  chain of, 64
Coordination, 172
  of dimensions, 136
  in tolerancing, 141
Copy exactly, 323
Cordless appliances, 338
Cordless screwdriver, 338
Cost, 489–493
  of assembly, 322, 382, 454
  of an assembly resource, 455
  of capital, 494, 501
  to disassemble, 403
  of labor, 322
  of materials, 322
  of a product, 26
Cost Drivers, 492
Cost–performance tradeoff, 22
Costs of accommodating variety, 354
C_p, 129
C_p and C_{pk}, 126
C_{pk}, 128, 129, 132, 152, 163
  for assembly workstation, 476
Cross-threading, 278
Cuisinart
  DFC example, 231
Custom products, 2
Cut set method for finding assembly
    sequences, 184
Cycle time, 427, 441, 451,
    477, 479

D

Daimler-Chrysler, 492
Data model of assembly, 51
Datum, 147, 213
Datum A, 120
Datum B, 120
Datum C, 120
Datum coordination, 113
Datum feature, 120, 218, 226

Datum flow chain (DFC), 6, 36,
    185, 211, 342, 348, 360, 384
  for aircraft wing subassembly, 251
  for assembly workstation, 476
  as a carrier of specifications in a supply
    chain, 322
  to deliver each KC, 216, 245
  examples, 217
  of a product family, 361
  segment inside a part, 227
Datum hierarchy, 119
Datum shift, 156, 242
Datum surfaces, 66, 119
DC-3 airplane, 345
Decoupling point, 356, 357, 358, 374
Defining ribs, 70
Degree of freedom (dof), 62, 65, 68, 73,
    78, 89, 148, 215, 224, 320,
    342, 484
  in assembly workstation, 476
  of flexible automation, 431
  of hand or gripper motion, 256
  in a product platform, 370
Delayed commitment, 180, 358,
    359, 374
Delivery schedules for parts, 452
Dell, 322, 356, 365, 432
Delta wing aircraft shape, 366
Demand, unpredictable, 354
Demand pull, 319
Demand uncertainty, 364
Deming, W. E., 124
Denso, 6, 330, 381, 471
Denso alternator assembly line, 458
Denso DFA method, 406
Denso manufacturing technology
    roadmap, 460
Denso panel meter, 7, 356, 360, 361, 364,
    365, 396, 422, 424, 431, 445, 458
Denso roving robot line, 460
Denso variable capacity assembly
    system, 459
Depreciation, 445, 500
Derivative products, 350
Design decoupling point, 356
Design for assembly (DFA), 70, 182, 253,
    317, 324, 327, 365, 379, 422
  factors affecting assembly, 381
  general approach, 383
  goals, 380
  history, 380
  two phases, 382
Design for disassembly, 327
Design for manufacturing (DFM), 379, 396
Design improvements, 324, 327, 329
Designing quality in, 435
Design of assembly, 324

Design procedure for assemblies, 245
Design simplification, 15, 380
Desktop copier, 68, 105
Desktop stapler, 2, 212, 417
  assembly features, 44
  degrees of freedom and constraint, 63
  KC delivery chain, 5
  matrix transform model, 36
  variations, 123
DFx, 379
DFx in the large, 379, 392, 415
DFx in the small, 379, 385
Direct cost, 490
Directed graph, 215
Disassembly difficulty, 403
Disassembly sequence, 183, 205, 403
Disassembly to repair, 318
Discounting to present value, 493, 500
Discrete event simulation, 442,
    447, 452
Disk drives, 349
Displacement vector, 37, 83
Distribution, 358
Distribution chain, 11, 392
Distributive system, 34
Division of labor, 498
Documentation, 12
Dog point screw, 279, 281
Dominant design, 344, 345
Dot matrix, 320
Downtime, 420, 425, 432, 452
  scheduled, 421
  unscheduled, 422
Draft angle, 43
Draper, 194, 195, 199,
    201, 206
Dual in-line packages, 305

E

EADS
  successor to Airbus, 367
Economic analysis, 253, 317, 420, 424,
    449, 453, 489, 492
  of assembly workstation, 465
Edge-following, 258
Effector, 73
Efficiency
  of assembly station, 475
  contrasted with flexibility, 347
Electrical connectors, 293, 295, 302,
    305, 349
Electric circuit diagram
  analogy to DFC, 216
Electric drill, 331, 370, 395, 400, 417
Electric range, 9

Electric screwdrivers, 396
Engine block, 371
Engineered compliance, 266
Entities
    in a simulation, 447
Equipment cost, 423, 479
Error
    distinguished from mistake, 422
    random cause, 124
    repeatable cause, 124
Error accumulation
    rate of, if mean shift is zero, 130
    rate of, in worst case tolerancing, 125
Error analysis
    of assembly workstation, 465, 476
Escape direction, 190, 194
Escort memory, 469
Expert systems
    for DFA, 381
Exploded view drawings, 391
Exploded view method for finding
        assembly sequences, 182
Exploratory phase
    of an industry, 344

F

Fabrication, 423
    features, 43, 44
    operations, 420
    of parts at assembly station, 476
Fabrication-driven manufacturing, 364
Facets, 102
Facility constraints, 323, 423
Factory performance, 323
Factory's defect fraction, 402
Failure rate
    of different part presentation
        methods, 473
Fan motor
    DFC example, 226
Fastener, 182, 193, 194, 318, 324,
        328, 335, 384, 387, 395,
        400, 409
Fastener method for finding assembly
        sequences, 182
Fastening alternatives, 401
Fastening techniques
    relation to recycling, 403
Fatigue
    effect on assembly time, 388
Feasibility
    technical or economic, 330
Feature, 329
    assembly, 4, 6, 180, 187, 193, 199
    cost of molding, 398

fabrication, 43
functional, 199
in GD&T, 120
as objects, 43
operating or functional, 4
Feature-based design, 43
Feature control frame, 121
Feature interface transform, 47
Feature of size, 120, 149
Feature recognition, 43
Feedback gain, 257
Fine motion, 120, 181, 205, 253, 334, 407,
        409, 477
Fishbone assembly line, 432
Fitting, 114
Fixed automation, 13, 420, 422, 423, 431,
        443, 449, 451, 455, 497
    cost model, 495
    economic characteristics, 430
    technical characteristics, 429
Fixed cost, 428, 430, 489, 494
Fixture, 112, 116, 142, 152, 184, 197, 198,
        215, 216, 245, 332, 465
    for aircraft assembly, 240
    for automobile body assembly, 352
    for car door assembly, 239
    contribution to variation, 155
    cost, 190
    for mounting car doors, 238
    needed for Type 2 assemblies, 211
    part of transport system, 424
    powered, 186
    to provide missing constraint, 221
    providing constraint for sheet metal
        parts, 167
    for sheet metal part assembly, 165
    sources of variation in Type 2
        assemblies, 224
Fixturing features, 112, 232
Fixturing surface, 328, 329
Flash
    molding, 338
Flexibility, 15, 322, 341, 343, 425,
        431, 465
    assembly enabled by architecture, 354
    contrasted with efficiency, 347
    provided by platforms, 350
Flexibility requirements, 420
Flexible automation, 420, 423, 449, 497
    cost model, 495
    economic characteristics, 431
    technical characteristics, 431
Flexible parts, 216, 390
Floor layout, 423, 451
Flowdown
    of key characteristics, 23, 113, 141, 321
    of requirements, 16

Force closure, 74
Force feedback algorithm, 256
Force feedback matrix, 258
Force feedback strategy, 255
Forces
    during compliant part mating, 297
    in fine motion, 255, 256, 265, 274, 285
Force-torque sensor, 274
Ford, 113, 169, 236, 352, 432
Ford, Henry, 12, 114, 380, 385,
        443, 490
Ford Model T, 345, 353
Ford process for car doors, 239
Form closure, 74
Four bar linkage
    constraint and mobility analysis of, 67
Friction, 65, 74, 154, 273, 297,
        305, 329
    in fine motions, 255
    in force feedback, 256
    influence on assembly mechanics, 265
    influence on assembly mechanics of
        compliant parts, 293
    influence on jamming, 273
Friction cone, 301
    involved in wedging, 272
Friction force, 295, 304
    during assembly, 268
    role in jamming, 266
    role in wedging, 270
Friction stir welding, 346
Front wheel drive, 342
Fujitsu, 381
Functional build, 116, 168, 169, 170
Functions
    of assembly, subassembly, or part, 327
FX-1 assembly system, 458

G

Gages, 113
    go and no-go, 118
Gage tolerances, 120
Galileo, 74
Gantt charts, 449
Gas turbine blades, 26
Gaussian distribution, 126, 152, 157,
        159, 160
GD&T, 141, 177, 218, 221
    represented as matrix
        transformations, 147
    a worst case tolerancing method, 136
GD&T symbols, 122
Gear, 72, 263, 336, 351, 370
Gear assembly, 55, 280, 331
Gearbox, 336

GEC (UK), 381
General Motors, 236, 353, 432
Geometric Dimensioning and
        Tolerancing, 113, 118
Geometry
    influence on assembly mechanics of
        compliant parts, 293
Geometry of parts
    influence on assembly mechanics, 265
Global constraints, 187
Globalization, 345
GM process for car doors, 239
Goalposting, 124, 134
Gripping surface, 328, 329, 331
Gross motion, 181, 253, 334, 407, 408,
        468, 477
Group Technology, 380
Grübler criterion, 66, 209
Gyroscope, 9

H

Handling difficulty
    classification and coding, 385
Heisenberg's Uncertainty
        Principle, 75
Hertzian stresses, 72
Hewlett-Packard, 181
High-precision part fabrication, 114
High-risk areas
    in an assembly, 329
High-risk assembly, 330
High-volume assembly, 11, 423
High-volume production, 21
Hinges
    of car doors, 28, 157, 237
Hitachi, 277, 460
Hitachi Assembleability Evaluation
        Method, 388
Hitachi Assembly Reliability Evaluation
        Method, 389
Hi-Ti Hand, 277
Hoag, Michael, 454
Hole and slot compound feature, 49,
        143, 237
Hollow core molding, 395
Holmes-Cooprider method for assembly
        system design, 454
Honda, 169
Hub and spokes
    airline architecture, 365
    liaison diagram, 210
Human performance, 465
HVAC, 35
Hybrid mate-contact, 239, 240
Hydraulic actuator system, 362

I

IBM, 356
Igniter, 484
Impedance matching, 349
Independent rear axle for
        automobiles, 455
Indirect cost, 490
Industry standards, 351
Information processing, 348
    influence on architecture, 349
Information technologies, 320
Infrared detector, 318
Initial contact
    in compliant part mating, 300
Injection-molded plastic parts, 352
Injection molding, 388
Injection molds, 382
    fabrication time, 397
Ink jet, 320
Inner panel
    of car door, 28, 235
Insertion depth, 268, 269, 271, 273
Insertion directions
    statistics, 393
Insertion force, 265, 277, 293, 295, 300,
        303, 304, 390
    computer program, 288
    equations, 297, 305
    factors affecting, 296
Insertion force during chamfer
        crossing, 287
Insertion force during one-point
        contact, 287
Insertion force during two-point
        contact, 287
Insertion force experiments, 274
Insertion operations
    statistics, 393
Inspecting, 12
Inspection, 465
Instability
    in force feedback, 255
Instructions
    assembly, 469
Instrument cluster, 358
Instrument panel assemblies, 445
Integral and modular compared, 346
Integral architecture, 341, 345, 371, 392
Intel, 323
Interactions
    intended and unintended, 353
Interchangeable modules, 348, 354
Interchangeable parts, 75, 113, 115, 141
    goal of GD&T, 119
    via build to print strategy, 170
Interest, 493

Interest rate, 493, 494, 500, 501
Interfaces, 349, 354
    between parts, 322
    between product modules, 342
    standard, 349, 361
Interface standards, 349
Interference, 74
Interference fits, 263
Internal rate of return (IRoR), 501, 504
Internal rate of return method, 499, 501
International standards for
        tolerancing, 113
Intersection
    of screws, 84
    of twist matrices, 84, 86
    of wrench matrices, 86
Inventory, 322, 359, 374, 449, 492
Inventory holding cost, 360
Inventory management, 354

J

Jack, 477
Jamming, 265, 269, 293, 301, 329, 331,
        407, 436
    analysis of, 272
    conditions, 273
Jigging surface, 113
Jigless assembly, 7
Joining, 12
Juicer, 53, 55, 199
Juran, J. M., 124
Just-in-time production, 12, 114,
        443–444, 492

K

Kanbans, 364, 444
KC (key characteristic), 4, 19, 34, 72, 76,
        112, 116, 125, 133, 245, 321, 342,
        347, 348, 360
    affected by assembly sequence, 167
    for assembly success, 476
    of car doors, 28
    in control and capable, 163
    correlated, 224
    definition of, 21
    delivery of, 20, 23, 214
    desktop stapler, 4
    for missile seeker head, 53
    of optical storage disk drive, 26
    three-dimensional, 237
KC conflict, 25, 29, 214, 224, 246
    in aircraft fuselage assembly, 240
    in car doors, 31, 236

KC delivery chain, 5, 20, 119
  automobile engine, 116
  design steps for, 117
  DFC role as, 211
  including fixtures, 155
  length of, 156
KC flowdown, 23, 113, 141
KC priority, 25, 224, 243
KC proliferation, 26
Kinematically constrained assemblies,
  102, 114
Kinematic assembly, 62, 68, 74
  achieved by datum hierarchy in
    GD&T, 120
Kinematic design, 388
Kinematic realignment, 143
Kinematics, 247
Kits
  for part presentation, 469, 472
Kutzbach criterion, 66

L

Labor cost, 382, 423, 451, 491
Labor skill, 322, 323
Ladies inspecting fiber, 15
Large parts
  part presentation, 472
Lateral error, 152, 198, 255, 256, 265,
  266, 275
  influence on wedging, 271
Lateral stiffness, 267
Layout of a workstation, 469
Learning curve, 498
Least material condition (LMC), 122
Liaison, 203
  phantom, 199
Liaison diagram, 4, 21, 36, 182, 184, 187,
  191, 217, 245, 327
  for aircraft wing subassembly, 251
  of cockpit module, 371
  of juicer, 55
  loop closure rule, 185
  of missile seeker head, 53
  rules, 185
Liaison sequence, 184
Liaison sequence diagram, 188, 189,
  192, 199, 202, 203
Life cycle cost, 494
Line balance, 181
Line contact, 265
Local constraints, 187
Local content laws, 322
Locating features, 393
Location responsibility
  DFC defines, 214

Location, constraint, and stability
  distinguished, 73
Locator, 68, 71, 73
Locked-in stress, 71, 73, 166
Logistics, 354
Logistics cost, 491
Long lead time items, 358
Loop closure rule, 185, 191, 207
Loss function, 125
Lower control limit (LCL), 129
Lowering the water so you can see the
  rocks, 445
Lower natural tolerance limit (LNTL),
  127, 129
Lower specification limit (LSL),
  152, 163
Low-volume assembly, 11, 423
Low-volume production, 22
Low-wage regions
  assembly in, 16
Lucas, 381
Lucas/University of Hull DFA
  method, 399
Lucent Technologies, 338

M

Machines
  as assemblers, 14
Machining, 388
Main function carriers, 328
Makespan, 427, 451
Make to order, 180, 356
Make to stock, 180, 356
Manual assemblers, 428
Manual assembly, 13, 55, 181, 190, 197,
  199, 200, 323, 336, 370, 380, 383,
  385, 396, 407, 420, 422, 423, 436,
  449, 497
  cost model, 495
  igniter, 484
Manual assembly cost, 428
Manual assembly time, 386
Manual handling time, 385
Manual insertion time, 386
Manufacturing context, 323
Manufacturing engineer, 318, 348
Market segments, 342, 350
Market tiers, 350
Mason's algorithm, 258
Mass production, 68, 114
Material handling, 423, 424
Materials, 318, 327, 329, 336, 338, 343,
  344, 348, 400, 492
  cost component, 382, 490
  knowledge needed for DFA, 397

polymer, 395
  relation to recycling, 403
Mates, 5, 73, 212, 217
  in aircraft assembly, 213
  in automobile transmission, 230
  in car doors, 239
  in Cuisinart, 231
  defined using Screw Theory, 219
  between parts and fixtures, 222
  in sheet metal example, 219
  in wheel–axle example, 217
  incoming, 223, 251
Matrix transformations, 36, 75, 102, 141,
  214, 215
  chains or composition of, 40
  inverse of, 39
  to model variation, 42
  of nominal assembly, 37
  nominal and varied, 131, 147
  order of multiplication of, 41, 42
Mature phase
  of an industry, 344
Maximum material condition (MMC), 122
McCormick reapers, 114
Mean, 124, 126
  of a sum, 130, 139
Mean shift, 124, 133, 161, 168
  accounting for, 134
  result of goalposting, 125
Mechanisms, 35, 78
Microprocessors, 346, 347, 349
Microsoft, 351
Minimum attractive rate of return, 501
Minimum energy chamfers, 295
Missile seeker head, 53
Mistake
  distinguished from error, 422
Model mix, 420, 440, 443
Modular architecture, 341, 345, 371, 392
Modules, 342, 348
Molding, 328
Monte Carlo
  for calculating variation, 150, 177
Motion analysis, 86
  graphical technique for, 97
  results, 87
Motion and constraint analysis
  of assembly features, 86
  of multi-feature joints, 94
Motor, 333
  electric, 345, 369, 370
  pancake, 363
  in product family, 351
  on wheel automobile body
    architecture, 369
Motorola, 126
Mountain bike industry, 358

MRP (material requirements planning), 12
Multiple states
   in a product, 328
Munro and Associates, 397, 492
Mushroom product, 358, 360
Mystery features, 328, 329, 338

N

National Bureau of Standards, 114
Natural constraints, 259
Natural tolerance range, 127
Net build strategy, 114
Net present value (NPV) method, 499,
   501, 504
Network complexity factor, 208
Nissan, 385
Nominal assembly
   competent, 214
Nominal design, 22
Nominal design phase, 214, 245
Non-normal error distribution, 135
Normal distribution, 126, 149
   See also Gaussian distribution
Number
   of feasible assembly sequences,
     202, 208
   of liaisons per part, 208
   of people, 489
   of shifts, 489

O

Office copiers, 350, 367
Ohno, Taiichi, 443
One-point contact, 265, 268, 286
One shift operation, 451
One-sided constraints, 121
   in fixtures, 77
Onion skin method for finding assembly
   sequences, 184
Operator-dependent assembly, 388
   caused by overconstraint, 69, 113
Optical storage disk, 26
Ordo gear mating patent, 281
Outer panel
   of car door, 28, 235
Outsourcing, 2, 8, 10, 324, 342,
   345, 347
   of assembly system design, 422, 426
   percent by cost, 10
Outsourcing strategy, 324
Overconstraint, 197, 222, 348
   inside a feature, 89
   prevented by self-aligning bearings, 226

Overdimensioning, 152
Overhead, 322, 490, 492

P

Packaging, 35
Pallet, 353, 424, 433, 458, 478, 481
   for part presentation, 469, 472
Parallel workstations or operations,
   451, 454
Part complexity, 338
Part consolidation, 392, 395, 397, 398, 400
Part count reduction, 213, 324, 330, 365,
   382, 384, 388, 395, 400, 414
Part feeders, 465
Part feeding, 423
   design for manual and automatic, 384
Part feeding methods
   bulk, 470
   individual, 472
Parting lines, 328
Part mating, 12, 267
Part presentation, 12, 465, 470
Part size and thickness
   effect on assembly time, 388
Parts list, 327
Parts presentation cost, 407
Parts replenishment, 433
Part symmetry, 407
   effect on assembly time, 387
Part weight
   effect on assembly time, 388
   effect on workers, 391
Payback period method, 494, 499, 501
Peg and hole
   assembly, 264
   geometry, 266
Pentium, 330
People as assemblers, 13
Phantom liaison, 54, 55, 199
Phases of assembly, 264, 268
Piano mover's problem, 187
Pin–hole feature, 144, 155
Pin–hole joint, 90
Pin–slot feature, 89, 144, 155
Pins of electrical connectors, 293
Pitch circles of gears, 280
Plain vanilla box, 180, 360
Planetary gear sets, 227
Planetary gear train, 72, 74, 231
Plastic injection molded parts, 321, 336,
   338, 395
Platform strategy, 343
Plato, 119
Plug and play, 349
Plus-chord, 251

Poka yoke, 468
Polaroid cameras, 15
Polaroid Corporation, 431
Pontiac, 353
Poorly understood processes, 428
Poschmann Industrie-Plastic GmbH & Co.
   KG, 399
Powder metal parts, 396
Power
   influence on architecture, 349
Power line splice
   igniter, 484
   product architecture example, 371
Powershot Tool Company, 407
Power tool family, 351
Power tools
   product architecture example, 369
Precedence relation, 188, 199, 202, 244
   via Bourjault method, 191
Precision metal gage blocks, 114
Preloaded opposed bearing set, 72
Preplanning of assembly motions, 254
Present value, 500
Preventive maintenance, 421
Price-time product, 497
Print
   four ways to, 320
Printed circuit boards, 346, 347
Printers
   power supplies, 359
Process capability, 25, 126, 133
Process capability indices, 126, 128
Process capable, 126, 128
Process control charts, 127
   interpretation of, 128
Process design, 22
Process flow chart, 449
Process in control, 126, 128, 133
Process mean, 127, 129
Process standard deviation, 127
Process time
   variable, 330
Process yield, 420
Product architecture, 3, 253, 317, 321,
   324, 327, 328, 341, 379, 382,
   392, 415
   defined, 341
   influence on product life cycle, 341
   to manage variety and change, 354
   used to mitigate risks, 343
   used to mitigate uncertainty, 343
Product character, 318, 322, 399
Product design, 415
   influence on assembly system
     design, 423
Product development, 2, 319
   influenced by product architecture, 342

role of assembly in, 9, 317
web of decisions in, 318
Product families, 10, 321, 341, 349, 356
examples, 351
power tools, 369
Product functions, 342, 343, 351, 383
Production capacity required, 420
Production decoupling point, 357
Production rate, 489
Production smoothing, 444
Production volume, 322, 422, 423, 499
Product liability, 389
Product life cycle, 10, 392
Product performance, 324, 414
and DFA, 380
driver of integral architecture, 346
Product platform, 341, 350, 353, 354
Product redesign
compared to DFA, 399
related to assembly system design, 421
Product reliability, 389
Product simplification, 393
Product structure, 390
Product sub-functions, 342
Pro-forma cash flow analysis, 501
Properly constrained assemblies, 211
Pull process, 11
Pull system, 444, 445
Pump impeller, 197
DFC example, 232
Push process, 11
Push–pull boundary, 357
Push system, 444, 445

Q

Qualcomm, 351
Quality, 425, 435
Quality control, 123, 425
Quartz watch assembly, 473
Queues in a simulation, 447

R

Random errors, 136
Random events
in assembly system operation, 447
Random variables, 124
Rapid prototyping, 134
Rate of return, 499
desired, 493
R chart, 127

Rear axle, 201, 477
editing feasible sequences, 203
finding feasible sequences, 201
people near machines, 204
Rear wheel drive, 342
Rechargeable batteries, 338
Reciprocal of a screw, 82, 84
Recycling, 10, 342
design for, 403
Redundant locators, 72
Reliability, 347
Remote center compliance (RCC), 260, 274, 277
Remote center of rotation, 260
Removal processes, 134
Reorientation, 180, 190, 197, 396
Repeatable cause errors, 136
Replenishment
parts, 470
Requirements
on an assembly, 19
vague, 468
Resistors, 133
Resource utilization, 448
Responsiveness, 425
Reuse, 321, 329, 341, 350, 353
after disassembly, 403
of designs, 20
Reuse graph, 362
Revenues, 425
Rework, 435, 454
electric range, 9
Rework area
size of, 435
Rework loops, 438
RF cards, 469
Rho factor, 495
Rigid body, 65, 78, 82
Rigid locator, 73
Rigid part mating
equations, 285
Rigid parts, 65, 102, 165, 216, 253, 263, 293
compliantly supported, 263
Risk, 493
Risk premium, 494
Riveting, 165, 167
Robot, 181, 182, 186, 194, 195, 197, 201, 476, 484, 487
Robot assembly, 13, 14, 329, 388
of complex products, 15
Robot assembly station, 407
Robot dog example, 335, 338
Robotic automation, 423
Robots, 431, 447, 449, 451, 458, 460
Robust design, 23, 144, 421, 425
Robust design of DFC, 214, 245, 247

Root node of DFC, 215, 221
Root sum square tolerancing (RSS), 131
Rotary dial machine, 429–430
Rotation matrix, 37, 83
RSS analysis, 160
Rule #1, 122, 148–150, 177
Rules of thumb
for easy assembly, 385

S

Safety in assembly systems, 451
Safety stock, 12, 359
Sales volume, 322
Salvage value, 495
Sandia Laboratory, 194
Sanyo, 404
Savings, 425
Scheduling, 354
Screw, 336, 338, 390
definition of, 81
Screw Theory, 62, 77, 214, 247, 252
applied to basic surface contacts, 87
history, 77
Screw thread mating, 278
Seals, 328, 329
Sears, 407
Second shift, 451
Seeker head, 53
SelectEquip, 454, 457, 477, 495, 498, 499
applied to workstation design, 479
Selective assembly, 23, 73, 168, 229, 244
in automobile engine valve train, 118, 168
Self-tapping screws, 401
Semiconductor manufacturing, 323
Sensitivity
of assembly-level error to part-level error, 131
Sewing machine, 9
Shape
of chamfers, 304
of contacting surfaces, 296, 299
of mating surfaces (chamfers), 293
of pin on electrical connector, 295
Sheet metal, 279, 388
Sheet metal parts, 115, 155, 221, 293
DFC example, 219
variation in, 165
Shewart, W. A., 124
Shifts, 454
Shortest path algorithm, 190, 454, 479
Showstoppers, 330
Silma, Inc., 478
Simulation software
for assembly workstation design, 477

Singer sewing machines, 114
Single piece flow, 444
Skilled craftsmen, 9
Slip joints, 166, 167, 170, 215, 219, 221
Sloan, Alfred P., 353
Slope
    of contacting surfaces, 299, 305
Small parts
    part presentation, 470
Snap fits, 263, 336, 401
Social rate of return, 493
Socket, 295
    of electrical connectors, 293
Solder, 384
Solder joints, 126, 347
Soldering, 336
Sonic Cruiser, 501
Sony, 356, 359, 381, 431, 458
Sony APOS, 472, 481
Sony DFA method, 391
Sony Phenix 10 assembly station, 481
Sony robots, 15, 481
Sony Walkman, 15, 351, 359, 363, 392, 458, 481
Splice stringer, 251
Splines
    assembly of, 281
Sport fishing reel, 469
Spot welding, 221, 249
Spring-back, 165, 166
Springs, 328
    closed end and open end, 476
Stability
    of assembly or subassembly, 182, 193
    of force feedback algorithm, 257
Stack architecture, 370
Stack product structure, 392, 393
Stacks of clutch plates, 484
Stamped parts, 395
Stamping dies, 347, 382
    adjustment by hand grinding, 170
    for making sheet metal parts, 166
Standard deviation, 126
Standard interfaces, 351
Standardization, 380
Standard part, 327, 343, 350, 384, 447
Staple gun, 449, 483
    Gantt chart, 450
    manual assembly system, 451
    robot assembly system, 453
Staple gun, low-cost
    DFA example, 414
Staple gun, rugged
    DFA example, 407
Stapler, 205
Starved workstations, 440

States
    assembly sequence, 189
Statically determinate assemblies, 64
Statically indeterminate assemblies, 64
Statics, 63, 64, 75, 247
Statistical process control (SPC), 21, 25, 126, 168, 170, 435
Statistical tolerancing, 113, 123, 130, 141, 168
    example, 132
    in tolerance allocation, 162
    represented by matrix transformations, 149
    summary of conditions for, 133
Statistical variation analysis
    of car doors, 159
Stoppages of assembly equipment, 440, 441
Structure, 15, 35
    in planning assembly, 254
Structured bill of materials (BOM), 36
Subassemblies, 1, 214, 317, 322, 327, 329, 342, 350, 356
    of automobile body, 352
    containing only contacts, 244
    contrasted with modules, 348
    fully constrained, 244
    multiple occurrences, 362
    product with no, 9
    related to outsourcing, 324
Subset rule, 186, 207
Sun, 351
Superset rule, 186, 207
Supplier, 114, 118, 136, 156, 322, 324, 343, 356, 385, 445, 492
    dependency on, 321
    multiple, for the same subassembly, 323
    role in coordination, 170
Supply chain, 8, 9, 16, 322, 392, 445, 491
    cost distribution in, 382, 490
    KC delivery in, 11, 25
Surface-constrained assembly model, 46
Surface contacts, 87
System engineering, 16, 348

T

Taguchi, S., 124
Takt time, 190, 427, 484, 499
Tape recorder mechanism, 352, 458
Tape recorders, 350
Task assignment, 423
Taylor, Frederick, 443
Team assembly, 13, 385, 432
Technological change, 343

Technologies
    new, 342
Technology push, 319
Telemechanique, 363
Testable function, 324
Test an assembly, 180, 184, 204
    rear axle, 203
Testing, 12
    for correct assembly, 384
    a subassembly, 348
Testing an assembly, 338
Test of function, 324
Test strategy, 435, 436
Theoretical assembly time, 402
Theoretically eliminating parts, 393
Theory of Constraints, 442
Thermoplastics, 403
Thermoset plastics, 403
Third shift, 322, 451
33 MHz 486s, 133
Threaded joints, 199
Threaded parts, 263
Three sigma range, 127, 152, 159
3-2-1 design, 68
3-2-1 principle, 65
Throttle body
    DFC example, 234
Throughput, 465
Time value of money, 493, 501
Tolerance, 112
Tolerance allocation, 131, 162, 168, 247
    to achieve assembly C_{pk}, 163
    to minimize fabrication cost, 162
Tolerance analysis, 112, 213
Tolerance chains, 213
Tolerances, 329, 428
Tolerance synthesis, 112
Tolerance zone, 119, 122, 147
Tool changes, 427, 481
    time needed for, 384, 454, 475
Tool cost, 454
Tooling drawings, 31
Tool turrets, 481
Top-down design, 16, 20, 211, 213, 227, 317, 328
Torque versus turn angle
    in screw thread mating, 280
Toshiba, 356
Total cost, 494
Touch
    human sense of, 385
Tower Automotive, 169
Toyota, 6, 169, 353, 364, 381, 385, 440, 443, 445, 490
Toyota production system, 421, 443
Transitions
    assembly sequence, 189

Translational velocity vector, 80
Transport
    parts, 12, 469
Traveler, 469
True position, 122
True position tolerancing, 114, 118
TVAL (Toyota Verification of Assembly
        Line), 391, 467
Twist, 78
    definition of, 81
    general form of, 80
Twist matrix, 78
    of basic surface contacts, 87
    of a combination of features, 84
    of pin–hole joint, 79, 80, 91
    of pin–slot joint, 93
    of plate on plate joint, 93
Twist space, 78, 81, 219, 220
    of pin–slot feature, 89
Two-point contact, 265, 268, 286
Two-sided constraints, 121
Type 1 assembly, 211, 221,
        245, 409
    of bicycles, 358
Type 2 assembly, 211, 221, 245
    of bicycles, 358
Typewriter, 320
Typical assembly cycle, 466

U

Underconstrained assembly, 211
Underconstraint, 215
Unibody
    automobile body architecture, 368
Uniform distribution, 136, 157, 159
Unimation, Inc., 465
Union
    of screws, 84
Unit assembly cost, 423, 499
    fixed automation, 495
    flexible automation, 497
    manual assembly, 495
Upper control limit (UCL), 129
Upper natural tolerance limit (UNTL),
        127, 129
Upper specification limit (USL),
        152, 163
Uptime, 421, 442

U.S. Air Force Materials Laboratory, 324
U-shape assembly line, 434
Utilization, 421

V

Vacuum cleaner
    example, 338
Variable cost, 428, 429, 490, 499
Variable cost per unit, 494
Variance, 124, 126
Variance of a sum, 130
    derivation, 139
Variation, 4, 6, 34, 46, 112, 142, 348, 390
    assembly, 190
    in car doors, 239
    DFC role in, 214
    sequence-dependent, 190
    in sheet metal parts, 165
    sources of, 143
Variation analysis, 112, 211, 247
    to compare two candidate DFCs, 223
Variation design, 22
Variation design phase, 214, 245
Variation management, 168
Variation risk management, 22
Variety, 341, 354, 406, 445
    high, 7
    product, 16
Variety–change tradeoff, 355
VCR tape changing mechanisms, 458
Verification of assembly operation, 465
Versions of a product, 322, 343, 449
Vibration, 353
Vibratory part feeders, 380, 424, 429, 470
Virtual environments, 396
Vision
    to avoid gross motion errors, 254
    human sense of, 385
Vision systems
    for part feeding, 471
Volvo, 385, 432

W

Washing uncertainty, 167, 243
Waste, 445
    target of Toyota Production System, 443

Wedging, 152, 197, 265, 269, 293,
        329, 436
    analysis of, 271
Wedging angle, 270
Wedging conditions, 154, 272
Weighted average cost of capital (WACC),
        494, 501
Weight fit ratio
    metric for design for disassembly, 403
Welding, 155, 165, 167
Westinghouse DFA calculator, 391, 402
Whitney, David, 454
Whitney, Karl, 334
Withdrawal force, 297, 300
Work force characteristics, 323
Work in process inventory,
        441, 444
Workstation cost, 190
Workstation design, 424
World coordinate assembly
        model, 45
Worst-case tolerancing, 113, 123,
        125, 141, 168
    example, 132
    represented by matrix
        transformations, 148
    in tolerance allocation, 162
Worst-case variation analysis
    of car doors, 159
Wrench, 78
    definition of, 82
    general form of, 80
Wrench matrix, 78
    of pin–hole joint, 80
Wrench space, 78, 81, 219, 220
    of pin–slot feature, 89

X

X-bar and R statistics, 126
X-bar chart, 127
X-ray systems, 362

Z

Zero clearance, 74
Zero mean shift, 130
Zero present value, 501